Nanoengineering of Structural, Functional, and Smart Materials

Nanoengineering of Structural, Functional, and Smart Materials

Edited by
Mark J. Schulz, Ajit D. Kelkar,
and Mannur J. Sundaresan

CRC Taylor & Francis
Taylor & Francis Group
Boca Raton London New York

A CRC title, part of the Taylor & Francis imprint, a member of the Taylor & Francis Group, the academic division of T&F Informa plc.

Reprint, 2017

Published in 2006 by
CRC Press
Taylor & Francis Group
6000 Broken Sound Parkway NW, Suite 300
Boca Raton, FL 33487-2742

Transferred to Digital Printing 2006

Printed and bound in India by Bhavish Print Solutions Pvt.Ltd.,

International Standard Book Number-13: 978-1-138-09688-2

Library of Congress Cataloging-in-Publication Data

Catalog record is available from the Library of Congress

T&F informa

Taylor & Francis Group
is the Academic Division of T&F Informa plc.

Visit the Taylor & Francis Web site at
http://www.taylorandfrancis.com

and the CRC Press Web site at
http://www.crcpress.com

For sale in India, Pakistan, Nepal, Bhutan, Bangladesh and Sri Lanka only.

Preface

Need for the Book. In most areas of science and engineering, there is research underway related to nanotechnology. However, the research is in different disciplines and the basic and applied research is often not in step. The intent of this book is therefore to connect science and technology under the umbrella of nanoengineering in order to design and build practical and innovative materials and devices from the nanoscale upward. Nanoengineering is fast becoming a cross-cutting field where chemists, physicists, medical doctors, engineers, business managers, and environmentalists work together to improve society through nanotechnology. Nanoscale materials such as nanotubes, nanowires, and nanobelts have extraordinary properties and unique geometric features, but utilizing these properties at the nanoscale and bringing these properties to the macroscale are very challenging problems. The authors of the 24 chapters of the book explain these problems and have attempted to develop well integrated coverages of the major areas where materials nanotechnology has shown advances and where the potential to develop unique structural, functional, and smart materials exists.

Structural materials are defined as load bearing and are designed mainly based on mechanical properties. Examples where nanoscale materials can improve mechanical properties include polymer and metallic materials reinforced with nanoparticles and thin films to increase the surface hardnesses of the materials. Functional materials are designed to have special properties, and are not primarily used for their mechanical characteristics. Functional materials can have tailored or functionally graded physical attributes such as electrical and thermal conductivity, magnetic properties, gas storage, and thermoelectric properties, and sometimes graded mechanical properties such as hardness. Nanoscale functional materials can be used in high-tech applications including magnetic devices, electronics conducting thermoplastics, anisotropic polymer nanocomposites, surface coatings, biomaterials, sensor materials, catalysts, polymers, gels, ceramics, thin films, and membranes. Smart or intelligent materials have sensing or actuation properties such as piezoelectric or electrochemical transduction activities. Carbon nanotubes are smart materials because their electrochemical and elastic properties are coupled, and they have higher theoretical actuation energy densities than existing smart materials.

Scope of the Book. Our goal is to provide readers with background in the various areas of research that are needed to develop unique atomically precise multifunctional materials that may be the strongest, lightest, and most versatile materials ever made. The background needed to accomplish this encompasses synthetic chemistry, biotechnology, self-organization, supramolecular self-assembly, nanophased particles, films and fibers, chemical vapor deposition, oxide evaporation, and various approaches to develop extraordinary strength, toughness, and net shape processing of multifunctional materials and structures. In addition, molecular sensors, active nanocomposites, thin film skins, power generation,, high thermal and electrical conductivities, and biomimetics are all discussed with the aim of optimizing material systems to monitor their performance and maintain their integrity. These processes may exploit the large elastic and transduction properties of carbon nanotube materials for developing extraordinary multifunctional capabilities. Moreover, because the nanotube structure is not limited to carbon, the benefits of nanoscale inorganic fullerene-like materials and nanotubes for developing multifunctional and polar materials are also examined. Many elements and compounds are known to form stable two-dimensional sheets, and hence create many exciting possibilities for developing new types of nanoscale materials. Most materials that can be formed by physical or chemical vapor deposition have the potential to form nanotubes, nanobelts, nanowires, or some form of nanostructure, and many of these new materials are discussed in this book. One goal of nanoscale research is to produce synthetic analogic bionic materials that evolve their own nanostructures, sense and react to their environments, self-monitor their conditions, and have super-elastic and self-healing properties to provide enduring performance.

This book provides engineers and scientists the broad foundation needed to attack barrier problems and produce high-payoff nanotechnology. As you will see, this book contains quite a variety of research representing different approaches and viewpoints about nanotechnology. Nanoengineering is a new field, and this book serves as a focal point and reference that can be used to conceptualize and design new materials and systems. It was made possible by the generous contributions of scientists from around the world, and presents state-of-the-art nanoscience and nanotechnology including comments on future directions for research. The book will help researchers, students, managers, those working in industry, and investors understand where we are and where we are headed in the area of nanoscale, nanophase, and nanostructured materials and systems. Many figures and detailed descriptions of the synthesis, processing, and characterization of nanoscale materials are included so that the book serves as a learning tool for nanoengineering, and so that readers can reproduce the results presented. Problems are included at the end of each chapter to test understanding of the concepts presented and to provoke further investigations into the subject. This book is also meant to be used as a textbook for graduate level nanotechnology courses. It is hoped that the book will inspire students of all ages and disciplines to study nanotechnology and to think of different ways to use it to help humanity.

Multifunctional Materials. This book also explores the multifunctionality that is common in nature. Multifunctional materials have several important properties simultaneously, such as a structural material that can also sense and actuate. Presently, no smart material is also a structural material. Multifunctionality is actually a universal trait of biological materials and systems. Since the beginning of time, biological materials and systems have been designed by nature from the smallest components upward, and they have capabilities unmatched by man-made materials. Therefore, it makes sense to integrate biomimetics and nanoengineering; biomimetics provides the architecture for materials design, whereas nanoengineering essentially provides the route to build materials starting on the atomic scale, as in nature. Biologically inspired nanotechnology or bionanotechnology can be described as the process of mimicking the chemical and evolutionary processes found in nature to synthesize unique almost defect-free multifunctional material systems starting from the nanoscale up. Bionanotechnology is becoming a new frontier in the development of advanced biomedical, structural, and other materials. Bionanotechnology is exemplified through biological materials constructed in layered anisotropic and self-assembled designs that provide strength and toughness at the same time and biological systems in which sensing and actuation are performed using millions of identical parallel nerves and muscle fibers. This architecture allows billions of bits of sensory information to be processed in the neural, auditory, and visual systems in an efficient hierarchical order and millions of identical micro-actuators to work in harmony. In the book, initial concepts are discussed to mimic the basic functions of nerves and muscles using nanoscale materials. These concepts may someday lead to digitally controlled intelligent and enduring materials and structures.

Applications and Benefits. The socio-economic benefit of nanoengineering will be ubiquitous and lead to improved safety, security, and standard of living throughout the world. Future materials and structures will have vastly improved properties and durability. Smart machines will control their own performance, preserve their integrity, and partially self-repair when damaged, and when they are worn out or obsolete, they will be programmed to demanufacture and be recycled into new machines. Building without machining may be another outgrowth of nanoengineering. Nanoengineering will produce new launch vehicles, lightweight agile aircraft, and may allow the human exploration of space. Major areas of impact include future space missions that will use hybrid nanocomposites to provide a wholesale reduction in weight in space vehicle systems through material substitution, redesign, and integration; autonomous reconfigurable structures will increase speeds, reduce fuel consumption, reduce pollution, reduce noise, and provide lasting performance for aircraft; intelligent materials will provide structural health and performance monitoring to prevent degradation and failure of structures in all types of critical applications; nanocoatings, fillers, sprays, and films will provide protection from abrasion, EMI, heat, and provide artificial skins for materials. Commercial applications of nanocomposite materials potentially include all composite material products, brake disks, turbine engine shrouds, composite bushings, brake parts, metallic composites, smart materials, biosensing, and power harvesting. New applications will emerge as our knowledge increases. Nanoengineering is also important in fuel cells where functionalized nanotubes may store hydrogen safely for use in automobiles. Electronics, medicine, and computing are other areas where nanotechnology promises advances. Indeed, our vision of nanoengineering is to obtain nanoscale control over the synthesis of matter to build designer materials that can be used to solve the most difficult scientific and medical problems that face humanity.

Outline of the Book. The book is organized into an introduction and three parts that cover the major areas of focus in nanoscale materials development. The *Introduction to Nanoengineering* chapter gives an insightful overview of where we are in understanding nanoscale phenomena, and possible future directions for research. Part 1 of the book is focused on *Synthesis of Nanoscale Materials* and contains beautiful microscopy images of different synthetic materials, a comprehensive exposition of the properties and synthesis of carbon nanotubes and bismuth nanowires, unique methods of producing zinc oxide nanobelts, advanced methods for carbon nanotube synthesis, synthesis of self-assembled nanodots, and a ball milling-annealing method to synthesize boron nitride nanotubes.

Part 2 concentrates on *Manufacturing Using Nanoscale Materials* and includes a technique for functionalizing nanoscale materials for material property improvements, techniques for producing structural and metal–ceramic nanocomposites, the use of low-cost carbon nanofibers to form fibers and films, a comprehensive overview of techniques for producing macroscopic fibers from single-walled carbon nanotubes, a means of fabricating microdevices through self-assembled monolayers, using nanotubes to improve the strength of polymers, properties and applications of nanoscale intelligent materials, thermal properties of nanostructured polymers, and pultruded nanocomposite materials.

Part 3 of the book focuses on *Modeling of Nanoscale and Nanostructured Materials*. Nanomechanics and modeling of nanoscale particles and their vibration properties are discussed, along with methods of continuum and atomistic modeling of the nanoindentation of thin films. Modeling of thin film heterostructures, polarization in nanotubes, uneven stress distribution in nanocrystalline metals, carbon nanotube polymer composites, and multi-scale heat transport are also discussed.

In summary, this book provides a broad synopsis of the nanomaterials research conducted in university and government labs. Because the size of the book is limited, much of the important research in the field could not be included. Readers are therefore encouraged to use this book as a starting point from which to explore the literature on nanotechnology, which is becoming more exciting every day.

The editors thank Cindy Carelli and Yulanda Croasdale of Taylor & Francis, and Larry Schartman and Frank Gerner of the University of Cincinnati whose support was instrumental in producing this book.

Editors

Mark J. Schulz is an associate professor of mechanical engineering and co-director of the Cincinnati Smart Structures Bio-Nanotechnology Laboratory. This laboratory integrates nanotechnology and biomimetics to develop new smart materials and devices for structural and medical applications. The laboratory includes a nanotube synthesis lab, a processing lab for nanoscale materials, and a smart structures and devices lab. Research in the labs focuses on building structural neural systems using continuous piezoceramic and carbon nanotube neurons and electronic logic circuits for structural health monitoring, carbon nanotube array biosensors for cancer diagnostics, active catheters for diagnostics and surgery, reinforcing polymers using carbon nanotubes and nanofibers, and developing wet and dry carbon nanofiber hybrid actuators to control large structures. His contribution to the book is dedicated to the memory of his parents, Jeanne and Joseph.

Ajit D. Kelkar is a professor of mechanical engineering and associate director of the Center for Advanced Materials and Smart Structures (CAMSS) and the founding member of Center for Composite Materials Research (CCMR) at North Carolina Agricultural and Technical State University, Greensboro, NC. He is also a member of the National Institute of Aerospace (NIA). His research interests include finite element modeling, atomistic modeling, performance evaluation and modeling of thin films, and nanomechanics. Some of the projects he is presently involved with include continuum and atomistic modeling of thin films, nanoindentation studies of thin films, low-cost manufacturing of ceramic composites using a nanoparticle alumina matrix, and the effects of alumina nanoparticles on the mechanical behaviors of epoxy resins. In addition he is involved in the low-cost manufacturing of composite materials, damage characterization of thin and thick composite laminates subjected to low-velocity impact loading, fatigue behavior of textile composites, and finite element modeling of woven and braided textile composites.

Mannur J. Sundaresan is an associate professor of mechanical engineering and the director of the Intelligent Structures and Mechanisms Laboratory at North Carolina Agricultural and Technical State University. This laboratory is dedicated to developing novel sensors, instrumentation, and signal processing techniques applicable to smart structures and structural health monitoring. It also integrates the micromechanics of damage evolution in heterogeneous materials and structural health monitoring techniques for the life prediction of such materials. He has worked in the areas of micromechanics of damage evolution, development of novel processing techniques for carbon–carbon composite materials, and experimental mechanics.

Contributors

Christian V.D.R. Anderson
Department of Mechanical Engineering
University of Minnesota
Minneapolis, Minnesota

Debasish Banerjee
Department of Physics
Boston College
Chestnut Hill, Massachusetts

F. James Boerio
Department of Chemical and Materials Engineering
University of Cincinnati
Cincinnati, Ohio

Donald W. Brenner
Department of Materials Science and Engineering
North Carolina State University
Raleigh, North Carolina

Han Gi Chae
School of Polymer, Textile, and Fiber Engineering
Georgia Institute of Technology
Atlanta, Georgia

Ying Chen
School of Physical Sciences and Engineering
The Australian National University
Canberra, Australia

Richard O. Claus
Fiber and Electro-Optics Research Center
Virginia Polytechnics Institute and State University
Blacksburg, Virginia

Saurabh Datta
Smart Structures Bionanotechnology Laboratory
University of Cincinnati
Cincinnati, Ohio

Virginia A. Davis
Department of Chemical Engineering
Rice University
Houston, Texas

Mildred S. Dresselhaus
Massachusetts Institute of Technology
Cambridge, Massachusetts

Edward H. Glaessgen
Analytical and Computational Methods Branch
NASA Langley Research Center
Hampton, Virginia

Rahul Gupta
Department of Mechanical Engineering
North Carolina Agricultural and Technical State University
Greensboro, North Carolina

Peng He
Smart Structures Bionanotechnology Laboratory
University of Cincinnati
Cincinnati, Ohio

Yun-Yeo Heung
Smart Structures Bionanotechnology Laboratory
University of Cincinnati
Cincinnati, Ohio

David Hui
Department of Mechanical Engineering
University of New Orleans
New Orleans, Louisiana

Douglas Hurd
Smart Structures Bionanotechnology Laboratory
University of Cincinnati
Cincinnati, Ohio

Farzana Hussain
Department of Mechanical and Aerospace Engineering
Oklahoma State University
Tulsa, Oklahoma

Sachin Jain
Smart Structures Bionanotechnology Laboratory
University of Cincinnati
Cincinnati, Ohio

Ado Jorio
Federal University of Minas Gerais
Belo Horizonte, Brasil

Inpil Kang
Smart Structures Bionanotechnology Laboratory
University of Cincinnati
Cincinnati, Ohio

Ajit D. Kelkar
Department of Mechanical Engineering
North Carolina Agricultural and Technical State University
Greensboro, North Carolina

Goutham Kirkeria
Smart Structures Bionanotechnology Laboratory
University of Cincinnati
Cincinnati, Ohio

Joseph H. Koo
Department of Mechanical Engineering
University of Texas
Austin, Texas

Dhanjay Kumar
Department of Mechanical Engineering
North Carolina Agricultural and Technical State University
Greensboro, North Carolina

Satish Kumar
School of Polymer, Textile, and Fiber Engineering
Georgia Institute of Technology
Atlanta, Georgia

Young W. Kwon
Department of Mechanical Engineering and Energy Processes
Southern Illinois University
Carbondale, Illinois

Jingyu Lao
Department of Physics
Boston College
Chestnut Hill, Massachusetts

Kin-Tak Lau
Department of Mechanical Engineering
The Hong Kong Polytechnic University
Hong Kong, China

Hongbing Lu
Department of Mechanical and Aerospace Engineering
Oklahoma State University
Tulsa, Oklahoma

John F. Maguire
Air Force Research Laboratory
Wright–Patterson Air Force Base, Ohio

Hassan Mahfuz
Department of Ocean Engineering
Florida Atlantic University
Boca Raton, Florida

David B. Mast
Department of Physics
University of Cincinnati
Cincinnati, Ohio

Vincent Meunier
Computer Science and Mathematics Division
Oak Ridge National Laboratory
Oak Ridge, Tennessee

Atul Miskin
Smart Structures Bionanotechnology Laboratory
University of Cincinnati
Cincinnati, Ohio

Serge M. Nakhmanson
Department of Physics
North Carolina State University
Raleigh, North Carolina

Suhasini Narasimhadevara
Smart Structures Bionanotechnology Laboratory
University of Cincinnati
Cincinnati, Ohio

Jagdish Narayan
Department of Material Science and Engineering
North Carolina State University
Raleigh, North Carolina

Marco Buongiorno Nardelli
Computer Science and Mathematics Division
Oak Ridge National Laboratory
Oak Ridge, Tennessee

Sudhir Neralla
Department of Mechanical Engineering
North Carolina Agricultural and Technical State University
Greensboro, North Carolina

Gregory M. Odegard
Department of Mechanical Engineering
Michigan Technological University
Houghton, Michigan

Sri Laxmi Pammi
Smart Structures Bionanotechnology Laboratory
University of Cincinnati
Cincinnati, Ohio

Matteo Pasquali
Department of Chemical Engineering
Rice University
Houston, Texas

Dawn R. Phillips
Lockheed Martin Space Operations
NASA Langley Research Center
Hampton, Virginia

Louis A. Pilato
KAI, Inc.
Austin, Texas

Oded Rabin
Massachusetts Institute of Technology
Cambridge, Massachusetts

Zhifeng Ren
Department of Physics
Boston College
Chestnut Hill, Massachusetts

Samit Roy
Department of Mechanical and Aerospace Engineering
Oklahoma State University
Tulsa, Oklahoma

Erik Saether
Analytical and Computational Methods Branch
NASA Langley Research Center
Hampton, Virginia

Jagannathan Sankar
Department of Mechanical Engineering
North Carolina Agricultural and Technical State University
Greensboro, North Carolina

J. David Schall
Department of Material Science and Engineering
North Carolina State University
Raleigh, North Carolina

Mark J. Schulz
Smart Structures Bionanotechnology Laboratory
University of Cincinnati
Cincinnati, Ohio

Vesselin N. Shanov
Smart Structures Bionanotechnology Laboratory
University of Cincinnati
Cincinnati, Ohio

Donglu Shi
Nanoparticle Coating Laboratory
University of Cincinnati
Cincinnati, Ohio

Vishal Shinde
Smart Structures Bionanotechnology Laboratory
University of Cincinnati
Cincinnati, Ohio

Mannur J. Sundaresan
Department of Mechanical Engineering
North Carolina Agricultural and Technical State University
Greensboro, North Carolina

Lakshmi Supriya
Fiber and Electro-Optics Research Center
Virginia Polytechnic Institute and State University
Blacksburg, Virginia

Kumar K. Tamma
Department of Mechanical Engineering
U.S. Army High Performance Computing Research Center

Ashutosh Tiwari
Department of Material Science and Engineering
North Carolina State University
Raleigh, North Carolina

Tetsuya Uchida
School of Polymer, Textile, and Fiber Engineering
Georgia Institute of Technology
Atlanta, Georgia

Kalivarathan Vengadassalam
Department of Mechanical and Aerospace Engineering
Oklahoma State University
Tulsa, Oklahoma

Xudong Wang
School of Materials Science and Engineering
Georgia Institute of Technology
Atlanta, Georgia

Zhong Lin Wang
School of Materials Science and Engineering
Georgia Institute of Technology
Atlanta, Georgia

Cindy K. Waters
Department of Mechanical Engineering
North Carolina Agricultural and Technical State University
Greensboro, North Carolina

Jim S. Williams
School of Physical Sciences and Engineering
The Australian National University
Canberra, Australia

Vesselin Yamakov
National Institute of Aerospace
Hampton, Virginia

Sergey Yarmolenko
Department of Mechanical Engineering
North Carolina Agricultural and Technical State University
Greensboro, North Carolina

Contents

PART 2 Manufacturing Using Nanoscale Materials

PART 3 Modeling of Nanoscale and Nanostructured Materials

1 Introduction to Nanoengineering

John F. Maguire and David B. Mast

CONTENTS

Nanoengineering offers the very real promise of a veritable cornucopia of enabling new materials, devices, and products. Examples range from improved materials for everyday uses such as self-cleaning paints and bathroom surfaces and deicing surface treatments for aircraft and automobiles in northern climates to new forms of structural materials that might be stronger than steel yet lighter than Styrofoam. The technology may also enable the development of adaptive soft materials like foams and polymer composites that could enable fundamentally new sorts of products. For example, imagine computers in which a CPU the size of a sugar cube has vastly more computing power than all existing machines combined, or wallpapering a room using "paper" that acts as a very-large-screen television. Similarly, work is under way to produce very large mirrors from nanoengineered plastics so that huge, lightweight plastic mirrors can be launched into space and used to see nearly the beginning of time. Many major companies have recognized that this area of science and technology holds the key to new products, processes, technologies, and medicines, with multifaceted societal and economic benefits.

The above paragraph resonates with the kind of hyperbole that has become the lingua franca of much of the nanoscience and technology (NST) discussed in the popular press. Although many of these benefits may very well come true, and some are already here, it would be exceedingly naive to expect that such major innovations as quantum computing and advanced "smart" materials might be developed without substantial scientific and technical breakthroughs on the one hand, and concomitant major capital investment on the other.

Whereas some of the key enabling advances — monumental discoveries such as buckyballs, carbon nanotubes, and so forth — have already been made, some are still coming down the pipeline. In particular, it will be necessary to develop fundamentally new paradigms in nanoengineering if the great promise of nanoscience and technology is to translate into concrete societal benefits and the creation of wealth. It is of critical importance, therefore, to quantify as fully and realistically as possible the scale and scope of the scientific, engineering, and manufacturing challenges that must be met to make even a rough order of magnitude estimate of the return on investment.

What Maxwell[1] once said about the field of classical thermodynamics is also applicable to the development of the field of nanoengineering. Nanoengineering must meet three essential criteria: there must be a solid scientific foundation, there must be clear definitions, and there must be distinct boundaries. So, how about NST? How sound are the foundations and where might they be strengthened? How good are the definitions and standards and how might they be improved? And where indeed are the boundaries?

The remainder of this introduction will address these questions in the course of providing a general and brief review of the background of NST. We should point out that we make no effort to provide any kind of review of the field but simply point out a few salient works. Our discussion also must focus on the non-bio aspects of NST as related to the theme of this book.

In the broadest sense, NST represents the work of human minds expressed by human hands to add value to and to create wealth from the natural resources of the Earth. The subject is concerned with how to turn the very "dirt" of the Earth, the metal ores and the crude oil, for example, into the automobiles, aircraft, aircraft engines, computers, materials, and medicines of our advanced civilization. This cycle requires the accumulated interdisciplinary knowledge of generations of scientists and engineers and represents the real, or bedrock, "knowledge environment" on which our economy is based and against which progress should be measured.

The essential basis for the ongoing NST revolution is the development of materials and devices that operate over mesoscopic distance scales, where the material response depends ultimately on the behavior of matter in what are called thermodynamically small systems. The *small* is used here in a precise scientific sense that refers to a piece of matter (or material system) where the size of the system is of the same order as some relevant correlation length, ξ, such as the length scale related to interparticle interactions or magnetic moment orientation. Behavior similar to that of thermodynamically small granular systems often occurs near a critical point or phase transition or within a thin interface between different materials, where the thickness of the interface is approximately equal to ξ. The importance of developing a deeper understanding of these types of systems has long been recognized and as such has been demonstrated by the awarding of several recent Nobel Prizes in Physics and Chemistry.[2]

1.1 THERMODYNAMIC AND STATISTICAL FOUNDATIONS OF SMALL SYSTEMS

In his now famous talk, "There's plenty of room at the bottom," Richard Feynman[3] essentially posed the nanomanufacturing problem in terms of a challenge for miniaturization. Feynman was very careful to dispose of the "normal" miniaturization that had been ongoing, especially in the electronic industry from the early 1960s, and made very clear that he was referring to the manipulation of matter over atomistic distance scales. His foresight has captured the imagination of a generation of physicists, engineers, and more recently chemists. It should not be forgotten, however, that Professor Feynman's lifelong interest and particular scientific expertise was in the area of statistical mechanics of dense many-particle systems. One of his earliest achievements was to show that superfluidity in liquid helium was due to long-range quantum correlation effects. It can be presumed that Feynman was very well aware that a search for the ultimate in miniaturization would, of necessity, reveal a world of new physical phenomena in which the macroscopic physics of our everyday experience would require significant revision.

A number of early workers have made pivotal contributions to the foundation of NST. Indeed, Sir John S. Rowlinson[4] has pointed out that understanding the basic science of how forces between molecules result in the observable properties of matter, especially "soft" matter at interfaces,[5] really represents a research effort of almost 200 years in the making. However, there has been particularly significant progress in our understanding of intermolecular forces and phase behavior over the last 30 years with the award of the Nobel Prizes for the renormalization group theory and the work in soft matter and polymers, as well as the discovery of nanoparticles. For example, it is now well known that the correlation length is on the order of the system size in three common situations (i.e., near a critical point, within an interface, and in granular materials). Notice that it is the ratio of system size to the correlation length that is important so that "nano" has nothing to do with a particular subdivision of the meter, per se. In this regard, use of the term *meso* (in between) would seem preferable to *nano* (dwarf). It is the ability to organize the mesoscopic structure of matter and hold spatial coherence over macroscopic distance scales that lies at the heart of nanoscience and technology.

These considerations have a number of examples. When machines become very small, the van der Waals[6] forces become more significant, and the components tend to undergo a jump to adhesion when they come close together.[7] Similarly, if materials are sheared over nanoscopic distance scales in low dimensionality they do not wear by stripping individual atoms from the surface but set up complex long-ranged vortex patterns in which solid matter tends to flow like a cold plastic solid.[8] Even though one might make a nanomachine, how would one lubricate it, and for how long would it run? The alkanes that form the basis of everyday oils and lubricants crystallize into two-dimensional structures at nanoscopic interfaces.[9] Although it is straightforward to build simple computer models of molecules that look like turbine engines, gears, or trucks, it must never be forgotten that matter over these distance scales obeys the laws of statistical mechanics (the bridge between classical mechanics and quantum mechanics). It is the free energy (perhaps local free energy) that matters, and if an attempt were made to make such a molecular machine, it would certainly show a tendency to spontaneously jump into reverse and, at worst, might even separate into two or more liquid phases. Similarly, imagine that one deposits a feature that is 10 atoms wide (~2 nm) and 5 atoms thick using a metal deposited on a semiconductor surface. Would such a structure tend to diffuse over the surface, would it remain stable as deposited, or would it tend to rearrange into some other nanostructure?

Questions of thermodynamic and kinetic stability in nanoscopic structures are critically important in this regard and will be ignored at our peril.

To achieve the promise of nano, it will be necessary to develop new nanomanufacturing technologies that can actually produce materials and devices in large quantities and at low cost in a reasonable time. Here it is important to recognize that while it is possible at the moment to make nanoscopic structures and make pictures (the famous IBM logo made from atoms is an excellent example), those techniques that involve direct atomistic manipulation will be far too slow to produce bulk products. If we could pick up and place an atom, perhaps with the tip of a scanning microscope, at a rate of one per second, it would take $\sim 10^{23}$ s or longer than the age of the universe to make a single mole of product. While this form of nanoassembly might be appropriate for some applications involving atto-moles of material, clearly, it will be necessary to manipulate phase transitions or *directed* assembly techniques to arrange macroscopic matter in the required nanoscopic and mesoscopic structures. The chemical physics of small nanoscopic and mesoscopic systems is dominated by phase transitions that are largely driven by local entropic contributions to the free energy. There has been a good deal of excellent theoretical work in the chemical physics of small systems over the last 10 years, but little of it has yet found its way into the current nano literature.

1.2 DEFINITIONS

The important point here is that before engaging in major real-world manufacturing enterprises, it is absolutely essential that proper design tools be developed and tested. The design tools in this case are the well-founded models (statistical physics–based) and measurement techniques that will generate the database on which reliable design and manufacturing plans can be established. There are few areas of technical endeavor in which recent fundamental theory and experiment bear so directly on emerging manufacturing practice as in NST.

To produce these nanomaterials and devices in the required quantities, quality, and cost, it will be necessary to do the following:

(a) Develop new nanomaterials characterization and metrology techniques that probe the relevant structural property of matter over an appropriate domain of energy and momentum.
(b) Explore theoretical and computer-modeling approaches that can help rationalize observed behavior and response and, more importantly, predict materials response in situations where direct measurement would be difficult or impossible.
(c) Integrate the experimental and theoretical knowledge gathered in (a) and (b) above and provide a seamless transition of this knowledge into the engineering and nanomanufacturing environment using advanced techniques such as computational methods in artificial intelligence.[10]

It is, therefore, essential that methods be developed to predict and measure the structure and properties of materials that are organized through directed assembly techniques. These might be achieved by "templating" using a surface or near-surface field or possibly by some combination of electric, magnetic (possibly of multipolar symmetry), and flow fields. These and related approaches offer the possibility of directing the nanoscopic and mesoscopic phase behavior and structure of matter in ways that hold the promise of producing new forms of matter in quantities that will be commercially viable.

1.2.1 Characterization and Metrology Needs

There has been tremendous growth in the past 15 years in the development of tools for the imaging and manipulation of nanometer-scale materials, even atoms. In truth, the ability to image individual atoms with a Scanning Tunneling Microscope (STM) was one of the defining developments that gave birth to the nanoscience and nanoengineering age. The STM, and its cousin, the Atomic Force Microscope (AFM), have since become the grandparents of whole families of different Scanning Probe Microscopes (SPM) that are now known by their initials: LFM (Lateral Force), SCM (Capacitance), MFM (Magnetic Force), and so forth. Also of importance is the ability to manipulate nanometer-scale objects and even single atoms. The previously mentioned IBM logo spelled out by moving individual xenon atoms on a nickel surface and the building of "quantum corrals"[11] for trapping atoms are well-known examples of using an STM to manipulate single atoms. Equally impressive has been the use of optical methods such as "laser tweezers"[12] for the manipulation of biological samples such as DNA and "optical traps"[13] for the confinement and cooling of atoms, for example in Bose-Einstein Condensation.[14] However, there are still critical metrology requirements that need to be met and tools that need to be developed before effective nanomanufacturing can take place. Some examples of such requirements and tools are as follows:

- *Rapid detection of individual functional groups over large-scale surface areas and interfaces.* Spectroscopic techniques have long been of critical importance in many areas of science and engineering, and it is to be expected that the use of such techniques will play a central role in the continuing development of NST. For example, consider Raman spectroscopy, a powerful tool for the determination of the vibronic properties of solids and liquids. The usefulness of Raman spectroscopy was further extended to small and even single molecular groups with the use of Surface Enhanced Raman Spectroscopy (SERS) techniques.[15] With SERS, these molecular groups can be studied by attaching them to or near the surface of nanometer-sized gold particles and using laser-induced surface plasmon modes to greatly enhance the local Raman scattering. Unfortunately, SERS is not a spatially scanning technique like the STM and AFM previously discussed. For scanned optical spectroscopy at spatial resolutions on the order of 50 nm, various near-field techniques such as a Near-field Scanning Optical Microscope (NSOM) or total internal reflection using a Solid Immersion Lens have been widely used. Work to develop scanning, nanometer-scale Raman probes that will further reduce the size of the scattering volume so that individual functional groups or moieties can be investigated with spatial resolution on the order of 1.0 nm will be of key importance to many areas of NST, both wet and dry. Recent work by Novosky[16] using a scanned sharp metal tip, and Pettinger et al. and Mast,[17] using a STM, have begun the development of such a scanning nano-Raman system. As well as providing the ultimate in surface spectroscopy, it will be recognized that such technology would also allow, for example, the very rapid mapping of genomes using massively parallel fiber-optic nanoprobes. This will clearly be of central importance in the health-care industry.

A cartoon depicting the STM-Raman approach is shown in Figure 1.1, with a photograph of such a system in the authors' (DBM) lab shown in Figure 1.2. Figure 1.3(a) shows the Raman spectrum of a thin sample of *p*-nitrobenzoic acid taken with a conventional micro-Raman system, and Figures 1.3(b) and (c) show the Raman spectra taken with the STM tip positioned over the sample and retracted back from the sample.

- *Determination of local and interfacial mechanical properties using non-contact and light-scattering measurements.* Of critical importance for many areas of NST is controlling the mechanical properties at the interface of composite materials. Central to obtaining this desired control is having the capability of accurately measuring these properties. For example, if accurate Rayleigh-Brillouin studies could be conducted using near-field interfacial scattering at this level of spatial resolution, it may well prove

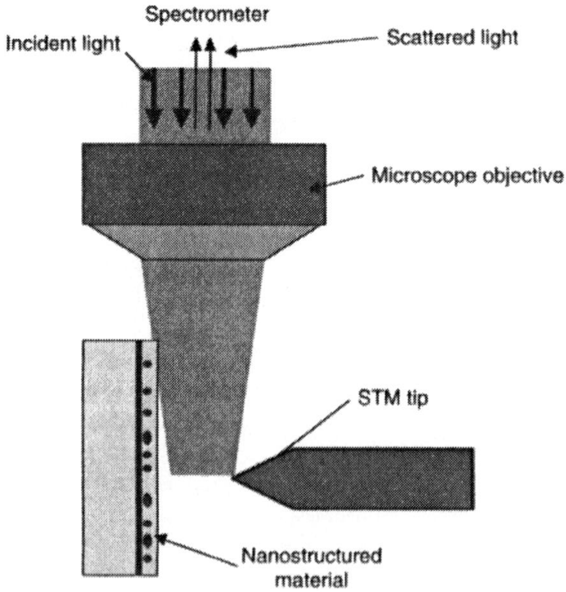

FIGURE 1.1 Cartoon of combined Raman-STM System.

FIGURE 1.2 (Color figure follows p. 12.) Photograph of miniature STM mounted on the Raman microscope stage.

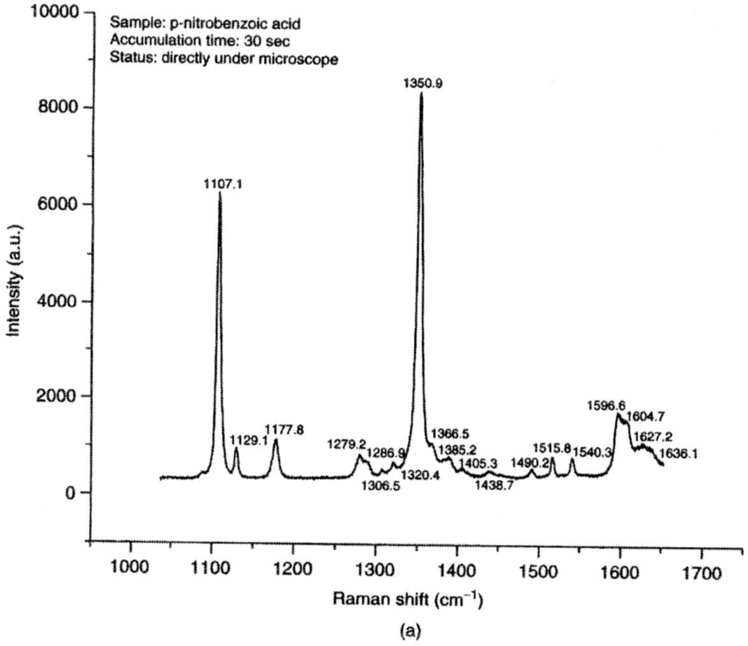

FIGURE 1.3(a) The Raman spectrum of the *p*-nitrobenzoic acid sample with the laser focused directly onto the sample; the sample plane is perpendicular to the illumination/detection direction.

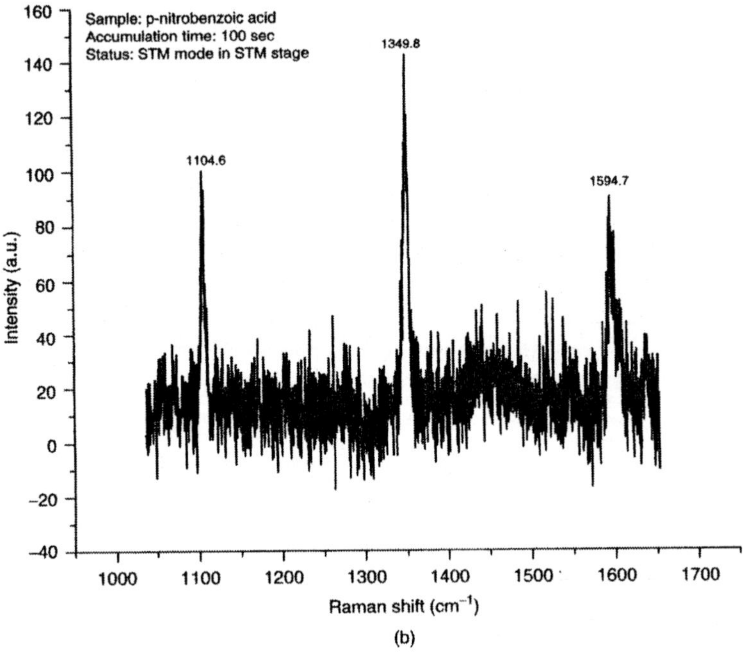

FIGURE 1.3(b) The Raman spectrum of the *p*-nitrobenzoic acid sample with the laser focused directly onto the STM tip when the STM tip is close enough to the sample for actual STM operation. The sample plane is parallel to the illumination/detection direction.

possible to measure experimentally the *local* free energy as a function of position (at nanometer resolution) through an interface. Such experimental information is absolutely vital if we are to understand materials transport and reaction in interfaces and thin films, including the cell wall.

- *Multiplexed sensor arrays for massively parallel detection.* It is clear that nanoengineered materials will be used in the near future to produce very large structures. For example, if suitable nanoparticles are finely dispersed in polymer films, it will be possible to engineer the microstructure such that the films will conform to a particular shape on application of a stress. Huge lightweight "inflatable" mirrors could be produced using this approach. How does one measure such microstructure over meters and even tens of meters in length? Although the single probe metrology tools previously described are essential to characterize the initial development of nanoengineered materials, they are many orders of magnitude too slow for use on large systems. Tools will need to be developed that contain arrays of large numbers of individual sensor probes for massively parallel operation. In addition, software will be needed that will allow these massively parallel tools to intelligently adapt their spatial resolutions for the detection of defects at different spatial resolutions.

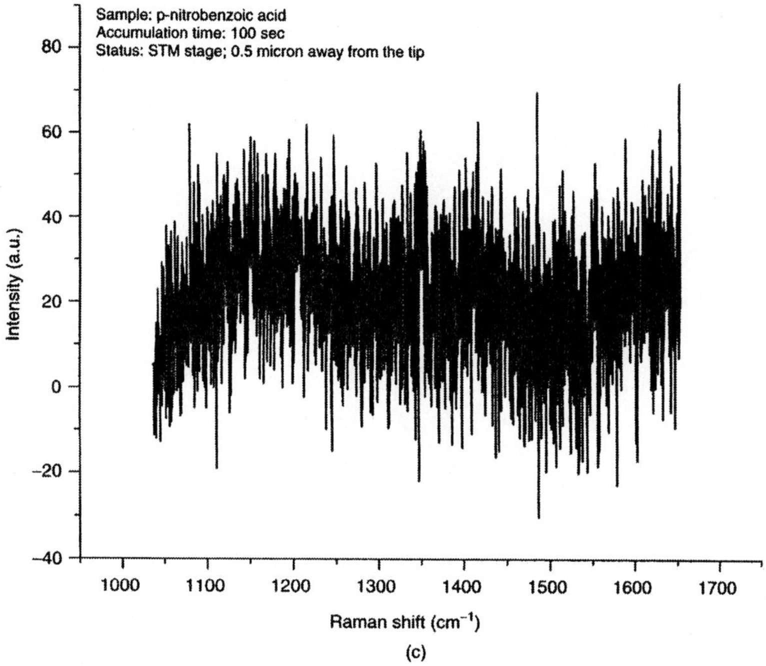

FIGURE 1.3(c) The Raman spectrum of the *p*-nitrobenzoic acid sample with the laser focused directly onto the STM tip when the STM tip moved a fraction of a micron away from the sample at the same point above the sample as in Figure 1.3(b). The sample plane is parallel to the illumination/detection direction.

1.2.2 Computer Modeling Needs

On the modeling front, our understanding of the fundamental nature of intermolecular and interparticulate forces has increased considerably over the last 20 years. It can now be said with some justification that the essential physics of simple dense systems with rapid ($\sim 10^{-12}$ s) relaxations are reasonably well understood. This is, however, not the case for complex molecular fluids or polymers. Here the relaxation times can be relatively slow ($>10^{-3}$ s), and the phenomena may be highly cooperative and long range. Prior to the development of recent methods based on the application of artificial intelligence in statistical mechanics,[18] such systems were quite outside the realm of *exact,* albeit numeric, machine calculations. With the development of new machine simulation techniques,[19] such systems are now amenable to more rigorous treatment, and many interesting results will undoubtedly follow. The coupling of artificial intelligence techniques with massively parallel machines brings the solution of a number of realistic nanomaterial simulations within reach if not yet quite within grasp. This should allow, for example, the first direct simulations of nucleation and crystal growth from equilibrium along the melting curve.

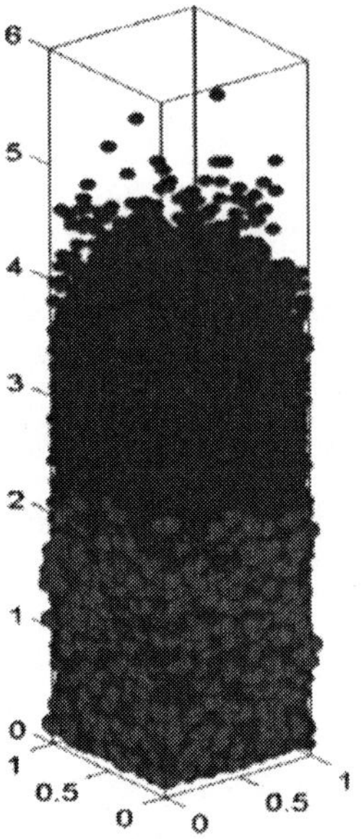

FIGURE 1.4 Showing the results of a computer simulation on the "thermodynamics" of fluidized powders investigating gravitational effects on mixing and segregation. This simulation is on a system comprising two dissimilar hard spheres under vibration. These simulations show that the system forms an ordered two-component solid near the bottom of the column, followed sequentially upward by a two-component fluid, an ordered single-component "solid," a single-component fluid, and finally a single-component "gas." (Image courtesy of L.V. Woodcock.)

Similarly, early work has shown the potential importance of external fields on surfaces and how the presence of a surface affects the conformation and adsorption of individual molecules.[20] In addition, the mechanical properties and functionality of advanced nanocomposite materials critically depend on the processes that produce the mesoscale structure at the nanoparticle–host matrix interface. This is especially true for granular nanomaterials.

Mesoscale simulations can be used to predict process dependence of properties for granular nanomaterials and will enable electronic prototyping of new continuous production processes to tailor and characterize the properties of nanomaterials. For example, Figure 1.4 shows the results of a numerical simulation, of a two-component, hard, spherical, granular material under the combined influence of gravity and a sinusoidal vibration. This figure shows the positions of each particle in the ensemble after a large number of vibration cycles and reveals the wide variation of "order" that has developed.[21] Again, new methods will allow the study of larger ensembles where the fully coupled nature of the phenomena, possible near phase transitions, will undoubtedly enable serious investigation in systems ranging from catalysis to the nature of protein interaction on cell walls or the formation of mesoscale entropically driven structures.

1.3 BOUNDARIES FOR NANOSCIENCE AND TECHNOLOGY

One of the very important issues that scientists and engineers need to keep in mind is that, like the well-known proverb, "All that glitters is not gold," "not all that is small is nano." Scientists and engineers, as individuals and as part of groups and institutions, feel a strong pull to join in, to be a part of all exciting fields of endeavor; this has been especially true for all things nano. The National Nanotechnology Initiative, directed funding from NSF, DOD, and DOE, and national, regional, and local nanoresearch centers have all contributed to the rush to add the word *nano* to many well-established, existing fields of research and engineering. We must be very selective with the use of *nano* lest we dilute its meaning and over-hype expectations of the benefits of what nanoscience and nanoengineering can contribute to society. If we are not careful, this saturation of the field with inappropriate uses of *nano* may likely result in a dilution of research funding below the amounts necessary for success.

1.4 SOME FINAL THOUGHTS

The interdisciplinary nature of nanoscience technology and manufacturing is a recurring theme. As evidence of the fruitfulness of the interdisciplinary approach, we can cite the spectacularly successful work that has resulted at the interface between chemistry and biology, called *molecular biology*. Work in this field has given rise to the new discipline of molecular science, which has resulted in tremendous advances in medicine, genetics, and agriculture. At the interface between chemistry and physics there has also been a fusion of new ideas in an area variously called *materials chemistry, soft matter,* and sometimes *surface and interfacial science*. Here the focus is to understand the forces between atoms and molecules in a dense medium and to use this knowledge to design new and useful forms of matter. As Philip Ball writes in *Made to Measure* (quoted by Dr. Rita Colwell, Director, National Science Foundation (NSF) in her address to the Materials Research Society), "We can make synthetic skin, blood, and bone. We can make an information superhighway from glass. We can make materials that repair themselves, that swell and flex like muscles, that repel any ink or paint, and that capture the energy of the sun." There is hardly a better example of how the old barriers have crumbled. The power and momentum of NST lie in no small measure in its porous boundary between physics, chemistry, and biology. Thus in the real world of today we routinely turn chemists into physicists and physicists into materials scientists and nanomanufacturing engineers, no longer recognizing divisions that are not nature-made but man-made. It is this "nanoengineering," done by chemists, physicists, materials scientists, and engineers "to put nanoscience into technology," that is the theme of this book.

REFERENCES

1. Gibbs, J.W., *Collected Works,* Vol. 2, Longmans, New York, 1928, p. 262.
2. For example, Steven Chu, Claude Cohen-Tannoudji, and William D. Phillips, Physics; 1997; Ernst Ruska, Gerd Binnig, and Heinrich Rohrer, Physics 1986; Robert F. Curl Jr., Sir Harold Kroto, and Richard E. Smalley, Chemistry; 1996.
3. Feynman, Richard P., "There's Plenty of Room at the Bottom," Talk at the annual meeting of the American Physical Society at the California Institute of Technology, December 29, 1959
4. Rowlinson, J.S., Faraday Lecture, "The Molecular Theory of Small Systems," *Chemical Society Reviews,* 12, 3, 1983; Rowlinson, J.S. and Widom, B., *Molecular Theory of Capillarity,* Dover Publications, Mineola, NY, 1982 (2002 ed.).
5. Maguire, J., Talley, P., and Lupkowski, M., *Journal of Adhesion,* 45, 269–290, 1994.
6. van der Waals, J.D., *Z. Phys. Chem.,* 13, 657, 1894; English translation in *J. Stat. Phys.,* 20, 197, 1979; also Lord Rayleigh, *Phil. Mag.,* 33, 209, 1892.
7. Lupkowski, M. and Maguire, J., *Phys. Rev. B,* 45, 23, 1992.
8. Maguire, J. and Leung, C.P., *Phys. Rev. B,* 43, 7, 1991.
9. Lupkowski, M. and Maguire, J., *Composite Interfaces,* 2, 1, 1–14, 1994.
10. Maguire, J., Benedict, M., LeClair, S., and Woodcock, L.V., *Proc. Mater. Res. Soc.,* 700, 241, 2002.
11. http://www.almaden.ibm.com/vis/stm/atomo.html;http://www.almaden.ibm.com/vis/stm/corral.html
12. Hirano, K. et al., *Appl. Phys. Lett.,* 80, 515, 2002.
13. For example, Myatt, C.J., Newbury, N.R., Ghrist, R.W., Loutzenhiser, S., and Wieman, C.E., *Opt. Lett.,* 21, 290, 1996.
14. For example, Ensher, J.R., Jin, D.S., Matthews, M.R., Wieman, C.E., and Cornell, E.A., *Phys. Rev. Lett.,* 77, 1996, 1996.
15. For example, McFarland, A.D. and Van Duyne, R.P., *Nano Lett.,* 3, 1057–1062, 2003.
16. Sanchez, E.J., Novotny, I., and Xie, X.S., *Phys. Rev. Lett.,* 82, 4014, 1999.
17. Pettinger, B., Rin, B., Picardi, G., Schuster, R., and Ertl, G., *Phys. Rev. Lett.,* 92, 096101, 2004; D. Mast, Final Report, SBIR Phase I, Contract Number: #504624-Sub Air Force, Advanced Adaptive Optical Coating Process Technologies, 01/04/02.
18. Saksena, R., Woodcock, L.V., and Maguire, J., *Mol. Phys.,* 102, 3, 259–266, 2004; Maguire, J. and Benedict, M., *Phys. Rev. B,* 70, 174 112, 2004.
19. See "A New Approach to Monte Carlo Simulation: Wang-Landau Sampling," D.P. Landau, S.H. Tsai, and M. Exler, *Amer. J. Phys.*
20. de Gennes, P.G., *Phys. Lett.,* 42, L-377, 1981; de Gennes 1991, in *Nobel Lectures, Physics, 1991–1995,* Gösta Ekspong, Ed., World Scientific, Singapore, 1997.
21. Knight, T.A. and Woodcock, L.V., "Test of the equipartition principle for granular spheres in a saw tooth shaker," *J. Phys. A Math & Gen.,* 29, 43654386, 1997.
22. http://www.nsf.gov/od/lpa/forum/colwell/rc81202.htm
23. Maguire, J.F., "Contributions to Materials Science and Engineering," D.Sc. thesis, University of Ulster, 2004; http://www.nsf.gov/od/lpa/forum/colwell/rc81202.htm.

FIGURE 1.2 Photograph of miniature STM mounted on the Raman microscope stage.

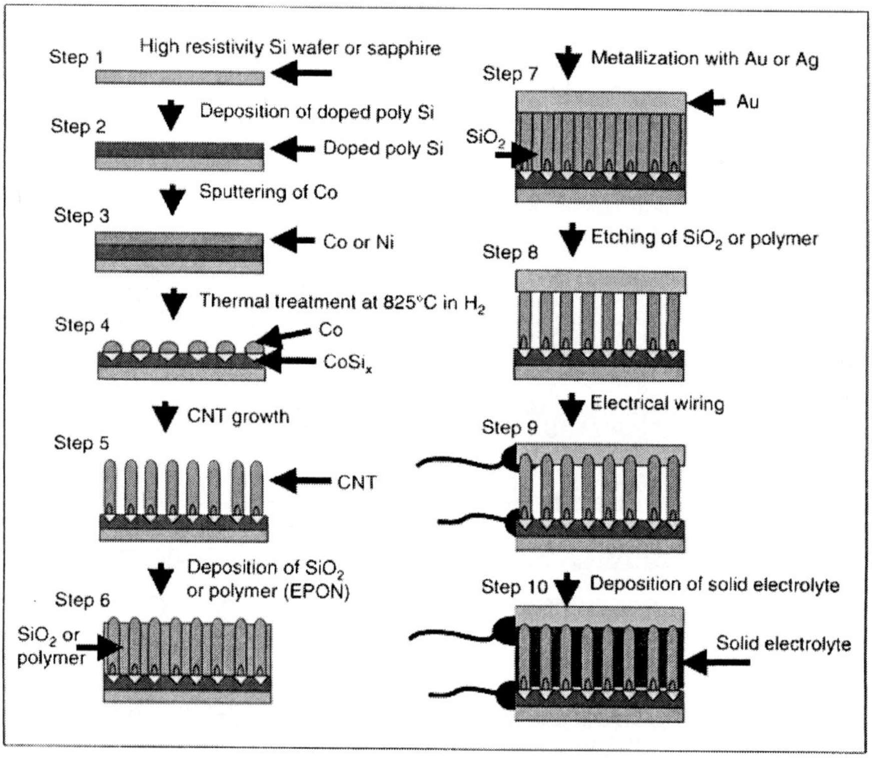

FIGURE 5.18 Schematic illustration of the fabrication process of CNT segment for actuator application.

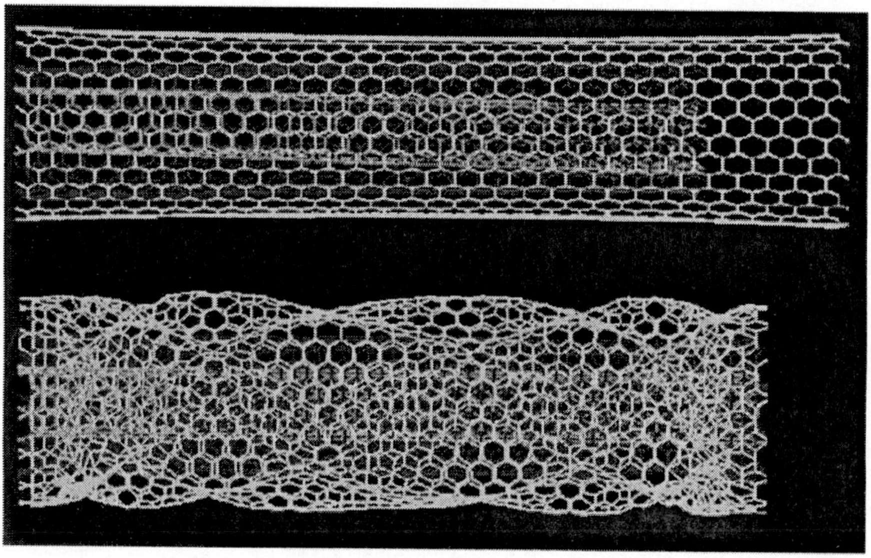

FIGURE 14.4 MD-simulated results of three-layer MWNTs. The inner two layers would not be affected by the surface layer when the MWNTs are subjected to tensile (top) and torsional (bottom) loads.

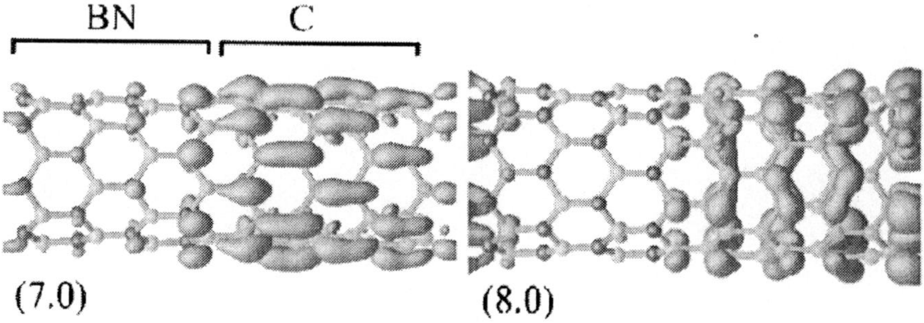

FIGURE 21.10 3-D plots of the valence state for (7,0) and (8,0) BN/C superlattices.

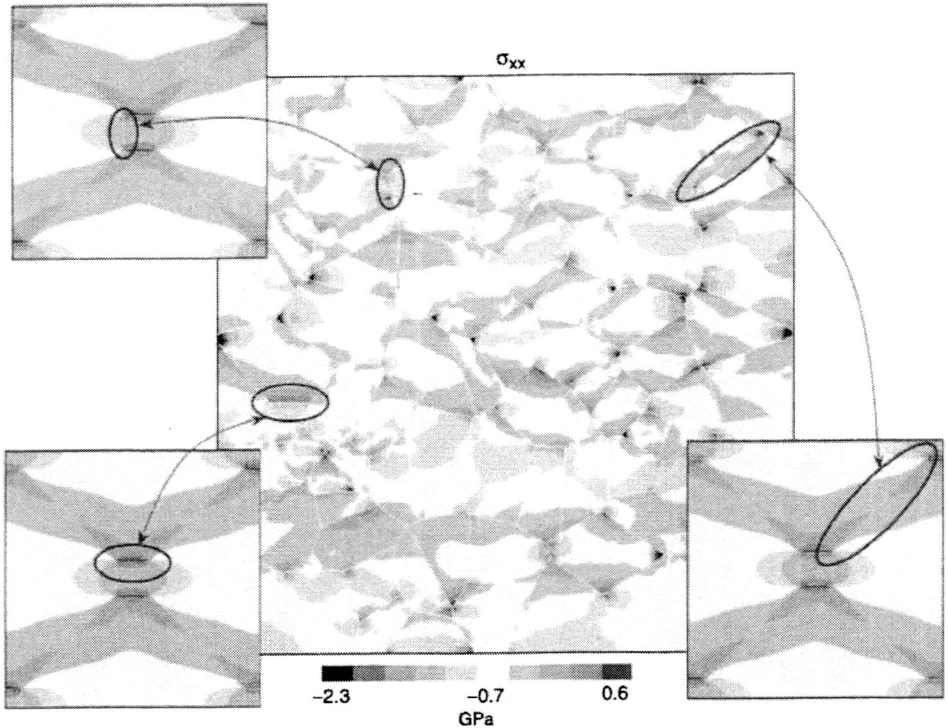

FIGURE 22.7 Stress contour $_{-xx}$ for 100-grain material subjected to 1.23% strain with similarities to the idealized model highlighted.

Part 1

Synthesis of Nanoscale Materials

2 Design of Nanostructured Materials

Debasish Banerjee, Jingyu Lao, and Zhifeng Ren

CONTENTS

2.1 INTRODUCTION

Downsizing of existing microstructures to nanostructures (those structures in which at least one dimension lies between 1 and 100 nanometers[1] (nm)) is found in widely varied research fields. It possesses the potential to become one of the most fruitful and farsighted scientific and technological innovations of recent decades. The journey toward nanoscale science began in 1905, when Albert Einstein published a paper estimating the diameter of a sugar molecule to be about 1 nm. In spite of several important developments and a vague understanding of small-scale science, this field remained dormant until critical technological progress was achieved. Progress includes developments such as significant improvements in high-magnification electron microscopy, the development of digital electronics, and greatly improved chemical and physical control and manipulation capabilities. Although development of the electron microscope, which enabled subnanometer imaging, emerged around 1931, interest in nanotechnology was stimulated only later on the occasion of Richard Feynman's[2] remarks in his 1959 landmark talk on aspects of miniaturization entitled, "There's Plenty of Room at the Bottom." Feynman's prediction of an age of nanoscience served as a lightning rod for innovative thinking and set the pace for the evolution of today's nanotechnology. Among the early milestones, the design in 1981 by Binnig and Rohrer of the tunneling scanning microscope (STM), which can image an individual atom, provided further encouragement to researchers everywhere. This development led to accelerated studies in small-scale science, stimulated further by the discovery of C_{60} by Curl, Kroto, and Smalley[3] and the discovery of the carbon nanotube by Iijima[4] in the mid-1980s and early 1990s, respectively.

Contemporary scientists, engineers, futurists, and investors foresee nanoscale science as a pathway to a vast field of smaller, faster, more energy-efficient technology through the expansion of their successes in downsizing microelectronics and other cutting-edge developments. The unique properties[5] of miniaturization have already demonstrated benefits in information storage,[6] in energy storage,[7] in chemical storage,[8] in chemical and biological sensing devices,[9] and over the entire areas of electronics, photonics, and communication.[10] These technological efforts have simultaneously enriched our understanding of physical phenomena through direct observation of novel behavior in low-dimensional systems, including size dependence, size-dependent photon emission (or photo-excitation),[11] coulomb blockage (or single electron tunneling),[12] metal-insulator transitions,[13] and quantized (or ballistic) conductance.[14] Quantum confinement[15] of electrons in nanometer-sized structures may provide a powerful tool for controlling and functionalizing the electric, optical, magnetic, and thermoelectric properties of these advanced materials. Several unique nanostructures, including nanotubes,[16] nanowires, nanobelts,[17] and nanodots,[18] have been synthesized in laboratories worldwide. They serve as building blocks for prototypes of many high-performance and ultrafast devices.[5,17,18]

Here we provide a brief account of the strategies for growing various non-carbon-based nanostructures and a description of our contributions to their synthesis and characterization. We present the findings of recent measurements of some of the physical and chemical properties[22,23] of complex hierarchical nanostructures,[19–21] nanowalls,[22] nanowires,[21,23] and the self-assembly of nanocrystal chains and nanowire circuits[24] in various systems. In a separate section, we consider a strategy for the production of large quantities of these structures in freestanding versions to meet the challenges of practical applications.[25,26] Lastly, we consider some possible future applications.

2.2 MOTIVATION, BACKGROUND, AND STRATEGIES

Developments in contemporary microelectronics have reached a point where further progress is limited by requirements for faster computing capabilities and by fundamental issues with the conventional top-down approach to fabrication, as well as with its high cost. Significant advances toward improved production efficiency and precise control over the morphology and microstructure of nanoscale materials through self-assembly have been achieved. This opens a totally new, more cost-effective, more economical, and more versatile bottom-up approach to fabrication. Furthermore, self-assembly in nanoengineering has considerable potential for providing both high-density fabrication capabilities and broad flexibility in choosing superior materials for improved device functionality. Also, enhancement of the controllability and tuning capacity of material properties can be accomplished by downsizing, powered by quantum confinement effects.[30] In this area, several research groups have demonstrated prototype basic electronic devices such as field effect transistors (FET), bipolar junction transistors, p-n junctions, and resonant tunneling diodes[27] using nanowires or other nanotube-based systems.[28] This approach becomes particularly powerful when the ease and control offered by self-assembly is combined with the electronic, magnetic, or photonic properties of inorganic components.[29] Accelerated progress in nanotechnology beyond silicon will require substantial elevation of capability in designing and manufacturing various complex three-dimensional (3-D) hierarchical assembly nanostructures. We will comment further on our recent successes in synthesizing novel 3-D hierarchical semiconducting nanostructures.

The growth of crystalline structures in low dimension[31] has been studied for more than a century, but little of a quantitative nature has been accomplished. Furthermore, the details of the control processes have not yet been mastered. The general methodology is first, to concentrate the atomic or molecular building blocks to form crystal nuclei that behave as seeds for growth, and second, to provide a continuous and well-controlled environment in which the building blocks can assemble the crystalline product. Control over the growth processes raises a nontrivial issue. Perfection in high-quality crystals requires, in addition to a homogeneous supply of building blocks, a reversible pathway between building blocks in the solid phase surface and those in the liquid phase. These constraints provide for long-range homogeneous enhancement of the building block's ability to acquire the proper orientation for forming a perfect crystal lattice with the selected morphology. Over the past few years Lieber and coworkers, at Harvard University, and Yang et al., at the University of California, Berkeley, have invented ingenious methods[32–34] for growing one-dimensional nanowires, rods, and belts. Materials with a highly anisotropic crystal structure prefer one-dimensional growth because of the highly anisotropic bonding between the building blocks in a chosen direction. Some polymeric materials and biological systems, such as polysulfurnitride $(SN)_x$, cellulose, and collagen,[35] have high anisotropy and thus a preferred growth orientation. Other anisotropic inorganic substances, including selenium, tellurium,[36] and the relatively complex molybdenum chalcogenides,[37] which have a natural tendency to grow in a preferred direction, have been studied extensively. Induced symmetry breaking is required to achieve a preferred growth direction in isotropic crystals. Arc discharge, laser ablation, and solution- and template-based[33] processes have been employed for growing low-dimensional systems. The most general and convenient method, however, has proven to be growth by a vapor phase transport process followed by chemical vapor deposition.[34] Vapor phase epitaxy has thus become the most common method for growing nanostructures of all solid materials. In principle any bulk material can be converted to its one-dimensional nanostructure form through this approach. In the early 1920s, Volmer and colleagues were the first to grow mercury nanofibers by condensing mercury vapor on glass.[38] The fibers were 20 nm in diameter and 1 μm in length. Later, in the mid-1950s, a series of studies conducted by Sears and colleagues determined that one-dimensional growth occurs through a general mechanism based on axial screw dislocations.[39] This result appears to explain the fundamental growth kinetics. We are now convinced that supersaturation during growth is the key to acquiring diverse shapes and morphologies. The vapor-liquid-solid (VLS) approach, first proposed by Wagner and Ellis in 1964, is the most widely discussed process for growing one-dimensional nanowires. This procedure involves the absorption of source material from the gas phase into a liquid droplet of a catalyst[40] to form liquid alloy clusters. The liquid alloy clusters serve as preferential sites for absorption and supersaturation of the reactant, thus breaking the symmetry. This leads to the formation of one-dimensional nanowires. One-dimensional growth occurs as long as the catalyst remains liquid. The most important issue with this approach is to discover the perfect catalyst for a particular material, one that can coexist as a liquid at the eutectic temperature. Observation of catalyst particles on the ends of nanostructures is generally considered to be proof of VLS growth. Semiconducting oxide nanowires of ZnO,[23,25,41] In_2O_3,[42] SiO_2, Ga_2O_3,[43] and nanobelts of ZnO, SnO_2, In_2O_3, CdO, and PbO[44] have been synthesized in recent years using the vapor phase epitaxy approach. Lee and coworkers have produced semiconducting Si nanowires at relatively high yields using an oxide-assisted growth (OAG) mechanism, an approach claimed to be fundamentally different from the VLS method.[45] In this oxide-assisted growth, the external metal catalyst is not necessary. Semiliquid suboxide droplets, which have a lower melting point than the oxides, act as a catalyst for the preferential growth sites.

We have already discussed the utility of ordering nanostructures in useful and relatively complex forms. In the past few years, the nanoscience research community has achieved reasonable success in growing and applying one-dimensional nanostructures, but few studies on hierarchical 3D structures have appeared. This is most likely because of their complex mechanism and the difficulties in process control. One simple approach for growing multidimensional nanostructures is first to grow nanowires by the VLS method and then alter the synthesis conditions to grow different materials on the *surface* of the nanowires by chemical vapor deposition (CVD). Lauhon and coworkers demonstrated the synthesis of Si/Ge and Ge/Si coaxial nanowires[46] and have shown that the outer shell can be formed epitaxially on the outer core. Wang and colleagues[47] generated a mixture of coaxial and biaxial SiC-SiO_x nanowires via a catalyst-free high-temperature reaction of a mixture of amorphous silica and graphite. Earlier, Zhu and coworkers[48] synthesized 3D flower-like nanostructures consisting of SiO_x fibers radially attached to a single catalytic particle by solid–solid or gas–solid reaction. Using the same method, they also synthesized a MgO fishbone assembly and a fernlike fractal nanostructure. Several other nontrivial 3D structures, such as Bi_2S_3 skeletons[49] and SnO[50] nanoribbon networks, have also been demonstrated. Yang and colleagues showed ZnO nanocrystals with exotic shapes, including tetrapods, tripods, and nanocombs, synthesized by thermal evaporation of mixtures of ZnO and graphite.[51] In a similar way, Wang and coworkers produced nanocombs and nanowindmills of wurtzite, ZnS.[52]

2.3 EXPERIMENTAL SET-UP

We have used a vapor phase epitaxy–oriented growth process, which employs vapor transport and condensation to grow complex microstructures. The process is relatively simple and powerful for preparation of one-dimensional as well as more complex 3-D structures, even in large quantities. First, we thermally evaporate from either a mixed source or two different sources and then direct the vapor molecules (building blocks for nanostructure growth) toward the substrate, which is at a lower temperature. The vapor condenses on a suitable substrate, such as graphite foil or sapphire, to form the nanostructures. The morphologies of the nanostructures are controlled by varying the experimental conditions. We will describe the particular changes separately in the corresponding sections.

The basic experimental set-up is illustrated schematically in Figure 2.1.[53] A powder source is placed at the sealed end of an open quartz tube. The open end of the quartz tube is covered with a graphite foil to serve as the substrate. The entire assembly is inserted into a high-temperature horizontal ceramic tube furnace in which a steep temperature gradient is established. To maintain the desired internal atmospheric pressure between 1 militorr and 10 torr, one end of the furnace tube is attached to a rotary vacuum pump while the other end is closed by a silicon stopper connected to an adjustable needle valve. The temperature of the furnace is ramped from ambient to a temperature between 900 to 1100°C at a rate between 60 and 80°C/min. The quartz tube is positioned in the furnace so that it is exposed to the appropriate temperature gradient. The substrate temperature is also controlled by its location in the temperature gradient. Because of the temperature gradient and the pumping direction, vapor produced from the heated source condenses and deposits on the graphite foil. At the end of the process, a fine powder is usually found deposited on the graphite foil surface facing the vapor stream, as indicated by the coating visible in Figure 2.1. The powder is analyzed using a JEOL JSM-6340F scanning electron microscope (SEM), a Bruker Analytical X-ray System, and a JEOL 2010 transmission electron microscope (TEM).

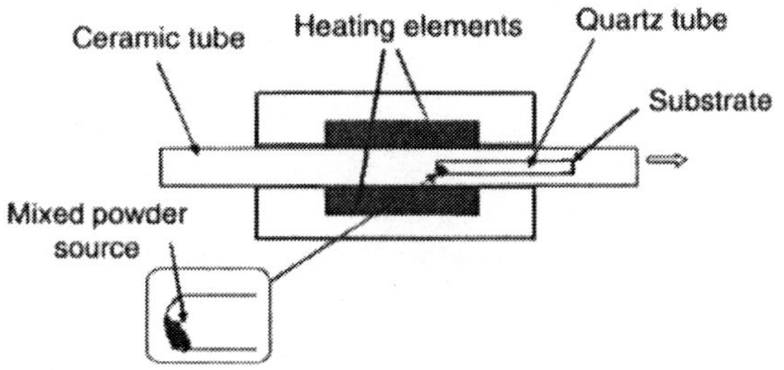

FIGURE 2.1 Experimental setup to grow nanostructures via vapor condensation and vapor transport process.

2.4 RESULTS AND DISCUSSION

Simple thermal evaporation and condensation of various mixtures of semiconductor oxides can effectively synthesize very complex and useful nanostructures. The precise growth mechanism of these nanocrystal formations is still obscure because of the complex transport characteristics of various building blocks and the intricacy of orientation-related hetero– and homo–epitaxial crystallographic matching among different materials. Consequently, control over morphology and size is far from satisfactory. In the sections that follow, we will concentrate on the interesting binary semiconducting materials zinc oxide (ZnO), indium oxide (In_2O_3), and tin oxide (SnO_2). ZnO is a direct wide band gap (3.37 eV) semiconducting material with significantly higher exciton binding energy (60 meV) than that of its competitors, GaN (25 meV) and ZnSe (22 meV). This fact deserves substantial attention because, to achieve efficient excitonic lasing action at room temperature, the excitonic binding energy must be higher than the thermal binding energy at room temperature (26 meV). ZnO naturally is a promising candidate to be a room-temperature short-wavelength laser.[10] Low-dimensional systems, such as nanostructures, have lower threshold lasing energy as a consequence of quantum effects that create a substantial density of states near the band edges. Carrier confinement enhances radiative recombination.[54] Thus ZnO nanostructures have a significant potential for opto-electronic applications. Apart from electronic and opto-electronic[19,41(a)] applications, ZnO has demonstrated several other optical and electro-optical applications in optical waveguides,[55] optical switches,[56] transparent ultraviolet protective conducting films,[57] and acousto-optic and surface–acoustic applications.[58] ZnO also is a well-established sensor material[59] that has demonstrated utility in solar cells[58] and photo-catalysts.[61] Among the other wide band gap transparent materials, the interesting optical and electronic properties of In_2O_3 have drawn considerable attention. In_2O_3 is also a direct wide band gap (3.6 eV) material with various applications in solar cells, organic light–emitting diodes, and gas sensors.[62]

2.4.1 ZnO Nanostructures and Their Properties

ZnO nanostructures can easily be synthesized with predetermined size and morphology through the application of vapor phase epitaxy. Chemical vapor deposition (CVD),[63] metal organic chemical vapor deposition (MOCVD),[64] physical vapor deposition (PVD),[65] and thermal evaporation[25,26,66] represent alternative methods for growing ZnO nanowires on various substrates. All of these have been explored in recent years. Control over orientation and size of the nanowires has also been achieved.[26,41(a)] A popular method for producing ZnO nanowires through CVD or PVD is to evaporate either pure Zn, a mixture of Zn and another material such as Se, or a mixture of ZnO and graphite in the proper proportion as a source material in a furnace similar to that shown in Figure 2.1. In this synthesis the vapor would be directed toward the substrate in a flow of argon or a mixture of oxygen, argon, and hydrogen. In the MOCVD technique, metal organic zinc precursors such as diethyl zinc ($Zn(C_2H_5)_2$) or zinc acetylacetonate hydrate ($Zn(C_5H_7O_2)_2 \times H_2O$) are normally used. ZnO nanowires can also be produced through the simple thermal evaporation of a mixture of ZnO and graphite while maintaining an appropriate oxygen partial pressure in the tube furnace and condensing the vapor on a selected substrate utilizing the temperature gradient along the tube. Alignment and size control of the nanowires can be achieved by employing an appropriate custom-designed substrate. Nanowires of certain materials have a preferred direction of growth. For example, Si nanowires prefer to grow along the ⟨110⟩ direction, whereas ZnO nanowires prefer the ⟨001⟩ direction. Therefore the proper strategy for growing vertically aligned nanowires is to select the appropriate thin film substrate with a suitable material to serve as a catalyst. When the substrate is heated to the melting temperature of the catalyst, it melts. Because of the high surface free energy, small particles appear that condense and form clusters of alloy. These supersaturate with the incoming vapor and become a one-dimensional nanostructure through the VLS mechanism as discussed previously. The diameter of the nanowires can be adjusted by controlling the thickness of the film. The preferred substrate for Si and ZnO nanowires is either a silicon or sapphire substrate with a thin Au coating as the catalyst. One may use an Si (111) wafer as a substrate for the growth of vertically aligned Si nanowire arrays,[45] whereas a-plane (110) sapphire can serve to grow epitaxial vertical ZnO nanowires.[41(a)] For a more comprehensive discussion of these procedures, please browse the respective references.

Epitaxial growth of vertically aligned ZnO nanowires on an a-plane sapphire substrate with Au coating is shown in Figure 2.2. In the initial stage of the formation of nanowires, the basis for the formation of aligned arrays of nanorods, zinc-gold alloy deposits in separated dots on the substrate. The SEM image of a large array of the aligned ZnO nanorods is shown in Figure 2.2(a), and Figure 2.2(b) is a top view at medium magnification. In this case, only the heads of the nanorods are visible. The nanorods are 50 to 120 nm in diameter, with a length between several hundred

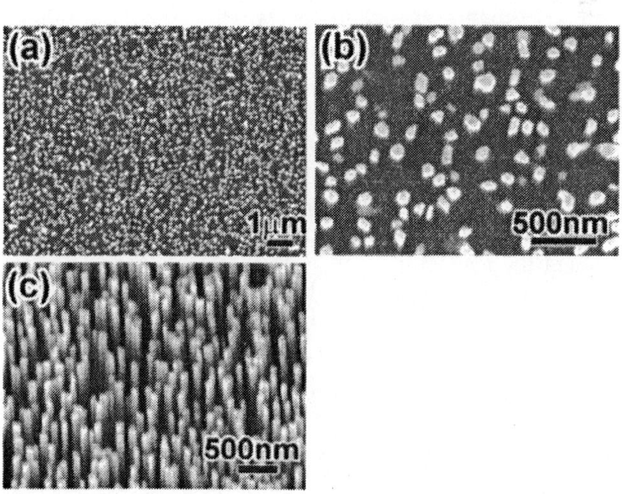

FIGURE 2.2 Aligned ZnO nanowires grown on a-plane sapphire substrate. Top view of aligned hexagonal ZnO nanowires at low (a) and medium (b) magnifications, and tilted view (c).

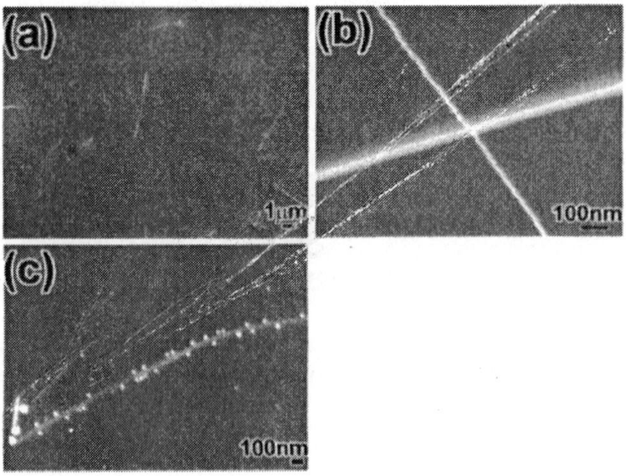

FIGURE 2.3 Formation of ZnO nanowires on an Si substrate coated with few nanometers of gold (Au). Low (a) and high (b) magnification SEM image of the small ZnO nanowires formed on the Si substrate where gold droplets acts as catalysts, and shows aggregation problem of Au clusters at high processing temperaturec(c).

nanometers and a few microns. Figure 2.2(c) is a tilted view of the aligned ZnO nanorods. For field emission and other electronic applications, the use of a sapphire substrate is not convenient because of its low conductivity. Crystalline Si substrate is preferable instead. Gold nanoparticles, 3 nm in diameter, are spread on the silicon substrate to serve as a catalyst. This procedure also presents a problem, because Au nanoparticles are very mobile at high temperatures, and aggregation may occur. A summary of the results we have obtained is shown in Figure 2.3. In Figure 2.3(a) a ZnO nanowire has grown on the Si surface. The nanowires usually have a diameter between 10 and 20 nm and a length of several microns. A high-magnification view of such nanowires is illustrated in Figure 2.3(b). Some larger nanowires are also visible. The formation of these large nanowires is caused by the aggregation of Au nanoparticles. Figure 2.3(c) shows the large aggregated Au nanoparticles and ZnO nanowires nucleated from these nanoparticles. Gold nanoparticles have even aggregated on the side of the nanowire.

ZnO nanowalls have been synthesized under similar conditions on 1- to 3-nm-thick Au coated a-plane sapphire substrate. In this case the substrate temperature was held between 875 and 950°C, with the tube pressure between 0.5 and 1.5 torr. Nanowalls grown at high temperature show a milky white-gray color, whereas structures grown at low temperature show a reddish color. The SEM images of the morphology of typical ZnO nanowalls are shown in Figures 2.4(a), through (c). The nanowalls are interconnected to form a network, with most of the nanowall flakes normal to the substrate. The pore size varies from 100 nm to 1 μm. These nanowall flakes do not exhibit a clearly ordered pattern but rather show a quasi-hexagonal pattern with most of the flakes forming angles that are multiples of 30°. The x-ray

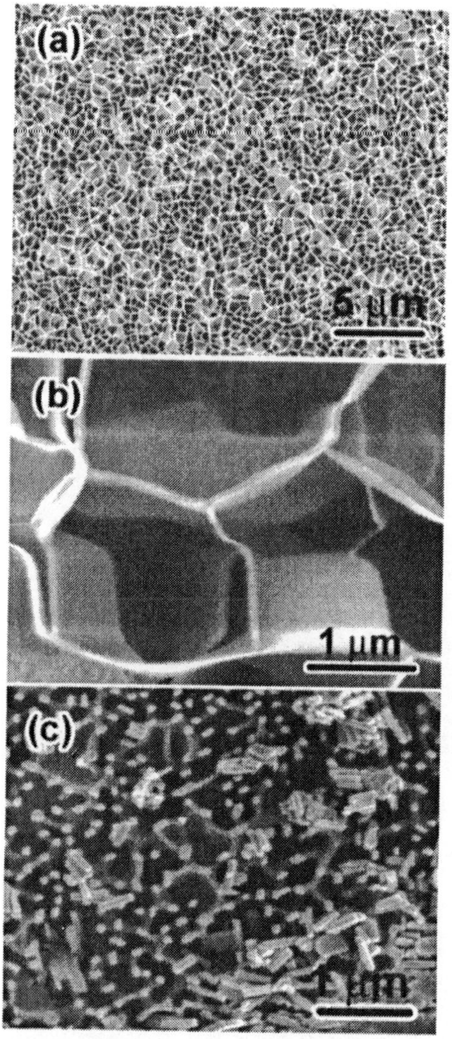

FIGURE 2.4 SEM images of the ZnO nanowalls synthesized by vapor transport and condensation method. (a) A medium magnification SEM image of the small size nanowalls. (b) A medium magnification SEM image of the large size nanowalls. (c) A high magnification SEM image of the large size nanowalls.

2–theta diffraction patterns of the nanowalls are shown in Figure 2.5(a). As a result of the good epitaxial relation between the *c*-plane of the ZnO nanowalls and the a-plane of the sapphire, only the ZnO (0002) and (0004) peaks can be seen. In the omega scan the peak splits with a full width at half maximum (FWHM) of 0.03° as displayed in Figure 2.5(b). This split could originate from the Si substrate itself. A similar split (with similar FWHM) in the bare Si substrate can be observed (Figure 2.5(c)). The phi scan of the (0112) peak of a ZnO nanowall on sapphire is shown

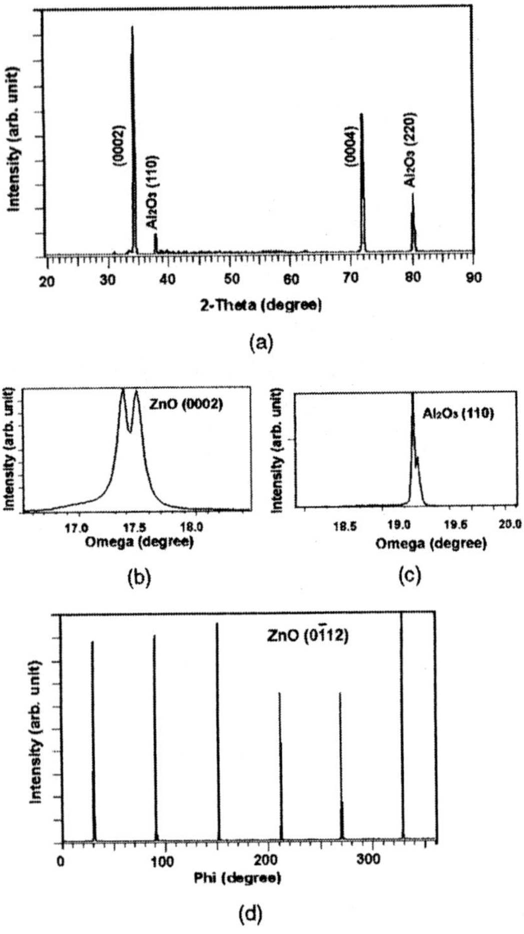

FIGURE 2.5 XRD spectra of the nanowalls structure. (a) Theta–2 theta scan. (b) and (c) Omega scan of the nanowalls and substrate, respectively. (d) Phi scan of the nanowalls.

in Figure 2.5(d). The six peaks, with equivalent distances of 60° demonstrate the in-plane epitaxial relation of the ZnO nanowalls with the a-plane single crystal sapphire substrate. The TEM images of nanowall flakes reveal that the atomic orientations are somewhat inconsistent with the presence of a significant number of dislocations. There is a low-magnification view of such a nanowall flake in Figure 2.6(a), and, in Figure 2.6(b), an associated electron diffraction pattern is shown. The appearance of extinction diffraction spots such as (0001), (0003), and so forth, is caused by double diffraction. From the SAD pattern, the nanowall flake is on the (1110) plane, and the direction from bottom to top of the plot is [0001], consistent with the XRD examination (Figure 2.5(a)). Dislocations on the surface of the flake can easily be identified by the parallel lines shown in Figure 2.6(c). These

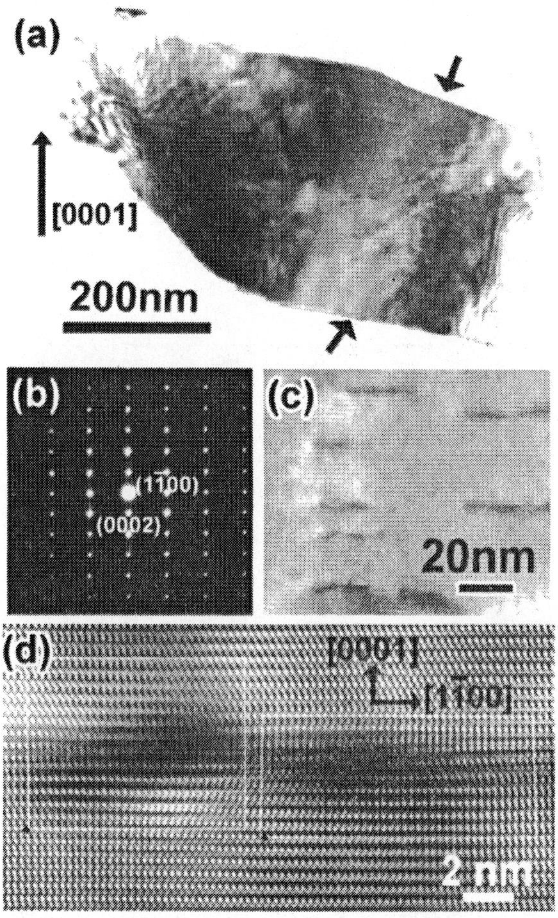

FIGURE 2.6 TEM micrographs of the structure of the nanowalls. (a) A low magnification TEM image of a nanowall flake. (b) An SAD pattern. (c) A high magnification phase contrast image showing the edge dislocation dipoles. (d) High-resolution TEM image of the dislocation dipole.

dislocations, with a length of about 20 nm, distribute periodically along the flake. The high-resolution image shown in Figure 2.6(d) indicates that each dislocation line is associated with two heavily strained areas. If one draws a Burgers circuit to enclose each heavily strained area, one can see that each strained area is associated with a perfect dislocation and a Burgers vector of either $1/3\langle 1\bar{1}\bar{1}0\rangle$ or $1/3\langle \bar{2}110\rangle$. The fact that each dislocation line is associated with two dislocations with opposite Burgers vectors suggests that these dislocations are actually dislocation dipoles. The Burgers vectors of these unit-perfect dislocations are parallel to the (0002) plane, typical mobile dislocations for a hexagonal structure. The reason for the existence of dislocations in the flakes is not yet clear.

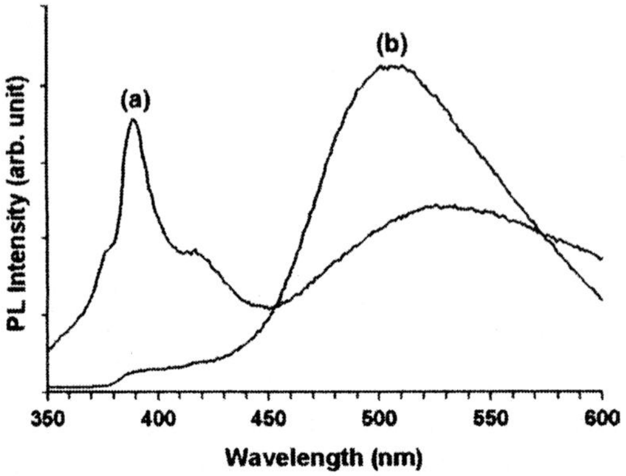

FIGURE 2.7 Photoluminescence spectra of the nanowalls. (a) Spectrum of white-gray nanowalls grown at high temperature and (b) spectrum of reddish nanowalls grown at low temperature.

Photoluminescence (PL) studies of ZnO nanowalls were conducted at room temperature using a dual scanning microplate spectrofluorometer (Molecular Devices Inc., SpectraMax Gemini XS) with an excitation wavelength of 325 nm. Emission scanning was conducted over the range between 340 and 600 nm. The PL spectra of ZnO nanowalls is shown in Figure 2.7. The white-gray nanowalls (curve a) grown at high temperature show significantly different PL spectra from those of the reddish nanowalls (curve b) grown at a lower temperature. Nanowalls grown at higher temperature exhibit a strong UV peak (380 nm), attributed to the band-edge emission of ZnO wurtzite, with a weak deep band green-yellow peak. Reddish samples grown at lower temperature have a weak UV peak with a large, broad, green-yellow peak. The green-yellow is attributed to defects (e.g., oxygen vacancies) that result in a radiative transition between shallow donors and deep acceptors.[67] It is widely accepted[68] that the green peak of ZnO is due to radiative recombination of photo-generated holes and electrons associated with oxygen vacancies. Annealing in an oxygen atmosphere at 600°C for 2 h suppresses the green-yellow emission completely, and vacuum annealing at 400°C suppresses both the UV and green-yellow emission by a small amount. The strong emission from the band edge suggests that the nanowalls grown at higher temperature have an excellent crystalline structure, consistent with the other reports on nanowires.[41(a)]

Because of the high aspect ratio in nanotubes and nanowires, a thin film of such nanostructures is found to be an excellent field emission electron source. Among other nanostructures, nanotube thin films have already been studied in great detail.[69] With better chemical stability and structural rigidity, ZnO nanowires are expected to provide a more stable and better field emission electron source than do nanotubes. For investigation of the field emission properties of ZnO nanowires, we used two separate methods for preparing the Au catalysts on a ⟨100⟩ substrate. One method was thermal evaporation of a 1 to 3 nm Au film, and the other was dispersion of 3 nm Au nanoparticles in toluene on the Si substrate. The production of Au nanoparticles

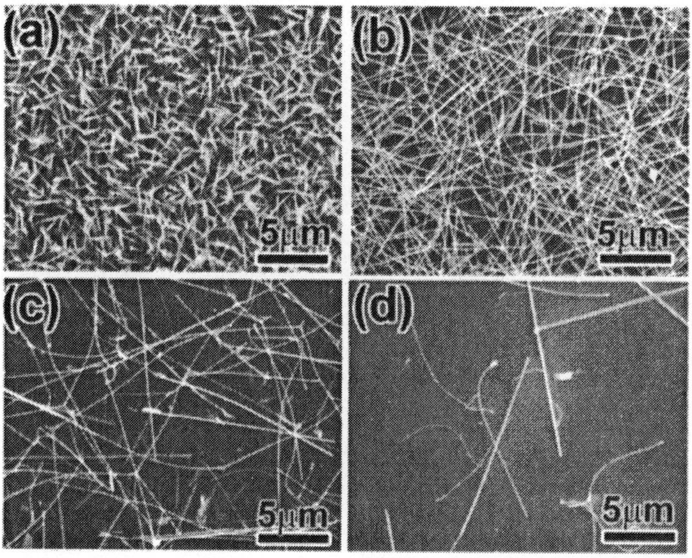

(e)

Sample ID	Areal density (/cm^2)
A	Continuous
B	Continuous
C	5.6×10^6
D	5.6×10^6
E	1.2×10^6
F	1.2×10^6
G	0.7×10^6
H	0.4×10^6

(f)

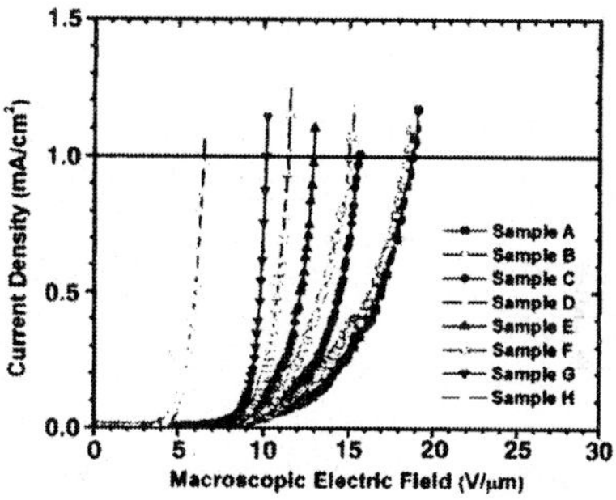

FIGURE 2.8 SEM micrographs of (a) sample B, with continuous Au film, (b) sample D, with density $5.6 \times 10^6/cm^2$, (c) sample F with density $1.2 \times 10^6/cm^2$, and (d) sample H, with density $0.4 \times 10^6/cm^2$. (e) Table showing different areal density of various substrates. (f) The measured current densities as a function of the macroscopic electric field for 8 samples. The measurements were carried out at a vacuum level of about 2×10^{-6} torr.

employed tetraoctylammonium bromide, as a transfer agent to move Au ions from an aqueous environment to toluene, and dodecylamine, as a capping molecule for the crystalline nanoparticles.[71] The suspension was diluted to different concentrations in toluene for different areal densities of Au nanoparticles on the Si surface. Eight samples with different Au catalyst densities, as described in Figure 2.8(e), were prepared in this study. Two of these samples were prepared using a continuous Au catalyst thin film, and the remaining six samples were prepared with four different concentrations of the Au catalyst nanoparticle suspension. A mixture of ZnO and

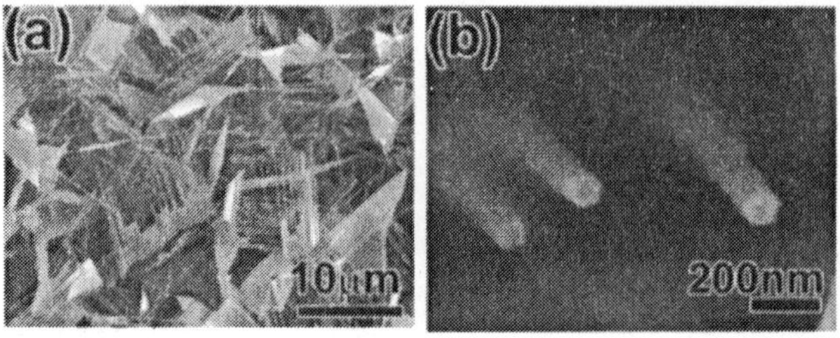

FIGURE 2.9 ZnO comb-like nanostructures. (a) Low magnification and (b) high magnification SEM images.

graphite was used as a source material. Some of our SEM images of these samples are shown in Figure 2.8. A ZnO nanowire thin film formed from the continuous Au film deposited on the silicon substrate is shown in Figure 2.8(a). Figures 2.8(b), 2.8(c), and 2.8(d) show the corresponding ZnO nanowire thin films generated from the Au nanoparticles with areal densities on the Si substrate of 5.6×10^6, 1.2×10^6, and $0.4 \times 10^6/cm^2$, respectively. The field emission current of all these samples was measured using a simple diode configuration.[71] A plot of the measured current densities as a function of the microscopic electric field is shown in Figure 2.8(f). The horizontal line corresponds to a current density of 1 mA/cm^2, and the values of electric field required to obtain this current density are 18.77, 18.50, 15.57, 14.96, 12.92, 11.43, 10.16, and 6.46 V/μm for samples A through H, respectively. The pressure was 2×10^{-6} torr during the measurement, much lower than is usually employed.[72] Therefore, it is expected that a much lower macroscopic electric field could have been achieved at a somewhat higher pressure if the same screening effect observed on carbon nanotube field emitters also affects the field emission from thin films of ZnO nanowires. Thin films with the lowest areal density of ZnO nanowires showed much stronger field emission characteristics, comparable to those of carbon nanotubes. The field emission characteristics of ZnO nanowire thin films were further improved on annealing in hydrogen.

Under experimental conditions similar to those in the nanowire synthesis, we found some nanocomb-like structures at a relatively higher temperature, as shown in Figure 2.9. This is consistent with other previous reports[51] of comb-like ZnO nanostructures that appear when pure Zn is used as the evaporation source. The large amount of such structures is seen in Figure 2.9(a) with a high-magnification image in Figure 2.9(b). Interestingly, in Figure 2.9(b), the nanostructure grows along the [1120] direction, which is different from the usual [0110] direction common in the hierarchical structure (discussed in Section 2.4.3) formed from binary sources. We explored some other types of nanostructures, where nanocrystals were attached to nanowires (see Figure 2.10). The small ZnO nanowires of tens of nanometers in diameter are decorated with nanocrystals (Figures 2.10(a) and 2.10(b)). Some of the nanocrystals are attached epitaxially to the nanowire. An unusual structure appears where a nanorod meets the nanobelt during growth and penetrates through it (Figure 2.10(c)).

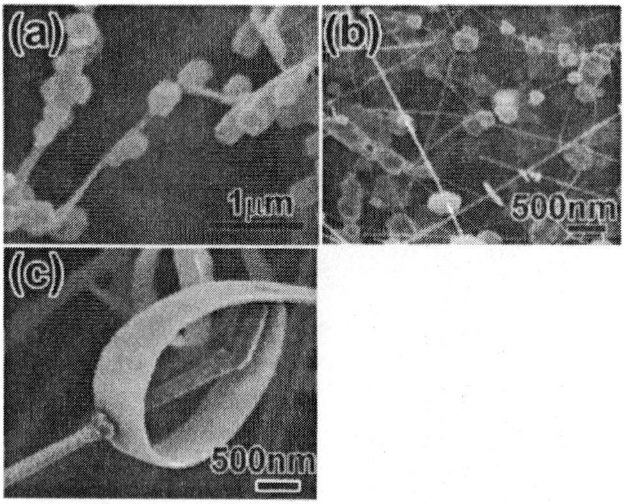

FIGURE 2.10 Figure 2.6 (a) and (b) ZnO nanocrystal decorated nanorods and nanowires. (c) A ZnO nanorod penetrated nanobelt.

2.4.2 The In_2O_3 Nanocrystal Chain and Nanowire Circuit

Transition from a top-down to a bottom-up approach in nanoelectronics through self assembly, as we have mentioned, requires a significant research effort and a profound understanding of the influences of external parameters on the growth of nanostructures. When air was introduced into the tube furnace, we found that In_2O_3 nanostructures grown by vapor transport and condensation are greatly influenced by the atmospheric pressure in the growth zone and by the indium or indium suboxide vapor pressure. By controlling the internal atmospheric pressure, it is possible to substantially change the morphology from a nanowire structure to a nanocrystal or a nanowire chain in circular form. We used a mixture of In_2O_3 and graphite placed near the sealed end of the quartz tube as the source material. Native SiO_2 coated Si (SiO_2/Si), on which a 1 to 3 nm Au thin film was deposited, formed on the substrate, which was kept close to the open end of the quartz tube and loosely sealed. Finally, the entire assembly was introduced into a quartz tube furnace, pumped, and heated to a temperature between 1000 and 1030°C, and held for 15 to 30 min. The pressure inside the tube during growth was controlled by a gauge valve. When the substrate temperature is between 875 and 950°C and the pressure is between 0.1 and 1.0 torr, In_2O_3 nanostructures grow on the SiO_2/Si substrate.

In_2O_3 nanowires can be grown in this setup at a pressure of 1.0 torr maintained from the beginning of heating. Scanning electron microscopy (SEM) images of the In_2O_3 nanowires on the SiO_2/Si substrate usually obtained from the lower temperature end of the substrate are shown in Figure 2.11(a). The In_2O_3 nanowires, with a typical body-centered cubic structure, have diameters ranging from 15 to

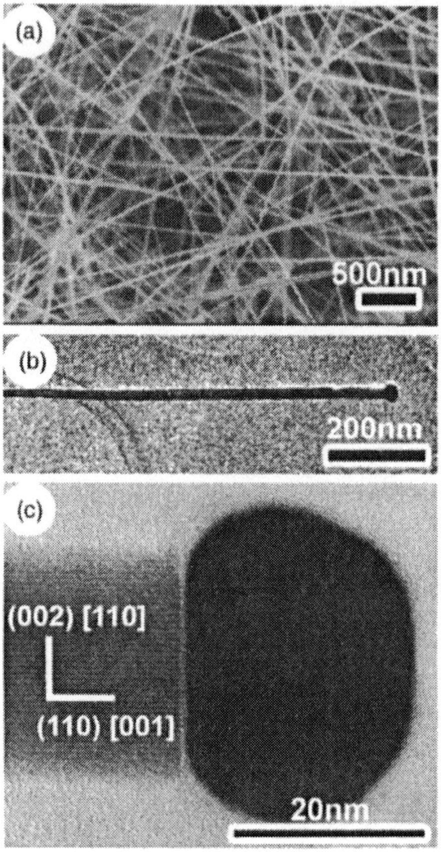

FIGURE 2.11 SEM and TEM microscopic images of the In_2O_3 nanowires. (a) Medium magnification SEM image of the nanowires. (b) TEM image of the nanowire. (c) HRTEM image showing the nanowire with an Au catalyst on the tip.

60 nm and lengths of tens of micrometers. A TEM image and a high-resolution transmission electron microgram (HRTEM) image of a nanowire having a [001] growth direction are seen in Figures 2.11(b) and 2.11(c). The Au particle that serves as a catalyst is clearly visible on the tip of the nanowire. We observed an increase of pressure from 0.05 torr to a peak of about 0.13 torr with continuous pumping in a tight system (no air inlet). This occurred because of the generation of indium vapor followed by a decline due to exhaustion of the In_2O_3 source. If air is introduced at the time when the Indium vapor pressure has already declined from a peak of 0.13 to about 0.06 torr, we obtained various complex self-assembled crystal chains and circuit networks (Figure 2.12). In this case we introduced an air pressure of 1.0 at this low indium vapor pressure (about 0.06 torr). The nanocrystal chains have diameters 20 to 100 nm and lengths of a few micrometers (Figure 2.12(a)).

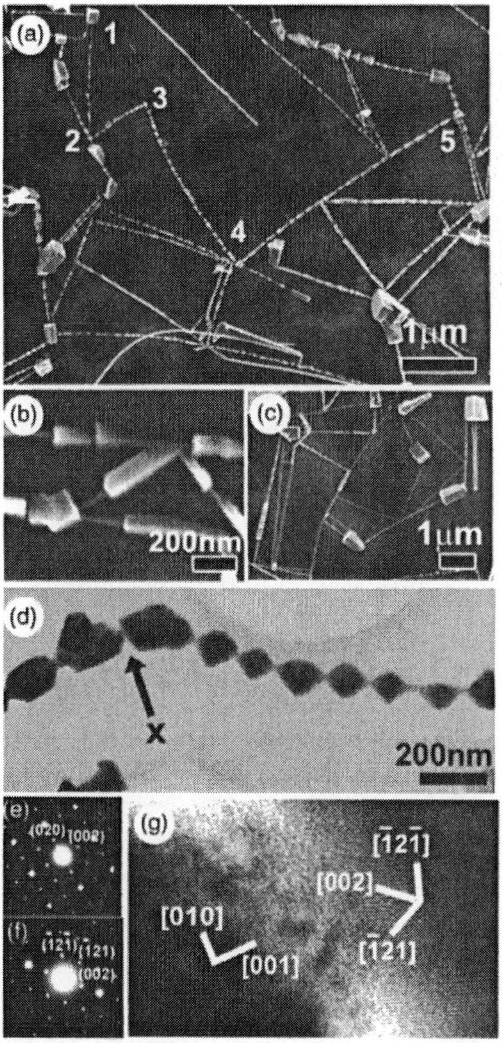

FIGURE 2.12 SEM and TEM microscopic images of the In_2O_3 nanowire and nanocrystal chain circuits. Big crystals are part of the circuit. (a) SEM image showing the nanocrystal chain circuits. (b) SEM image showing the circuit junctions. (c) SEM image showing the nanowire and nanocrystal circuits. (d) TEM bright field image of part of a nanocrystal chain. (e) SAD pattern corresponding to the nanocrystal on the left of point X and (f) on the right of point X. (g) HRTEM showing the domain boundary at point X.

The nanocrystal chains are connected together in a complex circuit network. The network junctions can be either nanocrystals or larger crystals. Based on our knowledge of crystal growth and our observation of the detailed morphologies, we find the reasonable growth direction of this particular nanocrystal chain must be 1-2-3-4-5, as shown in the figure. On the way from point 4 to point 5, two additional nanochains were bounced off of it. The large crystals are purely cubic and have a solid connection (see Figure 2.12(b)) with another crystal. Hence they are available to be used as leads for a nanoelectronic circuit. Figure 2.12(c) illustrates some networks where there are solid nanowires instead of crystal chains. The TEM image of part of a nanocrystal chain is visible in Figure 2.12(d). The chain is formed by the interconnection of an individual nanocrystal with a growth direction of [001] with the nanocrystals epitaxial to one another. The structures of these nanocrystals are perfect, and no amorphous layer is found on the surface. The chain can change geometric direction when it meets with an obstacle. It is found that the change of the chain's geometric direction, however, is caused by a change in crystal orientation, not by meeting an obstacle. The turning point, X, is indicated by an arrow in Figure 2.12(d). Electron diffraction patterns from the right and left of point X are shown in Figures 2.12(e) and 2.12(f), respectively. The chain crystal growth direction is always along the [001] direction. Figure 2.12(g) is an HRTEM image taken from point X. The domain boundary and change of lattice orientation can be clearly seen between two nanocrystals.

The nanowire circuits where few large crystals exist after a 0.3 torr pressure of air was introduced into the low indium vapor pressure of 0.06 torr are seen in a SEM image. Figures 2.13(a) and 2.13(b) show the SEM image of such circuits. These nanowires, having a diameter of 20 to 100 nm and a length of tens of micrometers, grow in a zigzag fashion. After running into another nanowire, the initial nanowire changed direction and continued growth. Figure 2.13(c) shows an interesting hexagonally shaped nanowire circuit with other nanowires growing inside

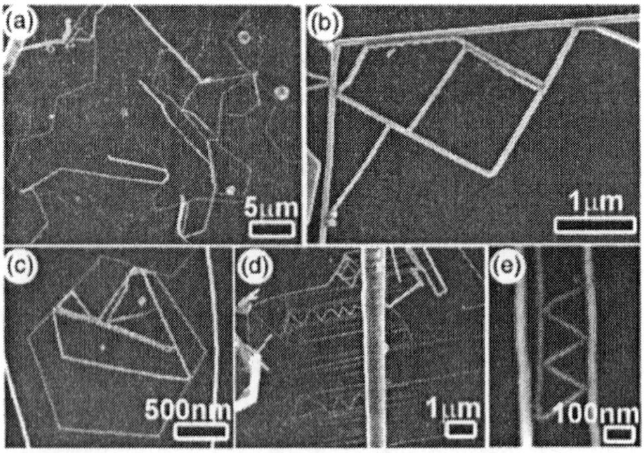

FIGURE 2.13 SEM images of the In_2O_3 nanowire circuits grown at 0.3 torr pressure. (a) Low magnification SEM image showing the circuit. (b) Medium SEM image showing the junctions. (c) A hexagonally shaped circuit. (d) SEM image showing the parallel nanowires from a big nanofiber. Some nanowires with zigzag growth direction can be seen. (e) High magnification SEM image showing the zigzag growth direction of a nanowire confined between the two parallel nanowires.

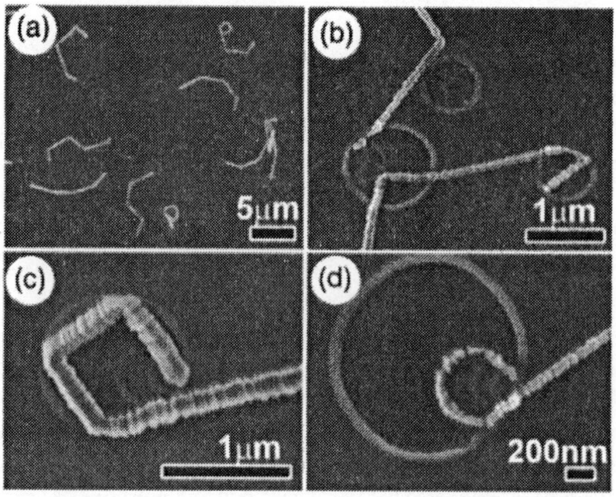

FIGURE 2.14 SEM images of the wavy In_2O_3 nanowires grown without Au catalyst. (a) Low magnification SEM image. All the nanowires are started from the edge of the holes. (b) Wavy nanowire crossed between two holes. (c) A wavy nanowire changed direction three times in a hole. (d) A wavy nanowire ring formed along the edge of a hole.

the hexagon. Many small nanowires are seen growing along the side of an In_2O_3 large fiber (Figure 2.13(d)). These nanowires are parallel to each other at the start and change growth direction after a certain length. Small particles are visible at the tip of many nanowires. A small zigzag nanowire confined between two nanowires is illustrated in Figure 2.13(e).

Under the same experimental conditions, but without the Au catalyst layer, many In_2O_3 microcrystals were formed on the substrates. A few interesting In_2O_3 nanocrystal chains were also found on some areas with an In/In_2O_3 film deposited on the SiO_2/Si substrate. Figures 2.14(a) and 2.14(b) show some wavy nanocrystal chains on the In/In_2O_3 film. All the nanocrystal chains are nucleated from the edge of the holes in the film. One In_2O_3 nanocrystal chain is seen growing directly out of the hole (Figure 2.14(b)). Another chain changed direction after meeting with the hole edge and then grew out of the hole and ran into another hole. It then changed direction after meeting with the edge of the smaller hole and finally grew out of the hole. One nanocrystal chain changed growth direction three times at the hole edge before growing out of the hole (Figure 2.14(c)). A chain is seen forming into a circle confined to the edge of a smaller hole before growing out (Figure 2.14(d)).

The nanostructures grown using the vapor-liquid-solid (VLS) mechanism[73] usually have either a constant diameter or a periodically changing diameter[74] along the growth direction, whereas here we see nanostructures composed of many nanocrystals with different sizes. Therefore, it is likely that the nanocrystal chains shown in Figure 2.12 are created through a vapor-solid (VS) mechanism. The higher atmospheric pressure over the indium oxide leads to these crystal chains, which is also consistent with the VLS mechanism. It is important to emphasize that a gold catalyst is necessary for the formation of these circuits. Formation of nanowires instead of nanochains is probably a consequence of the temperature gradient in the furnace. It is fascinating to find that the force behind the nanowire/nanocrystal chain formation is so strong that, in most cases, the nanowire continues to grow albeit with a direction change after meeting obstacles. It is generally expected in consideration of the VS mechanism that the deposited molecules on the side of the nanowire will move to the growth front and contribute to the one-dimensional nanostructure growth. Because of the limited diffusion length of the molecules deposited on the side of the nanocrystals, not every nanocrystal will exhibit direct materials exchange. These nanocircuits have immense potential as building blocks for nanoelectronic devices.

2.4.3 Zn-In-O Hierarchical Nanostructures

Hierarchical nanostructures with 6-, 4-, and 2-fold symmetries can be grown using the experimental set-up shown in Fig. 2.1 as described in Section 2.3. ZnO nanorods can be grown on an In_2O_3 nanowire core with various symmetries and orientations by vapor transport and condensation. ZnO and In_2O_3 have high melting points, (i.e., 1,975°C and 1,931°C respectively). We therefore mixed in graphite powder as a reducing agent, which brings down the reaction temperature to below 1,000°C. A mixture of ZnO, In_2O_3, and graphite powders can be used as a coevaporation source to be placed at the sealed end of a single-ended sealed quartz tube (shown in Fig. 2.1). The nanostructures can grow on many different collectors, including graphite foil, single crystal silicon, and $LaAlO_3$. In our experiment, mixed powders were heated between 950 and 1000°C for 30 min. in the ceramic tube under a pressure between approximately 0.5 and 2.5 torr. The vapor was transported after evaporation to the open end of the inner quartz tube, where it was oxidized in the presence of air and condensed on the graphite foil. (Other substrates, such as silicon and $LaAlO_3$, also have been used.) The oxidation rate was controlled by the amount of air leaked into the tube furnace through a needle valve connected to the open end of the ceramic tube. A region of relatively sharp temperature gradient was chosen so that the condensation temperature could be controlled between 820 and 870°C. The flux ratio of In alloying agent to Zn in the vapor phase was controlled by mixing the desired amount of In_2O_3 with ZnO powder in the original source. It has been determined that the In:Zn flux ratio is the key to control of the preferred symmetry in the hierarchical nanostructure. For example, a higher In:Zn flux ratio tends to grow six 4-fold symmetric nanostructures, whereas medium and lower In:Zn flux ratios grow 2-fold symmetric or nano ribbon structures.

The SEM images of the ZnO nanostructures with In_2O_3 as the alloying agent at low and medium magnifications are illustrated in Fig. 2.15. Large quantity hierarchical ZnO nanostructures are shown in Fig. 2.15(a). Under medium magnification, Fig. 2.15(b) clearly shows all the major 6-, 4-, and 2-fold structural symmetries. The length of the major In_2O_3 core is usually on the order of tens of microns, and the diameter is between 50 and 500 nm. The secondary ZnO nanorods on the core surfaces are about 0.2 μm in length and possess a diameter between 20 and 200 nm. They are oriented either normal to or at an angle to the core surface. This path branches out to a few more subsymmetries associated with the

FIGURE 2.15 SEM images of the ZnO nanostructures synthesized by vapor transport and condensation technique. (a) Low magnification SEM image of the ZnO nanostructures to show the abundance. (b) Medium magnification SEM image of ZnO nanostructures to show the various structural symmetries. Three major basic symmetries of 6-, 4-, and 2-fold were clearly seen. (c) XRD observation of the ZnO hierarchical nanostructures.

major symmetries. Three subsymmetries, namely 6S-, 6M-, and 6M*, have been observed for 6-fold major symmetries. Similar subsymmetries can also be found with 4-fold (4S-, 4S*1-, 4S*2-, 4M-, and 4M*) and 2-fold (2S-, 2S*, and 2M-) symmetries respectively. Powder x-ray diffraction (XRD) reveals that the sample is a mixture of hexagonal wurtzite ZnO and cubic In_2O_3, as shown in Figure 2.15(c). The lattice constants for ZnO have been calculated to be a = 3.249 Å and c = 5.206 Å, and for cubic In_2O_3, a = 10.118 Å. In both cases these values are in agreement with the values for their bulk forms.[76] Detailed SEM and TEM observations of 6-fold nanostructures associated with various subsymmetries are found in Figure 2.16. Figures 2.16(a) and 2.16(c) show, at medium and high magnification, the majority of 6S-fold nanostructures possessing smaller core diameters. As the nomenclature suggests, they have a single row of secondary ZnO nanorods on the core surface. If the available surface area is large enough, they can grow in multiple rows with 6M-fold symmetry as shown in Figures 2.16(b) and 2.16(d), in medium and high magnification, respectively. It is reasonable to conclude that the hexagonal symmetric surface of the major In_2O_3 nanowire core is responsible for the exotic secondary growth of ZnO nanorods. The hexagonal surface of the core is illustrated in Figures 2.16(e) and 2.16(f). Clearly it can be assumed, depending on the available surface area on the each segment of the hexagonal core, that single or multiple rows may grow. The formation of a 6M*-fold structure, where the rows of secondary ZnO nanorods are at an angle to the In_2O_3 major core, is visible in Figure 2.16(g). An energy dispersive x-ray spectroscopy (EDS) study of the sample demonstrates that the major core axis is pure In_2O_3, whereas secondary nanorods are pure ZnO. This is further confirmed by the clearly evident contrast in the TEM image in Figure 2.16(i). Orientation relationships between major core axes (In_2O_3) and secondary nanorods (ZnO) can be studied by selected area diffraction patterns (SAD). In each diffraction pattern, two separate sets of features are prominent, where one is from the core and the other, is from the secondary rods. The diffraction pattern in Figure 2.16(j) has been indexed using the [110] zone axis of In_2O_3 and the $[11\bar{2}0]$ zone axis of ZnO. Hence the crystallographic relationship comes out to be $[110]_{In_2O_3}$, $[11\bar{2}0]_{ZnO}$ for the 6-fold symmetry. When In_2O_3 is along the [110] direction, the core nanowire is enclosed by $\pm[1\bar{1}2]$, $\pm[1\bar{1}\bar{2}]$, and $\pm[1\bar{1}0]$ facets. The angle between each of these adjacent facets is very close to 60°, so a quasi 6-fold symmetry is often observed when an In_2O_3 nanowire grows along the [110] direction. In addition to the [001] and [110] directions, In_2O_3 nanowires also grow along the [111] direction, as shown in Figures 2.16(e) and 2.16(f) (and illustrated in Figure 2.16(h)), where hexagon end planes are clearly observable at the end of major cores and on secondary ZnO. A detailed conclusion on the orientation relationship of 6-fold symmetry in contrast to 4-fold symmetry is given later.

The various aspects of 4-fold symmetric nanostructures are found in Figure 2.17. The majority formation is in 4-fold symmetry, as seen in Figure 2.17(a). As already discussed, there are at least five possible varieties of 4-fold structures: 4S, 4S*1-, 4S*2-, 4M-, and 4M*. A closer look at 4S symmetry, where a single row of secondary nanorods are normal to the core, is given in Figure 2.17(b), whereas in

* Definitions of the symbols are found in Reference 75.

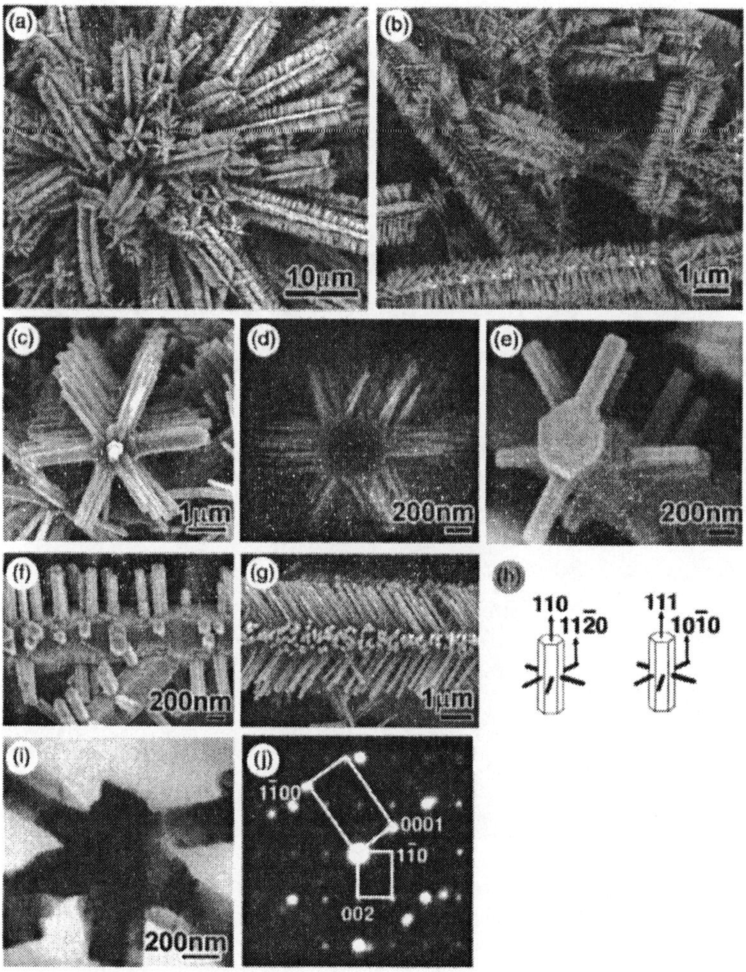

FIGURE 2.16 SEM and TEM images, selected area diffraction patterns, and the schematic growth models of the 6-fold ZnO nanostructures. (a) SEM image showing the abundance of the 6S-fold symmetry. (b) SEM image showing the 6M-fold symmetry. (c) High magnification SEM image of the 6S-fold symmetry. (d) High magnification SEM image of the 6M-fold symmetry. (e) Head-on view of a 6S-fold symmetry to show the hexagonal nature of the major core nanowire. (f) Side view of the structure in (e) to show the hexagonal nature of all the secondary ZnO nanorods and their same growth orientations with the major In_2O_3 core nanowire. (g) 6M*-fold symmetry, where the nanorods are not perpendicular to the major core. (h) Schematic diagram of orientation relationships between the major In_2O_3 core nanowire and the secondary ZnO nanorods with the core along [110] direction (h left) and along [111] direction (h) right). (i) Cross-sectional bright-field TEM image of 6-fold symmetry showing the six facets of the central core. (j) Selected-area electron diffraction pattern of (i) corresponding to the major In_2O_3 core and the secondary ZnO nanorods.

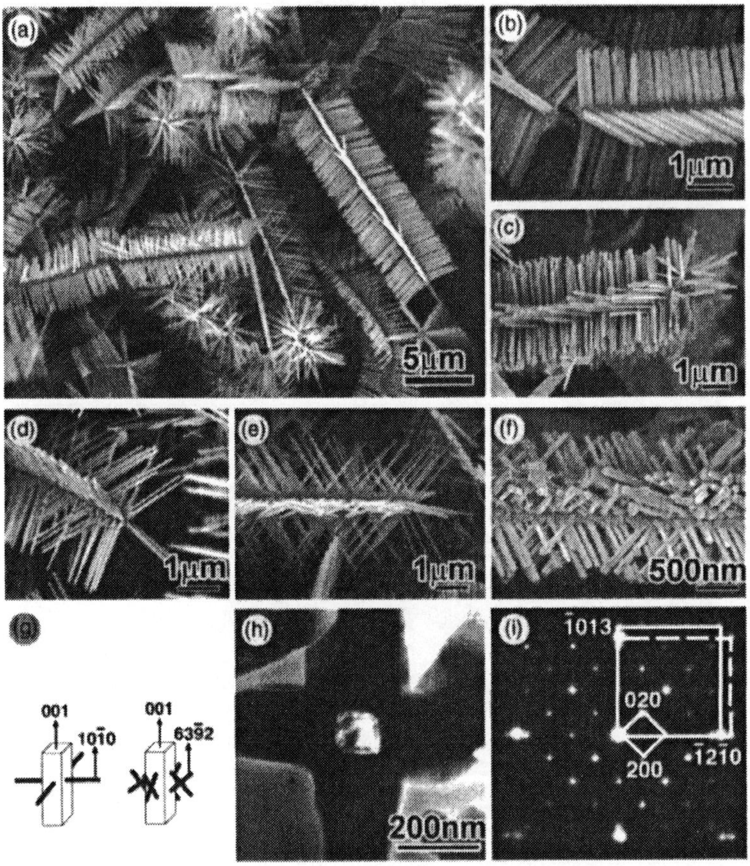

FIGURE 2.17 SEM and TEM images, selected area diffraction patterns, and the schematic growth models of the 4-fold ZnO nanostructures. (a) Medium magnification SEM image showing the abundance of the 4-fold nanostructures. Scale bar, 5 μm. (b) High magnification SEM image showing the 4S-fold symmetry. (c) High magnification SEM image showing the 4M-fold symmetry. (d) High magnification SEM image of the $4S^{*1}$-fold symmetry. (e) High magnification SEM image of the $4S^{*2}$-fold symmetry. (f) High magnification SEM image of the $4M^{*}$-fold symmetry. (g) Schematic model of the 4S- and $4S^{*}$-fold symmetry. (h) Cross-sectional bright-field TEM image of 4-fold symmetry to show the four facets of the central core. (i) Selected-area electron diffraction pattern of (h) corresponding to the major In_2O_3 core and the secondary ZnO nanorods. The diffraction consists of two sets of patterns; the small rectangle corresponds to [001] zone axis of In_2O_3 while the large solid one corresponds to [63$\bar{9}$2] zone axis of ZnO. The dashed rectangle is from another arm perpendicular to the solid rectangle.

Figure 2.17(c), multiple rows of perpendicular nanorods (4M) can be seen for the same reason discussed for its 6-fold counterpart. Similar to what was seen in the 6-fold case, in 4-fold nanostructures, secondary ZnO nanorods are not always perpendicular to the major In_2O_3 core nanowire but rather grow at discrete angles to the In_2O_3 core nanowire. In Figure 2.17(e), a single row of secondary ZnO nanorods grow at the same angle in all four directions. In Figure 2.17(d), the growth is seen to exist only in the two opposing directions (parallel to the page). The other two opposing directions are perpendicular to the major In_2O_3 core nanowire (into and out of the page). We define these two nanostructures as 4S*1- and 4S*2-fold, respectively. When the major In_2O_3 core nanowire is large enough, we can even see multiple rows of ZnO nanorods growing at an angle in all four directions (4M*-fold symmetry) as shown in Figure 2.17(f). Figures 2.17(h) and 2.17(i) show TEM bright field images and selected area diffraction patterns, which reveal, in contrast to 6-fold symmetry (with hexagonal core), that the major core is of cubic In_2O_3 nanowires, and the secondary rods are hexagonal ZnO nanorods for the 4-fold case. In a bright field TEM image, Figure 2.17(f) shows four facets of the central 4-fold symmetric In_2O_3 core. The diffraction patterns in Figure 2.17(g) consist of two sets. The small rectangle corresponds to [001] zone axis of In_2O_3, whereas the large solid one corresponds to the $[63\bar{9}2]$ zone axis of ZnO. The dashed rectangle is from a different arm perpendicular to the solid rectangle. Figure 2.17(g) is the schematic model of 4S- and 4S*-fold growth. The orientation relationship in the basic 4-fold symmetry is as follows: $[001]_{In_2O_3}//[0\bar{1}10]_{ZnO}$, $[001]_{In_2O_3}//[\bar{1}\bar{1}20]_{ZnO}$. When an In_2O_3 nanowire is along the [001] direction, the core nanowire is enclosed by ± (100) and ± (010) facets. In this case, all ZnO nanorods in the four arms grow perpendicular to the core nanowire because ZnO nanorods grow along the [0001] direction. They can be expressed symbolically as $[001]_{In_2O_3} \perp [0001]_{ZnO}$. In the case of the tilted secondary nanorod with 4-fold symmetry, the angle between a ZnO nanorod and a core nanowire (around 60°) is equal to the angle between $[63\bar{9}2]_{ZnO}$, and $[0001]_{ZnO}$. Because there is no difference between + and − $[63\bar{9}2]_{ZnO}$ for the growth of a nanorod on a [001] core, In_2O_3 nanowire nanorods can grow at an angle of either 60 or 120°. This results in all the variations of tilted growth. The heteroepitaxial nature of ZnO nanorods from In_2O_3 cores provides many possible crystal orientation relations among the cores and nanorods. This is the source of so many different ZnO nanorod orientations with respect to the core. Thus the symmetry of these hierarchical nanostructures is dependent on the crystallographic orientation of the In_2O_3 core nanowires. The orientation of an In_2O_3 nanowire along the [110] or [111] direction creates all the 6-fold symmetries, whereas along the [001] direction, all the 4-fold symmetries are produced. The orientation relationships demonstrated previously in Figures 2.16(h) and 2.17(g) are listed in Table 2.1 and can be understood through the theory of coincidence site lattices. For example, the In_2O_3 a-plane has 4-fold symmetry with a = 10.18 Å. The ZnO c-plane has 6-fold symmetry with a = 3.24 Å. This results in a lattice mismatch of about 3.7% (a factor of 3 for the In_2O_3 a-axis to the ZnO a-axis), a reasonable value for epitaxial growth.

In addition to 4-fold and 6-fold nanostructures, 2-fold symmetry has also been observed under similar experimental conditions. SEM images of 2-fold symmetric nanostructures appear in the In/Zn system. It was mentioned earlier that a medium In : Zn flux ratio in the source material enhances the growth of the 2-fold symmetric structures. Typical 2S-fold symmetric structures, where single perpendicular row nanorods grow on the two sides of the core, are shown in Figures 2.18(a) and 2.18(b). As previously stated, multiple rows may also be grown depending upon the

TABLE 2.1
Observed Orientation Relationships between the Major In_2O_3 Core Nanowire and the Secondary ZnO Nanorods

Core Axis	Orientation Relationship
$[110]_{In_2O_3}$(Figure 2.16(h))	$[110]_{In_2O_3}//[11\bar{2}0]_{ZnO}$ $[222]_{In_2O_3}//[0001]_{ZnO}$

$[111]_{In_2O_3}$(Figure 2.16(h))	$[111]_{In_2O_3}//[10\bar{1}0]_{ZnO}$
	$[11\bar{2}]_{In_2O_3}//[1\bar{2}10]_{ZnO}$
	$[1\bar{1}0]_{In_2O_3}\perp[0001]_{ZnO}$

$[001]_{In_2O_3}$(Figure 2.17(g))	$[001]_{In_2O_3}//[10\bar{1}0]_{ZnO}$
	$[100]_{In_2O_3}//[\bar{1}2\bar{1}0]_{ZnO}$
	$[001]_{In_2O_3}\perp[0001]_{ZnO}$

$[001]_{In_2O_3}$(Figure 2.17(g))	$[001]_{In_2O_3}//[63\bar{9}2]_{ZnO}$
	$[110]_{In_2O_3}//[\bar{1}2\bar{1}0]_{ZnO}$
	$[\bar{1}10]_{In_2O_3}//[\bar{1}013]_{ZnO}$

availability of sufficient surface area on the core. Figures 2.18(c) and 2.18(d) illustrate typical 2M-fold symmetric structures. 2S*-fold structures are also possible, where secondary nanorods make an angle with the core, as demonstrated in Figures 2.18(e) and 2.18(f). If the In : Zn flux ratio is further reduced, wool-like nanoribbons, with lengths ranging from 10 to 20 μm and widths from 50 to 200 nm, have been observed. A HRTEM study [18(b)] reveals that these In-alloyed ZnO nanoribbons consist of a central nanowire core with two adjacent nanoribbons flanking it, in contrast to the previously observed [15(c)] pure ZnO nanoribbon specimen. The TEM image of a single ribbon, where the central core and the two adjacent ribbons can easily be distinguished, is seen in Figure 2.18(h). With increasing In : Zn flux ratio, additional growth of nanorods perpendicular to the plane of the ribbons (in the [0001] direction) can be found. As in Figure 2.18(i), the SEM image indicates that these nanowires grow only from the central core of the nanoribbons and then self-align to form a straight row of nanorods. An EDS study indicates that the nanorods and nanoribbons grew epitaxially on the nanowire core.

We have already suggested that the variation of In : Zn flux ratio is crucial in controlling the morphology of the Zn-In-O hierarchical nanostructures. By reducing the In_2O_3 content in the source, we found a few more exotic nanostructures in the Zn-In-O system, including nanobridges, nanonails, and nanopins. The experimental conditions were the same except that we set the atmospheric pressure in the furnace to 2.0 torr for nanobridges. For nanonails it was held between 0.5 and 1.0 torr. The temperature of the furnace was 1000°C, and the condensation temperature was approximately between 950 and 970°C. A summary of the SEM and TEM observations of nanobridges defined as geometrical analogs to ancient stone bridge structures is shown in Figure 2.19. A typical nanobridge can be tens of microns long and up to a few microns in width and height (Figure 2.19(a)). The rows of nanorods grow at the edge of both sides of the nanobelts with varying density and are usually directed perpendicular to the belt surface (Figure 2.19(b)). It is even possible to

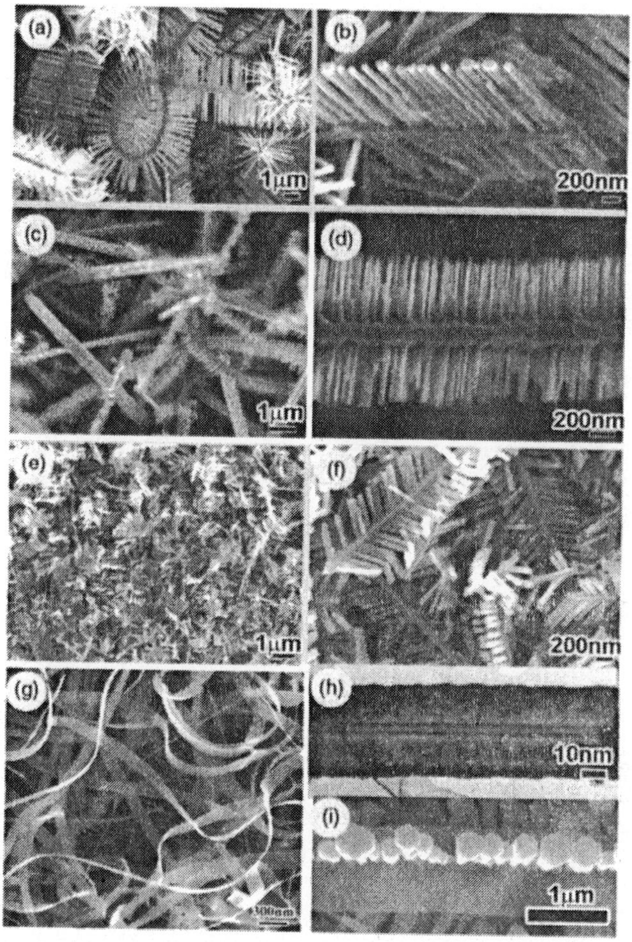

FIGURE 2.18 SEM and TEM images of the 2-fold nanostructures. (a) Medium SEM image to show the abundance of the 2S-fold symmetry. (b) High magnification of 2S-fold symmetry. (c) Low magnification SEM image to show the abundance of the 2M-fold symmetry. (d) High magnification SEM image of 2M-fold symmetry. (e) Low magnification SEM image to show the abundance of the 2S*-fold symmetry. (f) High magnification SEM image of 2S*-fold symmetry. (h) SEM image of In-alloyed ZnO nanoribbons grown at low In : Zn ratio (i). TEM image of a nanoribbon with showing dark contrast between the core and the side ribbons. (j) SEM image of a nanowire core with both nanoribbons and nanorods, where nanorods are well aligned and normal to the growth of side nanoribbons.

grow multiple rows of nanorods. The nanorods are 50 to 200 nm wide and 2 μm long. Figure 2.19(c) is a side view of the TEM image of a nanobridge, with the selected area diffraction pattern (SAD) illustrated in the inset. The nanobelt has a diffraction pattern similar to that of the nanorods except that its high-index diffraction spots are split. It is well established that the individual hexagonal nanorod grows

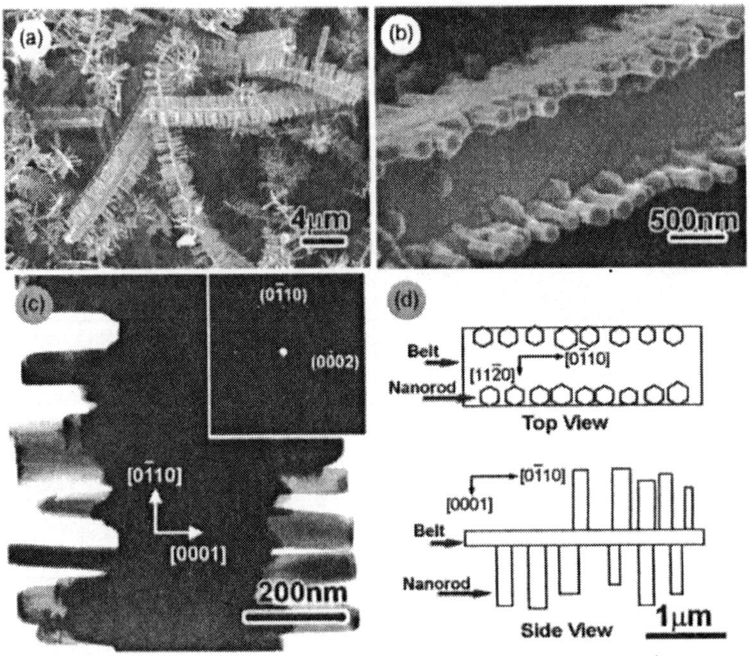

FIGURE 2.19 SEM images of the ZnO nanobridges synthesized by vapor transport and condensation method. (a) Low magnification image showing the abundance of the nanobridges. (b) Medium magnification image of top view of a nanobridge. (c) TEM image of the side view of a nanobridge. The insert is the electron diffraction pattern of a nanorod, with zone axis of [$\bar{2}$110] direction. Scale bar = 200 nm. (d) Schematic drawing of top and side view of part of a nanobridge. The page planes for the top view and side view are (0001) and (11$\bar{2}$0), respectively.

epitaxially along the [0001] direction on the (0001) plane and/or along the [000$\bar{1}$] direction on the (000$\bar{1}$) plane of the belt. The nanorod rows are parallel to the [0$\bar{1}$10] direction of the belt. The zone axis of the diffraction pattern in the insert is [$\bar{2}$110], as illustrated in Figure 2.19(d). Explanations of the various symmetries and of the versatility of the nanobridge growth we have found thus far are given in Figure 2.20. In Figure 2.20(a) the belt is shown to form a central ring, with the nanorods growing along the belt edges exactly perpendicular to the belt surface. This behavior is due to the strong epitaxial relation between the nanobelt and nanorods. Figure 2.20(b) shows a rare situation where two belts join together perpendicular to each other, with the associated nanorods growing according to their original direction perpendicular to their own belt. The availability of an edge surface on the belt determines the nanorod's growth habit. For example, in Figure 2.20(c) nanorods grew on all four sides of the belt. In total, eight rows of nanorods can be observed with 4-fold symmetry. Figure 2.20(d) shows another variation. The density of the nanorods (perpendicular to the plot) on the belts varies considerably. One edge has high density, whereas the other edge has low density. In addition, there is

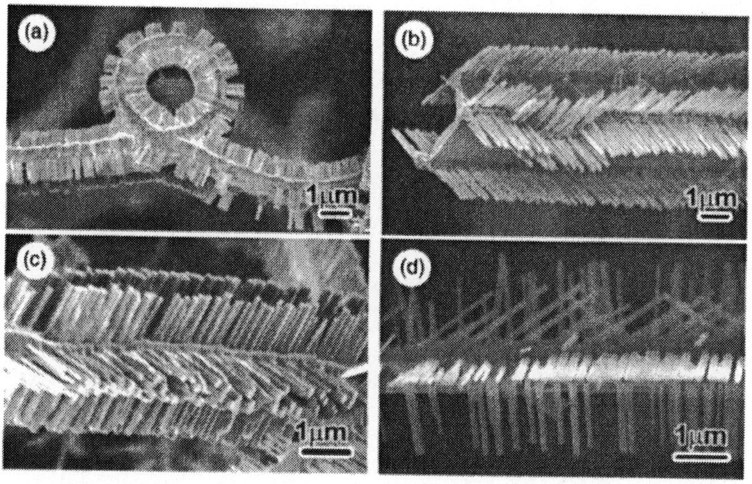

FIGURE 2.20 SEM images of ZnO nanobridge variations. (a) Roller coaster-like and (b) joined twin nanobridges; (c) and (d) are both combinations of a nanobridge and 4-fold symmetry.

another row of nanorods growing parallel to the belt surface at an angle of about 36° to the belt growth direction.

We found additional interesting structures, such as nanonails and nanopins, by slightly varying the experimental conditions. The SEM and TEM studies of such nanonails are described in Figure 2.21. A low-magnification view of the aligned nanonails, together with some nanonail flowers, can be found in Figure 2.21(a). The inserted XRD pattern shows the strong (002) peak of the wurtzite ZnO structure and much weaker (100) and (101) peaks due to the imperfect vertical growth of the nanonails. Figure 2.21(b) shows a side view of the nanoflowers to illustrate their hemispheric shape. Figures 2.21(c) and 2.21(d) are medium- and high-magnification views of the aligned nanonails. These nanonails are several microns in length, with a shaft diameter of 150 to 200 nm. They form roughly perpendicular to the surface. The hexagonal cap is approximately 1 μm in diameter and 50 to 100 nm thick. The change of diameter from shaft to cap is abrupt. The density of nanonails is high. They form quasi-perpendicular to the substrate, a habit that is probably a consequence of the epitaxial growth of ZnO along the c-axis of the graphite foil substrate. A TEM side view displays an image of a nanonail that exhibits dark contrast because it is too thick to be transparent to the electron beam (Figure 2.21(e)). The inserted SAD pattern confirms that the nanonail grows along the c-axis. No splitting of the diffraction spots has been found here. A high-resolution TEM (HRTEM) image taken from the cap of the nanonail demonstrates the perfect lattice structure of the nanonail and also confirms that the nanonail possesses wurtzite structure (Figure 2.21(f)). The EDS spectrum shows that the nanonails are composed only of zinc and oxygen, which means that no indium contamination has been found. Similar nanonails of a much larger size were made previously[66] by vaporizing a mixture of only ZnO and graphite powder in a similar

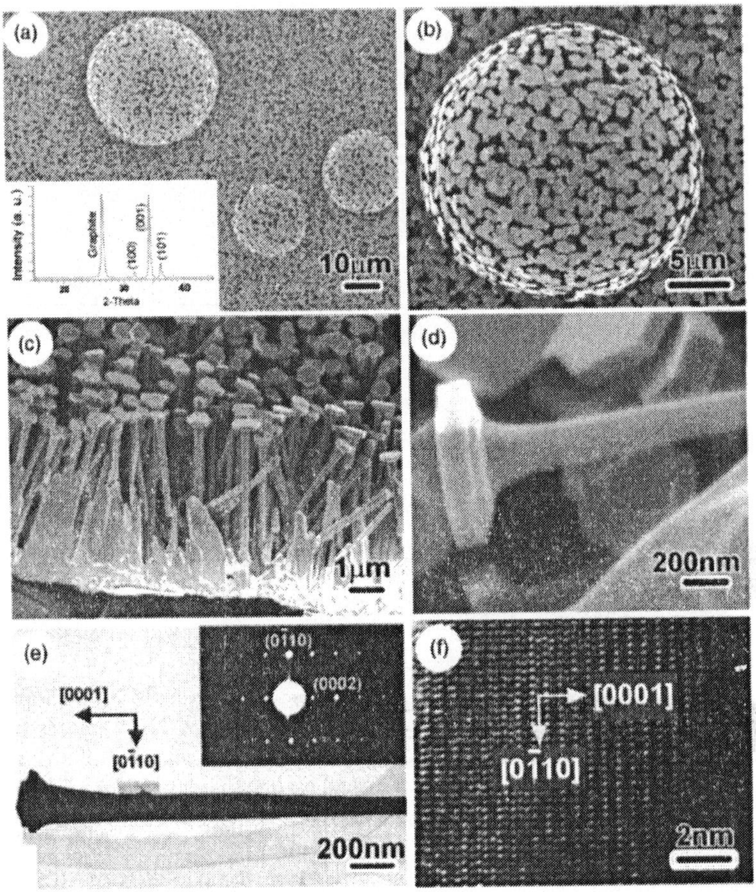

FIGURE 2.21 (a) Low-magnification SEM image of the ZnO nanonails synthesized by vapor transport and condensation method showing the aligned growth of nanonails and the nanonail flowers. Insert is the x-ray diffraction pattern. (b) Medium-magnification view of nanonail flower. Scale bar = 5 μm. (c) Side view showing the vertical growth of nanonails. (d) High-magnification side view SEM image of a nanonail. (e) TEM image of the nanonail. The insert is the electron diffraction pattern, with zone axis of [$\bar{2}$110]. (f) HRTEM image taken from the cap of the nanonail along [$\bar{2}$110] direction.

experimental set-up. In a different region we found very small (100 nm cap diameter with 1 μm length) nanonails (or nanopins). The SEM images of such nanonails in low and high magnification are illustrated in Figures 2.22(a) and 2.22(b). The diameter decreases to 40 nm at the end. Figures 2.22(c) and 2.22(d) illustrate SEM images of another type of nanonail where the size reduction from bottom to top is much larger. In this case the diameter is continuously reduced from 800 nm at the top to 20 nm at the bottom with a perfect hexagonal cap at the top. In yet another set of nanonails, we find that the caps are not perfectly

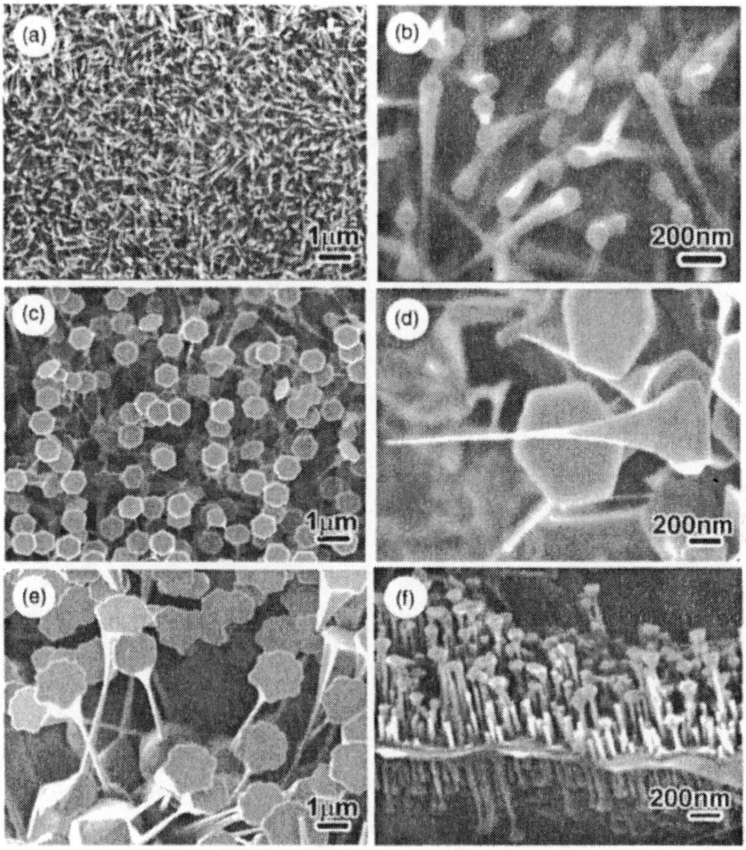

FIGURE 2.22 SEM images of several different ZnO nanonail structures. (a) Low and (b) medium magnification images of small nanonails. (c) Medium and (d) high magnification images of thin shaft nanonails. (e) Non-hexagon shape nanonails on ZnO rod bases. (f) Nanonails on ZnO sheet.

hexagonal, although they have hexagonal symmetry as indicated by the six facets (Figure 2.22(e)). These nanowires generally grow on ZnO nanorod bases (Figure 2.22(f)). There is also evidence that nanonails can grow on a thin ZnO sheet (Figure 2.22(g)).

The growth process of these Zn-In-O hierarchical nanostructures can be extremely complex because of their binary transport mechanism and complicated crystallographic orientations. The reduction of oxides, vapor transport, vapor oxidation, and condensation at the low-temperature collector are also involved. Because we have not used any catalyst in this process, the growth mechanism cannot be described as being of VLS origin, but rather it may follow a vapor-solid route.[77] In the ZnO-In_2O_3-graphite system, first both ZnO and In_2O_3 are probably reduced to suboxides (ZnO_x or InO_x [$x < 1$]), which have much lower melting points.

Between these two suboxides, InO_x vaporizes first and then, because of the temperature gradient along the furnace, condenses on the collector, where it deposits hexagonal, tetragonal, or rectangular In_2O_3 nanowires. ZnO vaporizes later and then condenses on the existing In_2O_3 nanowire core to form a layer of ZnO with 6-, 4-, and 2-fold symmetries. These provide the basis for further growth of ZnO nanorods through the vapor-solid mechanism. Cubic In_2O_3 prefers to grow along the [001] direction to form in 4-fold symmetry, while it chooses the [110] or [111] direction for 6-fold symmetry. Because we observed nanorods prematurely deposited on the In_2O_3 core surface, we conclude that it is very unlikely that both the core and secondary structures grow simultaneously. Because premature growth of nanorods on the In_2O_3 core surface in the early stage of nucleation occurs, we believe random fluctuations in the early stages of nucleation might initiate growth. Formation of ZnO nanoribbons with a central core has been studied in detail as previously reported.[20] A TEM study reveals the existence in the ribbons of a $(ZnO)_{11}In_2O_3$ nanowire core. The presence of this alloying element is necessary for the growth of side branching, which suggests that, during growth of the central nanowire, the alloying element continuously migrates to the facets of the nanowire core. The alloy-rich layers of the facets become nucleation sites for subsequent growth of the side branches by forming 2-fold symmetric structures. Zinc suboxide and liquid zinc metal, both of which have much lower melting points than ZnO, dominate the formation of nanobridges and nanonails. From the EDS and HRTEM studies of the nanobridges, it is seen that nanorods are composed of pure ZnO, and the belt is ZnO dominated, contaminated with only 0 to 3% atomic indium. This suggests that the growth of a ZnO nanorod from the nanobelt is homoepitaxial. We believe formation of the Zn-In-O eutectic phase significantly contributes to the nanobridge structure, because it has been observed that without In_2O_3, the evaporation rate becomes very low. The reason that nanorods grow on the edges is somewhat unclear, but a high surface free energy probably plays a role. For nanonails, ZnO_x and liquid Zn metal dominate. Following nucleation, small nanorods sprout epitaxially from the graphite foil substrate or from ZnO_x structures. The incoming ZnO_x vapor deposits epitaxially on both the shaft and the cap of the nanorod sprout. The nanorod cap, having a higher surface area, has the better chance to absorb ZnO_x vapor than does the bottom region. The absorbed ZnO_x vapor increases as the diameter of the hexagonal cap increases. Further, ZnO_x vapor is absorbed at the bottom. This causes the entire structure to be elevated. The diameter often continuously decreases from top to bottom. This is not always the case, however, because other factors, such as vapor dynamics and the oxidation rate, occasionally play a critical role.

2.4.4 Zn-Sn-O Hierarchical Nanostructures

As in the Zn-In-O system, hierarchical nanostructures can also be observed in the Zn-Sn-O system. A mixture of ZnO, SnO_2, and graphite powder has been used as a source for synthesizing Sn doped ZnO nanostructures. The experimental conditions and set-up are similar to those previously described (see Figure 2.23 for SEM images of these structures). The majority of the product is nanobelt-like, as shown

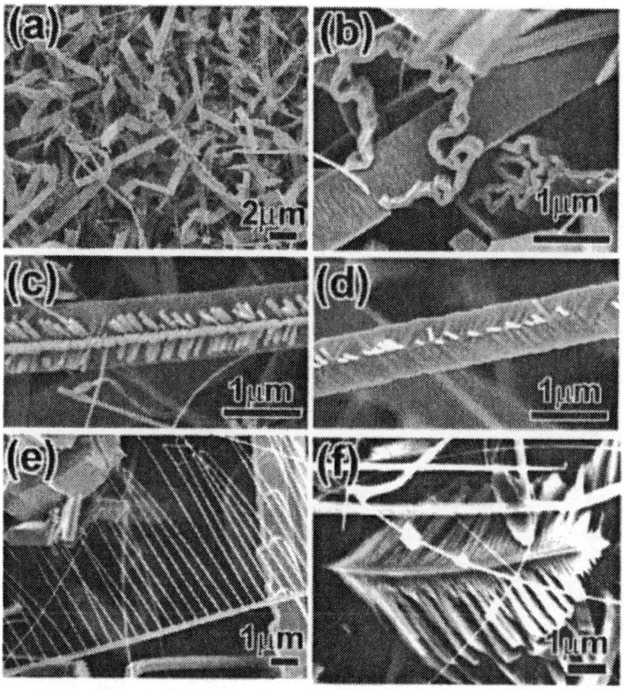

FIGURE 2.23 SEM image of the hierarchical Zn-Sn-O nanostructures. (a) Low-magnification and (b) high-magnification image of the nanobelts. (c) An 8-fold structure. (d) A 4-fold structure. (e) A 2-fold structure. (f) A rare leaf-like structure.

in Figure 2.23(a). On closer examination we find that it consists of either straight or twisted nanorods welded together, as shown in Figures 2.23(a) and 2.23(b). Small nanorods occasionally grow from the middle core onto the surface of the nanobelts, which are also the junctions of the elemental nanorods. The elemental nanorods can either appear perpendicular to the nanobelt growth direction as shown in Figure 2.23(d) or form at other angles. Figure 2.23(c) shows a nanobelt with three rows of nanorods growing on one side of the nanobelt surface. The middle row of the nanorods is perpendicular to the nanobelt surface, and the other two rows appear to form at 45°. We speculate that such a structure would possess 8-fold symmetry if secondary nanorods also grow on the other side of the nanobelt surface. Unlike the Zn-In-O nanobridge structure, no growth of secondary nanorods has been observed on the edges of the nanobelts. Two-fold symmetric structures, which may be the premature stage of the nanobelt formations, are shown in Figure 2.23(e). In Figure 2.23(e), only one side of the core nanorods has secondary nanorods, whereas there appears in Figure 2.23(f) a very rare tree leaf-like structure. It is likely that the Zn-Sn-O nanostructures are formed in two steps. The Zn-Sn-O core nanobelt forms first, and then the secondary nanorods grow epitaxially on the nanobelt under

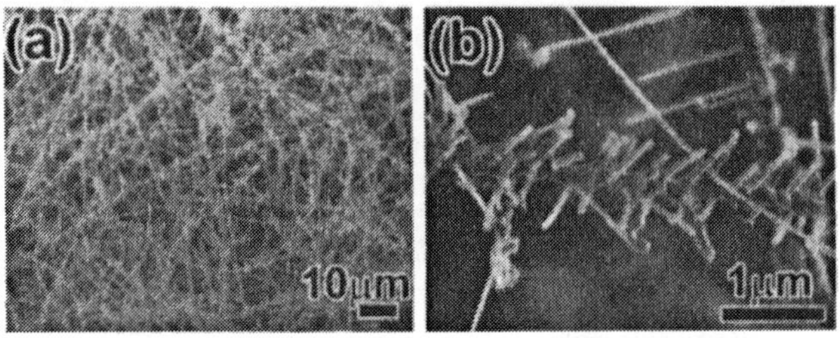

FIGURE 2.24 Microscopy studies of Zn-Ge-In-O hierarchical nanostructure. (a) Low-magnification and (b) high-magnification SEM images.

the right conditions. Most likely all the structures are based on homoepitaxials due to the dominant number of bare nanobelts and the similarity of the core nanobelt to hierarchical nanostructures with major bare nanobelts. Based on these observations, we believe there are several other missing hierarchical structures to be found in the Zn-Sn-O system.

2.4.5 Zn-In-Ge-O Quaternary Hierarchical Nanostructures

Success with binary and ternary systems encouraged us to explore some quaternary systems such as Zn-In-Ge-O. A mixture of ZnO, GeO_2, In_2O_3, and graphite powder was used as a source. Initial experiments produced some complicated ordered hierarchical nanostructures as shown in Figures 2.24(a) and 2.24(b). The secondary nanorods grew disordered on the major core nanorod. A higher magnification image of the structure is shown in Figure 2.24(b). The structures usually are tens of microns in length, and the core nanorods are in the range of 100 nm in diameter. The secondary nanorods are smaller in diameter, with the length varying between 200 nm and 2 μm.

2.5 LARGE QUANTITY NANOSTRUCTURES

From the preceding sections of this discussion, it is quite apparent that significant progress has been achieved in producing several exotic and useful nanostructures for various systems. It is important for practical purposes, as well, to have the ability to supply large quantities in a freestanding form. Most of the nanostructures we have discussed were grown on a substrate such as graphite foil, single crystal silicon, or sapphire. Growth of nanostructures on any two-dimensional substrate, however, does not yield large quantities. Rather, one must devise a process that will produce freestanding nanostructures in gram quantities at least. In this section we shall describe our recent success in producing such large quantity, freestanding ZnO nanowires. The method we have employed can easily be extended to any of the other nanostructures discussed in preceding sections.

The set-up we used is similar to that shown in Figure 2.25 except that the quartz tube was replaced by a quartz boat. The entire boat was loosely covered by a quartz

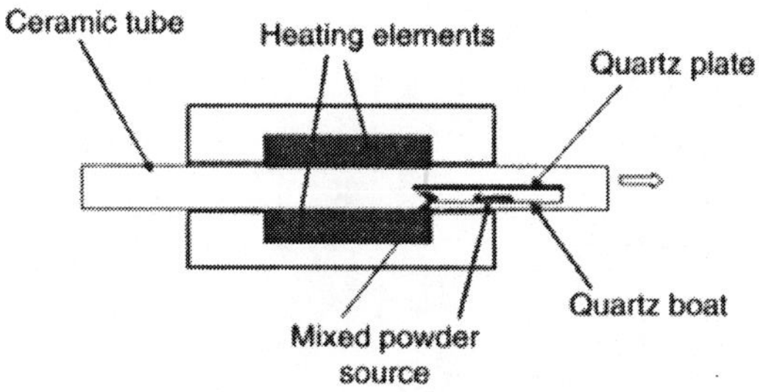

FIGURE 2.25 Experimental set-up to grow large quantity nanostructures via vapor transport and condensation method.

plate with a nozzle hole at the front to introduce air flow. One end of the furnace was connected to a rotary pump to regulate the air pressure between 0.1 and 10 torr, according to the specific requirements of the process. The other end was closed by a silicone stopper connected to an adjustable needle valve. A mixture of ZnO and graphite powder was placed at one end of the boat close to the nozzle hole, while between 2 and 3 cm from the source material, micron-size graphite flakes were spread uniformly along the direction of air flow to serve as a medium for collecting the nanowires. The boat was positioned within the steep temperature gradient of the furnace. At the end of the process, a substantial quantity of nanowires were found to have grown on the high surface area of the graphite flakes. The yield was substantial. The weight of the graphite flakes was more than twice their original weight, and 40% of the original ZnO was converted to nanowires. A SEM view of the product is shown in Figure 2.26. Nanowires as long as 15 μm formed in the relatively high temperature zone of the furnace near the source, whereas at a lower temperature, far from the source, the nanowires were only 2 to 5 μm long, with diameters between 70 and 100 nm. Figures 2.26(a), (b), (c), and (d) show such long nanowires under various magnifications. They possess smooth round surfaces with very sharp tips. The shorter nanowires have blunt hexagonal heads, with a hexagonal surface (Figures 2.26(e), (f), (g), and (h)). Under XRD analysis, these nanowires were proven to be entirely of the wurtzite hexagonal structure, similar to that of bulk ZnO. Their unit cell constants are as follows: a = 3.248 Å and c = 5.206 Å. These results were later confirmed by selected area diffraction (SAD). Further, an HRTEM study confirms that the nanowires are perfect crystals without dislocations, slaking, or other defects. The nanowires were separated from the graphite flakes to acquire freestanding nanowires by oxidizing the sample in flowing O_2 at 700°C. The XRD study shown in Figure 2.27 confirms that total removal of the graphite content was achieved by this procedure. More convincing proof of removal is seen in the SEM image of the oxidized sample, which shows voids that mimic the shape of the flakes (Figure 2.28(c)). Under TEM observation, the surface of the nanowires is nearly always wrapped in amorphous carbon or turbostratic graphite (Figure 2.28(a)) which was completely removed through the oxidation process as can be seen in Figure 2.28(b). Occasionally the oxidation

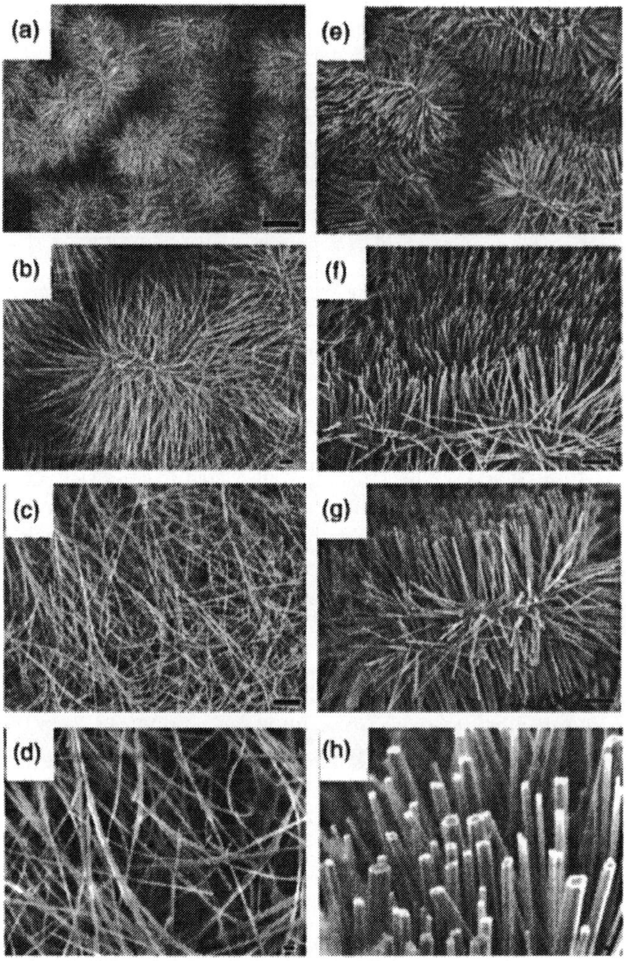

FIGURE 2.26 Morphology of as-made ZnO nanowires shown using SEM images. (a), (b), (c) and (d) show long (10–15 μm) and thin (30–60 nm) nanowires grown at the higher temperature region of the furnace whereas (e), (f), (g), and (h) show short (1–2 μm) and fat (60–100 nm) nanowires grown at the lower temperature region of the furnace. The scale bar for (a) is 10 μm, for (b), (c), (e), (f), and (g) is for 1 μm and for (d), and (h) is 100 nm.

process causes the nanowires to become less electrically conductive. Fortunately, the conductivity can be restored by further annealing at 500°C in a vacuum.

The high surface area of the graphite flakes is the key to large quantity growth of ZnO nanowires. Other nanopowders, such as Al_2O_3, have not yielded similar quantities. Success in producing large quantity ZnO nanowires raises the prospect of synthesizing the other previously observed nanostructures in large quantities as well. Our initial experiments in that direction have already begun with interesting positive results. A detailed study of the variation of yield with different growth

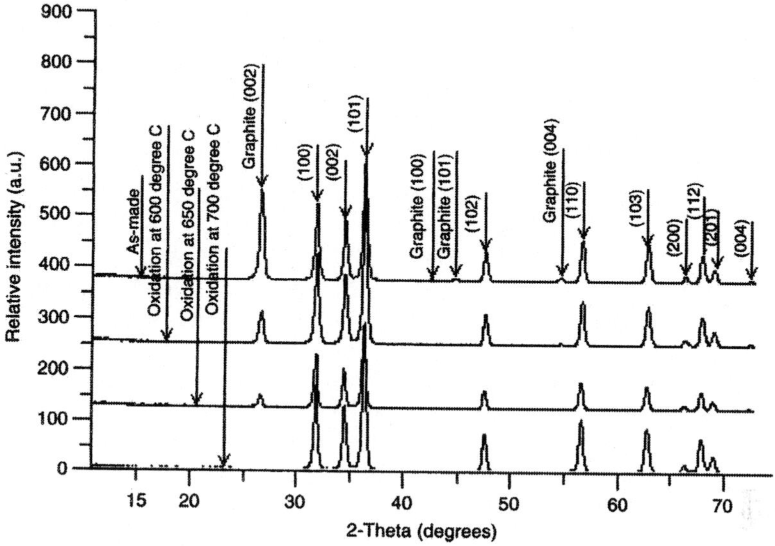

FIGURE 2.27 XRD pattern recorded from ZnO nanowires grown via vapor transport and condensation method.

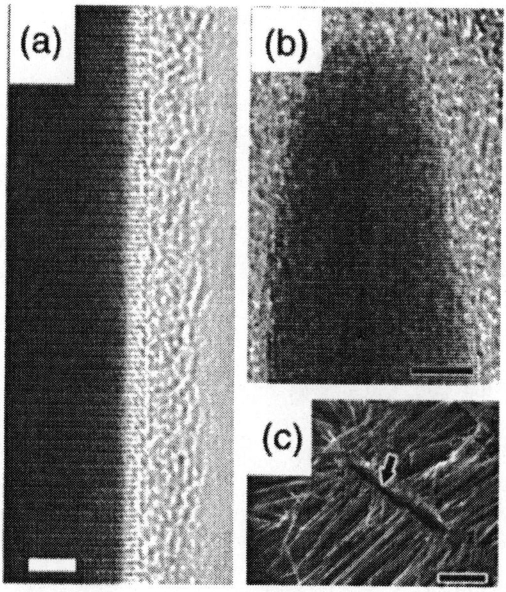

FIGURE 2.28 Reduction of amorphous graphite layer through annealing process. (a) TEM image of the surface of an as-made nanowire with amorphous graphite shell. (b) TEM image of an oxidized ZnO nanowire with no amorphous graphite shell on it. (c) SEM image of the oxidized ZnO nanowires, voids mimicking the shape of the graphite flakes. Scales bars are (a) 2 nm, (b) 5 nm, (c) 1 μm.

parameters, such as pressure and temperature in the tube furnace, the amount of graphite in the original mixture of the source, and the source-to-collector ratio, has been conducted and will be reported shortly.[26]

2.6 CONCLUDING REMARKS

We have discussed the synthesis and analysis of nanostructures (nanowires, nanowalls, nanocircuits, nanochains, and others) composed of binary systems, notably ZnO and In_2O_3; ternary systems, especially Zn-In-O and Zn-Sn-O; and significant modifications of both, including exotic ternary hierarchical systems and quaternary Zn-In-Ge-O hierarchical structures. We have described their morphology and structure as well as their electrical and optical properties. At this point our detailed understanding of their chemical and physical fundamentals is far from complete. Complex epitaxial relationships, with numerous possible orientations and relations among different components of these mixed systems, complicate access to a straightforward understanding. We do believe, however, that we are making progress. Our results make us certain that the vapor transport and condensation process is a powerful tool for the production of a large variety of nanoscale self-assemblies aimed at meeting the challenges of a bottom-up approach to nanoengineering. We also described our initiative growing these nanostructures in freestanding gram quantities. We find that the important matter of gaining precise control over size and morphology for practical applications requires a significant effort toward a more profound, detailed understanding. All the results we have discussed are very recent. This inspires us, in pursuit of direct practical applications, to continue close observation of our results as we go forward. We are certain that these nanostructures have great potential in a variety of fields including field emission; photovoltaics; transparent EMI shielding; super capacitors; fuel cells; high-strength, multifunctional composites; and many other branches of nanoelectronics and optoeletronics. Last, we must recognize that this is a truly interdisciplinary field encompassing physics, chemistry, materials science, and engineering. Our immediate goals will be to fully master the synthesis of the assemblies with reliably predictable morphology and physical properties.

PROBLEMS*

1. What are the key advantages of the bottom-up approach to top-down approaches in nanoscale fabrication?
2. Discuss the strategies one can adopt to grow nanostructures out of their bulk form.
3. Discuss the vapor phase epitaxy approach for nanostructure growth.
4. How does the vapor-solid-liquid (VLS) growth mechanism of nanostructures work? What are its advantages? What is the key difference between VLS and OAG (oxide-assisted growth)?
5. What are the strategies for growing nanostructures in complex hierarchical form? Explain and give examples.

* Readers are also advised to read corresponding references to answer the questions.

6. Why is ZnO superior to other oxide semiconductors for shortwave lasing applications? What are the advantages of ZnO in its nanostructure over its bulk form?
7. Discuss applications of ZnO and In_2O_3 in various fields.
8. What are the various techniques of producing nanostructures? Explain the vapor transport and condensation technique. What are the useful characterization techniques for nanostructures?

REFERENCES

1. H. Gleiter, Nanostructure materials: Basic concepts and microstructures, *Acta Mater.*, 48, 1, 2000.
2. R.P. Feynman, *Miniaturization,* Reinhold, New York, 1961.
3. H.W. Kroto, J.R. Health, S.C. O'Brian, R.F. Curl, and R.E. Smalley, *Nature,* 318, 6042, 1985.
4. S. Iijima, *Nature,* 354, 56, 1991.
5. a) A.S. Edelstein, and R.C. Cammarata, Eds., *Nanomaterials: Synthesis, Properties, and Application,* Institute of Physics, Philadelphia, PA, 1996;
   b) M. Grundmann, Ed., *Nano-optoelectronics,* Springer-Verlag, 2002; Miture, S., *Nanomaterials,* Pergamon Press, New York, 2000;
   c) Lojkowsky, W., Ed., *Interfacial Effects and Novel Properties of Nanomaterials,* Scitec Publications, 2003;
   d) H.S. Nalwa, Ed., *Hand-book of Nanostructured Materials and Nanotechnology,* Academic Press, New York, 2000.
6. a) H. Chopra, and S. Hua, *Phys. Rev.*, B 66, 020403, 2002; b) Y.C. Kong, D.P. Yu, B. Zhang, W. Fang, and S.Q. Feng, *Appl. Phys. Lett.*, 78, 4, 2001; c) R. O' Barr, S.Y. Yamamoto, S. Schultz, W.H. Xu, and A. Scherer, *J. Appl. Phys.*, 81, 4730, 1997.
7. a) R.H. Baughman, A.A. Zakhidov, W.A. de Heer, *Science,* 297 (5582), 787–792, 2002; b) S.A. Chesnokov, V.A. Nalimova, A.G. Rinzler, R.E. Smalley, J.E. Fischer, *Phys. Rev. Lett.*, 82 (2): 343–346, 1999; c) P.A. Gordon and P.B. Saeger, *Industrial & Engineering Chemistry Research,* 38 (12): 4647–4655, 1999; d) K.H. An, W.S. Kim, Y.S. Park, Y.C. Choi, S.M. Lee, D.C. Chung, D.J. Bae, S.C. Lim, Y.H. Lee, *Adv. Mater.*, 13 (7), 497, 2001.
8. a) S.M. Lee and Y.H. Lee, *Appl. Phys. Lett.*, 76 (20): 2877–2879, 2000; b) R.H. Baughman, A.A. Zakhidov, and W.A. de Heer, *Science,* 297 (5582): 787–792, 2002.
9. a) F. Favier, E.C. Walter, M.P. Zach, T. Benter, and R.M. Penner, *Science,* 293 (5538): 2227–2231, 2001; b) Y. Cui, Q. Wei, H. Park, and C.M. Lieber, *Science,* 293, 1298, 2001.
10. a) R.F. Service, *Science* 1997, 276, 895; b) R.H. Baughman, A.A. Zakhidov, W.A. de Heer, *Science,* 297 (5582): 787–792, 2002.
11. A.P. Alivisator, *Science,* 1996, 271, 1993.
12. a) C. Schonenberger, H. Vanhouten, H.C. Donfersloot, *Euro. Phys. Lett.*, 20 (3): 249–254, 1992; b) Y. Tokura, D.G. Austing, S. Tarucha, *J. Phys-Cond Matter,* 11 (31): 6023–6034, 1999; c) N. Asahi, M. Akazawa, Y. Amemiya, IEEE Transaction on Electron Devices, 44 (7): 1109–1116, 1997; d) D. Ali, H. Ahmed, *Appl. Phys. Lett.,* 64 (16): 2119–2120, 1994.

13. a) Z. Zhang, X. Sun, M.S. Dresselhaus, J.Y. Ying, *Phys. Rev.* B, 61, 4850, 2000; b) G. Markovich, C.P. Collier, S.E. Henrrichs, F. Remacle, R.D. Levine, J.R. Heath, *Acc. Chem. Res.*, 32, 415, 1999.
14. J.M. Krans, J.M. van Rutenbeek, V.V. Fisun, I.K. Yanson, L.J. de Jongh, *Nature*, 375, 767, 1995.
15. D.J. Lockwood, Z.H. Lu, J.M. Baribeau, *Phys. Rev. Lett.*, 76 (3): 539–541, 1996.
16. a) S. Iijima, *Nature*, 354, 56, 1991; b) T.W. Ebbesen, P.M. Ajayan, *Nature*, 358, 220, 1992; c) Z.F. Ren, Z.P. Huang, J.W. Xu, J.H. Wang, P. Bush, M.P. Siegal, and P.N. Provencio, *Science*, 282, 1105–1107, 1998.
17. a) M.S. Dresselhaus, Y.M. Lin, O. Robin, M.R. Black, G. Dresselhaus, *Nanowires*, 2003 (available the Mildred S. Dresselhaus's Group Web site: http://eecs-pc-05.mit.edu/x986.pdf; b) Z.L. Wang, Ed., *Nanowires and Nanobelts: Materials, Properties and Devices*, Kluwer Academic Publishers, 2003; c) Z.W. Pan, Z.R. Dai, Z.L. Wang, *Science*, 291, 1947–1949, 2001.
18. S. Bandyopadhyay, H.S. Nalwa, Eds., *Quantum Dots and Nanowires*, American Scientific Publication, 2003.
19. J.Y. Lao, J.G. Wen, Z.F. Ren, *Nano Lett.*, 2(11), 1287, 2002.
20. a) J.Y. Lao, J.Y. Huang, D.Z. Wang and Z.F. Ren, *Nano Lett.*, 3(2), 235, 2002; b) J.G. Wen, J.Y. Lao, D.Z. Wang, T.M. Kyaw, Y.L. Foo, Z.F. Ren, *Chem. Phys. Lett.*, 372, 717–722, 2003.
21. J.Y. Lao, J.Y. Huang, D. Banerjee, S.H. Jo, D.Z. Wang, J.G. Wen, D. Steeves, B. Kimball, W. Porter, R.A. Farrer, T. Baldacchini, J.T. Fourkas, Z.F. Ren. SPIE Proceedings, 5219, 99, 2003; J.Y. Lao, J.Y. Huang, D.Z. Wang, J.G. Wen, Z.F. Ren, Novel Hierarchical Oxide Nanostructures, *J. Chem. Mater.*, 14, 770, 2004.
22. J.Y. Lao, J.Y. Huang, D.Z. Wang, Z.F. Ren, D. Steeves, B. Kimball, W. Porter, *Appl. Phys.* A, 78, 539, 2004.
23. S.H. Jo, J.Y. Lao, and Z.F. Ren, R.A. Farrer, T. Baldacchini, J.T. Fourkas, *Appl. Phys. Lett.*, 82, 3520, 2003.
24. J.Y. Lao, J.Y. Haung, D.Z. Wang, Z.F. Ren, *Adv. Mater.*, 16, 65, 2004.
25. D. Banerjee, J.Y. Lao, D.Z. Wang, J.Y. Huang, Z.F. Ren, D. Steeves, B. Kimball, and M. Sennett, *Appl. Phys. Lett.*, 83, 2062, 2003.
26. D. Banerjee, J.Y. Lao, D.Z. Wang, J.Y. Huang, D. Steeves, B. Kimball, and Z.F. Ren, *Nanotechnology*, 15, 404, 2004.
27. a) X.F. Daun, Y. Huang, Y. Cui, J.F. Wang, and C.M. Lieber, *Nature*, 409, 66, 2001; b) Y. Huang, X. Daun, Y. Cui, L.J. Lauhon, K.H. Kim, C.M. Lieber, *Science*, 294, 1313, 2001; c) D.H. Cobden, *Nature*, 409, 32, 2001; d) G.Y. Tseng, J.C. Ellenbogen, *Science*, 2001, 294, 1293, 2001.
28. R.H. Baughman, A.A. Zakhidov, W.A. de Heer, *Science*, 297 (5582): 787–792, 2002.
29. a) C.T. Black, C.B. Murray, R.L. Sandstrom, S. Sun, *Science*, 290, 1131–1134, 2000; b) T. Thurn-Albrecht, *Science*, 290, 2126–2129 (2000); c) C. Sanchez, B. Lebeau, *Mater. Res. Soc. Bull.*, 26, 377–387, 2001.
30. a) B.B. Li, D.P. Yu, S.L. Zhang, *Phys. Rev.* B, 59 (3), 1645–1648, 1999; b) S.Q. Feng, D.P. Yu, H.Z. Zhang, Z.G. Bai, Y. Ding, *J. Crys. Growth*, 209 (2–3): 513–517, 2000; c) A.M. Rao, E. Richter, S. Bandow, B. Chase, P.C. Eklund, K.A. Williams, S. Fang, K.R. Subbaswamy, M. Menon, M.A. Thess, R.E. Smalley, G. Dresselhaus, M.S. Dresselhaus, *Science*, 275 (5297): 187–191, 1997.

31. a) R.J. Davey and J. Garside, *From Molecules to Crystallizers,* Oxford Chemistry Primers, No. 86, Oxford University Press, Oxford, 2001; b) P. Knauth, J. Schoonman, Eds., *Nanostructured Materials: Selected Synthesis Methods, Properties, and Applications* (Electronic Materials: Science & Technology) Kluwer Academic Publishers, 2002.
32. P. Yang, Y. Tu, R. Fan, *International Journal of Nanoscience,* 1(1) 1, 2002.
33. a) Y. Yu, R. Fan, P. Yang *Nano Lett.,* 2, 83, 2002; b) Y.C. Choi, W.S. Kim, Y.S. Park, S.M. Lee, D.J. Bae, Y.H. Lee, G.S. Park, W.B. Choi, N.S. Lee, J.M. Kim *Adv. Mater.,* 12, 746, 2000; c) X.F. Duan, C.M. Lieber, *Adv. Mater.,*12, 298, 2000; d) T.J. Trentler, K.M. Hickman, S.C. Geol, A.M. Viono, P.C. Gibbons, W.E. Buhro, *Science,* 270, 1791, 1995.
34. a) Y. Wu, P. Yang, *Chem. Mater.,* 12, 605, 2000; b) Y. Wu, B. Messer, P. Yang, *Adv. Mater.,* 13, 1487, 2001; c) Y. Wu, P. Yang, *J. Am. Chem. Soc.,* 123, 3165, 2001; d) M. Yazara, M. Koguchi, A. Muto, M. Ozawa, K. Hiruma, *Appl. Phys. Lett.,* 61, 2051, 1992.
35. a) J. Stejny, J. Dlugosz, A. Keller, *J. Mater Sci.,* 14, 1291, 1979; b) K.H. Meyer, H. Mark, *Helv. Chim. Acta,* 61, 1932, 1937; c) L. Stryer, *Biochemistry,* 3rd ed., W.M. Freeman and company, New York, p. 261, 1998.
36. A.A. Kundryavtsev, *The Chemistry and Technology of Selenium and Tellurium,* London, pp. 1–8, 1974.
37. P. Davidson, J.C. Gabriel, A.M. Levelut, P. Batail, *Europhys. Lett.,* 21, 317, 1993.
38. M. Volmer and I. Estermann, *Z. Phys.,* 7, 13, 1921.
39 a) G.W. Sears, *Acta. Metall.* 3, 367, 1955; b) G.W. Sears, *Acta. Metall.,* 1, 427, 1953.
40. J. Hu, T.W. Odom and C.M. Lieber, *Acc. Chem. Res.,* 32, 435, 1999.
41. a) P. Yang, H. Yan, S. Mao, R. Russo, J. Johnson, R. Saykally, N. Morris, J. Pham, R. He, H.J. Choi, *Adv. Funct. Mater.,* 12(5), 2002; b) Y.C. Kong, D.P. Yu, B. Zhang, W. Fang and S.Q. Feng, *Appl. Phys. Lett.,* 78, 4, 2001.
42. C.H. Liang, G.W. Meng, Y. Lei, F. Phillipp, L.D. Zhang, *Adv. Mater.,* 13, 1330, 2001.
43. a) Z.L. Wang, R.P. Gao, J.L. Gole, J.D. Stout, *Adv. Mater.,* 12, 1938, 2000; b) C.H. Liang, G.W. Meng, G.Z. Wang, Y.W. Wang, L.D. Zhang, S.Y. Zhang, *Appl. Phys. Lett.,* 78, 3202, 2001.
44. a) Z.W. Pan, Z.R. Dai, Z.L. Wang, *Science,* 291, 1947, 2001; b) Z.W. Pan, Z.R. Dai, Z.L. Wang, *Appl. Phys. Lett.,* 80, 309, 2002; c) Z.R. Dai, Z.W. Pan, Z.L. Wang, *Solid State Commun.,* 118, 351, 2001.
45. R.-Q. Zhang, Y. Lifshitz, and S.T. Lee, *Adv. Mater.,* 15(7–8), 635, 2003.
46. L.J. Lauhan, M.S. Gudiksen, D. Wang, C.M. Leiber, *Nature,* 420, 57, 2002.
47. Z.L. Wang, Z.R. Dai, Z.G. Bai, R.P. Gao, J.L. Gole, *Appl. Phys. Lett.,* 77, 3349, 2000.
48. Y.Q. Zhu, W.K. Hsu, W.Z. Zhou, M. Terrones, H.W. Kroto, D.R.M. Walton, *Chem. Phys., Lett.,* 347, 337, 2001.
49. D. Wang, M. Shao, D. Yu, W. Yu, and Y. Qian, *J. Crys. Growth,* 254, 487, 2003.
50. Z.L. Wang, Z. Pan, *Adv. Mater.,* 14(15), 2002.
51. H. Yun, R. He, J. Pham, P. Yang, *Adv. Mater.,* 15(5), 402, 2003.
52. C. Ma, D. Moore, J. Li Z.L. Wang, *Adv. Mater.,* 15 (3), 228, 2003.
53. J.Y. Lao, J.G. Wen, D.Z. Wang, Z.F. Ren, *Intl. J. Nanosci.,* 2, 149, 2002.
54. M.H. Huang, S. Mao, H. Feick, H. Yan, Y. Wu, H. Kind, E. Weber, R. Russo, P. Yang, *Science,* 292, 1897, 2001.
55. a) E.L. Paradis, A.J. Shuskus, *Thin Solid Films,* 83, 131, 1976; b) F.C.M. Van, D.E. Pol, *Ceram. Bull.,* 69, 1959, 1990.
56. a) K. Nashimoto, S. Nakamura, H. Moriyama, *Jpn. J. Appl. Phys.,* 43, 5091, 1995; b) T. Nagata, T. Shimura, A. Asida, N. Fujimura, T. Ito, *J. Crys. Growth,* 237, 533, 2002.

57. a) A.J. Freeman, K.R. Poeppelmeier, T.O. Mason, R.P.H. Chang, T.J. Marks, *MRS Bull.*, 2000, 25, 45, 2000; b) D.S. Ginley, C. Bright, *MRS Bull.*, 25, 15, 2000.
58. a) N. Chubachi, *Proc. IEEE*, 64, 772, 1976; b) M. Kadota, *Jpn. J. Appl. Phys.*, 36, 3076, 1997.
59. a) G. Heiland, *Sensors and Actuators*, 2, 343, 1982; b) G. Sberveglieri, S. Groppelli, P. Nelli, A. Tintinelli, G. Giunta, *Sensors and Actuators* B, 25, 588, 1995.
60. K. Hara et al., *Sol. Energy. Mater. Sol. Cells.*, 64, 115, 2000.
61. H. Yumoto, T. Inoue, S.J. Lee, T. Sako, K. Nishiyama, *Thin Solid Films*, 345, 38, 1999.
62. a) C.G. Granqvist, *Appl. Phys.* A, 57, 19, 1993; b) T. Takada, K. Suzukik, M. Nakane, *Sensor Actuat.* B, 13, 404, 1993; c) C. Li, D.H. Zhang, X.L. Liu, S. Han, T. Tang, J. Han, C.W. Zhou, *Appl. Phys. Lett.*, 82, 1613, 2003; d) D.H. Zhang, C. Li, S. Han, X.L. Liu, T. Tang, W. Jin, C.W. Zhou, *Appl. Phys. Lett.*, 82, 112, 2003.
63. a) M.H. Huang, Y. Wu, H. Feick, N. Tran, E. Weber, P. Yang, *Adv. Mater.*, 13, 113, 2001; b) Y.W. Wang, L.D. Zhang, G.Z. Wang, X.S. Peng, Z.Q. Chu, C.H. Liang, *J. Crys. Growth*, 234, 171, 2003.
64. a) J.J. Wu, S.C. Liu, *Adv. Mater.*, 14, 215, 2002; b) S.C. Lyu, Y. Zhang, H. Ruh, H.J. Lee, H.W. Shim, E.K. Suh, C.J. Lee, *Chem. Phys. Lett.*, 363, 134, 2002.
65. Y.C. Kong, D.P. Yu, B. Zhang, W. Fang, S.Q. Feng, *Appl. Phys. Lett.*, 78, 4, 2001.
66. B.D. Yao, Y.F. Chan, and N. Wang, *Appl. Phys. Lett.*, 81, 4, 2002.
67. H.-J. Egelhaaf and D. Oelkrug, *J. Cryst. Growth*, 161, 190, 1996.
68. K. Vanheusden, C.H. Seager, W.L. Warren, D.R. Tallant, and J.A. Voigt, *Appl. Phys. Lett.*, 68, 403, 1996.
69. a) L. Nilsson, O. Groening, C. Emmenegger, O. Kuettel, E. Schaller, L. Schlapbach, H. Kind, J.-M. Bonard, and K. Kern, *Appl. Phys. Lett.*, 76, 2071, 2000; b) S.H. Jo, Y. Tu, Z.P. Huang, D.L. Carnahan, D.Z. Wang and Z.F. Ren, *Appl. Phys. Lett.*, 82, 3520, 2003.
70. D.V. Leff, L. Brandt, and J.R. Heath, *Langmuir*, 12, 4723, 1996.
71. S.H. Jo, Y. Tu, Z.P. Huang, D.L. Carnahan, D.Z. Wang, Z.F. Ren, *Appl. Phys. Lett.*, 82, 3520, 2003.
72. C.J. Lee, T.J. Lee, S.C. Lyu, Y. Zhang, H. Ruh, and H.J. Lee, *Appl. Phys. Lett.*, 81, 3648, 2002.
73. R.S. Wagner, W.C. Ellis, *Appl. Phys. Lett.*, 4, 89, 1964.
74. H. Kohno, S. Takeda, *Appl. Phys. Lett.*, 73, 3144, 1998.
75. Definition of the symmetry symbols: for example, $4S^{*1}$-, the first letter such as 6 or 4 or 2 means the basic 6-, 4-, and 2-fold symmetry of the major core nanowires, respectively. The second letter S or M indicates the single or multiple rows of the secondary ZnO nanorods. If there is nothing after S or M, all the ZnO nanorods are perpendicular to the major In_2O_3 core nanowire. The third symbol * means that the secondary nanorods have an angle with the major core nanowire. The number 1 or 2 after the symbol * means that not all the secondary ZnO nanorods, branches have an angle with the core nanowire.
76. Powder Diffraction File Release 2000, PDF Maintenance 6.0 (International Center for Diffraction Data, Philadelphia).
77. A.P. Levitt, Ed., *Whisker Technology*, Wiley-Interscience New York, 1970.

3 Carbon Nanotubes and Bismuth Nanowires

Mildred S. Dresselhaus, Ado Jorio, and Oded Rabin

CONTENTS

3.1 INTRODUCTION

Carbon nanotubes and bismuth nanowires are both model systems for describing the one-dimensional properties of nanotubes and nanowires and will be discussed in this context in this chapter. For the case of carbon nanotubes, their one-dimensional aspects are best seen when they are prepared as single-wall carbon nanotubes (cylinders with walls one carbon atom in thickness) with approximately 1-nm diameters and lengths typically of about 10 μm. Their one-dimensional (1-D) properties are best exemplified in their 1-D electronic density of states (DOS), because their special physical properties are closely connected with their unique 1-D electronic DOS. The special properties associated with bismuth nanowires are closely connected to their unique semimetal-to-semiconducting transition, which occurs at approximately 50-nm wire diameter, making available the special properties of bismuth in the form of a semiconductor with highly anisotropic constant energy surfaces, effective mass components, and physical properties. This chapter focuses on the preparation and characterization of these materials.

For both of these systems, simple model calculations can be carried out to explain most of the unusual properties that are observed in their nanostructured form. These model calculations can be used to modify the conditions of their physical preparation to produce desired properties. Characterization of these materials is important to establish what their intrinsic properties are and how these basic properties can be used, controlled, or modified for both scientific studies and practical applications.

3.2 CARBON NANOTUBES

3.2.1 Overview

The special interest in carbon nanotubes stems from their unique structures and properties, their very small size (as small as ~0.42 nm in diameter), the possibility for carbon nanotubes to be metallic or semiconducting depending on their geometric structure, their exceptional properties of ballistic transport, extremely high thermal conductivity, extremely high optical polarizability, and possibilities of high structural perfection. Single-wall carbon nanotubes (SWNTs) have only one atomic species (carbon) and a relatively simple structure (a sheet of regular hexagons of carbon atoms rolled in a seamless way into a cylinder one atom thick). Single-wall nanotubes thus provide a benchmark for studying and characterizing new 1-D phenomena originating from their unique molecular density of states, whereby every structurally distinct (n,m) nanotube has its own unique and characteristic 1-D density of states, as discussed below. Many unexpected phenomena that do not occur in the parent graphite material have been discovered in carbon nanotubes and these discoveries have energized not only nanotube research, but also carbon and nanoscience research. Major gaps in basic knowledge remain, with the major obstacle confronting the carbon nanotube field being the lack of a detailed understanding of the nanotube growth mechanism. Such understanding is needed so that nanotubes of desired diameter and chirality can be grown in a monodispersed and controlled way because of the unusually close connection between nanotube properties (such as their metallicity, i.e., whether they are metallic or semiconducting) and their geometric structure. For this reason, much current research effort is directed toward the synthesis and characterization of SWNTs.

There are basically three different methods used to grow single-wall carbon nanotubes (SWNTs): the arc discharge, laser vaporization, and chemical vapor deposition (CVD) methods.[1–3] The synthesis methods will be briefly described subsequently, after the uniqueness of each (n,m) nanotube is further clarified; (n,m) denotes the structural indices used to characterize SWNTs.

Double-wall carbon nanotubes (DWNTs), consisting of two coaxial cylindrical SWNTs, one inside the other as in a Russian doll arrangement, provide a prototype for studying the structures and properties of multiwall carbon nanotubes (MWNTs) in a quantitative way. Because of the increased stability and durability of DWNTs and MWNTs relative to SWNTs, the more robust DWNTs and MWNTs are expected to have more potential for applications, exploiting the exceptional mechanical strength and stiffness and the very high thermal conductivity of carbon nanotubes.[4] Before discussing the synthesis and characterization of SWNTs and DWNTs, the unique aspect of (*n*,*m*) nanotubes is presented so that the synthesis and characterization can be discussed in a proper framework.

Since every structurally distinct (*n*,*m*) nanotube can be considered as a distinct molecule, it is important to provide the framework used to distinguish one nanotube from another. As illustrated in Fig. 3.1, a single-wall carbon nanotube is geometrically only a rolled-up graphene sheet. Its structure can be specified or indexed by its circumferential periodicity ($\vec{C}_h$), as described using the chiral vector (*AA′* in Fig. 3.1), which connects two crystallographically equivalent sites (*A* and *A′*) on a graphene sheet. In this way, the geometry of each SWNT is completely specified by a pair of integers (*n*,*m*) denoting the relative position $\vec{C}_h = n\vec{a}_1 + m\vec{a}_2$ of the pair of atoms on a graphene strip that, when rolled onto each other, form a tube. Here $\vec{a}_1$ and $\vec{a}_2$ are basis vectors of the hexagonal honeycomb lattice, as shown in Fig. 3.1. This chiral vector $\vec{C}_h$ also defines a chiral angle θ, which is the angle between $\vec{C}_h$ and the $\vec{a}_1$ direction of the graphene sheet. For the tubes with (*n*,0), called zigzag tubes, and the tubes with (*n*,*n*), called armchair tubes,[5] the translation and rotation symmetry operations are independent (and these tubes form symmorphic space groups), while for all other (*n*,*m*) tubes, the rotations and translations are coupled (nonsymmorphic space groups).[5] The assembly of MWNTs follows a Russian doll arrangement with the stacking of tubes coaxially at an interlayer distance greater than about 0.34 nm, which is the interlayer separation of turbostratic graphene layers that lack the ABAB . . . interlayer stacking order found in 3-D graphite. There does not seem to be a correlation between the (*n*,*m*) on one layer and the (*n′*,*m′*) of the adjacent layers beyond a preference for minimizing

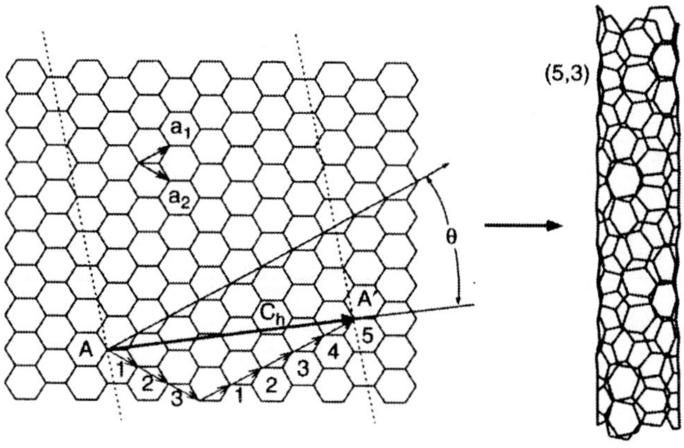

FIGURE 3.1 Schematic diagram showing a possible wrapping of a two-dimensional graphene sheet into a tubular form. In this example, a (5,3) nanotube is under construction and the resulting tube is illustrated on the right.[6]

the interlayer separation to be as close as possible to the turbostratic interplanar distance to maximize the coupling between adjacent SWNTs.

Before the first SWNT was ever synthesized in the laboratory, theoretical calculations[7–9] showed that the electronic properties of SWNTs are very sensitive to their geometric structures. Although graphene (2-D graphite) is a zero-gap semiconductor,[5] theory predicted that carbon nanotubes can serve as metals or semiconductors with different-size energy gaps, depending very sensitively on the indices (n,m) or equivalently on the diameter and helicity of the tubes. In addition, each chiral (n,m) nanotube ($0 < \theta < 30°$) also has a distinct handedness that further defines its uniqueness.[10]

The physics behind this sensitivity of the electronic properties of carbon nanotubes to their structures can be understood by relating the nanotube electronic structure to that of the graphene sheet that is rolled to produce the SWNT. The unique band structure of a graphene sheet shown in Fig. 3.2 has states crossing the Fermi level at only one point in k-space, denoted by the point K in Fig. 3.2. The quantization of the electron wave vector along the circumferential direction resulting from the periodic boundary conditions leads to the formation of energy subbands associated with cutting lines $k_{\perp} = \mu K_1$ separated from one another by a distance $K_1 = 2/d_t$. The d_t is the nanotube diameter, μ is an integer denoting each quantum state in the circumferential direction for each cutting line, and a quasicontinuum of states $k_{||}$, described by a reciprocal lattice vector K_2 occurs along the length of such cutting lines.[11] The K_1 and K_2 reciprocal lattice vectors form the basis vectors of the nanotube Brillouin zone.

An isolated sheet of graphite is a zero-gap semiconductor whose electronic structure near the Fermi energy is given by an occupied π band and an empty π^* band.

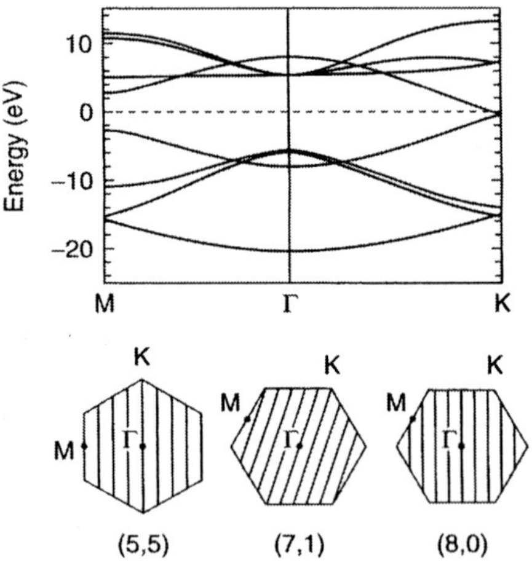

FIGURE 3.2 (Top) Tight-binding band structure of graphene (a single basal plane of graphite), showing the π and σ bands in the main high symmetry directions. (Bottom) Allowed $\vec{k}$-vectors of the (5,5), (7,1), and (8,0) tubes (solid lines) mapped onto the graphite Brillouin zone.[6]

These two bands have a linear dispersion $E(k)$ around the K point in the Brillouin zone and, as shown in Fig. 3.2, they meet at the Fermi level at the K point. The Fermi surface of an ideal graphite sheet consists of the six corner K points. The allowed set of k's in SWNTs, indicated by the lines in Fig. 3.2, depends on the diameter and helicity of the tube. Whenever the allowed k's include the point K, the system is a 1-D metal with a nonzero density of states at the Fermi level, resulting in a one-dimensional metal with two linear dispersing bands. When the point K is not included in the set of allowed states, the system is a semiconductor with different-size energy gaps depending on both d_t and θ. The d_t dependence comes from the $K_1 = 2/d_t$ relation, and the θ dependence comes from the fact that the equi-energies around the K point are not circles and exhibit trigonally warped shapes[5]. It is important to note that the states near the Fermi energy for both metallic and semiconducting tubes are all from states near the K point, and hence their transport and other electronic properties are related to the properties of the states on the cutting lines, with the conduction band and valence band states of a semiconducting tube coming from states along the cutting line closest to the K point.

The general rules arising from symmetry considerations tell us that SWNTs come in three varieties: armchair (n,n) tubes, which are always metals; (n,m) tubes with $n - m = 3j$, where j is a nonzero integer — very tiny-gap semiconductors, that behave like metals at room temperature; and large-gap semiconductors for which $n - m = 3j \pm 1$ (Fig. 3.3). As the tube diameter d_t increases, the band gaps of the large-gap and tiny-gap varieties decrease with a $1/d_t$ and $1/d_t^2$ dependence, respectively. The $1/d_t^2$ dependence for the tiny gap is due to a curvature effect and depends on chiral angle, the tiny gap being a maximum for zigzag and zero for armchair SWNTs. Thus, in Fig. 3.2, a (7,1) tube would be metallic at 300 K, whereas an (8,0) tube would be semiconducting, while the (5,5) armchair tube would always be metallic, consistent with whether a cutting line goes through the K point of the 2-D Brillouin zone. The corresponding electronic density of states for these three SWNTs are shown in Fig. 3.3. The main point to be emphasized is that the energy eigenvalues for every (n,m) SWNT are unique, and therefore each (n,m) nanotube is considered a distinct molecule with potentially different properties. Van Hove singularities (vHSs), characteristic of the density of states for 1-D systems, are found for both semiconducting and metallic SWNTs. To the extent that a specific application, such as for an electronic device, depends on the distinction between different (n,m) tubes, the synthesis of SWNTs depends on a high degree of control of the synthesis process discussed in the next section.

3.2.2 Nanotube Synthesis

All three synthesis methods [arc discharge, laser vaporization, and chemical vapor deposition (CVD)] produce carbon atoms in a hot gaseous form by evaporation from a condensed phase and these highly energetic carbon atoms then reassemble themselves to form carbon nanotubes. All methods use nanosize catalytic particles to induce the synthesis process and to control the diameter of the nanotubes that are synthesized. For the arc and laser methods, carrier gases are used for transporting nanotubes from where they are formed in the reactor to where they are collected,

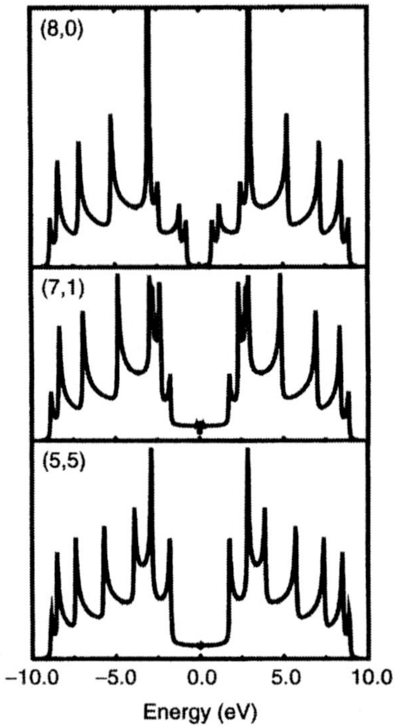

FIGURE 3.3 Electronic densities of states for the (5,5), (7,1), and (8,0) SWNTs showing singularities characteristic of 1-D systems. The (5,5) armchair nanotube is metallic for symmetry reasons, with a constant energy-independent density of states between the first singularities in the valence and conduction bands. The (7,1) chiral tube displays a tiny band gap due to curvature effects, but will display metallic behavior at room temperature. The (8,0) zigzag tube is a large-gap semiconductor.[12]

clearing the way for the subsequent growth of SWNTs in the more active reaction region of the reactor.[13] The arc discharge[14,15] and laser ablation[16] methods were the first to allow gram quantities of SWNTs to be synthesized, and these methods are still used today as sources of high-quality samples for specific fundamental studies. However, the chemical vapor deposition (CVD) method is the method most amenable for large-scale production of SWNTs. The HiPco (high pressure catalytic decomposition of carbon monoxide) adaptation of the CVD approach uses floating catalytic particles and is a common commercial source of SWNTs for both research and applications use.[17] However, the adaptation of the CVD process based on alcohol[18] produces relatively clean SWNT bundles with fewer residual catalytic particles, and when operated at relatively low temperatures (e.g., ~650°C), smaller diameter tubes are favored.[19] Common catalytic particles are Mo, Co, and Fe, but alloys of Co—Mo and Fe—Co have been found to produce high SWNT yields. Use of Co—Mo bimetallic catalysts and a fluidized bed CVD reactor in a Co–Mo CAT process[20,21] has been especially successful in the efficient production of small-diameter SWNTs.

Rapid progress is being made on nanotube synthesis to increase the control of the process, steadily narrowing the diameter and chirality ranges of the nanotubes that are produced, decreasing their defect and impurity content, increasing their production efficiency and yield, while expanding their functionality. The main directions in the pursuit of controlled nanotube synthesis include the synthesis of molecular catalytic clusters with atomically well defined sizes and shapes, the development of mild catalytic synthesis conditions at reduced temperatures, the development of patterned growth with a high degree of control in nanotube location and orientation, and the synthesis of complex and organized networks or arrays of nanotubes on substrates.[13]

Much attention is also being given to the synthesis of multiwall carbon nanotubes (MWNTs) and double-wall carbon nanotubes (DWNTs) because of their attractiveness for specific applications where mechanically robust nanostructures are needed. DWNTs can be considered model systems because of their similarity in physical properties to MWNTs and their relative simplicity. The synthesis of double-wall nanotubes can be accomplished either in a direct synthesis process or from a SWNT whose core has been previously filled with C_{60} molecules by a gas phase reactor at ~400°C on an acid-purified open-ended SWNT bundle. The transformation of the C_{60}-filled SWNTs (called peapods) to DWNTs is accomplished by heat treatment in the temperature range of 800 to 1200°C.[4]

Typically the growth of single-wall carbon nanotubes usually involves a catalytic process, with nanotubes grown in the presence of catalytic particles, but these growth conditions typically produce a variety of other carbon species, and many other unwanted constituents. Therefore, much attention has also been given to nanotube purification from other species, techniques for the characterization of nanotube purity, and for the preparation and sorting of nanotubes by length, diameter, chirality, and metallic or semiconducting properties and perhaps other attributes.[22] Chemical methods are generally used for the removal of metal particles and other unwanted impurities[23–26] and high-temperature heat treatment in vacuum can be used to remove structural defects such as heptagon–pentagon pairs.[26,27]

Nanotube synthesis and purification are closely connected from two points of view. First an improvement in the synthesis process regarding purity and monodispersity in tube diameter and chirality would reduce the need for extensive post-synthesis purification and separation. Second, if efficient and precise purification and separation techniques can be developed, the stringent limit that the growth process should produce a single (n,m) species could perhaps be relaxed. The convergence or combination of the two approaches, that is, the controlled or preferential growth of a particular (n,m) nanotube, and an effective and easy purification and separation process[4] could prove powerful in producing nanotubes with well defined diameters and chiralities.

It is well known that SWNTs can be either semiconducting or metallic, depending on their geometry (see Section 3.2.1). At present, all known synthesis processes yield a mixture of semiconducting and metallic SWNTs within a single SWNT bundle. For DWNTs or MWNTs, the various shells will contain random collections of semiconducting (S) and metallic (M) constituents with an approximate 2:1 ratio expected for $S:M$ tubes, assuming equal *a priori* probability for occurrence, which is realized when many different (*n*,*m*) constituents are present in the sample. Nanotube applications in some cases, such as electronics, require control of the synthesis process to discriminate between *S* and *M* SWNTs, and for circuit applications, control of diameter and chirality could also be required. Such control of the synthesis process will be a research challenge of the future.

For the present, effort is also being given to the separation of SWNTs into *M*-enriched and *S*-enriched samples by a variety of processes that preferentially distinguish *S* tubes from *M* tubes. One of these separation processes is attributed to the enhanced chemical affinity of the octadecylamine (ODA) surfactant for *S* SWNTs dispersed in tetrahydrofuran (THF), rendering *M* SWNTs more prone to precipitation when the THF is partially evaporated, leaving behind a supernatant enriched in *S* SWNTs.[28] Another separation procedure for enriching *M* or *S* SWNTs is based on ion exchange liquid chromatography of DNA-assisted SWNT solubilization, relying on the lower negative charge density of DNA-*M* SWNT hybrids. In this process, *M* SWNT-enriched fractions elute before *S* SWNT-enriched fractions.[29,30] More recently the separation of *M* from *S* SWNTs using alternating current dielectrophoresis in an aqueous SWNT suspension was achieved, making use of the higher induced dipole moments of *M* SWNTs.[31] More recently selective functionalization of *M* SWNTs against *S* SWNTs after addition of 4-chlorobenzenediazonium tetrafluoroborate to an aqueous suspension of SWNTs was reported, based on the higher activity of covalent aryl bonds of diazonium with *M* SWNTs than with *S* SWNTs.[32–34] *M*/*S* separation has also been reported using centrifugation of a surfactant-stabilized aqueous suspension of SWNTs with the addition of diluted bromine. Here *M* SWNTs form more highly structured charge-transfer complexes with bromine than *S* SWNTs, thereby yielding an excess of *M* SWNTs in the supernatant and of *S* SWNTs in the sediment.[35]

The development of many different separation procedures demands that a simple and reliable method be developed for the quantitative evaluation of separation efficiency. Several optical techniques widely used in nanotube science can provide such characterization tools. Optical absorption spectroscopy[36] is one suitable candidate for the characterization process of the separated samples once the individual peaks in the absorption spectra are assigned to *M* and *S* SWNTs,[37] as discussed in Section 3.2.5. Such measurements are difficult to quantify, because the most intense optical peak is strongly influenced by exciton effects and because the intensities of all of the optical absorption peaks are not only influenced by the number density of the *M* and *S* species but by their oscillator strengths for each of the optical transitions. Resonance Raman spectroscopy (RRS)[38] provides an alternative for the characterization of *M*- and *S*-enriched aliquots, since *M* and *S* SWNTs yield distinct features in the resonance Raman spectra, as discussed in Section 3.2.4.[39] With the RRS technique, the intensities of the radial breathing mode peaks are likewise influenced by the strength of the matrix elements for the Raman intensity. The pertinent matrix elements for the Raman effect include the electron–photon and electron–phonon matrix elements. In addition, Raman spectra have to be taken at a sufficient number of laser excitation energies to get a good sampling of the SWNTs present in the sample. Photoluminescence would be useful in making a determination of how the diameter distributions change during the separation process, but it is less useful in evaluating the *M* and *S* separation, since *M* SWNTs show no luminescence and, furthermore, the presence of *M* SWNTs in a nanotube sample quenches the luminescence from *S* SWNTs by sharply reducing the lifetimes of the excited states of the *S* SWNTs through nonradiative decay pathways.[40]

3.2.3 Nanotube Characterization

There are a number of methods used to characterize the structure and properties of both individual nanotubes and nanotube bundle samples. In this section, general characterization approaches are discussed while in Section 3.2.4 the characterization of SWNTs by Raman spectroscopy is discussed in some detail.

One obvious method to determine whether a SWNT is metallic or semiconducting would be to measure its electrical conductivity. Such measurements can be done, but they are not easy to carry out quantitatively because of the small SWNT diameter and the difficulty in making reproducible electrical contacts. Transport measurements, however, are not useful for specific (n,m) identification.

In practice, electron microscopy has been widely used for structural characterization. Scanning electron microscopy (SEM) is routinely used to evaluate the morphology and the number of remaining catalytic particles and other carbon forms present in an as-prepared SWNT sample, while an EDAX (energy dispersive x-ray analysis) attachment is typically used for the chemical detection of the noncarbon species present in the sample and their relative concentrations. Transmission electron microscopy (TEM) is used to image the lattice planes to evaluate the structural perfection of the sidewalls of an individual SWNT. Imaging is also used to study the cap structures of SWNTs and MWNTs, as well as to study the bamboo structures of the inner tubes of MWNTs. For SWNTs, the TEM is commonly used to obtain electron diffraction patterns, and from these patterns to determine the (n,m) indices for an individual SWNT.

Scanning probe microscopy (SPM) has been used to characterize isolated SWNTs spread on surfaces. Atomic force microscopy (AFM) has been widely used to measure the diameters of individual SWNTs, while low temperature, atomic resolution scanning tunneling microscopy (STM) images can be used to measure the chiral angles of individual SWNTs. In the scanning tunneling spectroscopy (STS) mode, a scanning probe instrument can be used to measure (I/V) (dI/dV) for individual SWNTs, thereby yielding scans of the electronic density of states from which the (n,m) indices for individual SWNTs can be determined.

Optical techniques, including optical absorption, photoluminescence, and Raman scattering have been widely used to characterize carbon nanotubes by studying both their electronic and phonon (lattice vibrational) spectra. These methods have the advantage of not requiring contacts and in being a very weakly perturbing probe. In Section 3.2.4, Raman spectroscopy, which is the most widely used spectroscopic tool for nanotube characterization, is discussed in some detail, while in Section 3.2.5 other photophysical techniques are briefly reviewed.

3.2.4 Raman Spectroscopy

The Raman spectra of SWNTs have been particularly valuable for providing detailed information for characterizing samples,[38] and several of the features in the Raman

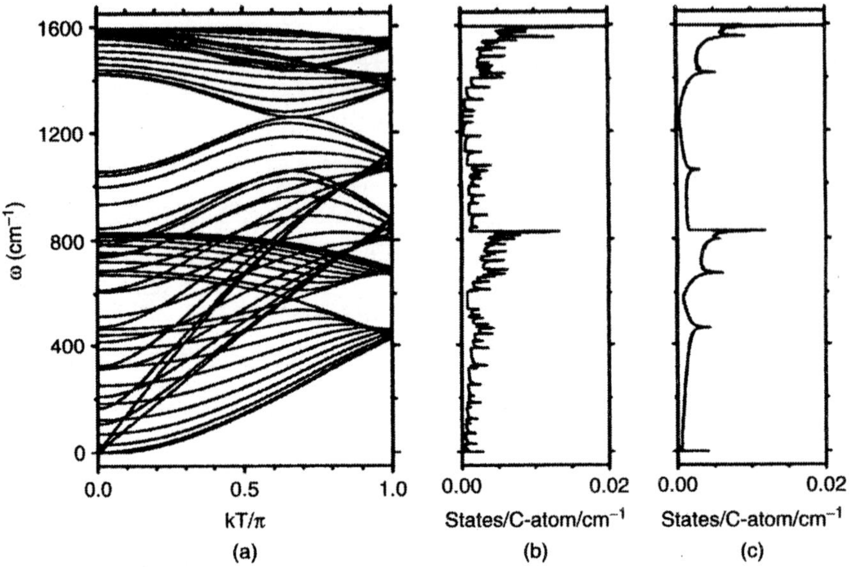

FIGURE 3.4 (a) The calculated phonon dispersion relations of an armchair carbon nanotube with $(n,m) = (10,10)$, for which there are 120 degrees of freedom and 66 distinct phonon branches.[5] (b) The corresponding phonon density of states for a (10,10) nanotube. (c) The corresponding phonon density of states for a 2-D graphene sheet, for purposes of comparison.[5]

spectra have been used for SWNT characterization, as indicated below. In the first report of Raman spectra from SWNTs,[38] despite the large number of branches in the SWNT phonon dispersion relations (Fig. 3.4(a)), the Raman spectra for a SWNT bundle (Fig. 3.5) exhibited only two dominant features, namely the radial breathing mode (RBM), which in this trace appears at 186 cm^{-1} for a laser excitation energy $E_{\text{laser}} = 2.41$ eV, and the tangential band for vibrations along the surface of the SWNT appears in the range of 1520 to 1620 cm^{-1}. Because of the strong connection of this tangential band to the corresponding mode in 2-D graphite, this higher frequency band for SWNTs is commonly called the *G*-band. Other lower intensity features, seen in Figs. 3.5 and 3.6 and discussed below, also provide important and unique information about SWNTs.

In this first paper on the Raman effect in SWNTs,[38] the strong and nonmonotonic dependence of the SWNT Raman spectra on the laser excitation energy E_{laser} established the Raman scattering to be a resonance process occurring when E_{laser} matches the optical transition energy E_{ii} between van Hove singularities in the valence band and conduction bands (Fig. 3.3). Raman scattering in SWNTs is very important for the characterization of SWNTs because it relates the photon excitation directly to both phonon and electron processes. Since the E_{ii} value for a particular SWNT is dependent on its diameter d_t, so are the Raman spectra that are observed. Because of the very small diameters of SWNTs (~1 nm), the joint density of states exhibits very large singularities, with associated large enhancements in Raman intensity, allowing the

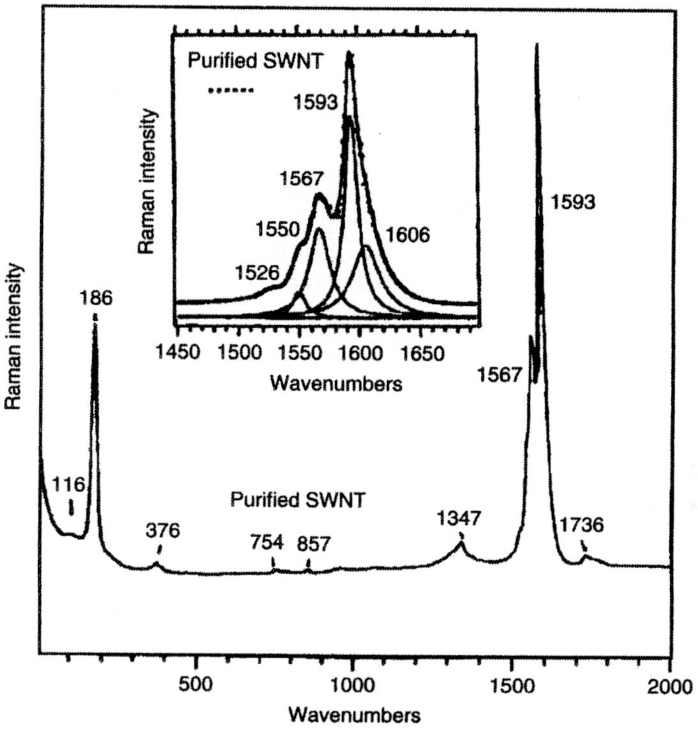

FIGURE 3.5 Experimental Raman spectrum taken with 514.5 nm (2.41 eV) laser excitation from a SWNT bundle sample with a diameter distribution $d_t = 1.36 \pm 020$ nm. The inset shows an expanded version of the spectra in the 1450 to 1700 cm^{-1} range, fit by a sum of Lorentzians.[38]

observation of spectra from an individual SWNT that is in strong resonance with E_{laser}.[41] From the point of view of nanotube characterization, the resonance Raman effect allows the selection for detailed examination of only the few SWNTs in a sample that are in resonance with a given E_{laser}. For example, the RBM for which all carbon atoms in a given nanotube are vibrating in phase in the radial direction has a frequency ω_{RBM} that is highly sensitive to the nanotube diameter, as discussed below, thereby providing information on both the mean diameter and the diameter distribution of a particular SWNT sample.[42]

Figure 3.5 indicates that the G-band feature consists of a superposition of two dominant components, shown at 1593 cm^{-1} (G^+) and at 1567 cm^{-1} (G^-) (see the inset to Fig. 3.5), although the G^- component exhibits a weak diameter dependence. The G^+ feature is associated with carbon atom vibrations along the nanotube axis and its frequency W_{G^+} is sensitive to charge transfer from dopant additions to SWNTs (upshifts in W_{G^+} for acceptors, and downshifts for donors). The G^- feature, in contrast, is associated with vibrations of carbon atoms along the circumferential direction of the nanotube, and its line shape is highly sensitive to whether the SWNT is metallic

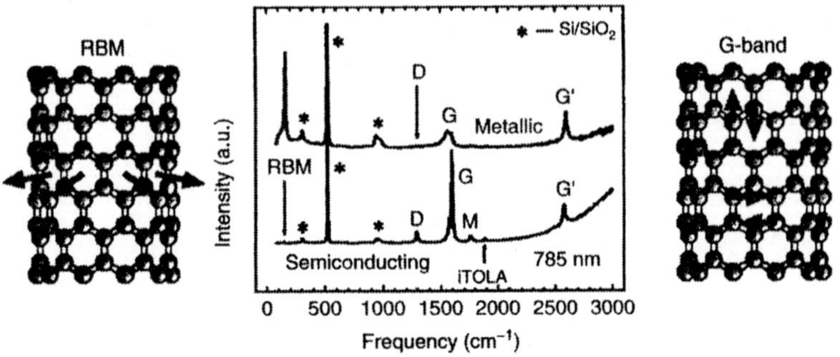

FIGURE 3.6 Raman spectra from metallic (top) and semiconducting (bottom) SWNT at the single nanotube level using 785-nm (1.58-eV) laser excitation, showing the radial breathing mode (RBM), *D*-band, *G*-band, and *G′* band features, in addition to weak double resonance features associated with the *M*-band and the iTOLA second-order modes. Insets on the left and the right show, respectively, atomic displacements associated with the RBM and *G*-band normal mode vibrations. The isolated carbon nanotubes are sitting on an oxidized silicon substrate that provides contributions to the Raman spectra denoted by asterisks and are used for calibration purposes.

(Breit–Wigner–Fano line shape) or semiconducting (Lorentzian line shape).[43,44] All in all, there are six modes contributing to the *G*-band, with two each having A, E_1, and E_2 symmetries, and each symmetry mode can be distinguished from the others by its behavior in polarization-sensitive Raman experiments.[45] The features of highest intensity are the G^+ and G^- features of A symmetry, and these components are normally used to characterize carbon nanotubes with regard to their metallicity and involvement in charge transfer. Polarization studies on SWNTs are best done at the single-nanotube level.[46]

Also commonly found in the Raman spectra in SWNT bundles are the *D*-band (with ω_D at 1347 cm^{-1} in Fig. 3.5), stemming from the disorder-induced mode in graphite, and its second harmonic, the *G′* band (Fig. 3.6) occurring at ~$2\omega_D$. Both the *D*-band and the *G′*-band are associated with a double resonance process,[47] and are sensitive to the nanotube diameter and chirality. The large dispersions of the *D*-band and *G′*-band features have been very important in revealing detailed information about the electronic and phonon properties of SWNTs as well as of the parent material graphite.[48–50]

Because of the sharp van Hove singularities occurring in carbon nanotubes with diameters less than 2 nm, the Raman intensities for the resonance process can be so large that it is possible to observe the Raman spectra from one individual SWNT,[41] as shown in Fig. 3.6, where the differences in the *G*-band spectra between semiconducting and metallic SWNTs can be seen at the single-nanotube level. Because of the trigonal warping effect, whereby the constant energy contours in reciprocal *k* space for 2-D graphite are not circles but show three-fold distortion, every (n,m) carbon nanotube has a different electronic structure and a unique density of states (Fig. 3.3). Therefore the energies E_{ii} of the van Hove singularities in the joint density

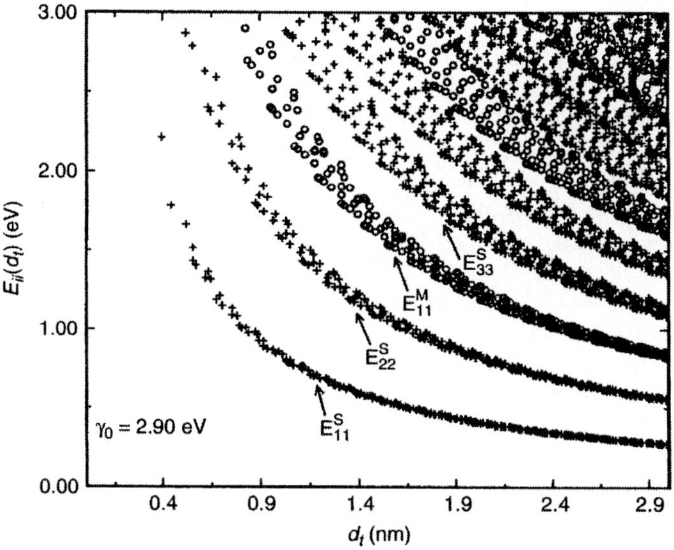

FIGURE 3.7 Calculated[51] energy separations E_{ii} between van Hove singularities i in the 1-D electronic density of states of the conduction and valence bands for all possible (n,m) values versus nanotube diameter in the range $0.4 < d_t < 3.0$ nm, using a value for the carbon–carbon energy overlap integral of $\gamma_0 = 2.9$ eV and a nearest neighbor carbon–carbon distance $a_{C-C} = 0.142$ nm for making this plot.[49,58] Semiconducting (S) and metallic (M) nanotubes are indicated by crosses and open circles, respectively. The subscript $i = 1$ denotes the index of the lowest energy of a singularity in the joint density of states.

of states for each SWNT (as shown in Fig. 3.4(b) and calculated, for example, by the tight binding approximation) are different, as shown in Fig. 3.7, where the E_{ii} (i = 1, 2, 3, ...) values for all (n,m) nanotubes in the diameter range up to 3.0 nm are shown for E_{ii} up to 3.0 eV.[51] In this so-called Kataura plot (Fig. 3.7), we see that for small-diameter SWNTs (d_t < 1.7 nm) and for the first few electronic transitions for semiconducting and metallic SWNTs, the E_{ii} values are arranged in bands, whose widths are determined by the trigonal warping effect[49] so that if there were no trigonal warping, the spread in energy at constant diameter would go to zero. A general chiral semiconducting SWNT will have van Hove singularities in the $E^S_{11}, E^S_{22}, E^S_{33}, \ldots$ bands, while a chiral metallic SWNT will have two vHSs in each of the $E^M_{11}, E^M_{22}, \ldots$ bands. The tight binding approximation provides predicted values for the E_{ii} energies for the vHSs for SWNTs with diameters in the 1.1-2.0-nm range (to an accuracy of better than 20 meV).[52] This Kataura plot can be used to estimate the appropriate laser energy that could be used to resonate with a particular (n,m) SWNT.

The Raman effect furthermore provides a determination of E_{ii} values, either by measurement of the relative intensities of the radial breathing mode for the Stokes (phonon emission) and anti-Stokes (phonon absorption) processes or by measurements of the RBM Raman intensity for the Stokes process for a particular (n,m) SWNT relative to the intensity for many SWNTs measured under similar conditions.[39] The frequency of the RBM has been used to determine the diameter of an isolated SWNT sitting on an oxidized Si surface, using the relation ω_{RBM} (cm^{-1}) = $248/d_t$ (nm), which is established by measurements on many SWNTs. Moreover, from a knowledge of the (E_{ii}, d_t) values for an individual SWNT (Fig. 3.7), the (n,m) indices for that SWNT can be determined from the Kataura plot. Resonance Raman spectroscopy at the single-nanotube level can not only determine the (n,m) indices from study of the RBM frequency and intensity, but can also determine E_{ii} values for individual SWNTs to about 10-meV accuracy, using a single laser line or to about 5-meV accuracy using the Stokes–anti-Stokes intensity analysis method.[39] Resonance Raman measurements with a tunable laser system can give E_{ii} values with an accuracy of ±3 meV.[53,54] The precision for this assignment using Stokes versus anti-Stokes Raman measurements depends on the determination of the shape of the resonance window (Raman intensity as a function of excitation laser energy E_{laser}). The resonance window could change for samples with different environments and with different sample preparation methods, and systematic work is still needed to increase the precision in determining E_{ii}. The ability to use a gate,[55] an externally applied potential,[56] or a tunable laser[53,54,57] to move the van Hove singularity for an individual SWNT into and out of resonance with the laser offers great promise for future detailed studies of the 1-D physics of single-wall carbon nanotubes using resonance Raman spectroscopy. The ability to characterize individual SWNTs for their (n,m) identification allows the determination of the dependence on nanotube diameter and chirality of many physical phenomena in SWNTs that can be measured at the single-nanotube level.

Near-field microscopy has also proven to provide a powerful tool for examining the Raman spectra on a spatially resolved basis (to 20-nm resolution), thus providing detailed information about the effects of specific defects on the vibrational spectra. The disorder-induced *D*-band provides a sensitive probe of symmetry-breaking agents such as defects in carbon nanotubes.[59,60]

Raman spectra of carbon nanotubes, particularly at the single-nanotube level, have been especially rich. Because of the simplicity of the geometrical structure of nanotubes, detailed analysis of the Raman spectra have yielded much information about the phonon dispersion relations, such as information about the trigonal warping effect for phonons. Such information was not yet available for 2-D graphite, but can now be studied in nanotubes because of their one-dimensionality.[50] Because of the close coupling between electrons and phonons under resonance conditions, Raman spectra have also provided valuable and detailed information about the electronic structure (such as an evaluation of the magnitude of the trigonal warping effect for electrons[48]), thus yielding valuable information about the electronic structure that can be used in the characterization of nanotube samples. At present, the use of many laser lines (over 50) provides the capability of carrying out Raman spectroscopy measurements with a quasicontinuous excitation energy source.[54,57] This capability promises to provide a means to access essentially all the SWNTs in a given sample, thereby providing a powerful new characterization technique for SWNTs.

3.2.5 Other Photophysical Techniques

Optical absorption measurements on carbon nanotube samples generally show broad peaks[61] corresponding to E_{11}^S, E_{22}^S, and E_{11}^M with decreasing intensity as the energy increases, as expected from the f-sum rule for optical matrix elements.[62] These peaks are characteristic of SWNTs and provide one of the most important characterization techniques to distinguish nanotubes from other carbonaceous species in an unpurified sample containing nanotubes in the presence of other carbonaceous species. The characterization measurement is carried out by comparing the absorption intensity of the E_{22}^S band relative to the background absorption in the same range of the infrared spectra. The E_{22}^S peak is selected because it is relatively strong and is not expected to be strongly perturbed by excitonic effects, as is the E_{11}^S peak. Optical absorption spectra are not only used for the characterization of such mixed samples but as an aid in the purification of such samples to isolate the nanotubes from other species.[22,63]

Although optical absorption techniques can be used to characterize almost any type of sample containing carbon nanotubes, optical emission as a characterization tool requires specially prepared nanotube samples with sufficiently long carrier lifetimes in the excited state, so that the emission process has sufficient intensity to be observable,[40] as discussed in the next paragraph.

Time-domain photoemission studies using fast optics provide information about the lifetime and electron–phonon matrix elements for the excited states. Time-domain studies on SWNT bundles have been performed, giving a measure of the lifetime of photoexcited charge carriers in SWNT ropes as a function of electron energy (Fig. 3.8).[64,65] The lifetimes of electrons excited to the π^* bands were found

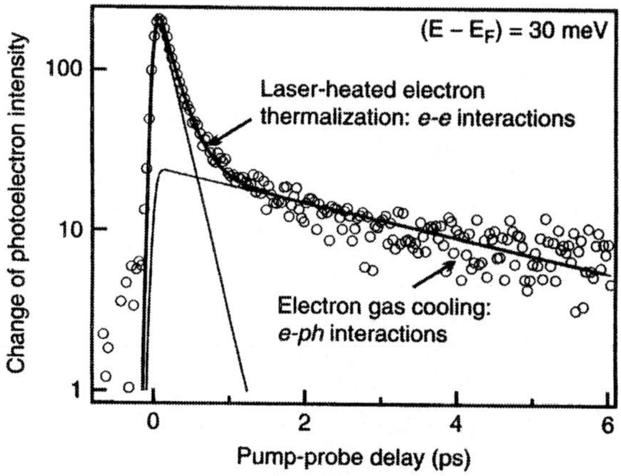

FIGURE 3.8 Time dependence of the photoelectron intensity during and after femtosecond laser generation of hot electrons in SWNT bundles.[65] Electrons monitored by this trace are 30 meV above the Fermi level. The fast and slow components, respectively, correspond to internal thermalization of the laser-heated electron gas and to its equilibration with the lattice, i.e., by cooling.

to decrease continuously from 130 fs at 0.2 eV down to less than 20 fs at energies above 1.5 eV with respect to the Fermi level.[64] These short lifetimes lead to a significant lifetime-induced broadening of the characteristic vHSs in the density of states, so that the optical effects related to the confinement of electrons into vHSs are stronger for the lower energy vHSs in the joint density of states. Experimental results show very similar excited state decay times for graphite and SWNT bundles, suggesting that electron–electron scattering of photoexcited carriers in SWNT bundles may lead to a rapid charge transfer between different tubes, thus allowing excited electron hole pairs in semiconducting SWNTs to relax almost as rapidly as those in metallic tubes. This result probably explains why photoluminescence is not observed for SWNT bundles.[40] The presence of conduction electrons in metallic tubes provides an effective nonradiative decay channel for the decay of photoexcited electrons. Thus to obtain photoluminescence spectra, the SWNTs contained in SWNT bundles need to be isolated from one another to reduce the nonradiative decay channel, and this has been done by the use of various species to wrap around individual SWNTs within a bundle to isolate individual SWNTs from each other. Some wrapping agents that have been effectively used include sodium dodecyl sulfate (SDS)[40,66] and single-stranded guanine–thyamine selected DNA chains (ss-GTDNA).[29]

The observation of photoluminescence (PL) from isolated SWNTs has made possible the observation of the E_{11}^S energy gap for semiconducting SWNTs, and interesting 3-D plots can be constructed (Fig. 3.9) showing strong intensity peaks for excitation of the electrons at E_{22}^S and emission at E_{11}^S, where each peak is related to one specific (n,m) SWNT. From this intensity plot in Fig. 3.9, a plot of the E_{11} and E_{22} electronic transition energies for many (n,m) SWNTs can be constructed.[66] By using an empirical expression for E_{ii} as a function of nanotube diameter and chiral angle for isolated SWNTs grown by the HiPco process, wrapped with SDS surfactant[67] and dispersed in aqueous solution, an assignment of the PL peaks with specific (n,m) values has been made. This assignment is based on the identification of families with $(2n + m)$ = constant, which becomes increasingly evident as the SWNT diameter decreases and curvature effects become more important. The widths of the PL peaks are ~25 meV, corresponding to room temperature thermal energy.[37,40,66,67] The strength of the PL characterization technique is that all the (n,m) semiconducting tubes in the sample are probed at once so that, in principle, their relative concentrations could be determined (if the matrix elements and lifetimes for the (n,m) emissions were known theoretically) from the measured intensities of each PL peak.

The limitation for the photoluminescence method is related to systems where nonradiative electron-hole recombination can occur, so that light emission from metallic SWNTs or SWNT in bundles cannot be observed. For such samples, resonance Raman experiments at the single-nanotube level could be alternatively used. Raman measurements on SDS-wrapped SWNTs in solution yield good agreement with E_{22}^S values obtained by PL measurements[54] within the accuracy of the PL measurements, which can only be made to 25 meV because of the broad line widths. In contrast, by using many laser lines, the accuracy of the E_{22}^S values is much greater (to 5 meV) as determined by resonance Raman spectroscopy, through measurement of both the Stokes and anti-Stokes RBM features, as discussed in Section 3.2.4.

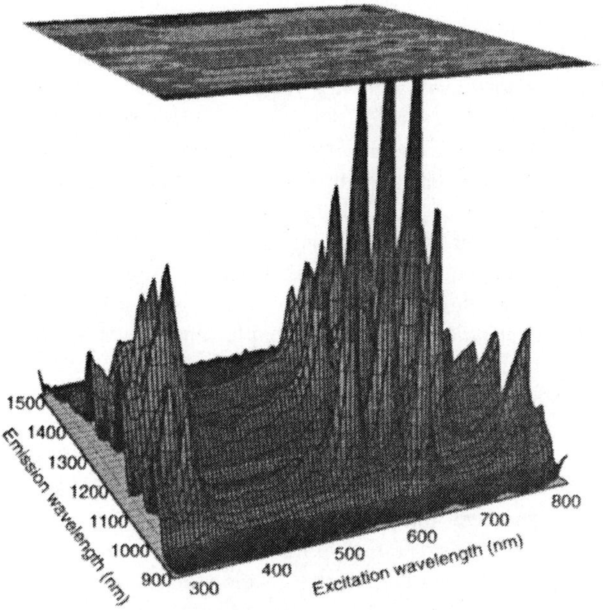

FIGURE 3.9 Fluorescence intensity versus excitation and emission wavelengths for SWNTs in an SDS suspension.[67] Since each peak corresponds to the absorption/emission of a single (n,m) SWNT, the intensity of each peak provides a measure of the relative concentrations of each (n,m) SWNT in the sample.

The discrepancies observed among various types of nanotube samples for the E_{22}^S and E_{11}^S transition energies (up to 100 meV) show that the electronic energy levels that we use to distinguish one type of (n,m) SWNT from another are, in fact, sensitive to the different wrapping agents used, and to the different environments of the SWNTs (freely suspended, isolated SWNTs on various oxide substrates, in aqueous or organic solvents, etc.).[54] Such phenomena are currently under investigation.

3.2.6 Future Directions for Nanotube Development

A large amount of information at the isolated SWNT level has provided accurate data concerning the structures and properties of single-wall carbon nanotubes and contributed in an important way to the development of more complete theoretical models to describe nanotube physics. Much effort must now focus on using these photophysics studies to characterize carbon nanotube systems in detail, because for such a nanometer scale system, the massless photon turns out to be a powerful probe for nanotube characterization without strongly disturbing this nanoscale system. For this reason we can expect spectroscopy to continue to serve as an important tool for characterizing nanotube properties reliably.

Although sample purity can be investigated using the *D*/*G* band intensity ratio in the Raman spectra from SWNTs, the development of systematic work is needed for quantitative information about sample purity using Raman spectroscopy. A quantitative procedure for the evaluation of the carbonaceous purity of bulk quantities of as-prepared SWNT soot can be obtained by the utilization of solution-phase near-infrared spectroscopy.[68] However, for defect characterization at the single-nanotube level, development of the physics related to the defect-induced Raman *D*-band feature is needed. Such a development will be important for the characterization of more advanced nanotube-based devices, where any junction, kink, doping additive, or wrapping agent will be seen by the Raman effect as a defect, and will affect device performance differently. The challenge will be to see how each of these different types of defects can provide a different and distinguishable footprint in Raman spectroscopy. By using polarized light and exploiting selection rules at the single-nanotube level and by using near-field Raman spectroscopy, a variety of challenging and promising experiments can be carried out for the development of the next important phase of carbon nanotube photophysics characterization.

For applications of SWNTs as electronic devices it is essential to have a source of inexpensive, pure single-phase material, preferably with the same (n,m) indices. Although progress has been made in both the synthesis of a more uniform carbon nanotube product and in the separation of metallic from semiconducting nanotubes,[22] the synthesis technology is not sufficiently developed for large-scale nanotube electronics applications for commercial products. Development of (n,m) selective synthesis is the most challenging present goal for the development of carbon nanotube science and applications.

3.3 BISMUTH NANOWIRES

Nanowires are especially attractive for nanoscience studies as well as for nanotechnology applications. Nanowires, compared to other low-dimensional systems, have two quantum confined directions, while still leaving one unconfined direction for electrical conduction. This allows nanowires to be used in applications where electrical conduction, rather than tunneling transport, is required. Because of quantum confinement effects, nanowires in the limit of small diameters are expected to exhibit significantly different optical and electrical properties from their bulk 3-D crystalline counterparts and this is very much the case for bismuth nanowires. Their increased surface area, diameter-dependent band overlap and bandgap, and increased surface scattering for electrons and phonons are only some of the ways in which nanowires differ from their corresponding bulk materials. However the diameters of typical bismuth nanowires are large enough (>5 nm) to have crystal structures closely related to their parent material, thereby allowing theoretical predictions about their properties to be made on the basis of an extensive literature relevant to bulk bismuth properties. Not only do bismuth nanowires exhibit many properties that are similar to, and others that are distinctly different from, those of their bulk counterparts, but they also have the advantage from an applications standpoint that some of their materials parameters that are critical for certain physical properties can be independently controlled in their nanowire form, but not in bulk bismuth, such as, for example, thermal and electrical conductivity. Certain properties can also be enhanced nonlinearly in small-diameter nanowires by exploiting the singular aspects of the 1-D electronic density of states. Bismuth nanowires provide a promising framework for applying the bottom-up approach[69] for the design of nanostructures for nanoscience investigations and for potential applications as, for example, for thermoelectricity. It is the aim of this section to focus on bismuth nanowire properties that differ from those of bulk bismuth, with an eye toward possible applications that might emerge from the unique properties of these nanowires and from future discoveries in this field.

3.3.1 Bismuth Nanowire Synthesis

Bismuth (Bi) nanowires in principle can be prepared by many techniques[70] and it is expected that the synthesis of Bi nanowires by various novel techniques will become an even more active current research area than it has been. Most of the presently available physical property studies of bismuth nanowires have been made on nanowires grown in porous anodic alumina templates. A variety of other templates are available for use in growing bismuth nanowires, such as those prepared from nanochannel glass, block copolymers, and ion track-etched polymer and mica films. Using porous anodic alumina templates (Fig. 3.10), highly crystalline bismuth nanowires with diameters in the 7- to 200-nm range and lengths up to 10 μm have been prepared with parallel alignment and a high packing density (~10^{11} nanowires/cm^2), which is favorable for a number of applications.[71] Nanochannel Vycor glass has been used to prepare Bi nanowires down to 4-nm diameter,[72] and the nanowires thus prepared have been useful for establishing the value of the wire diameter below which localization effects become important in electron transport. However, the random organization of the pores, and consequently of the wires filling the template pores, would limit the range of applications of such nanowire arrays. Polymer-based templates might be desirable because of their favorable properties for certain applications (such as low thermal conductivity of the host material, which is desirable for thermoelectric applications), but because of their limitations for use

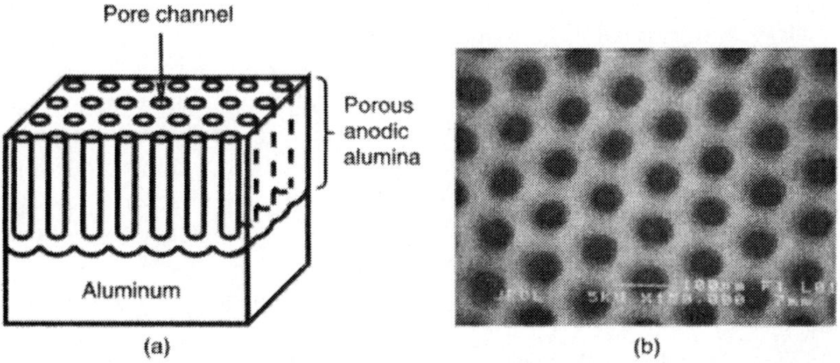

FIGURE 3.10 Porous anodic alumina templates showing (a) a schematic and (b) an SEM picture of the top surface.[75] The pores of these templates are filled to form nanowires.

at elevated temperatures, the filling of polymer-based templates would likely have to be done near room temperature by electrochemical means.[73,74]

The filling of the porous anodic alumina templates (Fig. 3.10) to form Bi nanowire arrays has been accomplished by at least three methods, including pressure injection, vapor growth, and electrochemical methods. Most of the properties measurements have been made with Bi nanowires prepared by pressure injection and vapor growth, although the electrochemical method is by far the most versatile and is the most logical candidate for scale-up for commercial applications.

The pressure injection technique can be employed for fabricating bismuth nanowires because of the low melting point of this material (271.4°C). In the high-pressure injection method, the nanowires are formed by pressure injecting the desired material in liquid form into the evacuated pores of the template. Due to the heating and the pressurization processes, the templates used for the pressure injection method must be chemically stable and be able to maintain their structural integrity at about 50 K above the melting point of bismuth at the injection pressure. Anodic aluminum oxide films and nanochannel glass are two typical host materials that have been used as templates in conjunction with the pressure injection filling technique.[76–78]

The pressure P required to overcome the surface tension for the liquid material to fill the pores with a diameter d_W is determined by the Washburn equation, $P = -4\gamma\cos\theta/d_W$, where γ is the surface tension of the liquid, and θ is the contact angle between the liquid and the template.[79] To reduce the required pressure and to maximize the filling factor, some surfactants are used to decrease the surface tension and the contact angle. For example, it is found that the introduction of a small amount of Cu in the Bi melt can facilitate the filling of the anodic alumina pores with liquid Bi and increase the number of nanowires that are prepared.[80] A typical pressure used for the preparation of bismuth nanowires is 0.3 to 1.5 kbar.[80,81] Nanowires produced by the pressure injection technique usually possess high crystallinity and a preferred crystal orientation along the wire axis.

The electrochemical deposition technique has attracted increasing attention as a promising alternative for fabricating nanowires because of its versatility and ability to produce bismuth nanowires of smaller diameter. Traditionally electrochemistry has been used to grow thin films on conducting surfaces. Since electrochemical growth is usually controllable in the direction normal to the substrate surface, this method can be readily extended to fabricate both quantum wire- and quantum dot-based nanostructures, if the deposition is confined within the pores of an appropriate template. In the electrochemical methods, a thin conducting metal film is first applied on one side of the porous membrane to serve as the working electrode for electroplating. The length of the deposited nanowires can be controlled by varying the duration of the electroplating process. This method has been used to synthesize a wide variety of electrically conducting nanowires including bismuth nanowires.[82]

In the electrochemical deposition process, the chosen template must be chemically stable in the electrolyte during the electroplating process. Cracks and defects in the templates are detrimental to nanowire growth because electrochemical deposition will then occur primarily in the more accessible cracks, thus leaving most of the nanopores unfilled. Particle track-etched mica films and polymer membranes are typical templates used in the simple DC electrolysis preparation method. To use anodic aluminum oxide films for DC electrochemical deposition, the insulating barrier layer that separates the pores from the bottom aluminum substrate has to be removed, and a metal film is then evaporated onto the back of the template membrane.[83] The electrochemical filling of anodic alumina templates has been carried out on both Bi and Bi_2Te_3.[84–86]

It is also possible to employ an AC electrodeposition method in filling anodic alumina templates without the removal of the barrier layer by utilizing the rectifying properties of the oxide barrier. In AC electrochemical deposition, although the applied voltage is sinusoidal and symmetric, the current is greater during the cathodic half-cycles, making deposition dominant over the etching, which occurs in the subsequent anodic half-cycles. Since no rectification occurs at defect sites, the deposition and etching rates are equal, and no material is deposited. Hence the difficulties associated with cracks are avoided. Pulse current electroplating techniques have been used for the synthesis of Bi nanowires[87,88] and they appear to be advantageous for the growth of crystalline wires generally because the metal ions in the solution can be regenerated between the electrical pulses and, therefore, more uniform deposition conditions can be produced for each deposition pulse.

One advantage of the electrochemical deposition technique is the possibility of fabricating multilayered structures within nanowires. By varying the cathodic potentials in the electrolyte, which contains two different kinds of ions, layers of different compositions can be controllably deposited. This electrodeposition method provides a low-cost approach to preparing multilayered 1-D nanostructures. One disadvantage of the electrochemical deposition approach for applications is that the Bi nanowires fabricated by the electrochemical process are often polycrystalline, with no preferred crystal orientations. Progress has been made in improving the crystallinity of Bi nanowires prepared by the electrochemical route by careful control of the deposition parameters.[75,89–91]

Vapor deposition is a convenient method for filling the template pores with bismuth and is usually capable of preparing smaller diameter ($\leq$20 nm) nanowires than pressure injection methods produce because it does not rely on high pressure to overcome the surface tension involved in inserting the bismuth into the pores of the template. In the physical vapor deposition technique, the bismuth is first heated to produce a vapor, which is then introduced through the pores of the template, and the template is subsequently cooled to solidify the bismuth nanowires. Using a specially designed experimental setup,[92] nearly single-crystal Bi nanowires in porous anodic alumina templates with pore diameters as small as 7 nm have been synthesized, and these Bi nanowires were found to possess a preferred crystal growth orientation along the wire axis, similar to the Bi nanowires prepared by pressure injection.[92]

Once the bismuth nanowires are grown in a template, it is possible to etch away the template by appropriate chemical treatments to gain access to the individual nanowires. An example of such free-standing bismuth nanowires on a silicon substrate is shown in Fig. 3.11.

Although porous alumina templates provide a convenient method for preparing arrays of nanowires for study and demonstration purposes, these arrays have some

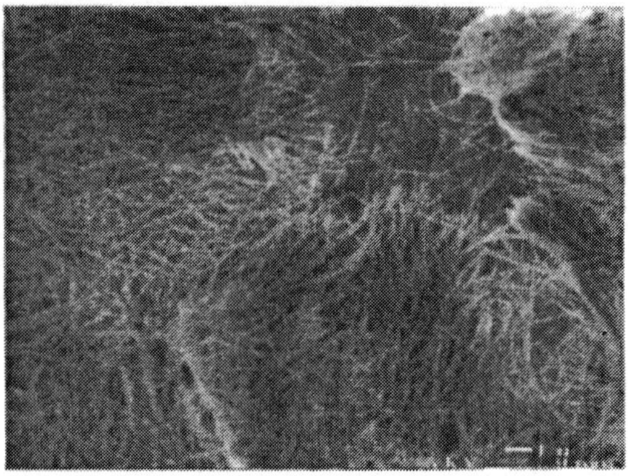

FIGURE 3.11 Bismuth nanowires on a silicon wafer exposed by etching a porous anodic alumina film.

severe disadvantages for large-scale applications. Porous anodic alumina is a brittle ceramic film grown on a soft aluminum metal substrate. Great care needs to be exercised in the preparation of the aluminum substrate and in the manipulation of the anodic film to produce pure defect-free porous anodic alumina films that are required in order to achieve uniform filling of the pores with the nanowire material. Between the porous layer and the aluminum substrate lies a continuous and dense alumina barrier layer that prevents the direct physical and electrical contact between the pore (or nanowire) and the substrate. It is therefore of great interest to develop methods to grow porous anodic alumina films on rigid substrates of technological relevance and to establish contact of both ends of the pores (or of the nanowires) while the film is still attached to the substrate. Such technology would allow integration of addressable arrays of parallel nanowires as active components in electronic, thermoelectric, optical, field emission, and sensing devices.

Techniques have recently been developed to grow arrays of bismuth nanowires on the surface of a silicon wafer coated with a porous anodic alumina film.[93] A conducting layer under the porous film serves as the working electrode for the electrochemical growth of the nanowire array. An example of bismuth nanowires grown electrochemically on the surface of a silicon wafer is shown in Fig. 3.12.[75,93] Unlike free-standing porous alumina films containing bismuth nanowires, the films grown on a silicon wafer are robust enough to allow growth on large areas of the silicon wafer. Furthermore, patterned porous films and patterned electrodes can be used to control the location and the shape of the nanowire array using conventional microelectronics fabrication techniques.

3.3.2 Structural Characterization

Thus far, most of the emphasis on the structural characterization of bismuth nanowires has been in terms of degree of crystallinity and the crystalline orientation of the nanowire axis. The reason for this focus on crystallinity stems from the

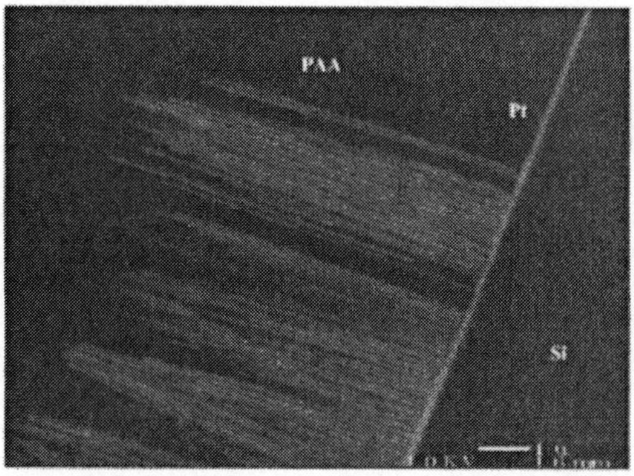

FIGURE 3.12 Cross-section of an array of bismuth nanowires grown on a platinum-coated silicon wafer, allowing more robust growth of bismuth nanowires in porous anodic alumina films over large areas on a silicon wafer.[93]

extraordinarily large anisotropy of the effective mass components and consequently of the constant electronic energy surfaces of crystalline bismuth.[71] This anisotropy is responsible for the exceptional electronic properties of bismuth and for the interest in bismuth nanowires both as model systems for studying 1-D physics phenomena and for possible future practical applications.

Scanning electron microscopy (SEM) produces images of bismuth nanowires down to length scales of ~10 nm, and provides valuable information regarding the structural arrangement, the spatial distribution of the nanowires, the filling factor of the nanowires in space, the morphological features of individual nanowires, and arrays of nanowires. Since many of the studies of bismuth nanowires focused on arrays of bismuth nanowires in templates, SEM has also provided valuable characterization of the templates used to grow the nanowires to determine the pore diameters and diameter distribution in the template arrays, the packing fraction, and the filling factor, as illustrated in Fig. 3.10, in which the top view of a typical empty anodic alumina template is shown. After the template has been filled with bismuth (by either liquid phase pressure injection, vapor phase injection, or electrochemical filling), the filling factor can be determined utilizing the very high contrast of the secondary electron emission (electron absorption) of the bismuth nanowire as compared to the very low interaction of the electron probes with the empty pores of the template.

Transmission electron microscopy provides a wide range of characterization information about bismuth nanowires including their crystal structure, crystal quality, grain size, and crystalline orientation of the nanowire axis. When operating in the diffraction mode, selected area electron diffraction (SAED) patterns can be used to determine the crystal structures of nanowires. The amount of crystallinity of individual nanowires is evaluated from lattice fringe images, such as the one shown in Fig. 3.13(a). Here we can see from the regularity of the lattice fringes that pressure injection produces bismuth nanowires with a high degree of crystallinity. The crystalline quality of bismuth nanowires obtained by vapor phase growth is also very high.

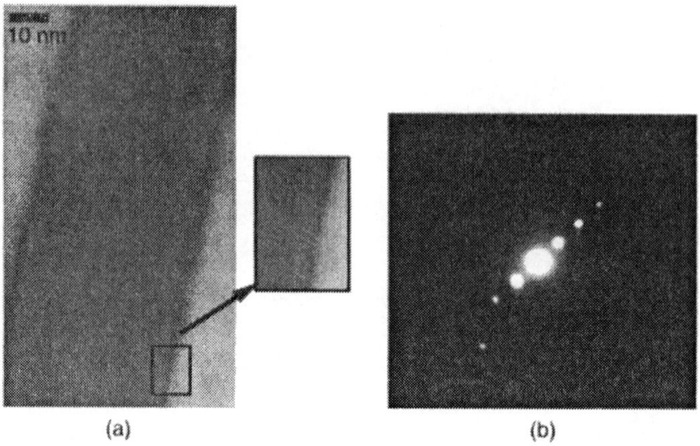

FIGURE 3.13 (a) HRTEM image of a free-standing individual Bi nanowire with a diameter of 45 nm. The inset, shown in an enlarged form on the right, highlights the lattice fringes of the nanowire, indicating high crystallinity. (b) Selected area electron diffraction (SAED) pattern of the same nanowire in the TEM image.[94]

To establish the crystalline direction of the nanowire axis, SAED measurements are made at the individual nanowire level (Fig. 3.13(b)). Such SAED patterns can also be used to establish the degree of crystallinity along a single nanowire by taking multiple SAED spectra at different points along the nanowire axis, and SAED measurement of several nanowires provides information about the distribution of the crystalline orientations.

X-ray diffraction (XRD) measurements on an array of aligned bismuth nanowires inside their templates can be used to acquire information about the crystallographic orientation of the nanowires within a single template, as shown in Fig. 3.14. Note the high degree of crystallinity with >80% crystalline orientation of the nanowire along a specific nanowire direction. The large diameter (>95 nm) nanowires show alignment predominantly along a (101) axis, while the smaller diameter nanowires (≤40 nm) show predominant alignment along an (012) axis, suggesting a wire diameter-dependent crystal growth mechanism. On the other hand, 30-nm Bi nanowires produced using a much higher pressure of >1.5 kbar show a different crystal orientation of (001) along the wire axis,[81] indicating that the preferred crystal orientation may also depend on the applied pressure, with the most dense packing direction along the wire axis for the highest applied pressure. The crystallinity and orientation of vapor-grown bismuth nanowires in the alumina templates are similar to those of nanowires prepared by pressure injection. Although early work on the electrochemical preparation of Bi nanowires produced fewer crystalline nanowires than the vapor-grown or pressure injection methods, subsequent efforts[89–91] have produced aligned and single-crystal nanowires, with crystal orientations along the wire axes depending on the preparation conditions. X-ray energy dispersion (EDS) analysis also provides information on the impurity content of the bismuth nanowires.

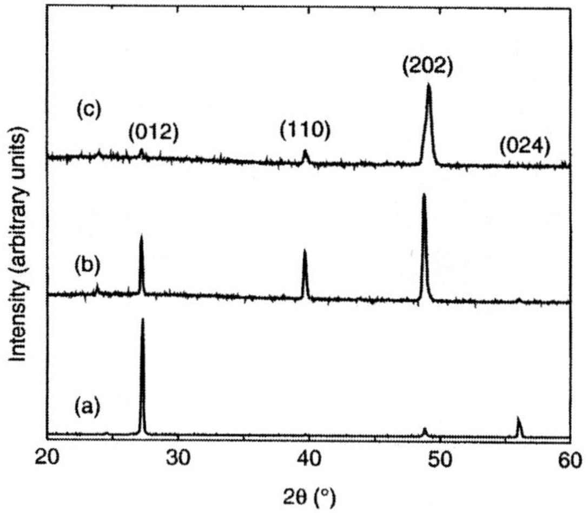

FIGURE 3.14 XRD patterns of bismuth/anodic alumina nanocomposites with average bismuth wire diameters of (a) 40 nm, (b) 52 nm, and (c) 95 nm.[78] The Miller indices corresponding to the lattice planes of bulk Bi are indicated above the individual peaks. The majority of the Bi nanowires are oriented along the $[10\bar{1}1]$ and $[01\bar{1}2]$ directions for $d_W \geq$ 60 nm and $d_W \leq 50$ nm, respectively.[78] The existence of more than one dominant orientation in the 52-nm Bi nanowires is attributed to the transitional behavior of intermediate diameter nanowires as the preferential growth orientation is shifted from $[10\bar{1}1]$ to $[01\bar{1}2]$ with decreasing d_W.

XRD, EDS and TEM characterizations of $Bi_{1-x}Sb_x$ alloy nanowires for $x \leq 0.15$ prepared by the pressure injection method[94] show good compositional homogeneity along the nanowire axis and good crystallinity of the alloy nanowires, with the same preferred crystalline alignment of the nanowire axis, as is the case of pure bismuth nanowires. On the other hand, a gradient in alloy composition along the nanowire axis was determined by EDS in 200-nm Sb-rich $Bi_{1-x}Sb_x$ nanowires electrochemically deposited in porous alumina templates.[75] It is significant that the SAED patterns (such as in Fig. 3.13(b)) noted for individual nanowires by electron diffraction are in good agreement with the x-ray diffraction patterns taken on large arrays of nanowires, as in Fig. 3.14.

The high resolution of the TEM also allows for the investigation of the surface structures of bismuth nanowires that form native surface oxides when exposed to oxidative conditions (i.e., atmospheric air). Knowledge of the surface composition and structure is critical for transport applications of bismuth and related nanowires, as, for example, for thermoelectric cooling or thermal management applications.[71] In an attempt to control the formation of these oxide surface layers, the surface structures of bismuth nanowires have been studied by using an *in situ* environmental TEM chamber that allowed TEM observations to be made while different gases were introduced or as the sample was heat treated at various temperatures, as illustrated in Fig. 3.15. This figure shows the high resolution TEM images of a Bi nanowire with an oxide coating

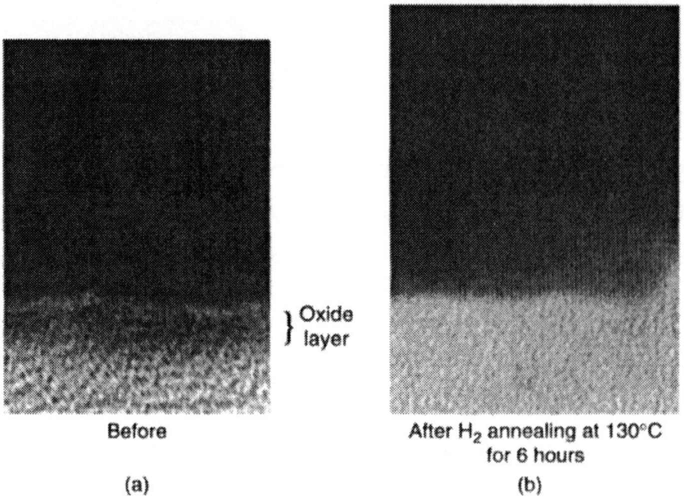

FIGURE 3.15 High-resolution transmission electron microscope image of a Bi nanowire (a) before and (b) after annealing in hydrogen gas at 130°C for 6 h within the environmental chamber of the instrument to remove the oxide surface layer.[95]

and the effect of a dynamic oxide removal process carried out within the environmental chamber of the TEM.[95] The amorphous bismuthoxide layer coating of the nanowire (Fig. 3.15(a)) is removed by exposure of the wire to hydrogen gas within the environmental chamber of the TEM, as indicated in Fig. 3.15(b).

By coupling the powerful imaging capabilities of TEM with other characterization tools, such as an electron energy loss spectrometer (EELS) or an energy dispersive x-ray spectrometer within the TEM instrument, additional properties of the nanowires can be probed with high spatial resolution. With the EELS technique, the energy and momentum of the incident and scattered electrons are measured in an inelastic electron scattering process to provide information on the energy and momentum of the excitations in the nanowire sample. Figure 3.16 shows the dependence on nanowire diameter of the electron energy loss spectra of Bi nanowires.[96] The spectra were taken from the center of the nanowire, and the shift in the energy of the peak position (Fig. 3.16) is used to monitor the plasmon frequency in the bismuth nanowires as the nanowire diameter is varied. The results show changes in the electronic structures of Bi nanowires as the wire diameter decreases. Such changes in electronic structure as a function of nanowire diameter are related to quantum confinement effects that are also observed in their transport and optical properties, as discussed in the next section.

3.3.3 Electronic Characterization

Bismuth in bulk form is a semimetal, and therefore when undoped, bulk bismuth has an equal number of electrons and holes, each with highly anisotropic constant energy surfaces. The corresponding highly anisotropic Fermi surfaces make bismuth

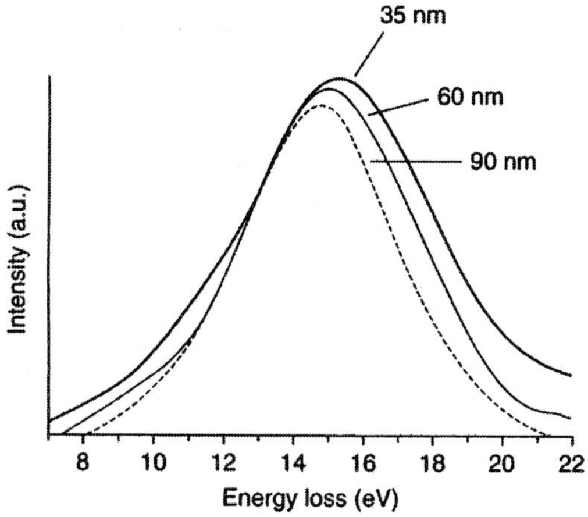

FIGURE 3.16 Electron energy loss spectra taken from the centers of bismuth nanowires with diameters of 35, 60, and 90 nm. The shift in the volume plasmon peaks is due to the wire diameter effects on the electronic structure.[96]

very interesting from an applications standpoint, because this aspect of bismuth makes the electronic properties of bismuth nanowires so different from bulk bismuth. Because of the very small effective mass components of bismuth, the subbands formed under quantum confined conditions are widely spaced in energy. Thus, as the nanowire diameter decreases, the lowest conduction subband increases in energy, while the highest valence subband decreases in energy, thereby decreasing the band overlap between the valence and conduction band, which is 38 meV for bulk bismuth and for Bi nanowires of very large diameter (Fig. 3.17). The figure further shows that at a nanowire diameter of about 50 nm, the band overlap at 77 K goes to zero and bismuth nanowires become semiconducting. As the nanowire diameter further decreases, the bandgap of the semiconducting nanowires increases, as is also seen in Fig. 3.17. The figure also shows that the diameter where the semimetal–semiconductor transition occurs is dependent on the crystalline orientation of the nanowire along its axis, with the transition occurring at the largest diameter (55 nm) for wires oriented along the highest symmetry direction, which is the three-fold trigonal symmetry direction.

What is exceptional about bismuth is that because of the highly anisotropic constant energy surfaces, the effective mass components normal to the light mass component can be quite heavy. This means that it is possible to achieve very high carrier mobility along the nanowire axis by having conduction along a high mobility, low effective-mass crystallographic direction, while at the same time having a reasonably high density of states, associated with the heavy carrier masses. Such an unusual situation is very beneficial for applications such as thermoelectricity.

It would seem that the best way to study these anisotropic properties would be through measurements of the temperature dependence of the resistivity of individual bismuth nanowires. However, the incorporation of bismuth nanowires into test device

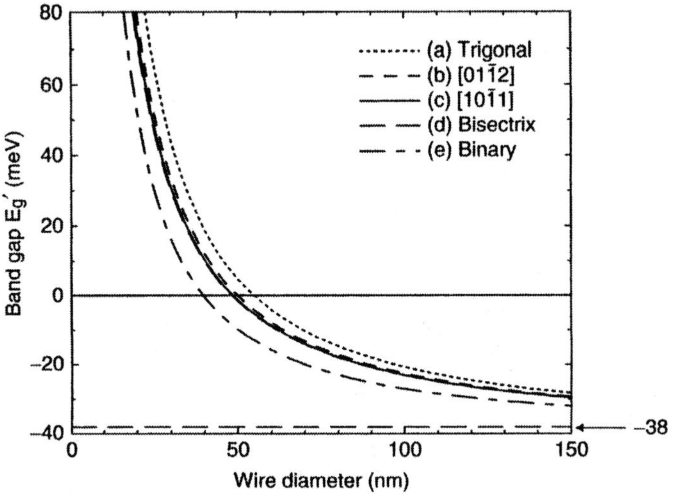

FIGURE 3.17 Calculated bandgap energy between the lowest electron subband and the highest hole subband of Bi nanowires oriented along the (a) trigonal, (b) [01 $\bar{1}$2], (c) [01$\bar{1}$1], (d) bisectrix, and (e) binary crystallographic directions as functions of wire diameter. The nonparabolic effects of the *L*-point electron pockets have been taken into account, and cylindrical wire boundary conditions were used in the calculations. The values of d_c, the critical wire diameter, where the semimetal-to-semiconductor transition occurs, are 55.1, 50.0, 49.0, 48.5, and 39.8 nm for nanowires oriented along the trigonal, [01 $\bar{1}$ 2], [01 $\bar{1}$ 1], bisectrix, and binary directions, respectively. The curves for the bisectrix and [01 $\bar{1}$1] directions are too close to one another to be resolved.[94]

structures for such measurements has not yet been explored extensively. Two- and four-point resistance test structures have been prepared (see Fig. 3.18) by photolithography (or e-beam lithography), and metal deposition of contacts has been achieved on individual bismuth nanowires dispersed on an oxidized silicon wafer substrate.[97,98] The challenge in studying the temperature dependence of individual bismuth nanowires has been the fabrication of reliable contacts to these nanowires, because of the native oxide coating that forms on their surfaces. Progress thus far has been made in developing reliable ohmic contacts that are stable over the 2 to 400 K temperature range. A method for avoiding the burnout of bismuth nanowires due to electrostatic discharge has also been developed.[97]

The experimental verification of the transition from semimetallic bulk bismuth to a semiconducting nanowire can be observed in transport measurements of the temperature-dependent resistance of nanowire arrays, as shown in Fig. 3.19(a). Here the temperature-dependent resistance is normalized to its value at 300 K because the number of nanowires in the template to which electrical contact is made is not known. Figure 3.19(a) shows that the temperature-dependent normalized resistance $R(T)/R(300\ K)$ changes radically from the curve for the bulk sample to that for a 200-nm diameter nanowire sample, where a peak in the normalized

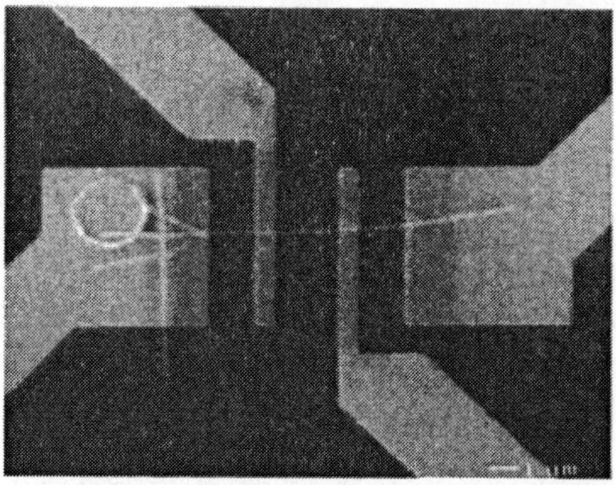

FIGURE 3.18 SEM image of four contacts to a single bismuth nanowire used for transport measurements.[98]

$R(T)/R(300\text{ K})$ curve appears. The peak gains prominence as fewer and fewer subbands contribute to the carrier conduction, as shown by the experimental results for the 70-nm diameter nanowire. This behavior is further confirmed by a model calculation of the normalized resistance $R(T)/R(300\text{ K})$ (Fig. 3.19(b)). Finally when the wire diameter enters into the semiconductor regime below 50 nm (Fig. 3.17), the experimental

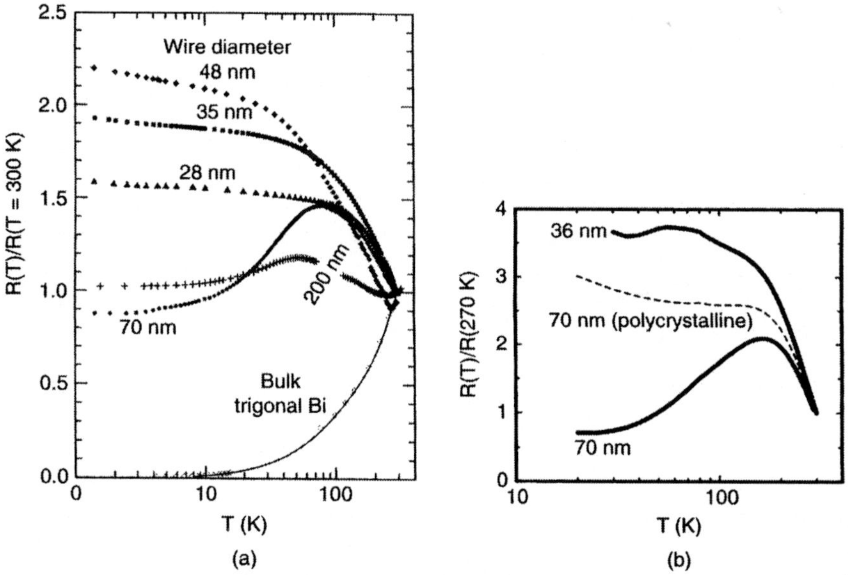

FIGURE 3.19 (a) Measured T dependence of the normalized resistance $R(T)/R(300\text{ K})$ for highly crystalline Bi nanowires of various d_w.[92] (b) Calculated $R(T)/R(300\text{ K})$ for 36- and 70-nm Bi nanowires.[78] The dashed line refers to a 70-nm polycrystalline wire with increased grain boundary scattering.

curves for $R(T)/R(300\text{ K})$ in Fig. 3.19(a) show a different monotonic functional form, as seen by comparing the experimental curves in this lower diameter range with results shown for wire diameters of 200 nm and 70 nm.[92] Model calculations for nanowires in the semiconducting regime (Fig. 3.19(b)) reproduce the observed form of $R(T)/R(300\text{ K})$.[78] To further verify the validity of the model calculations, a curve for $R(T)/R(300\text{ K})$ for polycrystalline bismuth wires was calculated (Fig. 3.19(b)) and these results were also in good agreement with measurements of Hong et al.[99]

Further evidence for the semimetal–semiconductor transition in bismuth nanowires comes from measurements made on randomly oriented bismuth nanowires of very small diameter prepared in the pores of silica, Vycor glass, and alumina templates.[72] The wire diameters in this case were sufficiently small so that the bandgaps were large enough to allow large resistance changes of several orders of magnitude over the measured temperature range, so that a fitting of the experimental points could be made to the functional form $R(T) = R_0\exp(-E_g/kT)$, as shown in Fig. 3.20(a), in order to obtain a measure of the bandgap. The results thus obtained for the semiconducting bandgap are shown in Fig. 3.20(b),[72] and good agreement was obtained for these experimental results with theoretical predictions[100] for the

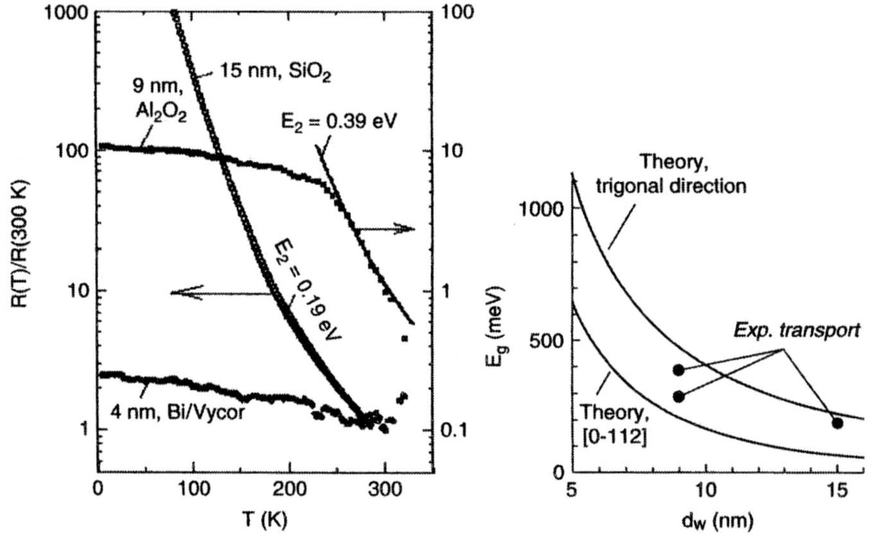

FIGURE 3.20 (a) Temperature dependence of the normalized resistance $R(T)/R(300\text{ K})$ for randomly oriented Bi nanowires with various diameters in the range $d_w \leq 15$ nm in silica, alumina, and Vycor glass templates, as indicated in the figure. The points are experimental data, while the lines are fits to $R(T) = R_0\exp(-E_g/kT)$.[72] (b) Calculated[100] values for the energy gaps of Bi nanowires as functions of wire diameter d_w for two different crystallographic orientations of the nanowire axis (Fig. 3.17). The points are the experimental values of E_g obtained from fits to $R(T) = R_0\exp(-E_g/kT)$ fitted to three different samples as shown in (a).[72]

temperature dependence of the semiconducting bandgap. Using these very thin bismuth nanowires allowed exploration for carrier localization effects at low temperature.[75]

The semimetal–semiconductor transition in nanowires can also be probed by examination of the effect of doping. In the case of bismuth nanowires, doping with isoelectronic antimony allows variation of the nanowire diameter at which the semimetal-to-semiconductor phase transition occurs. As a 3-D system, $Bi_{1-x}Sb_x$ alloys form a substitutional 3-D solid-state system over the entire range of compositions x. As shown in Fig. 3.21(a) the band overlap of bismuth decreases upon Sb addition and the alloys become semiconductors for x in the range $0.07 < x < 0.22$ with a bandgap that depends on x. For $0.09 < x < 0.16$, Fig. 3.21(a) shows that the $Bi_{1-x}Sb_x$ bulk alloy is a direct gap semiconductor, with both electron and hole pockets located at the L points in the Brillouin zone (see inset of Fig. 3.21(b)). In the range $0.07 < x < 0.09$, the $Bi_{1-x}Sb_x$ bulk alloy is an indirect gap semiconductor with a hole pocket at the T point, while in the range $0.16 < x < 0.22$, the alloy is an indirect gap semiconductor with holes forming around the six H-points shown in the inset of Fig. 3.21(b). Therefore by antimony addition, the electronic band structure of $Bi_{1-x}Sb_x$ can be modified over a wide range of electronic behaviors, suggesting that $Bi_{1-x}Sb_x$ alloy nanowires should constitute an interesting 1-D system in which the electronic band structure and related properties can be varied over an even wider range of behaviors than for 3-D $Bi_{1-x}Sb_x$ by combining the quantum confinement effect with the Sb alloying effect.

Figure 3.21(b) displays the phase diagram of $Bi_{1-x}Sb_x$ nanowires as a function of wire diameter and Sb concentration.[101] This phase diagram is calculated[101] assuming the same unit cell geometry as occurs in the parent bulk $Bi_{1-x}Sb_x$ alloy, consistent with x-ray diffraction measurements. As can be seen from Fig. 3.21(b), there are two regions where a semimetal-to-semiconductor phase transition occurs, one involving L-point electrons and T-point holes (upper left of Fig. 3.21(b)), out to nearly $x = 0.05$ for a 100-nm nanowire, and the second region involving L-point electrons and H-point holes (upper right of Fig. 3.21(b)), extending to $x = 0.24$ for a 100-nm nanowire, but providing a semiconducting phase for small-diameter wires over a much larger Sb concentration range than in the bulk alloy.[101] The inset in Fig. 3.21(b) shows the Brillouin zone of the $Bi_{1-x}Sb_x$ alloy system and the locations of the L, H, and T points in the rhombohedral Brillouin zone. The semimetal-to-semiconductor phase transition involving the T-point holes and the L-point electrons, as indicated by the phase boundary line in Fig. 3.21(b), have been probed experimentally by temperature-dependent normalized resistance measurements, as discussed subsequently. The critical wire diameter d_c for the semimetal-to-semiconductor transition is about 50 nm for pure bismuth nanowires at 77 K, but d_c increases with increasing Sb concentration for $Bi_{1-x}Sb_x$ alloy nanowires. We note from Fig. 3.21(b) that the x range of the semiconducting phase for $Bi_{1-x}Sb_x$ nanowires broadens as the wire diameter decreases.

The semimetal-to-semiconductor phase transition was determined in the $Bi_{1-x}Sb_x$ nanowire system by measurement of the normalized temperature-dependent resistance $R(T)/R(270\text{ K})$ for 65-nm diameter nanowires of different Sb alloy composition x, as shown in Fig. 3.22. In this figure the measured normalized resistance of pure bismuth shows a non-monotonic T dependence with a maximum at ~70 K,

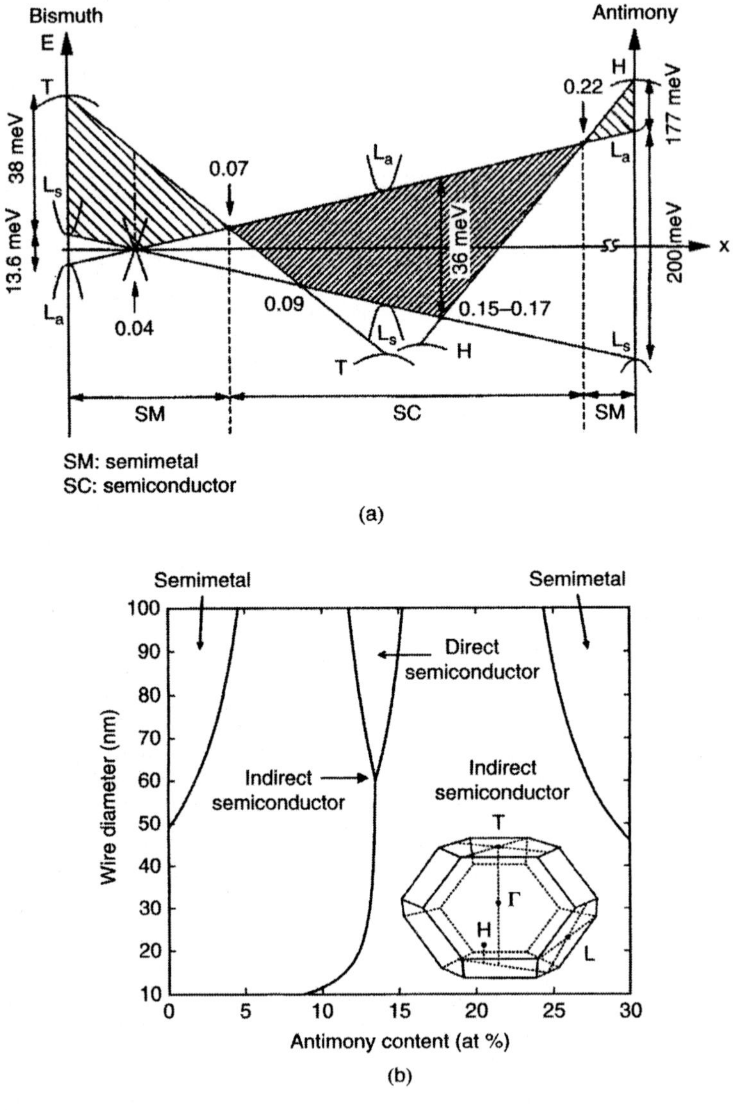

FIGURE 3.21 Diagrams for the electronic phases for (a) 3-D bulk $Bi_{1-x}Sb_x$ plotted as energy versus composition[75,102] and for (b) $Bi_{1-x}Sb_x$ nanowires as a function of wire diameter and Sb concentration.[101]

while the resistivities of the alloy nanowire samples (with 5 at.% and 10 at.% Sb) decrease monotonically with increasing *T*, characteristic of the semiconducting phase (Fig. 3.19(a)). Furthermore, *R*(*T*)/*R*(270 K) for the $Bi_{0.95}Sb_{0.05}$ nanowires displays a stronger *T* dependence than that of the $Bi_{0.90}Sb_{0.10}$ nanowires. The saturation of *R*(*T*)/*R*(270 K) at low *T* for all the three nanowire compositions is attributed to carriers associated with uncontrolled impurities that are estimated to

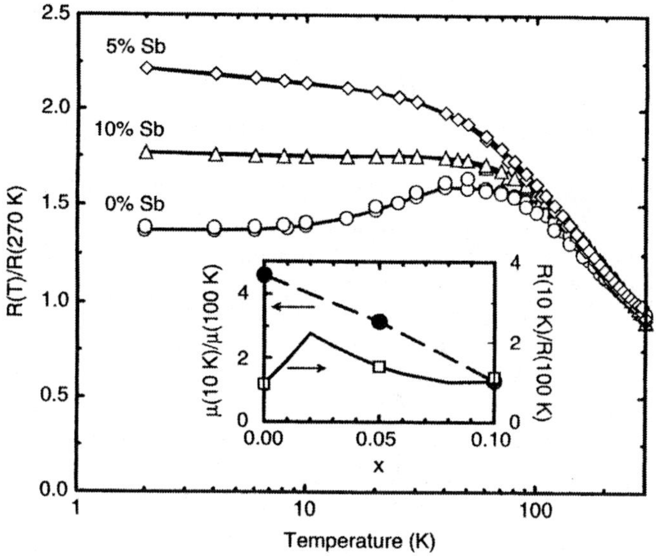

FIGURE 3.22 Measured temperature dependence of the 270 K normalized resistance of 65-nm $Bi_{1-x}Sb_x$ nanowires with different Sb contents. The inset shows ratios of the mobility $\mu(10\ K)/\mu(100\ K)$ between 10 K and 100 K for the three samples (•) and the ratios of the resistance $R(10\ K)/R(100\ K)$ (□) as a function of Sb content, where the points are experimental values and the curves are predictions.[103]

have a concentration of about 1 to 4 × $10^{16}/cm^3$. The carriers associated with the uncontrolled impurities become increasingly important as the bandgap of the semiconductor increases (Fig. 3.17). Based on the calculated temperature-dependent carrier concentration and the measured normalized temperature-dependent resistance $R(T)/R(270\ K)$ shown in Fig. 3.22, the temperature-dependent average mobility ratio $\mu(T)/\mu(100\ K)$ was obtained for the three samples. These calculations were then used to predict the Sb concentration at which the semimetal to semiconductor transition occurs, as indicated in Fig. 3.22, and the result was found to be in very good agreement with the theoretical prediction for this transition at $x \simeq 0.02$ for a 65-nm $Bi_{1-x}Sb_x$ nanowire, as shown in Fig. 3.21(b). Confirmation of the phase diagram shown in Fig. 3.21(b) allows the diagram to be used for guidance in designing nanowires in the semiconducting phase with some control of the bandgap of semiconducting bismuth as well as some control of other nanowire properties.

Measurements of the temperature-dependent Seebeck effect using the smallest available thermocouples for temperature sensors, as shown in Fig. 3.23(a), provide further support for the phase diagram shown in Fig. 3.21(b). The data in Fig. 3.23(b) show an increase in the magnitude of the Seebeck coefficient either as the wire diameter of bismuth nanowires decreases or as Sb is added to form a $Bi_{1-x}Sb_x$ alloy. Further enhancement of the magnitude of the Seebeck coefficient is achieved when both the nanowire diameter decreases and Sb is added, as shown in Fig. 3.23(b).[103,104]

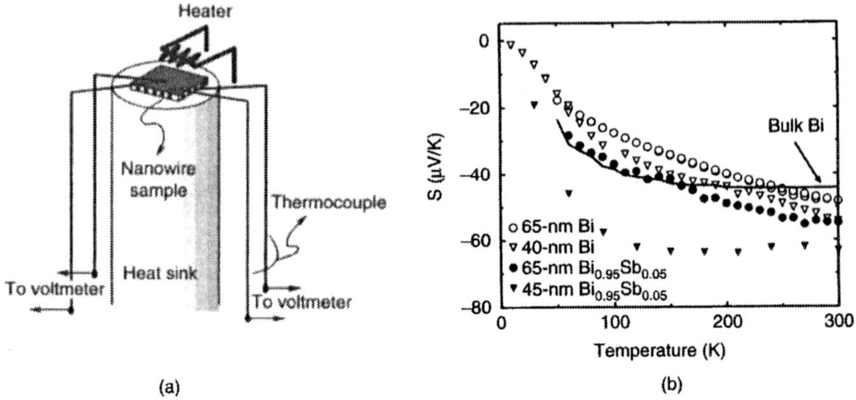

FIGURE 3.23 (a) Experimental setup of Seebeck coefficient measurement for nanowire arrays. (b) Measured Seebeck coefficient as a function of T for Bi and $Bi_{0.95}Sb_{0.05}$ nanowires with different diameters. The solid curve denotes the Seebeck coefficient for bulk Bi.[103]

3.3.4 Optical Characterization

Even though the bismuth nanowire diameter is typically an order of magnitude smaller than the wavelength of light, optical reflectivity and transmission measurements have nevertheless been used successfully to characterize bismuth nanowires.[105,106] What is surprising is that the intense and sharp absorption peak that dominates the infrared absorption spectrum in bismuth nanowires,[105,106] is not observed in bulk bismuth. The energy position E_p of this strong absorption peak increases with decreasing diameter. However, the rate of increase in energy with decreasing diameter $|\partial E_p/\partial d_W|$ is an order of magnitude less than that predicted for either a direct interband transition or for intersubband transitions in bismuth nanowires. On the other hand, the magnitude of $|\partial E_p/\partial d_W|$ agrees well with that predicted for an indirect L-point valence to T-point valence band transition (see inset to Fig. 3.21(b)). Since both the initial and final states for the indirect L-T point valence band transition downshift in energy as the wire diameter d_W is decreased, the shift in the absorption peak results from a *difference* between the effective masses for these two carrier pockets and is less sensitive to the actual value of either of the masses. Hence the diameter dependence of the absorption peak energy is an order of magnitude less for a valence-to-valence band indirect transition than for a direct interband L-point transition. Furthermore the band-tracking effect for this indirect transition gives rise to a large value for the joint density of states, thus accounting for the high intensity of this feature. This indirect transition arises through the gradient of the dielectric function, which is large at the bismuth–air or bismuth–alumina interfaces. In addition, the intensity can be quite significant because abundant initial state electrons, final state holes, and appropriate phonons are available for making an indirect L-T point valence band transition at room temperature. Interestingly the polarization dependence of this absorption peak is such that the strong absorption is present when the electric field is perpendicular to the wire axis, but is absent when the electric field is parallel to the wire axis, contrary to a traditional polarizer, such as a carbon nanotube where the optical E field is polarized by the nanotube itself to be aligned along the carbon nanotube axis. The observed polarization dependence for bismuth nanowires is consistent with a surface-induced effect that increases the coupling between the L-point and T-point bands throughout the full volume of the nanowire. The indirect L to T point valence band transition mechanism[105,106] is also consistent with observations of the effect on the optical spectra of a decrease in the nanowire diameter and of n-type doping of bismuth nanowires with tellurium.[105]

3.3.5 Future Directions for Bismuth Nanowires

Bismuth nanowires are likely to provide a model system for future nanowire studies because of their unique properties, namely their highly anisotropic constant energy surfaces that allow these nanowires to simultaneously have high mobility carriers at a reasonable high electronic density of states and to exhibit quantum effects at relatively high nanowire diameter. The anisotropy associated with the nanowire geometric structure breaks the symmetry so that the wire axis can be chosen to be along a low effective mass–high carrier mobility direction simultaneously. In this way, the highly anisotropic constant energy contours of bismuth can be explicitly exploited to yield the quantum confinement of the carriers that is responsible for the various semimetal-to-semiconductor electronic phase transitions. This phase transition allows the special electronic properties to be utilized in a semiconducting form of pure bismuth and over an extended range of wire diameter and of chemical composition in $Bi_{1-x}Sb_x$ alloy nanowires. Using the same dopants used in bulk bismuth to add electrons (from column IV in the periodic table) and to add holes (from column VI), semiconducting devices based on this nanowire system should be possible, for example, for thermoelectric applications.

Future progress in this field will be significantly enhanced by developing better bismuth nanowire synthesis methods yielding nanowires that are highly crystalline (so that the anisotropic electronic energy band structure can be exploited) and to which good electrical contacts can be made (circumventing the problems caused by surface oxidation when Bi nanowires are exposed to ambient conditions). If templates are used for nanowire self-assembly, electrical isolation and crystallographic alignment of nanowire arrays, more rugged template materials are needed, and for thermoelectric applications, template materials with lower thermal conductivity would be needed. The ability to control material properties that are linked together in bulk bismuth but are not so closely linked in bismuth nanowires requires further exploration through model-based selection of wire diameter and $Bi_{1-x}Sb_x$ alloy composition.

3.3.6 Concluding Remarks

When considering the one-dimensionality of bismuth, some comment is in order about the ability of bismuth (and many other materials systems for that matter) to form nanotubes and nanowires by a variety of synthesis approaches, including growth in templates initiated at the pore interfaces and using very short growth times.[107] One interesting aspect of a bismuth nanotube is that the wall thickness can be a single basic structural unit consisting of two atomic layers. The diffraction pattern for bismuth nanotubes reveals lattice constants similar to those of 3-D bismuth. Research on the physical properties of bismuth nanotubes is at an early stage, and the electronic structure remains to be explored. Study of the electronic structures of bismuth nanotubes in relation to those of bismuth nanowires with similar local atomic arrangements and similar outer diameters should prove to be fascinating. Also intriguing should be the comparison of the electronic structure of a bismuth nanotube to that of a carbon nanotube. Although bismuth and graphite are both semimetals as 3-D crystals, their electronic structures as 1-D nanowires and nanotubes, respectively, are very different, and it is expected that the electronic structure of a bismuth nanotube will differ significantly from its nanowire counterpart, as well as from that of a carbon nanotube.

From these considerations we can conclude that we are now within reach of studying the rich variety of electronic structures that 1-D systems might exhibit, learning how to realize this variety of structures experimentally and how to exploit these different electronic structures to yield interesting materials properties not yet achieved in any system, and someday be able to utilize these special properties for practical applications.

ACKNOWLEDGMENTS

A.J. acknowledges financial support by PRPq-UFMG, and the Instituto de Nanociências (Millennium Institute Program), CNPq, Brazil. O.R. and M.S.D. acknowledge support under NSF Grant DMR 04-05538.

PROBLEMS

1. Based on the cutting lines concept, explain why for an (n,m) single-wall carbon nanotube (SWNT) the nanotube is a large bandgap semiconductor, when $n - m$ is not a multiple of three, with the band gap increasing as the diameter decreases ($1/d_t$ dependence), while the nanotube shows metallic behavior (at room temperature) for $n - m = 3j$ (j = integer). Compare the electronic density of states near the Fermi level for semiconducting and metallic SWNTs and for a 2-D graphene sheet.
2. How are single-wall carbon nanotubes formed in the laboratory? Which mechanism for SWNT growth (catalytic root growth or tip growth) is most probable for the synthesis of very long (~cm) tubes? What are the major challenges that need to be addressed in advancing carbon nanotube synthesis?
3. List the common methods for nanotube characterization, and their advantages and limitations. What is it that each characterization method probes and why is it important to gain this information?
4. Why is Raman spectroscopy a particularly valuable tool for the study and characterization of carbon nanotube systems? What are its advantages and limitations?

5. Using the tight binding model, find an expression for the electronic energy levels of an armchair nanotube. Make an explicit application of your expression to a (10,10) nanotube.
6. For very small-diameter nanotubes, at what diameter does the tube curvature impose sp^3 bonding rather than sp^2 bonding. Explain in what sense we can consider this cross-over point to indicate that the graphitic nanotube becomes a nanodiamond.
7. Find the electronic energy levels for a free electron nanowire with a square cross section (a^2) and a length $10a$. At what temperature would you expect quantum confinement effects to become important at room temperature?
8. Find the electronic energy levels of a silicon nanowire of diameter d_w with its wire axis along (100) and then repeat for its wire axis along (111). Use the effective mass approximation and effective mass tensor components for 3-D Si. Explain how to find the effective mass that governs transport along the nanowire axis and how to find the threshold for optical transitions.
9. Assume that the carriers of a nanowire of diameter d_w have an effective mass of $0.1m_0$ in the plane of the nanowire and $0.05m_0$ along the nanowire axis. At what magnetic field value would the electron scattering rate for carriers in a nanowire of diameter d_w decrease rapidly for a magnetic field parallel to the nanowire axis? Repeat for the case of a magnetic field perpendicular to the wire axis.
10. Consider a silicon nanowire to be the same as in Problem 7. At what magnetic field along the nanowire axis is the spacing of the Landau levels (magnetic energy levels) equal to the subband spacing? Explain why carrier localization is expected as the nanowire cross-sectional area becomes very small.

REFERENCES

1. Dresselhaus, M.S., Dresselhaus, G., and Eklund, P.C., *Science of Fullerenes and Carbon Nanotubes,* Academic Press, New York, 1996.
2. Dresselhaus, M.S., Dresselhaus, G., and Avouris, Ph., *Carbon Nanotubes: Synthesis, Structure, Properties and Applications,* Vol. 80 of *Springer Series in Topics in Applied Physics,* Springer-Verlag, Berlin, 2001.
3. Harris, P., *Carbon Nanotubes and Related Structures: New Materials for the Twenty-First Century,* Cambridge University Press, Cambridge, UK, 2001.
4. Bandow, S., Hiraoka, K., Chen, G., Eklund, P.C., and Iijima, S., *Bull. Mater. Res. Soc., 29,* 260–264, 2004.
5. Saito, R., Dresselhaus, G., and Dresselhaus, M.S., *Physical Properties of Carbon Nanotubes,* Imperial College Press, London, 1998.
6. Dresselhaus, M.S., Dresselhaus, G., Charlier, J.C., and Hernández, E., *Phil. Trans. Roy. Soc. A, 362,* 2065–2098, 2003.
7. Hamada, N., Sawada, S., and Oshiyama, A., *Phys. Rev. Lett., 68,* 1579–1581, 1992.

8. Saito, R., Fujita, M., Dresselhaus, G., and Dresselhaus, M.S., *Appl. Phys. Lett., 60,* 2204–2206, 1992.
9. Mintmire, J.W., Dunlap, B.I., and White, C.T., *Phys. Rev. Lett., 68,* 631–634, 1992.
10. Samsonidze, Ge.G., Grüneis, A., Saito, R., Jorio, A., Souza Filho, A.G., Dresselhaus, G., and Dresselhaus, M.S., *Phys. Rev. B, 69,* 205402, 1–11, 2004.
11. Samsonidze, Ge.G., Saito, R., Jorio, A., Pimenta, M.A., Souza Filho, A.G., Grüneis, A., Dresselhaus, G., and Dresselhaus, M.S., *J. Nanosci. Nanotechnol., 3,* 431–458, 2003.
12. Chico, L., Crespi, V.H., Benedict, L.X., Louie, S.G., and Cohen, M.L., *Phys. Rev. Lett., 76,* 971–974, 1996.
13. Liu, J., Fan, S., and Dai, H., *Bull. Mater. Res. Soc., 29,* 244–250, 2004.
14. Bethune, D.S., Kiang, C.H., de Vries, M.S., Gorman, G., Savoy, R., Vazquez, J., and Beyers, R., *Nature* (London), *363,* 605, 1993.
15. Journet, C., Maser, W.K., Bernier, P., Loiseau, A., Lamy de la Chapelle, M., Lefrant, S., Deniard, P., Lee, R., and Fischer, J.E., *Nature* (London), 388, 756–758, 1997.
16. Thess, A., Lee, R., Nikolaev, P., Dai, H., Petit, P., Robert, J., Xu, C., Lee, Y.H., Kim, S.G., Rinzler, A.G., Colbert, D.T., Scuseria, G.E., Tománek, D., Fischer, J.E., and Smalley, R.E., *Science, 273,* 483–487, 1996.
17. Nilolaev, A.V., Bronikowski, M.J., Bradley, R.K., Rohmund, F., Colbert, D.T., Smith, K.A., and Smalley, R.E., *Chem. Phys. Lett., 313,* 91–97, 1999.
18. Maruyama, S., *Physica B, 323,* 193–195, 2002.
19. Hayashi, T., Kim, Y.A., Matoba, T., Esaka, M., Nishimura, K., Endo, M., and Dresselhaus, M.S., *Nano Lett., 3,* 887–889, 2003.
20. Kitiyanan, B., Alvarez, W.E., Harwell, J.H., and Resasco, D.E., *Chem. Phys. Lett., 317,* 497, 2000.
21. Alvarez, W.E., Kitiyanan, B., Borgna, A., and Resasco, D.E., *Carbon, 39,* 547, 2001.
22. Haddon, R.C., Sippel, J., Rinzler, A.G., and Papadimitrakopoulos, F., *Bull. Mater. Res. Soc., 29,* 252–259, 2004.
23. Liu, K., Chien, C.L., and Searson, P.C., *Phys. Rev. B, 58,* R14681–R14684, 1998.
24. Rinzler, A.G., Liu, J., Dai, H., Nikolaev, P., Huffman, C.B., and Rodriguez-Marcias, F.J., *Appl. Phys. A, 67,* 29–37, 1998.
25. Chattopadhyay, D., Galeska, I., and Papadimitrakopoulos, F., *Carbon, 40,* 985–988, 2002.
26. Endo, M., Kim, Y.A., Fukai, Y., Hayashi, T., Terrones, M., Terrones, H., and Dresselhaus, M.S., *Appl. Phys. Lett., 79,* 1531–1533, 2001.
27. Endo, M., Hayashi, T., Kim, Y.A., Terrones, M., and Dresselhaus, M.S., *Phil. Trans. Roy. Soc. 362*, 2223–2238, 2004
28. Chattopadhyay, D., Galeska, I., and Papadimitrakopoulos, F., *J. Am. Chem. Soc., 125,* 3370–3375, 2003.
29. Zheng, M., Jagota, A., Semke, E.D., Diner, B.A., McLean, R.S., Lustig, S.R., Richardson, R.E., and Tassi, N.G., *Nature Mater., 2,* 338–342, 2003.
30. Zheng, M., Jagota, A., Strano, M.S., Barone, P., Chou, S.G., Diner, B.A., Dresselhaus, M.S., McLean, R.S., Onoa, G.B., Santos, A.P., Semke, E.D., Usrey, M., and Walls, D.J., *Science, 302,* 1545–1548, 2003.
31. Lebedkin, S., Arnold, K., Hennrich, F., Krupke, R., Renker, B., and Kappes, M.M., *New J. Phys., 5,* 140, 2003.
32. Strano, M.S., *J. Am. Chem. Soc., 125,* 16148–16153, 2003.
33. Strano, M.S., Miller, M.K., Allen, M.J., Moore, V.C., O'Connell, M.J., Kittrell, C., Hauge, R.H., and Smalley, R.E., *J. Nanosci. Nanotechnol., 3,* 81–86, 2003.
34. Strano, M.S., *J. Am. Chem. Soc., 125,* 16148–16150, 2004.

35. Chen, G., Sumanasekera, G.U., Pradhan, B.K., Gupta, R., Eklund, P.C., Bronikowski, M.J., and Smalley, R.E., *J. Nanosci. Nanotech., 2,* 621–626, 2002.
36. Kataura, H., Kimura, A., Ohtsuka, Y., Suzuki, S., Maniwa, Y., Hanyu, T., and Achiba, Y., *Jpn. J. Appl. Phys., 37,* L616–L618, 1998.
37. Hagen, A., and Hertel, T., *Nanoletters, 3,* 383–388, 2003.
38. Rao, A.M., Richter, E., Bandow, S., Chase, B., Eklund, P.C., Williams, K.W., Fang, S., Subbaswamy, K.R., Menon, M., Thess, A., Smalley, R.E., Dresselhaus, G., and Dresselhaus, M.S., *Science, 275,* 187–191, 1997.
39. Dresselhaus, M.S., Dresselhaus, G., Jorio, A., Souza Filho, A.G., and Saito, R., *Carbon, 40,* 2043–2061, 2002.
40. O'Connell, M.J., Bachilo, S.M., Huffman, X.B., Moore, V.C., Strano, M.S., Haroz, E.H., Rialon, K.L., Boul, P.J., Noon, W.H., Kittrell, C., Ma, J., Hauge, R.H., Weisman, R.B., and Smalley, R.E., *Science, 297,* 593–596, 2002.
41. Jorio, A., Saito, R., Hafner, J.H., Lieber, C.M., Hunter, M., McClure, T., Dresselhaus, G., and Dresselhaus, M.S., *Phys. Rev. Lett., 86,* 1118–1121, 2001.
42. Dresselhaus, M.S., and Eklund, P.C., *Advan. Phys., 49,* 705–814, 2000.
43. Pimenta, M.A., Marucci, A., Empedocles, S., Bawendi, M., Hanlon, E.B., Rao, A.M., Eklund, P.C., Smalley, R.E., Dresselhaus, G., and Dresselhaus, M.S., *Phys. Rev. B Rapid, 58,* R16016–R16019, 1998.
44. Brown, S.D.M., Jorio, A., Corio, P., Dresselhaus, M.S., Dresselhaus, G., Saito, R., and Kneipp, K., *Phys. Rev. B, 63,* 155414, 2001.
45. Jorio, A., Dresselhaus, G., Dresselhaus, M.S., Souza, M., Dantas, M.S.S., Pimenta, M.A., Rao, A.M., Saito, R., Liu, C., and Cheng, H.M., *Phys. Rev. Lett., 85,* 2617–2620, 2000.
46. Jorio, A., Pimenta, M.A., Souza Filho, A.G., Samsonidze, Ge.G., Swan, A.K., Ünlü, M.S., Goldberg, B.B., Saito, R., Dresselhaus, G., and Dresselhaus, M.S., *Phys. Rev. Lett., 90,* 107403, 2003.
47. Thomsen, C., and Reich, S., *Phys. Rev. Lett., 85,* 5214, 2000.
48. Souza Filho, A.G., Jorio, A., Samsonidze, Ge.G., Dresselhaus, G., Dresselhaus, M.S., Swan, A.K., Ünlü, M.S., Goldberg, B.B., Saito, R., Hafner, J.H., Lieber, C.M., and Pimenta, M.A., *Chem. Phys. Lett., 354,* 62–68, 2002.
49. Saito, R., Dresselhaus, G., and Dresselhaus, M.S., *Phys. Rev., B, 61,* 2981–2990, 2000.
50. Samsonidze, Ge.G., Saito, R., Jorio, A., Souza Filho, A.G., Grüneis, A., Pimenta, M.A., Dresselhaus, G., and Dresselhaus, M.S., *Phys. Rev. Lett., 90,* 027403, 2003.
51. Kataura, H., Kumazawa, Y., Maniwa, Y., Umezu, I., Suzuki, S., Ohtsuka, Y., and Achiba, Y., *Synthet. Met., 103,* 2555–2558, 1999.
52. Souza Filho, A.G., Chou, S.G., Samsonidze, Ge.G., Dresselhaus, G., Dresselhaus, M.S., Lei An, J. Liu, Anna Swan, K., Ünlü, M.S., Goldberg, B.B., Jorio, A., Grüneis, A., and Saito, R., *Phys. Rev. B, 69,* 115428, 2004.
53. Jorio, A. , Souza Filho, A.G., Dresselhaus, G., Dresselhaus, M.S., Saito, R., Hafner, J.H., Lieber, C.M., Matinaga, F.M., Dantas, M.S.S., and Pimenta, M.A., *Phys. Rev. B, 63,* 245416, 2001.
54. Fantini, C., Jorio, A., Souza, M., Mai Jr., A.J., Strano, M.S., Dresselhaus, M.S., and Pimenta, M.A., *Phys. Rev. Lett., 93,* 147406, 2004.
55. Cronin, S.B., Barnett, R., Tinkham, M., Chou, S.G., Rabin, O., Dresselhaus, M.S., Swan, A.K., Ünlü, M.S., and Goldberg, B.B., *Appl. Phys. Lett., 84,* 2052–2055, 2004; *Virtual J. Nanoscale Sci. Technol.,* 2004.
56. Lin, Yu-Ming, and Dresselhaus, M.S., *Appl. Phys. Lett., 83,* 3567–3569, 2003.

57. Fantini, C., Jorio, A., Souza, M., Ladeira, L.O., Pimenta, M.A., Souza Filho, A.G., Saito, R., Samsonidze, Ge.G., Dresselhaus, G., and Dresselhaus, M.S., *Phys. Rev. Lett., 93,* 087401, 2004.
58. Dresselhaus, G., Pimenta, M.A., Saito, R., Charlier, J.-C., Brown, S.D.M., Corio, P., Marucci, A., and Dresselhaus, M.S. in Tománek, D. and Enbody, R.J., Eds., *Science and Applications of Nanotubes,* 275–295, Kluwer Academic, New York, 2000.
59. Hartschuh, A., Sanchez, E.J., Xie, X.S., and Novotny, L., *Phys. Rev. Lett., 90,* 95503, 2003.
60. Hartschuh, A., Pedrosa, H.N., Novotny, L., and Krauss, T.D., *Science, 301,* 1354–1356, 2003.
61. Saito, R., and Kataura, H., *Carbon Nanotubes: Synthesis, Structure, Properties and Applications,* Dresselhaus, M.S., Dresselhaus, G., and Avouris, P. Eds., pp. 213–246, Springer Series in Topics in Applied Physics, Vol. 80, Springer-Verlag, Berlin, 2001.
62. Kuzmany, Hans, *Solid-State Spectroscopy: An Introduction,* Springer-Verlag, Berlin, Vol. 50, p. 115. 1998.
63. Sen, R., Rickard, S.M., Itkis, M.E., and Haddon, R.C., *Chem. Mater., 15,* 4273, 2003.
64. Hertel, T., and Moos, G., *Chem. Phys. Lett., 320,* 359–364, 2000.
65. Hertel, T., and Moos, G., *Phys. Rev. Lett., 84,* 5002–5005, 2000.
66. Bachilo, S.M., Strano, M.S., Kittrell, C., Hauge, R.H., Smalley, R.E., and Weisman, R.B., *Science, 298,* 2361–2366, 2002.
67. Weisman, R.B., and Bachilo, S.M., *Nanoletters, 3,* 1235–1238, 2003.
68. Niyogi, S., Hamon, M.A., Hu, H., Zhao, B., Bhowmik, P., Sen, R., Itkis, M.E., and Haddon, R.C., *Acc. Chem. Res., 35,* 1105–1113, 2002.
69. Feynman, Richard P. (1959). From a talk reported at http://www.zyvex.com/nanotech/feynman.html.
70. Dresselhaus, M.S., Lin, Y.-M., Rabin, O., Black, M.R., and Dresselhaus, G., in Bhushan, B., Ed., *Springer Handbook of Nanotechnology,* Springer-Verlag, Heidelberg, 2004, pp. 99–145.
71. Dresselhaus, M.S., Lin, Yu-Ming, Cronin, S.B., Rabin, O., Black, M.R., Dresselhaus, G., and Koga, T. in *Semiconductors and Semimetals: Recent Trends in Thermoelectric Materials Research III,* Tritt, T. M., Academic Press, San Diego, 2001 Ed., 1 pp. 1–121.
72. Heremans, J., in *Thermoelectric Materials 2003—Research and Applications: MRS Symposium Proceedings, Boston, December 2003,* Nolas, G.S., Yang, J., Hogan, T.P., and Johnson, D.C., Eds., Materials Research Society Press, Pittsburgh, 2004, pp. 3–14.
73. Whitney, T.M., Jiang, J.S., Searson, P.C., and Chien, C.L., *Science, 261,* 1316, 1993.
74. Piraux, L., Dubois, S., Duvail, J.L., and Radulescu, A., *J. Mater. Res., 14,* 3042–3050, 1999.
75. Rabin, Oded, *Bismuth Nanowire and Antidot Array Studies Motivated by Thermoelectricity,* Ph.D. thesis, Massachusetts Institute of Technology, Cambridge, June 2004.
76. Huber, C.A., Huber, T.E., Sadoqi, M., Lubin, J.A., Manalis, S., and Prater, C.B., *Science, 263,* 800–802, 1994.
77. Zhang, Z., Ying, J.Y., and Dresselhaus, M.S., *J. Mater. Res., 13,* 1745–1748, 1998.
78. Lin, Y.-M., Cronin, S.B., Ying, J.Y., Dresselhaus, M.S., and Heremans, J.P., *Appl. Phys. Lett., 76,* 3944–3946, 2000.
79. Adamson, A.W., *Physical Chemistry of Surfaces,* Wiley, New York, 1982, p. 338.
80. Zhang, Z., Gekhtman, D., Dresselhaus, M.S., and Ying, J.Y., *Chem. Mater., 11,* 1659–1665, 1999.

81. Huber, T.E., Graf, M.J., Foss, Jr., C.A., and Constant, P., *J. Mater. Res., 15,* 1816–1821, 2000.
82. Liu, K., Chien, C.L., Searson, P.C., and Kui, Y.Z., *Appl. Phys. Lett., 73,* 1436–1438, 1998.
83. Almawlawi, D., Liu, C.Z., and Moskovits, M., *J. Mater. Res., 9,* 1014–1018, 1994.
84. Sapp, S.A., Lakshmi, B.B., and Martin, C.R., *Adv. Mater., 11,* 402–404, 1999.
85. Prieto, A.L., Sander, M.S., Martin-Gonzalez, M.S., Gronsky, R., Sands, T., and Stacy, A.M., *J. Am. Chem. Soc., 123,* 7160–7161, 2001.
86. Sander, M.S., Gronsky, R., Sands, T., and Stacy, A.M., *Chem. Mater., 15,* 335–339, 2003.
87. Yin, A.J., Li, J., Jian, W., Bennett, A.J., and Xu, J.M., *Appl. Phys. Lett., 79,* 1039–1041, 2001.
88. Li, L., Zhang, Y., Li, G., and Zhang, L., *Chem. Phys. Lett. 378,* 244–249, 2003.
89. Jin, C.G., Jiang, G.W., Liu, W.F., Cai, W.L., Yao, L.Z., Yao, Z., and Li, X.G., *J. Mater. Chem., 13,* 1743–1746, 2003.
90. Toimil Molares, M.E., Chtanko, N., Cornelius, T.W., Dobrev, D., Enculescu, I., Blick, R.H., and Neumann, R., *Nanotechnology, 15,* S201–S207, 2004.
91. Rabin, O., Chen, G., and Dresselhaus, M.S., *The Electrochemical Society Meeting Abstracts, 2003-02,* Abs. 25, 2003.
92. Heremans, J., Thrush, C.M., Lin, Yu-Ming, Cronin, S., Zhang, Z., Dresselhaus, M.S., and Mansfield, J.F., *Phys. Rev. B, 61,* 2921–2930, 2000.
93. Rabin, O., Herz, P.R., Lin, Y.-M., Akinwande, A.I., Cronin, S.B., and Dresselhaus, M.S., *Adv. Funct. Mater., 13,* 631–638, 2003.
94. Yu-Ming, Lin., *Thermoelectric Properties of* $Bi_{1-x}Sb_x$ *and Superlattice Nanowires,* Ph.D. thesis, Massachusetts Institute of Technology, Cambridge, June 2003.
95. Cronin, S.B., Lin, Y.-M., Rabin, O., Black, M.R., Dresselhaus, G., Dresselhaus, M.S., and Gai, P.L., *Microscopy Microanalysis, 8,* 58–63, 2002.
96. Sander, M.S., Gronsky, R., Lin, Y.-M., and Dresselhaus, M.S., *J. Appl. Phys., 89,* 2733–2736, 2001.
97. Cronin, S.B., Lin, Y.-M., Rabin, O., Black, M.R., Ying, J.Y., Dresselhaus, M.S., Gai, P.L., Minet, J.-P., and Issi, J.-P., *Nanotechnology J., 13,* 653–658, 2002.
98. Cronin, Stephen B., *Electronic Properties of Bi Nanowires,* Ph.D. thesis, Massachusetts Institute of Technology, Cambridge, June 2002.
99. Hong, K., Yang, F.Y., Liu, K., Reich, D.H., Searson, P.C., and Chien, C.L., *J. Appl. Phys., 85,* 6184–6186, 1999.
100. Lin, Y.-M., Sun, X., and Dresselhaus, M.S., *Phys. Rev. B, 62,* 4610–4623, 2000.
101. Rabin, O., Lin, Yu-Ming, and Dresselhaus, M.S., *Appl. Phys. Lett., 79,* 81–83, 2001.
102. Lenoir, B., Cassart, M., Mickenand, J.P., and Scharrer, H., Scharrer, S., *J. Phys. Chem. Solids 57,* 89–99, 1996.
103. Lin, Y.-M., Rabin, O., Cronin, S.B., Ying, J.Y., and Dresselhaus, M.S., *Appl. Phys. Lett., 81,* 2403–2405, 2002.
104. Lin, Y.-M., Rabin, O., Cronin, S.B., Ying, J.Y., and Dresselhaus, M.S., in Caillat, T., and Snyder, J. *21st International Conference on Thermoelectrics: ICT Symposium Proceedings, Long Beach, CA*, 2002, pp. 253–256.
105. Black, Marcie R., *The Optical Properties of Bi Nanowires,* Ph.D. thesis, Massachusetts Institute of Technology, Cambridge, June 2003.
106. Black, M.R., Hagelstein, P.L., Cronin, S.B., Lin, Y.-M., and Dresselhaus, M.S., *Phys. Rev. B, 68,* 235417, 2003.
107. Dresselhaus, M.S., Lin, Y.-M., Rabin, O., Jorio, A., Souza Filho, A.G., Pimenta, M.A., Saito, R., Samsonidze, Ge.G., and Dresselhaus, G., *Mater. Sci. Engin. C, 23,* 129–140, 2003.

4 Nanobelts and Nanowires of Functional Oxides

Xudong Wang and Zhong Lin Wang

CONTENTS

4.1 INTRODUCTION

Nanotechnology can be defined as the creation and utilization of materials, devices, and systems through the control of matter on the nanometer (nm) scale, that is, at the level of atoms, molecules, supermolecular structures, and with at least one characteristic dimension measured in nanometers (1 ~ 100 nm). Nanomaterials and systems are rationally designed to exhibit novel, unique, and significantly improved physical, chemical, and biological properties, phenomena, and processes because of their small sizes. Depending on the number of nanosized dimensions, nanomaterials can be classified as two-dimensional (2-D) thin/nanolayers, one-dimensional (1-D) nanowires, and zero-dimensional quantum dots. Quasi-1-D nanostructures[1] have recently attracted considerable research interest due to their unique properties and wide range of applications. They show great potential both for understanding size-dependent electrical, optical, thermal, and mechanical properties and for fabricating nanosized electrical junctions and optoelectronic and electromechanical devices. The nanobelt,[2] an important new member in the 1-D nanomaterial family, has a well-defined structure and is expected to be a good candidate as a 1-D nanoscale sensor, transducer, and resonator. This chapter provides a review of current progress in oxide-based nanowires and nanobelts, focusing on growth methods, growth mechanisms, properties, and potential applications. Some unique morphologies of 1-D nanostructures, such as ultra-narrow nanobelts, mesoporous nanowires, and hexagonal-patterned aligned nanorods, are presented.

4.2 THE NANOBELT: WHAT IS IT?

Several terminologies are defined for 1-D nanostructures, including nanobelt, nanowire, nanorod, and nanotube. These terminologies appear frequently in the literature owing to increasing research interests in the area. The term *nanowire,* which means a 1-D nanostructure that grows along a specific axial direction whose side surfaces may not be well defined, is fairly popular. The cross section of a nanowire can be round, hexagonal, or a polyhedron according to the crystallography of the material. The length of a nanowire varies from a few hundred nanometers to microns or even millimeters, and its thickness is always negligibly small compared to its length (Fig. 4.1(a)). A nanorod can be defined as a short nanowire of length in the range from tens to hundreds of nanometers (Fig. 4.1(b)).

Since the first discovery by our group in 2001, nanobelts, sometimes called nanoribbons, have attracted more and more research interest due to their unique morphology and shape-induced physical, electronic, and optical properties. The nanobelt represents a quasi-1-D nanostructure that not only has a specific growth direction but whose top/bottom surfaces and side surfaces are also well defined crystallographic facets (Fig. 4.1(c)). For example, a typical ZnO nanobelt[2] has a growth direction along [0001], the top/bottom surfaces are $\pm(2\bar{1}\bar{1}0)$, and the side surfaces are $\pm(01\bar{1}0)$, as shown in the of Fig. 4.1(c) insert. The as-synthesized metal oxide nanobelts exhibit a rectangular-like cross section and are single crystalline and structurally uniform. Nanobelts are nanowires that have well-defined structure. This is an important characteristic because the properties of nanobelts are strongly shape and structure dependent, especially when their sizes approach a few nanometers.

4.3 TECHNIQUES FOR GROWING NANOBELTS/NANOWIRES

Synthesis of nanomaterials is one of the most active fields in nanotechnology. Various 1-D nanostructures have been fabricated using a number of approaches, including thermal evaporation,[2] laser ablation,[3] template-assisted growth,[4,5] arc discharge,[6]

FIGURE 4.1 Quasi-1-D nanostructures. (a) ZnO nanowires; (b) ZnO nanorods; (c) ZnO nanobelts. The bottom right insert shows a transmission electronic microscopy image of ZnO nanobelts and the corresponding electron diffraction pattern.

lithography,[7] and sol-gel method.[8] Thermal evaporation and laser ablation, which have similar experimental setups and mechanisms, are usually the first choices of synthesis due to their simple setups and acceptable controllability of the self-assembly process. Both methods will be discussed in detail in this chapter.

4.3.1 Thermal Evaporation

Thermal evaporation is one of the simplest, most popular, and most versatile synthesis methods, and it has been very successful in fabricating nanobelts and nanowires with various characteristics. The basic process of this method is sublimating source material(s) in powder form at high temperature, and a subsequent deposition of the vapor in a certain temperature region to form desired nanostructures.

A typical experimental system is shown in Figure 4.2. The synthesis is performed in an alumina or quartz tube, which is located in a horizontal tube furnace. High-purity

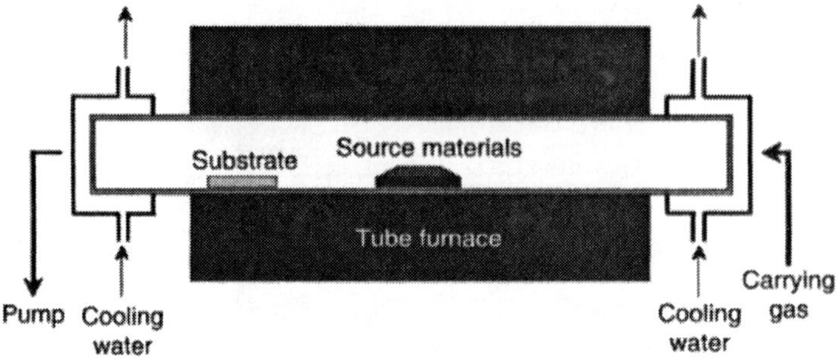

FIGURE 4.2 Thermal evaporate deposition system for synthesis of 1-D nanostructures.

oxide powders contained in an alumina boat are loaded in the middle of the furnace, the highest temperature region. The substrates for collecting the desired nanostructures are usually placed downstream following the carrier gas. The substrates can be silicon wafer, polycrystalline alumina, or single crystal alumina (sapphire). Both ends of the tube are covered by stainless steel caps and sealed with O-rings. Cooling water flows inside the cover caps to achieve a reasonable temperature gradient in the tube.

During the experiments, the system is first pumped down to around 10^{-2} torr. Then the furnace is turned on to heat the tube to the reaction temperature at a specific heating rate. An inert carrying gas, such as argon or nitrogen, is then introduced into the system at a constant flow rate to bring the pressure in the tube back to 200–500 torr. (Different pressures are required by different source materials and final deposited nanostructures.) The reaction temperature and pressure are held for a certain period of time to vaporize the source material and achieve a reasonable amount of deposition. Zinc oxide, for example, can be vaporized in two different processes:

$$2\text{ZnO}(s) \rightarrow 2\text{Zn}(v) + \text{O}_2(v) \tag{4.1}$$

$$\text{ZnO}(s) \rightarrow \text{ZnO}(v) \tag{4.2}$$

By controlling the growth kinetics it is possible to control the decomposition process. In Reaction 4.1, the zinc vapor and oxygen are transported by the carrier gas to the lower temperature region, and their recombination forms ZnO nanostructures. When the reaction is completed, the furnace is turned off and the system is cooled down to room temperature with flowing inert gas. Normally the system pressure is kept at the reaction pressure to prevent subsequent decomposition of the grown nanostructures.

The thermal evaporation process is basically a physical vapor deposition process that has been successfully used for synthesizing a variety of oxide and non-oxide nanobelts and nanowires.[9] Moreover, this system can also be used for chemical vapor deposition (CVD) by simply applying reaction gases instead of the carrier gas and placing substrates in the middle of the tube. For example, multiwall and single-wall carbon nanotubes have been successfully fabricated in this system using hydrogen and methane/acetylene as reactants.[10]

A disadvantage of the thermal evaporation method is the high temperature for controlling growth kinetics. Some modifications have been made to lower the growth temperature. Mixing ZnO with carbon powder, which is so-called carbon-thermal evaporation,[12] can reduce the vaporization temperature from 1300 to 800°C,

$$ZnO(s) + C(s) \xrightarrow{\sim 800^\circ C} Zn(v) + CO(v) \quad (4.3)$$

Metal catalysts, such as gold,[11–13] tin,[14,15] and copper,[16] have also been used to grow aligned nanostructures with reasonably controlled size.

4.3.2 Laser Ablation–Assisted CVD

Laser ablation–assisted CVD growth has been demonstrated as an effective technique for growing 1-D nanostructures, especially for semiconductors. Unlike the 2-D laser-ablation deposition system, in which the substrate is normally on the top of the target, this laser ablation system is a combination of a pulse laser and a thermal evaporation system with a few modifications, as shown in Fig. 4.3. A quartz tube has to be used and a fused silica window is assembled on the end cap for inputting and aiming the laser beam. The source materials are highly pressed into a small alumina crucible and face the laser beam at an angle of about 45°. The location of the source materials can be either in the middle or outside of the furnace depending on their temperature sensitivity. A focus lens with a long focal length is located in the front of the tube window to converge the laser beam onto the target. A pulsed laser vaporizes the source materials.

A typical experimental process is similar to that of the thermal evaporation process. The furnace temperature is always lower and ranges from 500 to 800°C. After the desired temperature and pressure are achieved, the laser beam is shot onto the target and vaporizes the source materials stoichiometrically. The vapor mixture subsequently deposits on the substrate with the help of the metal catalyst. If the local deposition temperature is sufficiently low and the distance from the

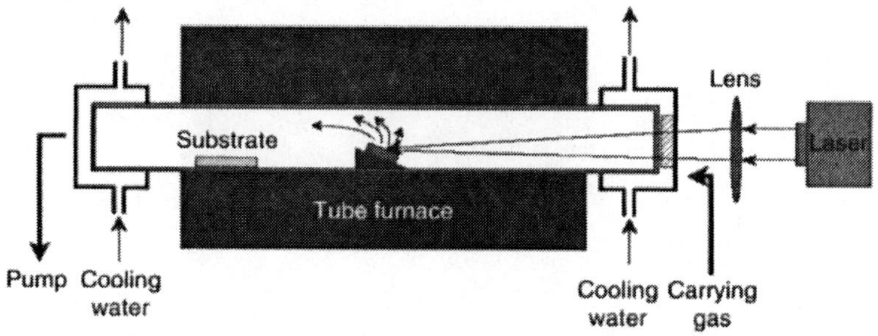

FIGURE 4.3 Laser ablation deposition system for synthesis of 1-D nanostructures.

substrate to the source is long, the vapor can also experience a homogeneous nucleation and growth during the transportation process, and the final nanostructures are collected by the substrate.[17]

The laser ablation technique can be used to fabricate nanostructures with more diverse and complex chemical composition. Due to the high power of the laser beam, the source materials could be sublimated stoichiometrically at a relatively low temperature. Without changing the tube or deposition temperature, the vapor density can be easily modified by varying the power and frequency of the laser beam so that size and growth control are achieved.

4.4 GROWTH MECHANISMS

One-dimensional nanostructures are formed by a vapor-solid or vapor-liquid-solid (VLS) process. When there is no catalyst used in the growth, the formation of 1-D nanostructures is dominated by a vapor-solid process, such as the growth of many oxide nanobelts (ZnO, SnO_2, Ga_2O_3, In_2O_3, etc.).[2,18] If a metal catalyst is introduced into the reaction system, 1-D nanostructures will grow through a VLS process.[19] The metal nanoparticles are liquid droplets at the growth temperature, and they act as nucleation sites for absorbing the incoming molecular vapor. Nanobelts or nanowires then grow out of the particles when the liquid droplets are supersaturated. Once the growth temperature drops below the eutectic temperature of the particle and/or the reactant vapor is no longer available, the growth is terminated.

Figure 4.4 shows the major steps of a VSL process using gold catalyst for growing ZnO nanowires/nanobelts. A thin layer of gold or presynthesized gold nanoparticles is used as catalyst (Fig. 4.4(a)). The gold catalyst forms liquid alloy droplets once the deposition temperature rises above the eutectic temperature of Au and ZnO (Figure 4.4(b)). With the constant incoming ZnO vapor, the percentage of ZnO component in the droplet increases and ultimately supersaturates. ZnO is precipitated out at the solid-liquid interface and forms wire- or belt-like nanostructures (Fig. 4.4(c)).

Typically the 1-D nanostructures obtained through a VLS mechanism have a solid metal ball at the growth tips. When the substrate has a crystalline surface, such as

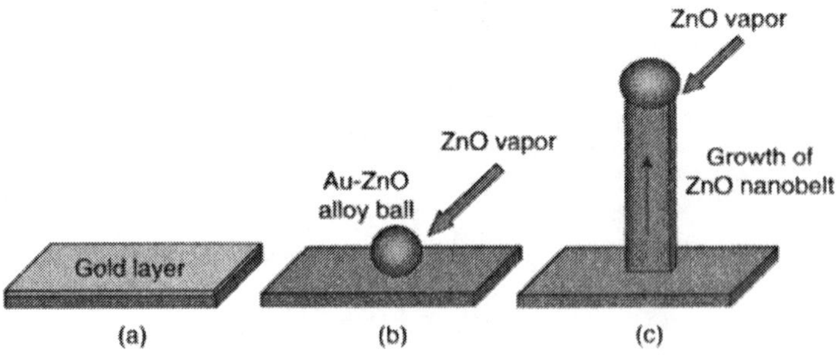

FIGURE 4.4 Schematic steps of a VLS growth process in a Au/ZnO system.

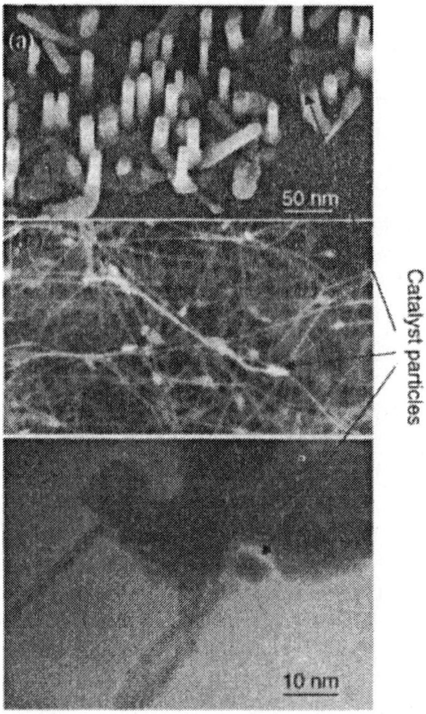

FIGURE 4.5 One-dimensional nanostructure grown by the VLS process. (a) Aligned ZnO nanorods growing on a sapphire substrate with gold catalyst at the tips; (b) ZnO nanowires growing on a silicon substrate with gold catalyst at the tips; (c) ZnO nanobelts rooted at tin catalysts.

sapphire,[20] the as-synthesized nanowires normally have roots at the substrate and point upward with the catalyst balls on the tips (Figure 4.5(a)), which may be a result of epitaxial growth of ZnO on alumina. If the surface of the substrate is disordered, such as silica or polycrystalline alumina, the catalyst balls are still at the growth fronts, but the nanowires are randomly oriented[12] (Figure 4.5(b)). However, once the metal layer is too thick, the local temperature is too low to form fully molten tiny catalyst balls, or the metal is wetted well onto the substrate, and the catalyst remains at the root of the nanowires. The ultra-small ZnO nanobelts reported recently are examples of this case[21] (Figure 4.5(c)). The structure is catalyzed by a thin layer of tin that was precoated on a silicon substrate. Using the VLS growth, random distributed, freestanding 1-D nanostructures can also be obtained by evaporizing a mixture of the catalyst and source materials. In this process, the formation of alloy droplets and the growth of 1-D nanostructures occurred during vapor transport.[22,23] ZnO nanowires, nanobelts, and nanorods have been grown using Sn as a catalyst, but Sn metal was not introduced directly, but rather by reducing SnO_2 mixed in the source material.

In VLS growth, the metal particles are usually believed to be in a liquid state during the growth and its solid-state structure may have no influence on the nature of the grown nanowires and nanobelts. Our recent study on the ZnO nanowire and nanobelts using Sn as a catalyst indicates that the crystalline orientation of the Sn particle can determine the growth direction and the side surfaces of the nanowires and nanobelts.[23] Studies on the interface relationship between catalyst tin particles and their guided ZnO 1-D nanostructures indicate that tin catalyst can guide not only [0001] growth of nanowires, but also $[01\bar{1}0]$ and $[2\bar{1}\bar{1}0]$ growth of nanobelts. The key is to minimize the interface mismatch energy. The results reveal that the interfacial region of the tin particle with the ZnO nanowire/nanobelt could be partially crystallized during the VLS growth, although the local growth temperature is much higher than the melting point of tin. This result may have an important impact on understanding the physical chemical process in the VLS growth.

4.5 THE NANOBELT FAMILY

Since the discovery of nanobelts in our laboratory in 2001, a wide range of interesting nanostructures of functional oxides have been synthesized. Nanobelts represent a unique group of nanostructures from semiconducting oxides to II-VI and III-V compounds to structural ceramics. The family of nanobelts is now experiencing a flourishing development and will be a new focus area in nanoscience and nanotechnology.

4.5.1 Binary Oxide Nanobelts

Belt morphology was first discovered in semiconducting oxide materials. Table 4.1 summarizes the nanobelt structures of function oxides that have been discovered experimentally thus far, including ZnO,[2] SnO_2,[24] Ga_2O_3,[25] In_2O_3,[26] CdO,[2] PbO_2,[27]

TABLE 4.1
Crystallographic Geometry of Functional Oxide Nanobelts

Nanobelt	Crystal Structure	Growth Direction	Top Surface	Side Surface
ZnO	Wurtzite	[0001] or $[01\bar{1}0]$	$\pm(2\bar{1}\bar{1}0)$ or $\pm(2\bar{1}\bar{1}0)$	$\pm(01\bar{1}0)$ or $\pm(0001)$
Ga_2O_3	Monoclinic	[001] or [010]	±(100) or ±(100)	±(010) or $\pm(10\bar{1})$
t-SnO_2	Rutile	[101]	$\pm(10\bar{1})$	±(010)
o-SnO_2	Orthorhombic	[010]	±(100)	±(001)
In_2O_3	C-rare earth	[001]	±(100)	±(010)
CdO	NaCl	[001]	±(100)	±(010)
PbO_2	Rutile	[010]	±(201)	$\pm(10\bar{1})$
MoO_3	Orthorhombic	[001]	±(010)	±(100)
MgO	Face Center Cubic	[100]	±(001)	±(010)
CuO	Monoclinic	$[\bar{1}10]$	±(110)	±(001)
Al_2O_3	Hexagonal	[001]	$\pm(\bar{1}10)$	±(110)

MoO_3,[28,29] MgO,[30,31] CuO,[32] and Al_2O_3.[33] Although these materials belong to different crystallographic families, they do have a common faceted structure, which is the nanobelt structure. Each type of nanobelt is defined by its crystallographic structure, the unique growth direction, and specific top/bottom surfaces and side surfaces. Some of the materials can grow along different crystal directions, which can be controlled by choosing specific experimental conditions.

4.5.2 Compound Semiconductor Nanobelts

In addition to the oxide nanobelts that are studied intensively, more and more function non-oxide nanobelts have been fabricated using various approaches. They are mostly II-VI or III-V compounds that have shown interesting physical or chemical properties in bulk or other nanomorphologies. The structure and structural characteristics are summarized in Table 4.2. ZnS[34] and CdSe nanobelts,[35] II-VI compounds with the wurtzite crystal structure, have been recently synthesized. Analogous to ZnO nanobelts, these nanobelts exhibit polarized side surfaces that induce anisotropic growth, forming a one-side structure with teeth, a so-called nanosaw.[36] Some other important III-V semiconducting materials, such as GaN,[36–38] GaP,[39] and AlN,[40] have also been reported to exhibit belt-like morphology. These nanobelts are single crystal and have specified growth directions and side facets.

4.5.3 Multielement Nanobelts

Because of the high requirements of stoichiometry, the formation of multielement nanobelts is restricted by the evaporation method, which is the traditional pathway used to grow a nanobelt. However, through some chemical routes, a limited number of three-component nanobelts can be fabricated in solution. For example, bismuth oxide bromide[41] and $BaCrO_4$[42] nanobelts have been prepared through a hydrothermal route and a reverse micelles directed synthesis, respectively.

Another type of nanobelt is achieved by doping. Using nanobelts as a template, doping with another type of element results in a new type of nanostructure. Manganese-doped ZnO nanobelts have been fabricated by ion implanting Mn^{2+} into as-synthesized ZnO nanobelts, which have very interesting applications in spintronics.[43] Nitrogen-doped GaP nanobelts have been achieved by growing GaP nanobelts in an NH_3 atmosphere,[39] which shows interesting optical properties.

TABLE 4.2
Crystallographic Geometry of Functional Non-Oxide Nanobelts

Nanobelt	Crystal Structure	Growth Direction	Top Surface	Side Surface
ZnS	Wurtzite	$[01\bar{1}0]$	$(2\bar{1}\bar{1}0)$	±(0001)
CdSe	Wurtzite	$[01\bar{1}0]$	$(2\bar{1}\bar{1}0)$	±(0001)
GaN	Hexagonal	$[10\bar{1}0]$	$\pm(01\bar{1}0)$	±(0001)
GaP	Zinc blende	[111]	±(100)	$\pm(0\bar{1}1)$
AlN	Face Center Cubic	[001]	±(100)	±(010)

4.6 ULTRA-NARROW ZnO NANOBELTS

In general, it is believed that nanostructures less than 10 nm in size will have novel and unique physical and chemical properties due to quantum confinement. Nanowires of a few nanometers in diameter have been synthesized for InP[44] and Si[45] by using nanoparticles as catalysts or using single-wall carbon nanotubes as templates. However, due to the fact that solid-vapor phase synthesis typically occurs at high temperature (between 700 and 1300°C), the synthesis of small-size (<10 nm) ZnO nanobelts requires an innovative approach.

Using thin films of metallic tin as a catalyst, ultrasmall single-crystalline ZnO nanobelts have recently been synthesized in large quantity by a carbon-thermal evaporation process.[21] Instead of using presynthesized nanoparticles as catalysts, a uniform thin film (~10 nm) of tin is coated on the silicon substrate. Since tin has a very low melting point (232°C), it is molten and slowly oxidized[46] before the Zn/ZnO vapor is formed in the reaction tube. During a VLS process, ultra-small ZnO nanobelts grow out from the partially oxidized tin balls and cover the entire substrate with a reasonable yield, as shown in Figure 4.6(a). Without using catalyst nanopar-

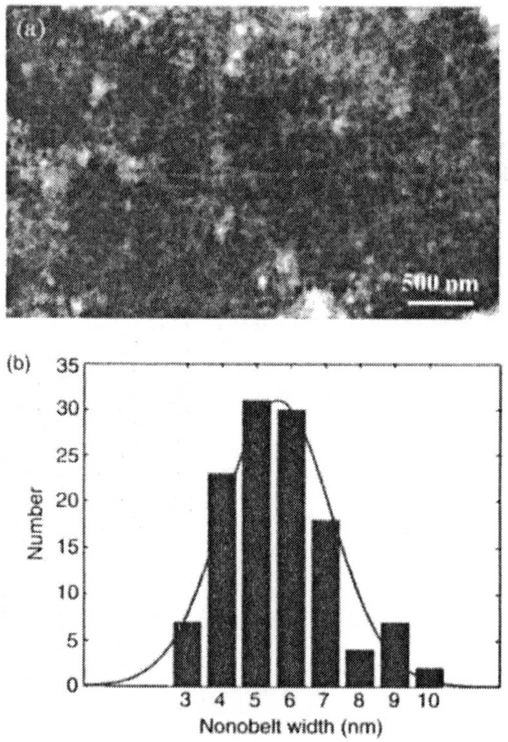

FIGURE 4.6 (a) Scanning electronic microscopy image of ultra-thin ZnO nanobelts grown on tin coated on silicon substrate. (b) Size distribution of the as-synthesized ZnO nanobelts measured from TEM images.

ticles to control the size of the nanobelts as in conventional VLS growth, this technique provides a much simpler way to fabricate ultra-small ZnO nanobelts at lower cost and in large quantity.

The nanobelts grow along the [0001] direction, with $(2\bar{1}\bar{1}0)$ top/bottom surfaces and $(01\bar{1}0)$ side surfaces. The average width of the nanobelts is only 5.5 nm with a narrow size distribution of ±1.5 nm. As shown in Fig. 4.6(b), most nanobelts are in the range of 4–7 nm, indicating very good size uniformity. All of the nanobelts display a single crystalline structure with the (0001) fringes of spacing 5.1 Å. Unlike larger ZnO nanobelts that are dislocation free, the ultra-narrow nanobelts have dislocations, which are revealed by high-resolution TEM. It is assumed that the formation of dislocations is due to a large local strain introduced by the bending of the nanobelts. This is demonstrated by the fact that a nanobelt of 10 nm in width is straight (Fig. 4.7(a)), while a thin nanobelt of 5 nm in width shows a curly shape (Fig. 4.7(b)). The strain in the thicker nanobelt is released by creating stacking faults perpendicular to the growth direction of the nanobelts (Fig. 4.7(c)). For the ~5-nm nanobelt, besides stacking faults, distortion in the orientation of the (0001) atomic planes (Fig. 4.7(d)) and edge dislocations (Fig. 4.7(e))

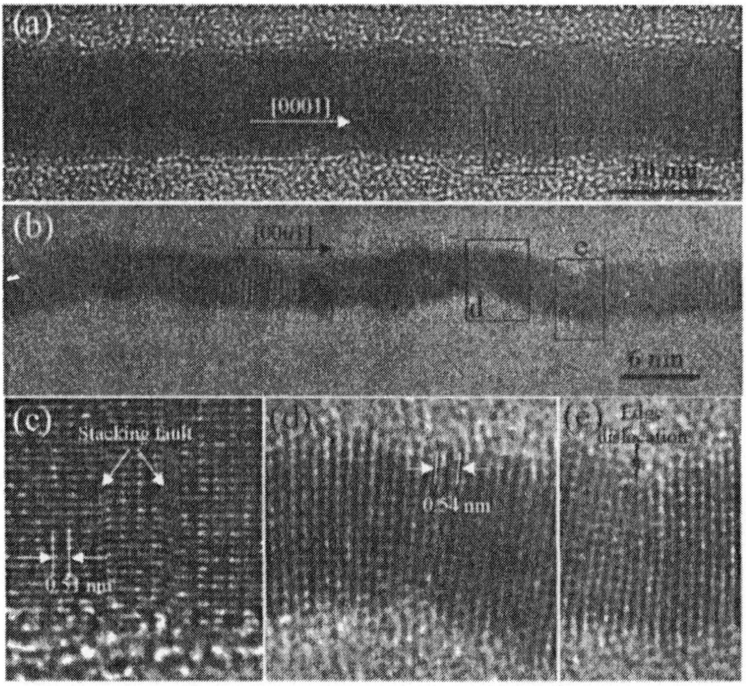

FIGURE 4.7 High-resolution TEM images of (a) 10 nm wide and (b) 5 nm wide ZnO nanobelts. (c–e) Enlarged images from areas c, d, and e marked in (a) and (b) showing the stacking fault, lattice distortion, and edge dislocation, respectively.

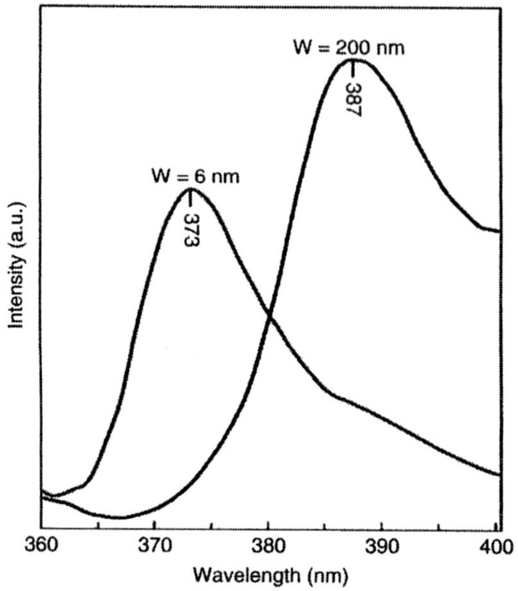

FIGURE 4.8 Photoluminescence spectra acquired from the ~200 nm-wide ZnO nanobelts (blue) and the 6 nm-wide ZnO nanobelts (pink).

are introduced to accommodate the local deformation. The interplanar distance can be expanded to ~5.4 Å at the exterior arc and compressed to ~4.8 Å at the inner arc to accommodate the local strain. This is possible for narrow nanobelts. The creation of edge dislocations in oxide nanobelts is rare.

Photoluminescence (PL) measurements are performed to examine the size-induced quantum effect in the ultra-thin ZnO nanobelts. Figure 4.8 compares the PL spectra recorded from the ZnO nanobelts that have an average width of ~200 nm and the PL from the ultra-thin ZnO nanobelts with an average width of 5.5 nm. The 387-nm peak corresponds to a 3.2-eV spontaneous emission of the normal ZnO nanobelts. This emission lies at an energy considerably below that of the band gap (3.37 eV) of ZnO, and is probably a result of exciton-exciton collision.[47] From this model, the energy of the resulting photon can be given by[48]

$$P_n = E_{ex} - E_b^{ex}\left(1 - \frac{1}{n^2}\right) - \frac{2}{3}kT \quad (n = 2, 3, 4, \ldots)$$

where P_n is the photon energy, E_{ex} is the free-exciton emission energy, E_b^{ex} is the binding energy of the exciton, n is the quantum number of the envelope function, and kT is the thermal energy. However, the ultra-thin ZnO nanobelts exhibit a near band edge emission at 373 nm (3.32 eV), which was measured under identical experimental conditions as for the larger nanobelts. It is believed that the increase in photon energy is due to an increase in E_{ex}, which indicates quantum confinement due to the reduced size of the nanobelts. The ultra-narrow ZnO nanobelts can be used in solid-state quantum computing, nanometer-size memory devices, UV laser diodes, etc. They could be a new type of quantum nanowire.

4.7 MESOPOROUS ZnO NANOWIRES

Porous materials have been used in a wide variety of applications in bioengineering, catalysis, environmental engineering, and sensor systems owing to their high surface-to-volume ratio. Normally most mesoporous structures are composed of amorphous materials[49,50] and the porosity is achieved by solvent-based organic or inorganic reactions.[51] Recently, a new wurtzite ZnO nanowire structure that is single crystal but composed of mesoporous walls/volumes was fabricated in our laboratory.[52] The porous nanowires have a hexagonal cross section with a variety of pore sizes from 10 to 50 nm. The typical length varies from 100 μm to 1 mm and the diameter is in the range of 50–500 nm (Figure 4.9(a)).

The growth of this porous nanostructure is catalyzed by a thin film of metallic tin as well, but the deposition temperature, which dictates the final morphology, is much higher than that used for growing ultra-small nanobelts. Condensation nucleation is

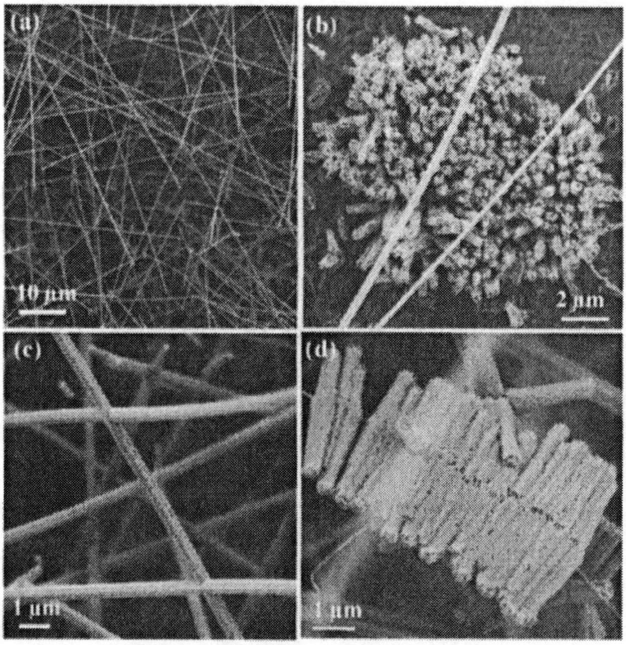

FIGURE 4.9 Mesoporous single-crystal ZnO nanowires. (a) Porous nanowires; (b) partially aligned porous nanorods; (c) junctions between porous nanowires; (d) rib-like porous nanorod arrays.

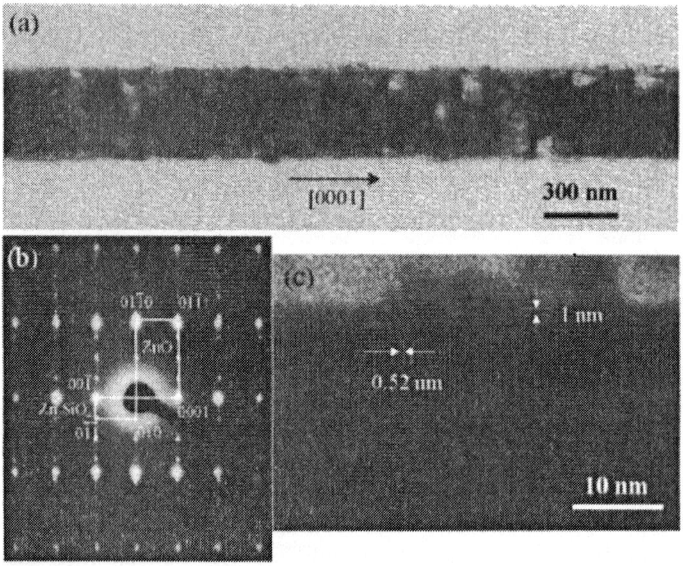

FIGURE 4.10 (a) Low-magnification TEM image of a porous ZnO nanowire and (b) corresponding diffraction pattern; (c) high-magnification TEM image taken on the edge of the porous ZnO nanowire.

also highly activated by this elevated local substrate temperature. Thus, a few more complex structures are formed, such as partially aligned ZnO sprouts that formed coral-like structures (Figure 4.9(b)), crossed structures between nanowires (Figure 4.9(c)), and rib-like short nanorod arrays (Fig. 4.9(d)). Each of these structures exhibits a high porosity surface.

TEM studies are performed to determine the mechanism of the formation of the porous structures. An electron diffraction pattern recorded from a porous nanowire (Figure 4.10(a)) is shown in Figure 4.10(b), which presents two sets of patterns. The brighter spots come from ZnO, which is the $[2\bar{1}\bar{1}0]$ zone axis pattern, displaying the nanowire axial direction as [0001]. The weaker diffraction spots correspond to a new Zn_2SiO_4 phase that is sheathed on the surface of the nanowire and has an epitaxial orientation relation with ZnO, which is clearly shown in a high-resolution TEM picture in Fig. 4.10(c). The diffraction pattern from the Zn_2SiO_4 phase can be indexed to be the [100] zone axis. The orientation relationship between ZnO and Zn_2SiO_4 is $(0001)_{ZnO} \parallel (001)_{Zn_2SiO_4}$, $[2\bar{1}\bar{1}0]_{ZnO} \parallel [100]_{Zn_2SiO_4}$, where structurally Zn_2SiO_4 has an orthorhombic crystal structure with lattice constants of $a = 0.9085$ nm, $b = 1.0625$ nm, and $c = 0.5962$ nm.

Therefore it is believed that the formation of this porous structure is due to the appearance of the Zn_2SiO_4 out-layer. In a high temperature region, ZnO nanowires can redecompose into Zn vapor and O_2, which could happen at the local substrate temperature of ~600°C. In the meantime, Si-O vapor sublimated from the silicon substrate can be quickly deposited on the nanowire surface and diffuse into the ZnO lattice, resulting in the formation of a new phase Zn_2SiO_4. The electronegativity of Si is 1.9 and the electronegativity of Zn is 1.65, which are quite close. The atomic sizes of Si and Zn are 0.117 and 0.133 nm, respectively, which are comparable. Zinc oxide and Si-O vapor can quickly combine to form a new phase:

$$2ZnO + SiO_2 \xrightarrow{\sim 600°C} Zn_2SiO_4$$

The newly formed Zn_2SiO_4 layer tends to have an epitaxial relationship with the ZnO in order to reduce the interface lattice mismatch. On the other hand, the lattice mismatch between $(0001)_{ZnO}$ and $(001)_{Zn_2SiO_4}$ is 14%, and between $(01\bar{1}0)_{ZnO}$ and $4(010)_{Zn_2SiO_4}$ is 9%, and Zn_2SiO_4 tends to form textured islands on the surface of ZnO, but they cannot form a continuous single-crystal film due to large mismatch, resulting in the growth of epitaxial Zn_2SiO_4 islands on the ZnO surface. The newly formed Zn_2SiO_4 may be more stable than ZnO at the local growth temperature. Therefore, there are open areas on the ZnO surface that are not covered by the Zn_2SiO_4 network; an evaporation of Zn-O from the open areas may form the pores in the volume of the nanowire. The sublimation of ZnO and growth of Zn_2SiO_4 proceed simultaneously and finally end with a high-porosity ZnO nanostructure.

This porous structure not only exhibits a high surface-to-volume ratio, but also provides numerous holes that are big enough to store other functional nanoparticles. Meanwhile the mesoporous ZnO nanowires still retain the crystallized structure, and their unique electronic and optical properties are preserved. The high surface-to-volume ratio guarantees a large activation area and the conductivity enables the signal to be easily transported. Therefore, the mesoporous ZnO nanowires are good candidates for sensor systems. The mesoporous ZnO nanowires can also be used in electrochemical reactions as electrodes that hold catalysts and provide a large reaction area.

4.8 PATTERNED GROWTH OF ALIGNED ZnO NANOWIRES

Aligned growth of ZnO nanorods/nanowires has been successfully achieved on a solid substrate via a VLS process with the use of gold[53,54] and tin[14] as catalysts. The catalyst initiates and guides the growth, and the epitaxial orientation relationship between the nanorods and the substrate leads to the aligned growth. Other techniques that do not use any catalyst, such as metalorganic vapor-phase epitaxial growth,[55] template-assisted growth,[4] and electrical field alignment,[56] have also been employed for the growth of vertically aligned ZnO nanorods. Based on a gold catalyst template produced by a self-assembled monolayer of submicron spheres and guided VLS growth on a single crystal alumina (sapphire) substrate, aligned ZnO nanorods have been hexagonally patterned in a large area.[57]

The synthesis process involves three main steps. First a 2-D, large-area, highly ordered monolayer of submicron polystyrene (PS) spheres was self assembled on

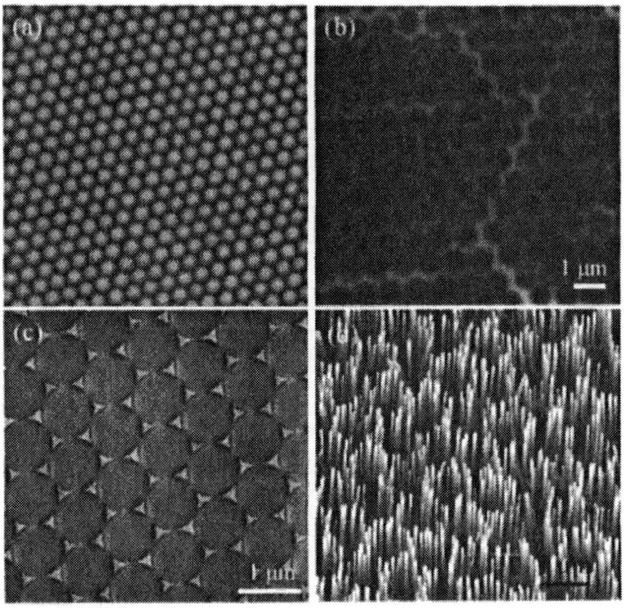

FIGURE 4.11 (a) Self-assembled monolayer of polystyrene spheres; (b) gold catalyst pattern achieved by sputtering; (c) gold catalyst pattern achieved by thermal evaporation; (d) aligned ZnO nanorods grown on substrate.

a single crystal Al_2O_3 substrate (Figure 4.11(a)). This can be achieved in many different ways.[58–59] Second, the self-assembled arrays of PS spheres were then used to pattern the gold catalyst. For this process, gold particles were either sputtered or thermally evaporated onto the self-assembled monolayer structure; as a result, a honeycomb-like hexagonal gold pattern and a highly ordered hexagonal array of gold spots pattern were obtained, respectively (Figures 4.11(b), (c)). Finally, using a carbon-thermal evaporation method, aligned ZnO nanorods are obtained through a VLS process. Following the patterned gold catalyst, the aligned ZnO nanorods exhibit a hexagonal arrangement on the sapphire substrate (Figure 4.11(d)).

Clearly the ZnO nanorods are oriented perpendicular to the sapphire substrate, with a gold particle at the tip. All of the ZnO nanorods have about the same height, of about 1.5 μm, and their diameters range between 50 and 150 nm. By changing the growth time, the height of the ZnO nanorods could be varied from a few hundred nanometers to a few micrometers.

The patterned and aligned 1-D ZnO nanostructures hold huge promise for applications as sensor arrays, piezoelectric antenna arrays, optoelectronic devices, and interconnects. Angular-dependent photoluminescence spectra showed that the luminescence was emitted mainly along the axis of the ZnO nanorods. As shown in Figure 4.12, by increasing the detection angle, the luminescence intensity dropped dramatically. Moreover the luminescence peak shifted very slightly from 384 to 383 nm

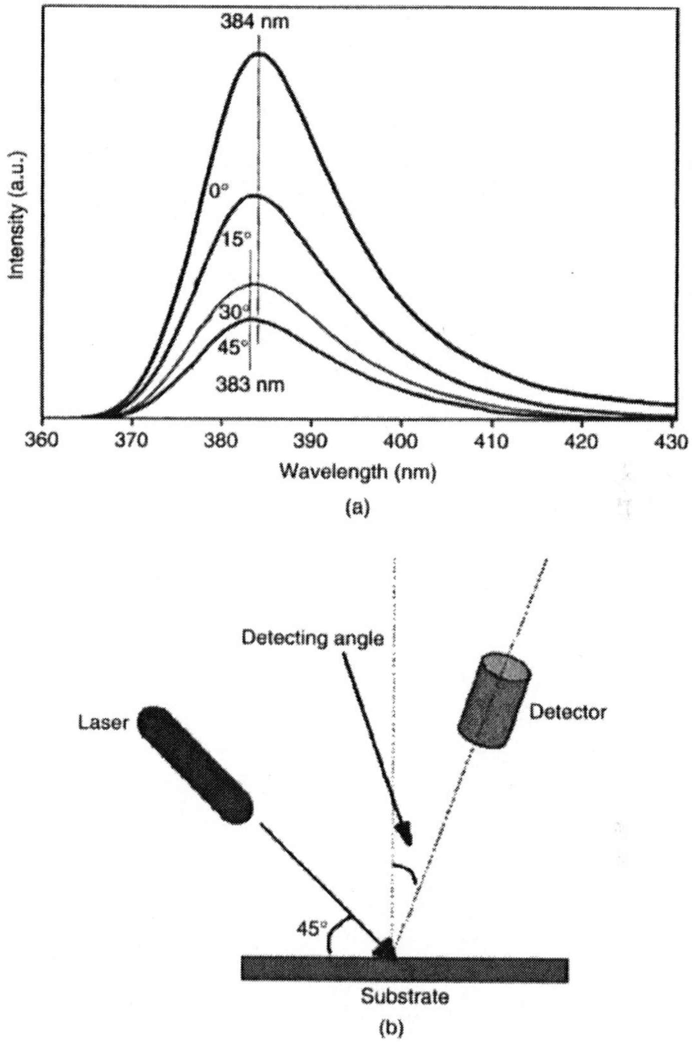

FIGURE 4.12 (a) Photoluminescence (PL) spectra acquired from an aligned ZnO nanorod array as a function of the angle between the detector and a direction normal to the substrate; (b) the experimental setup.

when the detection angle increased from 0° to 45°. This may be caused by the polarization of the emitted light from the aligned nanorods.[60] Thus a very tiny change of the optical or electrical property can be magnified by the aligned structure instead of canceling out by a random distribution so that the sensitivity could be highly increased and could be applied in a macrosensing system.

4.9 SELECTED APPLICATIONS OF NANOBELTS

4.9.1 NANOCANTILEVERS

Cantilever-based scanning probe microscopy (SPM) is a powerful imaging, manipulating, and measuring technique in the nanoscale. An ideal SPM cantilever to detect any forces between the tip and sample would be a robust 1-D structure that is small and rigid. The SPM cantilevers are conventionally made from Si, SiC, and Si_3N_4, with a typical size of ~125, ~35, and ~4 μm in length, width, and thickness, respectively. Recently upon the discovery of semiconducting oxide nanobelts, it was found that nanobelts could be ideal candidates for cantilever applications.

Structurally, most of the nanobelts are single crystals and dislocation free, which provides them excellent mechanical properties. A ZnO nanobelt of 56 nm in width can be bent for more than 60° without breaking, thus demonstrating extremely high mechanical flexibility. The elastic modulus of ZnO nanobelts has been measured by an *in situ* TEM technique based on electromechanical resonance. Geometrically, depending on the synthesis conditions, the dimensions of nanobelts vary from ~6 to 300 nm, which is about 35 to 1800 times smaller than the conventional cantilevers. The greatly reduced cantilever sizes will significantly increase the physical, chemical, and biological sensitivity of SPM cantilevers.

There are feasible methods of manipulating the nanobelts so as to realize the cantilever applications.[61] Due to the capillary force between the atomic force microscope (AFM) probe and the nanobelt, an individual ZnO nanobelt can be picked up selectively by an AFM probe and cut into specific lengths if the technician keeps bending the nanobelt using the AFM tip. Combining these techniques, individual ZnO nanobelts with different lengths have been aligned horizontally onto a silicon wafer (Figure 4.13(a)). This exemplifies our ability to tune the resonance frequency of each and to modify cantilevers for different applications such as contact, noncontact, and tapping mode AFM.

The nanocantilever can be applied to detect molecules (Figure 4.13(b)). When a small number of target molecules are absorbed onto the cantilever, the total mass of the cantilever will be changed. This small change can be magnified by shifting the resonance frequency, which can be applied to detect molecules at the sensitivity of a single molecule depending on the quality and size of the cantilever.

4.9.2 FIELD-EFFECT TRANSISTORS

By contacting the nanostructure to metal electrodes, a field-effect transistor (FET) can be produced, which allows the exploration of the electrical properties of the nanostructures. Based on single SnO_2 and ZnO nanobelts, FETs have been fabricated using two different means.[62] A SnO_2 FET is formed by depositing a nanobelt onto a SiO_2/Si(P^+) substrate, which is followed by making Ti electrodes using e-beam lithography. A ZnO FET is made by directly applying ethanol-dispersed nanobelts on predefined gold electrode arrays (Figure 4.14(a)). In both cases, a gold layer evaporated on the Si (P^+) side of the substrate is used as the back gate electrode, the control of which reveals the current-voltage (I-V) characteristics of the FETs.

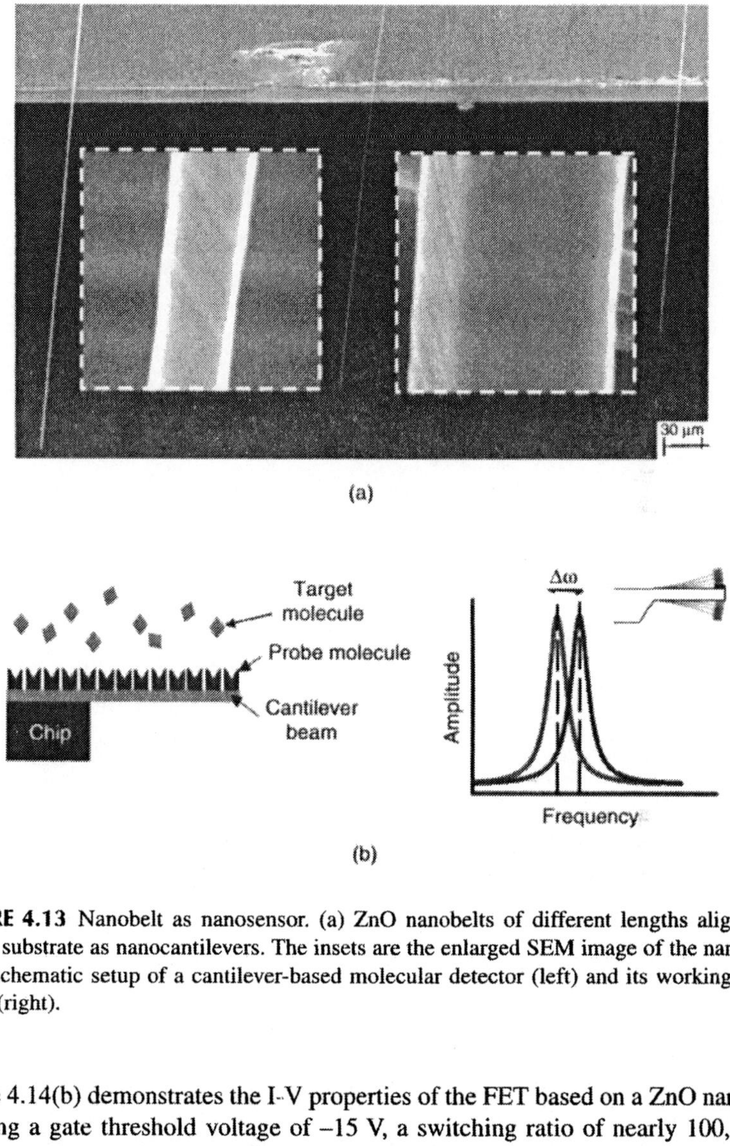

FIGURE 4.13 Nanobelt as nanosensor. (a) ZnO nanobelts of different lengths aligned on silicon substrate as nanocantilevers. The insets are the enlarged SEM image of the nanobelts. (b) A schematic setup of a cantilever-based molecular detector (left) and its working mechanism (right).

Figure 4.14(b) demonstrates the I-V properties of the FET based on a ZnO nanobelt, showing a gate threshold voltage of –15 V, a switching ratio of nearly 100, and a peak conductivity of 1.25×10^{-3} $(\Omega \cdot cm)^{-1}$. Furthermore a ZnO nanobelt–based FET can be controlled by UV illumination as well. When a UV light with a wavelength shorter than the band gap of ZnO is introduced, an immediate increase in the source-drain current can be observed, indicating a switch of status from OFF to ON. According to the measurements, the nanobelt FETs have switching ratios as large as six orders of magnitude, conductivity as high as 15 $(\Omega \cdot cm)^{-1}$, and electron mobility as large as 125 $cm^2/V \cdot s$.

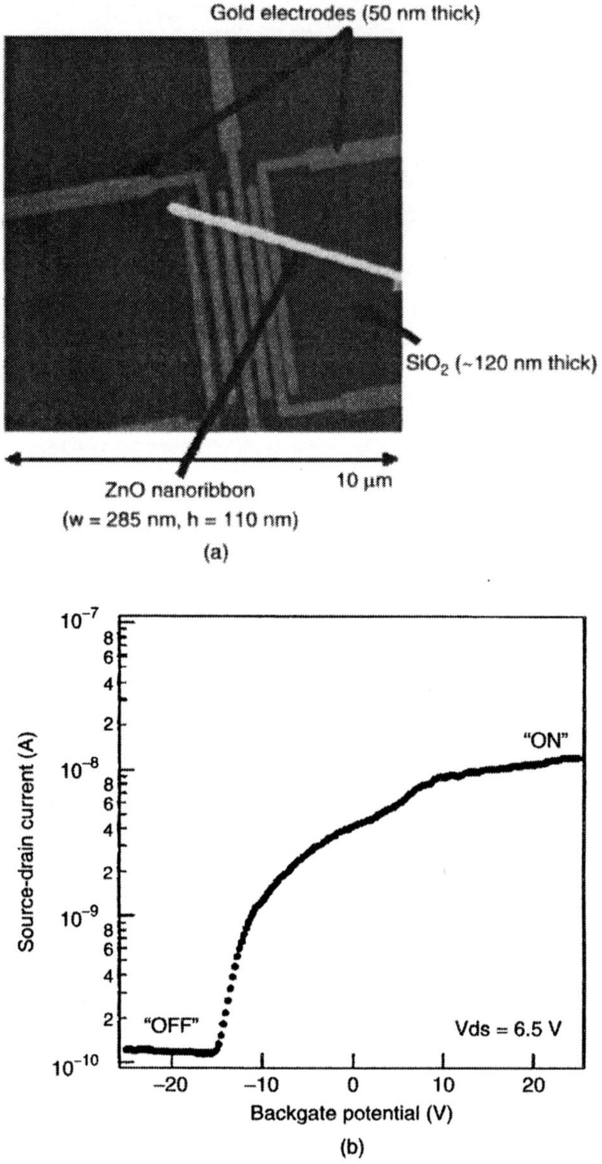

FIGURE 4.14 Field effect transistor using a nanobelt. (a) The experimental setup for measuring the electronic properties of ZnO nanobelts; (b) source-drain current vs. gate bias for a ZnO field-emission transistor.

4.9.3 Nanosensors

Semiconductor thin films are the most promising devices among solid-state chemical sensors due to their low cost, low power consumption, and high compatibility. However, in a thin-film device, which is normally polycrystalline, only a small fraction of the species absorbed near the grain boundaries is active in modifying the conductivity. Semiconducting metal oxide nanobelts are defect-free single crystals with a very high surface-to-volume ratio due to their small sizes. These features provide a possible high sensitivity because the faces exposed to the gaseous environment are always the same and the small size is likely to produce a complete depletion of carriers into the nanobelt, which will change the electrical properties.

SnO_2 nanobelts have been successfully fabricated as gas sensors[63] by placing the nanobelts on platinum electrodes, where the contact is confirmed to be ohmic by I-V measurements. Prompt responses of the current change are observed when this gas sensor is exposed to a gas flow with a small amount of certain species such as CO, NO_2, and ethanol. As shown in Figure 4.15, the current jumps up immediately

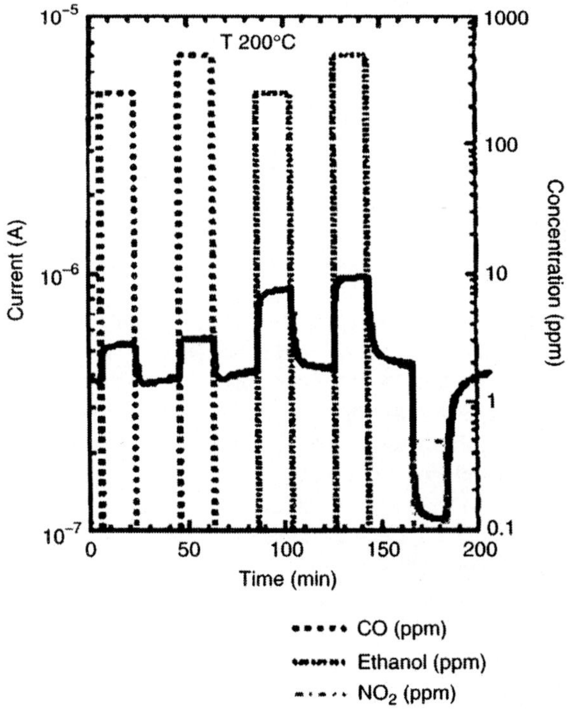

FIGURE 4.15 Nanosensors using nanobelts. Response of SnO_2 sensors to CO, ethanol, and NO_2 gases at 200°C.

once a few hundreds ppm CO or ethanol is introduced into the system, while only 0.5 ppm NO decreases the conductivity of the SnO nanobelts, obviously. When the gas species are depleted, the current recovers to the original level indicating a good repeatability of the SnO_2 nanobelts' sensitivity. The results demonstrate the feasibility for fabricating highly sensitive nanosized gas sensors using the integrity of a single nanobelt.

4.10 SUMMARY

This chapter provides an overview of the synthesis, structure, and potential applications of functional oxide nanobelts. Due to their structurally controlled morphology, nanobelts represent a new group of 1-D nanostructures. The most popular technique for synthesizing nanobelts is by vapor deposition with or without metal catalysts through the vapor-solid or vapor-liquid-solid (VLS) process, respectively. By using Sn and Au as catalysts, some new 1-D ZnO nanostructures have been synthesized. Ultra-narrow ZnO nanobelts with an average width of 5.5 nm exhibit a clear blue shift in photoluminescence due to size effect. Mesoporous ZnO nanowires with a pore size range from 10 to 50 nm could be good candidates for chemical and environmental sensors. By patterning gold catalyst through self-assembled submicron sphere arrays, aligned ZnO nanorods with a hexagonal arrangement are fabricated that exhibit great potential applications for sensor systems and optoelectronic devices. Owing to their well-controlled structures and unique chemical, physical, and electrical properties, semiconducting nanobelts could be applied in a broad area, including nanocantilevers, sensors, transducers, FETs, resonators, and more. We expect research in functional oxide to reach a new horizon in the next few years.

PROBLEMS

1. How can a nanobelt structure be defined? What are the differences between nanowires, nanorods, and nanobelts?
2. List at least five methods to synthesize 1-D nanostructures. What are the basic procedures for the thermal evaporation technique? What are the advantages and disadvantages of the thermal evaporation technique? How can the disadvantages be overcome?
3. Describe how 1-D nanostructures form through a vapor-liquid-solid (VLS) process. How does a metal catalyst affect the final morphologies of 1-D nanostructures in a VLS process?
4. Describe the mechanism of the formation of the porous ZnO 1-D nanostructure.
5. What are the three main steps for fabricating the patterned and aligned ZnO nanorods?
6. List the possible applications of nanobelts. What are the advantages of using a nanobelt as an SPM cantilever?

ACKNOWLEDGMENTS

The results in this chapter were partially contributed by our group members and collaborators. We are grateful to Yong Ding, William Hughes, M.S. Arnold, Ph. Avouris, E. Comini, G. Faglia, and G. Sberveglieri. Research was supported by NSF, NASA, and DARPA.

REFERENCES

1. Wang, Z.L., *Nanowires and Nanobelts, Vol. I: Metal and Semiconductor Nanowires, and Vol. II: Nanowire and Nanobelt of Functional Oxide*, Kluwer Academic Publishers Dordrecht, the Netherlands, 2003.
2. Pan, Z.W., Dai, Z.R., and Wang, Z.L., Nanobelts of semiconducting oxides, *Science, 291,* 1947–1949, 2001.
3. Morales, A.M. and Lieber, C.M., A laser ablation method for the synthesis of crystalline semiconductor nanowires, *Science, 279,* 208–211, 1988.
4. Liu, C., Zapien, J.A., Yao, Y., Meng, X., Lee, C.S., Fan, S., Lifshitz, Y., and Lee, S.T., High-density, ordered ultraviolet light-emitting ZnO nanowire arrays, *Adv. Mater., 15,* 838–841, 2003.
5. Wang, X.D., Gao, P.X., Li, J., Summers, C.J., and Wang, Z.L., Rectangular porous ZnO-ZnS nanocables and ZnS nanotubes, *Adv. Mater., 14,* 1732–1735, 2002.
6. Choi, Y.C., Kim, W.S., Park, Y.S., Lee, S.M., Bae, D.J., Lee, Y.H., Park, G.S., Choi, W.B., Lee, N.S., and Kim, J.M., Catalytic growth of β-Ga_2O_3 nanowires by arc discharge, *Adv. Mater., 12,* 746–750, 2000.
7. Ma, C., Berta, Y., and Wang, Z.L., Patterned aluminum nanowires produced by electron beam at the surface of AlF_3 single crystal, *Solid State Commun., 129,* 681–685, 2004.
8. Krumeich, F., Muhr, H.J., Niederberger, M., Bieri, F., Schnyder, B., and Nesper, R., Morphology and topochemical reactions of novel vanadium oxide nanotubes, *J. Am. Chem. Soc., 121,* 8324–8331, 1999.
9. Dai, Z.R., Pan, Z.W., and Wang, Z.L., Novel nanostructures of functional oxides synthesized by thermal evaporation, *Adv. Funct. Mater., 13,* 9–24, 2003.
10. Pan, Z.W., Xie, S.S., Chang, B.H., Wang, C.Y., Lu, L., Liu, W., Zhou, W.Y., Li, W.Z., and Qian, L.X., Very long carbon nanotubes, *Nature, 394,* 631–632, 1998.
11. Cao, H., Xu, J.Y., Zhang, D.Z., Chang, S.H., Ho, S.T., Seelig, E.W., Liu, X., and Chang, R.P.H., Spatial confinement of laser light in active random media, *Phys. Rev. Lett., 84,* 5584, 2000.
12. Huang, M.H., Wu, Y., Feick, H., Tran, N., Weber, E., and Yang, P., Catalytic growth of zinc oxide nanowires by vapor transport, *Adv. Mater., 13,* 113–116, 2001.
13. Gudiksen, M.S., Wang, J., and Lieber, C.M., Synthetic control of the diameter and length of single crystal semiconductor nanowires, *J. Phys. Chem. B., 105,* 4062–4064, 2001.
14. Gao, P.X., Yong, D., and Wang, Z.L., Crystallographic-orientation aligned ZnO nanorods grown by tin catalyst, *Nano Letters, 3,* 1315–1320, 2003.
15. Gao, P.X. and Wang, Z.L., Self-assembled nanowire-nanoribbon junction arrays of ZnO, *J. Phys. Chem. B., 106,* 12653–12658, 2002.
16. Li, S.Y., Lee, C.Y., and Tseng, T.Y., Copper-catalyzed ZnO nanowires on silicon (1 0 0) grown by vapor–liquid–solid process, *J. Crystl. Growth, 247,* 357–362, 2003.

17. Duan, X. and Lieber, C.M., Laser-assisted catalytic growth of single crystal GaN nanowires, *J. Am. Chem. Soc., 122,* 188–189, 2000.
18. Dai, Z.R., Pan, Z.W., and Wang, Z.L., Gallium oxide nanoribbons and nanosheets, *J. Phys. Chem. B, 106,* 902–904, 2002.
19. Wagner, R.S. and Ellis, W.C., Vapor-liquid-solid mechanism of single crystal growth, *Appl. Phys. Lett., 4,* 89–90, 1964.
20. Huang, M.H., Mao, S., Feick, H., Yan, H., Wu, Y., Kind, H., Weber, E., Russo, R., and Yang, P. Room-temperature ultraviolet nanowire nanolasers, *Science, 292,* 1897–1899, 2001.
21. Wang, X.D., Ding, Y., Summers, C.J., and Wang, Z.L., Large-scale synthesis of six-nanometer-wide ZnO nanobelts, *J. Phys. Chem. B*, 108, 8773–8777, 2004.
22. Duan, X. and Lieber, C.M., General synthesis of compound semiconductor nanowires, *Adv. Mater., 12,* 298–302, 2000.
23. Ding, Y., Gao, P.X., and Wang, Z.L., Catalyst-nanostructure interfacial lattice mismatch in determining the shape of VLS grown nanowires and nanobelts: A case of Sn/ZnO, *J. Am. Chem. Soc., 126,* 2066–2072, 2004.
24. Dai, Z.R., Gole, J.L., Stout, J.D., and Wang, Z.L., Tin oxide nanowires, nanoribbons, and nanotubes, *J. Phys. Chem. B., 106,* 1274–1279, 2001.
25. Dai, Z.R., Pan, Z.W., and Wang, Z.L., Gallium oxide nanoribbons and nanosheets, *J. Phys. Chem. B, 106,* 902–904, 2002.
26. Kong, X.Y. and Wang, Z.L., Structures of indium oxide nanobelts, *Solid State Commun., 128,* 1–4, 2003.
27. Pan, Z.W., Dai, Z.R., and Wang, Z.L., Lead oxide nanobelts and phase transformation induced by electron beam irradiation, *Appl. Phys. Lett., 80,* 309–311, 2002.
28. Li, Y.B., Bando, Y., Golberg, D., and Kurashima, K., Field emission from MoO_3 nanobelts, *Appl. Phys. Lett., 81,* 5048–5050, 2002.
29. Li, X.L., Liu, J.F., and Li, Y.D., Low-temperature synthesis of large-scale single-crystal molybdenum trioxide (MoO_3) nanobelts, *Appl. Phys. Lett., 81,* 4832–4834, 2002.
30. Li, Y.B., Bando, Y., and Sato, T., Preparation of network-like MgO nanobelts on Si substrate, *Chem. Phys. Lett., 359,* 141–145, 2002.
31. Liu, J., Cai, J., Sun, Y.C., Gao, Q.M., Suib, S.L., and Aindow, M., Magnesium manganese oxide nanoribbons: Synthesis, characterization, and catalytic application, *J. Phys. Chem. B, 106,* 9761–9768, 2002.
32. Wen, S.G., Zhang, W.X., and Yang, S.H., Synthesis of Cu(OH)2 and CuO nanoribbon arrays on a copper surface, *Langmuir, 19,* 5898–5903, 2003.
33. Fang, X.S., Ye, C.H., Peng, X.S., Wang, Y.H., Wu, Y.C., and Zhang, L.D., Temperature-controlled growth of α-Al_2O_3 nanobelts and nanosheets, *J. Mater. Chem., 13,* 3040–3043, 2003.
34. Ma, C., Moore, D., Li, J., and Wang, Z.L., Nanobelts, nanocombs and nano-windmills of wurtzite ZnS, *Adv. Mater., 15,* 228–231, 2003.
35. Ma, C., Ding, Y., Moore, D., Wang, X.D., and Wang, Z.L., Single-crystal CdSe nanosaws, *J. Am. Chem. Soc., 126,* 708–709, 2004.
36. Wang, Z.L., Kong, X.Y., and Zuo, J.M., Induced growth of asymmetric nanocantilever arrays on polar surfaces, *Phys. Rev. Lett., 91,* 185502, 2003.
37. Xiang, X., Cao, C.B., and Zhu, H., Catalytic synthesis of single-crystalline gallium nitride nanobelts, *Solid State Commun., 126,* 315–318, 2003.
38. Bae, S.Y., Seo, H.W., Park, J., Yang, H., and Song, S.A., Synthesis and structure of gallium nitride nanobelts, *Chem. Phys. Lett., 365,* 525–529, 2002.
39. Seo, H.W., Bae, S.Y., Park, J., Yang, H., Kang, M., Kim, S., Park, J.C., and Lee, S.Y., Nitrogen-doped gallium phosphide nanobelts, *Appl. Phys. Lett., 82,* 3752–3754, 2003.

40. Wu, Q., Hu, Z., Wang, X.Z., Chen, Y., and Lu, Y.N., Synthesis and optical characterization of aluminum nitride nanobelts, *J. Phys. Chem. B, 107,* 9726–9729, 2003.
41. Wang, J. and Li, Y., Synthesis of single-crystalline nanobelts of ternary bismuth oxide bromide with different compositions, *Chem. Comm., 18,* 2320–2321, 2003.
42. Shi, H.T., Qi, L.M., Ma, J.M., Cheng, H.M., and Zhu, B.Y., Synthesis of hierarchical superstructures consisting of $BaCrO_4$ nanobelts in catanionic reverse micelles, *Adv. Mater., 15,* 1647–1651, 2003.
43. Ronning, C., Gao, P.X., Ding, Y., Wang, Z.L., and Schwen, D., Manganese doped ZnO nanobelts for spintronics, *Appl. Phys. Lett., 84,* 783–785, 2004.
44. Gudiksen, M.S., Wang, J., and Lieber, C.M., Synthetic control of the diameter and length of single crystal semiconductor nanowires, *J. Phys. Chem. B, 105,* 4062–4064, 2001.
45. Ma, D.D.D., Lee, C.S., Au, F.C.K., Tong, S.Y., and Lee, S.T., Small-diameter silicon nanowire surfaces, *Science, 299,* 1874–1877, 2003.
46. Huh, M., Kim, S., Ahn, J., Park, J., and Kim, B., Oxidation of nanophase tin particles, *NanoStructured Mater., 11,* 211–220, 1999.
47. Bagnall, D.M., Chen, Y.F., Zhu, Z., Yao, T., Koyama, S., Shen, M.Y., and Goto, T., Optically pumped lasing of ZnO at room temperature, *Appl. Phys. Lett., 70,* 2230–2232, 1997.
48. Klingshirn, C., Luminescence of ZnO under high one-quantum and 2-quantum excitation, *Phys. Status Solidi B, 71,* 547–556, 1975.
49. Kramer, E., Forster, S., Goltner, C., and Antonietti, M., Synthesis of nanoporous silica with new pore morphologies by templating the assemblies of ionic block copolymers, *Langmuir, 14,* 2027–2031, 1998.
50. Hornebecq, V., Mastai, Y., Antonietti, M., and Polarz, S., Redox behavior of nanostructured molybdenum oxide-mesoporous silica hybrid materials, *Chem. Mater., 15,* 3586–3593, 2003.
51. Cooper, A.I., Porous materials and supercritical fluids, *Adv. Mater., 15,* 1049–1059, 2003.
52. Wang, X.D., Summers, C.J., and Wang, Z.L., Mesophorous single-crystal ZnO nanowires epitaxially sheathed with Zn, SiO_4, *Adv. Mater.,* 16, 1215–1218, 2004.
53. Ng, H.T., Chen, B., Li, J., Han, J., Meyyappan, M., Wu, J., Li, X., and Haller, E.E., Optical properties of single-crystalline ZnO nanowires on *m*-sapphire, *Appl. Phys. Lett., 82,* 2023–2025, 2003.
54. Zhao, Q.X., Willander, M., Morjan, R.R., Hu, Q.H., and Campbell, E.E.B., Optical recombination of ZnO nanowires grown on sapphire and Si substrates, *Appl. Phys. Lett., 83,* 165–167, 2003.
55. Park, W.I., Kim, D.H., Jung, S.W., and Yi, G.C., Metalorganic vapor-phase epitaxial growth of vertically well-aligned ZnO nanorods, *Appl. Phys. Lett., 80,* 4232–4234, 2002.
56. Harnack, O., Pacholski, C., Weller, H., Yasuda, A., and Wessels, J.M., Rectifying behavior of electrically aligned ZnO nanorods, *Nano Lett., 3,* 1097–1101, 2003.
57. Wang, X.D., Summers, C.J., and Wang, Z.L., Large-scale hexagonal-patterned growth of aligned ZnO nanorods for nano-optoelectronics and nanosensor arrays, *Nano Lett., 4,* 423–426, 2004.
58. Kempa, K., Kimball, B., Rybczynski, J., Huang, Z.P., Wu, P.F., Steeves, D., Sennett, M., Giersig, M., Rao, D.V.G.L.N., Carnahan, D.L., Wang, D.Z., Lao, J.Y., Li, W.Z., and Ren, Z.F., Photonic crystals based on periodic arrays of aligned carbon nanotubes, *Nano Lett., 3,* 13–18, 2003.
59. Garno, J.C., Amro, N.A., Mesthrige, W., and Liu, G.Y., Production of periodic arrays of protein nanostructures using particle lithography, *Langmuir, 18,* 8186–8192, 2002.

60. Johnson, J.C., Yan, H., Yang, P., Saykally, R.J., Optical cavity effects in ZnO nanowire lasers and waveguides, *J. Phys. Chem. B, 107,* 8816–8828, 2003.
61. Hughes, W. and Wang, Z.L., Nanobelt as nanocantilever, *Appl. Phys. Lett., 82,* 2886–2888, 2003.
62. Arnold, M.S., Avouris, P., Pan, Z.W., and Wang, Z.L., Field-effect transistors based on single semiconducting oxides nanobelts, *J. Phys. Chem. B, 107,* 659–663, 2003.
63. Comini, E., Faglia, G., Sberveglieri, G., Pan, Z.W., and Wang, Z.L., Stable and high-sensitive gas sensors based on semiconducting oxide nanobelts, *Appl. Phys. Lett., 81,* 1869–1871, 2002.

5 Advances in Chemical Vapor Deposition of Carbon Nanotubes

Vesselin N. Shanov, Atul Miskin, Sachin Jain, Peng He, and Mark J. Schulz

CONTENTS

5.1 THE CVD TECHNIQUE FOR GROWTH OF CNT

In recent years there has been an enormous number of publications on carbon nanotube (CNT) synthesis. Despite this information flow many challenges still need to be addressed. One such challenge is the production of large-scale and low-cost single-wall (SW) and multiwall (MW) carbon nanotubes. Another field of interest is the pursuit of controlled CNT growth in terms of selective deposition, orientation, and preselected metallic or semiconducting properties. Our understanding of the CNT growth mechanism has been rapidly evolving, but more consideration is still required to explain the variety of the observed growth features and experimental results. A significant amount of information covering these aspects of CNT research, along with other related topics, can be found in References 1–5.

Currently there are three principal techniques to produce high-quality SWCNT: laser ablation,[6] electric arc discharge,[7,8] and chemical vapor deposition (CVD).[9,10] Laser ablation and arc discharge are modified physical vapor deposition (PVD) techniques that involve the condensation of hot gaseous carbon atoms generated from the evaporation of solid carbon. However, the equipment requirements and the large amount of energy consumed by these methods make them mostly suitable for laboratory research.[9,10] Both the laser ablation and arc-discharge techniques are limited in the volume of sample they can produce in comparison to the size of the carbon source. In addition, more impurities accompany the nanotubes in the form of amorphous carbon and catalyst particles because of the high-temperature nature of the heat source. Since the growth is difficult to control, the final product consists mostly of MWCNT with poor alignment.

Chemical vapor deposition has become the most important commercial approach for manufacturing of carbon nanotubes. CVD is known as irreversible deposition of a solid from a gas or a mixture of gases through a heterogeneous chemical reaction. This reaction takes place at the interface of gas–solid substrate, and depending on the deposition conditions, the growth process can be controlled either by diffusion or by surface kinetics. CVD is the preferred technique for fabrication of thin layers of metals, insulators, and semiconductors on different substrates.[11–13] This method can easily be scaled up to industrial production. CVD is a continuous process that is currently the best-known technique for high-yield and low-impurity production of CNT at moderate temperatures. It offers better growth control because of the equilibrium nature of the chemical reactions involved. In addition, CVD has the ability to control the size, shape, and alignment of the nanotubes.[10]

CVD of CNT uses carbon precursor gases such as benzene, methane, acetylene, carbon monoxide, or ethanol. The process usually involves high-temperature decomposition of hydrocarbons in hydrogen over the catalyst, which is predeposited on the solid substrate. Figure 5.1 illustrates the CVD process by showing a typical reactor and thermal decomposition of methane precursor in a hydrogen atmosphere. The decomposition reaction can be controlled by the temperature, pressure, or concentration of the reactants involved. This reaction generates carbon atoms, which reach the surface of the substrate, where they diffuse into tiny islands of melted catalyst that dissolve the carbon atoms, forming a metal-carbon solution. This high-temperature solution may become supersaturated. When it reaches this point, carbon precipitates, thus providing the "building bricks" for growth of CNT structures. Transition metals such as Fe, Co, and Ni are frequently used as catalysts. The metal catalyst can stay on top of CNT, forming a cap that encapsulates the CNT.[14] In some cases, it remains at the bottom of the CNT, which promotes a bamboo-shaped morphology.[15] The described growth mechanism is very simplistic; it postulates that carbon atoms from the gas phase are transported through the melted carbon-catalyst solution to the growth surface of the CNT. This process is based on the

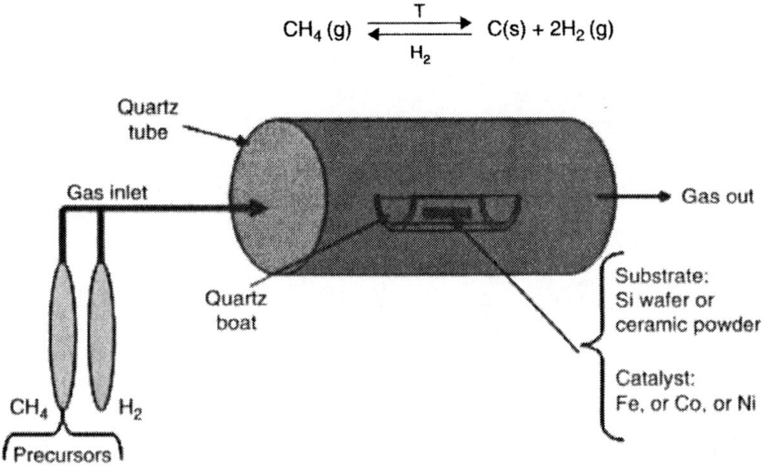

FIGURE 5.1 Schematic of a CVD reactor and the chemical reaction involved during the growth of CNT.

vapor–liquid–solid (VLS) mechanism. The VLS mechanism is used to explain crystal growth of whiskers from a gas phase through molten (liquid) media located at the gas–solid interface.[16] The selection and preparation of the substrate and catalyst are important aspects affecting the nanotube yield and the ability to purify the product of the synthesis.

The CVD synthesis of CNT can be divided into two major groups: bulk synthesis and surface synthesis on flat substrates.[10] The first group is based on ceramic particles such as Al_2O_3 or MgO, in some cases mixed with SiO_2 to form hybrid materials.[17–21] Catalysts of pure metals or alloys are deposited on the ceramics, which provide support and structural stability at elevated temperatures. High-surface-area catalysts with a porous structure of the supported ceramics are desirable for high yield of CNT grown by CVD. Alumina-supported Fe-Mo is a frequently used substrate/catalyst combination for synthesis of SWCNT.[9]

Surface synthesis of SWNT requires flat substrates such as polished or porous Si wafers, patterned with metal catalysts by various lithographic techniques.[22–27] One advantage of the surface synthesis is that the final CNT product is much cleaner than the one grown by bulk synthesis. In addition, this technique provides better control of the diameter and the alignment of the carbon nanotubes, which makes it attractive for device fabrication. The size and the pattern of the metal–catalyst islands on the Si substrate predetermine their orientation, geometric dimensions, and selective growth.

There are two different pathways to synthesize CNT by chemical vapor deposition. The first one, as previously described, involves a solid catalyst supported by a substrate. The second pathway requires the catalyst to be present in the vapor phase, where the growth of CNT takes place. This is known as the floating catalyst approach, which provides nanosize particles of the metal catalyst in the growth zone.[28–31] Metallocenes are used as a source of the catalyst precursor for the floating catalyst method. A typical floating catalyst compound that has been studied extensively is ferrocene.[29–31] Less popular are metallocenes such as cobaltocene.[31] Usually a solution of ferrocene in benzene is introduced into the CVD reactor, either as an aerosol or as a vapor along with a carrier gas. The reaction takes place at about 1100°C, depending on the stability of the hydrocarbon. The role of ferrocene is to provide the iron catalyst, while benzene delivers the carbon feedstock required for the CNT growth. Addition of sulfur in the form of thiophene has been known to increase the yield of carbon nanotubes and is used along with the catalytic metallocenes.[31]

5.2 CVD GROWTH SYSTEM

Conventional CVD reactors are frequently used for growing CNT from carbon precursors. An EasyTube™ Nanofurnace, shown in Fig. 5.2(a), is used at the University of Cincinnati (UC) for the synthesis of carbon nanotubes.[32] The furnace utilizes a thermally driven CVD process and consists of two main units: the control unit and the process unit. The process unit shown in Fig. 5.2(b) is composed of a furnace and a quartz tube reactor with a loader. The control unit contains the workstation of the operator; its display is shown in Fig. 5.2(c). This

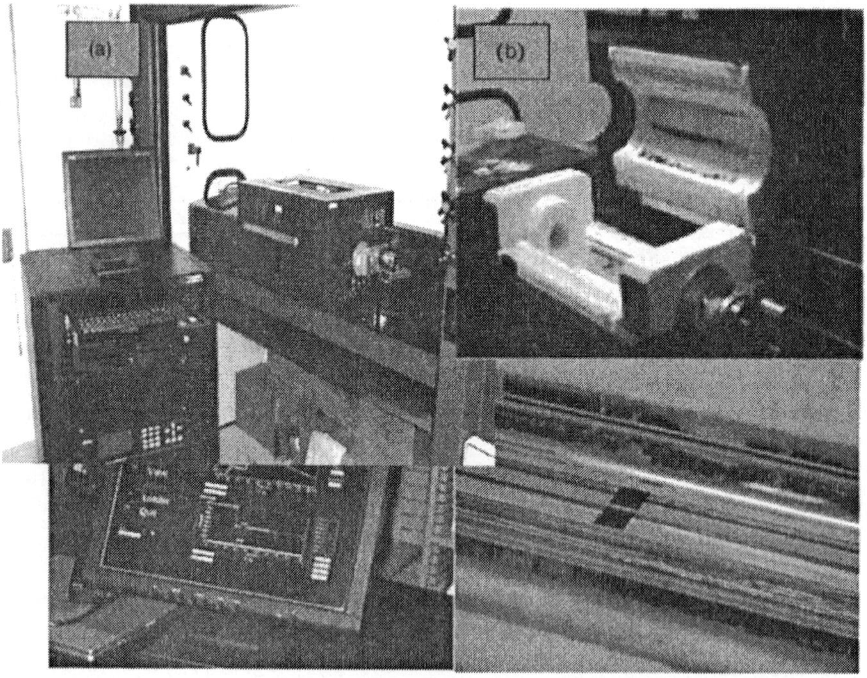

FIGURE 5.2 CVD system for CNT growth: (a) controller and furnace, (b) quartz tube in furnace, (c) display of process temperature and flow rates, and (d) two samples on the loader in the quartz tube.

module is a computer controller that regulates the furnace temperature and the gas flow. The computer uses the Microsoft Windows environment and LABVIEW as the programming language. The sample on the loader in the quartz tube is shown in Fig. 5.2(d). The system uses four process gases: methane, ethylene, hydrogen, and argon, which are involved in the synthesis of nanotubes. Nitrogen is used for operating the pneumatic gas valves in the furnace. Methane and ethylene are the carbon precursor gases. Hydrogen is used for moderating the chemical reaction of the hydrocarbon decomposition. Argon creates an inert atmosphere during the purging and cooling process steps. Safety is a concern when using the EasyTube™ System and combustible gases. Our CVD system is enclosed in a large fume hood with safety glass doors. In addition, leak detectors for hydrogen and hydrocarbons are installed in the gas cabinet and all along the gas supply lines to the furnace.

A gas supply system consisting of mass flow controllers and valves controls the flow rate and the sequence of the gas flows. The gases enter the quartz tube reactor where they are heated to a temperature of 900–1000°C. The reactor pressure is near atmospheric. The hydrocarbons decompose in the quartz tube over a catalyst, which is mounted on a substrate. The catalysts used in our experiments include Fe, Ni, and Co metals. Ceramic powdered materials such as alumina, silica, or magnesium oxide support them during the CVD growth. A quartz or molybdenum boat is used to accommodate a powder-supported catalyst for bulk nanotube growth. In some of our experiments, a Si substrate patterned with catalyst is used for surface synthesis of CNT.

In situ quadrupole mass spectroscopy (QMS) is a very useful technique for monitoring the species in the CVD reactor, and the related studies can improve the quality of the CNT. A dynamic mode of data acquisition can be employed to track the concentration of the major species with time. This mode easily reveals the changes in the partial pressure of the species when changing the concentration of methane or ethylene by pulsing of the carbon precursor flow. The pulsing flow approach is described later in this chapter. The acquired spectra usually illustrate change in the concentration of the major species such as C^+, H_2^+ CH_3^+, and other hydrocarbon radicals, which change with the carbon precursor pulses. This technique was successfully tested during plasma CVD growth of polycrystalline diamond films.[33]

To investigate the gas phase environment during the nanotube growth, the EasyTube™ system was modified by incorporating a quadrupole mass spectrometer, VISION 1000 P-M (300 amu), made by MKS Instruments. This instrument samples from the reactor during the deposition and provides quick feedback information about the species in the reaction zone. This approach helps optimize the process variables in terms of quality of the CNT. The modified CVD system with the mass spectrometer coupled to the reactor is illustrated in Fig. 5.3.

Sampling of the gas phase is done through a heated capillary inserted inside the reactor. The capillary is connected to the QMS sensor via a pressure-reducing system, allowing measurement at close to atmospheric pressure in the CVD reactor.

Pictures of the QMS incorporated into the CVD system are shown in Fig. 5.4. The extension of the heated capillary through a stainless steel probe is illustrated in

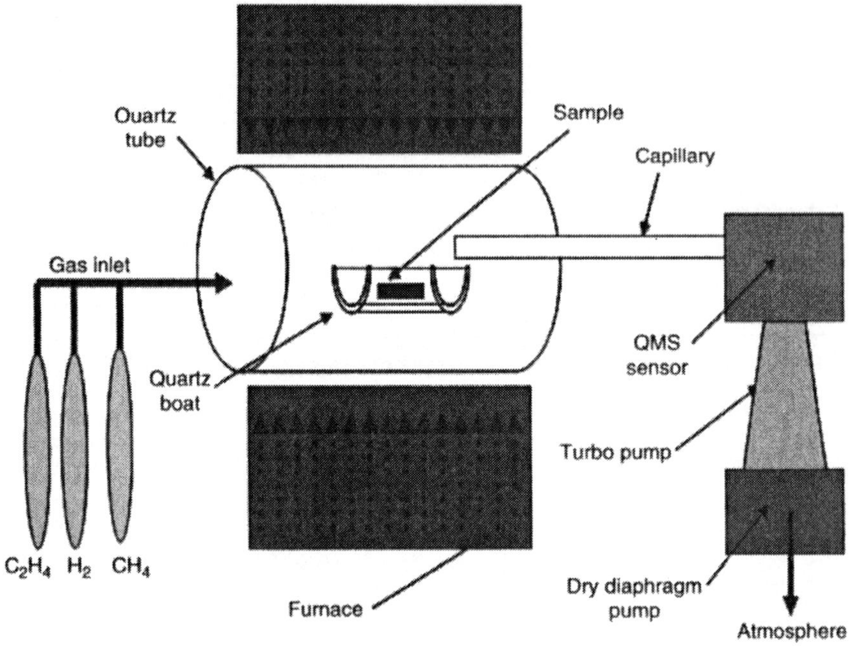

FIGURE 5.3 Schematic of the CVD system with the incorporated QMS.

Fig. 5.4(a). The capillary goes inside of the reactor and its inlet is close to the boat with the sample. This arrangement allows sampling of the gas phase within the vicinity of the growth zone. A 7 μm filter built up along the line between the capillary and the sensor protects the sensor from solid particles found in the reactor. Figure 5.4(b) shows the main components of the QMS.

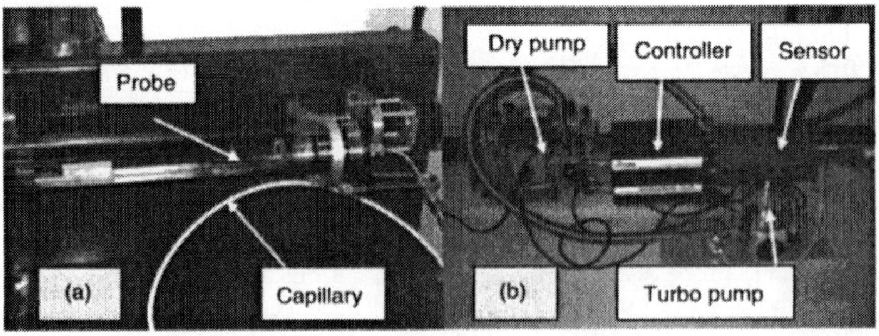

FIGURE 5.4 Pictures of the QMS incorporated into the CVD system: (a) heated capillary introduced into the reactor through a stainless steel probe and (b) major modules of the QMS.

5.3 CATALYST AND SUBSTRATE PREPARATION

One of the approaches tested in this work was to perform bulk synthesis of CNT using MgO powder as the supporting substrate for the metal catalyst. Our choice for MgO is based on previous experience with Al_2O_3 and SiO_2 as supporting ceramics that are difficult to remove during purification of the CNT. Magnesium oxide allows easy purification of CNT using hydrochloric acid. The substrate is prepared by thermal decomposition of $Mg(NO_3)_2$ or $MgCO_3$. The preparation of the metal catalyst on the MgO substrate is a multistep process. The ceramic powdered substrate is immersed into a $Fe(NO_3)_3$ aqueous or methanol solution and dispersed by high-power ultrasound. Later it is separated from the solution and dried at 100°C for 2 h.[20] In our experiments the preceding procedure was modified as follows: The ceramic powder was dispersed in water by high-power ultrasound. Then the dispersed powder was diluted by deionized water, and $Fe(NO_3)_3 \cdot 6H_2O$ was added to the MgO suspension. The mixture was treated for 30 min in an ultrasonic bath for further dispersion and dried at 100°C in air. The dried powder was ground in an agate mortar to reduce the grain size. In some experiments, Bis(acetylacetonato)-dioxymolybdenum (VI) is added to prepare a bimetal Fe-Mo catalyst.[34] As an alternative, Mo salt such as $(NH_4)_6\ Mo_7O_{24} \cdot 4H_2O$ can be used for the same purpose.[9] Further, the ceramic substrate impregnated with catalyst compound is calcinated in air at 500°C for 1.5 h. This step results in the formation of metal catalyst oxides. The last step in preparation of the catalyst takes place in the CVD reactor prior to the deposition of CNT. There the metal oxides undergo reduction with hydrogen and transform into metal particles, which cover the MgO powder.

Another approach we are exploring is the deposition of the catalysts on an oxidized Si substrate, patterned by photolithography, as shown in Fig. 5.5. This technique facilitates growth of high-purity nanotubes.[26] A suspension containing a catalyst salt such as $Fe(NO_3)_3$ and alumina particles in methanol was selectively applied by spin deposition on the Si substrate, followed by baking at 160°C for 5 min. Spin coating provides close control of the liquid film and of the catalyst thickness. It is important to note that for selective deposition of the catalyst, a photoresist mask of polymethylmethacrylate (PMMA) was used. In Fig. 5.5, the windows within the photoresist were filled with catalyst suspension. The windows can be square, circular, or cross-shaped. Prior to the high temperature growth, the photoresist was removed by a standard liftoff procedure using acetone. The substrate was then thermally treated in the CVD reactor to decompose the metal salt and to produce the metal catalyst by reduction with hydrogen. Alternatively E-beam lithography can be used to create islands of metal catalyst sputtered on silicon wafers for growth of aligned carbon nanotubes.

Photos of the catalysts, substrates, and substrate carriers for SWCNT growth are shown in Fig. 5.6. The substrate holder in Fig. 5.6(a) is made of quartz and is designed for surface synthesis of CNT. It allows carrying a Si substrate inside of the CVD reactor, which can be patterned on both sides. When the patterned side is facing down, the growth direction is the same as that of the gravitational forces. Catalysts and substrates used in our experiments are shown in Fig. 5.6(b). For bulk synthesis, the powdered catalyst is spread at the bottom of a Mo boat as a thin layer

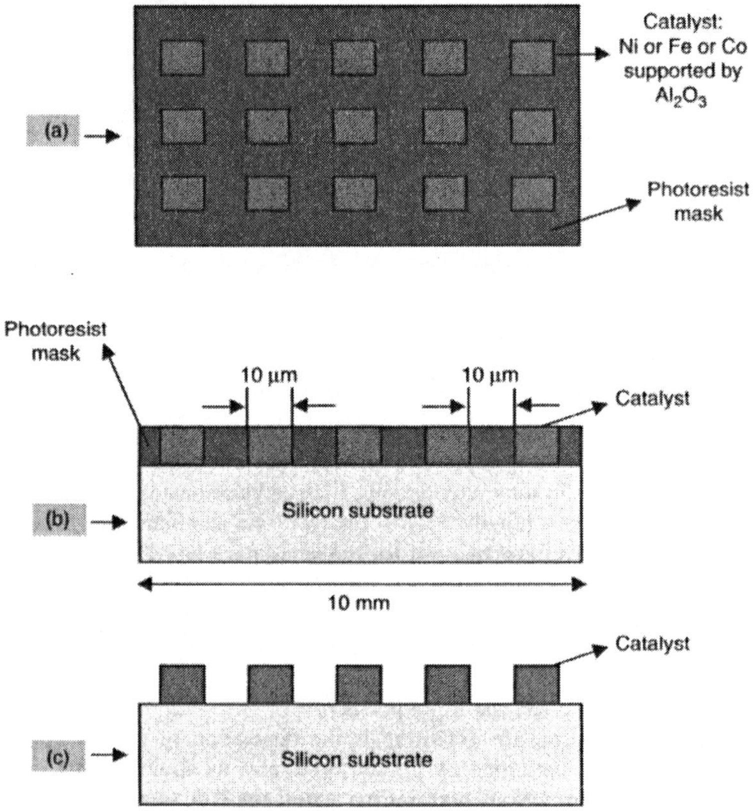

FIGURE 5.5 Patterned silicon substrate with catalyst before and after removal of the photoresist mask: (a) plane view before removal of the photoresist, (b) cross-section before removal of the photoresist, and (c) cross-section after removal of the photoresist.

in order to prevent any diffusion limitations during the growth of CNT. The Mo boat with the catalyst is shown in Fig. 5.6(c).

5.4 GROWTH OF CNT

For CVD growth of CNT, the substrate with the catalyst is inserted automatically inside the quartz reactor via a quartz loader tray driven by the computer. There are many recipes developed and currently in use at UC for synthesis of CNT at different conditions. The recipes are preprogrammed in a library on the computer that drives the CVD system. New recipes can be easily added. The standard recipe is capable of producing high-yield and high-purity SWCNT in three process steps: (1) maximum ramp, (2) deposition, and (3) cooling. In the maximum ramp step, the temperature of the furnace is raised from ambient to 900°C in 12 min. A continuous flow of

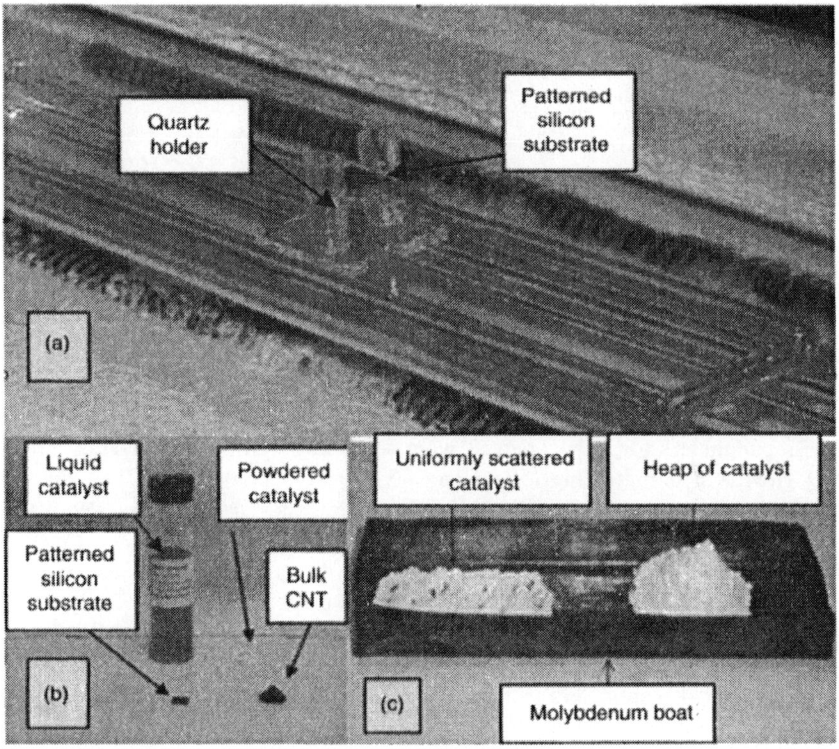

FIGURE 5.6 Photos of catalysts and substrates for CNT growth: (a) UC substrate holder for surface synthesis, (b) catalysts and CNT, and (c) molybdenum boat with catalyst for bulk synthesis.

argon is used to properly purge the reactor. Once the temperature reaches 900°C, it is maintained for 10 min. This is the deposition stage. Nanotube synthesis occurs during this stage when the supply of argon is stopped and the carbon source gases methane and ethylene are run for 10 min along with a flow of hydrogen. After the nanotubes are synthesized, they are cooled to ambient temperature, which completes the last process step. During cooling the methane, ethylene, and hydrogen flow is stopped, and the system is purged with argon. Apart from these three basic steps, a separate step called the timed ramp can be incorporated. Unlike maximum ramp in which the temperature rises as fast as possible, in the timed ramp the temperature is programmed to rise gradually. The variables responsible for nanotube growth include the temperature and duration of the process steps, the nature of the carbon precursor and its flow rate, the type and quantity of the catalyst, and the selection of the patterned substrate. These process variables determine the yield and quality of the nanotubes synthesized. Manipulating any one of them can lead to changes in size, shape, alignment, and the yield of the nanotubes. Many experiments are required to determine the optimum process parameters for fabrication of CNT with desired

TABLE 5.1
Standard Recipe for Growth of SWCNT, Experiment 1

		Temperature (°C)			Flow Rate (SCCM)			
Step	**Type**	**Start**	**End**	**Time (min)**	**Hydrogen**	**Methane**	**Ethylene**	**Argon**
1	Heat	Ambient	900	12	0	0	0	1000
2	Deposition	900	900	10	300	1200	30	0
3	Cool	900	Ambient	120	0	0	0	1000

qualities. Table 5.1 lists a set of process conditions for a standard recipe used in our experiments to grow SWCNT on MgO-supported Fe-Mo catalyst.

Another set of parameters designed and tested at UC is shown in Table 5.2. The process parameters for Experiment 2 were selected based on optimization of the CVD growth in terms of high yield and good quality of the CNT. The changes compared to Experiment 1 include reduced gas flow rates and increased deposition temperature and deposition time. These growth conditions produced good results, which are presented later in the chapter.

The deposition conditions for Experiment 2 are graphically illustrated in Fig. 5.7. The FirsNano CVD system allows on-line monitoring of the temperature and the gas flow profiles while the experiment is in progress.

A third set of parameters tested at UC is displayed in Table 5.3. The growth conditions for Experiment 3 were tuned to promote synthesis of MWCNT. Compared to Experiments 1 and 2 the flow rate of ethylene is increased and methane is not used. In addition, the hydrogen flow rate is reduced, which enhances the decomposition of the hydrocarbon precursor. This causes an increased carbon flux moving toward the substrate in the CVD reactor, which results in higher supersaturation of the carbon atoms within the growth zone, and especially in the metal catalyst-carbon solution. The overall effect is growth of MWCNT. Good-quality and high-yield MWCNT can be produced following the recipe in Table 5.3.

TABLE 5.2
New Recipe for Growth of SWCNT, Experiment 2

		Temperature (°C)		Time	Flow Rate (SCCM)			
Step	**Type**	**Start**	**End**	**(min)**	**Hydrogen**	**Methane**	**Ethylene**	**Argon**
1	Purge	Ambient	Ambient	10	0	0	0	1000
2	Timed Ramp	Ambient	500	10	0	0	0	1000
3	Purge	500	500	20	0	0	0	1000
4	Maximum Ramp	500	900	8	0	0	0	1000
5	Deposition	900	900	60	100	400	10	0
6	Cool	900	Ambient	120	0	0	0	1000

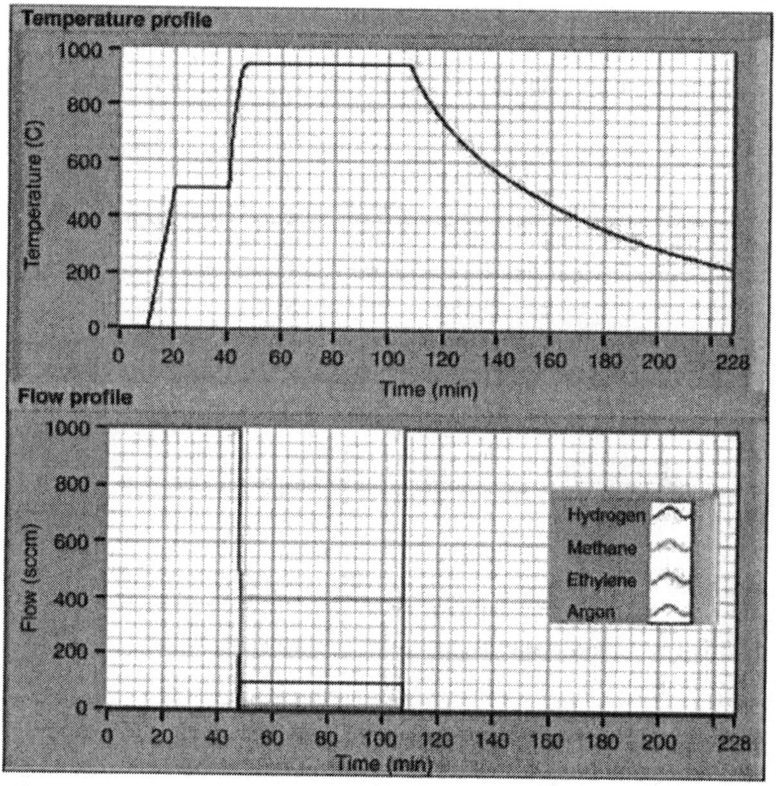

FIGURE 5.7 Graphical interface of operations for Experiment 2.

For bulk synthesis the catalyst is spread as a thin layer on the bottom of a molybdenum boat. The scanning electron microscope (SEM) photos of the CNT synthesized during this and other experiments are shown later in the characterization section of this chapter. The MgO supporting ceramic is easy to remove by purification

TABLE 5.3
New Recipe for Growth of MWCNT, Experiment 3

Step	Type	Temperature (°C)		Time	Flow Rate (SCCM)			
		Start	End	(min)	Hydrogen	Methane	Ethylene	Argon
1	Purge	Ambient	Ambient	5	0	0	0	1000
2	Timed Ramp	Ambient	400	10	0	0	0	1000
3	Purge	400	400	5	0	0	0	1000
4	Maximum Ramp	400	700	8	0	0	0	1000
5	Deposition	700	700	30	50	0	250	0
6	Cool	700	Ambient	120	0	0	0	1000

with acids. Different combinations of MgO-supported Fe-Mo catalyst are being tested to optimize the growth. With the long molybdenum boat and a thin coating of the substrate-catalyst spread on the bottom, approximately one-half gram of SWCNT can be produced in each experiment using our CVD system. The total processing time for our standard synthesis experiment is about 3 h. The purification of CNT, which includes acid treatment, thermal annealing, and washing, takes a couple of additional hours. Running simultaneously, synthesis and purification of different batches can produce 1–2 g of SWCNT per day. The use of more catalyst, a larger boat, and an appropriately modified CVD reactor can all increase the yield. New growth experiments are currently underway at UC that are expected to reach this objective. Other benefits may also be possible, such as increasing CNT length, and controlling the chirality and the alignment of the CNT.

Pulsed carbon precursor delivery during CVD growth of CNT is currently being developed at UC. It is known that high-quality CVD diamond films can be achieved by controlled manipulation of the growth parameters and particularly by changing the carbon precursor flow during the deposition process. Time-modulated CVD (TMCVD) is reported to produce smoother, harder, and better quality diamond films.[35,36] The results published in these papers show controlled diamond film growth by manipulating the flow of the carbon precursor. Another approach that is very similar to TMCVD is a two-step growth process. This is especially useful for low-temperature deposition, where it is difficult to obtain smooth films.[37] This process promotes secondary nucleation, which changes the product morphology. Recently, V. Shanov et al. reported pulsing of methane flow between a high and low level during the CVD process for producing small-grain-size diamond films.[33,38,39] The high methane concentration flow increased the carbon supersaturation on the substrate and enhanced secondary nucleation. This results in small grain size of the deposit, but a higher nondiamond (sp^2) phase in the film.

This previous experience is employed in pulsing of the carbon precursor delivery during the CVD process for better control of the quality and quantity of the CNT. For this purpose, the EasyTube™ furnace was computer-programmed to pulse the methane flow in the CVD reactor. This approach offers the experimental opportunity to manipulate the carbon supersaturation in the catalyst-carbon solution during the growth of the CNT. Another outcome of this process is expected to be the manufacturing of longer CNT grown in a vertical reactor. The study of the gas environment within the CNT growth zone is done based on measurements from the quadrupole mass spectrometer (QMS). The QMS allows *in situ* identification of the characteristic ionic and molecular species in the CVD reactor.

The magnesium oxide–supported iron catalyst for bulk growth, along with the patterned Si substrate for surface synthesis, described in Reference 40, was employed for these experiments.[40] Mixtures of CH_4 and H_2 are used as reactant gases for the CVD growth. The growth conditions for two experiments are shown in Table 5.4. These experiments were performed in two modes: (1) the continuous mode and (2) the pulsed mode. In the continuous mode the methane flow is maintained constant throughout the deposition process. During the pulsed mode, the methane flow is pulsed between a high and a low level. This procedure is

TABLE 5.4
Deposition Conditions for Continuous and Pulsed CVD Growth of CNT

Growth Time (min)	Temp. (°C)	Hydrogen Flow (sccm)	Methane Flow Mode	Methane Flow (sccm)
15	950	100	CM*	600 (high)
30	950	100	PM**	600-0 (high-low)

* Continuous mode.

** Pulsed mode.

automated by employing computer-controlled mass flow controllers. The duty cycle of the modulated methane flow is selected to be 50%, and the total time of one cycle is 10 min. This is illustrated in Fig. 5.8. The deposition time is maintained at 15 min for the nonpulsed flow experiments and is doubled for the pulsed flow experiments. During the growth step the hydrogen flow is maintained constant, but in future studies it will be manipulated along with the carbon precursor flow, and also with the floating catalyst flow.

The dynamic mode of the QMS was used to track the concentration of the major gaseous species with time. This mode conveniently reveals the change in the partial pressure of the species when the methane gas flow is pulsed from high to low and so on. The acquired spectra for continuous and pulse mode of the methane delivery are shown in Fig. 5.9. The four major species monitored in the CVD reactor during the CNT growth are H_2^+ (AMU = 2), C^+ (AMU = 12), CH_3^+ (AMU = 15), and CH_4^+ (AMU = 16). The partial pressure of CH_3^+ is close to that of CH_4^+, and both species follow the same chart trend. We selected to reveal in Fig. 5.9 only the CH_4^+ species

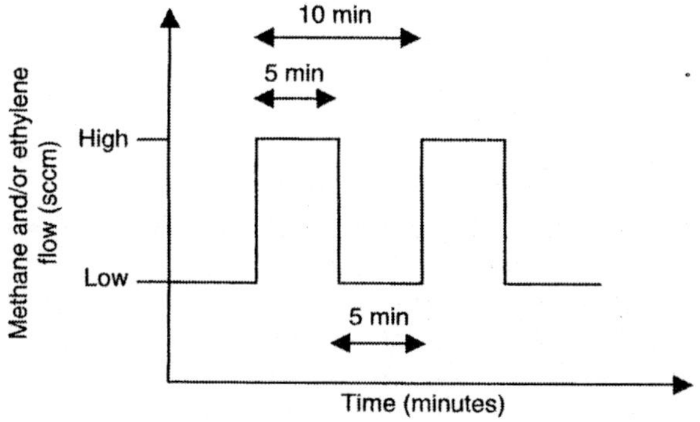

FIGURE 5.8 Modulated carbon precursor flow for growing of CNT.

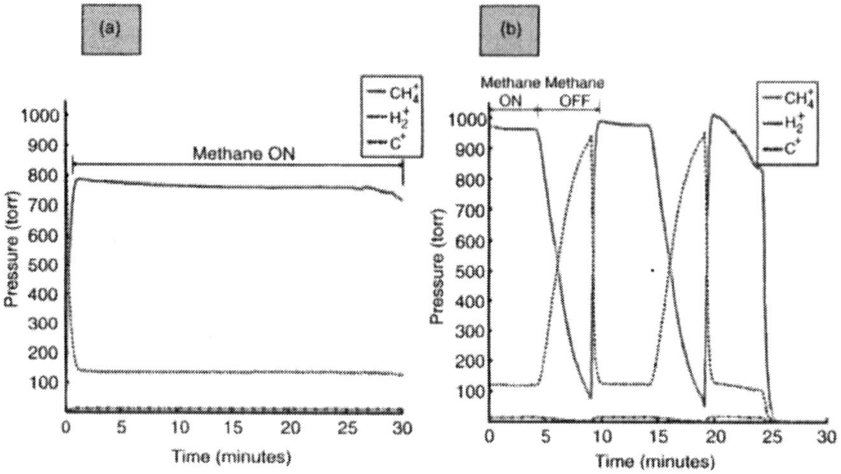

FIGURE 5.9 QMS spectra of CNT growth run in two different modes of methane delivery at conditions shown in Table 5.4: (a) continuous methane flow, and (b) pulsed methane flow.

along with H_2^+ and C^+. During the first growth experiment the methane flow rate was kept constant and the obtained QMS spectra are shown in Fig. 5.9(a). The concentrations of the three major species remain constant, which is an expected result. The amount of CH_4^+ dominates and C^+ reveals the lowest partial pressure. Figure 5.9(b) shows the results from the second experiment where methane flow pulses from 600 sccm to 0 and back to 600 sccm. The observed concentration of CH_3^+ and C^+ species first decreases and then increases accordingly. Because of residual methane in the reactor during the off period, the partial pressure of carbon-containing species does not reach zero value. The flow rate of hydrogen remains constant during the entire deposition time, but the partial pressure of H_2^+ increases with turning the methane flow off and vice versa.

The experiments conducted confirmed that the spectra of the monitored species are very sensitive to any flow changes and provide important information about the gas phase environment in the CNT growth zone.

These may be the first results reported using pulsed flow and residual gas analysis for CNT growth. Further studies are under way with more in-depth analysis of the CVD growth. Different duty cycles will be explored to refine the proposed pulsed carbon precursor flow technique. The obtained data will be correlated with Raman spectroscopy, SEM, and transmission electron microscope (TEM) studies of the CNT. Our preliminary results indicate that pulsed CVD with residual gas analysis will provide better control of the purity and morphology of the CNT in terms of diameter, growth alignment, and number of walls. This novel approach is more complex than standard CVD of CNT, and further research is needed to develop it.

5.5 PURIFICATION OF AS-GROWN CNT

As deposited CNT are frequently contaminated with metal catalysts, ceramic particles, amorphous carbon, and graphitic nanoparticles, purification of the as-grown CNT is a complex procedure. The first step of purification includes removal of the impurities by wet chemical oxidation. Hydrochloric and nitric acids are frequently used for this purpose. Additional steps include decanting, centrifugation, or filtration, followed by rinsing in deionized water. Gas or vapor phase oxidation is also used for deeper refinement. Finally gentle treatment in an ultrasonic bath helps to continuously break up agglomerated nanotubes. These approaches are frequently combined to achieve more comprehensive purification and a higher CNT yield.

Alumina/silica as a powder substrate was used in our initial experiments. These two oxides are hard to dissolve chemically, and HF acid was employed for removal of silica. This acid is difficult to handle due to its hazardous nature. Magnesium oxide[19,20] is easier to dissolve and is now routinely employed in our current research. The purification procedure involves sonication of the unpurified CNT sample in HCl acid with a concentration of 4 *M* for 1 h, and additional soaking of the sample in acid for 24 h. Washing in water and filtering of the CNT follows these steps.

A simple approach for purification of MWCNT is high temperature oxidation (325–425°C) in air. It is important to note that when using this technique the metal catalyst should be removed first by wet oxidation. The last step is required, since in the presence of oxygen the metal particles catalyze low-temperature oxidation of the CNT, which decreases the yield dramatically.[41] Purity evaluation of the CNT is also not a simple procedure. The metal catalyst weight in as-grown CNT is about 30%. Measuring this contaminant can be done by thermogravimetric analysis (TGA). Burning the CNT sample in air at 1000°C removes all the carbonaceous material and the TGA measurement gives the metal catalyst content in the original sample converted to oxides. Measuring the carbonaceous purity is a much more difficult procedure. Most of the qualitative and quantitative techniques have been based on spectroscopy. The ideal technique has to distinguish between graphitic and amorphous carbon phases in the CNT. A very promising approach for this is solution-phase near-infrared (NIR) spectroscopy. More details about this technique can be found in Reference 41.

Dry methods for purification of CNT represent another attractive approach, since they offer minimum loss during the treatment. Dry purification requires plasma and is more expensive than wet etching because of the cost of the plasma facility. Plasma etching for removal of amorphous carbon that covers aligned CNT has been tested using water plasma.[42] In addition, this treatment removes the end caps of the metal catalysts from the top of the CNT. However, a prolonged plasma procedure may disintegrate the graphitic structure of the nanotubes. Dry etching with H_2O and with H_2 plasma is a promising area that should receive further experimental investigation. Additional information on purification of CNT synthesized by different methods is provided in References 43–46.

5.6 CHARACTERIZATION OF CNT

Carbon nanotubes are extremely small objects and their characterization requires sophisticated instruments. The morphology of the CNT, their dimensions, and orientation can be easily revealed using scanning electron microscopes with high resolution.[6,10,17,18,23-26] Appropriate is the use of an environmental SEM (ESEM), which does not need preparation of the samples by conductive coating. Details on ESEM and its use for our research are given in a later section. Another advanced approach is atomic force microscopy (AFM), which provides the opportunity to study features and properties of CNT that are not available with other techniques.[10] High-resolution transmission electron microscopy (HRTEM) is widely used to characterize the overall shape and diameter of the CNT, the number of the walls, their thickness, and the distance between the walls.[8,9,15,17,18,20,47] In addition, the electron diffraction mode of the TEM helps to identify the nature of the cap on top of the CNT, which is usually composed of the metal catalyst.[14,23,24] Micro-Raman spectroscopy is frequently employed to study the quality of the carbon nanotubes.[20,21,27,48] The tube configuration can be investigated in detail using this technique.[8] It also provides information about the number of the walls and the presence of crystalline or amorphous carbon, including the diameter of SWCNT.

5.6.1 Scanning Electron Microscopy

A FEI XL-30 ESEM FEG made by Philips has been used to study the morphology of the CNT. A Shottky hot-field emission tip is employed as an electron source and has an ultimate resolution of 1.2–1.5 nm. A large specimen chamber housing a motorized stage with an internal CCD camera allows observation of fairly large specimens. Thanks to the variable pressure limiting apertures, the ESEM can be operated at a pressure chamber of 1 to 20 torr. Gaseous detection systems are used in imaging the samples. Partial ionization of the chamber gas causes charge neutralization at the sample surface, with the oppositely charged species being collected by the gaseous secondary detector through a cascade effect. The overall outcome is an improved image of samples regardless of the nature of the studied samples. Some of the samples discussed were observed using a Hitachi S4000 SEM.

Two groups of CNT samples grown at UC are studied by ESEM. The first group includes specimens synthesized on a flat Si substrate, and the second represents bulk synthesis of CNT on powdered ceramic carriers. The samples are directly mounted onto a standard aluminum fixture using double-sided carbon adhesive tape. No further sample preparation or metallization is carried out, allowing the nanotubes to be imaged in their natural condition. The following set of parameters is used to obtain high-resolution images of CNT: working distance of 8–10 mm, accelerating voltage of 10–30 KV, and a chamber pressure of 0.9–1.3 torr. The images in Fig. 5.10 were taken using the ESEM. They illustrate unpurified nanotubes grown in Experiment 1 on alumina particles coated with the Fe-Mo catalyst patterned on a Si substrate. The images reveal CNT dimensions of about 1 μm long and 2–3 nm in diameter, which suggests that the specimens are SWCNT. The nanotubes form a spaghetti-type pattern with tubes lying over each other. Figure 5.10(a) displays typical

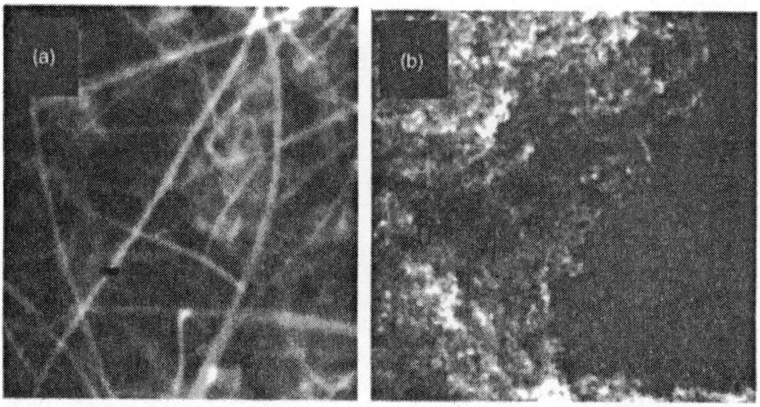

FIGURE 5.10 SEM images of purified SWCNT grown in Experiment 1 using alumina supported Fe-Mo catalyst on a patterned silicon substrate: (a) SWCNT at magnification of ×200,000, and (b) SWCNT grown between catalyst particles at magnification of ×50,000.

morphology of SWCNT grown on a catalyst island. Excessive shining catalyst is present below the nanotubes, which can be removed by purification with acids. Figure 5.10(b) shows growth of SWCNT between catalyst particles on a silicon substrate, indicating that CNT could be synthesized by bridging catalyst particles. It is observed that long SWCNT originate from one mound of catalyst and terminate at a different mound of catalyst running over the dark silicon substrate. This suggests the possibility for direct growth of a nanotube network between patterned islands on the substrate in two dimensions.

The images in Fig. 5.11 were taken using the Hitachi S4000 SEM. Figure 5.11(a) shows purified CNT filaments grown in the modified Experiment 2 using a powdered MgO substrate coated with Fe catalyst. In this experiment ethylene was not present and the methane gas flow rate was increased up to 420 SCCM to compensate for the absence of ethylene precursor. The growth time was 30 min. Figure 5.11(a) shows CNT filaments as well as isolated structures with smaller diameter. The samples have been further investigated by TEM and the results are presented later in this chapter. Figure 5.11(b) shows MWCNT grown in Experiment 3 using a powdered MgO substrate coated with Fe catalyst. The spaghetti-like bundles of MWCNT are further studied and reported later in the TEM section. The morphology and the dimensions of the CNT are very sensitive to the growth conditions and may change after purification.[17,18,42]

Figure 5.12 illustrates bulk nanoropes grown on powdered MgO substrate coated with Fe catalyst. The images contain bundles of CNT ropes with diameters of 40–80 nm. Further study of these structures was conducted by TEM and the results are reported in the next section.

The bundle-type morphology shown in Fig. 5.11 and Fig. 5.12 is similar to that obtained by other groups[17,18,34] and proves that bulk synthesis, although good for high-yield manufacturing, cannot provide aligned growth.

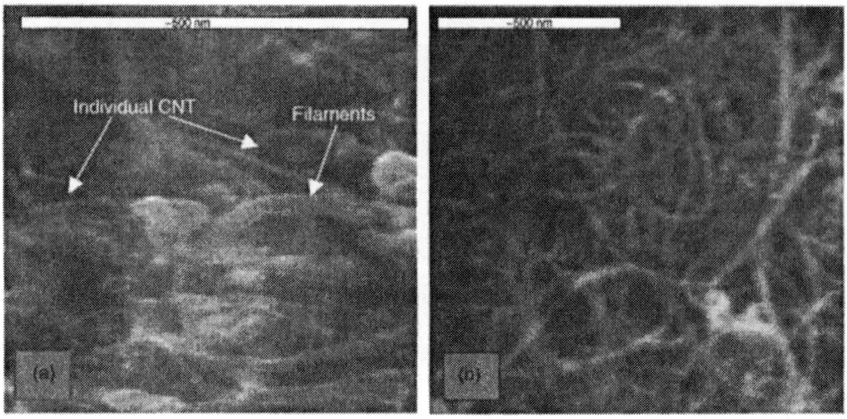

FIGURE 5.11 SEM images of purified CNT synthesized by bulk growth on MgO supported Fe catalyst: (a) CNT filaments grown in modified Experiment 2, magnification of ×100,000, and (b) MWCNT grown in Experiment 3, magnification of ×100,000.

5.6.2 Transmission Electron Microscopy

In transmission electron microscopy (TEM), a thin solid specimen (≤200 nm thick) is bombarded in vacuum with a highly focused, monoenergetic beam of electrons. The beam is of sufficient energy to propagate through the specimen. A series of electromagnetic lenses then magnifies this transmitted electron signal. Diffracted electrons are observed in the form of a diffraction pattern beneath the specimen. This information is used to determine the atomic structure of the material in the sample. Transmitted electrons form images from small regions of the sample that

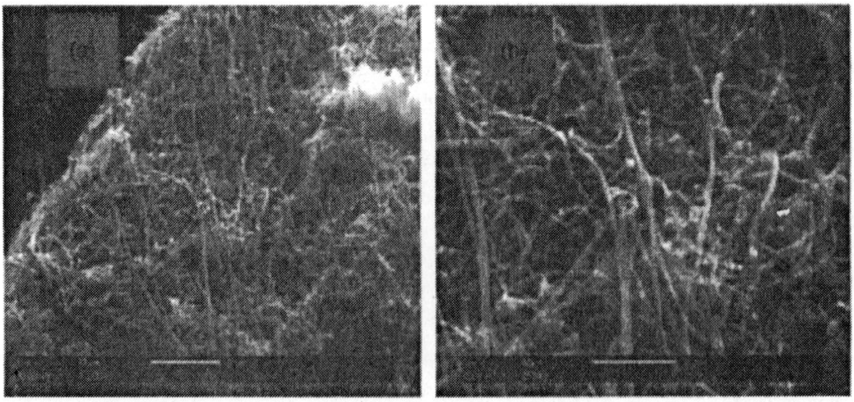

FIGURE 5.12 SEM images of bulk purified nanoropes grown on powdered MgO substrate coated with Fe catalyst at conditions of Experiment 2: (a) nanotubes in ropes, ×10,000, and (b) nanoropes at higher magnification, ×25,000.

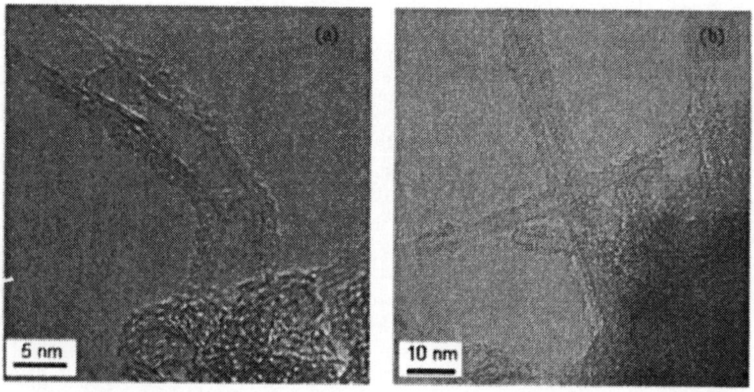

FIGURE 5.13 Purified SWCNT grown in modified Experiment 1 from MgO powder with Fe catalyst: (a) TEM image of SWCNT, and (b) TEM image of SWCNT at 2 times lower magnification than that used in Fig. 5.13(a).

contain contrast, due to several scattering mechanisms associated with interactions between electrons and the atomic constituents of the sample. Analysis of transmitted electron images yields information both about the atomic structure and defects present in the material. The lateral resolution is better than 0.2 nm on some instruments.[49]

A JEOL JEM 4000EX TEM instrument was used to acquire TEM images of CNT. The CNT samples are dispersed in water using a bath sonicator for 30 min. Two drops from the dispersed sample are dropped onto a supporting nickel grid and dried in air. The microscope is operated at 400 kV.

Figure 5.13 shows high-resolution TEM images of purified bulk SWCNT synthesized from MgO powder with Fe catalyst. The growth process was based on a modified Experiment 1, which was run at a longer deposition time of 60 min. The estimation of diameter of the SWCNT from the images shown in Fig. 5.13 is about 3 nm. The outer surface of the nanotubes looks rough which might be caused by an amorphous carbon coating. The high methane flow rate of 1200 sccm and longer deposition time may be the reason for this coating.

Figure 5.14 displays low-resolution TEM images of purified CNT filaments and bulk MWCNT synthesized on a MgO-supported Fe catalyst. Figure 5.14(a) shows CNT filaments grown in a modified Experiment 2, as described earlier. The filaments consist of bundles of CNT, which might consist of SWCNT. Analogous images of bundles of SWCNT are shown in References 8 and 10. Figure 5.14(b) exhibits MWCNT grown in Experiment 3. It reveals a typical bamboo-like structure of MWCNT with diameters ranging from 15–30 nm. Similar morphology is described in References 13, 17, and 18.

TEM images of nanoropes are shown in Fig. 5.15. The average diameter of the ropes is in the range of 40–60 nm. Figure 5.15(a) shows ropes bent in a U-shape. Figure 5.15(b) shows a TEM image of a CNT nanocoil. Similar carbon nanocoils with a 3-D spiral structure grown on a tungsten substrate using a nickel catalyst are described in Reference 50.

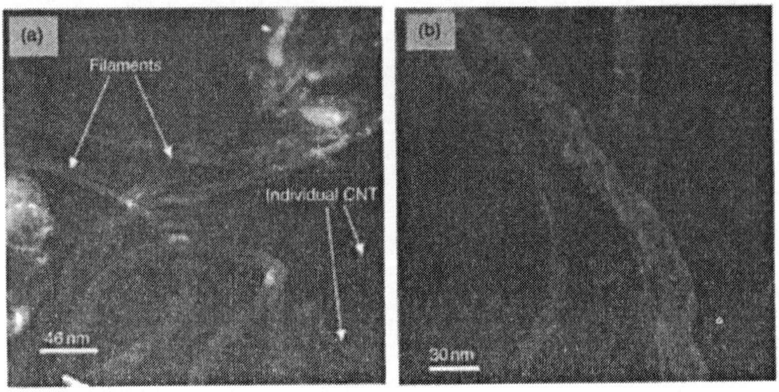

FIGURE 5.14 Low-resolution TEM images of purified CNT synthesized by bulk growth on MgO-supported Fe catalyst: (a) CNT filaments grown in Experiment 2, and (b) MWCNT grown in Experiment 3.

5.6.3 Raman Spectroscopy

The Raman technique is used in qualitative and quantitative analysis of carbon nanotubes. When a beam of light traverses a sample of a chemical compound, a small fraction of the light emerges in directions other than that of the incoming beam. Most of this scattered light is of unchanged wavelength. However, a small part has wavelengths different from those of the incident light, and their presence is a result of the Raman effect. The pattern of the Raman spectrum is characteristic of the particular molecular species, and its intensity is proportional to the number

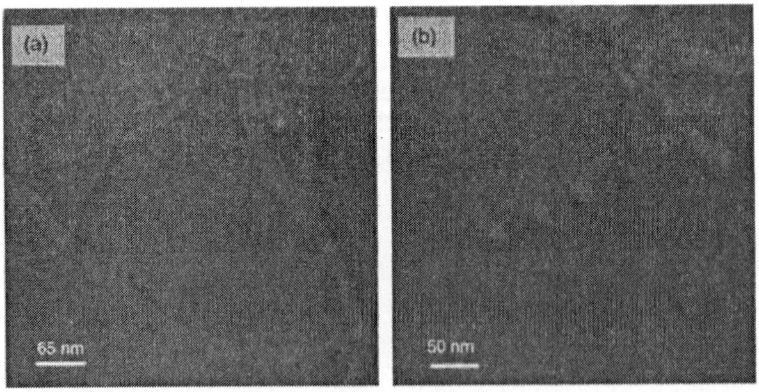

FIGURE 5.15 High-resolution TEM images of MWCNT grown at UC: (a) low-magnification image of dispersed CNT, (b) high-magnification image of a MWCNT separated from the array. Growth conditions: 200 sccm of H_2, 200 sccm of C_2H_4, deposition temperature of 750°C, and Fe catalyst patterned on oxidized Si substrate.

of scattering molecules in the path of the light. Resonance peaks are also observed in the spectrum, which symbolize the presence of a particular species type that is in abundance. Comprehensive information on the Raman technique and its application can be found in References 51–53.

Micro-Raman spectroscopy is used for characterization of CNT with a typical excitation wavelength of 632.8 nm (He-Ne laser) or of 514.5 nm (Ar ion laser). The diameter of single-wall CNT can be determined by SEM/TEM measurements and additionally verified by micro-Raman spectroscopy.

The characteristic spectrum of SWCNT exhibits three main zones at low (100–250 cm^{-1}), intermediate (300–1300 cm^{-1}), and high (1500–1600 cm^{-1}) frequencies.[8,20] These bands can be reduced by the presence of other carbon forms like multiwall nanotubes or amorphous carbon.[20,27] Other carbon materials produce other types of spectra.[54]

The high-frequency bands can be decomposed in one main peak around 1580 cm^{-1}, with a shoulder around 1540 cm^{-1}. This shoulder is more important in the case of SWCNT produced by arc discharge, where a separate peak can be observed.[20,27] In carbon-based materials, typically two main first-order peaks are present. One is the *D* peak, observed around 1300 cm^{-1} (for excitation with He-Ne laser) or at 1350 cm^{-1} when using an Ar ion laser. The *D* peak is related to the presence of defects.[21] The other one is the *G* peak at about 1580 cm^{-1}, which is associated with the in-plane vibrations of the graphene sheet.[48,55] In addition to the *G* line, the appearance of a side peak at about 1540 cm^{-1} indicates the existence of single-wall nanotubes with different diameters.[27,56] Ratios of the *D* peak to the *G* peak have been used as an indicator of the amount of disorder within nanotubes.[21,57] A small I_D/I_G ratio, typically in the range of 0.1–0.2, indicates that the defect level in the atomic carbon structure is low, which suggests reasonable crystalline quality.[48]

The high-resolution spectrum obtained in the low-frequency domain shows several components within the range of 100–250 cm^{-1}.[20,21] Prominent low-energy peaks around 191 and 216 cm^{-1} are the breathing modes of nanotubes vibrating along the radial direction[56] and can be clearly revealed by using a He-Ne laser.[27] The spectrum in the low-frequency domain reflects the SWCNT diameters and can be used to evaluate it. In general, the frequency increases with decreasing tube diameter (d). The frequency v of these modes is known to be inversely proportional to the diameter of the SWCNT. Based on the equation $v\ (cm^{-1}) = 223.75/d\ (nm)$ the diameter of the SWCNT can be determined.[58] The calculated (d) values reflect a distribution in the diameters of the CNT.

The peaks ranging from 400 to 1000 cm^{-1} are usually observed in single-wall nanotubes when investigated using as excitation wavelength of 632.8 nm and could be related to the finite length of the carbon nanotubes.[27,59]

In the case of MWCNT studied using excitation with a He-Ne laser, the *G* line reveals a small bump at the higher energy side around 1620 cm^{-1} and it is attributed to the multiwall structure. In addition, breathing modes are significantly diminished.[27] The Raman spectrum of nanotubes grown by thermal CVD on Co-coated Si substrates shows the *G*-line peak at 1587 cm^{-1}.[27] The observed broad peak near 1337 cm^{-1} indicates the existence of defective graphitic layers on the wall surfaces due to the relatively low growth temperature. Prominent peaks on the lower energy side cannot be observed in this case.[27] Another study claims investigation of low-frequency spectra of multiwalled CNT.[60] Low-frequency modes between 115 and 220 cm^{-1} are obtained after purification of the carbon nanotubes and excitation with a red laser line of 676 nm and with a green line of 514 nm. These radial-breathing modes originating from individual walls can be used for characterization of multiwalled CNT.[60]

The excitation energy of the laser sources can be used to explore the spectrum in the range 1400–1700 (cm^{-1}) and find variations in the metallic or semiconducting nature of the fabricated CNT. If the light excitation energy is between 1.7 and 1.9 eV, and resonance is observed in the 1400–1700 cm^{-1} wave number range, it is an indication of metallic nanotubes. If the light excitation energy is above 2.2 eV or below 1.5 eV, and resonance peaks are observed in the 1400–1700 cm^{-1} wave number range, it is evidence of the semiconducting nature of the CNT.[61]

Details on Raman spectroscopy of CNT are presented in Chapter 3 of this book.

An inVia Raman microscope was employed to obtain the CNT spectra synthesized in three runs based on Experiment 1. The Renishaw spectrometer is a 250-mm focal length system. For the 514.5 nm laser excitation, 1800 l/mm grating, and 50 μm slit width, the spectral resolution is about 5.5 cm^{-1}. This spectrometer is coupled with a RenCam high-performance cooled CCD detector, which is an all-in-one compact unit. It uses thermoelectric cooling to operate at −70°C. Its detector chip and electronics are carefully engineered to offer users ultra-low noise and low dark current.

One of the Raman spectra obtained in Experiment 1 is shown in Fig. 5.16. In this analysis only wave numbers 0–300 (cm^{-1}) are considered for calculation of diameter of SWCNT, and the results are presented in Table 5.5. The relationship between diameter and resonance wave number is $\nu\,(cm^{-1}) = 223.75/d\,(nm)$. The CNT diameters are almost the same in the three experiments and are in the range of 1.2–1.3 nm.

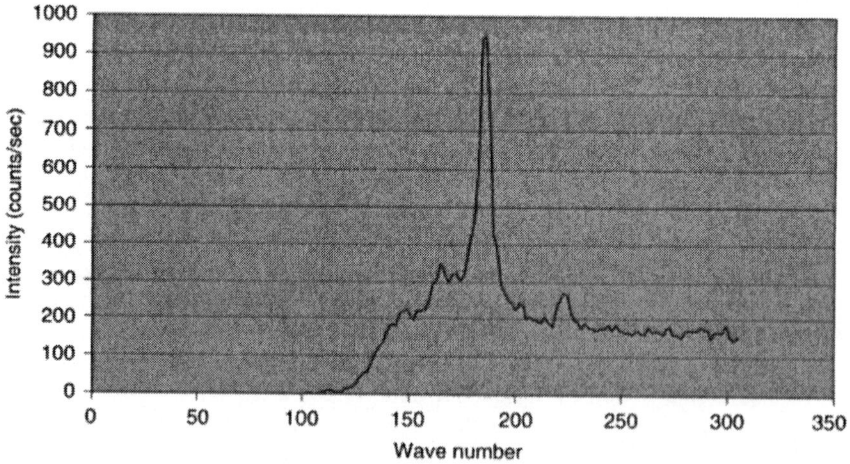

FIGURE 5.16 Raman spectrum of CNT grown in Experiment 1 from alumina-supported Fe-Mo catalyst on a patterned silicon substrate.

TABLE 5.5
Numerical Data Estimated from the Raman Spectra of SWCNT Grown on a Silicon Substrate

Experiment Number	Resonance Wave No. (cm^{-1})	Diameter (nm)
Experiment 1	192.7	1.2
Modified Exp. 1 (longer purging at elevated temperature)	183.9	1.2
Modified Exp. 1 (longer purging at elevated temperature and longer deposition time)	169.8	1.3

5.7 ADVANCED TOPICS AND FUTURE DIRECTIONS FOR CVD OF CNT

Our future work in the field of CVD synthesis of CNT includes three major directions:

1. The growth of long nanotubes is our long-term goal, which will have a positive impact on development of a new generation of composite materials and sensors. Increasing the length of nanotubes will improve the load transfer from the nanotubes to the matrix and thus the overall mechanical properties of CNT-based composites. This objective will be achieved by using our newly designed vertical CVD reactor, which will be fed with floating catalysts. *In situ* monitoring of the gas environment in the reactor by a quadrupole mass spectrometer is expected to provide better control of the process parameters. Pulsed carbon precursor delivery during the CVD deposition is currently being studied at UC, and the preliminary experiments show improved quality control of the grown CNT. These efforts are expected to enhance our abilities to synthesize longer CNT.
2. Controlled alignment of SWCNT, selectively grown on flat substrates, is another objective of our future research. This will open opportunities to create a "building brick" for CNT sensors and actuators. An array based on SWCNT for actuator applications has been designed and will be manufactured in the near future at UC. Semiconductor processing techniques will be used to fabricate the CNT array and devices.
3. Functionalization of the carbon nanotubes is another goal of our future research. It will help us change the nanotube surface properties by applying a very thin layer of functional materials or attaching selected functional groups. Different deposition techniques, including plasma CVD, are among the options we are currently exploring to achieve this objective. The controlled functionalization has a great potential for building advanced sensors and actuators with applications in new areas, including cancer diagnostics and therapy.

5.7.1 Synthesis of Long Carbon Nanotubes

An innovative aspect of nanomaterials synthesis involves the preparation of long carbon nanotubes in a vertical CVD reactor.[29,30] In its current horizontal configuration, the CVD system at UC can be used to make SWCNT that have a length of about several micrometers. We are modifying the horizontal furnace to a vertical configuration that it is hoped will produce SWCNT or strands with lengths of millimeters or even centimeters. The vertical furnace will be equipped with instruments to monitor the growth environment of SWCNT and to improve process control. The vertical CVD reactor will be fed with a floating catalyst. A schematic drawing of the CVD system with a vertical furnace is shown in Fig. 5.17. The vertical CVD system will have several unique features, including fiber optic control of temperature and individual preheat of all gases for fast growth. In addition, feedback from the

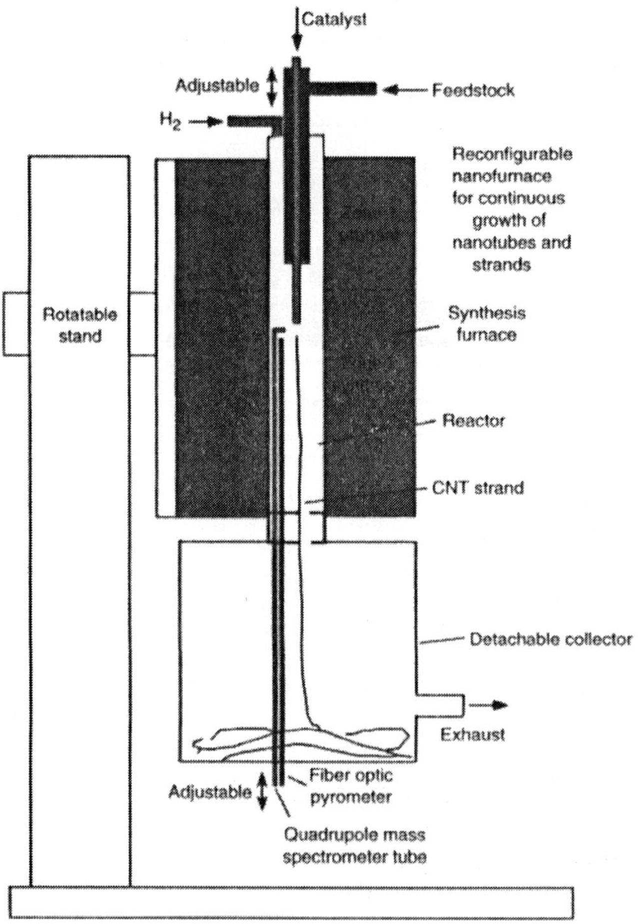

FIGURE 5.17 CVD system with a vertical furnace.

QMS to the mass flow controllers will provide precise control of the carbon and metal-catalyst concentration in the growth zone.

Floating catalysts will be introduced in the described CVD reactor for generating nanometer-size particles of the metal-catalyst from the gas phase. The advantages of this approach in a vertical CVD reactor include an uninterrupted downward growth environment and alignment of nanotubes in the gas flow direction. In addition, pulsed delivery of the carbon precursor will provide better control of the carbon supersaturation in the catalyst-carbon solution. This is critical for the CNT morphology and related properties.

5.7.2 Design and Fabrication of a Segment for CNT Actuator

A promising approach to making nanotube sensors and actuators is the fabrication of a single segment CNT array.[22,62] At UC we designed such a segment based on fabrication steps that are frequently used in semiconductor processing technology. Figure 5.18 illustrates the sequential steps in the fabrication process, which will result in producing a "building brick" for CNT actuators. Two types of segments are selected, depending on the electrolyte used for the actuator. The first one will

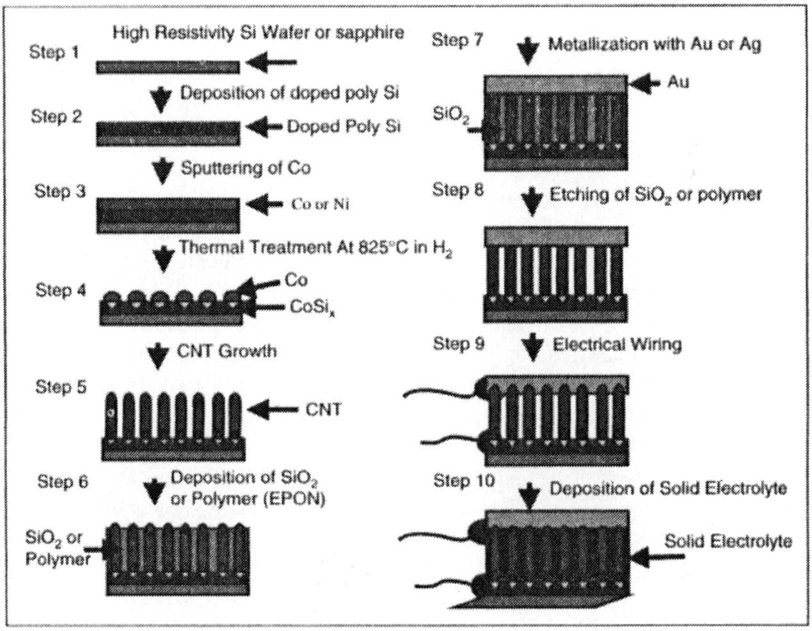

FIGURE 5.18 (Color figure follows p. 12.) Schematic illustration of the fabrication process of CNT segment for actuator application.

perform in a liquid electrolyte. The second segment type will be embedded in a solid electrolyte, which requires an additional step of processing, as revealed in Step 10 of Fig. 5.18. The starting substrate (Step 1) could be either a highly resistive Si single-crystal wafer or sapphire. This will provide an insulating base, which is required for electrical wiring of the segment. Step 2 involves deposition of highly conductive (doped) poly-Si, on which the CNT catalyst is planted. Doped Si will provide good electrical contact between the CNT and the substrate. CVD technique will be employed for depositing of poly-Si doped either with P or As. An alternative solution is to use an insulating Si wafer, which can be ion implanted for increasing the electrical conductivity. Steps 3 to 4 require sputtering of a thin metal film (20–40 nm) of catalyst such as Co and thermal annealing of the film in hydrogen at 825°C. The procedure is described in Reference 24. During the annealing, catalytic semi-spheres of Co islands are formed, driven by surface tension to lower the total energy. Simultaneously, a chemical reaction between the metal and the Si substrate takes place, producing highly conductive $CoSi_x$. This compound is expected to provide a strong bond between the CNT and the substrate and will also serve as a conductive electrical junction. The patterned substrate with catalyst provides aligned growth of MWCNT during Step 4. Metallization of the top portion of the CNT for good electrical contact requires several procedures. Step 6 fills up the gaps between the individual CNT within the block with either SiO_2 or a polymer, which can be done by standard deposition procedures for these materials.[22,62,63] This will give good mechanical support to the CNT for the next metallization. Step 7 illustrates metallization, performed by sputtering of Au on top of the CNT. Alternative metallization is possible by brushing the top surface with silver paste.

Step 8 removes the supporting SiO_2 material by etching that will liberate the CNT surfaces. Electrical wiring of the segment is shown in Step 9. Electrical signal can be provided to each individual block/column of aligned carbon nanotubes using the bottom wire.

In a new approach developed at UC, the top wire will be used to sense the strain in the nanotubes. Since the resistance of the nanotubes will change with strain, the array can become a self-sensing actuator. This will allow the actuator to have strain feedback and become smart. The strain feedback will allow precise feedback control of the actuator motion. More details on CNT actuators can be found in Chapter 15 of this book. Step 10 reveals the fabrication process for the second segment type. The CNT blocks need to be embedded in a solid electrolyte. Steps 1 to 9 are identical for both segment types and an additional Step 10 is added to the processing sequence. This step will provide the solid electrolyte, which will embed the CNT columns. The solid electrolyte will be deposited by plasma polymerization as discussed in Chapter 8 of this book and Reference 64. The array growth shown in Fig. 5.18 is feasible with our CVD system. Vertical MWCNT arrays grown on Si at the Universty of Cinncinati are shown in Fig. 5.19.

5.7.3 Oriented Growth of a Simple CNT Network

CVD is the most efficient method for synthesis of SWCNT with controlled orientation. Different approaches have been used to enhance CVD-oriented growth, including external gravitational, magnetic, and electric field. Electric field–directed growth has been

FIGURE 5.19 Long-array nanotubes grown at the University of Cincinnati: (a) a 4-mm MWCNT patterned array; and (b) SEM side view of a nanotube array at 1,000× magnification. The nanotubes were grown by Yun Yeo-Heung. The patterned substrate was provided by Yi Tu of First Nano Inc.

successfully used to deposit aligned SWCN showing two-dimensional network features.[65] Directed growth of free-standing SWCN on microtowers was demonstrated as well.[66] Results have been published about growth of millimeter-long SWCNT, and also about a two-dimensional nanotube network grown by a "fast heating" CVD process.[67] The controlled orientation during the growth process and deposition of a CNT network is extremely important for developing artificial neural microsystems.

Our future research objective is to synthesize a simple network of parallel SWCNT with a preset orientation, which is suitable for incorporation into a material neural system for sensing of microcracks, defects, and corrosion. The approach is based on two strategies: (1) to use a vertical CVD reactor for growth of long CNT, which has been described previously, and (2) to perform growth experiments where lateral electric fields will be applied across the prepatterned metal catalyst Si substrate during the CVD process. Different electric fields such as DC, pulsed DC, or AC field will be tested. The electrodes can be made of doped Si or other conductive materials. Alignment guidance by the field is expected to promote growth of the CNT network between parallel electrodes or between microposts on the Si surface. This concept is illustrated in Fig. 5.20. In addition, gravitational force experiments will be conducted to enhance oriented growth of CNT. A substrate holder for such experiments, shown in Fig. 5.6(a), has been made and is currently being tested. It will allow growth of the CNT in an upside-down direction, taking advantage of gravitational forces.

5.7.4 Functionalization of CNT

Functionalization of CNT is essential in changing their overall properties. It can be done by specific surface modification such as deposition of thin coating or attachment of selected functional groups. Different coatings can be tailored to alter the electrical and mechanical properties of the CNT due to a combination of dimensional changes and interfacial properties.

Low-temperature functionalization of CNT is described in Reference 68. A microwave plasma source is used in this study to generate hydrogen plasma, which is passed

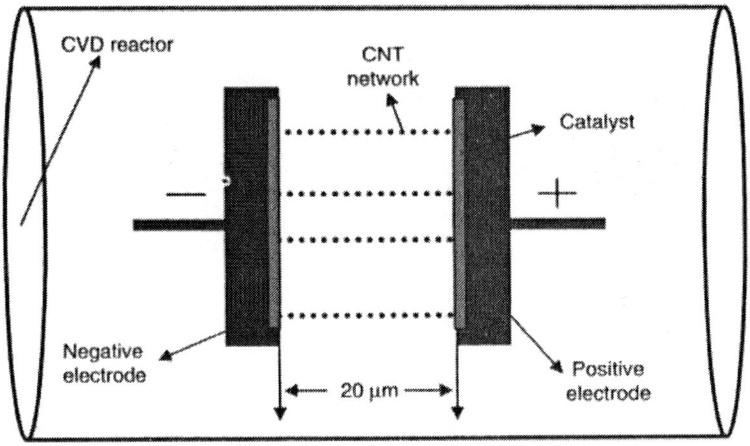

FIGURE 5.20 Schematic drawing of arrangement for oriented grow of parallel CNT under an electric field.

through a curved tube into the process chamber. The atomic hydrogen species reach the CNT and cause their hydrogenation. The process is dry, clean, and relatively simple.

Plasma polymerization coating of nanotubes is a novel approach used at UC to improve the surface properties of nanotubes. Polymerization is the process in which long-chain or network molecules are made from relatively small organic molecules. D. Shi and P. He at UC have developed a unique plasma polymerization process to coat CNT with different materials.[64] In a plasma environment, the surface atoms of the CNT are activated. At these active centers, a polymer film forms on the CNT surface. The polymerization conditions can be controlled so that the structure of the coating contains the desired functional groups. High-resolution TEM pictures of plasma-coated MWCNT are shown in Fig. 5.21. The coating thickness can be controlled by the processing time and it is in the range of 20 nm. The reader can refer to Chapter 8 of this book for more details on plasma coating of CNT.

Another approach for functionalization of CNT that we are currently investigating is based on deposition of a thin Pd coating. The goal of this research is to design and fabricate a hydrogen nanosensor. When palladium absorbs H_2, a new phase (palladium hydride) is produced, causing the metal lattice to expand by about 3.5%. As a result, the palladium grains within a thin Pd film swell.[69] The described effect is used in a hydrogen sensor demonstrated by R. Penner and coworkers.[70] They fabricated a miniature hydrogen sensor based on an array of Pd nanowires. Hydrogen absorption lowers the Pd wire electrical resistance by changing the Pd structure. When exposed to hydrogen, Pd wires expand because of H_2 absorption, and the nanoscopic gaps between the wires close. This allows more current to pass through the wires, which is proportional to the absorbed hydrogen.

We plan to coat columns of CNT, similar to those shown in Fig. 5.19, with 20–30 nm Pd by electroless deposition. Sputtering is also considered as a route for Pd metallization. The exposure of the Pd-coated CNT column to hydrogen will

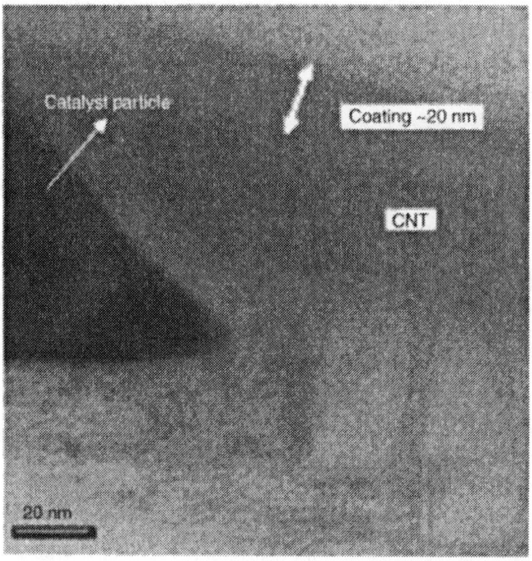

FIGURE 5.21 HRTEM images of coated MWCNT showing horizontal view of one nanotube with the catalyst particle on the top.

change the electrical and mechanical properties of the coated CNT. The column is expected to change the electrical resistance, the length, and also the direction of orientation. In addition, the nanogaps between the individual CNT will shrink. This process is reversible and the desorption of H_2 can be enhanced by slightly heating the CNT sensor. Multiple cycles of hydrogen adsorption and desorption from the Pd-coated CNT with predictable reproducibility are technically feasible.

5.8 CONCLUSIONS

The CVD technique is a powerful approach for synthesis of CNT with controlled properties. It has been demonstrated that the manipulation of growth parameters along with the right choice of catalyst can predetermine the deposition of either SWCNT or MWCNT. The use of MgO-supported catalyst results in easy removal of impurities by HCl acid.

Quadrupole mass spectroscopy can be employed for *in situ* monitoring of the CNT growth environment in the CVD reactor. It was observed that this technique successfully reveals changes in the concentration of the gas species during the pulsed hydrocarbon delivery. The synthesis of CNT using pulse flow of the carbon precursor is a promising approach. It is currently under investigation at UC, and new results will be published.

Advanced topics for CVD growth of CNT include (1) synthesis of long nanotubes by using floating catalyst and vertical CVD reactor and (2) controlled alignment of SWCNT selectively grown on flat substrate as arrays. These are the areas of interest to be pursued in order to fabricate high-quality CNT for sensor and actuator applications.

PROBLEMS

1. List the important benefits of the CVD approach as compared to other preparation techniques for synthesis of CNT.
2. Perform a literature search on CVD growth of CNT and reveal any new advances that have been made relative to the information provided in this chapter.
3. List the catalysts, the substrates, and the carbon precursors used for synthesis of CNT by CVD.
4. Describe the difference between the bulk synthesis and surface synthesis in terms of yield and alignment of CVD-grown carbon nanotubes.
5. List the impurities of as-grown CNT and describe the purification procedures for their removal.
6. Search the literature for new purification techniques that can decrease the loss of CNT during the removal of the impurities.
7. List the characterization techniques used for determination of CNT diameter. Comment on the limitations and advantages of each technique.
8. Describe the characterization techniques for studying the morphology and the structure of carbon nanotubes.

ACKNOWLEDGMENTS

The work described in this chapter was sponsored in part by the UC Summer Student Fellowship, the UC University Research Council, First Nano, Inc., and the Ohio Aerospace Institute. We thank S. Subrahmin and R. Tiwari for the ESEM and TEM images of CNT. The authors gratefully acknowledge C. Sloan, D. Adderton, and Y. Tu of First Nano, Inc. for their cooperation with the nanotube synthesis. We appreciate the high-resolution TEM images of SWCNT provided by L. Wang and J. Lian from the University of Michigan and the Raman spectroscopy of CNT done by D. Mast and R. Gilliland from UC and J. Busbee and J. Reber from AFRL/MLMT Wright-Patterson AFB.

REFERENCES

1. Sullivan, J., Robertson, J., Zhou, O., Allan, T., and Coll, B., Amorphous and nanostructured carbon, *MRS Symposium Proceedings, 593,* 2000.
2. Rao, A., Nanotubes and related materials, *MRS Symposium Proceedings, 633,* 2001.
3. Robertson, J., Friedmann, T., Geohegan, D., Luzzi, D., and Ruoff, R., Nanotubes, fullerenes, nanostructured and disordered carbon, *MRS Symposium Proceedings, 675,* 2001.
4. Bernier, P., Roth, S., Carroll, D., and Kim, G.T., Nanotube-based devices, *MRS Symposium Proceedings, 772,* 2003.
5. Dresselhaus, M. and Dai, H., Advances in carbon nanotubes, *MRS Bulletin, 29,* 4, 237, 2004.
6. Thess, A., Lee, R., Nikolaev, P., Dai, H., Petit, P., Robert, J., Xu, C., Lee, Y.H., Kim, S.G., Rinzler, A.G., Colbert, D.T., Scuseria, G.E., Tománek, D., Fischer, J.E.,

and Smalley, R.E., Crystalline ropes of metallic carbon nanotubes, *Science, 273,* 483, 1996.

7. Bethune, D., Kiang, C., DeVries, M., Gorman, G., Savoy, R., and Beyers, R., Cobalt-catalysed growth of carbon nanotubes with single-atomic-layer walls, *Nature, 363,* 605, 1993.
8. Journet, C., Maser, W.K., Bernier, P., Loiseau, A., Chapelle, M., Lefrant, S., Deniard, P., Lee, R., and Fischer, J.E., Large-scale production of single walled carbon nanotubes by the electric-arc technique, *Nature, 388,* 756, 1997.
9. Cassel, A., Raymakers, J., Kong, J., and Dai, H., Large scale synthesis of single-walled carbon nanotubes, *J. Phys. Chem. B., 103,* 6484, 1999.
10. Liu, J., Fan, S., and Dai, H., Recent advances in methods of forming carbon nanotubes, *MRS Bulletin, 29,* 4, 224, 2004.
11. Smith, D., *Thin-Film Deposition: Principles and Practice*, 1st ed., McGraw-Hill, New York, 1995.
12. Hitchman, M. and Jensen, K., *Chemical Vapor Deposition: Principles and Applications,* Academic Press, London, 1993.
13. Sherman, A., *Chemical Vapor Deposition for Microelectronics: Principles, Technology and Applications,* Noyes Publications, NJ. 1998.
14. Kuang, M., Wang, Z., Bai, X., Guo, J., and Wang, E., Catalytically active nickel {110} surfaces in growth of carbon tubular structures, *Appl. Phys. Lett., 76,* 10, 1255, 2000.
15. Lee, C. and Park, J., Growth model of bamboo-shaped carbon nanotubes by thermal chemical vapor deposition, *Appl. Phys. Lett., 77,* 21, 3397, 2000.
16. Tang, C., Ding, X., Huang, X., Gan, Z., Qi, S., Liu, W., and Fan, S., Effective growth of boron nitride nanotubes, *Chem. Phys. Lett., 356,* 254, 2002.
17. Li, W., Wen, J., and Ren, Z., Effect of temperature on growth and structure of carbon nanotubes by chemical vapor deposition, *Appl. Phys. A, 74,* 397, 2002.
18. Li, W., Wen, J., Tu, Y., and Ren, Z., Effect of gas pressure on growth and structure of carbon nanotubes by chemical vapor deposition, *Appl. Phys. A, 73,* 259, 2001.
19. Li, W., Wen, J., Sennett, M., and Ren, Z., Clean double-walled carbon nanotubes synthesized by CVD, *Chem. Phys. Lett., 368,* 299, 2003.
20. Colomer, J., Stephan, C., Lefrant, S., Tendeloo, G., Willems, I., Konya, Z., Fonseca, A., Laurent, C., and Nagy, J., Large-scale synthesis of single-walled carbon nanotubes by catalytic chemical vapor deposition (CCVD) method, *Chem. Phys. Lett., 317,* 83, 2000.
21. Geng, J., Singh, C., Shephard, D., Shaffer, M., Johnson, B., and Windle, A., Synthesis of high purity single-walled carbon nanotubes in high yield, *Chem. Commun.,* 2666, 2002.
22. Tu, Y., Lin, Y., and Ren, Z., Nanoelectrode arrays based on low site density aligned carbon nanotubes, *Nano Lett., 3,* 1, 107, 2003.
23. Wen, J., Huang, Z., Wang, D., Chen, J., Yang, S., Ren, Z., Wang, J., Calvet, L., Klemic, J., and Reed, M., Growth and characterization of aligned carbon nanotubes from patterned nickel nanodots and uniform thin films, *J. Mater. Res., 16,* 11, 3246, 2001.
24. Bower, C., Zhou, O., Zhu, W., Werder, D., and Jin, S., Nucleation and growth of carbon nanotubes by microwave plasma chemical vapor deposition, *Appl. Phys. Lett., 75,* 12, 2767, 2000.
25. Fan, S., Chaplin, G., Franklin, N., Tombler, T., Cassell, A., and Dai, H., Self-oriented regular arrays of carbon nanotubes and their field emission properties, *Science, 283,* 512, 1999.
26. Kong, J., Soh, H., Cassel, A., Quate, C., and Dai, H., Synthesis of individual single-walled carbon nanotubes on patterned silicon wafers, *Nature, 395,* 878, 1998.

27. Lee, C., Kim, D., Lee, T., Choi, Y.C., Park, Y., Kim, W., Lee, Y., Choi, W., Lee, N., Kim, J., Choi, Y.G., and Kim, J., Synthesis of uniform distributed carbon nanotubes on a large area of Si substrates by thermal chemical vapor deposition, *Appl. Phys. Lett., 75,* 12, 1721, 1999.
28. Chakrapani, N., Curran, S., Wei, B., Ajayan, P., Carrillo A., and Kane, R., Spectral fingerprinting of structural defects in plasma-treated carbon nanotubes, *J. Mater., Res., 18,* 10, 2515, 2003.
29. Zhu, H., Xu, C., Wu, D., Wei, B., Vajtai, R., and Ajayan, P., Direct synthesis of long single-walled carbon nanotube strands, *Science, 296,* 884, 2002.
30. Ci, L., Li, Y., Wei, B., Liang, J., Xu, C., and Wu, D., Preparation of carbon nanofibers by the floating catalyst method, *Carbon, 38,* 1933, 2000.
31. Singh, C., Quested, T., Boothroyd, C., Thomas, P., Kinloch, I., Kandil, A., and Windle, A., Synthesis and characterization of carbon nanofingers produced by the floating catalyst method, *J. Phys. Chem. B, 106,* 10915, 2002.
32. First Nano, Inc., EasyTube Users Manual, Preliminary Version — 10/16/02, Santa Barbara, CA, 2002.
33. Shanov, V., Ramamurti, R., and Singh, R.N., Effect of Pulsed Methane Flow on nanostructured diamond films synthesized by microwave plasma CVD, *Transactions, 154,* 283, 2003.
34. Hafner, J., Bronikowski, M., Azamian, B., Nikolaev, P., Rinzler, A., Colbert, D., Smith, K., and Smalley, R., Catalytic growth of single-wall carbon nanotubes from metal particles, *Chem. Phys. Lett., 296,* 195, 1988.
35. Fan, Q., Ali, N., Kousar, Y., Ahmed, W., and Gracio, J., Novel time-modulated chemical vapor deposition process for growing diamond films, *J. Mater. Res., 17,* 17, 2002.
36. Ali, N., Neto, V.F., and Gracio, J., Promoting secondary nucleation using methane modulations during diamond chemical vapor deposition to produce smoother, harder, and better quality films, *J. Mater. Res., 18,* 2, 2003.
37. Silva, F., Gicquel, A., Chiron, A., and Achard, J., Low roughness diamond films produced at temperatures less than 600°C, *Diamond Relat. Mater., 9,* 12, 1965, 2000.
38. Ramamurti, R., Shanov, V., and Singh, R.N., Synthesis of Nanocrystalline Diamond Films by Microwave Plasma CVD, in Proc. *Ceramic transactions,* Bansal, NP., and Singh, JP., Eds., The American Ceramic Society, OH, 135, 39, 2003.
39. Ramamurti, R., Shanov, V., and Singh, R. N., Processing of Nanocrystalline Diamond Films by Microwave Plasma CVD, in *Ceramic Engineering and Science Proceedings,* Kriven, W., and Lin, H.T., Eds., The American Ceramic Society, OH, 24(3), 15, 2003.
40. Jain, S., Kang, P., Yun,Y., He, P., Pammi, S., Miskin, A., Narsimhadevara, S., Hurd, D., Schulz, M., Chase, J., Subramaniam, S., Shanov, V., Boerio, F.J., Shi, D., Gilliland, R., Mast, D., and Sloan, C., Building Smart Materials Based on Carbon Nanotubes, in Proc., *SPIE,* Varadan, V.K., Eds., Society of Photo-Optical Instrumentation Engineers, Bellingham, WA, 5389, 167, 2004.
41. Haddon, R., Sippel, J., Rinzler, A., and Papadimitrakopoulos, F., Purification and separation of carbon nanotubes, *MRS Bulletin, 29,* 4, 252, 2004.
42. Huang, S. and Dai, L., Plasma etching for purification and controlled opening of aligned carbon nanotubes, *J. Phys. Chem. B, 106,* 3543, 2002.
43. Hou, P., Bai, S., Yang, Q., Liu, C., and Cheng, H., Multi-step purification of carbon nanotubes, *Carbon, 40,* 81, 2001.
44. Zheng, B., Li, Y., and Liu, J., CVD synthesis and purification of single-walled carbon nanotubes on aerogel-supported catalyst, *Appl. Phys. A, 74,* 345, 2002.
45. Hernadi, K., Fonseca, A., Nagy, J., Bernaerts, D., Riga, J., and Lucas, A., Catalytic synthesis and purification of carbon nanotubes, *Synth. Met., 77,* 31, 1996.

46. Dillon, A., Gennett, T., Jones, K., Alleman, J., Parilla, P., and Heben, M., A simple and complete purification of single-walled carbon nanotube materials, *Advanced Mater., 11*, 16, 1354, 1999.
47. Sloan, J., Luzzi, D., Kirkland, A., Hutchison, J., and Green, M., Imaging and characterization of molecules and one-dimensional crystals formed within carbon nanotubes, *MRS Bulletin, 29,* 4, 265, 2004.
48. Singh, C., Shaffer, M., Kozoil, K., Kinloch, I., and Wndle, A., Towards the production of large-scale aligned carbon nanotubes, *Chem. Phys. Lett., 372,* 860, 2003.
49. Loehman, R., *Characterization of Ceramics*, Butterworth-Heinemann, Boston, 993, 284.
50. Jiao, J., Einarsson, E., Tuggle, D., Love, L., Prado J., and Coia, G., High-yield synthesis of carbon coils on tungsten substrates and their behavior in the presence of an electric field, *J. Mater. Res., 18,* 11, 2580, 2003.
51. Lewis, I. and Edwards, H., *Handbook of Raman Spectroscopy: From the Research Laboratory to the Process Line*, Marcel Dekker, New York, 2001.
52. Szymanski, H., *Raman Spectroscopy: Theory and Practice*, Plenum Press, New York, 1967
53. McCreery, R., *Raman Spectroscopy for Chemical Analysis*, John Wiley & Sons, New York, 2000.
54. Bachmann, K. and Wiechert, D.U., Characterization and properties of artificially grown diamond, in Clausing, R., Horton, L., Angus, J., and Koidl, P., Eds., *Diamond and Diamond-like Films and Coatings*, Plenum Press, New York, 677, 1991.
55. Eklund, C., Holden, M., and Jishi, A., Vibrational modes of carbon nanotubes: Spectroscopy and theory, *Carbon, 33,* 7, 959, 1995.
56. Rao, A., Richter, E., Bandow, S., Chase, B., Eklund, P., Williams, K., Fang, S., Subbaswamy, K., Menon, M., Thess, A., Smalley, R., Dresselhaus, G., and Dresselhaus M.S., Diameter-selective Raman scattering from vibrational modes in carbon nanotubes, *Science, 275,* 5297, 187, 1997.
57. Tan, S., Zhang, P., Yue, K., Huang, F., Shi, Z., Zhou, X., and Gu, Z., Comparative study of carbon nanotubes Raman prepared by D.C. arc discharge and catalytic methods, *J. Raman Spectrosc., 28,* 369, 1997.
58. Bandow, S., Asaka, S., Sato, Y., Grigorian, L., Richter, E., and Ecklund, P., Effect of the growth temperature on the diameter distribution and chirality of single-wall carbon nanotubes, *Phys. Rev. Lett., 80,* 17, 3779, 1998.
59. Saito, R., Takeya, T., Kimura, T., Dresselhaus, G., and Dresselhaus, M.S., Finite-size effect on the Raman spectra of carbon nanotubes, *Phys. Rev. B, 59,* 31, 2388, 1999.
60. Benoit, J., Buisson, J., Chauvet, O., Godon, C., and Lefrant, S., Low-frequency Raman studies of multiwalled carbon nanotubes: Experiments and theory, *Phys. Rev. B*, 66, 073417/1, 2002.
61. Alvarez, L., Righi, A., Rols, S., Anglaret, E., and Sauvajol, J.L., Excitation energy dependence of the Raman spectrum of single-walled carbon nanotubes, *Chem. Phys. Lett., 320,* 441, 2000.
62. Lin, Y., Lu, F., Tu, Y., and Ren, Z., Glucose biosensors based on carbon nanotube nanoelectrode ensembles, *Nano Lett., 4,* 2, 191, 2004.
63. Bronikowski, M. and Hunt, B., Block copolymers as templates for arrays of carbon nanotubes, *NASA Tech Briefs,* 56–57, 2003.
64. Shi, D., Lian, J., He, P., Wang, L., van Ooij, W., Schultz, M., and Mast, D., Plasma deposition of ultrathin polymer films on carbon nanotubes, *Appl. Phys. Lett., 81,* 27, 5216, 2002.

65. Zhang, Y., Chang, A., Cao, J., Qian, Q., Woong, W., Li, Y., Yiming, Morris, N., Yenilmez, E., Kong, J., and Dai, H., Electric-field-directed growth of aligned single-walled carbon nanotubes, *Appl. Phys. Lett., 79,* 19, 3155, 2001.
66. Cassell, A., Franklin, N., Tombler, T., Chan, E., Han, J., and Dai, H., Directed growth of free-standing single-walled carbon nanotubes, *J. Am. Chem. Soc., 121,* 7975, 1999.
67. Huang, S., Cai, X., and Liu, J., Growth of millimeter-long and horizontally aligned single-walled carbon nanotubes on flat substrates, *J. Am. Chem. Soc., 125,* 19, 5636, 2003.
68. Khare, B. and Meyyappan, M., Low-temperature plasma functionalization of carbon nanotubes, *NASA Tech Briefs*, *45,* 2004.
69. Lewis, F., *The Palladium Hydrogen System*, Academic Press, New York, 1967.
70. Favier, F., Walter, E., Zach, M., Benter, T., and Penner, R., Hydrogen sensor and switches from electrodeposited palladium mesowire arrays, *Science, 293*, 2227, 2001.

6 Self-Assembled Au Nanodots in a ZnO Matrix: A Novel Way to Enhance Electrical and Optical Characteristics of ZnO Films

Ashutosh Tiwari and Jagdish Narayan

CONTENTS

6.1 INTRODUCTION

6.1.1 NANOSTRUCTURED MATERIALS

During the last few years nanostructured materials and devices have attracted a great deal of scientific and technological attention.[1–3] The properties of nanostructures are strongly influenced by quantum effects. They exhibit novel optical, magnetic, and electronic properties, and present the perspective of designing and tuning materials properties with exceptional flexibility and control. Nanostructured materials are of enormous interest, from the point of view of discovering new physical phenomena as well as their exploitation possibilities in novel devices. As the size and dimensionality of a material decrease, its electronic structure and properties change tremendously due to the reduction in the density of states and the effective length scale of electronic motion. Now the energy levels of material are determined mostly by the material's surface. This results in a transition from the bulk band structure to individual localized energy levels and hence the onset of quantum confinement effects. Because of the size-induced change in the electronic structure, optical properties of nanomaterials also change significantly. In the case of nanoparticles of noble metals, the surface plasmon resonance becomes very important and results in strong absorption of light in characteristic frequency range. Surface plasmon resonance occurs because of the coherent oscillations of the conduction band electrons induced by interaction with an electromagnetic field. The electric field of the incident light beam induces a polarization of the electrons with respect to the much heavier ionic core of the nanoparticles. A net charge difference is felt at the nanoparticle surface, which in turn acts as a restoring force. This creates a dipolar oscillation of all the electrons with the same phase. When the frequency of the electromagnetic field becomes resonant with the coherent electron motion, a strong absorption of incident radiation occurs. In the case of noble metals, particularly gold (Au) nanodots, a strong surface plasmon resonance occurs in the visible part of the electromagnetic radiation. Actual frequency and width of the surface plasmon resonance absorption depend on the size and shape of nanoparticles and the dielectric constant of metal and surrounding matrix. In this chapter we will show that how the phenomenon of surface plasmon resonance observed in gold nanodots can be utilized to enhance the optical and electrical characteristics of zinc oxide (ZnO).

6.1.2 TRANSPARENT AND CONDUCTING ZnO

Zinc oxide is one of the most promising materials in the fast-emerging optoelectronics industry.[4–6] It has a band gap of 3.27 eV at 300 K, which is very close to the band gap of 3.44 eV for GaN.[7] In addition, it has higher exciton binding energy of 60 meV, compared to 25 meV for GaN. High exciton binding energy results in less trapping of carriers, and therefore higher luminescent efficiencies are expected. Zinc oxide films are also being considered for ohmic contacts to GaN due to their high electrical conductivity and matching band gap with GaN. Therefore, one of the most challenging issues in ZnO-related research is to increase the electrical conductivity of ZnO without compromising its optical characteristics.[8,11] It has been realized that any effort to increase the conductivity of ZnO by doping or by creating oxygen deficiencies always results in the deterioration of the optical efficiency of the film.[8,9,12] These dopants and vacancies create defect levels, and hence a significant reduction in the band-edge PL peak is observed along with a very broad green band in PL spectrum.[13,14]

6.1.3 Au Nanodots in a ZnO Matrix

In this chapter we show that both the electrical and optical characteristics of ZnO can be improved by carefully embedding nanosized epitaxial Au particles in it. We have shown that ZnO-Au epitaxial composite films exhibit higher electrical conductivity and negligible green band emission in photoluminescence spectra. The choice of Au in this system was inspired by the following facts: (1) Au has negligible solubility, and therefore does not affect the chemical composition of the host ZnO matrix, and it grows epitaxially within the ZnO matrix; (2) Au has virtually no affinity for oxygen and hence retains its high conductivity, which results in overall enhancement in electrical conductivity of the composite film; and (3) the most important reason for this selection was that for Au, plasmon resonance occurs at around 520 nm,[15] which lies near the green band emission regime of ZnO. In the case of an assembly of nanosized metal particles with a size distribution, the plasmon resonance occurs over a range of frequencies. The exact value of this resonance frequency range is determined by the dielectric properties of the matrix and size distribution of nanoparticles. By carefully tailoring the size distribution of Au nanodots, it is possible to extend the plasmon resonance spectrum to the whole green band emission regime. Under this condition the green light radiation originating from the midgap impurity states of ZnO can be suppressed by Au nanodots, leaving behind pure excitonic emission of ZnO.

6.2 EXPERIMENTAL PROCEDURE

6.2.1 Pulsed Laser Deposition

We used pulsed laser deposition (PLD) technique to embed Au nanodots in a ZnO matrix. PLD is an efficient method to produce thin films and structures by utilizing the concept of laser-solid interaction. The principle of PLD is quite simple. A highly intense UV laser beam is focused on a target where the high energy density during the laser pulse (about 1 GW within 25 ns) ablates the material. The ablated material forms a plasma that is deposited on a substrate opposite the target. PLD technique can be used for almost any material, particularly for compounds that are difficult or impossible to produce in thin-film form by other techniques.

6.2.2 Experimental Setup

The experimental setup used in our method consists of a pulsed excimer laser, multitarget stainless steel chamber, and a turbomolecular pump (Fig. 6.1). A vacuum is maintained in the chamber by means of a port to the turbomolecular pump. Targets are positioned parallel to a substrate that is placed on a heater plate. The distance

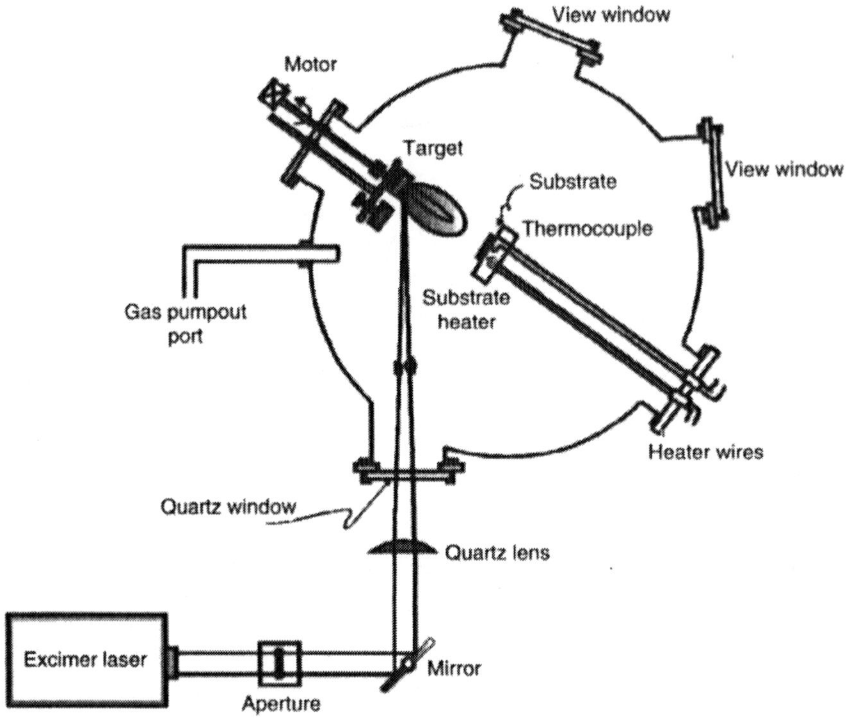

FIGURE 6.1 Schematic diagram of pulsed laser deposition system.

between the target and the substrate is kept at about 5 cm. We used a pulsed KrF excimer laser (wavelength 248 nm) to ablate the targets.

6.2.3 Target Arrangement

In order to grow nanocrystalline Au dots embedded in a ZnO matrix, high-purity Au and ZnO targets were alternately ablated by arranging the targets in a special geometrical configuration (Fig. 6.2). In this configuration, a small strip of Au was glued to a circular pallet of ZnO and a pulsed laser beam was made to hit this assembly at 45° from the normal. The whole assembly was rotated with a motor, which results in alternate ablation of ZnO and Au targets.

6.2.4 Growth Parameters

Key parameters for controlling the size of Au nanodots and distribution were (1) relative time of ablation of Au and ZnO (i.e., the relative area of ablation of these targets) and (2) energy density and repetition rate of the laser beam. These deposition parameters were carefully controlled to obtain the desired size and distribution of Au particles. Films with optimum electrical and optical performance were prepared

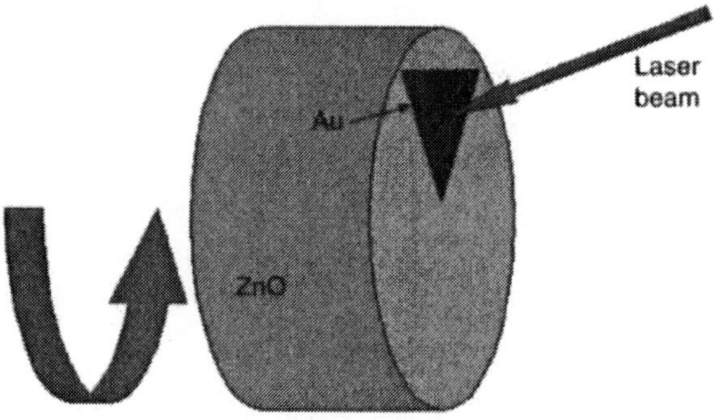

FIGURE 6.2 Geometrical arrangement of ZnO and Au targets for depositing ZnO-Au composite films.

by using the following combination of deposition parameters: relative area of ablation of Au and ZnO = 1:4, laser energy density = 1.5 J/cm^2, and repetition rate = 15 Hz. All the films were deposited on (0001) plane of sapphire at 650°C under the oxygen ambient of 1×10^{-4} torr. Under these conditions the deposition rate of ZnO-Au composite film was about 10 nm/min.

6.2.5 Characterization

Chemical composition, size distribution, and microstructure of nanodots and surrounding matrix were determined by using cross-sectional scanning transmission electron microscopy with atomic number (Z) sensitive contrast (STEM-Z). Photoluminescence measurements were performed using a Hitachi F-2500 spectrophotometer. Electrical resistivity of these films was measured by the four-point probe method in the temperature range of 12–300 K, using a Kiethely Current source (Model 220) and a nanovoltmeter (Model-182).

6.3 RESULTS AND DISCUSSION

6.3.1 Structural

Figure 6.3 shows the x-ray diffraction pattern of ZnO-Au composite film grown on sapphire substrate. In the 2θ range (20–80°), we observed only the peaks corresponding to (0002) and (0004) planes of ZnO, (0006) plane of sapphire, and (111) plane of Au.

Shown in Fig. 6.4(a) is a typical cross-sectional transmission electron microscopy image of one of the composite films. Uniformly distributed nanodots can clearly be seen. These nanodots are almost spherical in shape with a radius of ~2–6 nm. They are distributed in the matrix with an average spacing of ~4–12 nm. High-resolution TEM image of one typical Au nanodot is shown in Fig. 6.4(b). The Au nanodots grow epitaxially within the ZnO matrix via domain matching epitaxy,[16] where integral multiples of lattice planes match across the Au-ZnO interface.

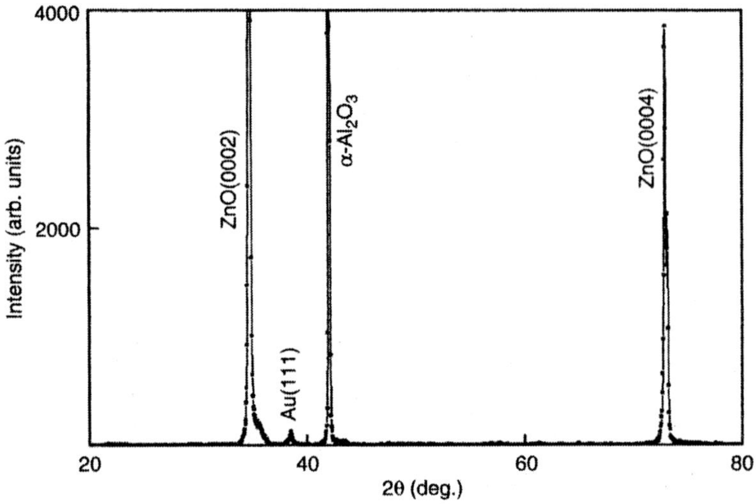

FIGURE 6.3 X-ray diffraction pattern of ZnO-Au composite film.

6.3.2 Optical

The transmission and absorption spectra of ZnO-Au composite film is shown in Fig. 6.5. Apart from the band-edge absorption of ZnO at ~375 ± 5 nm, we also observed a broad absorption peak corresponding to plasmon resonance spectrum of Au nanodots. The shape and the position of this broad absorption peak was found to be a function of particle size and its distribution. For the distribution of Au particles shown in Fig. 6.4(a), plasmon absorption peak lies in the wavelength range of 450–600 nm with a maxima at 522 nm.

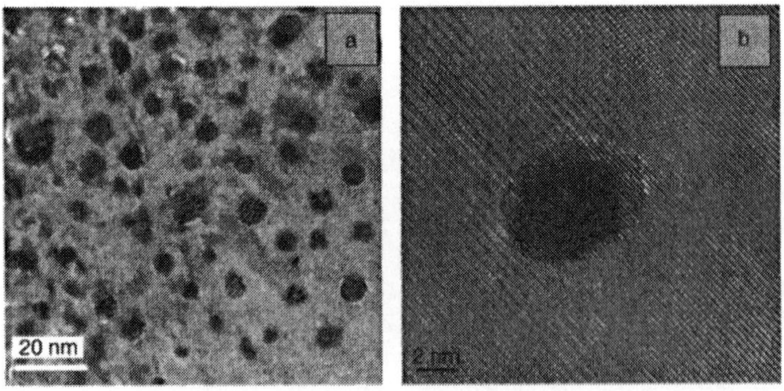

FIGURE 6.4 Cross-sectional TEM image of ZnO-Au composite film showing uniformly distributed Au nanodots (a), and high-resolution TEM image of one typical Au nanodot (b).

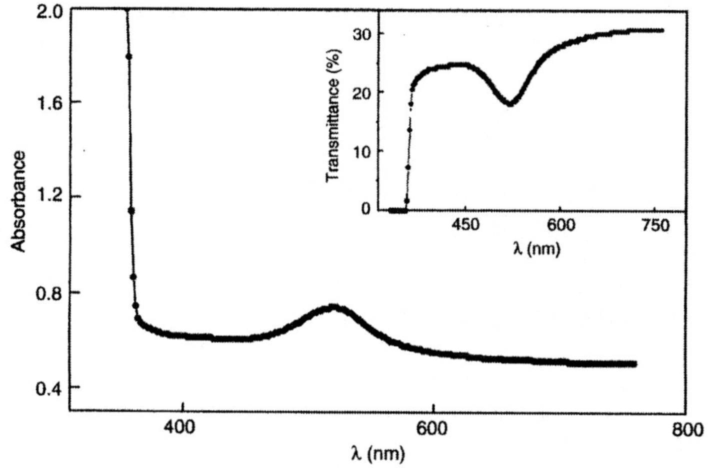

FIGURE 6.5 Transmission and absorption spectra of ZnO-Au composite film.

Because of this absorption peak, the intensity of any green band emission originating from the defect levels of ZnO is expected to be strongly damped. Shown in Fig. 6.6 is the room temperature photoluminescence of ZnO-Au composite films. This spectrum shows only one sharp peak at ~385 ± 5 nm. For a comparison, we have also shown the photoluminescence spectrum of ZnO films prepared under the

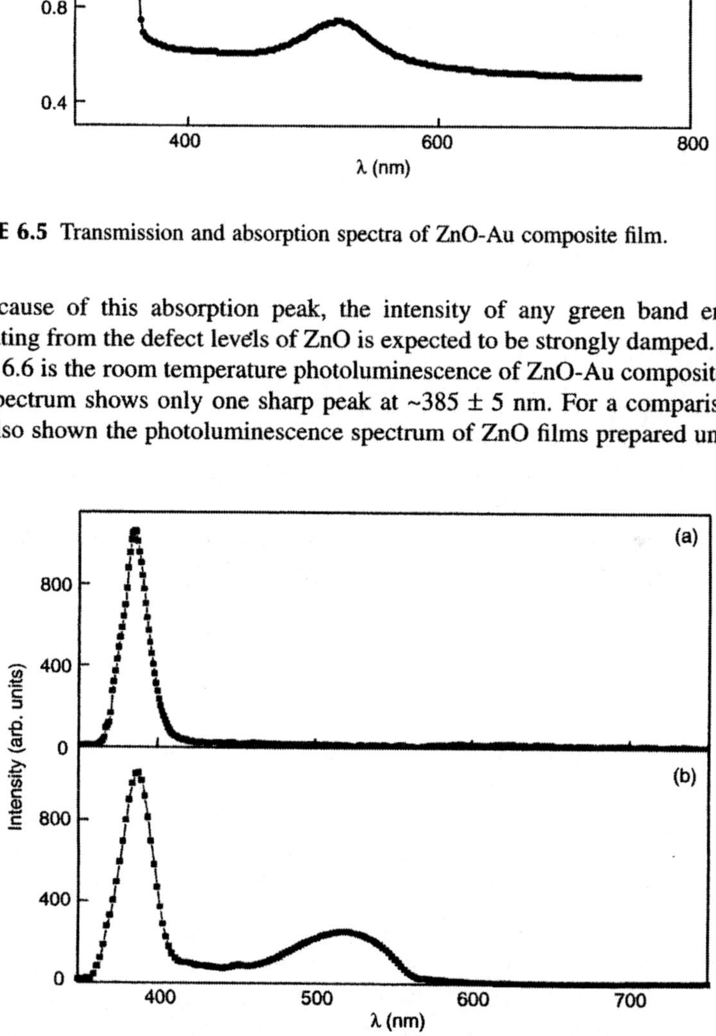

FIGURE 6.6 Photoluminescence spectra of ZnO-Au composite film (a), and photoluminescence spectra of ZnO film prepared under the exact same conditions as the composite film (b).

exact same conditions. Presence of a broad green band emission peak for bare ZnO film and its complete disappearance in the ZnO-Au composite film are clearly demonstrated.

6.3.3 Electrical

Figure 6.7(a) shows the variation of electrical resistivity of the ZnO-Au composite with temperature. At room temperature the value of electrical resistivity is 3.4 mΩ-cm, and it increases with the decrease in the temperature. The inset in this figure shows a plot of ln ρ vs. $1/T$. This graph exhibits a sharp change in the slope of ln ρ vs. $1/T$ at around 150 K. This suggests that the system exhibits two different values of excitation energies at low and high temperatures. Slope of ln ρ vs. $1/T$ at low temperature corresponds to an excitation energy of ~0.2 meV, while that at higher temperature ($T > 150$ K) corresponds to the excitation energy of ~1 meV. It is interesting to note that these energies are much smaller than the energies usually

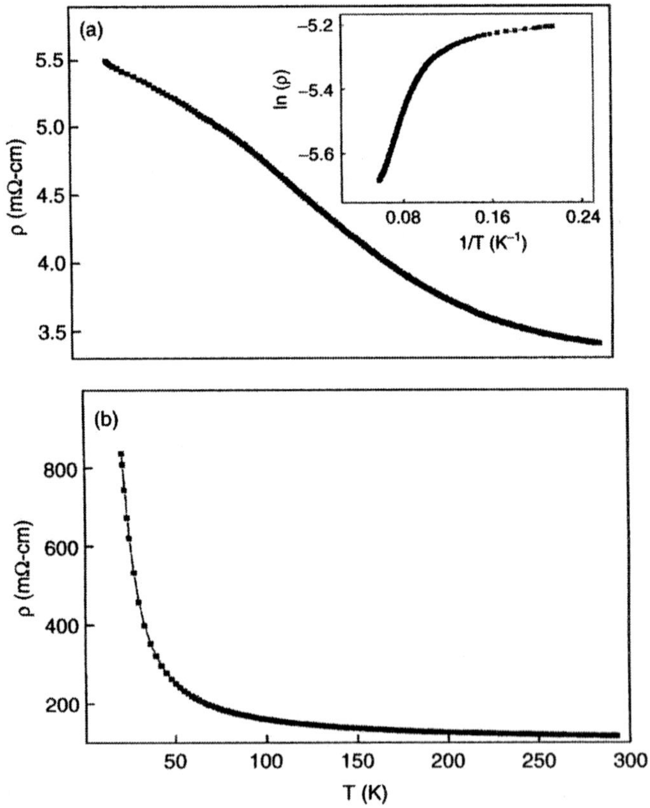

FIGURE 6.7 Electrical resistivity vs. temperature for ZnO-Au composite film. The inset shows the plot of ln vs. $1/T$ (a). Electrical resistivity versus temperature for pure ZnO film (b).

observed in ZnO systems, even with very shallow donor levels. For comparison we also measured the electrical resistivity of bare ZnO film prepared under the exact same conditions (Fig. 6.7(b)). This film exhibits activation energy of ~40 meV, which is in good agreement with previous reports. The extremely low values of activation energies for composite films suggest that the simple activated conduction mechanism cannot account for the observed behavior of electrical resistivity. However, the fact that at lower temperatures the estimated activation energy is smaller suggests that in this temperature regime electrostatic charging of nanodots provides another channel of electrical conduction.

6.4 CONCLUSION

In this chapter we have presented a novel approach to embed nanosized Au particles in epitaxial ZnO films via a pulsed laser deposition technique. Au-embedded epitaxial ZnO films were deposited on (0001) sapphire substrates. The crystalline quality of both the ZnO matrix and Au nanoparticles were investigated by x-ray diffraction and transmission electron microscopy. Composite films were characterized by photoluminescence, optical absorption, and low temperature electrical resistivity measurements. It has been shown that Au particles can suppress the green band emission of ZnO via the process of energy transfer from defect levels to the surface plasmon of Au. These ZnO-Au films exhibit high electrical conductivity with room temperature resistivity of 3.4 ± 0.2 mΩ-cm and a sharp excitonic peak at 3.22 ± 0.05 eV without any green band emission. Growth of highly conducting ZnO-Au composite films with improved optical characteristics is an important step in the fast-emerging optoelectronics industry.

PROBLEMS

1. Describe and discuss surface plasmon resonance.
2. What are the values of band gap and exciton binding energies for ZnO? Compare these numbers with corresponding values for GaN.

ACKNOWLEDGMENTS

The authors gratefully acknowledge financial support from the National Science Foundation the U.S. Army Research Laboratory, and the U.S. Army Research Office. We also thank Dr. C. Jin and Mr. Amit Chugh for useful discussion.

REFERENCES

1. Narayan, J. and Tiwari, A. "Methods of Forming Three-Dimensional Nanodot Arrays in a Matrix." U.S. Patent Application # 60/430, 210 (December 2, 2002).
2. Narayan, J. and Tiwari, A. "Novel Methods of Forming Self-Assembled Nanostructured Materials: Ni Nanodots in Al_2O_3 and TiN Matrices." *Journal of Nanoscience and Nanotechnology,* 4, 726, 2004.

3. Tiwari, A., Jin, C., Chugh, A., and Narayan, J. "Role of Self-Assembled Gold Nanodots in Improving the Electrical and Optical Characteristics of Zinc Oxide Films." *Journal of Nanoscience and Nanotechnology,* 3, 368 (2003).
4. Ohno, H. "Making Nonmagnetic Semiconductors Ferromagnetic." *Science,* 281, 951 (1998).
5. Yu, P., Tang, Z.K., Wong, G.K.L., Kawasaki, M., Ohtomo, A., Koinuma, H., and Segawa, Y. "Ultraviolet Spontaneous and Stimulated Emissions from Zno Microcrystallite Thin Films at Room Temperature." *Solid State Commun.,* 103, 459 (1997).
6. Tiwari, A., Jin, C., Kvit, A., Kumar, D., Muth, J.F., and Narayan, J. "Structural, Optical and Magnetic Properties of Diluted Magnetic Semiconducting Znl-Xmnxo Films." *Solid State Commun.,* 121, 371 (2002).
7. Viswanath, A.K., in *Handbook of Advanced Electronic and Photonic Materials and Devices,* edited by Nalwa, H.S., Academic Press, San Diego, CA (2001), Vol. 1, p. 109.
8. Hong, R.J., Jiang, X. Heide, G., Szyszka, B., Sittinger, V., and Werner, W. "Growth Behaviours and Properties of the ZnO: Al Films Prepared by Reactive Mid-Frequency Magnetron Sputtering." *Journal of Crystal Growth,* 249, 461 (2003).
9. Jin, B.J., Woo, H.S., Im, S., Bae, S. H., and Lee, S.Y. "Relationship Between Photoluminescence and Electrical Properties of Zno Thin Films Grown by Pulsed Laser Deposition." *Applied Surface Science,* 169, 521 (2001).
10. Exarhos, G.J., Rose, A., and Windisch, C.F. "Spectroscopic Characterization of Processing-Induced Property Changes in Doped ZnO Films." *Thin Solid Films,* 308, 56 (1997).
11. Bolzlee, B.J. and Exarhos, G.J. "Preparation and Characterization of Gold and Ruthenium Colloids in Thin Zinc Oxide Films." *Thin Solid Films,* 377, 1 (2000).
12. Reynolds, D.C., Look, D.C., and Jogai, B. "Find Structure on the Green Band in Zno." *Journal of Applied Physics,* 89, 6189 (2001).
13. Reynolds, D.C. Look, D.C., Jogai, B., and Morcoc, H. "Similarities in the bandedge and Deep-Centre Photluminescence Mechanisms of ZnO and GaN." *Solid State Commun.,* 101, 643 (1997).
14. Tiwari, A., Park, M., Jin, C., Wang, H., Kumar, D., and Narayan, J. "Epitaxial Growth of Zno Films on Si(111)." *Journal of Materials Research,* 17, 2480 (2002).
15. Averitt, R.D., Westcott, S.L., and Halas, N.J. "Linear Optical Properties of Gold Nanoshells." *Journal of Optical Society of America B-Optical Physics,* 16, 1824, 1999.
16. Narayan, J. and Larson, B.C. "Domain Epitaxy: A Unified Paradigm for Thin Film Growth." *Journal of Applied Physics,* 93, 278 (2003); J. Narayan, U.S. Patent 5, 406, 123 (4/11/1995).
17. Han, J., Senos, A.M.R., Mantas, P.Q., and Cao, W. "Dielectric Relaxation of Shallow Donor in Polycrystalline Mn-Doped ZnO." *Journal of Applied Physics,* 93, 4097, 2003.
18. Alivisatos, A.P. "Semiconductor Clusters, Nanocrystals, and Quantum Dots." *Science,* 271, 933, 1996.

7 Synthesis of Boron Nitride Nanotubes Using a Ball-Milling and Annealing Method

Ying Chen and Jim S. Williams

CONTENTS

7.1 INTRODUCTION

Boron nitride (BN) nanotubes have a structure similar to carbon (C) nanotubes but exhibit many interesting electronic and chemical properties. This chapter summarizes research in the synthesis of BN nanotubes using a new ball-milling and annealing method, which consists of high-energy ball milling and then isothermal annealing processes. The high-energy ball milling creates a precursor for BN nanotube formation during subsequent annealing. This review of BN nanotube synthesis under different milling and annealing conditions outlines new mechanisms of nanotube nucleation and growth involved in the milling-annealing method.

7.1.1 Boron Nitride Nanotubes

Boron nitride (BN) and carbon (C) share many common structures such as graphite and hexagonal BN, and diamond and cubic BN crystalline structures. After the discovery of C nanotubes in 1991,[1] BN nanotubes were first predicted theoretically[2] and then successfully synthesized using the same arc-discharge method by Chopra et al. in 1995.[3] Similar to C nanotubes, BN nanotubes can be simply regarded as a cylinder rolled up from a graphitic BN sheet with a diameter less than 100 nm and a length in the micrometer range.[4] However, there are two main differences between structures of BN and C nanotubes. First, in the nanotube tips a single-wall C nanotube is capped by a half-fullerene molecule (C_{60}) consisting of six-membered hexagons and five-membered pentagons, and thus C nanotubes have cone-shape tips (Fig. 7.1(a)).[5,6] However as shown in Fig. 7.1(b), most BN nanotubes have flat tips, which are formed with four-membered squares instead of five-membered pentagons.[7–9] Theoretical investigation reveals that five-membered pentagons in BN would require energetically unfavorable B—B or N—N bonds that destabilize the structure.[10,11] A second difference is in chirality. Chirality, the dominant BN nanotube structures, are so-called zigzag tubes.[12,13] Armchair and chiral tubes are less likely, possibly due to the special tip configurations.[10] On the other hand, C nanotubes with zigzag, armchair, and chiral structures are all particularly abundant.[14] In addition to typical parallel-walled cylindrical tubes, bamboo- and cone-like BN nanotubes can also be produced.[15,16] Similar to C nanotubes, BN nanotubes can be filled with metals (Fe, Ni, W, . . .)[3,8,17,18] and buckyballs.[19]

Due to the similar tubular structure and ultrastrong sp^2 B—N bonds, BN nanotubes have similarly superstrong mechanical properties as C nanotubes, with a Young's modulus of 1.22 ± 0.24 TPa.[20] Theoretical studies suggest that BN nanotubes are stable insulating materials with a uniform electronic band gap of about 5.5 eV, which does not depend on size and chirality of nanotubes as in the C

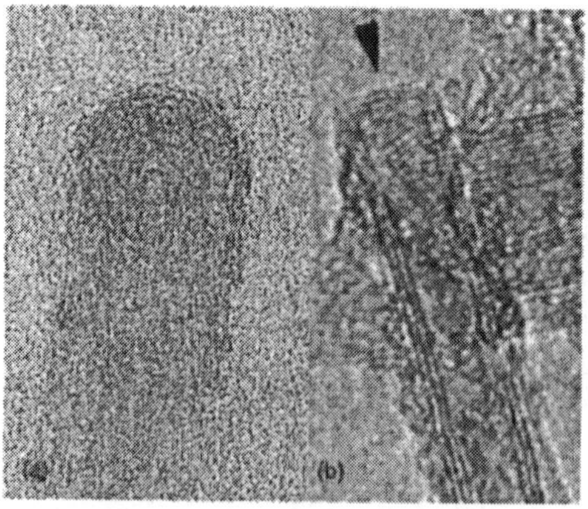

FIGURE 7.1 TEM images of (a) a multiwalled C nanotube with a cone-shape tip and (b) a BN nanotube with a flat tip, indicated by arrow.

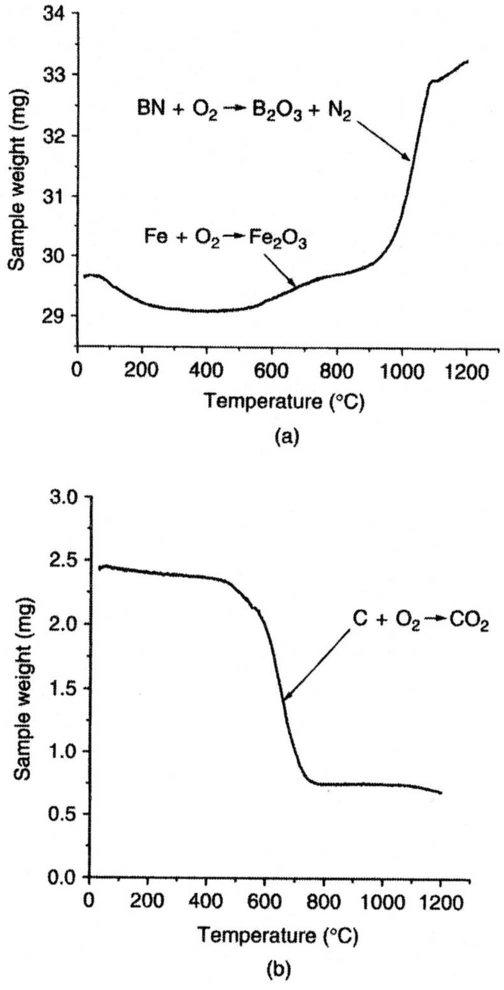

FIGURE 7.2 Thermal gravimetric curves of sample weight changes as a function of temperature for (a) BN nanotubes and (b) C nanotubes. The samples were heated at 20°C min^{-1} in an air flow of 10 ml min^{-1}.[22]

nanotube case.[21] BN nanotubes have a much better resistance to oxidation at high temperatures than C nanotubes.[22] C nanotubes readily oxidize in air at a temperature of 400°C, starting from the tips and defects on the outer layers. They burn completely at 700°C when sufficient oxygen is supplied.[23,24] We studied oxidation behavior of C and BN nanotubes using thermal gravimetric analysis. The results are illustrated in Fig. 7.2, which shows weight changes of nanotube samples as a function of heating temperature in an air flow.[22] In the case of BN nanotubes (Fig. 7.2(a)), a sharp weight increase from 800 to 1100°C, which is caused by the oxidation of BN with the formation of solid boron oxides, is observed. In the case of C nanotubes (Fig. 7.2(b)), a rapid weight loss due to oxidation with releasing carbon gaseous oxides indicates that the oxidation starts from 400°C and is complete at 700°C, consistent with literature reports.[23,24] Transmission electron microscopy (TEM) investigation confirmed that BN nanotubes with different nanostructures (cylinders and bamboos) and of a large range of sizes (2–200 nm) are stable in air at 700°C, at which temperature C nanotubes completely burn off. Thin BN nanotubes with perfect nanocrystalline structures can withstand temperatures of up to 900°C.[22] Thus, the superstrong mechanical properties and high resistance to oxidation make BN nanotubes ideal for composite materials applications and the stable electronic structure is useful in device applications.

The first successful synthesis of pure BN nanotubes was made by Zettl's group in 1995 using an arc-discharge process similar to that used for carbon nanotube production, where the tungsten anode contained an inserted BN rod rather than graphite.[3] Loiseau et al.[7] and Terrones et al.[8] confirmed the formation of BN nanotubes using the arc-discharge process with different metal catalysts. A high yield of BN nanotubes was obtained by using carbon nanotubes as templates, where C was substituted by boron and nitrogen atoms through a chemical reaction between boron oxide vapor and nitrogen gas at elevated temperatures.[25] A low-temperature oxidation is needed to remove residual C inside BN nanotubes.[26] Lee et al.[13] reported the production of gram quantities of single-walled nanotubes, and pure double-walled BN nanotubes were produced by Cunning et al.[27] However, all of the preceding deposition methods present difficulties in scaling up for an industrial process. Alternatively it was first demonstrated in 1999 that a milling-annealing process, in which boron powder was first milled in ammonia gas and followed by annealing in nitrogen atmosphere, had the potential to produce large quantities of BN nanotubes.[28] For example, up to 1 kg amounts of BN nanotubes are now possible using a laboratory-scale milling device, thus opening up studies of new properties and applications of BN nanotubes.[29] New formation mechanisms involved with such high-yield BN nanotube formation have been discovered, as outlined in the following sections.

7.1.2 High-Energy Ball Milling (HEBM)

Conventional ball milling is a traditional powder-processing technique that is widely used in mineral, pharmaceutical, and ceramic industries. Small grinding machines are popular tools in scientific laboratories for preparing fine powders for various analyses and applications. However, high-energy ball milling (HEBM), first reported by Benjamin et al.[30] in the 1960s, is fundamentally different from the conventional ball-milling technique. For example, a much more powerful attritor ball mill had to be developed to synthesize new oxide-dispersion-strengthened alloys since conventional ball mills could not provide sufficient grinding energy.[30] Intensive research in synthesis of new metastable materials by HEBM was stimulated by the pioneering work of amorphization of the Ni-Nb system conducted by Kock et al. in 1983.[31] Since then, a wide spectrum of metastable materials has been produced, including nanocrystalline,[32] nanocomposite,[33] nanoporous phases,[34] supersaturated

TABLE 7.1
Comparison of Milling Parameters for Conventional Ball Milling and High-Energy Ball Milling

	Conventional Ball Milling	High-Energy Ball Milling
Milling time	<1 h	20–200 h
Impact energy (W/g/ball)	0.001	0.2
Milling action	Rolling	Impact
Particle size	μm	Sub-μm (clusters)
Strucural changes	No	Yes
Chemical reactions	No	Yes
Milling atmosphere	No control	Vacuum, gases
Milling temperature	No control	Liquid nitrogen –700°C

solid solutions,[35] and amorphous alloys.[36] These new phase transformations induced by HEBM are generally referred to as mechanical alloying (MA). At the same time, it was found that HEBM can activate at room temperature chemical reactions, which are normally only possible at high temperatures.[37] This latter process is called reactive milling or mechanochemistry. Reactive ball milling has produced a large range of nanosized oxides,[38] nitrides,[39] hydrides,[40] and carbide[41] particles. The major differences between conventional ball milling and high-energy ball milling are listed in Table 7.1. The impact energy of HEBM is typically 1000 times higher than conventional ball milling energy. The particle fracturing and size reduction in the conventional ball milling is only the first stage of HEBM. A longer milling time is thus required for HEBM. In addition to the milling intensity, milling atmosphere and temperature controls of HEBM are crucial to create a desired structural change or chemical reaction.

Different types of high-energy ball mills have been developed for the HEBM process, including the Spex vibrating mill, planetary ball mill, rotating ball mill, attritors, and others.[42] In our nanotube synthesis research, two types of HEBM mills have been used: a rotating ball mill and a vibrating-frame ball mill. A vibrating-frame grinder (Pulverisette O, Fritsch) uses only one large ball (50 mm diameter). The media of the ball and vial can be stainless steel or ceramic (tungsten carbide). As shown in Fig. 7.3, the milling chamber has been modified, sealed with an O-ring. The atmosphere can be changed via a valve and the pressure monitored with an attached gauge.[43] The milling energy and frequency can be adjusted by varying vibrating amplitude and ball mass. By measuring the ball's movement, a milling intensity is defined as

$$I = M_b V_{max} f / M_p,$$

where M_b is the mass of the milling ball, V_{max} the maximum velocity of the vial, f the impact frequency, and M_p the mass of powder. The milling intensity is a very important parameter to MA and reactive ball milling. For example, an intensity threshold exists for full amorphization of Ni-Zr alloys.[44] When milling below this

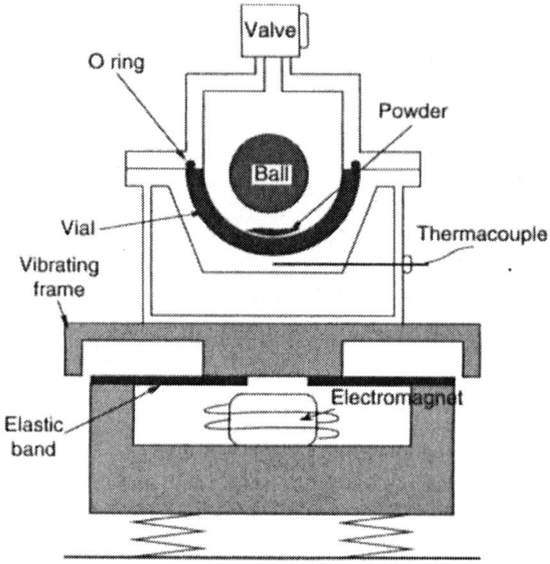

FIGURE 7.3 A schematic of a modified vibrating ball mill (Fritsch, Pulverisette O), showing impact actions and control of milling atmosphere and temperature.

threshold, a mixture of nanocrystalline and amorphous phases is produced. This threshold depends on milling temperature and alloy composition.

Figure 7.4 shows a schematic of a steel rotating ball mill, consisting of a rotating horizontal cell loaded with several hardened steel balls. As the cell rotates, the balls drop onto the powder that is being ground. An external magnet is placed close to the cell to increase milling energy.[45,46] Different milling actions can be realized by adjusting the rotation rate and magnet position. The rolling balls give a shear component of force to particles, while the falling balls provide head-on impacts with high kinetic or milling energy.[47]

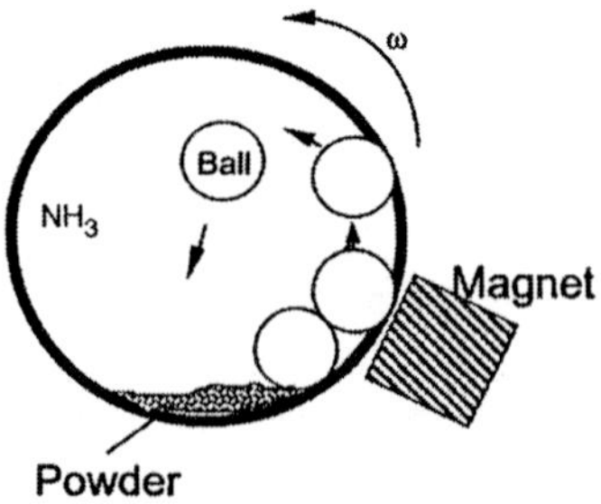

FIGURE 7.4 A schematic of a high-energy rotating ball mill, showing milling actions.

During HEBM, material particles are repeatedly flattened, fractured, and rewelded. Every time two steel balls collide or one ball hits onto the chamber wall, they trap some particles between their surfaces.[48] Such high-energy impacts severely deform the particles and create atomically fresh, new surfaces, as well as a high density of dislocations and other structural defects.[49] Depending on milling conditions and reaction kinetics, mechanochemical reactions may be confined to a very small volume during each impact due to an instantaneous injection of a huge amount of mechanical energy into material particles.[50] Therefore, a high number of impacts is required to complete a gradual transformation/reaction process. The repeated fracturing and welding of powder particles increases the contact areas between reactants and particles, allows fresh surfaces to come into contact repeatedly, and allows the reaction often to proceed to completion without the necessity for bulk diffusion of reactants through product layers.[51,52] Consequently, reactions that normally require high temperatures can occur at lower temperatures during HEBM. In addition, the high defect density induced by HEBM can accelerate the diffusion process.[53] Alternatively, the deformation and fracturing of particles cause continuous size reduction and can lead to reduction in diffusion distances. This can at least reduce the reaction temperatures significantly, even if they do not occur at room temperature.[54,55] Since newly created surfaces are most often very reactive and readily oxidize in air, the HEBM has to be conducted in an inert atmosphere.

It is now recognized that high-energy ball milling, along with other nonequilibrium techniques such as rapid quenching, irradiation/ion implantation, plasma processing, and gas deposition can produce a series of metastable and nanostructured materials that are usually difficult to prepare using melting or conventional powder metallurgy methods.[56,57] Detailed chemical reactions and phase transformations of B and BN induced by HEBM are presented in the following sections.

7.2 SYNTHESIS OF BN NANOTUBES FROM ELEMENTAL B

7.2.1 Milling with a Rotating Steel Mill

Boron nitride nanotubes were prepared from elemental B powder via the following two processes: high-energy ball milling and isothermal annealing. The milling experiment was conducted using a steel rotating ball mill. The milling cell was first loaded with several grams of B powder and four steel balls (diameter of 25.4 mm). The mill was then filled with reactive ammonia gas at a pressure of 300 kPa prior to milling. The pressure inside the milling cell was monitored regularly using a pressure gauge. The pressure changes recorded over the whole milling process are illustrated in Fig. 7.5. A rapid reduction of pressure from 300 to 190 kPa occurred during the first 36 h of milling, which was caused by absorption of ammonia gas onto fresh surfaces of B particles newly created by high-energy impacts.[28] The particle size reduction is the dominant event in the first stage of milling. Interestingly the pressure inside the mill started to rise during further milling and exceeded the starting pressure at the end of 150 h (315 kPa). This pressure increase is indicative of the occurrence of nitriding

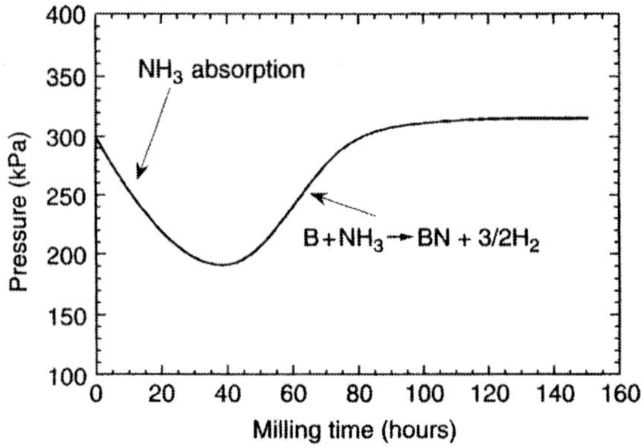

FIGURE 7.5 Pressure changes inside the milling chamber of a steel rotating mill during milling of B powder in NH_3 at room temperature.[28]

reactions between boron powder and the absorbed ammonia gas under HEBM. The possible reaction is $2B + 2NH_3 \rightarrow 2BN + 3H_2$. Hydrogen gas released from the ammonia leads to a pressure rise inside the mill. Similar gas–solid reactions induced by HEBM were also observed in Ti(Zr)-NH_3 systems.[50,58,59] To confirm the preceding reaction, chemical compositions of milled B samples were examined using x-ray energy dispersive spectroscopy (EDS) in a scanning electron microscopy (SEM), as well as combustion chemical analysis. Figure 7.6 shows the results obtained. The

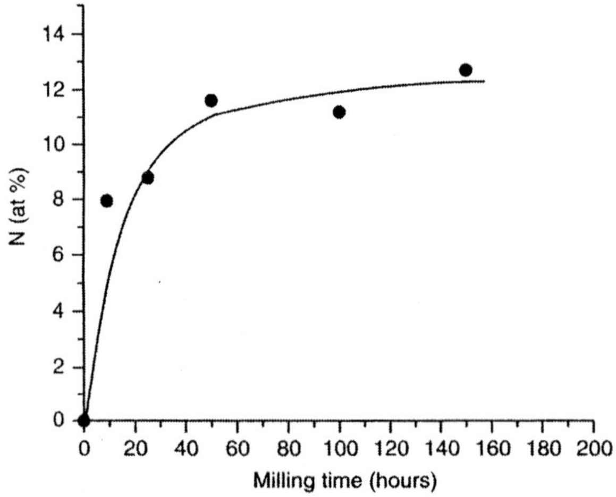

FIGURE 7.6 Nitrogen content of milled B samples in the rotating mill as a function of milling time.[29]

gradual increase of the N content over the whole milling process is consistent with

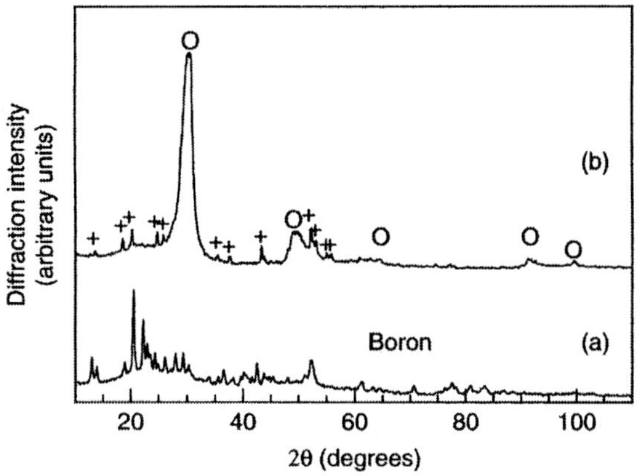

FIGURE 7.7 XRD patterns taken from (a) the 150-h milled B sample; (b) the milled sample subsequently annealed at 1300°C for 6 h in N_2. O = hexagonal BN; += Fe_xB_y.[28]

ammonia absorption and a nitriding reaction that results in the formation of BN. The high N content (10.1 wt%) and low H content (0.9 wt%) in the 150-h milled sample confirm the presence of nitriding reactions during further milling stages.

HEBM also created significant structural changes in B samples. The x-ray diffraction (XRD) patterns of the sample after 150 h of milling in Fig. 7.7(a) shows broadened peaks from unreacted B and a high background (a possible very large peak at around 25°) associated with amorphous B and newly formed nanostructural BN phases. The new diffraction peak at around 51° was found to be associated with Fe, which came from the steel milling balls and the cell as milling contamination. TEM (Philips 430 and CM300 instrument equipped with an EDS system) analysis of the milled B samples found a large number of Fe/Cr particles dispersed in a disordered B matrix. Figure 7.8 shows that the Fe and Cr contents increase gradually with increasing milling time and reached about 7 wt% for Fe and 1.4 wt% for Cr at the end of milling.[29] These milling contaminations actually have a positive role to play in nanotube formation. They might act as catalysts to help nanotube nucleation. TEM did not find any nanotubes, but rather thin layers of BN or B(N) solid solution on the surface of B particles.

Isothermal annealing was conducted in a tube furnace under nitrogen gas flow (0.5 l/min) so that the nitriding reaction could further proceed to full nitridation with the formation of BN nanotubes. The B sample milled for 150 h was heated at 1300°C for 6 h. The XRD pattern in Fig. 7.7(b) shows pronounced peaks of hexagonal BN and new diffraction peaks of an iron boride phase. No more B peaks are found, suggesting that residual boron reacted completely with nitrogen to form BN and some B formed Fe_xB_y. SEM observation found large aggregates of nanosized filaments and particles. Figure 7.9 shows a typical SEM image taken from a sample milled for 150 h and heated at 1200°C for 16 h in N_2. Filaments

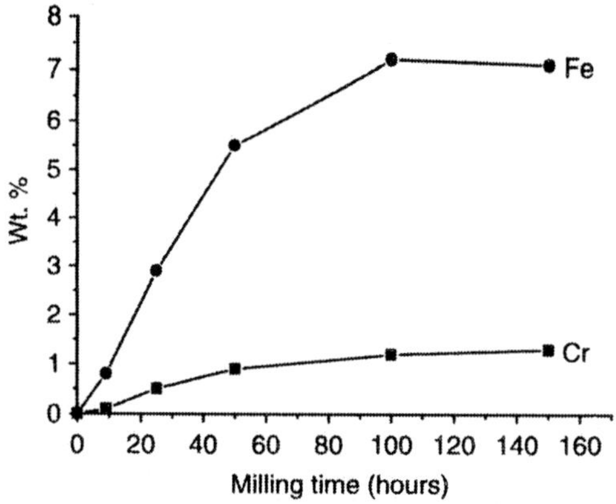

FIGURE 7.8 Iron and chromium contents of the milled B samples as a function of milling time.[29]

with a diameter of 20 to 200 nm are observed. Thinner filaments were difficult to observe clearly due to the instrument resolution limit. The filament yield was estimated by analyzing 25 SEM images taken from different aggregates. The ratios of the area occupied by the filaments to the whole sample area were calculated for each image. The formation yield was found to be in the range of 65% to 85%.[29]

TEM revealed that these filaments have tubular structure. A typical TEM image in Fig. 7.10 shows that thin (diameter less than 20 nm) nanotubes have typical

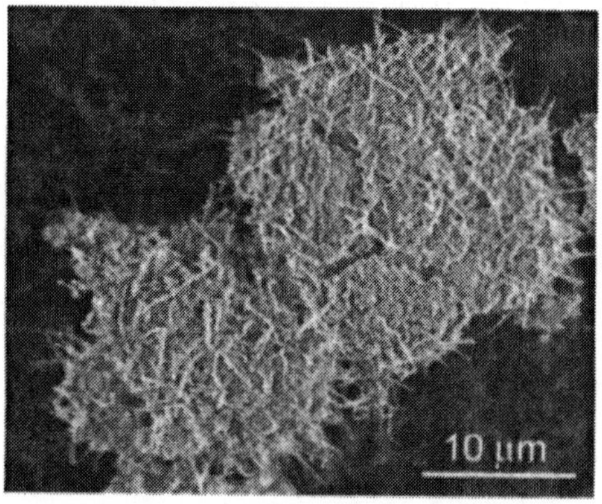

FIGURE 7.9 SEM image of a B sample first milled for 150 h in NH_3 and then heated at 1200°C for 16 h in N_2.

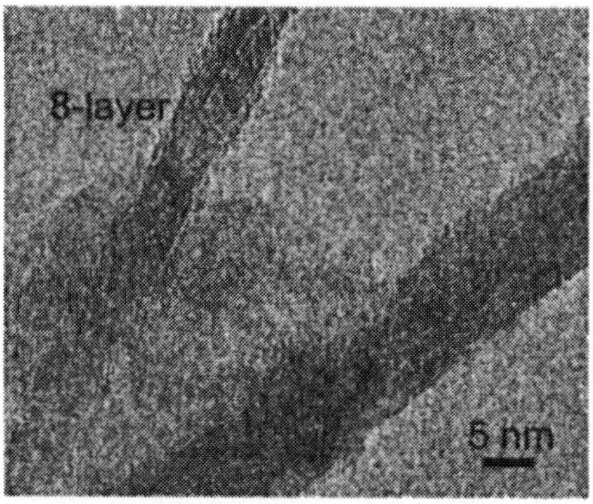

FIGURE 7.10 TEM image of two thin BN nanotubes with diameter less than 20 nm. Parallel fringes suggest a multiwalled, cylindrical structure.[29]

parallel-walled, cylindrical structures. The parallel fringes on the wall areas of the nanotube suggest eight atomic layers. Some thick nanotubes (diameter greater than 20 nm) exhibit a bamboo-like structure (Fig. 7.11). Nanosized polyhedral particles of BN are also found. The chemical nature of these nanotubes was investigated using electron energy loss spectroscopy (EELS) attached to a VG field-emission transmission electron microscopy. In order to avoid carbon contamination, a BN nanotube suspended over a hole in the carbon support film was analyzed, and EELS spectra are shown in Fig. 7.12. Two characteristic K-shell ionization edges (188 and 401 eV) of B and N are observed and the B/N ratio is 0.98. No carbon was detected from any BN nanotubes or particles.[28]

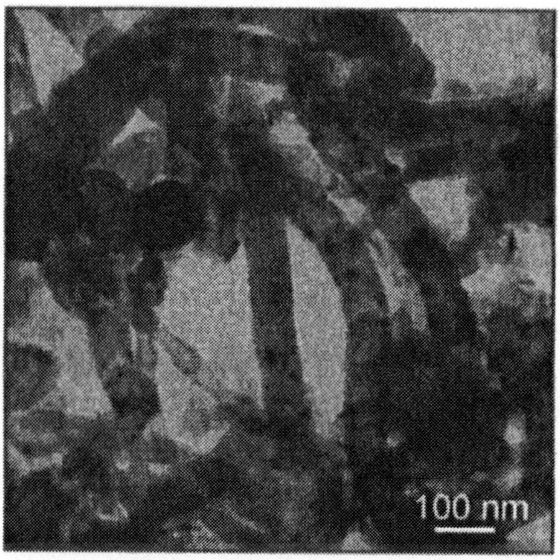

FIGURE 7.11 TEM image of BN nanotubes with bamboo-type structure.

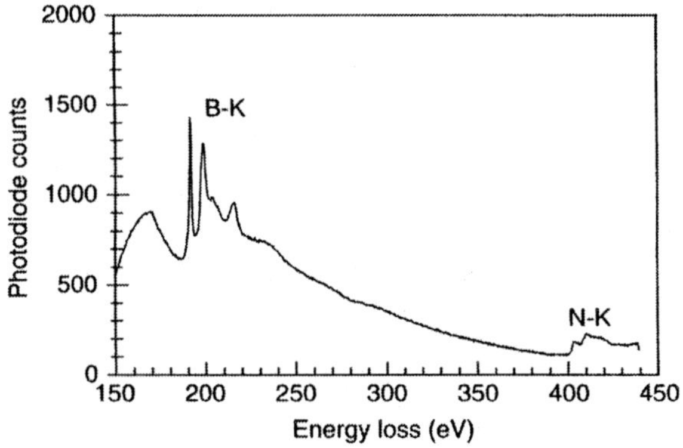

FIGURE 7.12 EELS spectra taken from a BN nanotube suspended over a hole in a carbon support film showing K-shell ionization edges of B and N. The calculated B/N ratio is 0.98 ± 0.16.[28]

The formation yield and sizes of nanotubes depend on both milling and annealing conditions. For the rotating ball mill, extensive milling treatment up to 150 h under high milling intensity seems to be required for a high formation yield. The yield increases with increasing milling time, which might be due to higher N and Fe contents in the prolonged milled samples, as shown in Fig. 7.6 and Fig. 7.8. The long milling time can be reduced by using a fast (higher milling frequency) milling device. The current rotating ball mill was designed for fundamental research purposes and involved a low (about 14 Hz) milling frequency, so that the evolution of chemical reactions or phase transformations induced by ball impacts could be examined over a reasonable milling time frame.[45,46]

Annealing temperatures are critical to BN nanotube formation. A temperature of 1000°C appears to be too low for full nitridation and therefore the yield is very low. The yield increases with increasing temperature up to 1300°C. At 1400°C, very thick nanotubes of diameter more than 200 nm, as well as large BN particles, were produced. By optimizing both milling and annealing conditions, a very high formation yield of about 80% can be achieved. Under controlled annealing conditions, different types and sizes of nanotubes can be produced selectively. For example, cylindrical nanotubes with a diameter less than 20 nm can be produced at a temperature below 1200°C; coarse bamboo or cone-structured tubes with diameter around 100 nm can be formed at temperatures above 1300°C.[29]

A large quantity up to 1 kg of BN nanotubes can be produced by running four large-sized milling cells at the same time.[29] The results obtained show that the HEBM and the presence of Fe particles each have important roles in nanotube formation. To understand the effect of HEBM and roles of Fe in nanotube formation, a different milling device (involving a different impact mode and using a ceramic vial and balls) was tested.

7.2.2 Milling with a Vibrating Ceramic Mill

As described previously, the vibrating ball mill uses only one large ball to "hammer crush" powders at the bottom of the vial. To avoid metal contamination, a tungsten carbide ball and vial were used. The high milling energy level ($I = 2.17$ ms^{-2})[44] was ensured using a heavy milling ball of 1 kg and a high vibrating amplitude of 1.5 mm. 2 grams of hexagonal B powder was milled in ammonia gas at a starting pressure of 300 kPa for up to 168 h. The pressure decreased from 300 kPa to 266 kPa after 60 h of milling and increased up to 297 kPa after milling for 168 h. Again a nitriding reaction was induced by high-energy ball milling.[16]

The milled samples were heated at 1200°C for different times up to 16 h. BN nanotubes with very different structures were obtained. Very thin nanotubes with only two or three atomic layers were found in the sample after heating for 10 h. Bundles of the thin nanotubes formed (Fig. 7.13). Abundant thick-walled BN nanotubes and some with conical structures were formed (Fig. 7.14(a)) in the sample after annealing for 16 h. Some nanotubes contain a W metal particle at their tips, implying different formation mechanisms to thin cylindrical tubes. The diameters of such nanotubes are around 100 nm with lengths of several microns. As shown in Fig. 7.14(b), the nanotubes have hundreds of crossed bands of BN, thus segmenting each tube into many chambers and resembling the structure of bamboo. Clearly annealing time has a very important effect on the final nanotube structures.

HEBM with the ceramic vibrating mill created similar disordered structure in B powder and the same nitriding reaction as the steel rotating mill. It confirms that HEBM has an important role to play in BN nanotube formation rather than a simple metal catalytic effect. No difference was found between pure impact milling actions in the vibrating mill and the mixture of actions involving both shearing force and impacts in the rotating mill.

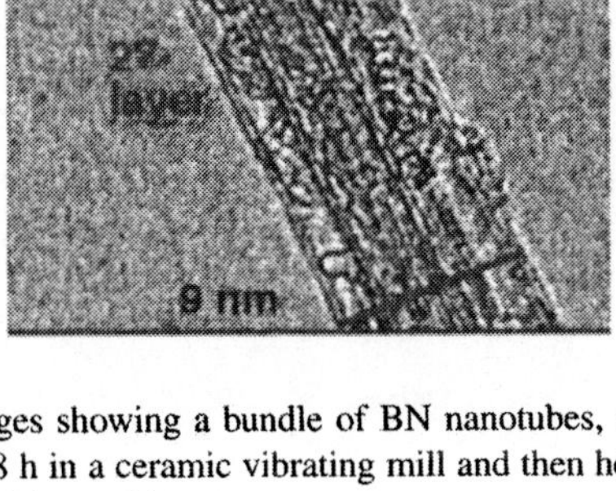

FIGURE 7.13 TEM images showing a bundle of BN nanotubes, which was taken from a B sample first milled for 168 h in a ceramic vibrating mill and then heated at 1200°C for 10 h in N_2. (Courtesy of Dr. John Fitzgerald.)

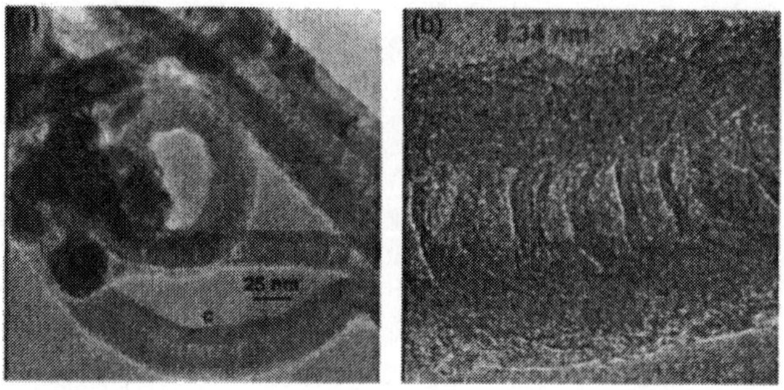

FIGURE 7.14 TEM images show thick-walled BN nanotubes with conical structure. The images were taken from a B sample first milled for 168 h in a ceramic vibrating mill and then heated at 1200°C for 16 h in N_2. (Image courtesy of Dr. John Fitzgerald.)

Although we did not expect any metal catalysts involved in the formation process, W reduced from WC appears to act as a catalyst in the formation of many coarse nanotubes, as seen in Fig. 7.14. The XRD patterns in Fig. 7.15 show that the milled sample has a similar disordered structure to the B samples milled in a steel mill. The weak WC peaks indicate the presence of milling contamination. The BN phase started to form after 4 h of annealing. WC was first reduced by B to W_2B_5. Some W_2B_5 was further reduced to W during longer heating for 10 and 16 h. An increased amount of BN phase was produced for these longer times, as suggested by the more pronounced peaks. The possible chemical reactions occurring during annealing are

$$WC + B \rightarrow WxBy + C$$

$$WxBy + N \rightarrow W + BN$$

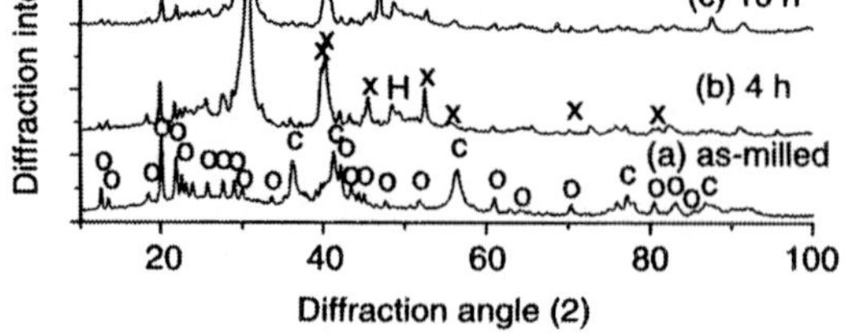

FIGURE 7.15 XRD patterns taken from samples: (a) B milled for 168 h in ceramic mill; (b–d) the milled sample subsequently annealed at 1200°C in N_2 for various times marked.[16]

In conclusion, BN nanotubes were produced successfully from two different milling devices and different milling media (stainless steel and ceramic tungsten carbide). The nanotubes obtained have the same nanostructure as BN nanotubes produced by other methods.[3,7–9] The aforementioned results suggest the following possible formation sequences:

1. Reactive ball milling: $B + NH_3 \rightarrow$ BN (nanoparticles/layers) + H_2 (gas)
   Annealing: BN nanoparticles/layers $\rightarrow$ BN nanotubes
2. Ball milling: $B + NH_3 \rightarrow$ B(nanoparticles) + B(N) (nanoparticles) + H_2(gas)
   Annealing: B(nanoparticles) + $N_2 \rightarrow$ BN(nanotubes)

In the first formation sequence, reactive milling produces BN particles, which transform to BN nanotubes during heating without further nitriding reaction. In the second sequence, the nanosized, disordered B particles created by HEBM react with N_2 during annealing and BN nanotubes are the end products of the reaction. The second sequence is confirmed by ball milling of B in Ar gas (rather than NH_3), so the milled B contains nanosized B particles but not BN phase. BN nanotubes formed after heating the milled sample at 1200°C for 6 h in N_2 gas. To clarify the possibility of the first formation processes, a further experiment with BN compound as the starting material for HEBM was carried out, and detailed results are introduced in the following section.

7.3 SYNTHESIS OF BN NANOTUBES FROM BN COMPOUNDS

Several grams of BN (purity > 98%) powder were milled in a steel rotating ball mill with four hardened steel balls. To prevent oxidation contamination, the milling cell was purged with nitrogen gas several times and a pressure of 300 kPa was established prior to milling. Because a nitridation reaction was not required, the more reactive ammonia gas was not used during the milling process in this case.

XRD analysis revealed that the starting hexagonal BN phase was transformed to a highly disordered BN phase after milling for 140 h. The XRD pattern in Fig. 7.16(a) shows two weak and broadened peaks (referred as $\alpha1$ and $\alpha2$), which can be associated with a turbostratic BN structure.[60] XRD patterns also show the presence of iron from the milling balls and the cell. The overall Fe level measured by XEDS is about 3.8 wt%. TEM did not find any nanotubes but aggregates consisting of particles below 50 nm, as shown in Fig. 7.17. The selected area diffraction pattern shows three diffuse rings associated with (002), (100), and (110) planes with lattice spacings of 3.61, 2.14, and 1.24 Å, respectively, which suggests a disordered nanocrystalline BN structure. There may also be some amorphous BN present. The larger (002) interlayer distance and the absence of a 3-D structure are features characteristic of a turbostratic structure, which is consistent with the XRD results.

Following milling, nanosized BN powders were heated under N_2 gas flow (1 ml min^{-1}) at 1200°C for about 10 h. The N_2 gas was used to prevent oxidation during the annealing rather than as a reactive gas. The TEM image in Fig. 7.18 shows the presence of clusters of BN nanotubes as well as nanoparticles of outer diameter in

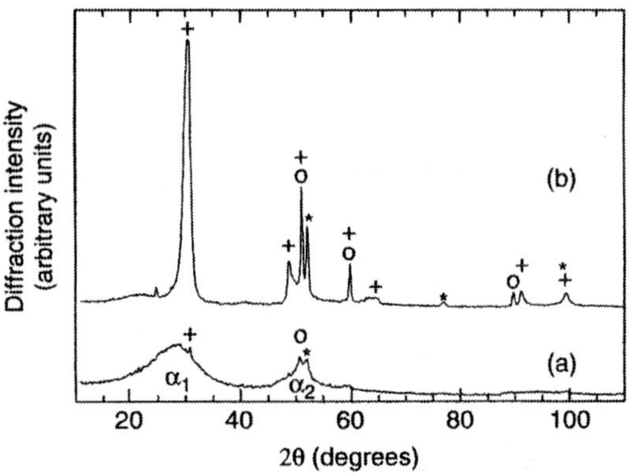

FIGURE 7.16 XRD patterns taken from sample: (a) BN milled for 140 h in N_2 in a steel mill; (b) the milled sample subsequently annealed at 1200°C for 10 h in N_2. α1 and α2: mixture of turbostratic and amorphous BN; + =hexagonal BN, ★ = Fe; O = Fe(C) austenite alloy.[15]

the range from 10 to 50 nm. These nanotubes have multiwalled, cylindrical structures. BN nanotubes with different tubular structures were produced at a higher annealing temperature, as shown in the TEM image in Fig. 7.19, which was taken from the sample annealed at 1300°C for 10 h. It shows three bamboo-like tubes. The biggest has a diameter of 280 nm and the smallest has a diameter of about 120

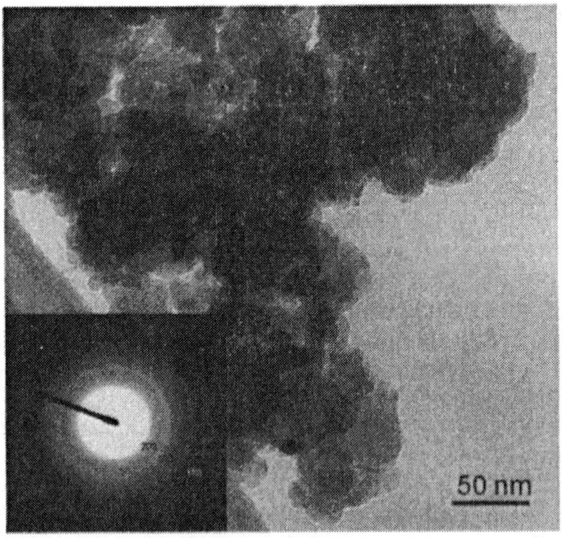

FIGURE 7.17 TEM image showing nanosized particles in the BN sample after milling for 140 h in a steel mill without heating treatment. The insert diffraction pattern suggests a disordered turbostratic structure.[15]

FIGURE 7.18 TEM image showing BN nanotubes and nanoparticles formed in the BN samples after first milling and then subsequently heated at 1200°C for 10 h.

nm. Each tube contains a metal particle at its tip and EDS reveals that these particles are mainly Fe with a little Ni and Cr. The outer diameter seems to be determined by the overall size of the metal particles and the inner diameter is close to the projected width of the edge of the Fe particle.

EELS analysis was conducted on a single nanotube. Figure 7.20 consists of a bright field (BF) image and three elemental maps of C, B, and N, which were produced using EELS imaging with inelastic electrons of C, B, and N *k*-edges, respectively. The BF image shows that a nanotube, containing a metal particle at its tip, appears across a hole between a large aggregate of BN and the C support film of the TEM grid. The brighter the region in the elemental map, the larger the elemental content. For the C map, compared with the bright C film region, the darker

FIGURE 7.19 TEM image showing BN nanotubes with bamboo-like structures, which formed in the BN samples after first milling, and then subsequently heated at 1200°C for 16 h. The dark particles at the tips are Fe.[15]

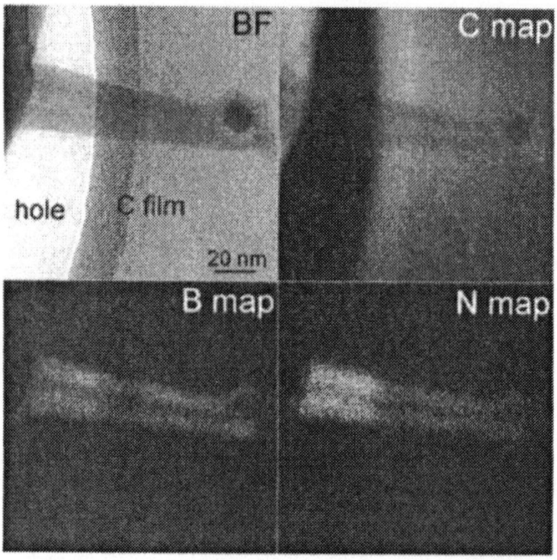

FIGURE 7.20 EELS elemental maps[3] of a BN nanotube lying across a hole between the BN particle and the C support file. The first BF image is a bright field image. The brighter contrast in the nanotube areas of the B and N maps suggest the nanotube observed is BN not C. (Image courtesy of Dr. Jin Zou.)

contrast in the nanotube areas suggests absence of C in the nanotube. In contrast, the B and N maps have light contrast only in the nanotube areas, clearly indicating strong B and N content in the nanotube.

The formation of BN nanotubes from the milled BN compound confirms that the first reaction sequence proposed in Section 7.3 is possible. It is interesting to explore how nanosized, disordered BN particles produced by HEBM are transformed into BN tubules during annealing. Comparing the XRD patterns of the milled BN (Fig. 7.16(a)) with those of the heated sample (Fig. 7.16(b)), the pronounced diffraction peaks of the BN in Fig. 7.16(b) indicate crystal growth of BN from the disordered BN during the annealing. Indeed, many two-dimensional (002) layers, the lowest free-energy layers for hexagonal BN,[61] were observed under TEM in the annealed samples. The (002) layers appear to extend into various tubular morphologies, with or without the catalytic effect of metal particles. At higher annealing temperatures (around 1300°C), the growth is faster and most tubes are both thicker and longer. It is interesting that some cylindrical tubes appear to have grown without the presence of a contaminant particle, while the bamboo-like tubes are most often capped with an Fe particle. Since the starting phase is BN, no chemical reactions are likely to be involved in the formation process of the nanotubes. BN vapor should not be present at such low annealing temperatures. Therefore, a new solid-state nucleation and growth process is indicated in the formation of BN, which is distinctly different from either the previous case of reactive milling and annealing of B or the alternative vapor phase (arc-discharge and laser ablation)[3,7–9] methods identified previously. Formation mechanisms are discussed in more detail in the following section.

7.4 FORMATION MECHANISM DISCUSSION

The previous sections demonstrate that ball milling and annealing produces a high-yield of BN nanotubes with the same nanotubular structures as those produced by other methods such as arc-discharge and laser ablation. How nanotubes are formed in such a process and what formation mechanisms are involved are the questions to be answered in this section.

7.4.1 SEPARATE NUCLEATION AND GROWTH

SEM and TEM observations found a high yield of BN nanotubes in heated samples but not in any milled samples, suggesting that BN nanotubes develop only during the thermal heating process following ball milling. What then is the role of high-energy ball milling in the formation process? Can we produce BN nanotubes directly from B or BN without the ball milling treatment? To examine this prospect, we heated B and BN powders without milling treatment under the same annealing conditions that produced nanotubes, but no nanotubes were obtained. This suggests that the milling process has an essential role to play in nanotube formation during subsequent annealing, which appears to be associated with the creation of suitable nucleation sites for nanotube growth. We suggest, therefore, that the milling and annealing processes actually correspond to separate nucleation and growth processes, respectively, that is,

Ball milling -> nucleation

Annealing -> growth

The apparent separate nucleation and growth processes afford us the advantage of investigating them separately.

7.4.2 NUCLEATION STRUCTURES

To explore possible nucleation structures, milled B and BN powders have been carefully characterized using a number of techniques. In both milled B and BN, the following common nanoscale structures were found:

1. *Amorphous phase.* Both XRD and TEM revealed that extended HEBM leads to amorphization of both crystalline B and BN in both steel and ceramic mills. As discussed previously, high-energy ball milling can transform ordered crystalline structures to disordered amorphous phases above certain milling intensity.[44] The milling intensities of the two mills used appear to be sufficiently high to create a predominantly amorphous structure in both B and BN, although a small amount of nanosized crystalline material or layers less than 100 nm was still found after milling. The amorphous phase is metastable and recrystallizes during heating.
2. *Nanosized metal particles.* A large number of metal particles were introduced into B and BN powders during ball milling with both steel and ceramic mills. The milling contamination is almost unavoidable by the mechanical milling process, especially under high milling intensity.

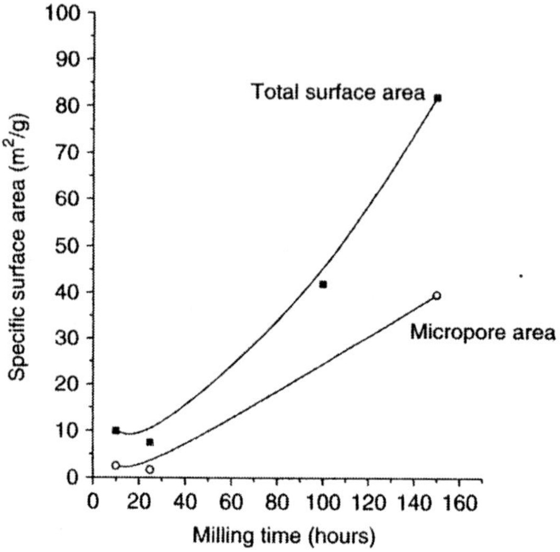

FIGURE 7.21 Variation of specific surface area of milled BN samples as a function of milling time. The micropore area was determined using the T-plot method.

In general, such milling contamination is undesirable, but, in the current case, metal particles (Fe, Ni, Cr from the steel mill and W from the ceramic mill) have been found to have positive catalytic roles in promoting nano tube formation. Actually, a high yield of BN nanotubes can only be obtained above a certain level of metal content (1 wt%). Such catalytic effects have been reported extensively in the literature for nanotube formations by other methods.[62,63]

3. *Microporous structures.* Both milled B and BN powders have high surface areas as a result of particle size reduction caused by mechanical grinding. Figure 7.21 illustrates the surface area of BN samples as a function of milling time. The total surface area of a BN sample has increased from 10 to 82 m^2/g after milling for 150 h. Analysis by the t-plot method found that almost half of the total surface area is actually contributed by micropores.[64] For example, the 150-h milled sample has a micropore area of 39.5 m^2/g. This result shows that the milled particles contain a large number of open micropores. Micropores presumably form during agglomeration and welding of BN particles under the ball impacts.[65] The nanoporous structure also could be created by severe plastic deformation of the (002) planes.[66] Indeed TEM analysis indicated a large number of severely deformed (200) basal layers in milled samples.

The formation of BN nanotubes has two important requirements: the presence of seeds or nucleation sites and the source of free atoms. Based on TEM observations, we presume that possible seeds are both nanoscale metal particles and micropores created by HEBM. Our experimental results have demonstrated clearly that Fe and W particles can act as seeds to catalyze nanotube growth. In addition, our observations suggest that micropores might also be responsible for the formation of nanotubes without requiring the presence of metal particles. It is now reasonably well accepted that multiwalled nanotubes can form without the help of metal particles, while the formation of single-wall nanotubes appears to depend on specific metal catalysts.[67] Our data indicate that micropores containing four-membered squares might serve as starting sites from which tubular structures are formed. Harris et al.[68] reported that multiwalled C nanotubes can be formed by heating an arc-discharge–produced carbon soot containing fullerenes and micropores. A porous, amorphous phase and highly disordered particles, observed in our ball milled samples, could be the source of free atoms, which are available to facilitate nanotube growth since the amorphous and disordered B or BN phases are metastable under heating. For example, XRD analysis (Fig. 7.7(b), Fig. 7.15, and Fig. 7.16(b)) has revealed that the amorphous phase crystallizes to form dominant (002) planes during heating. Therefore, milled B and BN samples appear to contain at least two possible nucleation structures, as well as a source of free atoms. These features make ball-milled material a perfect precursor for nanotube growth.

7.4.3 The Growth Process

BN nanotubes grow out from BN particle clusters during isothermal annealing in a temperature range of 1100–1300°C. This temperature range is above the bulk self-diffusion temperature of BN but well below the boiling point of B (2550°C) and the sublimation temperature of BN (3000°C).[69] We therefore conclude that vapor phases of B or BN are not possible during BN nanotube growth under our experimental conditions. In our cases, both nucleation and growth processes seem to occur in the solid state. At such a low growth temperature (1200–1400°C), surface diffusion is the most likely mechanism for nanotube growth.[70]

Noncatalytic and metal-induced catalytic growth appear to be the basis of two separate growth mechanisms. We now briefly review our results and those of others that affect such mechanisms.

1. Noncatalytic growth. Many BN nanotubes with perfect tip structures are not associated with any metal particles. They appear to form without the help of metal catalysts. For example, the TEM image in Fig. 7.22 shows two BN nanotubes with only three walls and flat tips. No metal particles are found either inside the tubes or at the tips. Although a large metal particle (dark spot) is present near the two tubes, it is completely covered with thick BN layers, and should not be responsible for the formation of the two small tubes. The small diameters of these nanotubes are close to the micropore sizes. We suggest that micropores might be the nucleation sites for these tubes. Indeed, carbon nanotubes have been successfully produced from nanoporous carbon containing pentagons and heptagons[71–73] without using metal catalysts, and catalyst-free synthesis of BN nanotubes has also been reported in a laser-ablation process.[13]

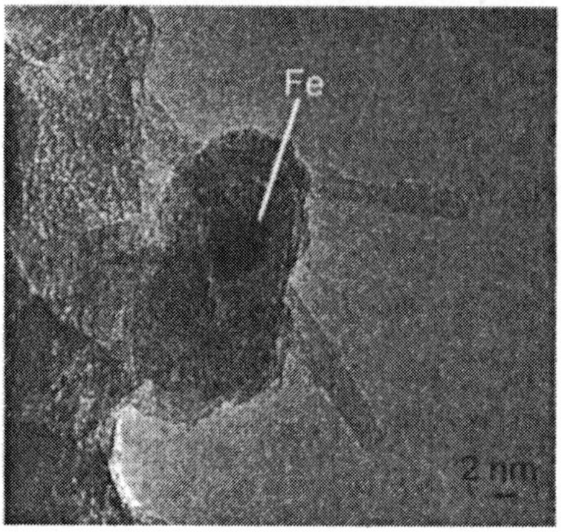

FIGURE 7.22 TEM image showing two thin BN nanotubes without containing any metal particles and an Fe particle covered by (002) basal planes nearby.

Thus, solid-state crystallization is considered the main process for non-catalytic growth in our case and surface diffusion may provide the driving force for atomic motion that feeds material to the nanotube during growth.

2. Metal catalytic growth. Metal catalytic growth has been found in both B (Fig. 7.11 and Fig. 7.14) and BN (Fig. 7.19 and Fig. 7.20) cases. The popular metal-catalytic growth model for C nanotubes, especially for the CVD process, is the vapor–liquid–solid (VLS) mechanism.[62,63] Carbon vapor dissolves into a metallic melt, forming a saturated solution at high temperature; when the temperature is reduced or when temperature or concentration gradients exist, carbon precipitates out in the form of nanotubes. Because of the lack of vapor phase in our case, the possible dissolution and precipitation process that results in BN nanotubes is most likely to be a solid–liquid–solid process.[74,75]

7.5 SUMMARY

BN nanotubes in large quantity and high formation yield have been prepared using a ball milling–annealing method consisting of high-energy ball milling of B or BN powders and then subsequent annealing. The method has the advantages of control in nanotube structures and sizes, high formation yield, and a much larger quantity compared with other methods. The available large quantity of BN nanotubes makes the studies of property and application possible. A stronger resistance to oxidation of BN nanotubes has been found, which makes them a better choice than C nanotubes for composite materials. The method can be scaled up to an industrial level and therefore commercialization of BN nanotube production is possible. Systematic research also found nanosized metal particles and micropores as possible nucleation structures and the amorphous/disordered phase as the source of free B and N atoms.

The growth process is via the solid state, driven by surface diffusion. In addition to the synthesis of BN, C,[73,76] and other oxide[77,78] nanotubes, the ball milling technique has also been used in processing of nanotubes, including opening of closed nanotubes,[79–81] the cutting of long nanotubes into short lengths,[81–84] and in assisting functionalization,[85,86] helping aligned nanotube growth,[87,88] demonstrating the important role of HEBM in nanotube research.

PROBLEMS

1. What are the structural differences between C and BN nanotubes?
2. Why do BN nanotubes have flat tips instead of cone-shaped tips?
3. What are the different properties of BN nanotubes compared with C nanotubes?
4. What is the role of high-energy ball milling in the milling-annealing process?
5. Is HEBM the same as conventional ball milling?
6. What are the possible applications of BN nanotubes based on their special properties?

ACKNOWLEDGMENTS

The authors gratefully acknowledge the contributions from all collaborators, especially Professor Lewis Chadderton, Dr. John Fitzgerald, Dr. Jin Zou, Mr. Martin Conway, Mrs. Jun Yu, Professor Stewart Campbell, and Professor Gerard Le Caer. They also thank Electron Microscopy Units of the Australian National University and of University of Sydney for their help with electron microscopes. This work was supported financially by several research grants from the Australian Research Council.

REFERENCES

1. Iijima, S., Helical microtubules of graphitic carbon, *Nature, 354,* 56, 1991.
2. Rubio, A., Corkill, J.L., and Cohen, M.L., Theory of graphitic boron nitride nanotubes, *Phys. Rev. B, 49,* 5081, 1994.
3. Chopra, N.G. et al., Boron nitride nanotubes, *Science, 269,* 966, 1995.
4. Dresselhaus, M.S., Dresselhaus, G., and Eklund, P.C., *Science of Fullerenes and Carbon Nanotubes,* Academic Press, New York, 1996, 756.
5. Ajayan, P.M., Carbon nanotubes: A new graphite architecture, *Condensed Matter News, 4,* 9, 1995.
6. Chen, Y. et al., The nucleation and growth of carbon nanotubes in a mechano-thermal process, *Carbon, 42,* 1543, 2004.
7. Loiseau, A. et al., Boron nitride nanotubes with reduced numbers of layers synthesized by arc discharge, *Phys. Rev. Lett., 76,* 4737, 1996.
8. Terrones, M. et al., Metal particle catalysed production of nanoscale BN structures, *Chem. Phys. Lett., 259,* 568, 1996.
9. Saito, Y., Maida, M., and Matsumoto, T., Structures of boron nitride nanotubes with single-layer and multilayers produced by arc discharge, *Jpn. J. Appl. Phys., 38,* 159, 1999.

10. Menon, M., and Srivastava, D., Structure of boron nitride nanotubes: Tube closing versus chirality, *Chem. Phys. Lett., 307,* 407, 1999.
11. Terrones, M. et al., Nanotubes: A revolution in materials science and electronics, *Fullerenes and Related Structures, 199,* 189, 1999.
12. Golberg, D. et al., Fine structure of boron nitride nanotubes produced from carbon nanotubes by a substitution reaction, *J. Appl. Phys., 86,* 2364, 1999.
13. Lee, R.S. et al., Catalyst-free synthesis of boron nitride single-wall nanotubes with a preferred zig-zag configuration, *Phys. Rev. B, 64,* 1214051, 2001.
14. Endo, M., Iijima, S., and Dresselhaus, M.S., *Carbon Nanotubes,* 1st Ed., Elsevier Science, Oxford, 1996, 27.
15. Chen, Y. et al., A solid state process for formation of boron nitride nanotubes, *Appl. Phys. Lett., 74,* 2960, 1999.
16. Fitzgerald, J.D., Chen, Y., and Conway, M.J., Nanotube growth during annealing of mechanically milled boron, *Appl. Phys. A, 76,* 107, 2003.
17. Chen, Y. et al., Mechanochemical synthesis of boron nitride nanotubes, *Mater. Sci. Forum, 312–314,* 173, 1999.
18. Bando, Y., Ogawa, K., and Golberg, D., Insulating "nanocables": Invar Fe-Ni alloy nanorods inside BN nanotubes, *Chem. Phys. Lett., 347,* 349, 2001.
19. Mickelson, W. et al., Packing C60 in boron nitride nanotubes, *Science, 300,* 467, 2003.
20. Chopra, N.G., and Zettl, A., Measurement of the elastic modulus of a multi-wall boron nitride nanotube, *Solid State Communications, 105,* 297, 1998.
21. Blase X. et al., Stability and band gap constancy of boron nitride nanotubes, *Europhys. Lett., 28,* 335, 1994.
22. Chen, Y., Zou, Y., Campbell, S.J., and Le Caer, G., Boron nitride nanotubes: Pronounced resistance to oxidation, *Appl. Phys. Lett., 84,* 13, 2430, 2004.
23. Tsang, S.C., Harris, P.J.F., and Green, M.L.H., Thinning and opening of carbon nanotubes by oxidation using carbon dioxide, *Nature, 362,* 520, 1993.
24. Ajayan, P.M., Opening carbon nanotubes with oxygen and implications for filling, *Nature, 362,* 522, 1993.
25. Han, W. et al., Synthesis of boron nitride nanotubes from carbon nanotubes by a substitution reaction, *Appl. Phys. Lett., 73,* 3085, 1998.
26. Han, W. et al., Transformation of BxCyNz nanotubes to pure BN nanotubes, *Appl. Phys. Lett., 81,* 1110, 2002.
27. Cumings, J., and Zettl, A., Mass-production of boron nitride double-wall nanotubes and nanococoons, *Chem. Phys. Lett., 316,* 211, 2000.
28. Chen, Y., Fitz Gerald, J., Williams, J.S., and Bulcock, S., Synthesis of boron nitride nanotubes at low temperatures using reactive ball milling, *Chem. Phys. Lett., 299,* 260, 1999.
29. Chen, Y., Conway, M., Williams, J.S., and Zou, J., Large-quantity production of high-yield boron nitride nanotubes, *J. Mater. Res., 17,* 8, 1896, 2002.
30. Benjamin, J.S., Dispersion strengthened superalloys by mechanical alloying, *Metall. Trans., 1,* 2943, 1970.
31. Koch, C.C. et al., Preparation of "amorphous" Ni60Nb40 by mechanical alloying, *Appl. Phys. Lett., 43,* 1017, 1983.
32. Suryanarayana, C., Nanocrystalline materials, *Inter. Mater. Rev., 40,* 41, 1995.
33. Debarbadillo, J.J., and Smith, G.D., Recent developments and challenges in the application of mechanically alloyed, oxide dispersion strengthened alloys, *Mater. Sci. Forum, 88–90,* 167, 1992.
34. Chen, Y., Fitz Gerald, J., Chadderton, L.T., and Chaffron, L., Nanoporous carbon produced by ball milling, *Appl. Phys. Lett., 74,* 19, 2782, 1999.

35. Tang, J. et al., Extended solubility and spin-glass behavior in a Ag-Gd solid solution prepared by mechanical alloying, *Phys. Rev. B, 52,* 12829, 1995.
36. Koch, C.C., Amorphisation by mechanical alloying, *J. Non-Crystalline Solids, 117–118,* 670, 1990.
37. Butyagin, P.J., The chemical forces in mechanical alloying, *Mater. Sci. Forum, 88–90,* 695, 1992.
38. Schaffer, G.B., and McCormick, P.G., Displacement reactions during mechanical alloying, *Metall. Trans. A, 21,* 2789, 1990.
39. Calka, A., and Williams, J.S., Synthesis of nitrides by mechanical alloying, *Mater. Sci. Forum, 88–90,* 787, 1992.
40. Chen, Y., and Williams, J.S., Formation of metal hydrides by mechanical alloying, *J. Alloys Compounds, 217,* 181, 1995.
41. Matteazzi, P., and Le Caer, G., Room-temperature mechanosynthesis of carbides by grinding of elemental powders, *J. Am. Ceram. Soc., 74,* 1382, 1991.
42. Suryanarayana, C., Mechanical alloying and milling, *Prog. Mater. Sci., 46,* 13, 2001.
43. Chen, Y., Le Hazif, R., and Martin, G., Amorphization in a vibrating frame grinder: An example of phase transition in driven systems, *Mater. Sci. Forum, 88–90,* 35, 1992.
44. Chen, Y. et al., Ball-milling-induced amorphization in Ni-Zr compounds: A parametric study, *Phys. Rev. B, 48,* 14, 1993.
45. Chen, Y., Halstead, T., and Williams, J.S., Influence of milling temperature and atmosphere on the synthesis of iron nitrides by ball milling, *Mater. Sci. Engin. A, 206,* 24, 1996.
46. Calka, A., and Radlinski, R.P. Formation of amorphous Fe-B alloys by mechanical alloying, *Appl. Phys. Lett., 58,* 119, 1991.
47. Calka, A. and Kikolov, J.I., The dynamics of magneto-ball milling and its effect on phase transformations during mechanical alloying, *Mater. Sci. Forum, 179–181,* 333, 1995.
48. Benjamin, J.S., and Volin, T.E., The mechanism of mechanical alloying, *Metall. Trans., 5,* 1929, 1974.
49. Galy, D., Chaffron, L., and Martin, G., Amorphization mechanism of $NiZr_2$ by ball milling, *J. Mater. Res., 12,* 688, 1997.
50. Chen, Y., and Williams, J.S., Investigation of gas-solid reactions realized by ball milling, *Mater. Sci. Forum, 225–227,* 545, 1996.
51. Suryanarayana, C., and Froes, F.H., Mechanical alloying of titanium-based alloys, *Adv. Mater., 5,* 96, 1993.
52. Chen, Y. et al., Study on mechanism of mechanical activation, *Mater. Sci. Engin. A, 226,* 95, 1997.
53. Bhattacharya, A.K., and Arzt, E., Plastic deformation and its influence on diffusion process during mechanical alloying, *Scripta Metall. Mater., 28,* 395, 1993.
54. Chen, Y., Hwang, T., and Williams, J.S., Ball milling induced low-temperature carbothermic reduction of ilmenite, *Mater. Lett., 28,* 55, 1996.
55. Chen, Y. et al., Mechanically activated carbothermic reduction of ilmenite, *Metall. Trans. A, 28A,* 1115, 1997.
56. Suryanarayana, C., Mechanical alloying and milling, *Prog. Mater. Sci., 46,* 1, 2001.
57. Gleiter, H., Nanostructured materials: State of the art and perspectives, *Z. Metallkd., 86,* 78, 1995.
58. Chen, Y., Li, Z.L., and Williams, J.S., The evolution of hydriding and nitriding reactions during ball milling of titanium in ammonia, *J. Mater. Sci. Lett., 14,* 542, 1995.
59. Chen, Y., and Williams, J.S., Competitive gas-solid reactions realized by ball milling of Zr in ammonia gas, *J. Mater. Res., 11,* 6, 1500, 1996.

60. Hirano, S.I., Yogo, T., Asada, S., and Naka, S., Synthesis of amorphous boron nitride by pressure pyrolysis of borazine, *J. Am. Ceram. Soc., 72,* 66, 1989.
61. Abrahamson, J., The surface energies of graphite, *Carbon, 11,* 337, 1973.
62. Tibbertts, G.G., Why are carbon filaments tubular? *J. Crystl. Growth, 66,* 632, 1984.
63. Laurent, C. et al., Metal nanoparticles for the catalytic synthesis of carbon nanotubes, *New J. Chem., 22,* 1229, 1998.
64. Seaton, W.G., A new analysis method for the determination of the pore size distribution of porous carbon from nitrogen adsorption measurements, *Carbon, 27,* 853, 1989.
65. Berlin, O., *Chemistry and Physics of Carbon,* Marcel Dekker, New York, 1989, 94.
66. Huang, J.Y., Yasuda, H., and Mori, H., Highly curved carbon nanostructures produced by ball-milling, *Chem. Phys. Lett., 303,* 130, 1999.
67. Charlier, J.C., and Iijima, S., Growth mechanisms of carbon nanotubes, in Dressehuas, M.S., Dresselhaus, G., and Avouris, Ph., *Carbon Nanotubes: Synthesis, Structure, Properties, and Applications,* Topics in Applied Physics 80, Springer-Verlag, New York, 2001, 55.
68. Harris, P.J.F. et al., High-resolution electron microscopy studies of a microporous carbon produced by arc-evaporation, *J. Chem. Soc. Faraday Trans., 90,* 2799, 1994.
69. West, R.C. et al., *Handbook of Chemistry and Physics,* 70th Ed., CRC Press, Boca Raton, Florida, 1990, B77.
70. Chadderton, L.T., and Chen, Y., Nanotube growth by surface diffusion, *Phys. Lett. A, 263,* 401, 1999.
71. Iijima, S., Growth of carbon nanotubes, *Mater. Sci. Eng., B19,* 172, 1993.
72. Setlur, A.A. et al., A promising pathway to make multiwalled carbon nanotubes, *Appl. Phys. Lett., 76,* 3008, 2000.
73. Chen, Y., Conway, M.J., and Fitzgerald, J.D., Carbon nanotubes formed in graphite after mechanically grinding and thermal annealing, *Appl. Phys. A, 76,* 633, 2003.
74. Chen, Y. et al., Solid-state formation of carbon and boron nitride nanotubes, *Mater. Sci. Forum, 343,* 63, 2000.
75. Chadderton, L.T., and Chen, Y., A model for the growth of bamboo and skeletal nanotubes: Catalytic capillarity, *J. Crystl. Growth, 240,* 164, 2002.
76. Chen, Y. et al., Investigation of nanoporous carbon powders produced by high energy ball milling and formation of carbon nanotubes during subsequent annealing, *Mater. Sci. Forum, 312–314,* 375, 1999.
77. Kholmanov, I.N. et al., A simple method for the synthesis of silicon carbide nanorods, *J. Nanosci. Nanotechnol., 2,* 5, 453, 2002.
78. Li, S.L. et al., Synthesis of TiTe2 nanotubes, *Acta Chimica Sinica., 62,* 6, 634, 2004.
79. Li, Y.B. et al., Transformation of carbon nanotubes to nanoparticles by ball milling process, *Carbon, 37,* 3, 493, 1999.
80. Gao, B. et al., Electrochemical intercalation of single-walled carbon nanotubes with lithium, *Chem. Phys. Lett., 307,* 153, 1999.
81. Pierard, N. et al., Production of short carbon nanotubes with open tips by ball milling, *Chem. Phys. Lett., 335,* 1, 2001.
82. Kim, Y.A. et al., Effect of ball milling on morphology of cup-stacked carbon nanotubes, *Chem. Phys. Lett., 355,* 279, 2002.
83. Shimoda, H. et al., Lithium intercalation into opened single-wall carbon nanotubes: Storage capacity and electronic properties, *Phys. Rev. Lett., 88,* 155021, 2002.
84. Liu. F. et al., Preparation of short carbon nanotubes by mechanical ball milling and their hydrogen adsorption behavior, *Carbon, 41,* 2527, 2003.
85. Breton, Y. et al., Functionalization of multiwall carbon nanotubes: Properties of nanotubes-epoxy composites, *Molecular Crystals and Liquid Crystals, 387,* 359, 2002.

86. Wang, Y., Wu. J., and Wei, F., A treatment method to give separated multi-walled carbon nanotubes with high purity, high crystallization and a large aspect ratio, *Carbon, 41,* 15, 2939, 2003.
87. Guo, X.M., Qi, J.F., and Sakurai, K., Mechanochemical formation of novel catalyst for preparing carbon nanotubes: Nanocrystalline yttrium aluminum iron perovskite, *Scripta Materialia, 48,* 8, 1185, 2003.
88. Chen, Y., and Chadderton, L.T., Improved growth of aligned carbon nanotubes by ball milling, *J. Mater. Res.*, 19, 10, 2791, 2004.

Part 2

Manufacturing Using Nanoscale Materials

8 Plasma Deposition of Ultra-Thin Functional Films on Nanoscale Materials

Peng He and Donglu Shi

CONTENTS

8.1 INTRODUCTION

Nanotechnology is the application of nanoscale materials. This has been the most important research topic in the world for the past few years, and it will change our lives in the very near future. One of the critical aspects of nanotechnology research is how to modify the surface of nanoparticles to make them more useful for different applications. This chapter describes a novel plasma polymerization technique for coating nanoparticles, nanotubes, and nanoplates. The coated nanoparticles/tubes/plates can be used to reinforce polymer materials and to provide special functional capabilities for materials.

In the current development of nanomaterials, it has become critical to modify the surfaces of the nanoparticles for both fundamental research and engineering applications.[1–8] Taking different film technologies into consideration, plasma polymerization coating as well as plasma surface treatment of powders are particularly promising because of the following unique features and advantages: (1) The starting feed gases used may not contain the type of functional groups, such as carbon double bond or triple bond, that are normally associated with conventional polymerization; (2) such films are often highly cohesive and adhesive to a variety of substrates, including conventional polymers, glasses, and metals; (3) polymerization may be achieved without the use of solvents; (4) plasma polymer films can be easily produced with thicknesses from several nanometers to 1 μm; and (5) through careful control of the polymerization parameters, it is possible to tailor the films with specific chemical functionality, thickness, and other chemical and physical properties.

Plasma-treated powders have a wide range of potential applications. The applications involve improving the following properties of powders: biocompatibility, sorption, wettability, triboelectricity, barrier and insulation, adhesion, diffusion, friction coefficient, corrosion, dispersion, flowability, and other properties. Plasma treatment has wide applications in medicine, pharmacy, the pigment industry, rubber chemistry, and others. The current research on plasma treatment of particles is mainly focused on the micron size or even larger. In the literature, not much research has been done investigating thin-film coating of nanosize particles. The long term main objective of the research reported here is to develop a broad set of techniques for plasma thin-film coating of nanoparticles and nanotubes. In order to develop this coating technology, different monomers such as styrene, pyrrole, and C_6F_{14}, and different substrates such as nanoparticles and carbon nanotubes have been used to form new material systems. The processing and characterization of these materials is discussed next.

8.2 THE PLASMA-COATING TECHNIQUE

The plasma-coating facility is a homemade system. The schematic diagram of the plasma reactor for thin-film deposition of nanoparticles is shown in Fig. 8.1. It consists mainly of a radio frequency (RF) source, a glass vacuum chamber, and pressure gauge. The vacuum chamber of the plasma reactor has a long Pyrex glass column about 80 cm in height and 6 cm in internal diameter.[9–11] The powder is vigorously stirred at the bottom of the tube and thus the surface of particles can be continuously renewed and exposed to the plasma for thin-film deposition during the plasma polymerization processing. A magnetic bar is used to stir the powders. The gases and monomers are introduced from the gas inlet during the plasma cleaning treatment and plasma polymerization. Before the plasma treatment, the basic pressure is pumped down to less than 50 mtorr and then the carrier gas (such as argon) or monomer vapors are introduced into the reactor chamber. The operating pressure is adjusted by the gas/monomer mass flow rate. During the plasma polymerization processing, the input power is 10–80 W, and the system pressure is 300–450 mtorr. The plasma treatment time varies from 15 to 120 min according to the different monomers used and the different film thicknesses desired.

After the plasma treatment, the treated powders are examined by using transmission electron microscopy (TEM), secondary ion mass spectrometry (TOFSIMS),

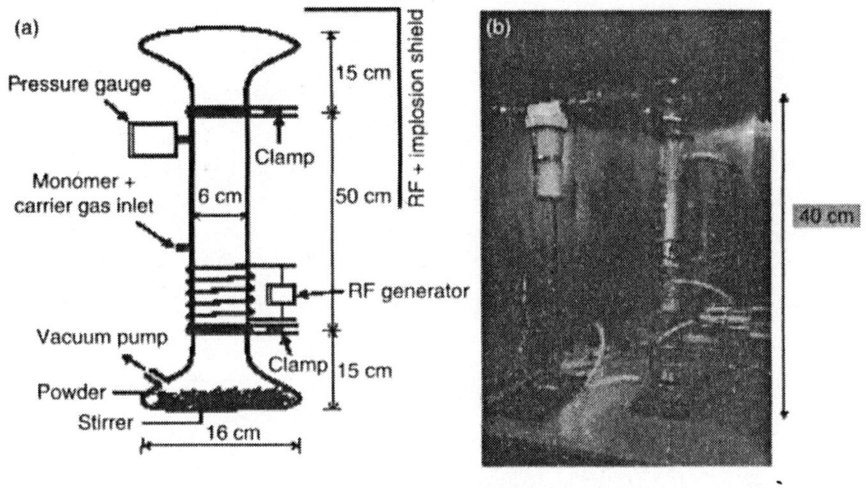

FIGURE 8.1 The plasma reactor for thin-film coating of nanoparticles: (a) schematic diagram, (b) fluidized bed reactor.

and Fourier transform infrared (FTIR). The high-resolution TEM (HRTEM) experiments are performed on a JEOL JEM 4000EX TEM.

In this research, the substrates used for plasma deposition include ZnO, $NiFe_2O_4$, $YYbErO_2S$, and carbon nanofibers. The ZnO powders are obtained from Zinc Corporation of America (Monaca, PA). The $NiFe_2O_4$ is obtained from Inframat Corporation (Willington, CT). The $YYbErO_2S$ is obtained from OraSure Technologies, Inc. (Bethlehem, PA). The carbon nanofiber Pyrograf-III PR-24 is obtained from Applied Science, Inc., (Cedarville, OH). The monomers, such as pyrrole, acrylic acid, C_6F_{14}, and styrene, were obtained from Alfa Aesar, a Johnson Matthey Company (Ward Hill, MA).

8.3 APPLICATIONS AND CHARACTERIZATION

Different applications of the plasma technique and the methods used for characterizing the new material systems produced are described below.

8.3.1 Plasma Polymer Films Deposited on ZnO Particles

In this experiment, a plasma polymer film is deposited on ZnO particles.[12] In transmission electron microscopy (TEM), the original and coated ZnO nanoparticles were dispersed onto a holy-carbon film supported by Cu grids for the TEM operated at 400 kV. Figure 8.2 shows the HRTEM image of the original, uncoated ZnO particles. As can be seen in this figure, the particle sizes are in the several hundreds of nanometers range. The particles also exhibit nonspherical shapes.

Figure 8.3 shows the HRTEM images of the coated particles. Compared with Fig. 8.2, a bright ring was found around the particles. From the HRTEM images, the bright ring was found to be one kind of amorphous layer outside the lattice

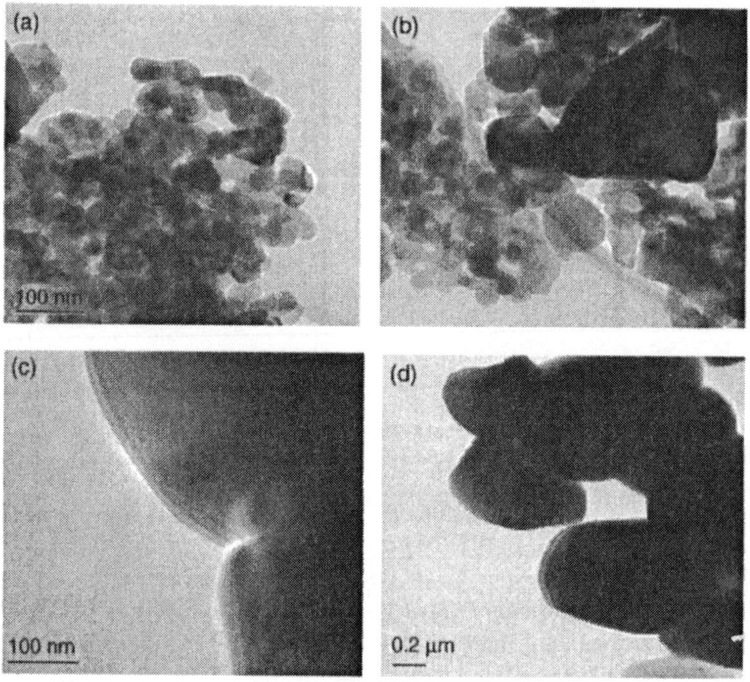

FIGURE 8.2 TEM image of uncoated ZnO at different resolutions (a), (b), (c), and (d).

structure. The ZnO particles have a crystalline structure. Thus the lattice structure in the images is the ZnO particle substrate. The amorphous layer that does not appear in the uncoated images (Fig. 8.2) is due to the coating. In Fig. 8.3, the low magnification images show a uniform coating on the particle surfaces. The coating thickness is approximately 10–20 nm thick over the entire particle surface. Although the shape of particles is nonspherical and at some places the shape is even very sharp, it is particularly interesting that the coating remains the same thickness, indicating that the plasma chamber produces a uniform coating on all the particles.

To confirm the TEM observations shown in Fig. 8.2 and Fig. 8.3, TOFSIMS was carried out to study the surface films of the particles. Figure 8.4(a) and (b) show the positive and negative TOFSIMS spectra of the coated ZnO particles. In Fig. 8.4(a), one can see that the spectrum of the positive ion from the coated ZnO have strong peaks of functional groups such as $C_4H_7^+$, $C_4H_9^+$, $C_6H_{13}O_4^+$, $C_7H_9COH^+$, and $C_7H_9COOH^+$, indicating a surface coating on the particles consistent with the HRTEM data presented in Fig. 8.2 and Fig. 8.3. In Fig. 8.4(b), where the spectra of the negative ions are presented, we also see the AA monomer, AA dimer, and AA dimer + C_2H_4 and AA dimer + C_3H_6, which are strong indications of these functional groups. These are the typical characteristic cluster patterns of a plasma-polymerized poly(acrylic acid) film.

FTIR was used to study the effect of the plasma power on the molecular structure of the AA films. As can be seen in Fig. 8.5(a), the strong C=O peak near 1700 cm^{-1}

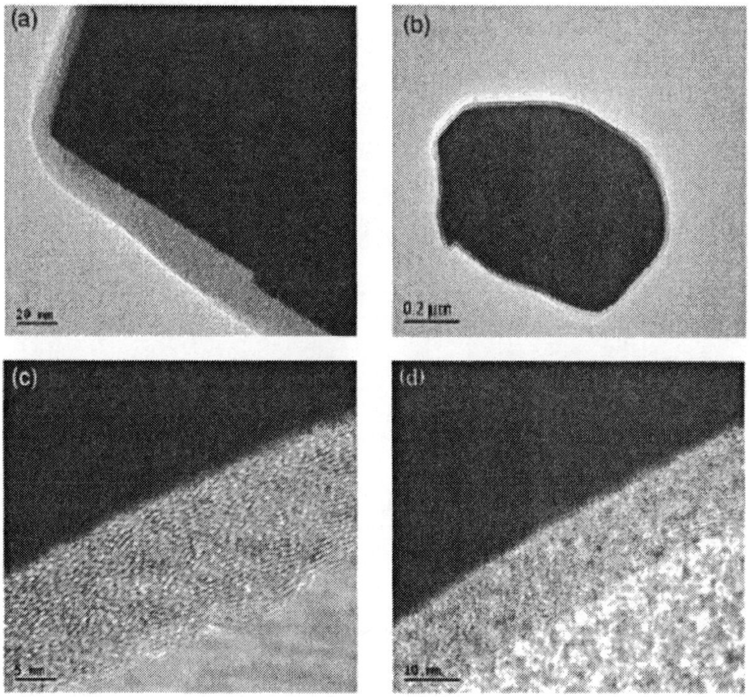

FIGURE 8.3 Bright-field HRTEM images of the AA-coated ZnO nanoparticles: (a) at low magnification; (b) at higher magnification; (c, d) showing the amorphous nature of the AA thin film and the crystal lattices of the ZnO structure.

indicates the surface coating of the nanoparticles, and it is consistent with the HRTEM data presented in Fig. 8.3. The peak of C=O intensifies as the plasma power increases up to 80 W, indicating a strong plasma power dependence. The intensification of this peak also indicates that an increased number of C=O function groups came from the acrylic acid.

In order to investigate the solubility of the plasma film, the coated ZnO powder was immersed into Ni^{+} solution. After 1 h, the sample was removed from the solution, dried, and the FTIR experiment was performed again on the nickel solution treated powder. The results are shown in Fig. 8.5(b), (c), and (d). In these figure parts, we can see that the AA film was entirely dissolved when the plasma power was 15 W. The FTIR spectra clearly show that the film was removed by the nickel solution after immersed for only 1 h. However, as the plasma power was increased to 60 W (Fig. 8.5(c)) and 80 W (Fig. 8.5(d)), the AA film remained after immersed in the nickel solution. This indicates that large cross links exist in the polymer, which prevented the AA film from dissolving in the nickel solution; even the polymer has the strong hydrophilic function groups COOH. Modified ZnO powder can be used in epoxy, PE, PP, or PS to improve the mechanical properties. At the same time, the ZnO powder can introduce special optical properties to composite materials.

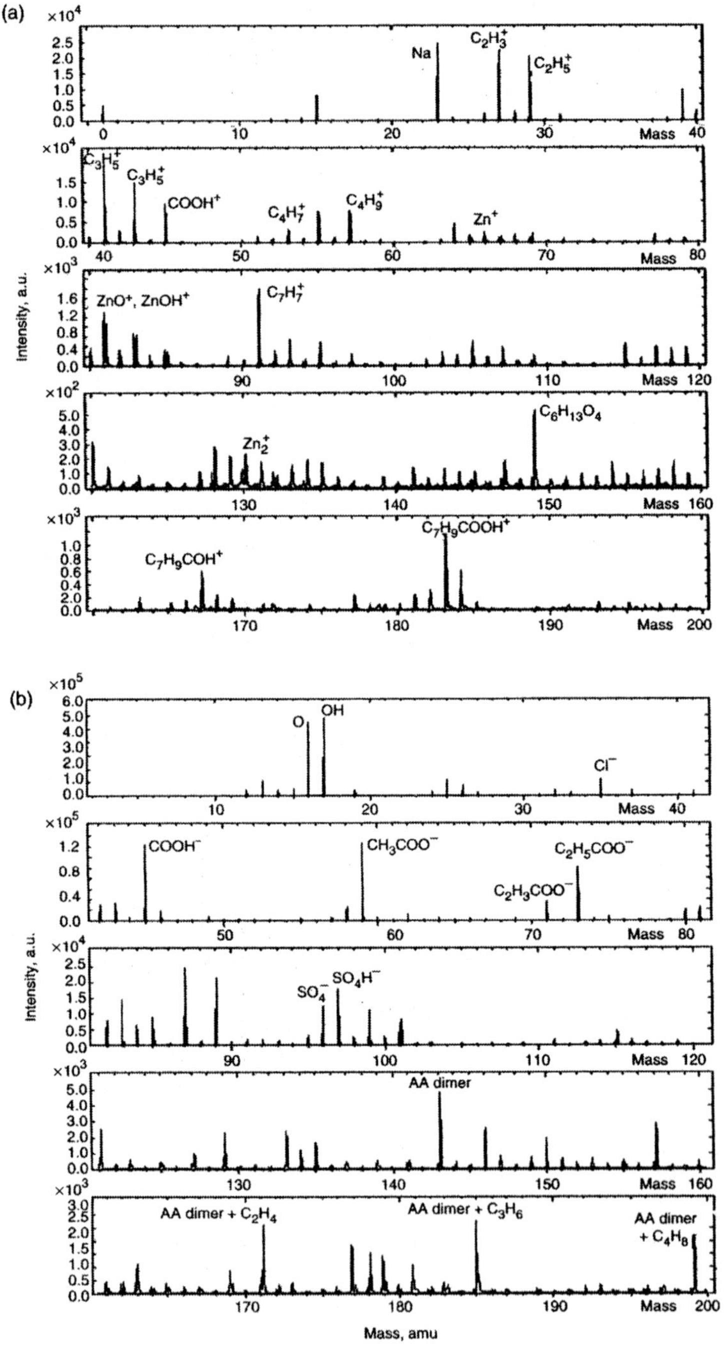

FIGURE 8.4 SIMS spectra of coated samples: (a) positive image (positive species); (b) negative image (negative species).

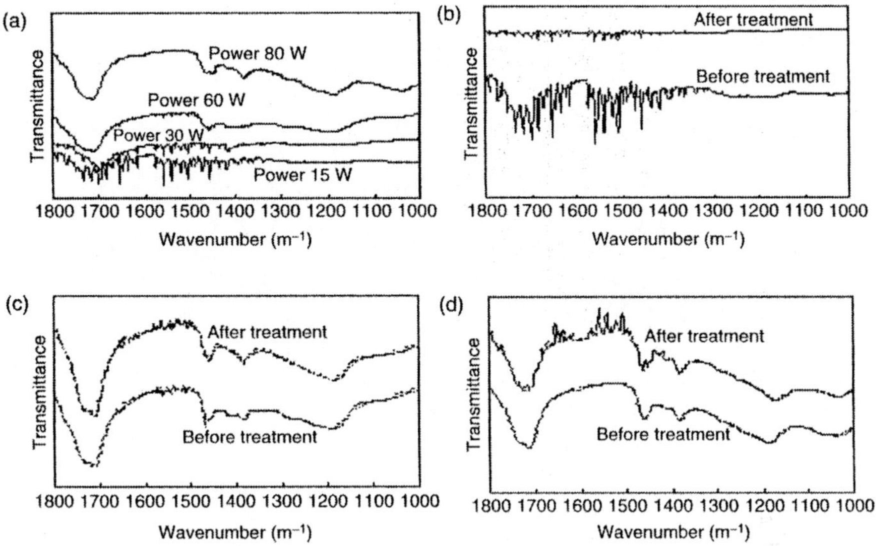

FIGURE 8.5 (a) FTIR spectra of a AA-coated ZnO at different plasma powers. The FTIR spectra of Ni solution treated, AA-coated ZnO nanoparticles at plasma power of (b) 15 W, (c) 60 W, and (d) 80 W.

8.3.2 Plasma Coating of Magnetic Nanoparticles

In this experiment, $NiFe_2O_4$ particles were coated using mechanical stirring instead of magnetic stirring because of the particles' magnetic properties. Figure 8.6 shows the HRTEM image of the original, uncoated $NiFe_2O_4$ particles. As can be seen in this figure, the particle sizes range from 10–50 nm, and the particles are severely aggregated. Figure 8.6(c) is the high-resolution image of the original $NiFe_2O_4$ nanoparticles. The lattice image further reveals the crystallographic features of the naked $NiFe_2O_4$ particles.

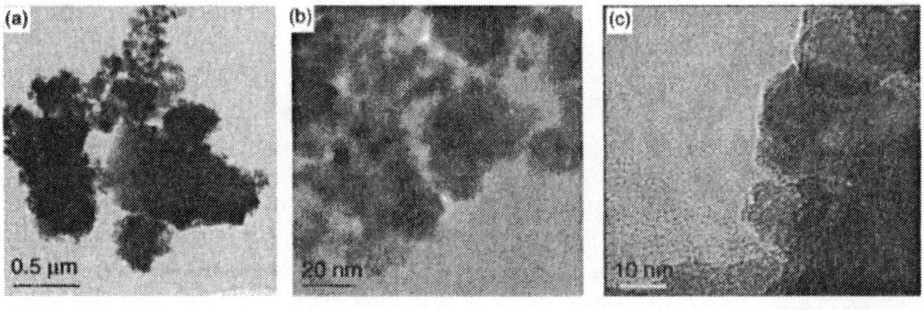

FIGURE 8.6 HRTEM images of the original $NiFe_2O_4$ nanoparticles (a, b) at different resolutions, with a particle size distribution ranging from 10 to 50 nm, and (c) crystal lattice and the uncoated nature of the nanoparticle surfaces.

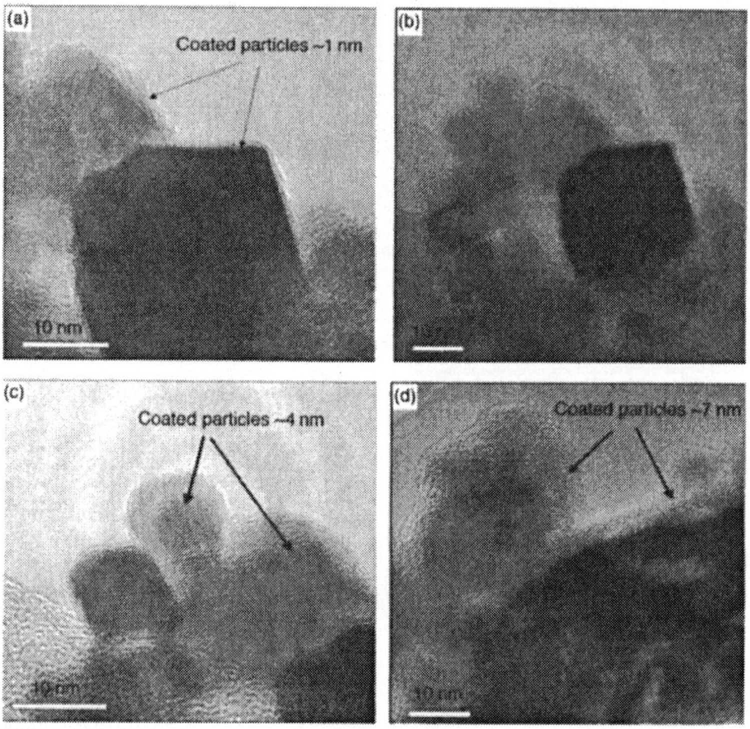

FIGURE 8.7 TEM image of the coated $NiFe_2O_4$ at different positions (a), (b), (c), and (d).

From Fig. 8.7, we can see the magnetic particles are still aggregated. The coating is only on the outside of the cluster, not exactly on the surface of each particle.

It is noted that the impurity particles such as TiOx and Cu (Fig. 8.8) have a very uniform coating on each individual particle. We can reach the conclusion that the mechanical stirring is satisfactory for many nanoparticles, but not for the magnetic particles. The natural magnetic properties of $NiFe_2O_4$ make them much more difficult to disperse than regular particles.

In Fig. 8.9, the spectrum exhibits several very strong peaks between 1400 cm^{-1} and 1600 cm^{-1}, which is due to the C—C vibrating special absorption of the benzene ring. These peaks confirm that the plasma polymer structure contains some benzene rings. In the FTIR absorption spectrum, the range from 690 cm^{-1} to 900 cm^{-1} belongs to the benzene C-H out of plane bending.

In Fig. 8.9, at 758 cm^{-1} and 700 cm^{-1} there are two peaks, which are the special absorption for one hydrogen atom on a benzene ring substituted by other function groups. Figure 8.9 is more complicated than the regular FTIR spectra of polystyrene, but the special absorption peaks for styrene are almost the same as those of regular polystyrene. This means that in plasma polymerization the molecular structure is destroyed by ion or electron bombardment, but the coating film still keeps some of the polystyrene molecular structure, which can be used to get good compatibility with a polystyrene matrix.

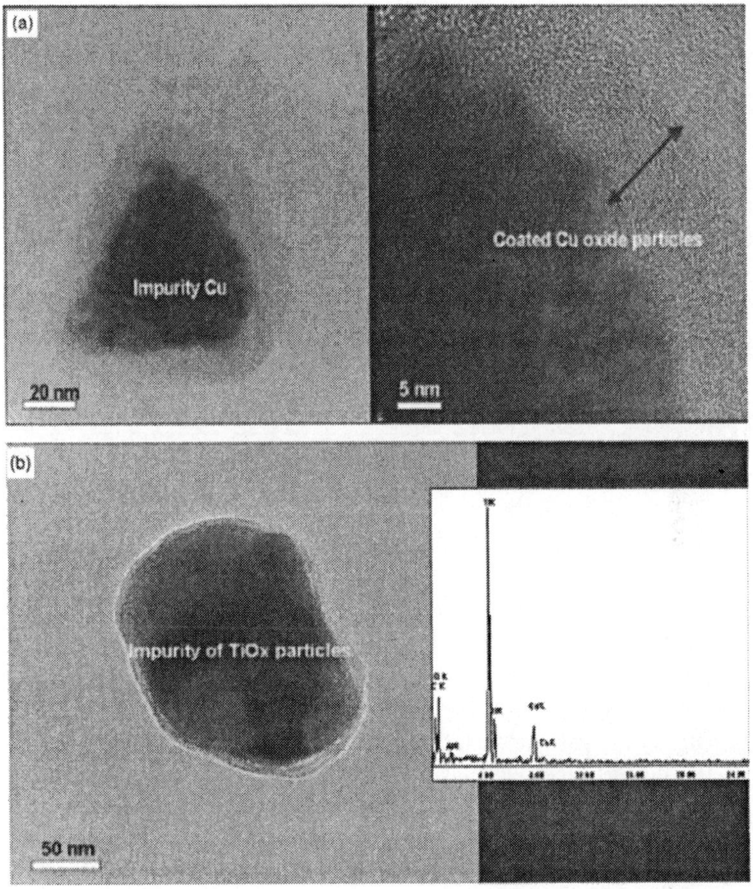

FIGURE 8.8 TEM image of coated impurity: (a) Cu; (b) TiOx in magnetic particles.

8.3.3 Copolymer Coating of $YYbErO_2S$

After using a single monomer to coat nanoparticles, several monomers are used together to modify the surface properties of nanoparticles. Different monomers can be used to introduce different function groups, which enable us to easily control the surface properties of nanoparticles and to easily modify the surface properties of nanoparticles for different purposes.

The particle used here is one type of luminescent particle. This particle will be used as the substrate for bacteria detection. Other applications with different coatings might be to use the nanoparticles in composites to detect cracks in structural materials, for radiation shielding, and other applications. For the current application, the outside coating needs to contain the C=O function group, which can be used to attract the bacteria. However, the concentration needs to be carefully controlled for different conditions. In order to get the C=O function group and control the

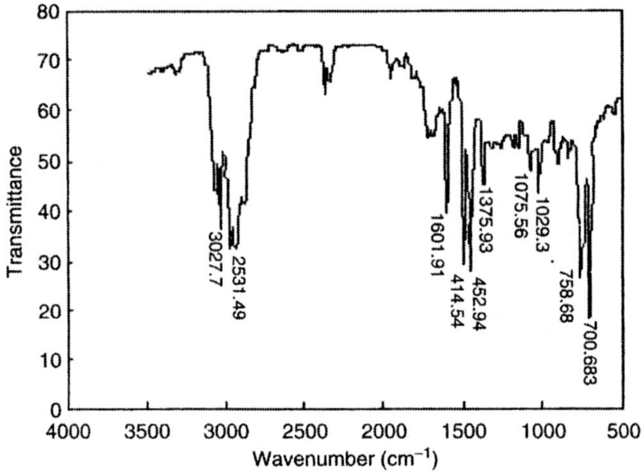

FIGURE 8.9 FTIR spectra of the coating of $NiFe_2O_4$ nanoparticles.

concentration, two kinds of monomers are used in this coating processing. One is methyl methacrylate (MMA), which used to introduce the C=O function group. Another one is styrene, which is used to control the concentration of C=O.

Figure 8.10 is the HRTEM image of the coated particle. In Fig. 8.10, there is a bright ring on the particle surfaces, which is the result of the polymer coating. Figure 8.10 contains high magnification images showing a uniform coating on the particle surfaces. The coating thickness is approximately 10 nm over the entire particle surface. The coating layer is amorphous based on the HRTEMs of different particles.

Figure 8.11 shows there are several strong peaks belonging to different function groups. The peak around 1700 cm^{-1} is the special absorption of C=O vibration. The peaks between 1450 cm^{-1} and 1700 cm^{-1} belong to the C-C vibration of the benzene ring.

From the molecular structure of the monomer, we can see that the benzene ring comes from styrene and the C=O function group comes from the MMA. Comparing Fig. 8.11(a), (b), and (c), the intensity of the peaks between 1450 and 1600 cm^{-1} and the peaks around 700 cm^{-1} and 760 cm^{-1} decreases as the ratio of styrene/MMA decreases, and the intensity of peaks changing means the concentration of different function groups has changed. With the MMA concentration increasing, the peak around 1700 cm^{-1} becomes stronger and stronger. This evidence means that as more MMA monomer is used in plasma coating, more C=O function groups occur in the coating films. The FTIR data strongly prove that through controlling the ratio of monomers, the molecular structure of the coating polymer can be modified as desired.

The molecular structure of the monomers styrene and methyl methacrylate (MMA) are shown in Fig. 8.12.

8.3.4 Plasma Coating of Carbon Nanofibers

After modifying nanoparticles, the so-called zero dimension substrate, the plasma-coating technique is applied to one-dimensional substrates. Commercial Pyrograf

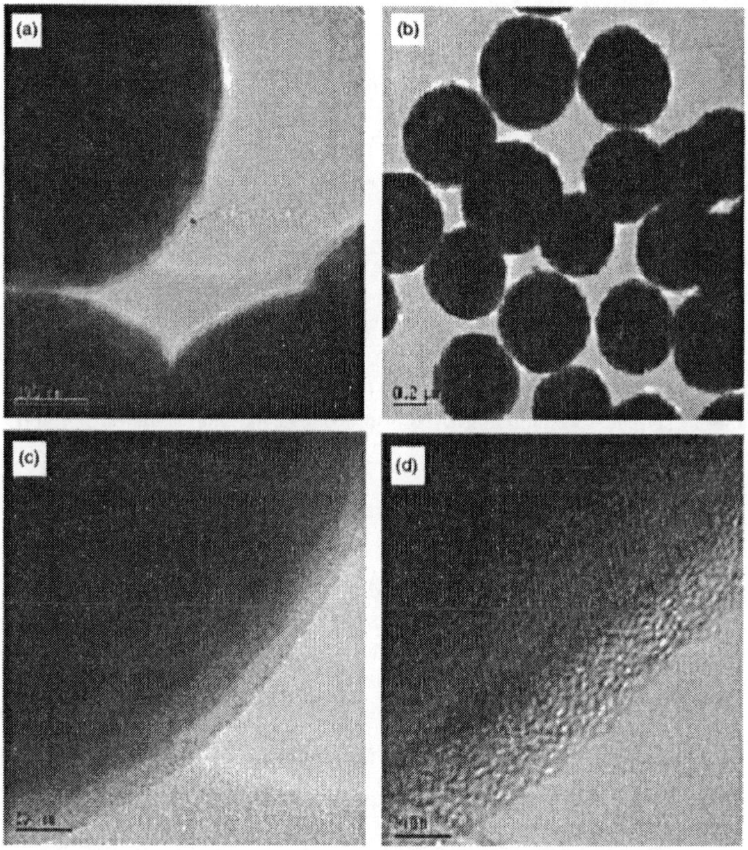

FIGURE 8.10 Bright-field HRTEM images of the coated $YYbErO_2S$: (a) image of the original, uncoated $YYbErO_2S$ nanoparticles with a particle size distribution ranging from 400 to 600 nm; (b), (c) image of the plasma coated $YYbErO_2S$ nanoparticles; (d) nanoparticles showing crystal lattice and the amorphous nanoparticle surfaces.

III carbon nanofibers were chosen as the substrate in this experiment.[13] The nanofibers are stirred in the sample container at the middle of the chamber, and thus the surface of the particles can be continuously renewed and exposed to the plasma for thin-film deposition during the plasma polymerization process. Two small magnetic bars were used to stir the powder. Because the density of the nanofiber is very small, a mesh cover was used to prove the nanofibers spray out from the sample container during the vigorous stirring. Before the plasma coating, the argon gas was introduced for a surface cleaning processing. After 30 min of ion bombardment, styrene and the C_6F_{14} gas with a controlled ratio were introduced from another gas inlet. At the same time, the argon gas was shut off. The operating pressure was adjusted by the gas/monomer mass flow rate. The total pressure was about 350 mtorr. During the plasma polymerization process, the input power was

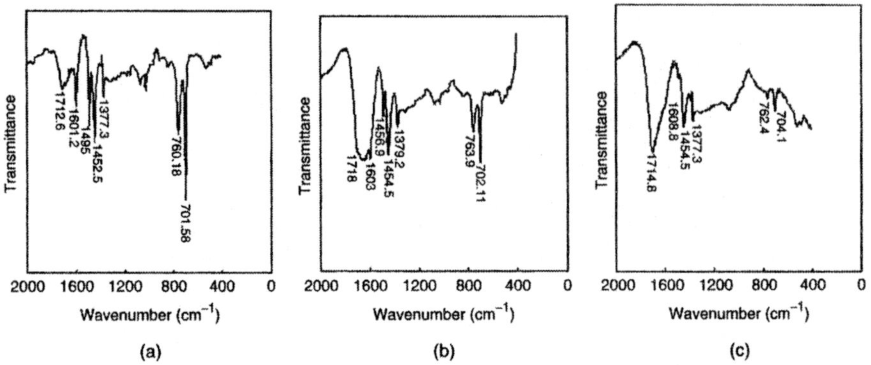

FIGURE 8.11 FTIR of the coated particle (a) condition: Styrene 200 mtorr, MMA 50 mtorr, (b) condition: styrene 150 mtorr, MMA 100 mtorr, (c) condition: styrene 100 mtorr, MMA 150 mtorr.

15 W. The time of plasma treatment was 10 min. Per batch, 0.1 g of powder were treated.

Figure 8.13 shows the bright-field TEM images of the original, uncoated Pyrograf III PR-24 carbon nanofibers. As can be seen in this figure, the carbon nanofibers have multiple walls with similar features to a multiwall carbon nanotube. However, the nanofibers have the tubes at a 20° angle to the fiber axis and thus the tubes terminate as shown in Fig. 8.14. The Pyrograf III PR-24 nanofibers have outside diameters averaging about 70 nm and are quite uniformly distributed. Some nanofibers become curved during their growth with the ends open. The HRTEM image (Fig. 8.13(a)) of the original Pyrograf III PR-24 carbon nanofibers shows the graphite structure with the interlayer spacing $d_{002} = 0.34$ nm. Based on the bright-field TEM and HRTEM images, the wall thickness of the nanofibers can be estimated to be about 20–30 nm. Nanofibers with axially parallel graphite layers and the nanofibers with axially parallel graphite layers oriented at an angle to the tube axis (Fig. 8.13(a)) are observed. The edge dislocations can be seen due to the disorder of the graphite layers (002). It can be seen that both outer and inner surfaces terminate at the graphite (002) layer without an addition of surface layer (the edge is very sharp) for originally uncoated nanofibers (Fig. 8.13(a)). The bright-field and high-resolution TEM images

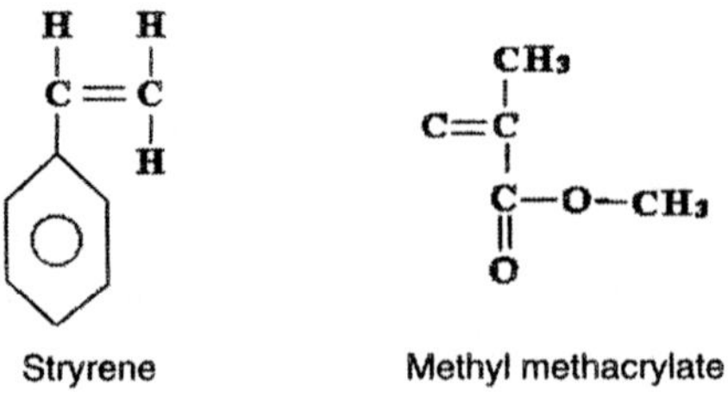

FIGURE 8.12 Molecular structure of the styrene and methyl methacrylate (MMA) monomers.

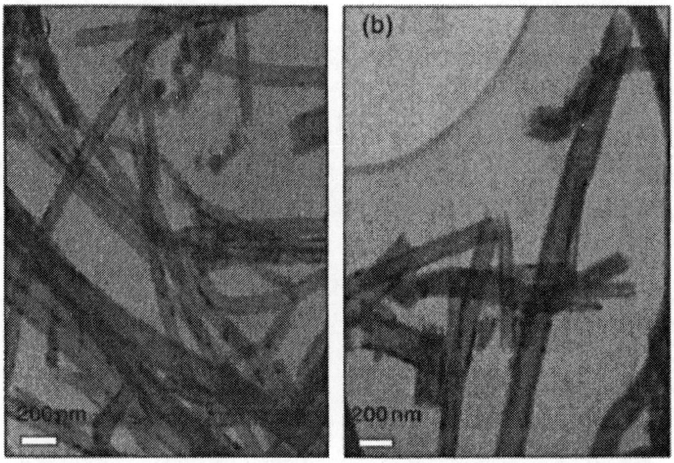

FIGURE 8.13 Bright-field TEM images of the original, uncoated carbon nanofibers.

of these nanofibers after plasma treatment are shown in Fig. 8.14(b) and Fig. 8.15, respectively. An ultrathin-film amorphous layer can be clearly seen over the surfaces of the Pyrograf III PR-24 nanofibers (Fig. 8.14(b)). The thin film is uniform on the surfaces of the nanofiber (Fig. 8.14(b)). The thickness of ultrathin film is approximately 2–7 nm all the way surrounding the nanofiber surfaces and is thicker and more uniform than the disturbance (<1 nm) on the outer surface of carbon nanofibers

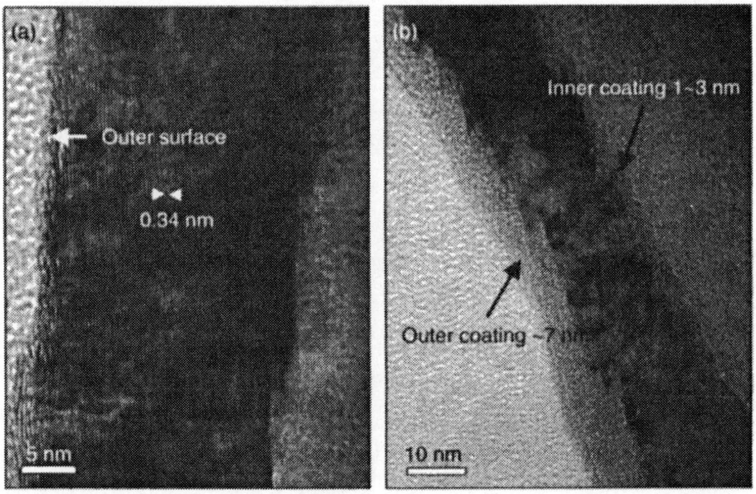

FIGURE 8.14 HRTEM images of nanofibers: (a) the fragments of the wall with inclined planes (002) showing lattice space on the outer and inner surfaces of uncoated nanofibers with slight disturbance (<1 nm) on the surface; (b) an ultrathin film can be observed on the surface of coated nanofibers.

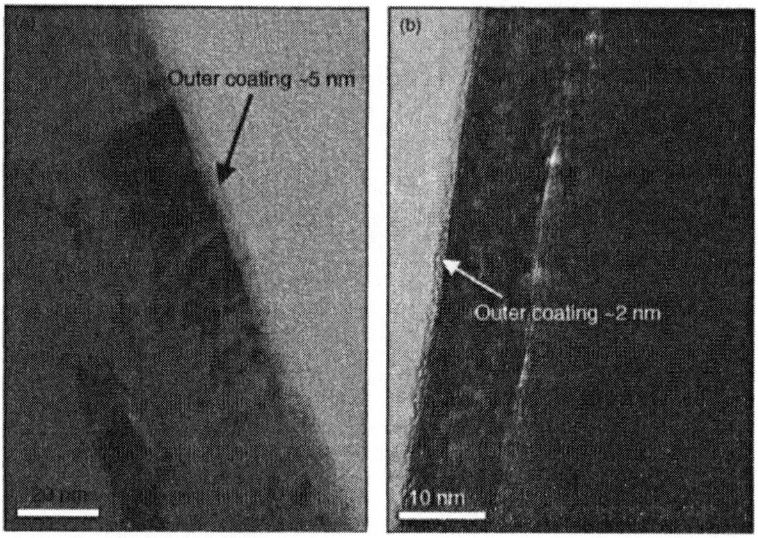

FIGURE 8.15 The bright-field TEM (a) and HRTEM images (b) showing the coating on the surface of the nanofibers. The lattice image of carbon can be clearly seen with an extremely thin layer of polymer film (~2 nm) on the outer surface of the coated Pyrograf III PR-24 carbon nanofiber.

(Fig. 8.14(a)). This coating can only be attributed to be a remarkably uniform layer of the plasma coating. In Fig. 8.14(b), we show the HRTEM image of a coated Pyrograf III PR-24 carbon nanofiber. The lattice image of graphite can be clearly seen with an extremely thin layer of polymer film on its surface.

To confirm the TEM observations shown in Fig. 8.14 and Fig. 8.15, TOFSIMS was carried out to study the surface films of the nanofibers. Figure 8.16 and Fig. 8.17 show the positive TOFSIMS spectra of coated and uncoated nanofibers. In Fig. 8.17, one can see that the spectra of the positive ion from the uncoated nanofibers has strong peaks of functional groups such as C_1, C_2, C_3, C_4, $C_7H_7^+$ and $C_{10}H_8^+$, indicating that the surface of nanofibers contains a hydrocarbon. Because the plasma coating of polystyrene contains only carbon and hydrogen, the hydrocarbon from the surface of nanofibers will cause a problem when trying to identify the coating from the nanofibers surface. In order to solve this problem, a small amount of C_6F_{14} was added to copolymerize with the styrene monomer. In Fig. 18, one can see that the spectrum of the positive ion from the coated nanofibers has strong peaks such as CF^+, C_2F^+, $C_4F_6^+$, $C_3F_7^+$, $C_4F_7^+$, and $C_5F_7^+$, indicating fluorine in the coating film. The fluorine can come only from the monomer, which strongly proves the existence of the coated film.

8.3.5 Plasma Coating of an Aligned Carbon Nanotube Array

Compared with other CNT materials, aligned carbon nanotube array (ACNT) has better organization and orientation. It will be easier to use in application areas. But CNT is

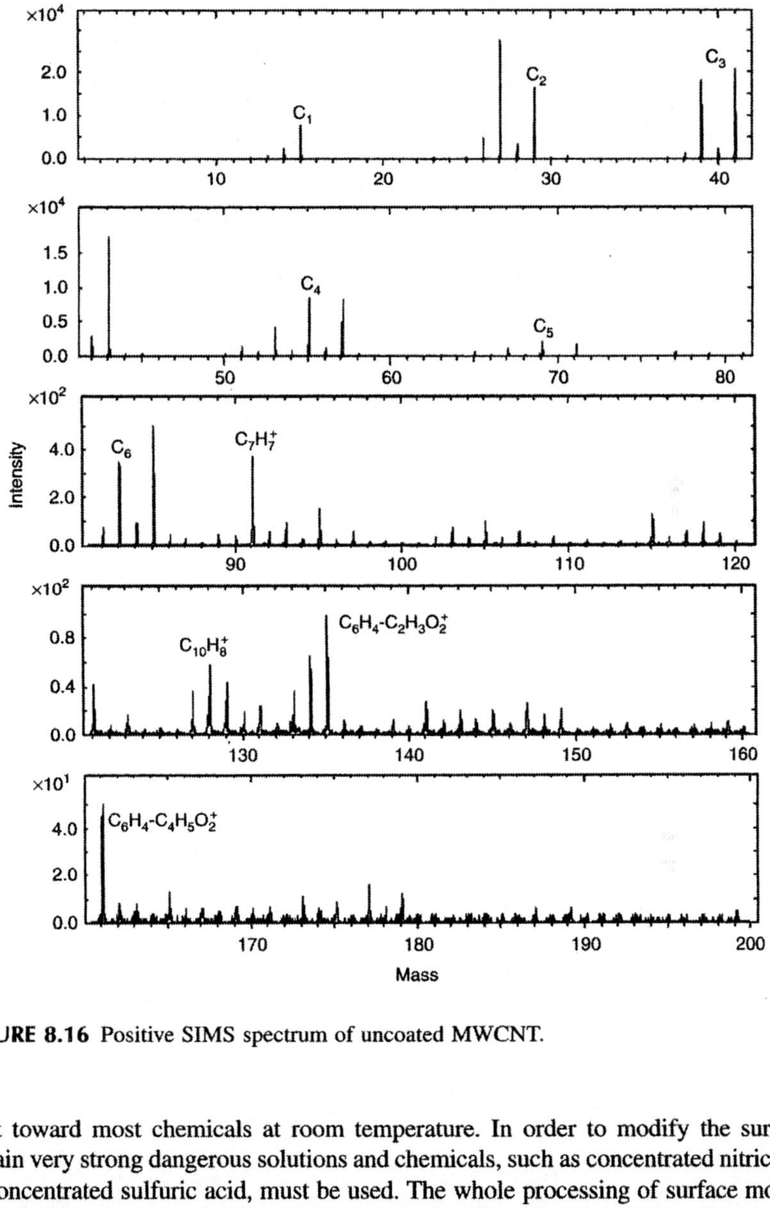

FIGURE 8.16 Positive SIMS spectrum of uncoated MWCNT.

inert toward most chemicals at room temperature. In order to modify the surface, certain very strong dangerous solutions and chemicals, such as concentrated nitric acid or concentrated sulfuric acid, must be used. The whole processing of surface modification will last about 10–20 h. Especially for ACNT, strong chemicals will not only modify the surface of the CNT, but also damage its unique forest structure. In order to find a better way to solve this problem, the plasma modification method was chosen. Compared with the nanofibers, ACNT is more suitable for plasma treatment. All the tubes grow along the same direction on the two-dimensional flat substrate. The gap between each individual tube can let the monomer gas easily pass through. The ACNT chip is simply put at the center of the plasma chamber. Other parameters used for

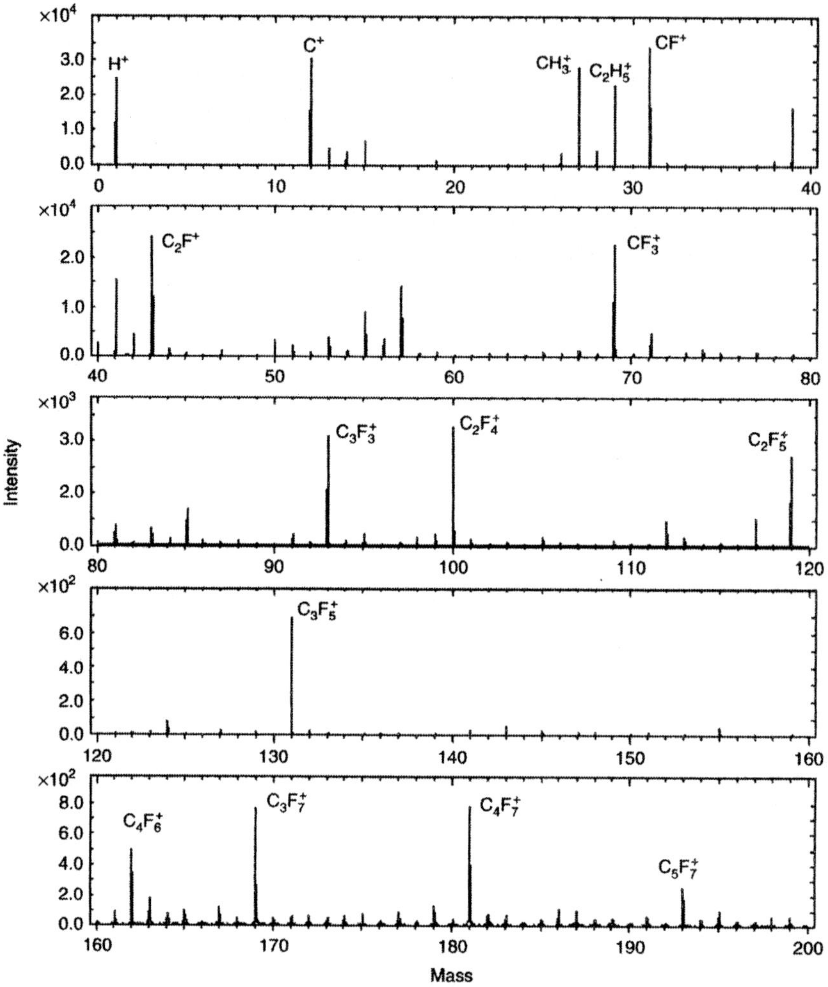

FIGURE 8.17 Positive SIMS spectrum of coated MWCNT.

ACNT processing are the same as for nanofiber processing, which were mentioned in Section 8.4.

Figure 8.18(a) is the SEM image of the coated ACNTs. Compared to uncoated ACNTs (Fig. 8.18(b)), one can see that the surface morphology exhibits smooth and round curvatures, indicating coated polymer films.

HRTEM was used to further identify the coating layer, Figure 8.19(a) is the HRTEM image of the uncoated ACNTs. In this figure, the silicon substrate, aligned CNTs, and the nickel catalyst used in the CNT growth can be well identified. Figure 8.19(b) shows the coating layer clearly on the surfaces of the aligned CNTs. The coating thickness is estimated to be about 10–30 nm.

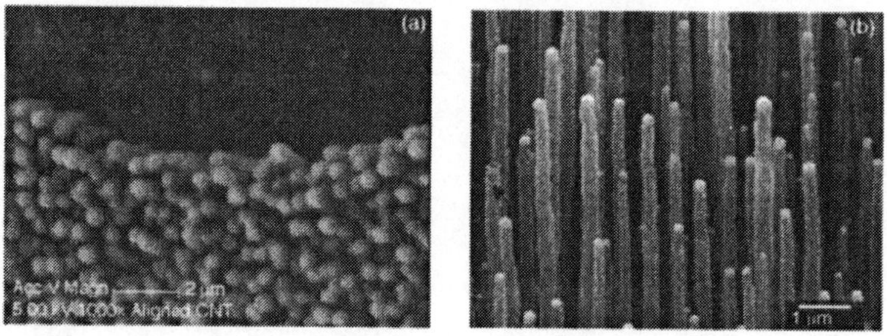

FIGURE 8.18 HRTEM images of aligned carbon nanotubes: (a) coated with polystyrene; (b) uncoated.

8.4 PROCESSING AND CHARACTERIZATION OF NANOCOMPOSITE MATERIALS

Plasma coating of carbon nanofibers for enhanced dispersion and interfacial bonding in polymer composites is discussed in this section. Carbon nanotubes can potentially be used in many applications because of their desirable bulk properties.[14–19] Recently it has been shown in laboratory scale tests that the physical properties and performance of polymer materials can be significantly improved by the addition of small percentages of carbon nanotubes and nanofibers.[20–23] However, there have not been many successful tests that have used larger percentages of nanofibers as fillers. This problem is associated with dispersing the nanofibers and creating a strong interface between the nanofiber and the polymer matrix.[24–25] The strong interface between the nanofiber and the polymer matrix is essential to transfer the load from the matrix to the nanofibers and thereby to enhance the mechanical properties of the composite. In addition, the as-produced nanofibers usually form as aggregates that behave differently in response to a load compared to individual nanofibers.[26,27] To maximize the advantage of nanofibers as reinforcing particles in high-strength composites, the aggregates need to be broken up and dispersed to prevent slippage.

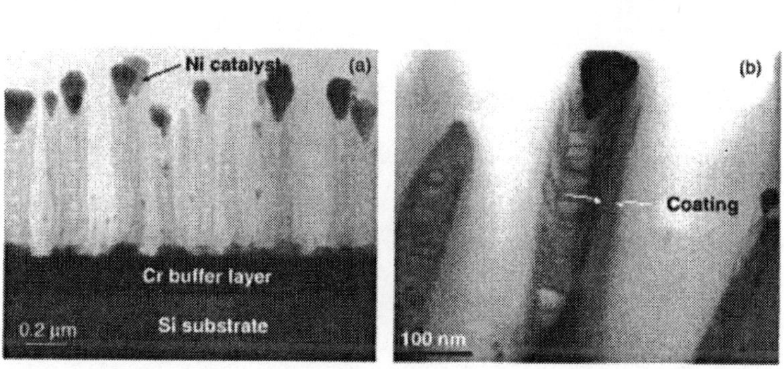

FIGURE 8.19 TEM images of (a) uncoated ACNT and (b) coated ACNT.

A key aspect of being able to manipulate the properties of the nanofibers is the surface treatment of the nanofibers using various processing techniques. In the previous section, we demonstrated the successful plasma deposition of a thin polymer film on the surfaces of carbon nanofibers. In this section, we present the results on the microstructure, dispersion, and mechanical properties of a polymer composite impregnated with coated carbon nanofibers. The fracture morphology of both coated and uncoated nanofiber composites have been identified by scanning electron microscopy (SEM) observation using a Philips XL30 FEG SEM. High-resolution transmission electron microscopy (HRTEM) images were acquired using a JEOL 2010F TEM to show the interface structures that are responsible for the improved properties. TEM samples of the composite were prepared by ultramicrotomy with a cutting thickness of 60 nm.

In this experiment, commercial Pyrograf III carbon nanofibers were used as substrates. The Pyrograf III nanofibers are 70–200 nm in diameter and 50–100 μm long. Polystyrene is used as the monomer for the plasma polymerization because of the good compatibility with the matrix polystyrene. The coating processing used here was the same as the method mentioned in Section 4.1.3.4.

The procedure to form the nanocomposite is described next. Two grams of polystyrene powder was weighed and then mixed mechanically with coated or uncoated nanofibers in appropriate proportions, i.e., 0 wt%, 1 wt%, 3 wt%, and 5 wt%. A solvent (50 ml toluene) was then added to the premixed powders and the powder was thoroughly dispersed ultrasonically using a low power bath–type sonicator. The solution was evaporated until its volume reduced to ~20 ml (the ultrasonic vibration was kept on during this process) and then poured into an 80-mm × 60-mm × 6.5-mm aluminum mold. The mold and the solution were kept at room temperature and dried for 7 days. After the sample was completely dried, it was sectioned into 50-mm × 6-mm × 0.4-mm samples for tensile testing according to ASTM D822-97, "Standard Test Method for Tensile Properties of Thin Plastic Sheeting."

An Instron mechanical testing machine, Model 2525-818, with a 1 mm/min cross head speed was used for the tensile test.[28]

Before the mechanical test, the first observation took place in the solution stage. After ultrasonic mixing, the solutions with coated and uncoated nanofibers were stored in small jars, as shown in Fig. 8.20. A stunning difference at this stage was

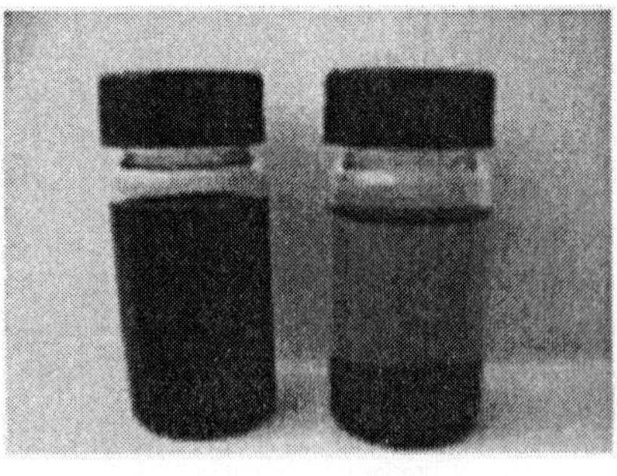

FIGURE 8.20 Solutions with coated (left) and uncoated (right) nanofibers.

that, as shown in this figure, the solution of uncoated nanofibers quickly separated within 24 h (Fig. 8.20, right, nanofibers are precipitated from the solvent and concentrated at the bottom of the jar), while the solution with coated nanofiber remained homogenously suspended as shown in Fig. 8.20 (left) for as long as several months. This phenomenon clearly indicates significantly modified surface behaviors due to plasma coating.

Figure 8.21 shows the strength as a function of nanofiber concentration for both coated and uncoated nanofiber composites. For the uncoated nanofiber composite, the strength of the composite decreases gradually as the nanofiber concentration

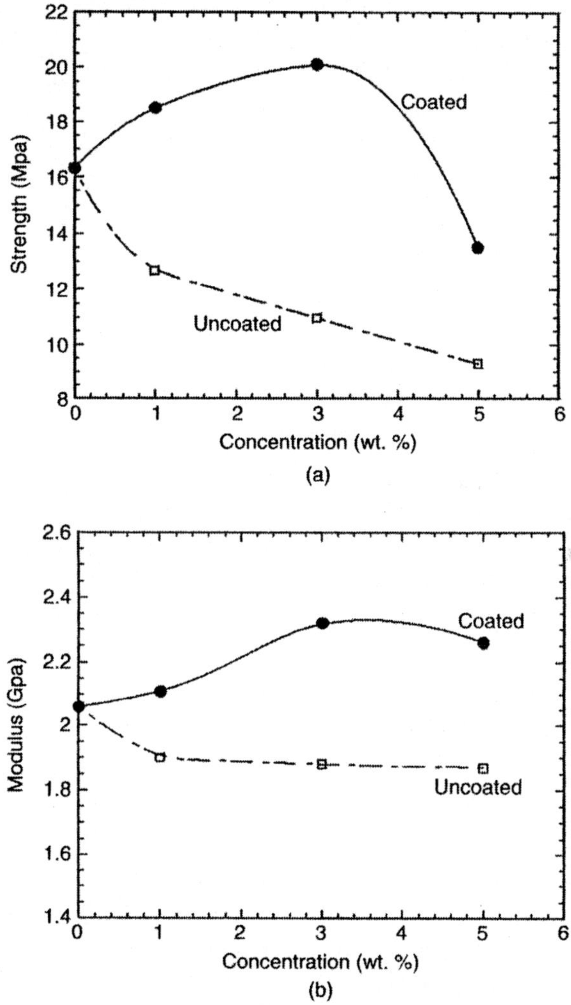

FIGURE 8.21 Properties of a polystyrene nanocomposite: (a) the strength of composite VS CNT concentration; (b) the modulus of composite VS CNT concentration.

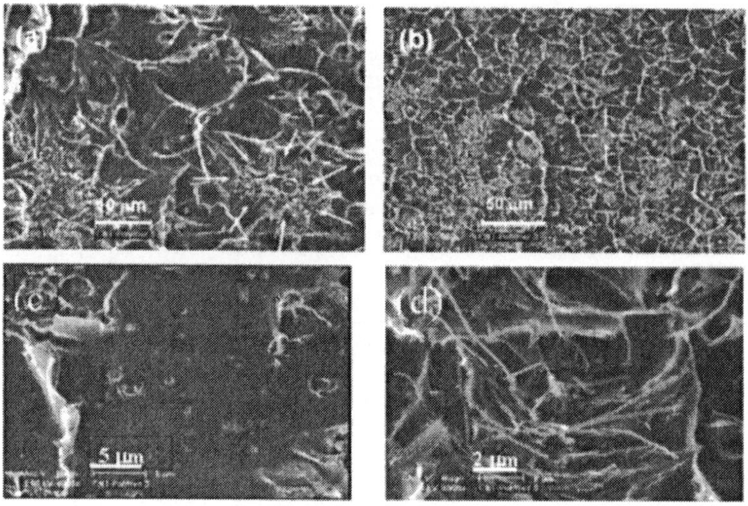

FIGURE 8.22 SEM images of the fracture surfaces of the 3 wt% uncoated sample at different resolution (a), (b), (c), and (d).

increases, while the coated counterpart showed a significant increase in strength. The maximum strength of the coated nanofiber composite takes place at 3 wt% and then decreases up to 5 wt%. The modulus value is shown for both composites in Fig. 8.19(b). A similar trend is seen, which is consistent with the strength values (Fig. 8.21(a)). The decrease in properties above 3 wt% loading may be due to the nanofibers not being initially well dispersed, and due to later agglomeration of the nanofibers in the matrix. It is anticipated that the composite properties will monotonically increase with the wt% loading of nanofibers if the dispersion can be improved and maintained.

Figure 8.22 shows the fracture surfaces of the 3 wt% uncoated sample. The nanofibers are highly clustered in the matrix with approximately a ~10-μm diameter (Fig. 8.22(a)), as indicated by the arrows. These clusters appear to be densely distributed with a small spacing of ~25 μm (Fig. 8.22(b)).

Another important characteristic of the uncoated nanofibers composite is the rather flat fracture surface (Fig. 8.22(c)), indicating the nature of brittle fracture. At these fracture surfaces, severe pullouts of nanofibers are also observed, as shown in Fig. 8.22(d). In sharp contrast, the dispersion is greatly improved in the coated nanofibers composite. Figure 8.23 shows the fracture surfaces of the 3 wt% coated nanofiber composite. The coated nanofibers are well dispersed (Fig. 8.23(a)) in the matrix with a wavy type of fracture surface morphology (Fig. 8.23(b)). The interface structure between the CNTs and polymer matrix was studied by HRTEM for both coated (Fig. 8.24(a)) and uncoated nanofiber (Fig. 8.24(b)) composite samples. The contrast in Fig. 8.24(a) clearly shows the coating layer between the carbon nanofiber and the matrix, whereas the uncoated carbon nanofiber surface is in direct contact with the matrix, as shown in Fig. 8.24(b).

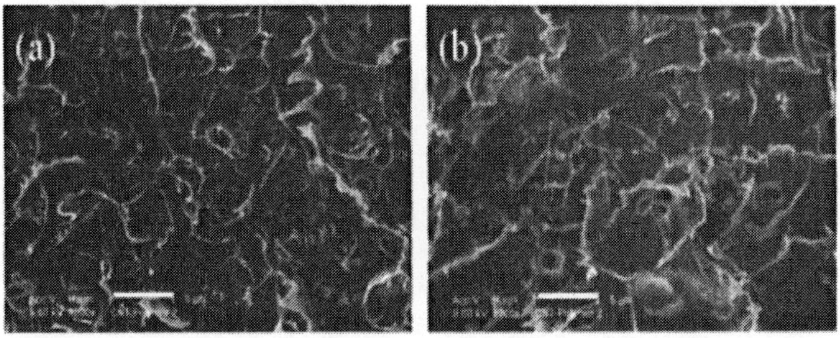

FIGURE 8.23 SEM images of the fracture surfaces of the 3 wt% coated sample (a) and (b).

The central focus of this study was the enhanced interfacial bonding due to plasma-coated thin films on nanofibers. The nature of strengthening in nanofiber-reinforced composites is dependent on the stress transfer between the matrix and nanofiber. For polymers, tensile loading can produce matrix cracking, nanofiber bridging, nanofiber rupture, nanofiber pullout, and debonding. In this experiment, pullouts of nanofibers were observed in the uncoated nanofiber composite, as indicated in Fig. 8.22(d), especially within the cluster regions. As the nanofibers are clustered, the interface area between the matrix and nanofiber is greatly reduced, leading to significantly lowered strength. Furthermore, these clusters act as large voids that are responsible for decreasing the strength of the composite as the nanofiber concentration increases (Fig. 8.21).

As the nanofiber surfaces are modified by plasma coating, the surface energy can be significantly lowered, which can enhance dispersion in the polymer matrix. The well-dispersed nanofibers in the matrix appear to have few clusters and pullouts. In addition, the adhesive film on the nanofiber surface, as shown in Fig. 8.24(a), can

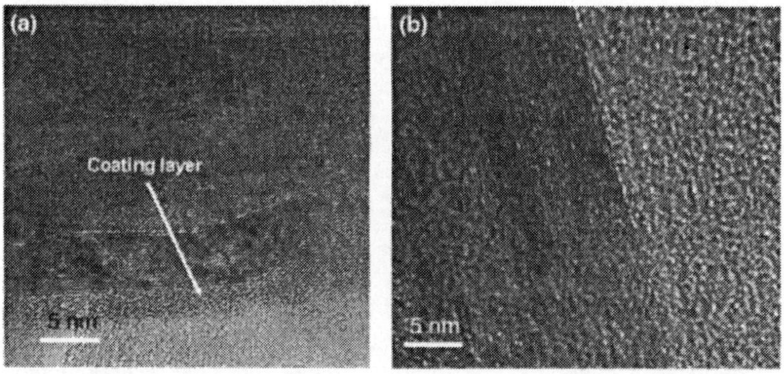

FIGURE 8.24 HRTEM image of the interface structure between the CNTs and polymer matrix: (a) coated; (b) uncoated.

provide enhanced bonding, and therefore contribute to a considerably increased strength in the coated-nanofiber composite. The efficiency of stress transfer is strongly dependent on the maximum value of the shear stress acting at the interface. This stress is also characterized as the interfacial shear strength that depends on the nature of bonding at the interface. As indicated by the interface HRTEM, there is clearly an interfacial adhesion layer due to the coated polymer film on the nanofiber surface. Although a quantitative measure of the interfacial shear strength has not been conducted, the effect of enhanced bonding is evident from the increased composite strength and fracture surface morphology.

Although polystyrene is brittle, a number of studies showed that polystyrene can be macroscopically toughened and manipulated to deform via shear yielding by controlling the microstructure. The improvement of toughness of heterogeneous polystyrene systems is mainly achieved by enhancing the strength of craze and thus the craze resistance, or decreasing the concentrated stress. In the coated nanofiber composite, the improved interfacial bond strength between the nanofiber and matrix due to an adhesive thin film on nanofiber surfaces could increase the strength of the craze. Furthermore, the increase of the interfacial adhesion may suppress the production of voids or flaws in the polymer matrix, which might grow into cracks. Thus, the fracture surface of the coated nanofiber composite exhibits typical shear yielding behavior. In contrast, in the uncoated nanofiber-polymer composite, a rather flat, brittle type of fracture surface occurs (Fig. 8.22(c)), similar to the fracture surface of pure polystyrene. This behavior suggests that the highly clustered nanofibers in the matrix do not contribute to shear yielding, and the uncoated nanofiber composite shows a brittle fracture feature. In addition, the uncoated nanofibers may cause the formation of voids (due to nanofiber pullouts) and defects. This could lower the stress required for craze initiation and thus decrease the craze resistance, consistent with the observation that the strength of the uncoated nanofiber-composite decreases with the increasing percentage of nanofibers (Fig. 8.21).

Overall a unique approach has been developed to enhance the dispersion and interfacial bonding of nanofibers in polymer composites. As a result of plasma coating, carbon nanofibers can be well dispersed in a polymer matrix. Both the fracture behavior and tensile strength data indicate that the well-dispersed nanofibers have contributed to enhanced interfacial shear strength, and therefore have increased the overall strength of the material. It is believed that the strength of the nanofiber composite will also be enhanced based on the identification of the bonding mechanisms between the nanofiber and coating, and the coating and the matrix material.

8.5 SUMMARY

Polymer thin films (such as acrylic acid, MMA, C_6F_{14}, and polystyrene) were deposited on the surfaces of nanoparticles and nanofibers by a plasma polymerization treatment. The average size of the nanoparticles was from several nanometers to several hundreds of nanometers in irregular shapes. HRTEM experiments showed that extremely thin films of the polymer (3–15 nm) were uniformly deposited on the surfaces of the nanoparticles and nanofibers. The HRTEM results were confirmed by time-of-flight secondary ion mass spectroscopy and IR. The deposition mechanisms and the effects of plasma treatment parameters were discussed. The experiments showed that the plasma processing is a powerful tool for surface modification of nanoparticles. However, there are still several challenges that must be faced before the plasma polymerization method can be widely used as a standard procedure for nanoparticle surface modification.

The first problem is to design a more efficient stirring method to improve the reproduction of the coating for each run. In plasma processing, if the substrate is a flat surface, controlling the plasma processing parameters will give good control of the film properties, including the thickness and the composition. This technique has been widely used in the semiconductor industry for the surface modification of wafers. But, for nanoparticles, the stirring method plays a very important role in the overall processing. The methods used here are magnetic stirring and mechanical stirring. The stirring parameters are difficult to control, which affects the control of the coatings. One approach that is being investigated is to disperse the nanoparticles in the monomer that is injected into the plasma system. In the future, an efficient method to stir the powder will be a key for improvement of the plasma technique when used for nanoparticle surface modification.

Another problem is standardization of the plasma system. The parameters for plasma processing are highly system dependent. The parameters obtained from one system may not be optimal when used in another system, even when the monomers and substrates are exactly the same. This is a disadvantage for data sharing and comparison.

Finally, to take full advantage of the plasma surface coating technique, it is suggested that many different monomers be investigated for use as coatings. With different monomers, different experimental conditions also need to be tried in order to get a uniform film on the nanosurfaces. Although polymers have been used widely in plasma treatment, metals or conducting materials can also be used as coatings to explore how the electrical conductivity and electrical properties of the coated nanoparticles change.

PROBLEMS

1. Briefly explain what plasma polymerization processing is and what the important parameters for this processing are.
2. Compared with other functionalization methods for carbon nanotubes, what are the advantages of using plasma processing?
3. A company has some amorphous carbon substrates. In order to modify the surface, plasma polymerization processing is suggested. In order to get better adhesion between the coating and substrates, which condition will you will try to use for this processing: (a) high power, low flow ratio or (b) low power, high flow ratio? If the substrate is polystyrene, which condition you will use for this case? (Assume that the monomer used here is styrene, and that other processing parameters are the same.)
4. A company wants to coat a glass substrate with poly-pyrrole to make the surface ionically conductive. If the regular plasma polymerization processing is used, could we achieve this objective? How could you modify the plasma processing to achieve the objective?

5. What are the advantages and limitations of plasma polymerization processing?
6. List as many parameters as you can think of that control the plasma processing. Explain what effect you think each parameter would have on the coating of nanoparticles. For example, faster or more vigorous stirring of the nanoparticles may expose the particle to the plasma longer and make the coating thicker and more uniform and prevent clumping of particles.

REFERENCES

1. Siegel, R.W., Nanostructured Materials-mind Over Matter, *Nanostruct. Mater.*, 3(1), 1–18, 1993.
2. Hadjipanayis, G.C. and Siegel, R.W., *Nanophase Materials, Synthesis-Properties Applications,* Kluwer Academic, Dordrecht, the Netherlands, 1994.
3. Whitesides, G.M., Mathias, J.P., and Seto, C.T., Molecular Self-Assembly and Nanochemistry—a Chemical Strategy for the Synthesis of Nanostructures, *Science,* 254, 1312–1319, 1991.
4. Stucky, G.D. and MacDougall, J.E., Quantum Confinement and Host/Guest Chemistry: Probing A New Dimension, *Science,* 247, 669–671, 1990.
5. Gleiter, H., Nanostructured Materials: State of the Art and Perspectives, *Nanostruct. Mater.,* 6(1), 3–14, 1995.
6. Wolde, A.T., Ed., *Nanotechnology,* STT Netherlands Study Center for Technology Trends, The Hague, the Netherlands, 1998.
7. Inagaki, N., Tasaka, S., and Ishii, K., Surface Modification of Polyethylene and Magnetite Powders by Combination of Fluidization and Plasma Polymerization, *J. Appl. Polym. Sci.,* 48(8), 1433–1440, 1993.
8. Bayer, C., Karches, M., Mattews, A., and Von Rohr, P.R., Plasma Enhanced Chemical Vapor Deposition on Powders in a Low Temperature Plasma Fluidized Bed, *Chem. Eng. Technol.,* 21(5), 427–430, 1998.
9. van Ooij, W.J., Luo, S., Zhang, N., and Chityala, A., in *Proceedings International Conference on Advanced Mfg. Technology,* Science Press, New York, 1999, p. 1572.
10. van Ooij, W.J. and Chityala, A., *Surface Modification of Powders by Plasma Polymerization,* Mittal, K.L., Ed., VSP, Utrecht, the Netherlands, 2000, pp. 243–253.
11. van Ooij, W.J., Zhang, N., and Guo, S., in *Fundamental and Applied Aspects of Chemically Modified Surfaces,* Blitz, J.P and Little, C.B., Ed., Royal Society of Chemistry, Cambridge, U.K. 1999, pp. 191–211.
12. Shi, D., He, P., Lian, J., Wang, L.M., and van Ooij, W.J., Plasma Deposition and Characterization of Acrylic Acid Thin Film on Zno Nanoparticles, *J. Mater. Res.,* 17(10), 2555–2560, 2002.
13. Shi, D., Lian, J., He, P., Wang, L.M., van Ooij, W.J., Schulz, M., Liu, Y.J., and Mast, D.B., Plasma Deposition of Ultrathin Polymer Films on Carbon Nanotubes, *Appl. Phys. Lett.,* 81(27), 5216–5218, 2002.
14. Iijima, S., Helical Microtubules of Graphitic Carbon, *Nature,* 354, 56–58, 1991.
15. Baughman, R.H., Cui, C., Zakhidov, A.A., Iqbal, Z., Barisci, J.N., Spinks, G.M., Wallace, G.G., Mazzoldi, A., De Rossi, D., Rinzler, A.G., Jaschinski, O., Roth, S., and Kertesz, M., Carbon Nanotube Actuators, *Science,* 284, 1340–1344, 1999.
16. Gao, M., Dai, L., Baughman, R.H., Spinks, G.M., and Wallace, G.G., *Electroactive Polymer Actuators and Devices,* SPIE Proceedings, 2000, p. 18–24.

17. Hadjiev, V.G., Iliev, M.N., Arepalli, S., Nikolaev, P., and Files, B.S., Raman Scattering Test of Single-Wall Carbon Nanotube Composites, *Appl. Phys. Lett.*, 78(21), 3193–3195, 2001.
18. Liu, C., Cheng, H.M., Cong, H.T., Li, F., Su, G., Zhou, B.L., and Dresselhaus, M.S., Synthesis of Macroscopically Long Ropes of Well-Aligned Single-Walled Carbon Nanotubes, *Adv. Mater.*, 12(16), 1190–1192, 2000.
19. Walters, D.A., Casavant, M.J., Qin, X.C., Huffman, C.B., Boul, P.J., Ericson, L.M., Haroz, E.H., O'Connell, M.J., Smith, K., Colbert, D.T., and Smalley, R.E., In-plane-aligned Membranes of Carbon Nanotubes, *Chem. Phys. Lett.*, 338(1), 14–20, 2001.
20. Zhu, J., Kim, J.D., Peng, H.Q., Margrave, J.L., Khabashesku, V.N., and Barrera, E.V., Improving the Dispersion and Integration of Single-Walled Carbon Nanotubes in Epoxy Composites through Functionalization, *Nano Lett.*, 3(8), 1107–1113, 2003.
21. Tostenson, E.T. and Chou, T.W., Aligned Multi-Walled Carbon Nanotube-Reinforced Composites: Processing and Mechanical Characterization, *J. Phys. D, 35*, L-77, 2002.
22. Cadek, M., Coleman, J.N., Barron, V., Hedicke, K., and Blau, W.J., Morphological and Mechanical Properties of Carbon-Nanotube-Reinforced Semicrystalline and Amorphous Polymer Composites, *Appl. Phys. Lett.*, 81(27), 5123–5125, 2002.
23. Franklanda, S.J.V., Harikb, V.M., Odegarda, G.M., Brennerc, D.W., and Gatesd, T.S., The Stress–Strain Behavior of Polymer–Nanotube Composites from Molecular Dynamics Simulation, *Composites Science and Technology*, 63(11), 1655–1661, 2003.
24. Lau, K.-T. *Chem. Phys. Lett.*, 370, 399, 2003.
25. Mamedov, A.A., Kotov, N.A., Prato, M., Guldi, D.M., Wicksted, J.P., and Hirsch, A., Molecular Design of Strong Single-Wall Carbon Nanotube/Polyelectrolyte Multilayer Composites, *Nat. Mater.*, 1, 190–194, 2002.
26. Ajayan, P.M., Nanotubes from Carbon, *Chem. Rev.*, 99(7), 1787–1800, 1999.
27. Ajayan, P.M. and Zhou, O.Z., in *Carbon Nanotubes*, Dresselhaus, M.S., and Avouris, Ph., Eds., Springer-Verlag, Berlin, 2001.
28. Shi, D., Lian, J., He, P., Wang, L.M., van Ooij, W.J., Schulz, M., Liu, Y.J., and Mast, D.B., Plasma Coating of Carbon Nanofibers for Enhanced Dispersion and Interfacial Bonding in Polymer Composites, *Appl. Phys. Lett.*, 83(25), 5301–5303, 2003.

9 Structural Nanocomposites

Hassan Mahfuz

CONTENTS

9.1 INTRODUCTION

The subject of nanocomposites is especially interesting because at least one of a nanocomposite's phases has one or more dimensions—length, width, or thickness—in the nanometer range, which is usually defined as 1 to 100 nm. This is the range where phenomena associated with atomic and molecular interactions strongly influence the macroscopic properties of the materials, but this is also the length scale where our knowledge of how to synthesize and process materials is the weakest. Nevertheless, it is very well known that the catalytic, mechanical, electronic, optical, and other properties of a material can significantly and favorably be altered when that material is fashioned from nanoscale building blocks.[1–6] For instance, nanocrystalline copper is up to five times harder than conventional copper, and ceramics, which normally are brittle, can be made more easily deformable if their grain size is reduced to low nanometer range. Such improved properties can also be incorporated into nanocomposites, in which the building blocks—say, nanoscale metal or carbon particles, or nanometer-thick filaments or sheets of a ceramic—are dispersed in a matrix of another material, such as a polymer. Because the building blocks of a nanocomposite are nanoscale, they have an enormous surface area, and therefore significantly create large interfaces between the two intermixed phases. The special properties of the nanocomposite arise from the interaction of its phases at these interfaces. The challenge, however, lies in fabricating these materials with the required atomic specifications.

One of the important areas for future investments in research identified by the National Nanotechnology Initiative (NNI) is called "beyond nano," which notes that the advances at the nanoscale will be meaningless if they cannot be interfaced well with the technology at larger material components, systems, and architectures to produce usable devices.[7] One of the primary objectives of the NNI is to integrate nano-objects and nanoscale phenomena into larger hierarchical systems. Development of large structural-level laminates from nanoinfused polymers is a step in the right direction.

With our continuing quest for lighter and stronger composites, the demand for new types of materials is increasing. No longer can traditional fibrous composites fulfill our stringent requirements, nor can they be engineered at the continuum level, which controls properties at the molecular or atomic level. However, it is well known that molecular forces and bonding, the interaction between the interfaces, and the physical phenomena occurring at this level will dictate the aggregate properties of materials. In this nanocrystalline state, the solids contain such a high density of defects that the spacing between them approaches interatomic distances and a large fraction of the atoms sits very close to a defect.[8–10] Consequently nanocrystalline materials are exceptionally strong, hard, and ductile at high temperatures, wear resistant, corrosion resistant, and chemically very active.[11,12]

It has been established in recent years that polymer-based composites reinforced with a small percentage of strong fillers can significantly improve the mechanical, thermal, and barrier properties of pure polymer matrix.[13–18] Moreover, these improvements are achieved through conventional processing techniques without any detrimental effects on processability, appearance, density, or aging performance of the matrix. These composites are now being considered for a wide range of applications including the packaging, coating, electronics, automotive, and aerospace industries. While nanoparticles have attractive attributes, their use in structural composites, which are relatively large in dimension, is almost nonexistent.

In recent years, studies have been conducted to see how traditional and nontraditional modifiers influence the thermal and mechanical properties.[19] A conventional modifier, microscale nylon particles, and nonconventional modifiers, nanoclays with different surface chemistries, were incorporated into an epoxy matrix used for prepeg-based, fiber-reinforced composites. The incorporation of particle modifiers resulted in changes in the interlaminar shear stress (ILSS) and char yield. In various other studies,[20–25] it has been established that the addition of a small amount of nanoparticles (<5 wt%) to a matrix can increase the strength, toughness, dimensional stability, and resistance to thermal degradation without compromising the weight or processability of the composite. This concept of infusion of nanoparticles into a polymer by low volume or weight fraction forms the foundation of the manufacturing of structural nanocomposites. Three routes are generally followed: (1) modification of the matrix through particle infusion and reinforcement with regular fibers;

(2) dispersion of acicular nanoparticles such as carbon whiskers, carbon nanotubes (CNT), or carbon nanofibers (CNF) into a textile polymer precursor, and melt extrusion into filaments; and (3) infusion of nanoparticles into liquid polymer foam and construction of sandwich panels using nanophased foam materials. A detailed description of the first route and a brief description of the other two routes are given in the following sections.

9.2 MATRIX MODIFICATION

A systematic study has been carried out to investigate matrix properties by introducing micro- and nanosized SiC fillers into an epoxy matrix. The study has revealed that with equal amount of loading, nanoparticle infusion brings about better thermal and mechanical properties to the matrix than what is usually given by the microfillers' infusion. The nanophased matrix is then utilized in a vacuum-assisted resin transfer molding (VARTM) set up with satin weave carbon fiber preforms to fabricate laminated composites. The resulting structural composites have been tested under flexural and tensile loads to evaluate mechanical properties. The fillers were nano- and micron-size silicon carbide particles that were mixed with the SC-15 epoxy resin using an ultrasonic processor. The amount of particle loading varied from 1.5 to 3.0% by weight of the resin. Ultrasonic mixing utilized high-energy sonic waves to force an intrinsic mixing of particles with the matrix via sonic cavitations. In parallel, control panels were also fabricated without particle infusion. It has been observed that nanoparticle infusion increases the thermal stability of the system by enhancing cross-linking in the polymer. Nanoparticles also tend to reduce void content of the as-fabricated composites and thus translate into increased mechanical properties. With 1.5 wt% loading, an average of 20–30% increase in mechanical properties has been observed. Fatigue tests were also performed under flexural loading, and the performance of the nanoinfused system was seen to be superior to that of the neat system.

9.2.1 Ultrasonic Mixing

Various techniques such as melt mixing, solution mixing, shear mixing, and mechanical stirring are employed to infuse nanoparticles into a polymer. In addition to these processes, acoustic cavitation is also one of the efficient ways to disperse nanoparticles into virgin materials.[26] In this case, the application of alternating acoustic pressure above the cavitation threshold creates numerous cavities in the liquid. Some of these cavities oscillate at a frequency of the applied field (usually 20 kHz), while the gas content inside these cavities remains constant. However, some other cavities grow intensely under tensile stresses, while yet another portion of these cavities, which are not completely filled with gas, start to collapse under the compressive stresses of the sound wave. In the latter case, the collapsing cavity generates tiny particles of debris and the energy of the collapsed one is transformed into pressure pulses. It is noteworthy that the formation of the debris further facilitates the development of cavitation. It is assumed that acoustic cavitation in liquid develops according to a chain reaction. Therefore, individual cavities on real nuclei develop so rapidly that within a few microseconds an active cavitation region is created close to the source of the ultrasound probe. The development of cavitation processes in the ultrasonically processed melt creates favorable conditions for the intensification of various physiochemical processes. Acoustic cavitation accelerates heat and mass transfer processes such as diffusion, wetting, dissolution, dispersion, and emulsification. Recently it has been reported that there is a clear acceleration of polymer reaction under ultrasound in both catalysed and uncatalysed reactions.[27] In the present investigation, ultrasonic mixing was employed to infuse SiC nanoparticles into part A of the SC-15 epoxy resin.

9.2.2 Manufacturing of Nanocomposites

The fabrication of nanophased carbon/epoxy composites was carried out in three steps. In the first step, spherical SiC nanoparticles of about 29 nm in diameter (manufacturer: MTI Corporation, Inc., Richmond, CA) were ultrasonically mixed with part A (mixture of diglycidylether of bisphenl A, 60 to 70%, aliphatic diglycidylether, 10 to 20%, and epoxy toughner, 10 to 20%) of SC-15 epoxy resin (manufacturer: Applied Poleramic, Inc., Benicia, CA). SC-15 is a two-phase toughened epoxy resin system that cures at room temperature and is extensively used in vacuum-assisted resin transfer molding (VARTM) processes. The loading of nanoparticles ranged from 1.5 to 3.0% by weight of the resin. The mixing was carried out in a sonics vibra cell ultrasonic liquid processor (Ti-horn, frequency = 20 kHz, intensity = 100 W/cm^2), as shown in Fig. 9.1. The mixing was carried out at 55% of the amplitude for about 30 min. At this time, the dispersion of nanoparticles seemed uniform through visual observation. In order to avoid rise in temperature during sonication, cooling was employed by submerging the mixing beaker in a liquid chamber where temperature was autocontrolled through a computer. In the next step, Part B (hardener, cycloaliphatic amine, 70 to 90% and polyoxylalkylamine, 10 to 30%) was added with the mixture at a ratio of 3:10 and the mixing was carried

FIGURE 9.1 Vibra-cell ultrasonic processing.

FIGURE 9.2 A typical VARTM setup.

out mechanically for about 10 min using a high-speed mechanical stirrer. In the final step, the reaction mixture was used in a VARTM set up[28–30] with satin weave carbon fiber preforms to fabricate carbon/epoxy nanocomposite panels, as shown in Fig. 9.2. In order to carry out studies with particle size, micron-sized SiC particles (average particle size ~1 μm; supplier: Atlantic Equipment Engineers, Bergenfield, NJ) were also infused with SC-15 parallel to nanoparticle infusion. Modified part A was then mixed with part B, and was cast into rectangular molds to cure to solid panels. Test coupons were extracted from each category of panels to conduct various chemical and mechanical tests.

9.2.3 Micron- and Nanosized SiC Particle Infusion

As stated earlier, the investigation began with the infusion of nano- and micron-sized particles into SC-15 epoxy matrix. Tensile response of these systems is shown in Fig. 9.3. As the figure shows, epoxy with nanoinfusion has the highest strength and stiffness when compared with the other three systems, namely neat epoxy, epoxy irradiated with ultrasonic cavitation, and epoxy with 1.5 wt% of micron-sized SiC particles. As stated earlier, a batch of epoxy matrix was also irradiated for equal length of time without any particle infusion to evaluate the effect of sonication on the epoxy. Figure 9.3 and Table 9.1 show that sonicated epoxy performs as well as the micron-SiC infused system, and much better than the neat epoxy. We see that sonication alone has contributed to about 18% and 3% increase in stiffness and strength, respectively. The reason for such change in stiffness of the matrix due to ultrasound irradiation comes from the fact that the ultrasound irradiation enhances the homogeneity of the reaction mixture part A of the SC-15 epoxy resin. As we mentioned earlier, part A of SC-15 epoxy resin contains the mixture of diglycidyl ether of bisphenol A, aliphatic diglycidyl ether, and epoxy toughner. The ultrasound irradiation helps in molecular mixing of these components and the formation of

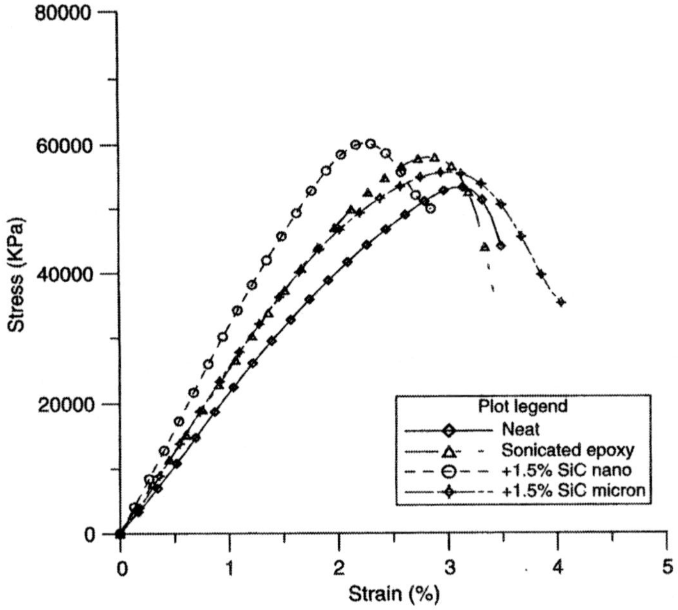

FIGURE 9.3 Tensile response of SiC infusion in SC-15 epoxy.

TABLE 9.1
Tensile Response of Neat and Nanoinfused SC-15 Epoxy Composites

Material	Tensile Modulus (GPa)		Gain/Loss Modulus (%)	Tensile Strength (MPa)		Gain/Loss Strength (%)
Neat Epoxy	2.16	2.14	—	57.30	55.00	—
	2.12			55.20		
	2.13			52.50		
Sonicated Epoxy	2.54	2.53	+18.22	57.13	56.57	+2.85
	2.50			55.46		
	2.55			57.14		
1.5% μm SiC-Epoxy	2.60	2.55	+19.15	56.52	56.21	+2.20
	2.5			55.91		
	2.54			56.21		
1.5% Nano SiC-Epoxy	3.10	3.10	+44.85	60.87	63.70	+15.81
	3.21			62.54		
	3.00			67.68		

reactive species, which ultimately leads to an increase in the cross-linking in the polymer when mixed with part B of SC-15. This effect is more prominent when each of the highly reactive surfaces of the nanoparticle is efficiently coated with part A of SC-15 resin and reacted with the hardener, part B of SC-15 resin. When the matrix is sonicated with SiC nanoparticles in it, it is observed that the enhancement in stiffness and strength significantly increases to around 45% and 13%, respectively, over the neat system. The corresponding improvement over the sonicated-epoxy system, however, is 22% and 12%, respectively, which we believe is entirely due to the nanoparticle infusion. On the other hand, the infusion of micron-sized particles with identical loading (1.5 wt%) did not contribute to any improvement of properties over the sonicated-epoxy system. Although we see some improvements in properties with respect to the neat system, the contribution is actually coming from the sonic cavitation effect rather than from the particle infusion. We also note from Fig. 9.3 that the percent strain to failure of the nanoinfused system reduces to 2.2 from around 3.0, which is typical with the other systems. A quick calculation based on 1.5 wt% loading shows that the particle volume fraction in the composite in both cases is around 0.53%. With such low particle volume fraction, it is very difficult to visualize a 44% increase in stiffness of the nanoparticle-infused system. The underlying phenomenon thus points to interatomic/molecular activities in the polymer that is taking place in the presence of the nanosized particles. The small size of these particles effectively increases the surface energy significantly, and subsequently enhances crystallinity and cross-linking of the polymer. This in turn translates into improved stiffness and strength of the resulting composites. The overall inference from the study was that with low amounts of loading in an epoxy matrix, nanoparticles offer significant benefits over the parallel microparticle infusion.

9.2.4 Thermal Analysis of Carbon/Epoxy Nanocomposites

In the next phase of the investigation, epoxy matrix infused with 1.5 wt% of SiC nanoparticles were reinforced with satin woven carbon fibers in a traditional composite manufacturing procedure (VARTM). Coupons extracted from the resulting fibrous composite laminates were then subjected to thermal and mechanical tests. Thermogravimetric analysis (TGA) of various specimens was carried out under nitrogen gas atmosphere at a heating rate of 10°C/min on a TA Instruments, Inc. apparatus. The real-time characteristic curves were generated by the Universal Analysis 200 (TA Instruments, Inc.) data acquisition system. Data from a typical thermogravimetric analysis of the fibrous composite are shown in Fig. 9.4(a).

In an attempt to determine the effect of the amount of particle loading, a separate batch of materials was manufactured with 3 wt% in addition to 1.5 wt% loading. The percent weight loss with temperature of these two systems, along with the neat composite, is shown in Fig. 9.4(a). In the present study, 50% of the total weight loss is considered the structural destabilization point of the system. It is also common practice to consider 50% weight loss an indicator for structural destabilization as given in References 31 and 32. Figure 9.4(b) clearly shows that the neat sample without SiC nanoparticles is stable up to 355°C, whereas in Fig. 9.4(c) shows the samples with 1.5% loading are stable up to 365°C. The reason for this increase in

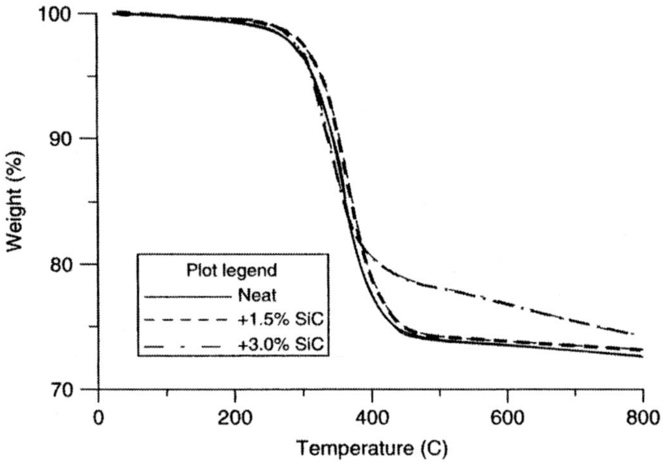

FIGURE 9.4(a) TGA comparison graph of neat vs. nanophased systems.

the thermal stability is due to the increase in cross-linking of the epoxy resin in the presence of SiC nanoparticles and having minimum particle-to-particle interaction. As the loading of SiC particles increases to 3%, as shown in Fig. 9.4(d), the thermal stability drops down to 342°C, which is lower than that of the neat carbon epoxy system shown in Fig. 9.4(b). The reason for the lowering of the temperature of the 3 wt% carbon/epoxy system may be explained macroscopically as a simple colligative thermodynamic effect of an impurity on a bulk solution. Microscopically it may be seen as the result of the perturbation that the SiC introduces to the three-dimensional structure of the polymer. This perturbation weakens the van der Waals interaction between the polymer chains. This affects the stability of the polymer, which is reflected in the lowering of the thermal stability. We believe this perturbation

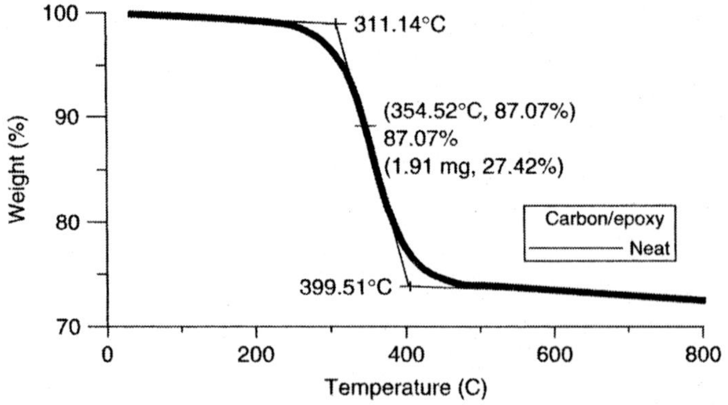

FIGURE 9.4(b) TGA response of neat carbon/epoxy.

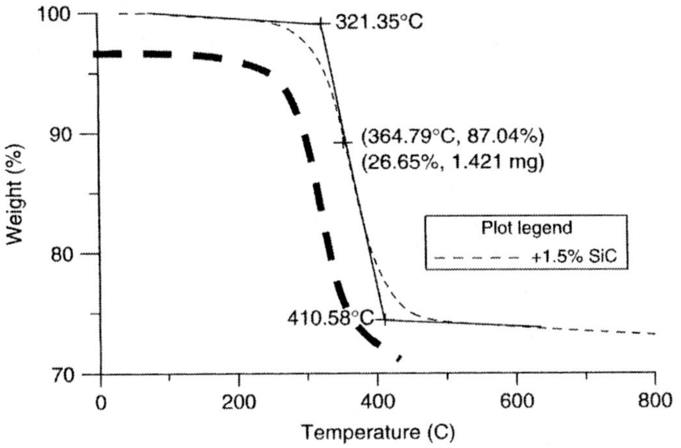

FIGURE 9.4(c) TGA response of carbon/epoxy with 1.5% SiC nanofillers.

begins at a point when the number of particle reaches a certain level and particle-to-particle interaction initiates leading to agglomeration of particles into lumps, which acts as an impurity in the system.

Thermal mechanical analysis (TMA) of the various systems is shown in Fig. 9.5. Figure 9.5 shows that the dimensional changes are more or less identical with the three systems until one reaches the Tg of the material. Beyond this temperature, dimensional change with 1.5 wt% system is quite lower than the neat and 3 wt% system. This dimensional change is also a measure of the coefficient of thermal expansion (CTE) of the material system being tested. Since the dimensional change was measured in the thickness direction of the laminate, the slope of the curves in

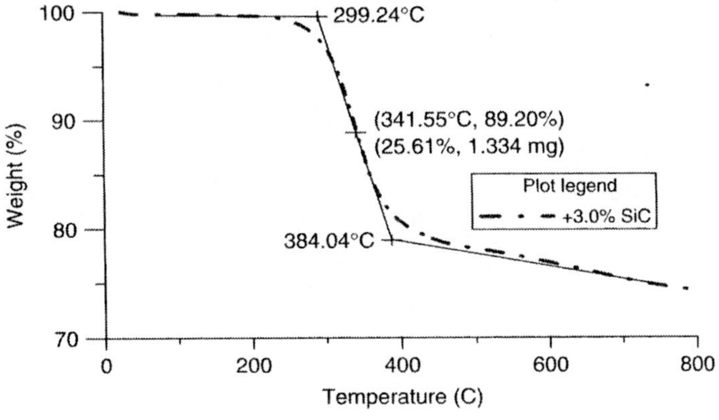

FIGURE 9.4(d) TGA response of carbon/epoxy with 3.0% SiC nanofillers.

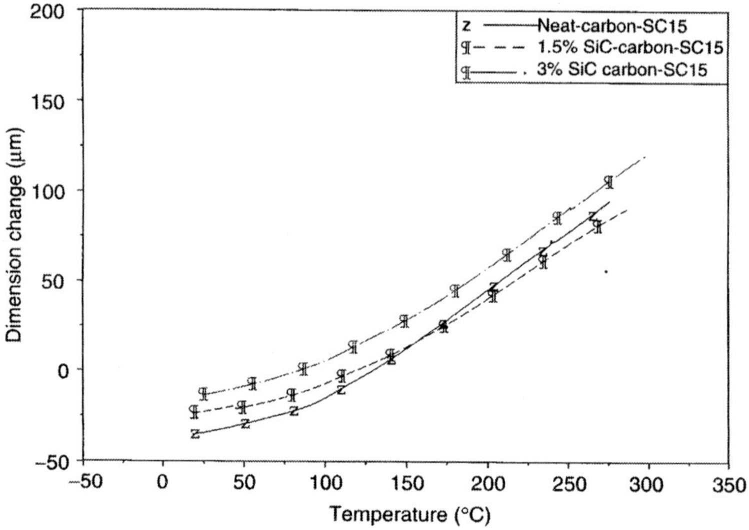

FIGURE 9.5 TMA of carbon/epoxy nanocomposites.

Fig. 9.5 accordingly indicates CTE in that particular direction. As Fig. 9.5 shows, there is a slight reduction in CTE of the 1.5 wt% system over the neat and 3 wt% systems.

9.2.5 Mechanical Tests

Two types of mechanical tests, namely flexure and tensile, were performed to evaluate the bulk stiffness and strength of each of the material systems. A typical stress strain behavior from the flexural test is shown in Fig. 9.6. It is observed that the system with

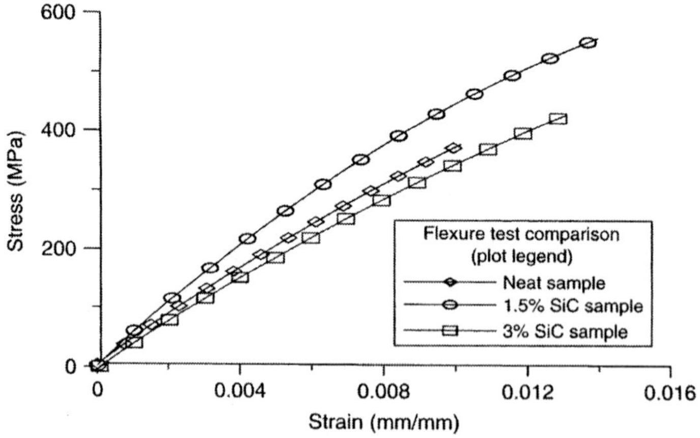

FIGURE 9.6 Flexural testing of carbon/epoxy nanocomposites.

TABLE 9.2
Flexural Test Data for Carbon/Epoxy Composites

Material	Flexural Strength (MPa)		Gain/Loss Strength (%)	Flexural Modulus (GPa)		Gain/Loss Modulus (%)
Neat	334.0	**381.8**	—	45.0	**45.6**	—
	390.0			46.0		
	395.0			39.0		
	390.0			44.0		
	400.0			54.0		
+1.5% SiC	550.5	**531.3**	**+39.2**	50.3	**51.26**	**+12.4**
	505.0			51.0		
	525.0			47.0		
	556.0			59.0		
	520.0			49.0		
+3.0% SiC	424.1	**399.8**	**+4.7**	37.5	**38.1**	**–16.4**
	390.0			36.0		
	395.0			39.0		
	390.0			41.0		
	400.0			37.0		

1.5% infusion has the highest strength and stiffness among the three systems indicated in Fig. 9.6. Gain in strength and stiffness of the 1.5% system is around 30% and 12%, respectively, over the neat, as shown in Table 9.2. Enhancement in strength by 30% during flexure was somewhat surprising due to the fact that previous studies with nanoclays[19,20] did not show such improvement. As Fig. 9.3 and Table 9.1 show, if the particle infusion is increased to 3%, there is no proportional improvement in properties. Rather, there is a very nominal increase in strength and a significant reduction in stiffness with the 3% wt system. Similar reflection in properties with somewhat different ratios is observed during tensile tests, as shown in Fig. 9.7 and Table 9.3. It is noted that the enhancements in strength and stiffness are consistent with those shown in Fig. 9.3. Large values of strength and modulus associated with Fig. 9.7 with respect to those in Fig. 9.3 are due to the reinforcement of fibers. The fiber reinforcement, however, did not affect the strain to failure of the 1.5% system, as it is seen to be around 2.25% (Fig. 9.7), which is almost identical to that indicated in Fig. 9.3. Possible reasons for such behavior may be that since the reinforcement is in cloth (satin woven) form, it did not contribute much to the elongation, and that the failure was mostly controlled by matrix- and delamination-related failure modes. Correspondingly, enhancement in the case of fibrous composites (Fig. 9.7) is very similar to that obtained with nanophased matrix (Fig. 9.3).

In an attempt to understand the gain in properties in both cases (flexure and tension), TEM micrographs were taken and investigated for the nanophased systems.

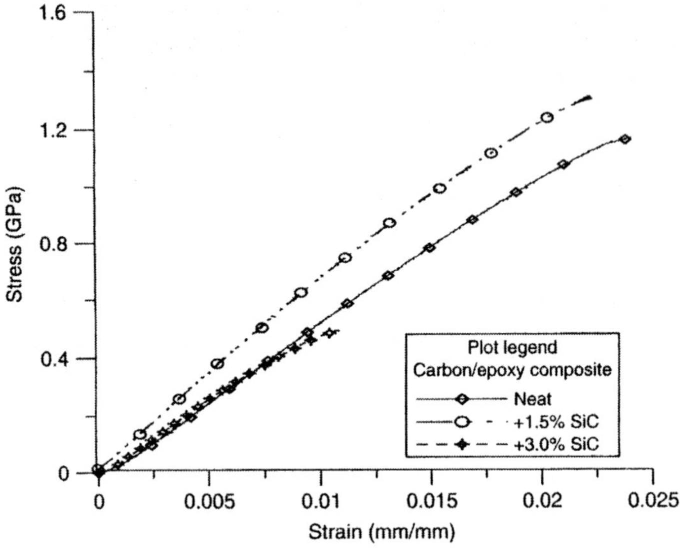

FIGURE 9.7 Tensile testing of carbon/epoxy nanocomposites.

TABLE 9.3
Tensile Test Data for Carbon/Epoxy Composites

Material	Tensile Strength (GPa)		Gain/Loss Strength (%)	Tensile Modulus (GPa)		Gain/Loss Modulus (%)
Neat	1.196	1.19	—	50.1	50.1	—
	1.192			51.5		
	1.193			39.0		
	1.194			53.9		
	1.194			55.9		
+1.5% SiC	1.330	1.33	+11.6	63.2	61.8	+23.5
	1.430			59.5		
	1.320			61.9		
	1.290			62.8		
	1.290			61.8		
+3.0% SiC	0.500	0.48	−53.9	53.1	54.0	+6.3
	0.500			51.3		
	0.450			61.0		
	0.380			50.4		
	0.550			54.0		

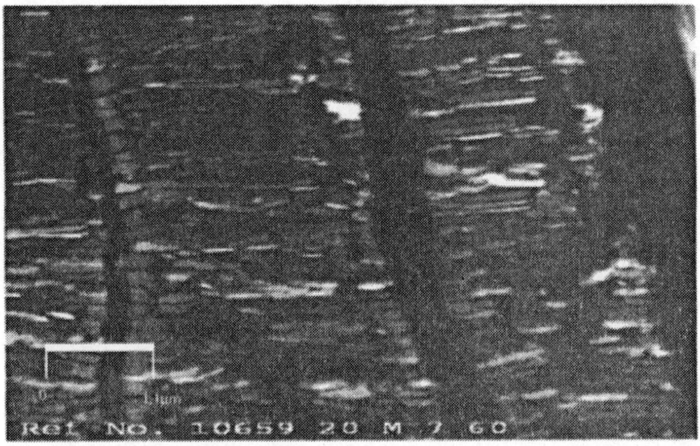

FIGURE 9.8 TEM micrograph of carbon/epoxy nanocomposites.

One such TEM of the fibrous composite with 1.5% infusion is shown in Fig. 9.8. It can be seen in Fig. 9.8 that the fiber matrix interface is very uniform, strongly bonded with the matrix, and is almost undisturbed by the infusion of nanoparticles. Nanoparticles are also uniformly dispersed over the entire body of the matrix. Micron-size voids are also seen in the figure, some of which are partially filled with nanoparticles. An enlarged view of these voids and nanoparticles is shown in Fig. 9.9. Void content tests performed on the nanophased composites revealed that there is a reduction in void content by about 30–40% due to 1.5% infusion. We believe this reduction in the void content of the matrix provided strength for the nanophased composites. When the infusion increased to 3%, as is observed in the TEM micrograph of Fig. 9.10, the particles started to form lumps that sometimes became larger

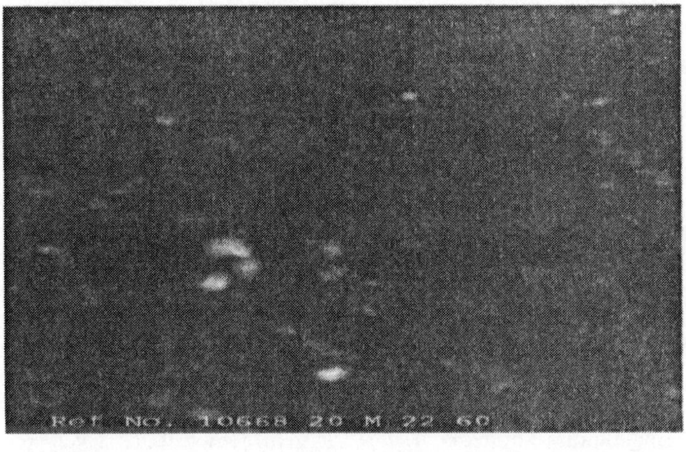

FIGURE 9.9 TEM Micrograph of nanoparticles and voids at 1.5% wt loading.

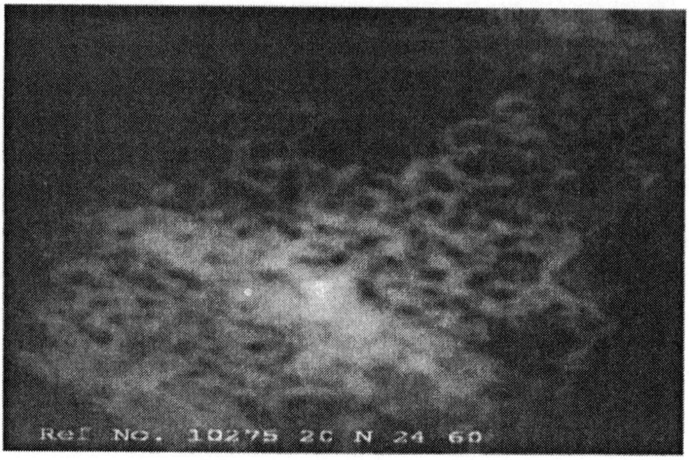

FIGURE 9.10 Nanoparticle agglomeration at 3 wt% loading.

than the size of the voids. Instead of filling voids, these lumps acted almost like impurities in the polymer. We believe with 3% loading, the number of nanoparticles are very high, which allows particle-to-particle interaction rather than the intended particle-to-polymer interaction. Once it reaches this state, the particles begin to agglomerate and form lumps that eventually affect the van der Waals interaction between the polymer chains, and reduce the cross-linking and increase void content in the nanocomposite. The resulting mechanical properties are hence degraded. This effect can be clearly visualized from careful observation of TGA data from 600°C to 800°C in Fig 9.4. Figure 9.4(a) and (b) shows that the percentage weight losses within the temperature range of 600°C to 800°C are about 1% and 0.5%, respectively, for neat carbon epoxy and 1.5% SiC carbon epoxy samples, whereas Fig. 9.4(c) indicates that the percentage weight loss at the same temperature range is about 2.5%. It is also seen in Fig. 9.4(c) that the curve continues to fall further beyond 800°C. The possible reasons for these relatively high percentage weight losses could be due to the continuous loss of void content in the system. In other words, this suggests the presence of higher void content in the 3% SiC carbon epoxy system than the neat and 1.5 wt% SiC carbon epoxy systems.

9.2.6 Fatigue Tests

Since property enhancement was positive from a quasistatic point of view, it was decided to test these materials under dynamic loading such as fatigue. Accordingly flexural fatigue tests were conducted at a stress ratio of 0.1 and a frequency of 3 Hz. S-N diagrams, shown in Fig. 9.11 were then generated for the three systems. We observe in general that above 60% load level, the neat system performs better than any of the nanoinfused systems. The 3% system demonstrates a very poor fatigue performance over the entire range of loading, which is similar to the 3 wt% system during quasistatic tension and flexure, as we saw earlier. During fatigue, once the load

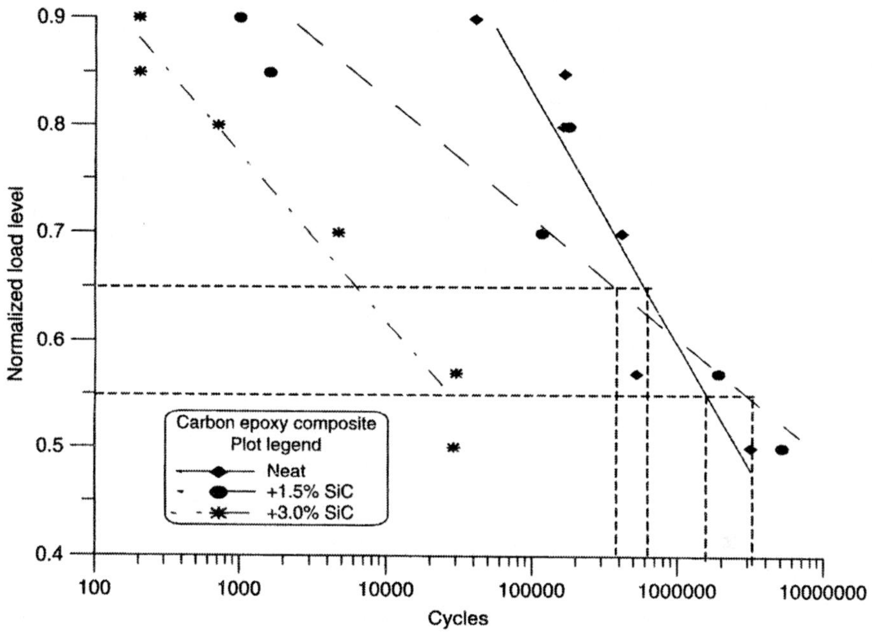

FIGURE 9.11 S-N diagram (flexural fatigue).

level goes below 60%, the fatigue performances of the neat and 1.5% system begin to reverse. For example, at 65% load level, the ratio of cycle numbers to failure between the neat and the 1.5% system is around 1.75, whereas this ratio changes to 0.56 at 55% load level. The reversal of the fatigue phenomenon is not yet fully understood. However, as shown in the S-N diagram, the slope of the neat system is much steeper that that of the 1.5% system, suggesting a lesser sensitivity of the neat system with respect to applied load. This indicates that for a change in the load level, the corresponding change in cycle numbers will be much smaller in the case of the neat system. Since this continued up to 50% load level, we were able to find an intersection point at around the 60% marker, which was defined as a threshold load level. Below this threshold stress level, we believe the fatigue failure mechanisms such as matrix cracks, filament splitting, and delamination[33–35] are significantly slowed down with the 1.5% system. And that is quite possible, since nanoinfusion (1.5%) certainly improves the matrix properties and matrix-dominated failure modes, as we saw earlier. In this investigation, we did not define any runout cycle numbers as is typical with fatigue studies. Data shown in Table 9.4 represent all failed samples. Although it took a good deal of time to conduct fatigue tests, especially at lower load levels, it was done intentionally to capture the threshold stress level. It should also be noted in Fig. 9.11 that the slopes of the two nanophased systems (1.5% and 3%) are similar, indicating that they are more or less equally sensitive to applied load with the exception that the 1.5 wt% system is outperforming the 3 wt% system by a large margin.

TABLE 9.4
Flexural Fatigue Response (Frequency = 3 Hz, R = 0.1)

Type	Load Level					
	50%	57%	70%	80%	85%	90%
Neat	3,109,800	496,090	621,000	66,260	69,260	38,824
	3,187,786	509,136	500,000	155,577	165,577	40,989
		560,050	244,900	37,503	37,503	41,860
			278,589	386,579	386,579	
Average	3,148,793	521,759	411,122	161,480	164,730	40,558
1.5% SiC	5,090,088	2,945,712	200,000	416,285	1284	1604
	5,146,479	120,281	39,350	55,327	1576	538
		2,607,053	166,393	63,490	608	1066
			105,525	52,744	3575	
				300,087	1400	
Average	5,118,284	1,891,015	127,817	177,587	1689	1069
3.0% SiC	29,897	4688	200,000	643	286	121
	27,765	5890	39,350	534	101	249
	28,578	3420	166,393	765		
	29,657		105,525	1065		
Average	28,974	4666	127,817	752	194	185

9.3 NANOPHASED FILAMENTS

In another route, nanoparticles with aspect ratios such as carbon nanowhiskers and carbon nanotubes were doped with textile polymer precursors. Two types of polymer precursors were used: linear low-density polyethylene and nylon-6. The fillers in each case were carbon nanowhiskers and multiwalled carbon nanotubes (MWCNT). The amount of loading varied 1–2% by weight of the base polymer. Nanoparticles were first mixed with the base polymer in powdered form. The mixture was then dried in a hopper and fed through a single screw extrusion machine. It passed through three stages of heating where the temperature was set slightly above the melting temperature of the polymer. After heating and at the end of the screw, the molten mass was forced through two stages of mixing. In the final stage, filaments were extruded through a micron-size orifice. The extruded filaments went sequentially through a cooling trough, a tension device, a heater, and eventually wound into a filament winder to form into spools. In one attempt, these filaments were cut into strands, laid out in 0° fashions in several layers, and consolidated in a compression molding machine. Test coupons were extracted from these laminates and were tested under tension and flexure. It was found that the improvement with carbon nanowhiskers was around 17%, while it increased to 34% with MWCNT. Out of the four systems (two fillers and two polymers) investigated, the system with nylon-6 infused with MWCNT yielded the most promising results. Tension tests on individual filament of this system, shown in Fig. 9.12, demonstrate that there is about 150–300%

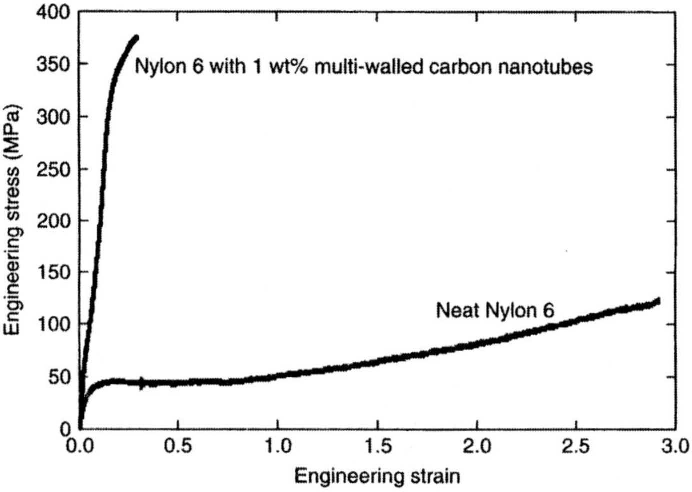

FIGURE 9.12 Stress versus strain for neat Nylon 6 and Nylon 6 with 1 wt% multiwalled carbon nanotubes with 102-mm gage length.

improvement in strength and stiffness with 1% MWCNT loading. TEM studies revealed that extrusion technique caused sufficient alignment of MWCNT along the length of the filament, which may have caused the gain in mechanical properties.[36]

9.4 CORE MODIFICATION

In the third route, an innovative technique to develop polyurethane foams containing nanoparticles was introduced. Polymethylene polyphenylisocyanate (part A) was mixed with nanoparticles such as SiC and TiO_2, and irradiated with a high-power ultrasound liquid processor. In the next step, the modified foams containing nanoparticles were mixed with part B (containing polyol resin systems, surfactant, and an amine catalyst) through a high-speed mechanical stirrer. The mixture was then cast into rectangular molds to make nanophased foam panels. Test coupons were then extracted from the panels to carry out thermal and mechanical characterizations. The as-prepared foams were characterized by scanning electron microscopy (SEM), x-ray diffraction, and thermogravimetric analysis (TGA). SEM studies showed that the particles were nonagglomerated and well dispersed in the entire volume of the foam. The foam cells were well ordered and uniform in size and shape. The TGA analyses indicate that the modified foams were thermally more stable than the parallel neat system.[37] Quasistatic flexure tests under three-point bend configuration were also conducted with both modified and neat foams. Test results showed a significant increase (approximately 50–70%) in the flexural strength and stiffness of the nanophased foams over the neat system. This enhancement in flexural properties was demonstrated repeatedly with multiple batches and with at least three specimens tested from each batch. The nanophased foam was then used with regular S-2 glass

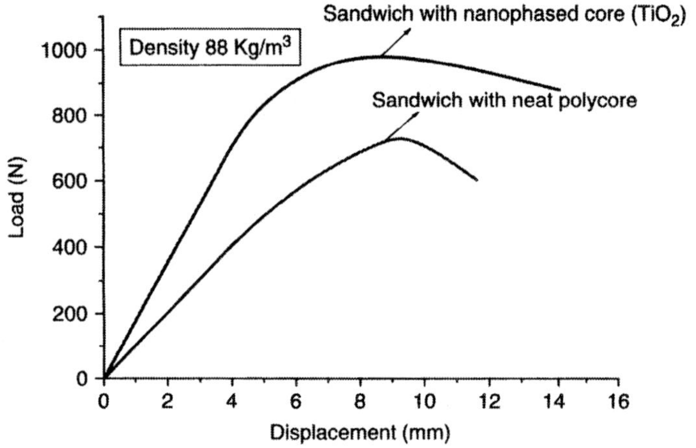

FIGURE 9.13 Flexural tests of neat and nanophased sandwich composites.

fiber preforms and SC-15 epoxy to manufacture sandwich composites in a VARTM setup. A parallel set of control panels was also made with neat polyurethane core materials. A significant improvement in flexural strength and stiffness was observed with 3% loading of TiO_2 nanoparticles in part A. Debond fracture toughness parameters (G_c) were also determined for both categories of sandwich constructions, and it was seen that nanoparticle infusion reduces the value of G_c by almost a factor of three. Despite this reduction, the strength of nanophased sandwich increased by about 53% over the neat system, as shown in Fig. 9.13.

9.5 SUMMARY

With identical loading fraction, nanoparticle infusion into an epoxy system is seen to be significantly superior to micron-sized particle infusion in enhancing the thermal and mechanical properties of the resulting composites. A low-cost but reliable manufacturing procedure can be used to fabricate large-scale laminated structural composites with a nanophased matrix. Nanoparticles loading into the base matrix are found to be optimal around 1.5% by weight to derive maximum gain in the mechanical and thermal properties of the structural composites. Nanoparticle infusion is seen to increase the thermal stability of the matrix by enhancing cross-linking in the polymer. The additional cross-linking seems to be due to the catalytic effects caused by the presence of the nanoparticles. Nanoparticles tend to act as fillers in the voids, which in turn reduces the void content of the composites and thus translates into enhanced mechanical properties. With 1.5% loading in the matrix, an average 20–30% increase in mechanical properties of the resulting fibrous composites has been observed both in tension and flexure. During flexural fatigue, a threshold load level around 60% of the ultimate flexural strength seems to exist. Below this threshold load level, the 1.5% system supersedes the neat system, whereas the situation is reversed if stress level is above this level. Throughout the entire loading range, the 3% system is seen to be inferior to the other two systems.

It is observed from our investigation that in textile precursors as well as in liquid foam there are significant improvements in chemical and mechanical properties of the resulting nanocomposites due to nanoparticle infusion. This improvement takes place with a very low loading of nanoparticle, and there seems to be an optimal loading corresponding to each system of nanoparticle and the polymer. But, in each route and in each system, the optimal loading of nanoparticles of either spherical or acicular shape is seen to be between 1 and 3% by weight. It is also noted that the gain in mechanical properties is much more significant if one goes through the fiber and the core modification routes.

PROBLEMS

1. Explain different methods that are commonly used in mixing nanosized particles into polymers.
2. Discuss the effects of micro- and nanosized SiC fillers in an epoxy matrix on the thermal properties of the composite.
3. Discuss the effects of micro- and nanosized SiC fillers in an epoxy matrix on the mechanical properties of the composite.
4. Discuss the advantages and disadvantages of nanosized SiC fillers on the fatigue performance of polymeric matrix composites.

REFERENCES

1. Lau, K. and Hui, D., The revolutionary creation of new advanced materials-carbon nanotube composites, *Composites: Part B*, 33, 263–277,.2002.
2. Becker, O., Varley, R., and Simon, G., Morphology, thermal relaxations and mechanical properties of layerd silicate nanocomposites based upon high-functionality epoxy resins, *Polymer, 43*, 4365–4373, 2002.
3. Agag, T., Koga, T., and Takeichi, T., Studies on thermal and mechanical properties of polyimide-clay nanocomposites, *Polymer, 42*, 3399–3408, 2001.
4. Gianellis, E. P., Polymer layered silicate nanocomposites, *Adv. Mater.*, 8, 29–35, 1996.
5. Edelstein, A.S. and Cammarata, R.C., *Nanomaterials: Synthesis, Properties and Applications,* Institute of Physics Pub., 1996.
6. Mahfuz, H., Islam, M., Rangari, V., Saha, M., and Jeelani, S., Response of Sandwich Composites with Nanophased Cores under Flexural Loading, *Composites Part B,* 35, 543–550, 2004.
7. Dagani, R., Putting the nano into composites, *Chemical & Engineering News, June 7, 1999.*
8. Birringer, R. and Gleiter, H., *Encyclopedia of Materials Science and Engineering,* Cahn, R.W. et al., Eds, Suppl. Vol. 1, Pergamon Press, Oxford, 1988, pp. 339.
9. Gleiter, H., Nanocrystalline materials, *Prog. Mater. Sci., 33,* 223–315, 1989.
10. Dagani, R., Nanostructural materials promise to advance range of technologies, *Chemical & Engineering News,* November, 18–24, 1992.

11. Uyeda, R., Studies of ultrafine particles in Japan: Crystallography. Methods of operation and technological application, *Prog. Mater. Sci.*, 35, 1–96, 1991.
12. Karch, J., Birringer, R., and Gleiter, H., Ceramics ductile at low-temperature *Nature*, 330, 556–558, 1987.
13. Teishev, A., Incardons, S., Migliaresi, C., and Marom, G., Polyethylene fibers-polyethylene matrix composites: Preparation and physical properties, *J. Appl. Polym. Sci., 50*, 503–512, 1993.
14. Sperling, L.H., *Introduction to Physical Polymer Science,* 2nd ed., John Wiley & Sons, New York, 1992.
15. Hlangothi, S.P., Krupa, I., Djokovic, V., and Luyt, A.S., Thermal and mechanical properties of cross-linked and uncross-linked linear low-density polyethylene–wax blends, *Polymer Degradation and Stability*, 79, 53–59, 2003.
16. Walter, R., Friedrich, K., Privalko, V., and Savadori, A., On modulus and fracture toughness of rigid particulate filled high density polyethylene, *J. Adhesion, 64*, 87–109, 1997.
17. Rothan, R.N., Mineral fillers in thermoplastics: Filler manufacture and characterization, *Advanced Polym. Sci., 139*, 67–107, 1999.
18. Wu, C., Zhang, M., Rong, M., and Friedrich K., Tensile performance improvement of low nanoparticles filled-polypropylene composites, *Composite Sci. Tech., 62*, 1327–1340, 2002.
19. Shiner, C., Timmerman, J., Eboneee, P.M., Williams and Seferis, J., Thermal and Mechanical Characteristics of Nano Modified Fiber-Reinforced Composites, paper presented at 48th International SAMPE Symposium, May 11–15, 2003, p. 2539.
20. Fukushima, Y., and Inagaki, S., Synthesis of an intercalated compound of montmorillonite and 6-polyamide, *J. Inclusion Phenomena*, 5, 473–482, 1987.
21. Garces, J., Moll, D., Bicerano, J., Fibiger, R., and McLeod, D., Polymeric nanocomposites for automotive applications, *Adv. Mater.* 12, 23, 1835–1839, 2000.
22. Gilman, J. and Kashiwagi, T., Literature review in Beall, G.W. and Pinnavaia, T., Eds., in *Polymer-Clay nanocomposites,* John Wiley & Sons, New York, 2000, p. 193.
23. Kojima, Y., Usukia, A., Kawasumi, M., Okada, A., Fukushima, Y., Kurauchi, T., and Kamigaito, O., Mechanical properties of nylon-6/clay hybrid, *J. Mater. Res., 8*, 1185–1189, 1993.
24. LeBaron, P., Wang, Z., and Pinnavaia, T., Polymer-layered silicate nanocomposites: an overview, *J. Appl. Clay Sci., 15*, 11–29, 1999.
25. Lee, A. and Lichtenhan, J., Thermal and viscoelastic property of epoxy-clay and hybrid inorganic-organic epoxy nanocomposites, *J. Appl. Polym. Sci.*, 73, 1993–2001, 1999.
26. Eskin, G., Broad prospects for commercial application of the ultrasonic (cavitation) melt treatment of light alloys, *Ulatrasonic Sonochemistry, 8*, 319–325, 2001.
27. Price, G., Recent developments in sonochemical polymerization, *Ultrasonics Sonochemistry*, 10, 277–283, 2003.
28. Mahfuz, H., Zaman, K., Hisham, M., Foy, Costee, Haque, A., and Jeelani, Fatigue life prediction of thick-section S2-glass/vinyl-ester composites under flexural loading, Transaction of ASME, *J. Eng. Mater. Tech., 122*, 402–408, 2000.
29. Mahfuz, H., Mamun, W., and Jeelani, S., High strain rate response of sandwich composites: Effect of core density and core-skin debond, *J. Adv. Mater., 34*, 1, 22–26, 2002.
30. Vaidya, U., Kamath, M., Hosur, M., Mahfuz, H., and Jeelani, S., Manufacturing and low velocity impact response of sandwich composites with hollow and foam filled Z-pin reinforced core, *J. Composites Tech. Res., 21*, 2, 84–97, 1999.

31. Vijaya Kumar, R., Koltypin, Y., and Gedanken, A., Preparation and characterization of nickel-polystyrene nanocomposite by ultrasound irradiation, *J. Appl. Polym. Sci.*, 86, 160, 2002.
32. Vijaya Kumar, R., Koltypin, Y., Cohen, Y.S., Cohen, Y., Aurbach, D., Palchik, O., Felner, I., and Gedanken, A., Preparation of amorphous magnetic nanoparticles embedded in polyvinyl alcohol using ultrasound radiation, *J. Mater. Chem.*, 10, 1125, 200.
33. Mahfuz, H., Maniruzzaman, M., Vaidya, U., and Jeelani, S., Fatigue damage and effects of stress ratio on the fatigue life of carbon-carbon composites, *Theor. Appl. Fracture Mech.*, 24, 21–31, 1995.
34. Mahfuz, H., Partha, S.D., Jeelani, S., Baker, D. M., and Johnson, S., Effect of missioncycling on the fatigue performance of SiC coated carbon-carbon composites, *Int. J. Fatigue*, 15, 4, 283–291, 1993.
35. Mahfuz, H., Maniruzzaman, M., Vaidya, U., and Jeelani, S., Response of SiC/Si3N-4 composites under cyclic loading—an experimental and statistical analysis, *J. Eng. Mater. Technol.*, 119, 186–193, 1997.
36. Mahfuz, H., Adnan, A., Rangari, V., and Jeelani, S., Carbon nanoparticles/whiskers reinforced composites and their tensile response, *Composites Part A, Appl. Sci. Manufacturing*, 35, 519–527, 2004.
37. Mahfuz, H., Rangari, V., Islam, M., and Jeelani, S., Fabrication, synthesis and mechanical characterization of nanoparticles infused polyurethane foams, *Composites Part A, Appl. Sci. Manufacturing*, 35, 453–460, 2004.

10 Synthesis and Characterization of Metal-Ceramic Thin-Film Nanocomposites with Improved Mechanical Properties

Dhanjay Kumar, Jagannathan Sankar, and Jagdish Narayan

CONTENTS

10.1 INTRODUCTION

Composite materials consisting of metallic clusters or crystals of nanometer dimension embedded in an insulator host can exhibit special optical, electrical, magnetic, and mechanical properties that may be used in various technological applications.[1–8] For example, metal nanocrystals embedded in insulating materials offer enhanced mechanical properties, making them promising candidates for use in machine tools.[9–14] The search for materials with enhanced hardness is driven by both the scientific curiosity of researchers to explore the possibilities of synthesizing a material whose hardness could approach or even exceed that of diamond and the technical importance of hard materials for wear protection, e.g., of machine tools. The importance of hard-wear protective coatings for machining applications is illustrated by the fact that today more than 40% of all cutting tools are coated by wear-resistant coatings and the market is growing fast. Wear-resistant hard coatings for high-speed dry machining would allow the industry to increase the productivity of expensive automated machines and to reduce the high costs presently associated with the use of environmentally hazardous coolants.

The metal nanocrystals embedded in insulating materials have been synthesized by quenching and heat treatments,[15] sol-gel processes,[16] sputtering,[17] and ion implantation.[18] Most of these processes are multistep processes, in which a postdeposition treatment is often needed to optimize the properties. However, these treatments can also alter the average grain size, size distribution, and spatial arrangement of the nanocrystals, with possible unfavorable effects on the mechanical properties of the nanocomposites. Unlike these methods, multiple-target sequential pulsed laser deposition (PLD) permits independent control of synthesis of the nanocrystals and the embedding matrix. PLD has shown particular success in stoichiometric thin-film deposition of complex oxides. Energetic (>100 eV) ions produced by ablation yield smooth, high-density films with good adhesion, especially desirable for mechanical applications. In this chapter, we report the synthesis of alternating-target PLD of Fe and Ni nanocrystals embedded in an Al_2O_3 matrix. The material produced is a thin-film composite consisting of metallic nanocrystals embedded in an insulator host. The structural quality of the thin-film composite was evaluated using high-resolution transmission electron microscopy and scanning transmission electron microscopy with atomic number contrast. This revealed the formation of a biphase system with thermodynamically driven segregation of Ni and alumina during pulsed laser deposition. This new nanocomposite material exhibits superior mechanical properties with enhanced hardness and Young's modulus. The improvement in hardness of Al_2O_3 thin films by embedding metal nanocrystals is related to the evolution of a microstructure that efficiently hinders the manipulation and movement of dislocations and the growth of microcracks. This is achieved by grain boundary hardening.

10.2 THEORY OF PULSED LASER DEPOSITION

The pulsed laser deposition technique has been used to prepare the thin-film nanocomposite samples discussed in the present chapter. Since this is a relatively new technique with respect to other thin-film deposition techniques such as evaporation, sputtering, and molecular beam epitaxy, it is appropriate to briefly describe the theoretical aspects of pulsed laser deposition. The pulsed laser deposition (PLD) process was first used more than two decades ago.[19] However, it gained prominence recently when it was found to be the most convenient and efficient technique for the synthesis of new high-temperature superconducting thin films.[20] The schematic of a pulsed laser deposition chamber is shown in Fig. 10.1. In a typical PLD of thin films, a pulsed laser strikes a solid bulk target. Some of the target materials are removed, escaping in the form of a plume. Part of the plume comes in contact with the surface of a heated substrate kept a few centimeters away from the target. The plume, consisting of the

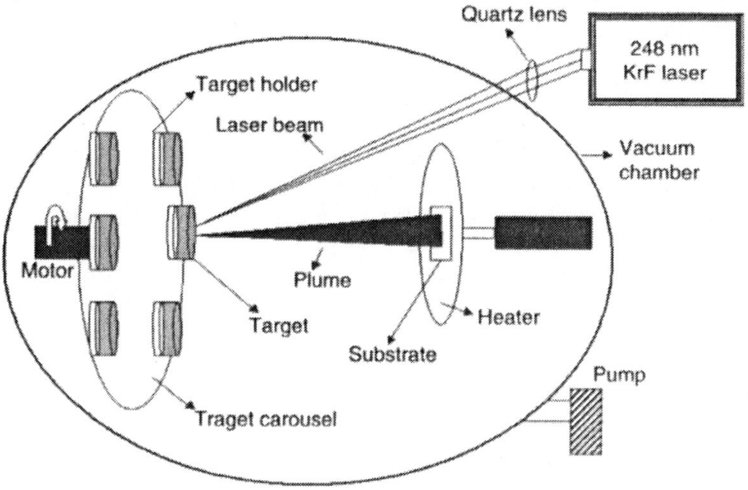

FIGURE 10.1 Schematic of a pulsed laser deposition system.

building block (lattice) of the material, covers the substrate. The result is the fabrication of a thin film of the given material with the same chemical structure as the target. PLD offers numerous advantages, including film stoichiometry close to the target, a low contamination level, a high deposition rate, and nonequilibrium processing. The relatively easily accessible experimental parameters in PLD make it very attractive for the synthesis of high-temperature superconducting thin films. These parameters are the substrate temperature, the energy of the atom flux, the relative and absolute arrival rate of atoms for compound films, and the pressure in the chamber. In order for any thin-film technique to be highly suitable for the growth of thin-film materials, the following conditions must be fulfilled: (1) the substrate temperature should be accurately controlled; (2) all the atoms in the deposition flux should have an energy of 5 eV to 10 eV to promote surface diffusion, nucleation, and high sticking probability without producing damage; (3) the atoms of the multicomponent material should arrive with the precise relative abundance required for the compound; and (4) it is useful to do all the processing in a high partial pressure of reactive gases such as oxygen to maintain stoichiometry of volatile species on the heated substrates and to control the energy of the deposited flux. The plasma generated by laser irradiation of the target can be used to assist the growth of very high quality thin films at relatively low processing temperatures. The processing temperature in PLD is generally 50–100°C lower than used in other thin-film growth techniques. Atomically sharp multilayer structures and superlattices can easily be fabricated by this technique. These advantages have made the PLD method one of the most popular methods to deposit thin films.

The basic experimental design for thin-film deposition by laser ablation is similar to any other physical vapor deposition process. The apparatus includes a vacuum chamber, a substrate holder with precise temperature control, and source materials (target). Usually an excimer laser that uses a mixture of Kr, F_2, He, and Ne generates pulses with wavelengths varying from 193 nm to 308 nm is employed for target ablation. Ultraviolet lasers have been found to result in congruent ablation that varies from 25 to 30 ns duration and the energy density ranges from 1.0 to 2.5 J/cm^2. An aperture is placed across the beam so that the nonuniform edge effects can be minimized. The laser fluence is varied by either varying the laser output energy or by focusing the beam. The optical imaging system consists of a mirror with reflective coating and a planoconcave lens. The laser beam is admitted to the chamber via a quartz window that is susceptible to damage if the energy density is too high or if the window is covered by dust particles. The sides of the chamber contain several quartz windows. One is used to admit the laser beam, while the others are used for *in situ* plasma diagnostics and for monitoring the growth. The chamber contains a sample heater block, a rotating target holder, and a shutter. The target block is mounted on a linear motion feedthrough that is used to vary the target-substrate distance. The target is rotated at about 1 Hz, which is incommensurate with the laser pulse rate but provides a more uniform erosion of the target. The oxygen pressure during deposition is regulated using a mass flow controller and a throttle valve.

Although PLD appears to be an evaporation process, the stoichiometry and the nature of species are quite different from standard evaporation processes. This is due to the complex interaction of the laser beam with the target and plasma and subsequent deposition of species. Depending on the type of interaction of the laser beam with the target, the PLD process can be classified into four separate regimes: (1) interaction of the laser beam with the target material resulting in ablation of the surface layers; (2) interaction of the ablated material with the incident laser beam resulting in an isothermal plasma formation and expansion; (3) anisotropic adiabatic expansion of the plasma leading to the characteristic nature of the laser deposition process; and (4) interaction of the plasma with the background gas. A schematic view of the interaction processes is shown in Fig. 10.2. The first two regimes (marked (a) and (b) in Fig. 10.2)

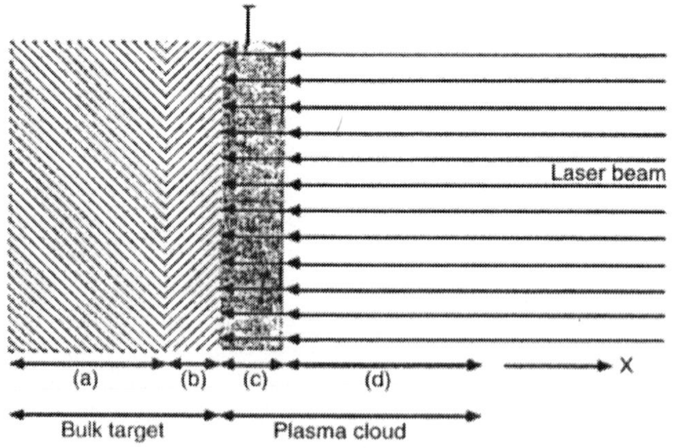

FIGURE 10.2 Schematic diagram showing the different phases present during laser irradiation of a target: (a) unaffected target, (b) evaporated target, (c) dense plasma absorbing laser radiation, and (d) expanding plasma outer edge transparent to the laser beam.

start with the laser pulse and continue throughout the laser pulse duration, while the next two regimes initiate after the laser pulse terminates. Under PLD deposition conditions, where the pulse energy density is in the range of 1 to 10 J/cm^2, the target ablation can be considered to be thermal in nature, while the interaction of the laser beam with the evaporated material gives rise to thermal characteristics of the species in the plasma.

The thermal effects of pulsed nanosecond laser irradiation of materials are determined by the laser pulse parameters (temporal power density $I(t)$, pulse duration t_p, wavelength, etc.), optical properties (reflectivity R, absorption coefficient α), and the thermal properties of the material (thermal conductivity K, latent heat per unit volume L_v, specific heat per unit volume C_v, ablation temperature T_v, etc.). The thermal diffusivity $D = K/C_v$ defines the thermal diffusion length $(2Dt_p)^{0.5}$, which determines the spread of the temperature profile during the laser pulses. The temperature in the target $T(x,t)$ during laser irradiation is controlled by the heat flow equation given by[23]

$$C_v(T)\frac{\partial T(x,t)}{\partial t} = \frac{\partial}{\partial x}\left[K(t)\frac{\partial T(x,T)}{\partial x}\right] + (1-R)I_0(x,t)\alpha e^{\alpha x} \tag{10.1}$$

with appropriate boundary conditions that take into account the formation and movement of the solid-liquid (or liquid-vapor) interface. Here, x refers to the direction in the plane perpendicular to the target, and t refers to the time. The second term on the right-hand side of the equation is the heat generation term due to the absorption of the incident laser beam by the target. If the surface of the material is highly absorbing ($\alpha \geq 10^6$ cm^{-1}) to the incident laser beam, the heat generation term can be removed from Equation 10.1, and applied to the front surface boundary condition.

The laser-target interactions play a critical role in the film quality deposited by the PLD technique. Depending on the nature of the heating and thermophysical properties of the materials, two separate regimes can be distinguished[24]: (1) a surface heating regime in which the optical absorption depth is much smaller than the thermal diffusion distance, $\alpha(2Dt_p)^{0.5} \gg 1$, and (2) a volume heating regime in which the optical absorption depth is much larger than the thermal diffusion distance, $\alpha(2Dt_p)^{0.5} \ll 1$. The two regimes are shown schematically in Fig. 10.3. In the first case, the laser energy is absorbed in the surface layer, and the thermal diffusivity controls the heating and ablation characteristics. During ablation of the target, a planar vaporization interface may initiate from the surface and propagate into the bulk of the target. However, at higher power densities, subsurface heating effects that may lead to explosive removal of material from the target become important, and thus lead to nonlinear ablation effects.[25] The transformation from surface to volume ablation regimes is dependent on several factors such as incident power density, absorption coefficient, and the thermophysical properties of the target. In the volume heating regime, the thermal conductivity plays an insignificant role in controlling the laser-target interactions, and the ablation depth is determined mainly by the optical absorption depth, which is given by the inverse of the absorption coefficient. In this regime, internal heating subsurface effects may predominate, leading to removal of particles from the target, and these particles may shield the

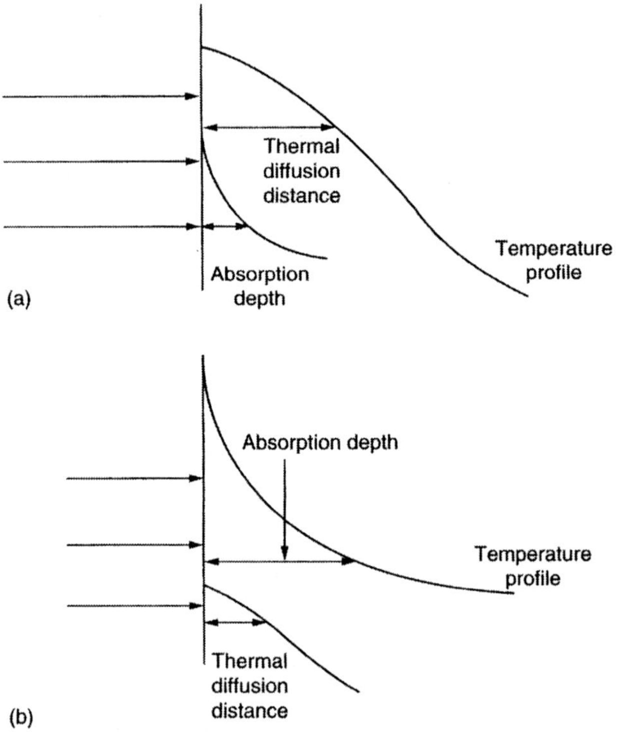

FIGURE 10.3 Schematic of optical absorption and thermal diffusion profiles for (a) the surface heating regime where the optical absorption length is very small and insignificant compared to the thermal diffusion length, and (b) the volume heating regime where the optical absorption length is much greater than the thermal diffusion length.

target from the incoming laser beam. The term *subsurface superheating* has been developed to describe temperature inversion on the surface during laser ablation. In the case of planar surface ablation when the vaporization interface moves rapidly into the target, the surface is constantly being cooled due to the latent heat of ablation.

However, the absorption of the laser beam in the target is characterized by a finite absorption depth in the material. The subsurface layers thus are heated directly by the laser beam where the heat dissipation mechanism is due to the thermal conduction losses in the target. As the energy density is increased, the subsurface temperatures may vastly exceed the surface temperature, which corresponds to the ablation point of the material (Fig. 10.4). These subsurface superheating heating effects may lead to explosive removal of material from the target in the form of particles. These particles may in turn shield the laser beam from the target, thus giving rise to nonlinear ablation characteristics.

According to Singh et al.,[26] the time after which the ablation front may change from a planar to volume type will depend on the degree of internal superheating and the time required for the nucleation to form a subsurface gaseous phase. Because of the number of unknowns, the nucleation of the stable gaseous phase near the

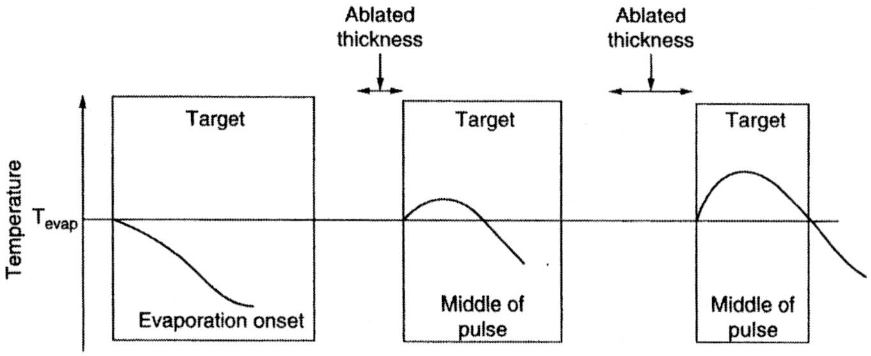

FIGURE 10.4 Schematic of subsurface heating of the target during laser ablation.

surface cannot be quantitatively analyzed using the basic nucleation theory. However, it is known that the nucleation rate is a strong function of the temperature, and thus a larger degree of superheating increases the probability of more rapid volume nucleation of the material. Using this approach, Singh et al.[26] analyzed the effect of various processing factors on subsurface superheating.

The high surface temperature induced by laser irradiation leads to emission of positive ions and electrons from a free surface. The thermionic emission of positive ions and electrons from hot surfaces has been widely recognized in the literature.[27] The flux of ions and electrons as a function of temperature can be predicted by Richardson's and Langmuir-Saha's equations, respectively. Both of these equations show an exponential increase in the fraction of ionized species with temperature; higher ionized fractions than predicted by the Langmuir-Saha equation have been observed in laser-irradiated targets.[28] This has been attributed to the higher temperatures induced by absorption of the laser beam by the evaporating material and electron impact ionization.

The physical mechanisms involved in the absorption and reflection of the laser energy by the evaporating material were identified in the early experiments as the sources for very high temperature (~1 KeV) plasma. The penetration and absorption of the laser beam by the plasma depend on the electron-ion density, temperature, and the wavelength of the laser light. For penetration (or reflection) of the incident laser beam, the plasma frequency (v_p) should be lower than the laser frequency. For excimer laser wavelengths ($\lambda = 193$–308 nm), the laser frequency varies from 9.74×10^{14} cm^{-1} to 1.5×10^{15} cm^{-1}. The plasma frequency is given by

$$v_p = 8.9 \times 10^3 n_e 0.5 \tag{10.2}$$

where n_e is the electron concentration of the plasma. Using Equation 10.2, the critical density for reflection of the laser beam is determined to be 1–3×10^{22}/cm^3. This critical density is approximately equal to the concentration of atoms in the solid or liquid. The presence of a diffuse plasma boundary and a gradual decrease in the plasma density away from the surface results in a lower electron density than the estimated value. Thus, energy losses due to the reflection of laser light from the plasma are insignificant.

The material that evaporated from the hot target is further heated by the absorption of the laser radiation. The primary absorption mechanism for ionized plasma is electron-ion collision. The absorption primarily occurs by an inverse Bremmstrahlung process, which involves the absorption of a photon by a free electron. The absorption coefficient (α_{plasma}) of the plasma can be expressed as

$$\alpha_p = 3.69 \times 10^8 \left(Z^3 n_i^2 / T^{0.5} v^3\right)\left[1 - \exp\left(-hv / kT\right)\right] \tag{10.3}$$

where Z, n_i, and T, are the average charge, ion density, and temperature of the plasma, respectively, and h, k, and v are the Planck constant, Boltzmann constant, and frequency of the laser light, respectively. The laser energy is highly absorbed if $(\alpha_p X) \sim 1$, where $X(t)$ is the dimension perpendicular to the target of the expanding plasma. This equation shows that the absorption coefficient of the plasma is proportional to n_i^2. Thus the plasma absorbs the incident laser radiation only at distances very close to the target where the densities of the charged particles are very high.

10.3 EXPERIMENTAL PROCEDURE

Nanocrystalline nickel and iron crystallites were embedded in an alumina matrix using a KrF excimer laser (252 nm, 30 ns full width at half maximum) focused alternately onto high-purity targets of nickel or iron and alumina. The depositions were carried out on silicon substrates in a high-vacuum environment ($\sim 5 \times 10^{-7}$ torr). The substrate temperature was approximately 500°C. The energy density and repetition rate of the laser beam used were 2 J/cm^2 and 10 Hz, respectively. The size distribution of metallic particles and the crystalline quality of both the matrix and metallic particles were investigated by cross-sectional scanning transmission electron microscopy with atomic number (Z) contrast (STEM-Z). Figure 10.5 is a schematic diagram showing the incoherent imaging under STEM conditions in which a high-angle annular detector is employed to collect the incoherent tail of the scattered electron distribution. The image is formed by focusing a very fine probe (2.2 Å at 100 KeV) on the sample. There is no imaging lens in this system. Instead, high-angle scattering is detected as a function of probe position and used to map the scattering power of the sample, which depends on the square of the atomic number. This is a dark-field technique, so that the atomic columns are seen as bright spots in the image; the heaviest columns are always brightest, allowing unambiguous discrimination of atoms with different atomic numbers. The diagram shows that this technique is able to resolve Ga and As in GaAs, even though these two atoms have a small atomic number difference between them.[21,22] It should be noted that the STEM-Z technique is capable of resolving the location of oxygen atoms at the interface. This is accomplished by simultaneously using the EELS mode in which a low-angle detector collects the oxygen signal. Thus, with the simultaneous use of the STEM-Z and EELS, both low- and high-atomic-number atoms can be identified at the particle/oxide interface.

Depth-sensing nanoindentation continuous stiffness measurements (CSMs) were made on the nanocomposite thin films. A MTS Nanoindenter-XP with a Berkovich

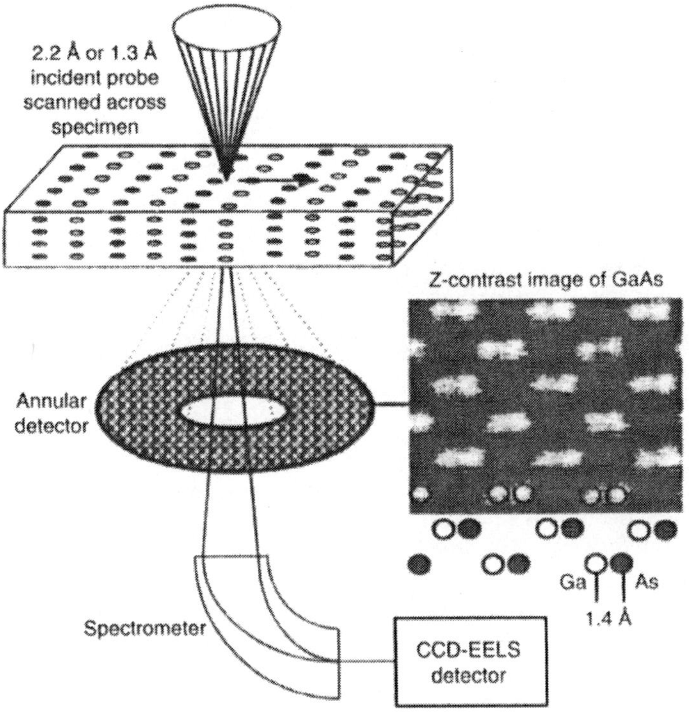

FIGURE 10.5 Schematic showing the optical arrangement for atomic imaging and analysis with the scanning transmission electron microscope. The inset shows the Z-contrast image of GaAs.

triangular pyramidal tip was used for these measurements. The instrument monitored and recorded the dynamic load and displacement values during the indentation. A series of arrays of indentations were made at a constant strain rate of 0.05 s^{-1} and at depths of 750 nm for hardness and modulus measurements. Four samples were made per deposition and three indentation tests were run on each sample. Each test comprised an array of six indents. Load displacement data collected from the indentations were used to calculate the hardness and Young's modulus using Oliver and Pharr analysis.

10.4 RESULTS AND DISCUSSION

10.4.1 Mechanical Properties

Shown in Fig. 10.6 is the variation of hardness of Ni-Al_2O_3 thin-film composites as a function of the size of the Ni particles embedded in the Al_2O_3 matrix. For the sake of comparison, we have shown in the figure the hardness data of pure alumina film deposited under identical conditions. It is clear from the figure that the hardness of the Ni-alumina composite is significantly higher than that of pure alumina films. However a further increase in the particle size of the embedded Ni particles resulted

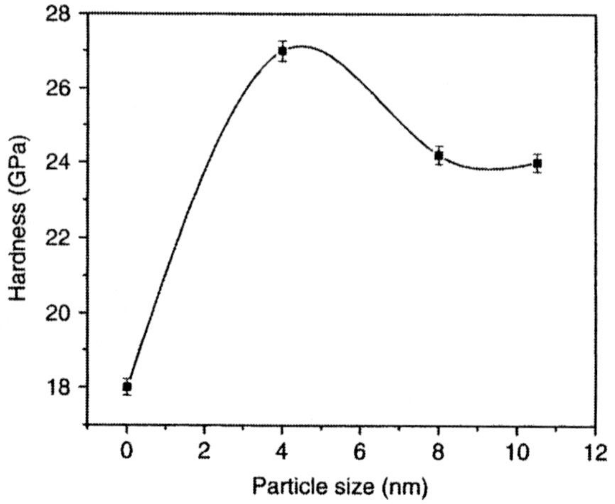

FIGURE 10.6 Hardness as a function of Ni nanoparticle size in different samples of the Ni-Al_2O_3 thin-film composite system.

in the deterioration of hardness, suggesting that there is an optimum size of metal particle that produces the greatest hardness in the composite. Supporting evidence in favor of this suggestion comes from hardness measurements performed on different thin-film composite systems consisting of Fe nanocrystals in alumina thin films. The results obtained are shown in Fig. 10.7. According to the results in this figure, the hardness of the first Fe-Al_2O_3 composite (size of Fe particles is 3 nm) is in fact lower than that of pure alumina film. However, the hardness of the Fe-Al_2O_3

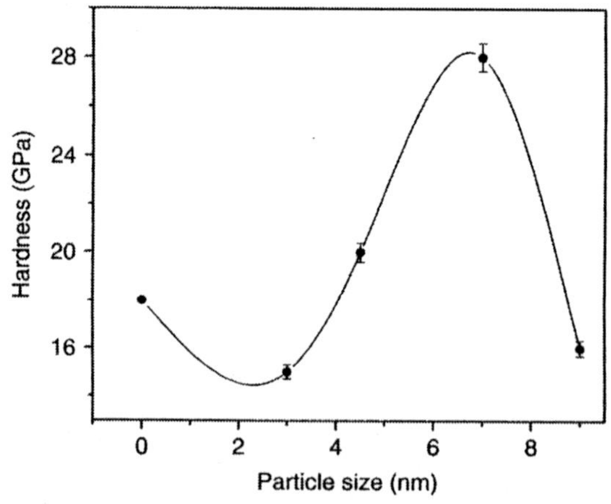

FIGURE 10.7 Hardness as a function of Fe nanoparticle size in different samples of the Fe-Al_2O_3 thin-film composite system.

composite becomes greater than that of pure alumina thin film when the size of the embedded Fe nanocrystals increases to 4.5 nm. The hardness reaches a peak at 7 nm, beyond which the hardness of the Fe-Al_2O_3 thin-film composite (Fe particle size is 9 nm) is again lower than that of pure alumina thin film.

The improvement in values of hardness of Al_2O_3 thin films by embedding metal nanocrystals is related to the evolution of a microstructure that efficiently hinders the manipulation and movement of dislocations and the growth of microcracks, which in turn is achieved by grain boundary hardening as described by the Hall-Petch relationship, which is valid down to a crystallite size of 20–50 nm[29–31]:

$$\sigma_c = \sigma_0 + \frac{k_{gb}}{\sqrt{d}} \tag{10.4}$$

Here σ_c is the critical fracture stress, d is the crystallite size, and σ_0 and k_{gb} are constants. With the crystallite size decreasing to this limit, the fraction of the material in the grain boundaries increases, which leads to a decrease in material strength and hardness due to increased grain boundary sliding.[32–36] A simple phenomenological model has been used to describe the softening in terms of an increasing volume fraction of the grain boundary material f_{gb} with the crystallite size decreasing below 1–6 nm[37]:

$$H(f_{gb}) = (1 - f_{gb})H_c + f_{gb}H_{gb} \tag{10.5}$$

with $f_{gb}\alpha(1/d)$. Due to the flaws present, the hardness of grain boundary materials h_{gb} is smaller than that of the crystallites H_c. Thus the average hardness of the material decreases with d decreasing below 10 nm, an effect commonly known as the reverse Hall-Petch effect.[38] Recent computer simulation studies confirm that the reverse Hall-Petch dependence in nanocrystalline materials is due to the grain boundary sliding that occurs due to a large number of small sliding events on atomic planes at the grain boundaries, without thermal activation. This will ultimately impose a limit on how strong the nanocrystalline metal may become.[39,40] Although many details are still not understood, there is little doubt that grain boundary sliding is the reason for softening in this crystallite size range. Therefore a further increase of the strength or hardness with decreasing crystallite size can be achieved only if grain boundary sliding is blocked by appropriate design of the material.[40,41] We have tried to achieve this condition by having a uniform distribution of isolated nanocrystals in an amorphous thin-film matrix. In such situations, the grain boundaries formed at the interface are very strong and avoid or greatly reduce the grain boundary sliding. We believe that this effect is primarily responsible for the higher values of hardness for metal-alumina thin-film composites composed of isolated nanocrystalline metal particles (size below 10 nm) in amorphous matrices.

10.4.2 Structural Characterization

Shown in Fig. 10.8 is the cross-sectional STEM-Z micrograph of Ni nanocrystallites embedded in an alumina matrix. The Z-contrast in the STEM provides an incoherent image in which changes in atomic structure and composition across an interface can be interpreted directly without the need for preconceived atomic structure models.

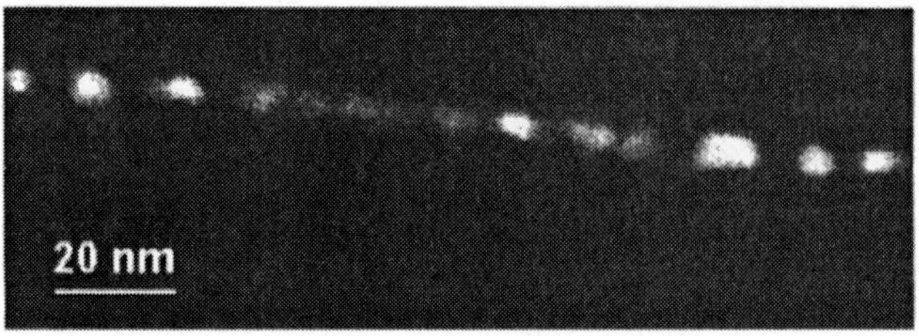

FIGURE 10.8 STEM-Z image of the Ni nanoparticles embedded in the alumina matrix.

This is true provided the incident electron probe is smaller than the lattice spacing (for a sample oriented along a major axis) or atomic columns. Since the probe is scanned, the resultant image is a map of the column scattering power, which in turn depends on the atomic number Z^2 of each column. The spatial resolution is limited primarily by the probe size, which is 1.2 Å in our field emission instrument. The Z-contrast image in the STEM is formed by electrons scattered through high angles. The average size of the Ni nanocrystals was determined to be ~5 nm. It is clear from the cross-sectional images that the Ni nanoparticles are mostly uniform in size and are well separated from each other. The average intralayer separation between the particles is estimated to be ~2–5 nm. In order to evaluate the crystalline quality of the metal particles and the alumina matrix, a high-resolution STEM-Z was recorded (Fig. 10.9) from a single metal particle and the alumina matrix in its immediate vicinity. It is clear

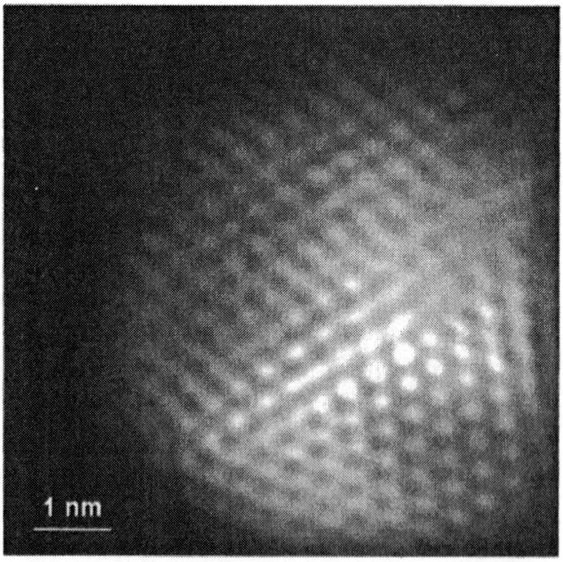

FIGURE 10.9 A STEM-Z image of a single Ni nanoparticle and the adjoining alumina matrix.

from this figure that the Ni nanoparticle is crystalline and the alumina matrix is amorphous in nature. Such a biphase system is formed as both materials are immiscible (i.e., they display thermodynamically driven segregation during deposition) and the cohesive energy at the interface between both phases is high. The amorphous phase of the matrix can provide high structural flexibility in order to accommodate the coherency strain without forming dangling bonds, voids, or other flaws as observed in Fig. 10.9. In this case, the nanostructure would be stable and grain boundary sliding is greatly reduced.

The improvement in the values of hardness of Al_2O_3 by embedding Ni and Fe nanoparticles may also be explained using the Koehler concept of multilayers for the design of strong solids.[42] According to this concept, under applied stress a dislocation that would form in a softer layer (metal in the present study) would move toward the metal-ceramic interface, and elastic strain in the second layer (alumina in the present study) with the higher elastic modulus would cause a repulsive force that would hinder the dislocation from crossing that interface. Therefore the hardness of Ni-Al_2O_3 and Fe-Al_2O_3 thin-film composites, which are essentially multilayered structures consisting of continuous layers of alumina thin films and discontinuous films of metals, is much stronger than expected from the rule of mixtures.

10.5 CONCLUSIONS

In summary, thin-film composite materials consisting of metallic nanocrystals embedded in an insulator were fabricated. These films exhibited significantly improved mechanical properties. The best hardness values measured using a nanoindentation technique were 20–30% larger than the hardness of pure alumina films fabricated under identical conditions. The improvement in the hardness of Al_2O_3 thin films by embedding metal nanocrystals is related to the evolution of a microstructure that impedes the manipulation and movement of dislocations and the growth of microcracks. The impedance in grain boundary movement is brought about by grain boundary hardening due to formation of well-separated metallic nanocrystallites in amorphous matrices.

PROBLEMS

1. Describe the role of the pulsed laser deposition method in the fabrication of thin-film nanocomposite materials.
2. Discuss the mechanism for the improvement in ceramic thin films by the addition of nanodimensional metal particles.
3. Explain how the mechanical properties of hard materials such as alumina can be improved by the addition of soft materials such as nickel or iron.

ACKNOWLEDGMENTS

The authors would like to thank Dr. S.J. Pennycook of Oak Ridge National Laboratory for collaboration on the STEM-Z characterization. Partial funding for this research was obtained from a U.S. National Science Foundation nanoscale exploratory research (NER) grant and from the Center for Advanced Materials and Smart Structures.

REFERENCES

1. Stamm, C., Marty, M., Vaterlaus, A., Weich, V., Egger, S., Maier, U., Ramsperger, U., Fuhrmann, H., and Pescia, D., *Science, 282,* 449, 1998.
2. Lu, L., Sui, M.L., and Lu, L., *Science, 287,* 1463, 2000.
3. Timp, G., Ed., *Nanotechnology,* Springer, New York, 1999.
4. Kumar, D., Zhou, H., Nath, T.K., Kvit, A.V., and Narayan, J., *Appl. Phys. Lett., 79,* 2817, 2001.
5. Kumar, D., Narayan, J., Kvit, A.V., Sharma, A.K., and Sankar, J., *J. Mag. Mater., 232,* 161, 2001.
6. Prinz, G.A., *Science, 282,* 1660, 1998.
7. Mehn, M., Punadjela, K., Bucher, J., Rousseausx, F., Decanini, D., Bartenlian, B., and Chappert, C., *Science, 272,* 1782, 1996.
8. Katiyar, P., Kumar, D., Nath, T.K., Kvit, A.V., Narayan, J., Chattopadhyay, C., Gilmore, W.M., Coleman, S., Lee, C.B., Sankar, J., and Singh, R.K., *Appl. Phys. Lett., 79,* 1327, 2001.
9. Ballesteros, J.M., Serna, R., Solis, J., Afonso, C.N., Petford-Long, A.K., Osborne, D.H., and Haglund, Jr., R.F., *Appl. Phys. Lett., 71,* 2445, 1997.
10. Hirono, S., Umemura, S., Tomita, M., and Kaneko, R., *Appl. Phys. Lett., 80,* 425, 2002.
11. Veprek, S., *J. Vac. Sci. Technol. A, 17,* 2401, 1999.
12. Cselle, T. and Barimani, A., *Surf. Coat. Technol., 76–77,* 712, 1995.
13. Rodeghiero, E.D., Tse, O.K., Chisaki, J., and Giannelis, E.P., *Mater. Sci. Eng. A, 195,* 151, 1995.
14. Sekino, T., Nakajima, T., Ueda, S., and Niihara, K., *J. Am. Ceram. Soc., 80,* 1139, 1997.
15. Lu, K., Wang, T., and Wei, W.D., *J. Appl. Phys., 69,* 522, 1991.
16. De, G., Tapfer, L., Catalano, M., Battaglin, G., Caccavale, F., Gonella, F., Mazzoldi, P., and Haglund, Jr., R.F., *Appl. Phys. Lett., 68,* 3820, 1996.
17. Liao, H.B., Xiao, R.F., Fu, J.S., Yu, P., Wong, G.K.L., and Sheng, P., *Appl. Phys. Lett., 70,* 1, 1997.
18. Mazzoldi, P., Arnold, G.W., Battaglin, G., Bertoncello, R., and Gonella, F., *Nucl. Instrum. Methods Phys. Res. B, 91,* 478, 1994.
19. Chrisey, D.B. and Hubler, G.H., *Pulsed Laser Deposition of Thin Films,* John Wiley & Sons, New York, 1994.
20. Singh, R.K. and Kumar, D., *Mater. Sci. Eng., R22,* 113–185, 1999.
21. Mcgibbon, A.J., Pennycook, S.J., and, Angelo, J.E., *Science, 269,* 519, 1995.
22. Pennycook, S.J., *Ann. Rev. Mater. Sci., 22,* 171, 1992.
23. Singh, R.K. and Narayan J., *Mater. Sci. Eng., B3,* 2923, 1989.
24. Bloemerben, N., in Ferris, S.D., Leamy, H.J., Poate, J.M., Eds., *Laser Solid Interactions,* American Institute of Physics, New York, 1979, p. 1.
25. Singh, R.K., Bhattacharya, D., and Narayan, J., *Appl. Phys. Lett., 57,* 2022, 1990.
26. Singh, R.K., *J. Non-Crys. Solids, 178,* 199, 1994.
27. Reddy, J.F., *Effects of High Power Laser Radiation*, Academic Press, New York, 1971.
28. Nakayama, T., Okigawa, M., and Itoh, N., *Nucl. Instru. Meth., B1,* 301, 1984.
29. Hertzberg, R.W., *Deformation and Fracture Mechanics of Engineering Materials,* 3rd ed., John Wiley & Sons, New York, 1989.
30. Hall, E.O., *Proc. Phys. Soc. London, Sect. B64,* 747, 1951.
31. Petch, N.J., *J. Iron Steel Inst.*, London, *174,* 25, 1953.
32. Siegel, R.W. and Fougere, E., *Mater. Res. Soc. Symp. Proc., 362,* 219, 1995.

33. Siegel, R.W. and Fougere, G.E., *Nanostruc. Mater, 6,* 205, 1995.
34. Hahn, H. and Padmanabhan, K.A., *Philos. Mag. B, 76,* 559, 1997.
35. Yip, S., *Nature, 391,* 532, 1998.
36. Schiotz, J., Di Tolla, E., and Jacobsen, K.W., *Nature, 391,* 561, 1998.
37. Carsley, J.E., Ning, J., Milligan, W.M., Hackney, S.A., and Aifantis, E.C., *Nanostruc. Mater., 5,* 441, 1995.
38. Chokshi, A.H., Rosen, A., Karch, J., and Gleiter, H., *Scr. Metall., 23,* 1679, 1989.
39. Nieman, G.W., Weertman, J.R., and Siegel, R.W., *J. Mater. Res., 6,* 1012, 1991.
40. Sanders, P.G., Youngdahl, C.J., and Weertman, J.R., *Mater. Sci. Eng. A, 234,* 77, 1997.
41. Youngdahl, C.J., Sanders, P.G., Eastman, J.A., and Weertman, J.R., *Scr. Mater., 37,* 809, 1997.
42. Koehler, J.S., *Phys. B, 2,* 547, 1970.

11 Macroscopic Fibers of Single-Walled Carbon Nanotubes

Virginia A. Davis and Matteo Pasquali

CONTENTS

11.1 INTRODUCTION

On the nano- to microscale, theoretical calculations and experimental measurements indicate that single-walled carbon nanotubes (SWNTs) have electrical conductivity and current-carrying capacity similar to copper,[1,2] thermal conductivity higher than diamond,[3,4] and mechanical strength higher than any naturally occurring or man-made material.[5,6] Although precise numbers are hard to pinpoint because of the limited accuracy of theoretical calculations and experiments on small bundles of nanotubes (NTs), Baughman et al. cite ballistic conductance in metallic SWNTs, a theoretical thermal conductivity of 2000 W/m K, an elastic modulus of 640 GPa, and a tensile strength of 37 GPa.[7] Some microelectronic and materials characterization applications (e.g., an atomic force microscopy tip) may be able to utilize the properties of individual and small groups of SWNTs. However, many important applications require low-cost synthesis of SWNTs and manufacturing methods that preserve some or all of the nanoscale properties in a macroscopic object.[8,9] Whereas at least one solution to the economic mass production of SWNTs is now available through the HiPco process[10] and potentially by other scalable routes,[11,12] technologies for processing SWNTs into macroscopic materials are at a much earlier stage. Some of the most promising ones for manufacturing SWNT fibers are reviewed here.

Macroscopic carbon nanotube fibers have the potential to form high-strength, lightweight, thermally and electrically conducting structural elements at lower cost than other forms of SWNTs such as unaligned bucky paper.[13] Whereas futuristic applications such as space tethers and the space elevator[14] will require ultra-strong SWNT fibers, many other applications will require supplementary multifunctional properties and not such high mechanical strength.[15] The electrical properties may be used for highly efficient transmission of electricity over long distances. Thermal properties could be exploited in microelectronic applications where thermal management is an increasing problem as miniaturization progresses.[14] The extremely high surface area of SWNTs (above 300 m^2/g) could be exploited in carbon nanotube supercapacitors, electrochemically driven artificial muscles, hydrogen storage, and electrical energy harvesting.[15] Carbon nanotube fiber actuators have already been tested at temperatures in excess of 350°C with a maximum observed actuator stress of 26 MPa, which is roughly 100 times higher than natural muscle.[7] Potential uses include robots for planetary exploration and controlling blade pitch in jet engine rotors at temperatures above 1000°C.[16]

Fibers containing SWNTs have been produced by four main types of processes: solid-state production or growth, electrophoretic processing, melt spinning, and solution spinning — i.e., liquid-state processing. Melt and solution spinning seem to be the most viable techniques for commercial scale-up; moreover, continuous fibers consisting predominantly or solely of SWNTs have been produced only by solution spinning. This chapter emphasizes liquid-state processing and only briefly summarizes other techniques.

11.2 FIBERS PRODUCED DIRECTLY FROM SWNT SYNTHESIS

Fibers of carbon nanotubes micrometers to millimeters long have been produced by variations of chemical vapor deposition (CVD).[8,17,18] For example, lengths up to 20 μm have been achieved using 1,3-dicyclohexylcarbodiimide to polymerize oxidized SWNTs into strands 50 to 150 nm in diameter.[19] Ribbons with preferential alignment (50–140 μm wide, 4–40 μm thick and 100 mm in length) have also been produced by heating oxidized acid-treated tubes at 100°C. Graphitization of the ribbons at 2200°C for 2 h under argon with a pressure of 0.5 MPa resulted in an increase of ribbon density from 1.1 to 1.5 g/cm^3 and an increase in Young's modulus from 24 to 60 GPa.[20] Electrical resistivity of as-grown ribbons was roughly 1 mΩ cm at room temperature.

On this microscale, alignment during tube growth can be enhanced by electric fields. Aligned SWNTs on the order of 10 μm can be grown under an electric field in the range of 0.5–2 V/μm.[21] Aligned CVD-grown nanotubes have been self-assembled into yarns up to 30 cm in length by drawing them out of arrays several hundred micrometers in height in a process similar to drawing silk out of a cocoon;[22] an array of 1 cm^2 area and 100 μm height could generate about 10 m of yarn. The diameter of the yarn can be controlled by the size of the drawing tool, with smaller tools generating smaller diameter yarns. The yarns were used to construct a light bulb filament and a polarizer suitable for use in the ultraviolet region.

Zhu et al. produced aligned strands up to 20 cm in length and several hundred micrometers in diameter using a floating catalyst method in a vertical furnace where n-hexane was catalytically pyrolyzed in the presence of a sulfur-containing compound (thiophene) and hydrogen. Typically, ferrocene-assisted CVD produced MWNTs at temperatures around 800°C and a mixture of SWNTs and MWNTs at temperatures above about 1000°C. The thiophene increased the yield of SWNTs; the temperature and flow rates were optimized to produce long SWNT strands in a continuous process.[23] The strands contained roughly 5 wt% iron and amorphous carbon impurities. These strands were tweezed into smaller ones 5 to 20 μm in diameter. SEM showed a structural hierarchy in the fibers, which consisted of thousands of well-aligned bundles of SWNTs (diameter 1.1 to 1.7 nm) arranged in a two-dimensional triangular lattice. The volume fraction of SWNTs in the strands was estimated to be less than 48% based on the spacing between ropes measured by SEM. Raman spectroscopy showed that the dominant SWNT diameter was 1.1 nm.[23] X-ray diffraction showed a full-width half-maximum (FWHM) of 44°.[24] Electrical resistivity on strands ranging in diameter from 50 μm to 0.5 mm was measured with a four-probe method. The crossover temperature (metallic to semiconducting) was approximately 90 K; from 90 to 300 K the behavior was metallic with resistivity $\rho = 0.55$ to 0.7 mΩ cm.[23] Tensile testing was complicated by difficulties in measuring the cross-section of the strand (and thus the true stress) during testing. The starting diameter was measured by SEM and was in the range of 5–20 μm. The estimated Young's modulus ranged from 49 to 77 GPa. The authors assert that if one bases the Young's modulus on a cross-section corresponding to a nanotube volume fraction of less than 48%, the result is 150 GPa, consistent with the modulus of SWNT bundles.[23]

Li et al. have demonstrated the ability to produce continuously SWNT and MWNT fibers during chemical vapor deposition.[25] This process entails continuously winding an aerogel of high-purity nanotubes formed in the furnace hot zone onto a rotating rod. By changing the angle of the rod relative to the furnace axis, fibers with different degrees of twist can be produced. As with any chemical vapor deposition process, the reactor conditions can be controlled to produce either SWNTs or MWNTs; thus either SWNT or MWNT fibers can be produced by varying the reactor conditions. However, the SWNT fibers contain approximately 50 vol% impurities compared to approximately 10% for the MWNT fibers. Both types of fibers showed high degrees of alignment. Electrical resistivities as low as 0.12 mΩ cm were achieved. Mechanical strength varied from 0.1 to 1 GPa, assuming a typical carbon fiber density of 2 g/cm^3.[25]

11.3 ELECTROPHORETIC SPINNING

Gommans et al. have spun fibers electrophoretically from purified laser vaporization-grown SWNTs dispersed in n,n-dimethylformamide (DMF) at concentrations of about 0.01 mg/ml.[26] A commercially available carbon fiber (8 μm in diameter, about 12 mm long) was connected to a conducting wire (0.5 mm diameter, 10 mm long) with silver conducting paint. The wire was attached to a motor-driven stage via a pin vise. The carbon fiber translated along its axis into the suspension at a depth of a few millimeters. The carbon fiber was turned into a positive electrode by applying voltage, causing the SWNTs to migrate toward it and to form a cloud around the carbon fiber. SWNTs migrate because they are negatively charged in DMF and move electrophoretically toward the positively charged carbon fiber. As the carbon fiber was slowly withdrawn from the suspension, another fiber, attached to its end, spontaneously formed from the SWNT cloud. The fiber length was limited by the travel distance of the translation stage, the size of the SWNT cloud, and the smoothness of the withdrawal from the solution. Fibers were typically several centimeters long with diameters between 2 and 10 μm. The mass of SWNTs below the bath surface and the surface tension in the meniscus promoted the coalescence and axial alignment of bundles of SWNTs. Alignment was measured by polarized Raman spectroscopy and the axial-versus-perpendicular ratio was in the range of 2 to 6.[26]

11.4 CONVENTIONAL FIBER SPINNING

There are three main types of commercial fiber spinning processes (Fig. 11.1): melt spinning, dry solution spinning, and wet solution spinning.[27] Melt spinning is used to produce many polymer fibers, such as nylon and polyethylene terephthalate fibers, for commodity applications such as residential carpeting; it can be applied to single-component as well as composite fibers. In melt spinning, the fiber-forming material is melted and extruded under tension, typically into cooled air; the rapid cooling induces the solidification of the fiber. Some polymer/SWNT composite fibers of which polymer is the major constituent have been produced using this method. However, SWNTs decompose without melting at approximately 750°C in air and at approximately 2000°C in an inert atmosphere (e.g., argon); thus melt spinning is not a viable option for fibers in which SWNTs are the sole components.

Most SWNT fibers have been produced by the solution spinning process. Solution spinning is more complicated than melt spinning because the solidification of the fiber involves additional steps. The fiber-forming material must be dissolved or finely dispersed into a solvent, and the solvent must be extracted after the extrusion to form the solid fiber. Therefore, solution spinning is typically used to produce fibers from materials that decompose before reaching their melting point or do not have a suitable viscosity for stable fiber formation. Solution spinning has been particularly effective for spinning fibers from stiff polymer molecules, e.g., poly(p-phenylene terephthalamide) (PPTA), that form liquid crystals when dispersed in a solvent.

Several research groups have used solution spinning to produce bulk quantities of SWNT fibers (continuous lengths of 1 m or longer). Solution spinning can be considered as a four-step process: (1) dispersion or dissolution of the fiber material

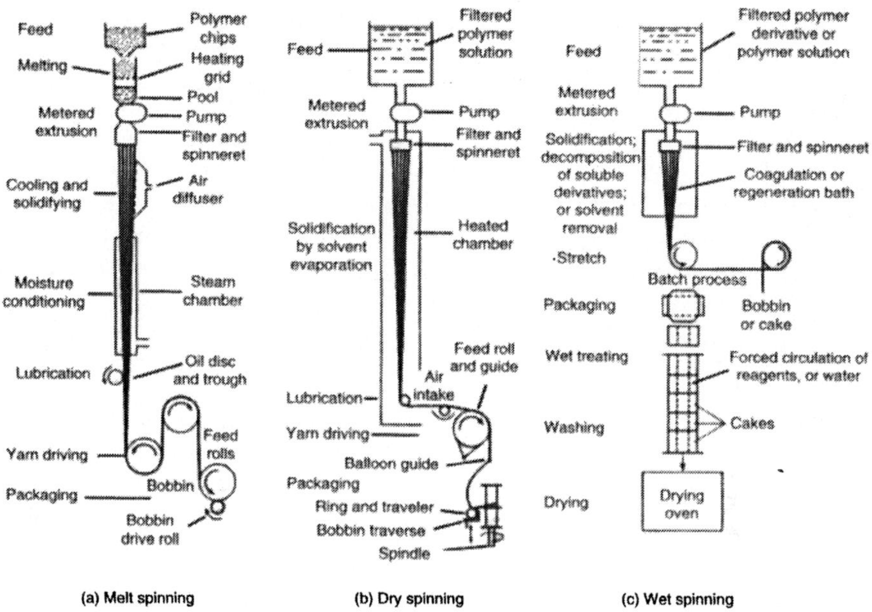

FIGURE 11.1 Three main types of spinning processes: melt spinning, dry solution spinning, and wet solution spinning.[27] (Reprinted with permission from Griskey, R.G., *Polymer Process Engineering,* Chapman & Hall, New York, 1995. Copyright CRC Press, Boca Raton, Florida.)

into a solvent, (2) mixing and spinning the dispersion, (3) coagulation and drawing into a solid fiber, and (4) postprocessing of the fiber through subsequent washing, drying, or annealing steps.

Solution spinning can be divided into two main categories: dry solution spinning and wet solution spinning (Fig. 11.1). In both cases, the spinning solution consists of the polymer dissolved in a solvent. The solution is extruded through one or more small orifices in the spinneret and the solvent is removed, solidifying the fiber. Typically the polymer content is less than 30 wt% and sometimes less than 10 wt%. In many cases, the solution is formed at the molecular level. Dry spinning is used for systems such as cellulose acetate in acetone where the solvent is sufficiently volatile that it can be evaporated rapidly from the fiber during its formation. Wet spinning is used when the polymer is dissolved in a nonvolatile solvent that must be extracted by using another liquid that is miscible with the solvent but cannot dissolve the fiber-forming material. A key example of a wet solution-spun fiber is DuPont Kevlar™ (Wilmington, DE), produced from poly(p-phenylene terephthalamide) (PPTA) dissolved in sulfuric acid. In wet solution spinning (Fig. 11.1), the solution typically goes through an air (or inert gas) gap and then enters the coagulation bath. In this case, the process is termed dry-jet wet spinning. The air gap allows elongation and cooling of the spinning solution prior to coagulation. The presence and length of the air gap depend on the ability of the polymer solution exiting the spinneret to withstand surface forces prior to entering the coagulation bath. Liquid filaments are unstable to surface waves, which are amplified by surface tension (Rayleigh instability);[28] higher surface tension and small filament diameters promote the growth of the waves and the eventual breakup of the filament into droplets, whereas viscous forces delay the development of the instability. High-viscosity fibers as well as thicker fibers can normally support a larger air gap, particularly if the polymer is extension-thickening, i.e., its viscosity in extension grows with strain rate. Typical air gaps used in PPTA spinning are less than 2 cm; a gap of 0.5 to 1.5 cm is considered optimal.[29] If larger gaps are used, the (liquid) fiber breaks before entering the coagulation bath. In contrast, higher viscosity poly(p-phenylene benzobisthiazole) (PBZT)/poly(phosphoric acid) and (p-phenylene benzobisoxazole) (PBO)/poly(phosphoric acid) spinning solutions can support air gaps of up to 20 cm without fiber line breakage.[29] If the fiber cannot support an air gap at all, the spinneret is immersed directly into the coagulation bath and the process is termed wet-jet wet spinning.

Upon entering the coagulation bath, mass transfer occurs between the coagulant (nonsolvent) in the bath and the solvent in the spinning solution, which can now be considered the protofiber.[30,31] The goal of the coagulation process is to freeze the structure of the protofiber and solidify the fiber. The final properties of the resulting fiber depend on the initial polymer solution, the coagulation process,[29,32–35] drawing of the fiber, and postprocessing steps such as secondary washing[29,34] or drying[36] of the fiber.

The first step of the solution spinning process, dispersal into a solvent, is far from trivial for SWNTs.[7] Due to high van der Waals interactions, it is typically only possible to disperse less than 0.5 wt% of SWNT in small molecule organic solvents such as dichlorobenzene or aqueous solutions of surfactants such as sodium dodecyl sulfate (SDS), although recent work has shown that superacids can disperse SWNTs at higher concentration.[37–39] Current research on SWNT functionalization and wrapping may provide new means for dissolving higher concentrations of SWNTs in these liquids for fiber production.[40–43]

Finding an appropriate dispersion for the SWNTs is not sufficient for producing high-quality fibers; understanding the phase behavior of the dispersion is critical.[15,37,44,45] The phase behavior and the related rheological properties of the dispersion profoundly influence the selection of the spinning solution concentration, spinning and coagulation process variables, and alignment in the as-spun fibers. Ultimate fiber properties are determined by these variables as well as subsequent washing, drawing, drying, and annealing processes.[46]

11.4.1 Melt-Spun Composite Fibers

Reinforcement of existing materials with nanomaterials including SWNTs is a topic of significant current interest because relatively low loadings of the costly component (SWNTs) could yield significant improvements in properties. Improvements in electrical and thermal conductivity are of particular interest, although mechanical reinforcement is also an important application. As with the development of other SWNT materials, this field is still in its infancy; current results should be viewed as preliminary and not indicative of the full potential of SWNT composite fibers.

Some of the first melt-spun composite SWNT fibers were produced by Andrews et al. using carbon pitch as the main component.[47] SWNTs from Carbolex, Inc. (Lexington, KY) were purified and mixed with isotropic petroleum pitch in quinoline. After sonication and solvent removal, the fibers were spun, drawn, and put through a two-stage heat treatment in air and nitrogen. Mixtures containing 8 and 10 wt% SWNT yielded fibers that were too weak to be collected; however, fibers produced with 1 wt% and 5 wt% SWNTs showed improved mechanical and electrical properties. Improvements were a 90% increase in tensile strength, a 150% increase in elastic modulus, and a 70% drop in electrical resistivity (Table 11.1).

Haggenmueller et al. produced SWNT polymethylmethacrylate (PMMA) films and fibers.[48] Three types of SWNT were investigated: raw and purified soot from laser ablation, and Tubes@Rice material. The SWNTs were dispersed in PMMA (M_w = 10,000) in dimethyl formamide (DMF) by prolonged sonication. After evaporation of the DMF, the resulting SWNT/PMMA pellets were melt mixed, spun from a 600-μm orifice, and drawn under tension by a windup spool at high speed. The maximum draw ratio (defined as $\lambda = (D_0/D)^2$) ranged from 60 to 3600, corresponding to fiber diameters between 75 and 10 μm, respectively. Increasing SWNT content dramatically increased the melt viscosity; this resulted in melt fracture evidenced by surface roughness, striations along the fiber axis, and nonuniform diameter. Increasing the SWNT weight fraction also decreased the attainable draw ratios and increased the frequency of fiber breaks; fibers were successfully produced from loadings up to 8 wt% SWNT. The 10 wt% SWNT mixture was too viscous to be extruded. Mechanical and electrical properties depended on SWNT concentration and draw ratio. For melt-pressed films, at a draw ratio of $\lambda = 4$, the electrical resistivity in the direction parallel to processing decreased approximately 100-fold from 850 Ω cm at 1.3 wt% SWNT to 8.7 Ω cm at 6.6 wt%. In the direction perpendicular to processing, the electrical resistivities were only slightly higher: 1280 and 14 Ω cm, respectively. The fibers were highly anisotropic; their conductivity was below the detection limit, in agreement with the fact that weight fractions of SWNTs for percolation increase as the nanotubes align. At a draw ratio of $\lambda = 70$, the elastic modulus G increased from 3.1 GPa in the absence of SWNT to 6 GPa for 8 wt% SWNT. The draw ratio had a small effect on the modulus but significantly affected the yield stress. In fibers containing 5 wt% purified SWNT soot, the yield stress doubled from 65 to 130 MPa as the draw ratio was raised from $\lambda = 40$ to 300. Similar results were observed with the other types of SWNTs. These improvements are consistent with the high degree of alignment measured by polarized Raman

TABLE 11.1
Properties of SWNT/Pitch Fibers[45,47]

Sample	Tensile Strength [MPa]	Elastic Modulus [GPa]	Electrical Resistivity [mΩ cm]
Pitch	480	33	8.4
Pitch/1 wt% SWNTs	620	41	2.5
Pitch/5 wt% SWNTs	850	77	2.2

spectroscopy at multiple angles relative to the fiber axis; a fiber containing 1 wt% purified soot had a full width half maximum (FWHM) of 4°.[48]

Efforts to incorporate SWNTs into polypropylene (PP) fibers have built on earlier success in incorporating carbon nanofibers into thermoplastics.[49–52] Bhattacharyya et al.[53] produced 0.8 wt% SWNT-PP fibers using HiPco SWNTs and polypropylene of 17 melt flow index (MFI). The SWNTs were purified (less than 1 wt% impurities) using methods found in the literature,[54] mixed with the PP in a Haake Rheomix 600 at 240°C, and filtered to remove the largest aggregates. This preparation procedure, however, failed to disperse satisfactorily the SWNTs that remained clumped into domains of tens of micrometers. The SWNTs were found to promote faster PP crystallization, probably by acting as nucleation sites, and to yield smaller PP crystallites. The SWNTs in the final fiber were better oriented axially than the PP. Mechanical properties were largely unaffected by the presence of the SWNTs as a result of the poor dispersion. The mechanical properties are expected to improve as better methods for dispersing the SWNTs in PP are found.

11.4.2 Solution-Spun SWNT Fibers

11.4.2.1 SWNT/Liquid Crystalline Polymer Composite Fibers

Improving the properties of high-performance fibers such as poly(p-phenylene benzobisthiazole) (PBO) through the incorporation of SWNTs is an important potential application of SWNTs. Commercial PBO fiber (Zylon™) is one of the strongest fibers with a tensile strength of 5.6 GPa.[55] Kumar et al.[56] synthesized PBO in poly(phosphoric acid) (PPA) in the presence of purified HiPco SWNTs at concentrations of 0, 5, and 10 wt%. Optical microscopy showed that the SWNTs were well dispersed in the liquid crystalline PBO/PPA solution. The liquid crystalline solution was dry-jet wet spun into a water coagulation gap with an air gap of 10 cm and draw ratios as high as 10. SWNT loading improved modulus, tensile strength, and elongation to break. For 10 wt% the improvements were 20, 60, and 40%, respectively (Table 11.2). Achieving similar improvements with commercial Zylon fibers (twice as strong as the laboratory control) could yield fibers with a tensile strength in excess of 8 GPa. Very importantly the incorporation of SWNTs improved the compressive strength of the fibers. The coefficient of

TABLE 11.2
Mechanical Properties of SWNT/PBO Fibers[56]

Sample	Tensile Modulus (GPa)	Strain to Failure (%)	Tensile Strength (GPa)	Compressive Strength (GPa)
PBO	138 ± 20	2.0 ± 0.2	2.6 ± 0.3	0.35 ± 0.6
PBO/5 wt% SWNTs	156 ± 20	2.3 ± 0.3	3.2 ± 0.3	0.40 ± 0.6
PBO/10 wt% SWNTs	167 ± 15	2.8 ± 0.3	4.2 ± 0.5	0.50 ± 0.6

thermal expansion dropped from −6 ppm/°C for the pure PBO fiber to −4 ppm/°C at 10% SWNT loading. Thermal degradation was not affected by the presence of SWNTs. The degree of alignment of the PBO in the fiber was also unaffected by the SWNTs. The electrical conductivity was below the limit of detection. The authors state that such low conductivity indicates both good dispersion and strong alignment; however, either one of these two morphological features would suffice to explain the lack of conductivity, because 10% is well below the percolation threshold of aligned slender anisotropic objects as well as randomly packed isotropic ones.[57]

Polyacrylonitrile (PAN) copolymers are used as carbon fiber precursors and for developing porous activated carbon that can be used in various applications (e.g., catalysis, electrochemistry, separations, and energy storage). Sreekumar et al.[58] produced PAN/SWNT fibers by dry-jet wet solution spinning. Composite fibers with 10 wt% SWNT had dramatically better properties than the pure PAN fibers. The tensile modulus nearly doubled from 7.9 GPa (pure PAN fiber) to 14.2 GPa (5 wt% SWNT fiber) and 16.2 GPa (10 wt% SWNT fiber). Tensile strength was not improved as much. The pure PAN fiber had a strength of 0.23 GPa and the 5% and 10% fibers had strengths of 0.36 GPa and 0.33 GPa, respectively. All fibers showed similar elongation at break (about 10%). The presence of SWNTs improved thermal stability. The glass transition temperature grew from 103°C (pure fiber) to 114°C (5 wt% fiber) to 143°C (10 wt% SWNT). The SWNT/PAN fibers retained their elastic modulus better than the pure PAN fibers at high temperature. (Interestingly the 5% SWNT fibers retained modulus better than the 10% fibers.) At 200°C, thermal shrinkage of the 10% fiber was nearly half that of the control. Supercapacitor films of SWNT/PAN were produced by a similar dispersion technique followed by film casting in vacuum.[59]

11.4.2.2 Fibers Produced from SWNT/Surfactant Dispersions

The first truly scalable continuous process for making fibers composed primarily or solely of SWNTs was developed by a team at the Centre National de la Recherche Scientifique (CNRS) and the Université Bordeaux.[60,61] This process produced fibers composed of a network of SWNTs and a polymer, typically polyvinyl alcohol (PVA). The SWNTs were dispersed in an aqueous solution with the aid of surfactants and then injected into a flowing polymer (PVA) solution. The polymer can be removed by washing the fiber repeatedly and by heating the fiber at temperatures between the decomposition temperature of the polymer and that of the SWNTs. The original process developed at CNRS employs a rotating polymer solution bath mounted on a turntable;[60–62] a modified process developed by the University of Texas at Dallas uses a coflowing stream of polymer solution.[63,64]

SWNT suspensions are analogous to solutions of rod-like polymers or anisotropic colloids.[44] Just as the nature of surfactants plays a critical role in the phase behavior of classical colloids, the same is true for nanotubes. In the CNRS process, SWNTs were sonicated in aqueous solutions of sodium dodecyl sulfate (SDS), an anionic surfactant that adsorbs at the surfaces of SWNT bundles. Similar adsorption behavior has also been observed with cationic surfactants such as tetratrimethylammonium bromide. Vigolo et al.[44,61] studied the phase behavior of SWNTs produced by electric arc, laser vaporization, and HiPco in SDS in order to identify the ranges of surfactant and SWNT concentration that yielded dispersions with optimal processability. The stabilization mechanism seems to be a balance between repulsive electrostatic forces induced by the charged groups on the surfactants and attractive forces induced by van der Waals forces between SWNTs (or bundles of SWNTs) and by micellar depletion.

At low surfactant concentration, the number of surfactant molecules adsorbed on the SWNTs present in small bundles after sonication was not sufficient to counterbalance the long-range van der Waals forces — van der Waals forces between parallel cylinders decay with $r^{-1.5}$, where r is the separation between the cylinders.[65] Therefore, the bundles stick back together after sonication and form large dense clusters. At high surfactant concentration, the excess surfactant that is not adsorbed on the SWNT bundles forms nanometer-sized micelles. As two SWNT bundles approach each other, they exclude micelles; this produces a difference of osmotic pressure,[65,66] which in effect brings the bundles closer together and causes the formation of flocs that coexist with well-dispersed thin bundles.[44] The proportion of clusters grows with surfactant concentration.

At intermediate surfactant concentration, the system forms an apparently homogeneous phase of well-dispersed small bundles. In the case of electric arc-produced SWNTs, this homogeneous phase exists up to approximately 0.35 wt% SWNT in 1 wt% SDS. Similar behavior was observed for SWNTs produced by all three production methods, but the exact phase boundaries were dependent on the production technique; the boundaries shifted to higher concentrations for the HiPco-produced SWNTs, probably because of differences in length and diameter distributions. The homogeneous dispersions with the highest content of SWNTs showed optimal processability, although fibers of HiPco SWNTs could be produced also from nonhomogeneous dispersions.[44] The origin of the SWNTs also influenced the ribbon/fiber morphology. For HiPco-produced tubes, a smaller weight fraction (approximately 0.15 wt%) was needed to make homogeneous fibers with a uniform and nearly cylindrical cross-section.[62] HiPco tubes also yielded stable fiber formation and good mechanical properties at higher velocities.[60]

Due to their anisotropy, SWNTs align in shear or elongational flow. Injecting the SWNT dispersion in a coflowing stream of polymer has two beneficial effects on the fiber formation. It aligns the SWNTs by imposing a small amount of extensional flow on the protofiber that can be controlled by the mismatch between the velocity of the SWNT dispersion being injected and the velocity of the polymeric bath. Moreover the rotating bath supports and transports the protofiber during the early stages of coagulation, while it is still too weak to support an applied tension. If the coagulation bath were not moving, the fiber would have to be drawn mechanically through the coagulation bath, which is not possible unless the liquid filament has a high enough viscosity and the ability to withstand tension along the streamlines (usually associated with extensional thickening). Most of the surfactant leaves the filament during the coagulation, while at the same time the PVA molecules penetrate the forming fiber and stabilize it by forming bridges with the SWNTs.[61,62] The SWNT/SDS dispersions were injected through a 0.5-mm diameter syringe needle into a rotating bath containing 2 to 6 wt% of PVA (M_w = 70,000, hydrolysis 89%, solution viscosity about 200 cP).[44,60,61] Amphiphilic polymers such as PVA are ineffective at stabilizing SWNTs; they stick to each other almost as soon as they enter the bath. This can be considered as coagulation through bridging flocculation.[62] Polarized optical microscopy of ribbons suspended in water showed that there is preferential alignment along the ribbon axis. SEM showed preferential axial alignment in the cleaned dry ribbons although the dried ribbons displayed weaker anisotropy than the ribbons suspended in water. Low-viscosity PVA solutions yielded fibers with lower alignment, possibly because they were less effective at rapidly stretching the extruded filament.

High strain rates in the processing flow yielded better aligned as-spun fibers.[60] The relevant quantities include the shear rate in the syringe tube, the ratio of the viscosity of the coagulant and SWNT dispersion, and the ratio between the coagulant velocity and the extrusion speed of the SWNT dispersion. The shear rate in the syringe tube should promote axial alignment of the SWNT bundles before they meet the PVA solution. A higher coagulant/SWNT dispersion viscosity ratio should accelerate the filament over a smaller distance, i.e., produce a higher rate of extension immediately after the syringe orifice. The ratio of the velocity of the rotating PVA bath and the extruded SWNT dispersion controls the total extensional strain that is experienced by the filament as it coagulates into a ribbon.[60] However, variations in polymer solution speed between 300 and 800 m/h and the SWNT dispersion injection rate (35–100 ml/hr) did not affect the orientation in the dried fibers. This suggests that alignment is fixed during the drying stage and not during the initial coagulation process.[62]

The fibers from the CNRS process possess a hierarchical structure with well-developed porosity and high surface area (160 m^2/g by the Brunauer-Emmett-Teller (BET) method, compared to 200 m^2/g for standard bucky paper).[67] This makes them of interest for nanofluidic devices, adsorbents and absorbents, catalysis support, and host systems for encapsulation of biomolecules for sensors, smart textiles, and other nanotechnologies in which wetting and sorption properties play an important role. Cross-sections of fibers dried under tension reveal a well-defined core consisting mostly of SWNTs and a shell that contains most of the carbon impurities; such impurities were randomly distributed in the initial fibers. Thus, separation occurred as the fibers collapsed and this may provide a means of large-scale SWNT purification. Chemical irradiative or thermal removal of the shell would result in purified SWNTs.[61]

Neimark et al.[67,68] made detailed studies of the fiber morphology and gas adsorption. Gas adsorption measurements were made on fibers that had been heated in air at 320°C for 3 h. This removed 95% of the organic species while keeping the fibers intact. The fibers have several levels of structural organization; 10- to 50-μm fibers are composed of straight elementary filaments 0.2 to 2 μm in diameter with a hairy surface.[67] These elementary filaments consist of densely aligned nanotube bundles 10–30 nm in diameter, which in turn consist of SWNTs 1–2 nm in diameter. The elementary filaments are aligned in a closely packed configuration in the fiber skin of 1–5 μm diameter, while the core consists of a nanofelt of loosely packed bundles; the skin-core morphology resembles that of polyacrylonitrile (PAN)-based carbon fibers. This was confirmed by comparing alignment measured by x-ray scattering and polarized Raman spectroscopy. In x-ray scattering, where the whole fiber is exposed to radiation, the FWHM was 75°. In polarized Raman spectroscopy, where a limited boundary layer of about 1 μm is probed, the FWHM was about 35°; this indicates that the surface alignment is much better than the alignment in the core.

The original CNRS process resulted in fibers that showed a plastic deformation before breaking and could be curved through 360° within a length of tens of micrometers, demonstrating that the fibers are more flexible and resistant to torsion than classical carbon fibers.[61] Four-probe electrical resistivity measurements gave a room temperature resistivity of approximately 0.1 Ω cm and a nonmetallic behavior when the temperature was decreased. This value was influenced by the presence of the polymer and carbon impurities; it was three orders of magnitude higher than those of small bundles of SWNTs.[61] In an improvement to the original process, the washed and dried fibers were rewetted, swollen, and dried under tensile load with a weight attached to one end of the fiber. They could be stretched up to 160% and did not disassemble even in solvents where the polymer was highly soluble. This suggests that the SWNTs and adsorbed polymer form an effectively cross-linked network that can be easily deformed.[62] The qualitative effect of stretching on fiber morphology and alignment can be seen in Fig. 11.2. This effect can be quantified by the change in FWHM — from 75 to 80° for raw fibers to less than 50° for fibers with draw ratios of 125% or greater. At higher draw ratios, the SWNTs seemed to slide and unbind from the PVA; both the applied stress needed to achieve the higher draw ratio and the FWHM remained constant. The improvement in alignment with stretching translated into improved mechanical properties: The Young's modulus increased four-fold from about 10 GPa for the raw fibers to 40 GPa for fibers stretched 145% in water with a load of 0.44 g. The tensile strength of these fibers increased from 125 MPa to 230 MPa, respectively. The effectiveness of the stretching process was found to be dependent on the liquid used to rewet the fiber; poorer PVA solvents yielded better alignment. For a draw ratio of 117%, increasing the concentration of acetone (poor PVA solvent) in water (good PVA solvent) from 0% to 70% decreased the FWHM from 65° to 50°. Vigolo et al. hypothesize that in a poorer solvent there is less sliding and the cohesion of the SWNT-polymer network is better.[62]

The improved CNRS process has also been used to spin fibers from DNA-stabilized SWNT dispersions.[69] Typical fiber dimensions are 50 cm in length and 20 to 30 μm in diameter. Young's modulus as high as 19 GPa, and tensile strength as high as 130 MPa were achieved. These values did not vary significantly with stretching; they are two to three times better than those the authors achieved for unstretched fibers spun from SWNT-SDS dispersions. However, these values are similiar to to those achieved by Vigolo et al. (2002) for unstretched SWNT-SDS fibers and lower than those Vigolo et al. achieved for stretched SWNT-SDS fibers. The electrical resistivity of fibers produced from the DNA-stabilized dispersions was more than 50% greater than resistivities of the fibers produced from the SDS-stabilized suspensions.[69]

Baughman and coworkers at the University of Texas at Dallas have built on the CNRS work and developed a process for producing SWNT fibers, primarily from HiPco SWNTs dispersed in aqueous lithium dodecyl sulfate with PVA as a coagulant.[7,63,64]

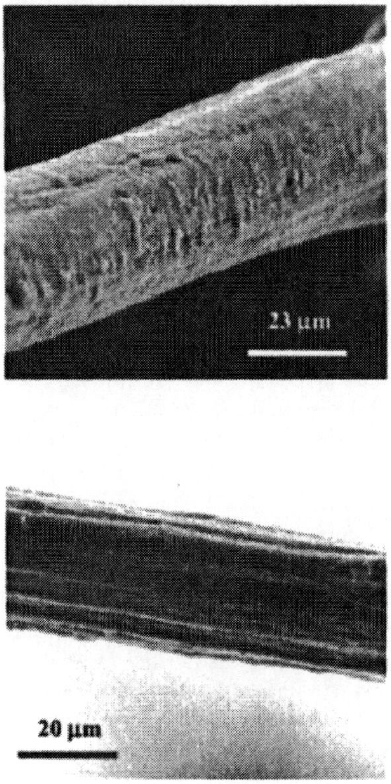

FIGURE 11.2 Effect of drying under load on alignment. (Reprinted with permission from Vigolo, B. et al. *Appl. Phys. Lett.*, *81*, 1210–1212, 2002. Copyright American Institute of Physics, Melville, NY.)

In this process, the spinning dispersion is injected into a stream of coagulant that is flowing in a pipe coaxial with the spinning needle. The gel protofibers are wound onto a mandrel and then passed through a series of acetone and water baths for drawing and solvent removal. The gel protofibers can be continuously processed at approximately 100 cm/min to produce reels of several hundred continuous meters of 50-μm diameter fibers consisting of approximately 60 wt% SWNT and 40% PVA.[70] Like Vigolo et al., Baughman et al. also found that HiPco tubes spin over a much broader concentration range and produced more mechanically robust ribbons.[63]

The fibers have impressive mechanical and electrical actuation properties.[64,70] The PVA forms a coating on the SWNTs and the mechanical properties are believed to result from internanotube stress transfer via the PVA.[70] Tensile strengths up to 3.2 GPa have been achieved; this is nearly twice that of spider silk, and about ten times higher than that of the CNRS fibers. The Young's modulus is 80 GPa, which equals that of the fibers grown by Zhu et al.[23] and is double that of fibers spun by Vigolo et al.;[62] however, this value is still roughly an order of magnitude less than that of high-performance graphite fibers or individual SWNTs. Normalized for density, the Young's modulus and tensile strength of these fibers are more than twice that of steel and the fibers are approximately 20 times as tough as steel.[64,70] Fiber toughness, the total mechanical energy a specified mass of fiber absorbs before breaking, surpasses that of spider silk, the benchmark for fiber toughness. The predrawn SWNT/PVA fibers match the energy absorption of spider silk with a 30% breaking strain and continue to absorb energy until reaching an energy to break of over 600 J/g, which is higher than the best spider drag line silk (165 J/g).[64] The high elongation required for toughness is achieved because the fiber does not neck during loading; testing showed that strain rate was linearly dependent on stress.[64]

Dalton et al.[64] anticipate extending their technology to coagulation polymers other than PVA, particularly those with higher modulus or ionic conductivity. Initial commercial applications are anticipated in niche (low volume/high value) applications such as actuators for microcatheters for microsurgery, moving pins on dynamic braille displays, three-dimensional optical fiber switches, and as minority components for electronic textiles.[64] The ability to make the fibers into supercapacitors and weave them into textiles has been demonstrated; 100-micron supercapacitors provide high life cycle and comparable energy storage to large aqueous-electrolyte supercapacitors. However, the slower discharge rates indicate that higher fiber electrical conductivity must be achieved.[70] Other potential applications for weavable and sewn SWNT/PVA fibers are distributed sensors, electronic interconnects, electromagnetic shields, antennas, and batteries.[64] Fiber toughness could be exploited in antimeteorite/antiballistic shields for satellites, antiballistic vests, and explosion-proof blankets for aircraft cargo bays. Longer term applications include actuators (artificial muscles) for prosthetics and severe environments such as those encountered in planetary exploration and aircraft engines.[70]

11.4.2.3 Fibers Produced from SWNT/Superacid Dispersions

Highly aligned fibers consisting solely of SWNTs have been produced by wet spinning SWNT/superacid dispersions.[37,45,71,72] The direct protonation of single-walled carbon nanotubes in superacids allows them to be dispersed at more than an order of magnitude higher concentration than that typically achieved in surfactants or organic solvents.[37–39,72] Rheology and microscopy show that SWNTs in superacids roughly parallel the phase behavior of rod-like polymer solutions (such as PPTA in sulfuric acid) used for solution spinning of high-performance fibers from nematic liquid crystals.[37]

The phase behavior of rod-like polymer solutions is depicted in Fig. 11.3. With increasing concentration, such a system transitions from a dilute solution where

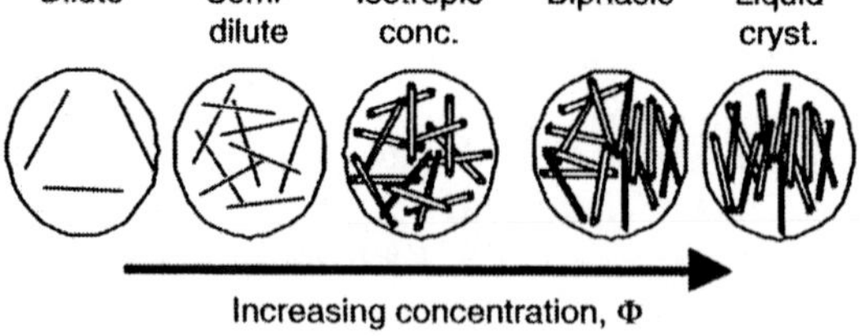

FIGURE 11.3 Phase behavior of rigid rods in solution.[37] Reprinted with permission from *Macromolecules,* 37, 154–160, 2004. Copyright 2004 America Chemical Society.

individual rods do not interact with each other, to a semidilute solution where rotation is inhibited, to an isotropic concentrated solution where both translation and rotation are inhibited. At concentrations above the percolation threshold, steric effects cause the system to phase separately into a liquid crystalline phase in equilibrium with the isotropic concentrated phase.[73,74] Raising the concentration further increases the proportion of the liquid crystalline phase until the system becomes fully liquid crystalline.

In superacids, SWNTs form charge-transfer complexes of positively charged nanotubes surrounded by acid anions.[38] Such charge-transfer complexes repel each other electrostatically at short distance; however, the charges are screened by the acid; thus, electrostatic repulsion is expected to decay rapidly with SWNT separation, and the SWNTs can attract each other at longer separation distances through van der Waals forces (which have long range in the case of long, finite-diameter cylinders[65] such as SWNTs). Such long-range forces, together with excluded volume, seem responsible[75] for the inception of the biphasic region at much lower concentration than would be predicted by the theory for noninteracting rods.[37] The biphasic region consists of aligned, mobile, thread-like SWNT/acid charge-transfer complexes surrounded by sulfuric acid anions in equilibrium with an isotropic phase. "The aligned phase has been dubbed SWNT spaghetti. At high enough concentration, the spaghetti form a single-phase nematic liquid crystal; this occurs at approximately 4 wt% SWNT for purified[54] HiPco SWNTs with an average aspect ratio *L/D* of 470 dispersed in 102% sulfuric acid." This liquid crystal has a birefringent polydomain structure and key rheological signatures such as a maximum in the viscosity versus concentration curve, long oscillatory transients, and the first normal stress difference changing sign from positive to negative and back to positive with increasing shear rate.[37] Moreover, a variety of aligned species can be formed from the dispersion. Liquid crystals of oxidized MWNTs[76] and SWNTs in isopropyl acrylamide gel[77] have also been reported.

Production of SWNT fibers from 102% H_2SO_4 requires mixing of the dispersion in an anhydrous environment. The introduction of moisture destabilizes the dispersion resulting in the formation of tactoid-shaped aggregates dubbed alewives, roughly 50 μm long and 2 μm wide.[37,38,78] The alignment in the dispersion facilitates the production of highly aligned neat SWNT fibers even in the absence of drawing. Simply pushing the dispersion through a 125-μm syringe needle into a diethyl ether coagulation bath results in the formation of fibers with 12 : 1 axial to radial alignment as measured by Raman spectroscopy.[45] The development of an integrated mixing and spinning apparatus for producing SWNT fibers from SWNT acid dispersions has resulted in fibers with better alignment and properties.[45,72] Fibers have been produced from 4 to 10 wt% purified SWNT in 102% sulfuric acid. The maximum achievable SWNT loading has been dictated by the ability to mix and extrude the dispersion. The viscosity of the 6 wt% dispersion is roughly 1000 Pa s at 0.1 s^{-1} (one million times that of water) at room temperature, and heating has little effect on the viscosity.

Typical coagulants for the SWNT/superacid dispersions are diethyl ether, dilute sulfuric acid solutions, and water. Diethyl ether is an effective coagulant for these fibers because it is miscible with sulfuric acid and it evaporates rapidly in air upon removal from the bath. During coagulation, the SWNTs do not have time to reorient and the fiber quickly collapses into a dense structure.[45] Both dilute sulfuric acid solutions and water facilitate mass transfer of the 102% sulfuric out of the fiber and result in a denser fiber of more uniform cross section. Water-coagulated fibers spun from 8 wt% SWNT in 102% H_2SO_4 had a density of 1.1 g/cc, approximately 77% of the theoretical close-packed density for 1.0-nm SWNTs (1.5 g/cc).[72] In all cases, annealing at high temperatures (on the order of 1100°C) is needed to remove residual sulfur.[13,72]

Scanning electron microscopy (SEM) of fibers produced from 6–10 wt% SWNT show significant alignment and a consistent, clean surface morphology generally free of impurities.[37,45] High-magnification images of the neat SWNT fibers reveal a fiber substructure of aligned SWNT bundles approximately 200–600 nm in diameter and a bundle substructure of ropes on the order of 20 nm.[45,72] The fibers possess a relatively uniform circular cross section, which indicates a homogeneous coagulation without brittle skin formation or structural collapse. Peeling of the fibers as well as cross-sectioning of the fibers by microtome and UV-ozone etching[45,72,79] show that the alignment and packing are consistent throughout the fiber. This differs from the SWNT/PVA fibers where the surface is more aligned and has a different morphology from the interior.[67]

Fiber alignment as high as 28 : 1 axial to perpendicular has been measured by polarized Raman spectroscopy; the fibers are the most highly aligned macroscopic neat SWNT material produced to date.[45,80] This alignment is achieved without any extensional drawing during fiber production; it is largely the result of the liquid crystalline nature of the SWNT/superacid dispersion. Zhou et al. determined that for 6 to 8 wt% purified SWNTs dispersed in 100% H_2SO_4 and spun into diethyl ether, decreasing extrusion orifice diameter from 500 to 125 μm improved alignment and electrical and thermal transport.[13] The improvement in alignment was quantified by the FWHM of wide-angle x-ray scattering peaks, decreasing from 63° to 45° and the aligned fraction as measured by Raman increasing from 0.83 to 0.94.[13] Subsequent improvements in fiber spinning have resulted in even more highly aligned fibers (FWHM of 31°).[72] Annealing the fibers at 1100°C in argon for 24 hours or at 1150°C in vacuum for 2 hours had little effect on the alignment. Annealing also had little to no effect on the thermal conductivity of the fibers. Thermal conductivities at 300 K were 17 W/mK and 5 W/mK for the most and least aligned fibers, respectively.[13] These low values, compared to that of graphite parallel to the layers, are likely the result of several factors. The cross-sectional area was based on the macroscopic dimensions and did not account for gross voids or porosity. Also the fibers are not perfectly aligned, and the gaps between bundles represent thermal barriers.[13]

Unlike alignment and thermal conductivity, electrical properties were significantly affected by annealing. The as-produced fibers were heavily p-doped with low resistivity and metallic behavior above 300 K; resistivity decreased from 0.64 to 0.25 mΩ cm at 300 K for the 500- and 125-μm orifices, respectively. Annealing the fibers removed the residual acid dopants causing an order of magnitude increase in resistivity at 300 K and a nonmetallic temperature dependence.[13] Thermoelectric power (S) measurements showed an unusual low-temperature behavior explained by a one-dimensional phonon-drag model.[81] The two characteristic features of a phonon-drag contribution are the upturn in the thermopower at $T_0 \approx 0.1\eta\omega_Q/k_B$, where $\eta\omega_Q$ is the energy of the relevant phonon mode, and a peak around $dS/dT \approx 2.3T_0$. The fibers from SWNT/acid show an upturn in thermopower at $T_0 = 26$ K and a differential temperature maximum at 49–55 K. Modeling the data by a sum of phonon-drag thermopower and diffusion thermopower resulted in the phonon energy $\eta\omega_Q = 0.02$ eV, diffusion thermopower coefficient $b = 0.04\ \mu V/K^2$, and $L_{ph}(Q) = 0.6$ and 0.4 μm at 10 and 300 K, respectively.

Fiber mechanical properties depend on SWNT concentration, orifice diameter (alignment), and coagulant. A Young's modulus of 120 GPa and tensile strengths of 120 MPa have been achieved in fibers on the order of 50 μm in diameter and tens of meters in length.[72] The tensile strength is limited by the presence of defects, voids, and rope-rope interfaces, and was improved by a factor of 30 over the course of one year.[45] Further optimization of the dispersion, spinning conditions, coagulation conditions, and postprocessing are expected to reduce fiber defects and improve mechanical properties.

11.5 CONCLUSION

The development of SWNT nanotube fibers is in its infancy, and the performance expected from the nanoscale properties of SWNTs has yet to be realized in macroscopic objects. However, preliminary composite and neat fibers show promising results. It is likely that spinning processes will evolve to meet specific applications' needs. SWNTs dispersed in a polymer matrix may be best suited for extending the potential applications of existing polymeric materials by enhancing thermal and electrical conductivity and compressive strength at a relatively low-cost differential. SWNT/surfactant dispersions may be solution spun for applications where toughness and other mechanical properties are most critical. SWNT/acid dispersions used to produce neat fibers may be best suited for applications with the most demanding thermal and electrical requirements. The development of these spinning technologies along with improvements in SWNT production, purification, and functionalization will undoubtedly result in a number of revolutionary fiber applications.

PROBLEMS

1. Describe a potential application for nanotube fibers. What properties would be required? What kind of nanotubes would you use? Which of the fiber production methods described in this section would you use? (At this stage, assume that cost is not an issue.)
2. How does the polydispersity in chirality, length, and diameter of SWNTs affect the different fiber spinning processes? How does this polydispersity affect fiber properties?
3. What are the biggest challenges in producing nanotube polymer composite fibers? How do these differ from the challenges in producing neat SWNT fibers?

ACKNOWLEDGMENTS

For discussions and for sharing ideas and experimental results we thank Lars Ericson, Sivarajan Ramesh, Rajesh Saini, Nicholas Parra-Vasquez, Hua Fan, Yuhuang Wang, Haiqeng Peng Wei Zhou, Carter Kittrell, Bob Hauge, Wade Adams, Ed Billups,

Boris Yakobson, Howard Schmidt, Wen-Fang Hwang, Rick Smalley, Satish Kumar, Karen Winey, Jack Fischer, Alan Dalton, Ray Baughman, Edgar Munoz, Philippe Poulin, and Erik Hobbie. We acknowledge funding from the DURINT initiative of the Office of Naval Research under Contract N00014-01-1-0789, from the Texas Advanced Technology Program through grant 003604-0113-2003, and from the National Science Foundation through Award EEC-0118007.

REFERENCES

1. Tans, S.J., Devoret, M.H., Dai, H.J., Thess, A., Smalley, R.E., Geerligs, L.J., and Dekker, C., Individual single-wall carbon nanotubes as quantum wires, *Nature, 386,* 474–477, 1997.
2. Hone, J., Llaguno, M.C., Nemes, N.M., Johnson, A.T., Fischer, J.E., Walters, D.A., Casavant, M.J., Schmidt, J., and Smalley, R.E., Electrical and thermal transport properties of magnetically aligned single wall carbon nanotube films, *Appl. Phys. Lett., 77,* 666–668, 2000.
3. Berber, S., Kwon, Y.K., and Tomanek, D., Unusually high thermal conductivity of carbon nanotubes, *Phys. Rev. Lett., 84,* 4613–4616, 2000.
4. Hone, J., Whitney, M., Piskoti, C., and Zettl, A., Thermal conductivity of single-walled carbon nanotubes, *Phys. Rev. B, 59,* R2514–R2516, 1999.
5. Krishnan, A., Dujardin, E., Ebbesen, T.W., Yianilos, P.N., and Treacy, M.M.J., Young's modulus of single-walled nanotubes, *Phys. Rev. B, 58,* 14013–14019, 1998.
6. Buongiorno-Nardelli, M., Yakobson, B.I., and Bernholc, J., Brittle and ductile behavior in carbon nanotubes, *Phys. Rev. Lett., 81,* 4656–4659, 1998.
7. Baughman, R.H., Zakhidov, A.A., and de Heer, W.A., Carbon nanotubes—the route toward applications, *Science, 297,* 787–792, 2002.
8. Zhang, X.F., Cao, A.Y., Li, Y.H., Xu, C.L., Liang, J., Wu, D.H., and Wei, B.Q., Self-organized arrays of carbon nanotube ropes, *Chem. Phys. Lett., 351,* 183–188, 2002.
9. Salvetat-Delmotte, J.P. and Rubio, A., Mechanical properties of carbon nanotubes: a fiber digest for beginners, *Carbon, 40,* 1729–1734, 2002.
10. Bronikowski, M.J., Willis, P.A., Colbert, D.T., Smith, K.A., and Smalley, R.E., Gas-phase production of carbon single-walled nanotubes from carbon monoxide via the HiPco process: a parametric study, *J. Vac. Sci. Technol. A Vac. Surf. Films, 19,* 1800–1805, 2001.
11. Dai, H.J., Nanotube growth and characterization, *Topics Appl. Phys. Carbon Nanotubes, 80,* 29–53, 2001.
12. Kitiyanan, B., Alvarez, W.E., Harwell, J.H., and Resasco, D.E., Controlled production of single-wall carbon nanotubes by catalytic decomposition of CO on bimetallic Co-Mo catalysts, *Chem. Phys. Lett., 317,* 497–503, 2000.
13. Zhou, W., Vavro, J., Guthy, C., Winey, K.I., Fischer, J.E., Ericson, L.M., Ramesh, S., Saini, R., Davis, V.A., Kittrell, C., Pasquali, M., Hauge, R.H., and Smalley, R.E., Single wall carbon nanotube fibers extruded from super-acid suspensions: preferred orientation, electrical, and thermal transport, *J. Appl. Phys., 95,* 649–655, 2004.
14. Yakobson, B.I. and Smalley, R.E., Fullerene nanotubes: $C_{1,000,000}$ and beyond, *Am. Scientist, 85,* 324–337, 1997.
15. Baughman, R.H., Materials science—putting a new spin on carbon nanotubes, *Science, 290,* 1310–1311, 2000.

16. Baughman, R.H., Materials science—muscles made from metal, *Science, 300,* 268–269, 2003.
17. De Zhang, W., Wen, Y., Liu, S.M., Tjiu, W.C., Xu, G.Q., and Gan, L.M., Synthesis of vertically aligned carbon nanotubes on metal deposited quartz plates, *Carbon, 40,* 1981–1989, 2002.
18. Thostenson, E.T., Ren, Z.F., and Chou, T.W., Advances in the science and technology of carbon nanotubes and their composites: a review, *Compos. Sci. Technol., 61,* 1899–1912, 2001.
19. Li, X.H., Zhang, J., Li, Q.W., Li, H.L., and Liu, Z.F., Polymerization of short single-walled carbon nanotubes into large strands, *Carbon, 41,* 598–601, 2003.
20. Li, Y.H., Wei, J.Q., Zhang, X.F., Xu, C.L., Wu, D.H., Lu, L., and Wei, B.Q., Mechanical and electrical properties of carbon nanotube ribbons, *Chem. Phys. Lett., 365,* 95–100, 2002.
21. Zhang, Y.G., Chang, A.L., Cao, J., Wang, Q., Kim, W., Li, Y.M., Morris, N., Yenilmez, E., Kong, J., and Dai, H.J., Electric-field-directed growth of aligned single-walled carbon nanotubes, *Appl. Phys. Lett., 79,* 3155–3157, 2001.
22. Jiang, K.L., Li, Q.Q., and Fan, S.S., Nanotechnology: spinning continuous carbon nanotube yarns—carbon nanotubes weave their way into a range of imaginative macroscopic applications, *Nature, 419,* 801–801, 2002.
23. Zhu, H.W., Xu, C.L., Wu, D.H., Wei, B.Q., Vajtai, R., and Ajayan, P.M., Direct synthesis of long single-walled carbon nanotube strands, *Science, 296,* 884–886, 2002.
24. Wei, B.Q., Vajtai, R., Choi, Y.Y., Ajayan, P.M., Zhu, H.W., Xu, C.L., and Wu, D.H., Structural characterizations of long single-walled carbon nanotube strands, *Nano Lett., 2,* 1105–1107, 2002.
25. Li, Y.L., Kinloch, I.A., and Windle, A.H., Direct spinning of carbon nanotube fibers from chemical vapor deposition synthesis, *Science, 304,* 276–278, 2004.
26. Gommans, H.H., Alldredge, J.W., Tashiro, H., Park, J., Magnuson, J., and Rinzler, A.G., Fibers of aligned single-walled carbon nanotubes: polarized Raman spectroscopy, *J. Appl. Phys.*, 88, 2509–2514, 2000.
27. Griskey, R.G., *Polymer Process Engineering,* Chapman & Hall, New York, 1995, chap. 11.
28. Levich, V.G., *Physiochemical Hydrodynamics,* Prentice Hall, Englewood Cliffs, 1962, chap. 8.
29. Rakas, M.A., *The Effect of Coagulants on the Microstructure and Mechanical Properties of Lyotropic Fiber-Forming Polymers,* Ph.D. Thesis, Department of Polymer Science and Polymer Engineering, University of Massachusetts, available through UMI, 1990.
30. Paul, D.R., Diffusion during the coagulation step of wet-spinning, *J. Appl. Polym. Sci., 12,* 383–402, 1968.
31. Ziabicki, A., *Fundamentals of Fibre Formation: The Science of Fibre Spinning and Drawing,* John Wiley & Sons, London, 1976.
32. Hancock, T.A., Spruiell, J.E., and White, J.L., Wet spinning of aliphatic and aromatic polyamides, *J. Appl. Polym. Sci., 21,* 1227, 1977.
33. Hancock, T.A., *An Experimental and Theoretical Study of the Wet Spinning Process,* Ph.D. Thesis, Department of Chemical Engineering, University of Tennessee, available through UMI, 1981.
34. Chenevey, E.C. and Wadhwa, L.H., Technical Report AFML-TR-82-4194, Air Force Materials Laboratory, Dayton, OH, 1982.

35. Berry, G.C., Wong, C.P., Venkatamen, S., and Chu, S.G., Technical Report: AFML-TR-79-4115, Air Force Materials Laboratory, Dayton, OH, 1979.
36. Pottick, L.A., *The Influence of Drying on the Structure and Mechanics of Poly (P-Phenylene Benzobisthiazole) Fibers*. Ph.D. Thesis, Department of Polymer Science and Polymer Engineering, University of Massachusetts, available through UMI, 1986.
37. Davis, V.A., Ericson, L.M., Parra-Vasquez, A.N.G., Fan, H., Wang, Y., Prieto, V., Longoria, J.A., Ramesh, S., Saini, R., Kittrell, C., Billups, W.E., Adams, W.W., Hauge, R.H., Smalley, R.E., and Pasquali, M., Phase Behavior and Rheology of SWNTs in Superacids, *Macromolecules, 37,* 154–160, 2004.
38. Ramesh, S., Ericson, L.M., Davis, V.A., Saini, R.K., Kittrell, C., Pasquali, M., Billups, W.E., Adams, W.W., Hauge R.H., and Smalley, R.E., Dissolution of pristine single walled carbon nanotubes in superacids by direct protonation, *J. Phys. Chem. B, 108,* 8794–8798, 2004.
39. Davis, V.A., Ericson, L.M., Saini, R., Sivarajan, R., Hauge, R.H., Smalley, R.E., and Pasquali, M., *Proceedings of the 2001 AIChE Annual Meeting,* Reno, NV, 351–359, 2001.
40. Bahr, J.L., Mickelson, E.T., Bronikowski, M.J., Smalley, R.E., and Tour, J.M., Dissolution of small diameter single-wall carbon nanotubes in organic solvents?, *Chem. Commun., 2,* 193–194, 2001.
41. Georgakilas, V., Voulgaris, D., Vazquez, E., Prato, M., Guldi, D.M., Kukovecz, A., and Kuzmany, H., Purification of HiPco carbon nanotubes via organic functionalization, *J. Am. Chem. Soc., 124,* 14318–14319, 2002.
42. Saini, R.K., Chiang, I.W., Peng, H.Q., Smalley, R.E., Billups, W.E., Hauge, R.H., and Margrave, J.L., Covalent sidewall functionalization of single wall carbon nanotubes, *J. Am. Chem. Soc., 125,* 3617–3621, 2003.
43. Ying, Y.M., Saini, R.K., Liang, F., Sadana, A.K., and Billups, W.E., Functionalization of carbon nanotubes by free radicals, *Org. Lett., 5,*1471–1473, 2003.
44. Poulin, P., Vigolo, B., and Launois, P., Films and fibers of oriented single wall nanotubes, *Carbon, 40,* 1741–1749, 2002.
45. Ericson, L.M., *Macroscopic Neat Single-Wall Carbon Nanotube Fibers,* Ph.D. Thesis Department of Applied Physics, Rice University, available through UMI, 2003.
46. Kiss, G.D., *Rheology and Rheo-Optics of Concentrated Solutions of Helical Polypeptides,* Ph.D. Thesis, Department of Polymer Science and Polymer Engineering, University of Massachusetts, available through UMI, 1979.
47. Andrews, R.J., Rao, D., Rantell, A.M., Derbyshire, T., Chen, F., Chen, Y., J. Haddon, R.C, Nanotube composite carbon fibers, *Appl. Phys. Lett., 75,* 1329, 1999.
48. Haggenmueller, R., Gommans, H.H., Rinzler, A.G., Fischer, J.E., and Winey, K.I., Aligned single-wall carbon nanotubes in composites by melt processing methods, *Chem. Phys. Lett., 330,* 219–225, 2000.
49. Van Hattum, F.W.J., Bernardo, C.A., Finegan, J.C., Tibbetts, G.G., Alig, R.L., and Lake, M.L., A study of the thermomechanical properties of carbon fiber-polypropylene composites, *Polym. Composites, 20,* 683–688, 1999.
50. Tibbetts, G.G. and McHugh, J.J., Mechanical properties of vapor-grown carbon fiber composites with thermoplastic matrices, *J. Mater. Res., 14,* 2871–2880, 1999.
51. Lozano, K., Bonilla-Rios, J., and Barrera, E.V., A study on nanofiber-reinforced thermoplastic composites (II): investigation of the mixing rheology and conduction properties, *J. Appl. Polym. Sci., 80,* 1162–1172, 2001.

52. Ma, H.M., Zeng, J.J., Realff, M.L., Kumar, S., and Schiraldi, D.A., Processing, structure, and properties of fibers from polyester/carbon nanofiber composites, *Compos. Sci. Technol., 63,* 1617–1628, 2003.
53. Bhattacharyya, A.R., Sreekumar, T.V., Liu, T., Kumar, S., Ericson, L.M., Hauge, R.H., and Smalley, R.E., Crystallization and orientation studies in polypropylene/single wall carbon nanotube composite, *Polymer, 44,* 2373–2377, 2003.
54. Chiang, I.W., Brinson, B.E., Huang, A.Y., Willis, P.A., Bronikowski, M.J., Margrave, J.L., Smalley, R.E., and Hauge, R.H., Purification and characterization of single-wall carbon nanotubes (SWNTs) obtained from the gas-phase decomposition of CO (HiPco process), *J. Phys. Chem. B, 105,* 8297–8301, 2001.
55. Kitagawa, T., Ishitobi, M., and Yabuki, K., An analysis of deformation process on poly-p-phenylenebenzobisoxazole fiber and a structural study of the new high-modulus type PBO HM+ fiber, *J. Polym. Sci. B, 38,* 1605–1611, 2000.
56. Kumar, S., Dang, T.D., Arnold, F.E., Bhattacharyya, A.R., Min, B.G., Zhang, X.F., Vaia, R.A., Park, C., Adams, W.W., Hauge, R.H., Smalley, R.E., Ramesh, S., and Willis, P.A., Synthesis, structure, and properties of PBO/SWNT composites, *Macromolecules, 35,* 9039–9043, 2002.
57. Balberg, I., Anderson, C.H., Alexander, S., and Wagner, N., Excluded volume and its relation to the onset of percolation, *Phys. Rev. B, 30,* 3933–3943, 1984.
58. Sreekumar, T.V., Liu, T., Min, B.G., Guo, H., Kumar, S., Hauge, R.H., and Smalley, R.E., Polyacrylonitrile single-walled carbon nanotube composite fibers, *Adv. Mater., 16,* 58–61, 2004.
59. Liu, T., Sreekumar, T.V., Kumar, S., Hauge, R.H., and Smalley, R.E., SWNT/PAN composite film-based supercapacitors, *Carbon, 41,* 2440–2442, 2003.
60. Poulin, P., Vigolo, B., Penicaud, A., and Coulon, C., U.S. Patent, 2002/0102585 A1, Method for Obtaining Macroscopic Fibres and Strips from Colloidal Particles and in Particular Carbon Nanotubes, 2003.
61. Vigolo, B., Penicaud, A., Coulon, C., Sauder, C., Pailler, R., Journet, C., Bernier, P., and Poulin, P., Macroscopic fibers and ribbons of oriented carbon nanotubes, *Science, 290,* 1331–1334, 2000.
62. Vigolo, B., Poulin, P., Lucas, M., Launois, P., and Bernier, P., Improved structure and properties of single-wall carbon nanotube spun fibers, *Appl. Phys. Lett., 81,* 1210–1212, 2002.
63. Lobovsky, A., Matrunich, J., Kozlov, M., Morris, R.C., Baughman, R.H., and Zakhidov, A.A., U.S. Patent, 2002/0113335 A1, Spinning, Processing, and Applications of Carbon Nanotube Filaments, Ribbons and Yarns, 2002.
64. Dalton, A.B., Collins, S., Munoz, E., Razal, J.M., Ebron, V.H., Ferraris, J.P., Coleman, J.N., Kim, B.G., and Baughman, R.H., Super-tough carbon-nanotube fibres—these extraordinary composite fibres can be woven into electronic textiles, *Nature, 423,* 703–703, 2003.
65. Israelchvili, J.N., *Intermolecular and Surface Forces,* 2nd ed., Academic Press, London, 1992, chaps. 11 and 14.
66. Hiemenz, P.C. and Rajagopalan, R., *Principles of Colloid and Surface Chemistry 2nd Edition,* Marcel Dekker, New York, 1997, chap. 11.
67. Neimark, A.V., Ruetsch, S., Kornev, K.G., and Ravikovitch, P.I., Hierarchical pore structure and wetting properties of single- wall carbon nanotube fibers, *Nano Lett., 3,* 419–423, 2003.
68. Nativ-Roth, E., Levi-Kalisman, Y., Regev, O., and Yerushalmi-Rozen, R., On the route to compatibilization of carbon nanotubes, *J. Polym. Eng., 22,* 353–368, 2002.

69. Barisci, J.N., Tahhan, M., Wallace, G.G., Badaire, S., Vaugien, T., Maugey, M., and Poulin, P., Properties of carbon nanotube fibers spun from DNA-stabilized dispersions, *Adv. Funct. Mater., 14,* 133–138, 2004.
70. Dalton, A.B., Collins, S., Razal, J., Munoz, E., Ebron, V.H., Kim, B.G., Coleman, J.N., Ferraris, J.P., and Baughman, R.H., Continuous carbon nanotube composite fibers: properties, potential applications, and problems, *J. Mater. Chem., 14,* 1–3, 2004.
71. Smalley, R.E., Saini, R., Sivarajan, R., Hauge, R.H., Davis, V.A., Pasquali, M., Ericson, L.M., U.S. Patent, 2003/0170166A1, Fibers of Aligned Single-Wall Carbon Nanotubes and Process for Making the Same, 2003.
72. Ericson, L., Fan, H., Peng, H., Davis, V., Zhou, W., Sulpizio, J., Wang, Y., Booker, R., Vavro, J., Guthy, C., Parra-Vasquez, A., Kim, M., Ramesh, S., Saini, R., Kittrell, C., Lavin, G., Schmidt, H., Adams, W., Billups, W., Pasquali, M., Hwang W., Hauge, R., Fischer J., and Smalley, R., Macroscopic, neat, single-walled carbon nanotube fibers, *Science, 305,* 1447–1450, 2004.
73. Onsager, L., The effects of shape on the interaction of colloidal particles, *Ann. N.Y. Acad. Sci., 51,* 627, 1949.
74. Flory, P.J., Phase equlibria in solutions of rod-like particles, *Proc. Royal Soc. London, Ser. A, 234,* 1956.
75. Lee, C. S. and Yakobson, B.I., in preparation.
76. Song, W.H., Kinloch, I.A., and Windle, A.H., Nematic liquid crystallinity of multiwall carbon nanotubes, *Science, 302,* 1363–1363, 2003.
77. Islam, M.F., Alsayed, A.M., Dogic, Z., Zhang, J., Lubensky, T.C., and Yodh, A.G., Nematic nanotube gels, *Phys. Rev. Lett., 92,* 2004.
78. Saini, R., Sivarajan, R., Hauge, R.H., Davis, V.A., Pasquali, M., Ericson, L.M., Kumar, S., Veedu, S., U.S. Patent, 2003/0133865A1, Single-Wall Carbon Nanotube Alewives, Process for Making and Compositions Thereof, 2003.
79. Wang, Y., *Seed Crystals and Catalyzed Epitaxy of Single-Walled Carbon Nanotubes,* Ph.D. Thesis, Department of Applied Physics, Rice University, available through UMI, 2004.
80. Walters, D.A., Casavant, M.J., Qin, X.C., Huffman, C.B., Boul, P.J., Ericson, L.M., Haroz, E.H., O'Connell, M.J., Smith, K., Colbert, D.T., and Smalley, R.E., In-plane-aligned membranes of carbon nanotubes, *Chem.Phys. Lett., 338,* 14–20, 2001.
81. Vavro, J., M. C. Llaguno, J. E. Fischer, S. Ramesh, R. K. Saini, L. M. Ericson, V. A. Davis, R. H. Hauge, M. Pasquali and R. E. Smalley, Thermoelectric power of p-doped single wall carbon nanotubes and the role of phonon drag, *Phys. Rev. Lett., 90,* 065503, 2003.

12 Carbon Nanofiber and Carbon Nanotube/Polymer Composite Fibers and Films

Han Gi Chae, Tetsuya Uchida, and Satish Kumar

CONTENTS

12.1 INTRODUCTION

"In the early nineteenth century, Herman Staudinger began to publish articles asserting that polymeric molecules are practically endless chains held together by ordinary chemical bonds".[1] Even though this idea of practically endless chains appeared foreign to most chemists of that time, these concepts ultimately led to the development of the synthetic polymer and fiber industry, resulting in revolutionary developments in the field, particularly during the period of 1930–1980.

Developments in nanoscale science and technology and their applications to polymers and fibers suggest that we are at the initial stages of another revolutionary period that will bring about polymeric and fibrous materials with properties not yet seen. The field of polymer/carbon nanotube composites alone is expected to open new vistas in materials properties. About 100 research papers were published on polymer/carbon nanotube composites in the 1991–2000 period. However, the number of publications in this field is growing at a rapid rate, with 80, 160, and over 300 publications in 2001, 2002, and 2003, respectively. The pace of this research activity is only going to grow, as carbon nanotubes, and particularly single-wall carbon nanotubes, become more affordable, and as methods are developed to synthesize or separate SWNTs of specific chirality and diameter. Due to variation in electronic structure, different types of nanotubes are likely to have different interactions with a given polymer. There are many different solvents or solvent combinations (surfactant/ water, organic solvents, inorganic acids, etc.) that can be used to disperse nanotubes in polymers. And there are literally unlimited opportunities for nanotube functionalization, and for chemical reaction with polymers. One can quickly see that there are thousands of combinations (polymer/nanotube/solvents/chemical reactions) on which to base experiments.

Despite the fact that nearly 500 research papers appeared on polymer carbon nanotube composites in 2002–2003, the field is still in its infancy and is at the same state where carbon fiber-based composites were in the 1960s when carbon fibers were just being developed. Carbon fibers and their composite manufacturing technology have matured after 40 years of development. In the past 40 years, carbon fiber cost has also come down dramatically. From almost no composites in aerospace structures in the 1960s, we have now reached the stage where 50% of the weight of the Boeing 7E7 to be produced in 2008 will be composite materials. There is sufficient scientific data to suggest that in 20+ years, carbon nanotube-enabled materials may reach similar application levels, with the added difference that carbon nanotube-based materials will be much more functional. This chapter summarizes recent developments in the field in the authors' laboratory at Georgia Institute of Technology.

12.2 VAPOR-GROWN CARBON NANOFIBERS (VGCNFS) AND POLYMER COMPOSITE FIBERS

Carbon fibers (typical diameter 7 ~ 10 μm) for composite applications were developed during the 1960s, while the development of carbon nanofibers (typical diameter 50 ~ 200 nm) dates back to the 1980s.[2] The primary interest of this section is the VGCNF, which is synthesized from the pyrolysis of hydrocarbons or carbon monoxide in the gaseous state in the presence of a catalyst.[2–4] VGCNFs distinguish themselves from other types of carbon fibers, such as polyacrylonitrile- or mesophase pitch-based carbon fibers, in their method of production, physical properties, and structure. The reported moduli of VGCNFs are varied in a wide range from less than 100 up to 700 GPa.[5–8] This noticeable variance in mechanical properties originates from the characteristic morphology of VGCNFs. High-resolution transmission electron microscopy (HRTEM) of VGCNFs revealed two types of morphologies.[9] In one case, a truncated cone microstructure was seen, with outer and inner diameters of 60 and 25 nm, respectively. In this type of VGCNF, graphite sheets were oriented at an average angle of about 15° with respect to the fiber axis. Figure 12.1 and

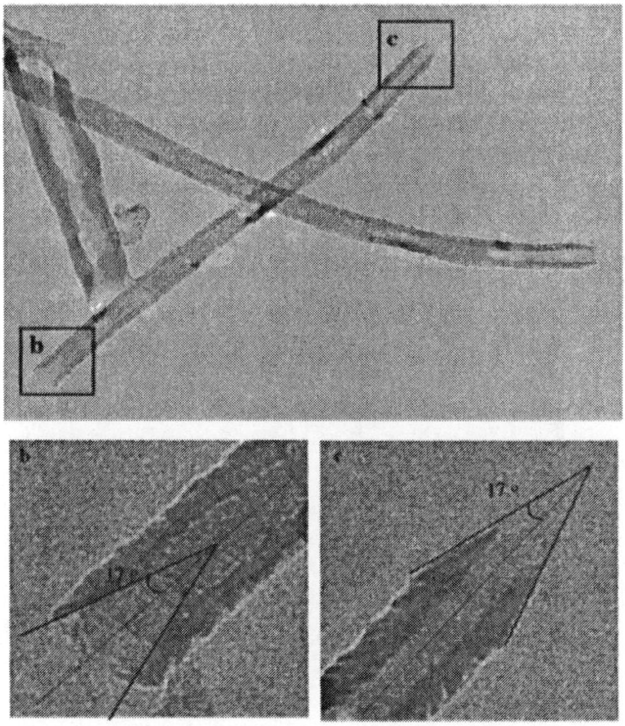

FIGURE 12.1 Transmission electron micrographs of single-layer carbon nanofibers.[9]

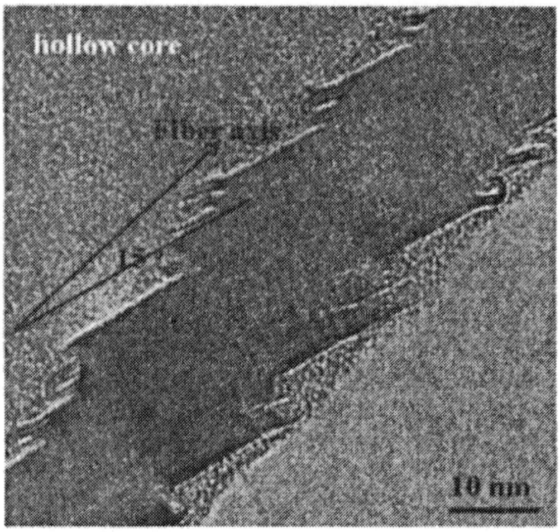

FIGURE 12.2 High-resolution transmission electron micrograph of single-layer carbon nanofiber.[9]

Fig. 12.2 show the TEM and HRTEM images of single-layer VGCNF. The HRTEM image shows graphite layer stacking as well as graphite layer folding.

The second type of VGCNF is a double-layer carbon nanofiber, with outer and inner diameters of 85 and 20 nm, respectively. Truncated cone structure is also observed in the double-layer VGCNF as shown in Fig. 12.3.

The focus of VGCNF-reinforced composites has been the engineering applications that require superior strength, stiffness, and electrical and thermal conductivities.

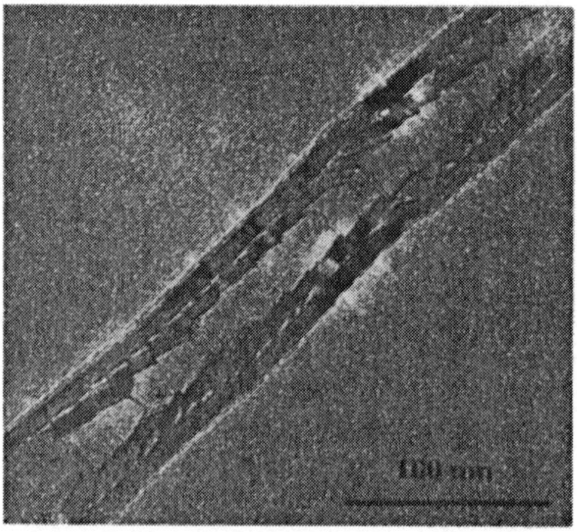

FIGURE 12.3 Transmission electron micrograph of double-layer carbon nanofiber.[9]

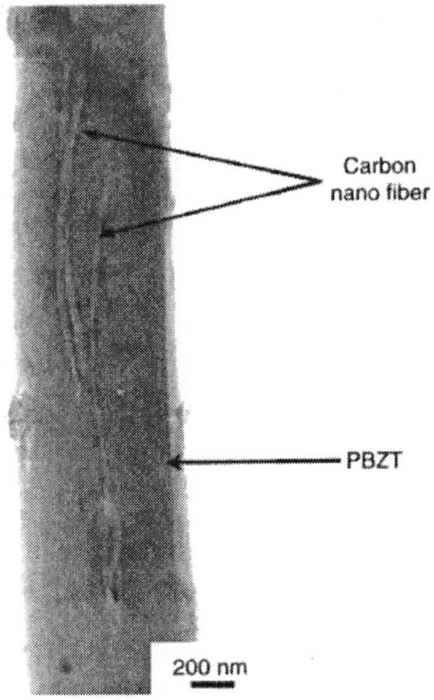

FIGURE 12.4 Transmission electron micrograph of PBZT-CNF composite fiber.[20]

The matrices have included a variety of commodity polymers, including polypropylene (PP),[10–12] polycarbonate,[13,14] nylon,[15] polymethyl methacrylate (PMMA),[16] poly(ether ether ketone,[17] polyamide,[18] and epoxies.[19] The next sections highlight the properties of a few VGCNFs containing polymers.

12.2.1 PBZT-VGCNF Composite Fibers[20]

A dope of poly(p-phenelyene benzobisthiazole) (PBZT)-VGCNF composite was prepared by *in situ* polymerization in polyphosphoric acid (PPA) in the presence of VGCNFs, and fibers were spun by dry-jet wet spinning. The degrees of dispersion and orientation of the VGCNFs were assessed using transmission electron microscopy (TEM). The TEM image in Fig. 12.4 shows that the VGCNFs were isolated without aggregation and were well oriented along the fiber axis. The lengths of the VGCNFs appeared to be in the order of a few micrometers, that is, nearly the same as pristine VGCNFs, which suggests that the PPA used in the *in situ* polymerization of PBZT did not affect the geometry of the VGCNFs.

12.2.2 PP-VGCNF Composite Fibers[21]

Carbon nanofibers (5 wt%) and polypropylene (PP) composite fibers were spun using conventional melt spinning.

TABLE 12.1
Mechanical Properties of PP and PP-CNF (95-5) Composite Fibers[21]

Sample	Tensile Modulus (GPa)	Tensile Strength (GPa)	Strain to Failure (%)	Compressive Strength (MPa)
PP	4.6 ± 0.7	0.49 ± 0.06	23 ± 5	25 ± 1
PP + CNF	7.1 ± 0.9	0.57 ± 0.07	16 ± 2	48 ± 10

The tensile and compressive properties of the spun fibers are shown in Table 12.1. The reinforcement effects of VGCNFs are evident from significant improvements in tensile modulus as well as in compressive strength. Scanning electron microscopy (SEM) and laser scanning confocal microscopy (LSCM) were employed to study the microstructures of the composite fibers. The SEM image in Fig. 12.5 shows good dispersion of VGCNFs in the matrix. In addition, the confocal microscope images (not shown here) revealed an interesting phenomenon; that is, the outer layer of the composite fiber contained fewer VGCNFs while having a higher degree of orientation, as compared to the core.

12.2.3 PET-VGCNF Composite Fibers[22]

Polyethylene terephthalate (PET)-VGCNF composite fibers were also processed by melt spinning. Figure 12.6 provides an SEM image of a PET-VGCNF composite before fiber spinning, showing good VGCNF dispersion.

The mechanical properties of the pristine and the 5 wt% VGCNFs containing composite fibers are given in Table 12.2. The tensile moduli of composite fibers are slightly higher than those of the pristine PET fiber. On the other hand, the tensile strengths of composite fibers are either comparable to or lower than the control fiber. A series of process steps, including mixing, melt blending, and fiber spinning tended to shorten the VGCNFs, which was detrimental to tensile properties. In addition,

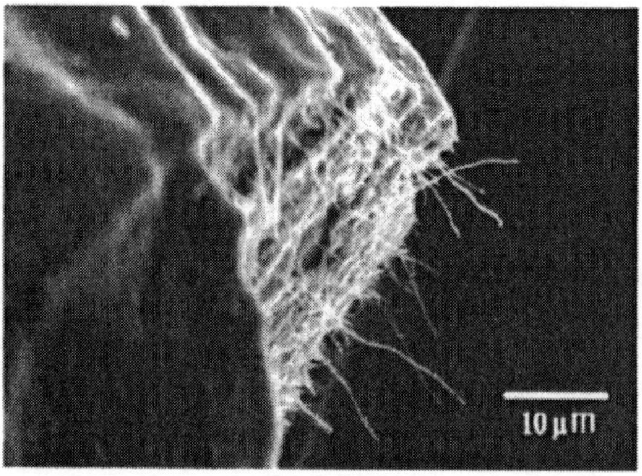

FIGURE 12.5 Scanning electron micrograph of PP-CNF composite fiber.[21]

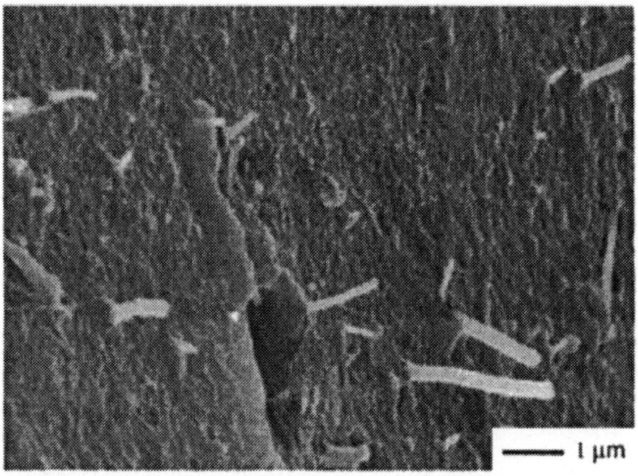

FIGURE 12.6 Scanning electron micrograph of PET-CNT composite before fiber spinning.[22]

the wide-angle x-ray diffraction (WAXD) study (not shown here) suggested that significant misalignment of graphitic planes in the VGCNFs with respect to the nanofiber axis direction may be another factor contributing to the relatively low modulus of the PET-VGCNF composite fiber. The low orientation of the graphite plane is consistent with the graphite plane misorientation (Fig. 12.2).[9] While only moderate improvement in tensile modulus was observed, the composite fiber properties in compression and torsion improved by as much as 50% compared to the control PET fiber (Table 12.2). Kozey et al.[23] reported that compressive failure in polymeric fibers occurs by yielding, which results in the formation and propagation of kinks. VGCNFs act as barriers for kink propagation, thus improving the compressive strength.

TABLE 12.2
Mechanical Properties of PET and PET-CNF (95-5) Composite Fibers[22]

Sample	Fiber Diameter (μm)	Tensile Modulus (GPa)	Tensile Strength (GPa)	Strain to Failure (%)	Compressive Strength (MPa)	Torsional Modulus (GPa)
PET	24 ± 3	10 ± 3	0.43 ± 0.08	26 ± 2	80	0.7
PET/PR-24-PS	27 ± 6	11 ± 2	0.33 ± 0.06	18 ± 7	120	1.0
PET/PR-21-PS	27 ± 5	11 ± 3	0.25 ± 0.1	13 ± 8	100	0.9
PET/PR-24-HT	25 ± 2	11 ± 2	0.42 ±0.06	20 ± 4	90	1.1

PR-24-PS, PR-21-PS, and PR-24-HT represent different fiber grades. PS: pyrolytically stripped; HT: heat treated.

TABLE 12.3
Mechanical Properties of PMMA and PMMA-CNF Composite Fibers[24]

Sample	Fiber Diameter (μm)	Tensile Modulus (GPa)	Tensile Strength (GPa)	Strain to Failure (%)	Compressive Strength (MPa)
PMMA	60 ± 4	4.7 ± 1.5	0.20 ± 0.04	16 ± 3	28 ± 2
PMMA/PR-21-PS (95–5)	61 ± 12	8.0 ± 1.2	0.17 ± 0.04	10 ± 6	73 ± 11
PMMA/PR-21-PS (90–10)	63 ± 10	7.7 ± 1.0	0.16 ± 0.04	10 ± 6	N/A
PMMA/PR-24-PS (95–5)	62 ± 5	7.5 ± 1.3	0.16 ± 0.03	10 ± 5	66 ± 20
PMMA/PR-24-PS (90–10)	63 ± 5	7.6 ± 0.9	0.15 ± 0.01	9 ± 4	N/A

12.2.4 PMMA-VGCNF Composite Fibers[24]

Polymethyl methacrylate (PMMA) nanocomposites were processed by melt blending, containing two different grades of VGCNF at two different VGCNF loadings (5 and 10 wt%). The PMMA-VGCNF composites were processed into 40-mm diameter rods and 60-μm diameter fibers. The respective microstructures and mechanical and thermal properties were characterized.

The SEM image captured perpendicular to the extrusion direction and the fractograph of the fracture surface (not shown here) of a PMMA-VGCNF composite fiber showed good dispersion and bonding of the VGCNFs in the matrix. The mechanical properties of pure PMMA and PMMA-VGCNF composite fibers are compared in Table 12.3. Tensile modulus increased by more than 50%, and compressive strength more than doubled by the addition of 5 wt% VGCNF. However, further incorporation of VGCNFs up to 10 wt% did not result in any further improvements in the mechanical properties. Tensile strength and elongation at break slightly decreased with the addition of VGCNFs.

A series of thermal analysis experiments showed significant improvements in the thermal behavior of the composite fiber. Thermogravimetric analysis showed that the onset degradation temperature increased by about 50°C by reinforcing PMMA with VGCNFs. In addition, dynamic mechanical analysis revealed that the glass transition temperature of the PMMA-VGCNF composite fiber is about 5°C higher than that of the control PMMA fiber. Finally thermomechanical analysis showed only 5% of thermal shrinkage in the PMMA-CNF (95-5) composite fiber at 110°C, while the control PMMA fiber shrank by 30%. The improvement in thermal shrinkage was especially dramatic compared to other thermal behavior because PMMA, a glassy polymer, tends to be subjected to considerable shrinkage when the oriented fibers are allowed to relax at elevated temperatures.

12.3 CARBON NANOTUBES (CNTs) AND POLYMER COMPOSITE FIBERS

Carbon nanotubes (CNTs), which were discovered in 1991,[25] are extremely versatile materials; they are very strong, yet highly elastic, highly conducting, and nanoscale in size, but stable and robust in most chemically harsh conditions.[26] As a result, many potential applications have been proposed for CNTs, including conductive and high strength composites, energy storage devices, sensors, field emission displays, nanoscale semiconductor devices, probes, and interconnects.[26–30] Significant research activity has focused on the development of a new class of high-performance polymer composites with nanotubes as the reinforcing fillers.

While carbon fibers, carbon nanofibers, multiwall carbon nanotubes (MWNTs), and single-wall carbon nanotubes (SWNTs) are being used for reinforcing polymer matrices, SWNTs are sometimes heralded as the ultimate in carbon reinforcement. Therefore, this section will mostly discuss SWNT-polymer composite fibers and their potential.

12.3.1 PBO-SWNT Fibers[31]

Poly(p-phenylene benzobisoxazole) (PBO) was *in situ* polymerized in the presence of single-wall carbon nanotubes (SWNTs) in polyphosphoric acid (PPA) using typical PBO polymerization conditions. PBO and PBO-SWNT lyotropic liquid crystalline solutions in PPA were spun into fibers using dry-jet wet spinning. The tensile strength of the PBO-SWNT fiber containing 10 wt% SWNT was over 50% higher than those of the control PBO fibers containing no SWNT (Table 12.4).

Synthesis and fiber spinning experiments were performed at three levels of impurity (30, 10, and 1 wt% metal catalytic impurity) in the SWNTs. At 30% catalytic impurity, continuous fiber could not be drawn. At 10 wt% impurity, fiber could be drawn; however, its mechanical properties were comparable to those of the PBO synthesized and spun under the same conditions. Dramatic improvements in tensile strength were obtained only when SWNTs with about 1 wt% impurity were used. Extensive high-resolution transmission electron microscopy in PBO-SWNT (90–10) has not revealed SWNT aggregates or bundles, suggesting that SWNTs

TABLE 12.4
Mechanical Properties of PBO and PBO-SWNT Composite Fibers[31]

Sample	Fiber Diameter (μm)	Tensile Modulus (GPa)	Tensile Strength (GPa)	Strain to Failure (%)	Compressive Strength (GPa)
PBO	22 ± 2	138 ± 20	2.6 ± 0.3	2.0 ± 0.2	0.35 ± 0.6
PBO-SWNT (95–5)	25 ± 2	156 ± 20	3.2 ± 0.3	2.3 ± 0.3	0.40 ± 0.6
PBO-SWNT (90–10)	25 ± 2	167 ± 15	4.2 ± 0.5	2.8 ± 0.3	0.50 ± 0.6

were indeed well dispersed in these samples. On the other hand, both oriented and unoriented nanotube bundles have been observed in PBO-SWNT samples containing 15 wt% nanotubes. Attempts are being made to achieve better SWNT dispersion at this and higher concentrations.

12.3.2 PAN-SWNT Composite Fibers and Films

12.3.2.1 PAN-SWNT Composite Fibers[32]

The tensile and storage moduli of polyacrylonitrile (PAN)-SWNT fibers containing 10 wt% SWNT increased by factors of two and ten at room temperature and at 150°C, respectively, when compared to the PAN fiber (Fig. 12.7(a)). In addition, the glass transition temperature increased by 40°C (Fig. 12.7(b)).

Mechanical properties of a two-phase composite, in which both the matrix and the reinforcement have different orientations,[33] require a full set of elastic constants for the anisotropic matrix that are currently not available for PAN. Therefore the composite fiber modulus (E_c) was estimated using the equation,

$$E_c = V_{NT}E_{NT} + V_{PAN}E_{PAN}$$

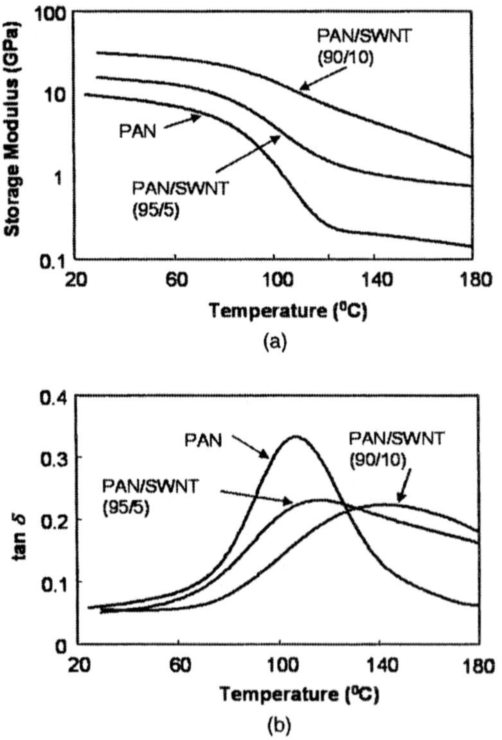

FIGURE 12.7 (a) storage modulus and (b) tan δ as a function of temperature for PAN and PAN-SWNT composite fibers.[32]

TABLE 12.5
Physical and Mechanical Properties of SWNT-PAN Composite Fibers[32]

	PAN	PAN-SWNT (95–5 wt%)	PAN-SWNT (90–10 wt%)
V_{NT}†	0	4.6	9.2
f_{PAN} (x-ray)	0.58	0.68	0.66
f_{NT} (Raman)	—	0.94	0.92
E_{PAN} (GPa)§	7.9	9.4	9.1
E_{NT} (GPa) (SWNT rope diameter > 20 nm)*	—	28.2	22.2
E_{NT} (GPa) (SWNT rope diameter ~ 4.5 nm)*	—	149.1	122.6
E_c (GPa) (SWNT rope diameter ≥ 20 nm)	—	10.3	10.3
E_c (GPa) (SWNT rope diameter ~ 4.5 nm)	—	15.8	19.5
E_{exp} (GPa)	7.9	14.2	16.2

† V_{NT} calculated assuming PAN and SWNT densities of 1.18 g/cm^3 and 1.30 g/cm^3, respectively. § E_{PAN} for the composite fibers was calculated from the control PAN fiber modulus to account for the difference in orientation. *Obtained from Reference 35 for the given f_{NT} and rope diameter.

where E_{NT} and E_{PAN} are the nanotube and PAN moduli along the fiber axis, and V_{NT} and V_{PAN} are their volume fractions, respectively. Based on the SWNT orientation factors for 5 and 10 wt% composite fibers and the orientation dependence of the SWNT modulus, E_{NT} values obtained from elsewhere[34,35] for two different rope diameters are listed in Table 12.5, along with values of E_{PAN}, V_{NT}, and V_{PAN}. Table 12.5 shows that the experimental moduli values of PAN-SWNT composite fiber are well within the range predicted for the two rope diameters. Rope diameter for the SWNT powder used in this study was measured to be 37 ± 8 nm from scanning electron microscopy.[36] However, the measured modulus is closer to the predictions based on smaller diameter ropes. This suggests that some SWNT rope exfoliation has occurred during mixing, fiber processing, and fiber drawing. Based on high resolution transmission electron microscopy study (Fig. 12.8(a) and (b)), average single-wall nanotube bundle diameter in PAN-SWNT (95-5) fiber was measured to be 11 nm, confirming partial exfoliation.[37] Further improvement in nanotube exfoliation and orientation is expected to result in further modulus increase.[32,35]

Composite fiber also showed significant reduction in thermal shrinkage, as well as in polymer solubility. For example, the shrinkage at nominal stress at 200°C was approximately 25% in PAN fiber and only about 10% in PAN-SWNT (90–10) composite fiber, as shown in Fig. 12.9. After extensive sonication and being subjected to elevated temperature, less than 50% of the PAN could be dissolved from the PAN-SWNT (90–10) composite fiber, thus making the composite fiber substantially more solvent resistant. These observations suggest good interaction between PAN and SWNTs.

SWNTs exhibited higher orientation than PAN as determined by Raman spectroscopy and x-ray diffraction. Infrared spectroscopy confirmed the anisotropic optical absorption behavior of SWNTs, as shown in Fig. 12.10.

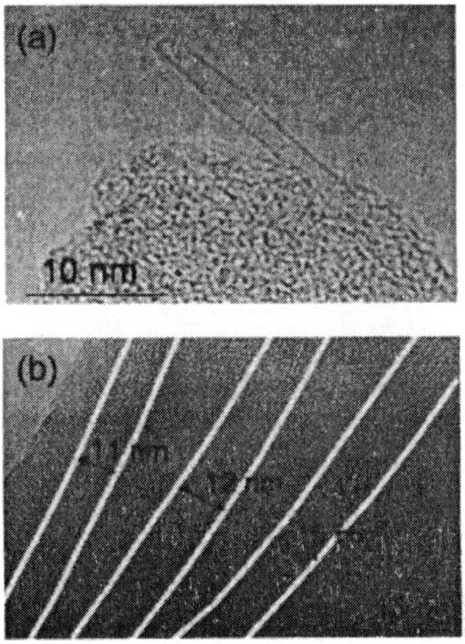

FIGURE 12.8 Transmission electron micrographs of SWNT ropes in PAN: (a) tip of nanotube; (b) intertube distances.[37]

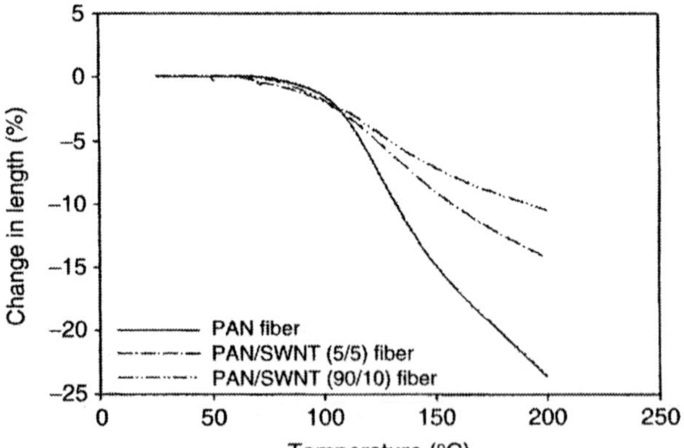

FIGURE 12.9 Thermal shrinkage in (A) PAN, (B) PAN-SWNT (95–5), and (C) PAN-SWNT (90–10) composite fibers as a function of temperature.[32]

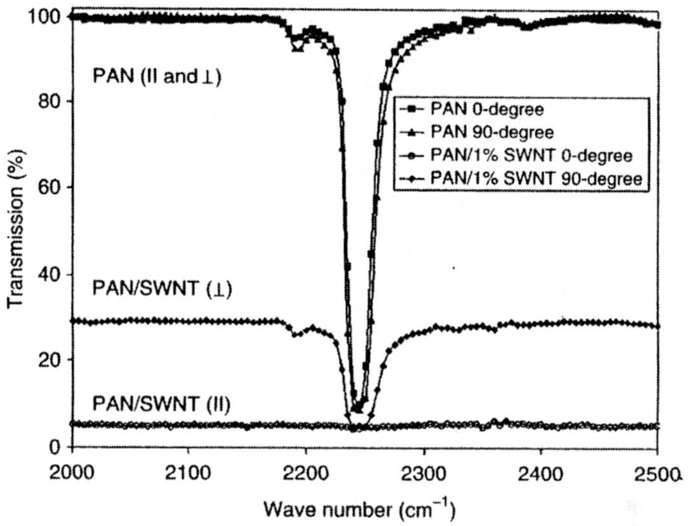

FIGURE 12.10 Polarized IR spectra of PAN and PAN-SWNT (99–1) composite fibers.[32]

12.3.2.2 Oxidative Stabilization on PAN-SWNT Composite Fibers[38]

The work on PAN-SWNT composites is also important because PAN is a carbon fiber precursor and PAN-SWNT composite fiber is likely to result in higher tensile strength and modulus carbon fibers for aerospace applications. Polyacrylontrile,[39,40] petroleum pitch, and cellulosic fibers have been used as carbon fiber precursors. Currently carbon fibers are predominantly made from polyacrylonitrile copolymer precursors.[41] For converting PAN to carbon fiber, thermo-oxidative stabilization typically in the 200 to 300°C range is a key step.

At 10 wt% SWNTs, breaking strength, modulus, and strain to failure of the oxidized composite fiber increased by 100, 160, and 115%, respectively (Table 12.6). Tensile fracture surfaces (not shown here) of thermally stabilized PAN and the PAN-SWNT fibers exhibited brittle behavior, and well-distributed SWNT ropes covered with the oxidized matrix can be observed in the tensile fracture surfaces of the fibers.

Figure 12.11 shows weight loss in PAN and PAN-SWNT composite fibers when isothermally oxidized at 250°C. Weight loss during oxidative stabilization of PAN is a well-known phenomenon, even though there is oxygen uptake.[42] One reason for the difference in weight loss between the control PAN and the PAN-SWNT composite fibers is that SWNTs are thermally stable at this temperature. However, the mass balance analysis suggests that in the oxidized PAN-SWNT (90–10) composite fiber, the weight residue is about 0.3 wt% more than the calculated value, assuming that weight loss in PAN in the composite fiber is the same as in the control PAN and that there was no weight loss in SWNT. This suggests that the presence of SWNT may alter PAN oxidative stabilization rate, mechanism, or both.

TABLE 12.6
Mechanical Properties of PAN and PAN-SWNT Composite Fibers before and after Oxidation[38]

	SWNT Content (wt%)	Tensile Modulus (GPa)	Tensile Strength (GPa)	Strain to Failure (%)
Before oxidation	0	7.9 ± 0.4	0.23 ± 0.03	11.6 ± 1.4
	5	14.2 ± 0.6	0.36 ± 0.02	11.9 ± 1.3
	10	16.2 ± 0.8	0.33 ± 0.02	9.7 ± 1.6
After oxidation	0	5.2 ± 0.5	0.06 ± 0.01	1.4 ± 0.3
	5	8.9 ± 1.0	0.10 ± 0.01	2.2 ± 0.5
	10	13.5 ± 0.9	0.13 ± 0.01	3.1 ± 0.7

The tensile strength and modulus of carbon fibers strongly depend on the structure and properties of the PAN precursor fiber. As listed in Table 12.6, the mechanical properties of the pristine and stabilized composite fibers were higher than those of the control PAN fiber. In addition, PAN had higher orientation in the composite fiber than in the control PAN fiber, and composite fiber exhibited reduced stress during stabilization. In the 100 to 1060 GPa modulus range, carbon fiber modulus depended very strongly on the orientation. With a misorientation factor of less than 5°, graphite in plane modulus dropped from 1060 GPa to below 200 GPa.[43] Carbon fibers used in the aerospace industry today have a tensile modulus in the 200 to 300 GPa range

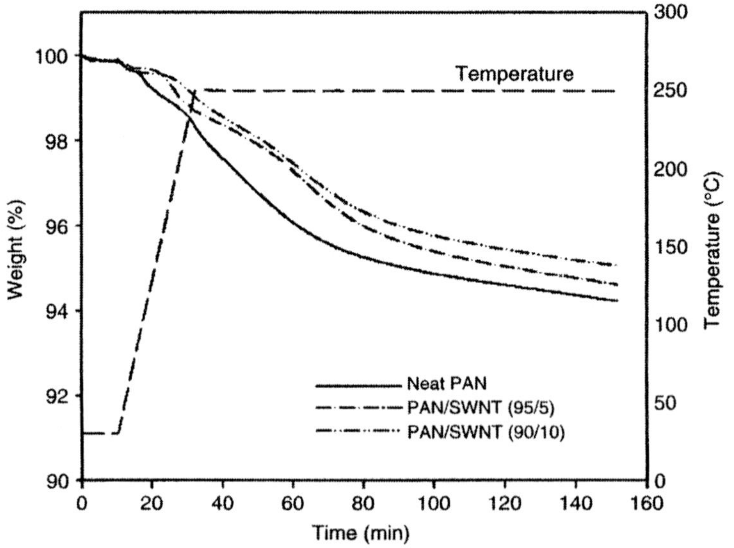

FIGURE 12.11 Weight loss in PAN and PAN-SWNT composite fibers when isothermally heated at 250°C under air flow.[38]

and are carbonized at about 1500°C. Although PAN-based carbon fibers with moduli higher than 500 GPa can be produced, these fibers are carbonized at ~2500°C and have low compressive strength (~1 GPa). By comparison, PAN-based carbon fibers carbonized at ~1500°C with a modulus in the 200 to 300 GPa range have high compressive strength values (~3 GPa).[44] All pitch-based carbon fibers have low compressive strength (less than 1 GPa).[44,45] Significant improvement in carbon fiber modulus without sacrificing the compressive strength will represent a major development in the carbon fiber composite materials technology. Fiber compressive strength decreases with increasing carbonization temperature and with increasing graphitic order. The key to increasing the carbon fiber modulus without sacrificing compressive strength is to achieve higher orientation at moderate carbonization temperatures (~1500°C). Since PAN-SWNT composites exhibit lower stress during stabilization and have higher PAN orientation, PAN-SWNT fibers may yield higher modulus carbon fibers while maintaining compressive strength. In PAN-SWNT fibers reported here, SWNT ropes were partially exfoliated. The rope diameter was reduced from 30 nm to 11 nm. Exfoliation to individual nanotube will result in fiber property improvements at lower SWNT concentration than have been used in the current work, thus reducing the nanotube requirements, which should make these composite systems a commercial reality.

12.3.2.3 PAN-SWNT Composite Films[46]

The mechanical properties of PAN-SWNT composite films listed in Table 12.7 show that the initial modulus of the composite film is up to 2.7 times the modulus of PAN film, and more than eight times the modulus of the bucky paper. Tensile strength of the composite film is up to twice the strength of the PAN film, and ten times the strength of the bucky paper. This clearly demonstrates the reinforcing effect of PAN in SWNTs and vice versa. The tensile strength and modulus values of the composite film with the unpurified SWNT are larger than those using the

TABLE 12.7
Mechanical and Electrical Properties of PAN, PAN-SWNT, and SWNT Films[46]

Sample	SWNT Content (wt%)	Tensile Modulus (GPa)	Tensile Yield Strength (MPa)	Strain to Failure (%)	Conductivity (S/m)
PAN	0	2.7 ± 0.4	57 ± 4	22.3 ± 6.5	N/A
PAN/SWNT (80/20)[a]	20	6.4 ± 0.7	78 ± 10	1.9 ± 0.4	2.4×10^3
PAN/SWNT (60/40)[a]	40	7.2 ± 1.0	65 ± 22	2.0 ± 0.6	6.3×10^3
PAN/SWNT (60/40)[b]	40	10.9 ± 0.3	103 ± 18	1.6 ± 0.6	1.5×10^4
SWNT bucky paper	100	0.8 ± 0.1	10 ± 2	5.6 ± 0.3	3.0×10^4

[a] Purified HiPCO SWNTs were used to prepare composite film.

[b] Unpurified HiPCO SWNTs were used to prepare composite film.

purified SWNTs at the same SWNT loading. Due to the presence of 2–4 nm diameter catalytic particles, unpurified SWNTs are easier to disperse in dimethylformamide (DMF). Better mechanical properties of the composite film processed using unpurified SWNTs are attributed to the difference in dispersion quality between unpurified and purified films. The mechanical properties of PAN-SWNT composites are significantly higher than the SWNT bucky paper and control PAN film. In addition, the electrical conductivities of PAN-SWNT films exhibited fairly comparable values, which implies that PAN-SWNT composite films maintain the degree of percolation as well as reinforce each other.

12.3.3 Polyvinyl Alcohol (PVA)/SWNT Composite Fibers and Films

12.3.3.1 PVA-SWNT Composite Fibers[47]

The concept of producing high-performance fibers from flexible polymers is based on arresting polymer orientation and preventing chain folding by gel spinning, which induces crystallization and minimizes chain entanglements. High mechanical drawability is the key for obtaining high orientation. In order to obtain high drawability, entanglement control between the chains is crucial. A certain degree of entanglement is needed to endure drawing, but too much entanglement would reduce drawability. In other words, the concentration of the polymer solution should be near the polymer overlap concentration. But, in reality, much higher concentration is chosen as fibers cannot be drawn near the overlap concentration. For gel spinning of PVA or other polymers, crystallization-induced gelation is the key step in the process. These crystals act as physical cross-links for mechanical drawing. For good drawability, the crystal size should be as small as possible.

The PVA-SWNT dispersions in DMSO and DMSO-H_2O solvents were prepared and successfully spun into composite fibers through gel spinning. Mechanical properties of gel-spun PVA and PVA-SWNT composite fibers are listed in Table 12.8.

12.3.3.2 PVA-SWNT Composite Films[48]

The ultimate reinforcement of fillers can be achieved only when the applied stress to matrix polymer is transferred into dispersed fillers. Intensive study for Raman spectroscopy of CNTs indicated that D^* (disorder) band of CNTs can represent the deformation of CNTs from external stresses. Figure 12.12 and Fig. 12.13 show the

TABLE 12.8
Mechanical Properties of PVA and PVA-SWNT Gel-Spun Fibers[47]

Sample	Fiber Diameter (μm)	Tensile Strength (GPa)	Tensile Modulus (GPa)	Strain to Failure (%)
PVA	26.5 ± 1.6	0.9 ± 0.1	25.6 ± 2.6	7.5 ± 1.6
PVA/SWNT (3 wt%)	27.0 ± 2.0	1.1 ± 0.2	35.8 ± 3.5	8.8 ± 1.7

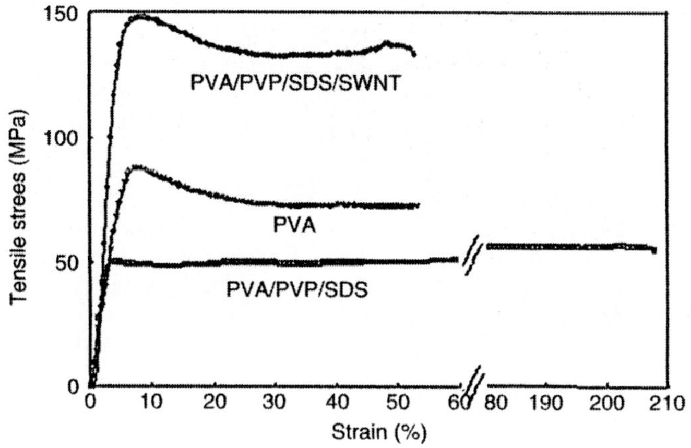

FIGURE 12.12 Stress-strain curves for various films.[48]

stress-strain curves of PVA-SWNT films and D^* band shift versus strain, respectively. The shifting behavior of the D^* band looks similar to that of stress variation with strain, which implies the load transfer to CNTs from polymer matrix.

12.3.4 PMMA-SWNT Composite Fibers[49]

Liu et al.[50] showed the wrapping behavior of polymer around SWNT. Figure 12.14 shows the SEM image of electrospun PMMA-SWNT composite fiber. SWNT is electrically conductive, while PMMA is an insulating matrix around SWNT. Therefore this

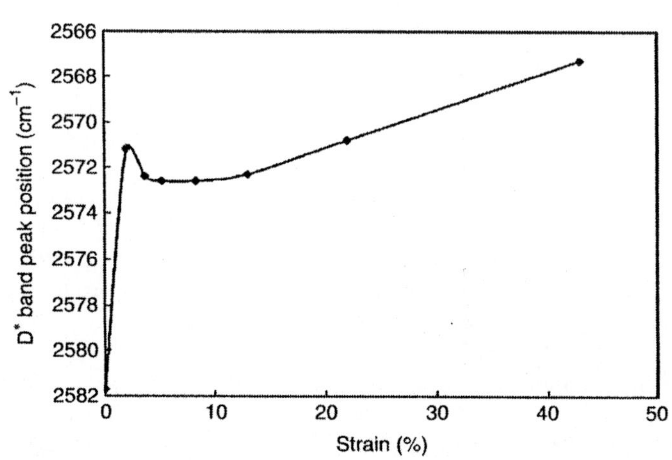

FIGURE 12.13 Raman D^* band peak position as a function of strain for PVA-PVP-SDS-SWNT composite films containing 5 wt% SWNT. Excitation laser wavelength was 785 nm, and the laser power was ·1.5 mW.[48]

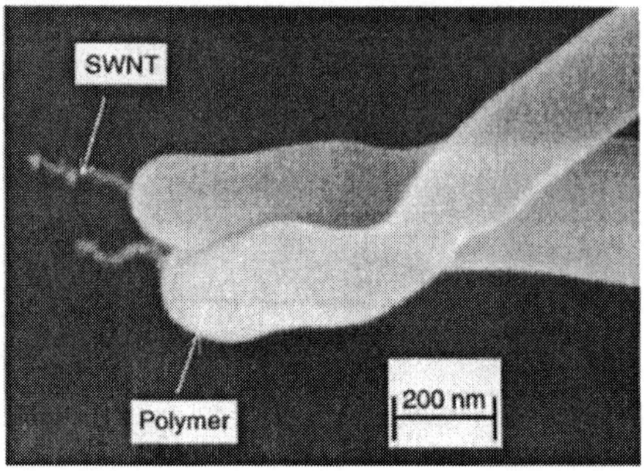

FIGURE 12.14 Scanning electron micrograph of electrospun PMMA-SWNT fiber.[49]

type of fiber can be used as nanowire in nanoscale electrical devices. Such fibers can also be used as atomic force microscope (AFM) tips.[51]

12.4 ADDITIONAL ASPECTS OF CNT-POLYMER COMPOSITES

12.4.1 Cross-linking of SWNTs by Oxidation and Effect of Nitric Acid Treatment[52,53]

It has been proposed that mechanical properties of SWNT products can be improved by covalent cross-linking.[54–56] Oxidation of CNTs is one possible way to achieve cross-linking. Functional groups such as carboxylic acid, quinine,[57] phenol,[58] ester,[59] amide,[60] and zwitterions[61] have been reported on the oxidized SWNTs. Oxidation can also occur during SWNT purification in strong acid.[62,59] It is understood that oxidation occurs at end caps and at the defect sites.[63] The presence of polycyclic aromatic amorphous carbon has been reported in the oxidized SWNT.[53,64] Small-diameter SWNTs have been reported to be destroyed during nitric acid treatment,[53,65] and the formation of continuous phase morphology has been observed in HNO_3-purified SWNTs, which is attributed to amorphous carbon formed by the decomposition of SWNTs.[66]

As listed in Table 12.9, the nitric acid-treated bucky paper showed higher mechanical properties than oleum-treated bucky paper, which may be evidence of cross-linking by oxidation of SWNTs. However, the in-plane electrical conductivity decreases with nitric acid treatment, which is a result of SWNT functionalization leading to imperfections and defects along the SWNT wall.

The storage modulus of SWNT films is fairly constant (Fig. 12.15) in the measured temperature range (there is slight up-turn above 150°C) and the tan δ value

TABLE 12.9
Mechanical and Electrical Properties of Various SWNT Films[53]

Sample	Tensile Modulus (GPa)	Tensile Strength (MPa)	Strain to Failure (%)	Conductivity (S/m)
Control SWNT film*	0.8 ± 0.1	10 ± 2	5.6 ± 0.3	3.0×10^4
3M HNO_3 SWNT film	1.4 ± 0.1	16 ± 1	1.4 ± 0.2	2.3×10^4
6M HNO_3 SWNT film	2.9 ± 0.2	71 ± 5	3.4 ± 0.2	2.4×10^4
10M HNO_3 SWNT film	5.0 ± 0.2	74 ± 2	3.0 ± 0.1	1.2×10^4

* Control SWNT film is bucky paper processed from aqueous solution without treating nitric acid.

is very low (~0.02), suggesting that these films are fairly elastic in this temperature range.

It also has been reported that the most important parameter, that determines the chemical oxidation attack of the nanotubes is their diameter. Small diameter tubes, due to the stress induced by the curvature, are first attacked and destroyed. Some reactivity is also reported for the larger diameter metallic nanotubes.[67] From the quantitative analysis of CNT diameter with nitric acid treatment, one can see that the relative fractions of small diameter SWNTs (0.88 and 0.89 nm) significantly decreased with increasing nitric acid concentration, as shown in Fig. 12.16. The percentage of these small diameter tubes decreased from 70% in the control film to less than 20% in 10 M samples, confirming the selective degradation of the small

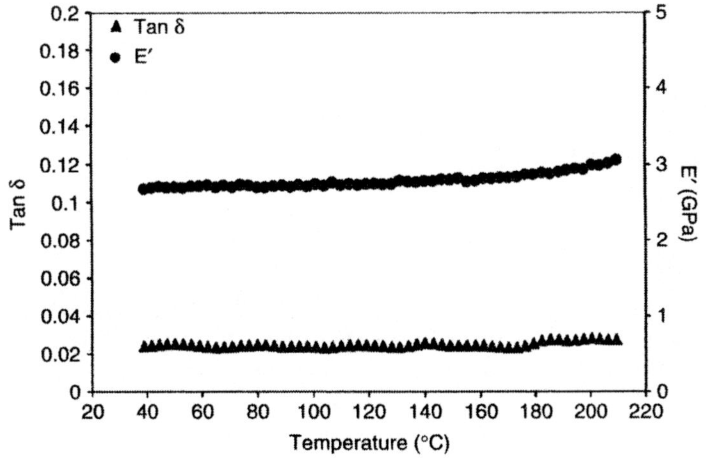

FIGURE 12.15 Dynamic mechanical behavior of SWNT film processed from 6 molar nitric acid.[53]

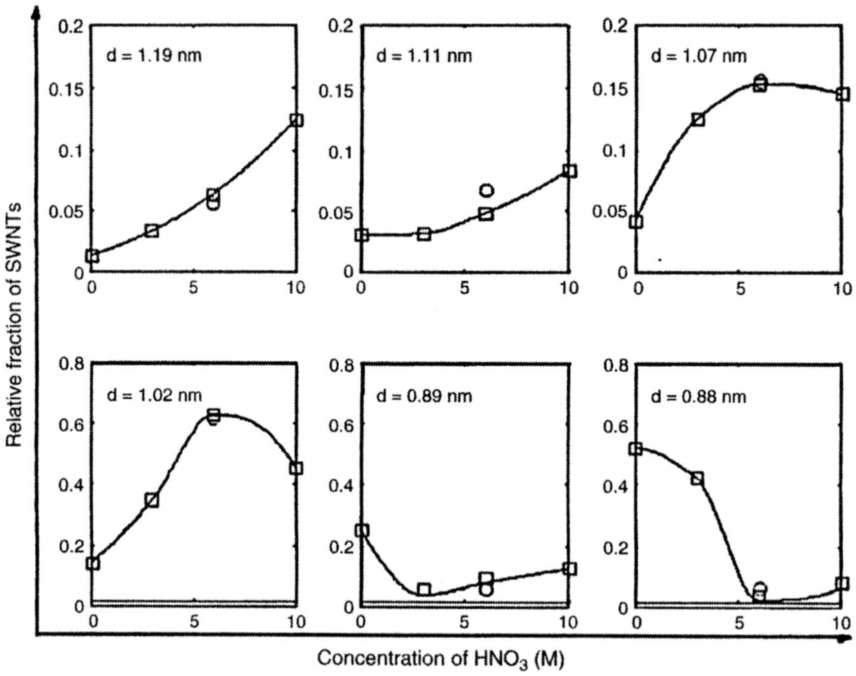

FIGURE 12.16 Relative fractional changes of different diameter SWNTs as a function of HNO_3 concentration.[53]

diameter SWNTs by HNO_3. As a result, the relative fraction of the large diameter SWNTs (1.19 and 1.11 nm) increased monotonically with nitric acid concentration. It has been pointed out that the data in Fig. 12.16 are based on the Raman radial breathing mode using 785-nm wavelength laser. Therefore metallic tubes have not been observed in this study.

12.4.2 Crystallization Behavior of Polymer with SWNT[67]

The spherulite size in PP is much larger than in PP-SWNT, suggesting that SWNTs act as nucleating sites for PP crystallization (Fig. 12.17(a) and (b)). Based on the half-crystallization time as a function of isothermal crystallization temperature, it has been shown that the addition of 0.8 wt% SWNT increases the crystallization rate by as much as an order of magnitude or higher (Fig. 12.18). In the drawn PP-SWNT composite fibers, SWNT orientation was monitored using Raman spectroscopy and that of PP using x-ray diffraction. These studies show that the orientation of SWNTs (orientation factor 0.95 assuming Gaussian distribution) was higher than that of PP (orientation factor 0.86).

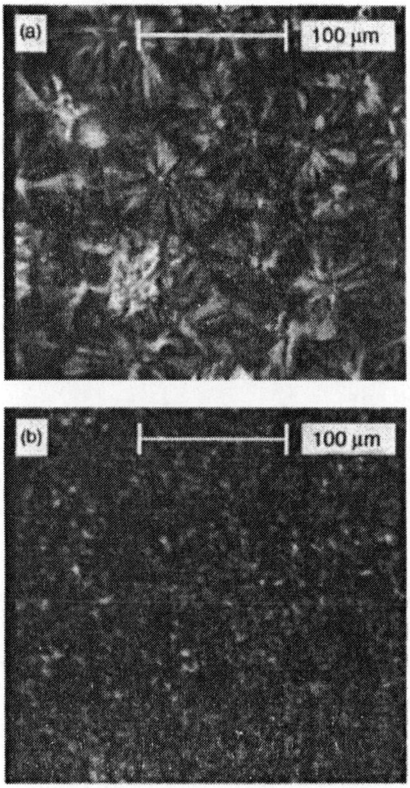

FIGURE 12.17 Optical micrographs (with cross-polars) of (a) polypropylene (PP) and (b) PP-SWNT composite with 0.8 wt% SWNT content.[67]

12.4.3 Effect of SWNT Exfoliation and Orientation[35,77]

Modulus of single-wall carbon nanotube (SWNT) films and fibers, calculated using continuum mechanics and the SWNT rope elastic constants, is consistent with the experimentally measured moduli of these products. Shear modulus of 20-nm diameter ropes is about 1 GPa, and that of 4-nm diameter ropes is about 6 GPa. The axial tensile modulus of large diameter ropes drops precipitously with orientation, while the smaller diameter rope modulus exhibits lower orientation dependence (Fig. 12.19).[35] Therefore, in addition to orientation, for the ropes composed of SWNTs with varying diameters and helicity, exfoliation also appears to be critical in achieving high modulus SWNT films and fibers.

Due to the strong resonance-enhanced Raman scattering,[69–71] polarized Raman spectroscopy is commonly used for SWNT orientation determination.[72–76] By fitting the polarized Raman scattering intensity as a function of angles that the sample makes with respect to the polarization direction of the incident exciting laser source,

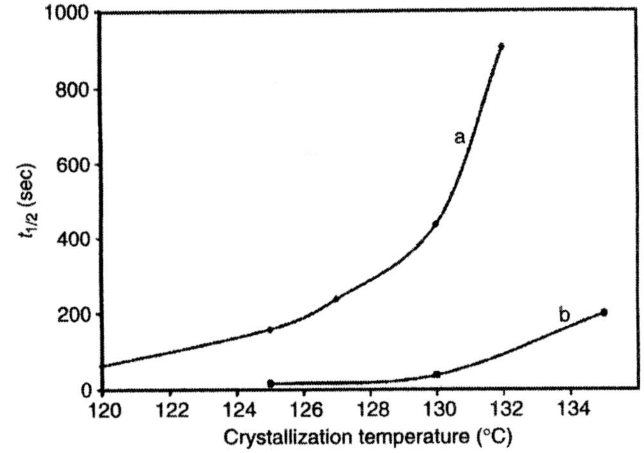

FIGURE 12.18 Crystallization half-time ($t_{1/2}$) of (a) PP and (b) PP-SWNT (99.2–0.8) as a function of crystallization temperature.[67]

the full width at half-maximum (FWHM) of Gaussian or Lorentzian distribution function is generally taken as the measure of orientational order. While this approach is mathematically simple, the assumed distribution function may not truly reflect the orientation distribution of SWNTs. Therefore the orientational order of SWNTs would

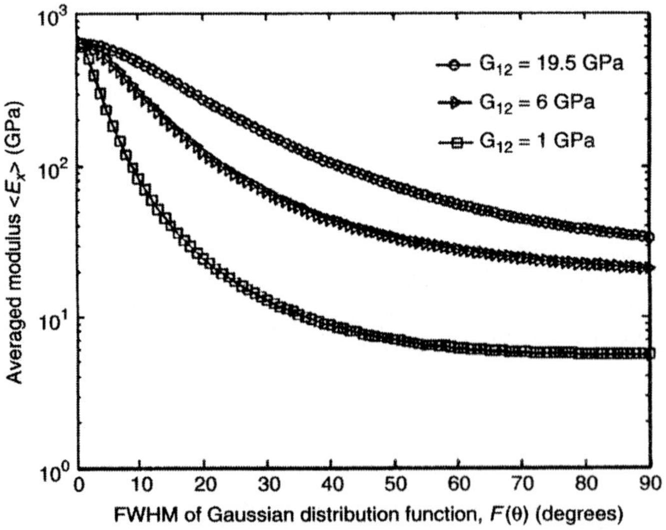

FIGURE 12.19 Effect of SWNT rope orientation on the axial tensile modulus of SWNT fiber.[35]

not be quantified correctly. To rectify this problem, a generalized spherical harmonics expanded orientation distribution function (ODF) has been used to determine the SWNT orientation.[77] A relationship between the second and the fourth order SWNT orientation parameters and the Raman scattering intensity and depolarization ratio has been established. By determining the VV and VH (direction of polarizer and analyzer; V and H indicate vertical and horizontal, respectively) intensities at 0 and 90° to the orientation direction, the second and the fourth order SWNT orientation parameters can be determined. From the second and the fourth order orientation parameters, orientation distribution function can be constructed using the maximum entropy formalism.

12.5 POLYMER/SWNT APPLICATION (SUPERCAPACITOR)[36]

The diameter of the as-produced HiPco SWNT ropes used in this study was 35 ± 7 nm and that for the as-prepared SWNT-PAN composite film was estimated to be 57 ± 8 nm (Fig. 12.20(a) and (b)). The increased rope diameter indicates PAN copolymer adsorption by the SWNT ropes. Physical or chemical activation of PAN or its copolymers is generally used for producing activated carbon with high specific surface area and porosity, and it is expected that similar activation treatment of SWNT-PAN composite film can also be used for developing SWNT activated carbon composite films.

The specific capacitance of the SWNT activated carbon film is significantly higher than that of the SWNT bucky paper (Fig. 12.21). The specific capacitance of SWNT activated carbon at 0.001 ampere discharging current is strongly dependent on the discharging voltage, a result of its nonlinear discharging behavior. When the dis-

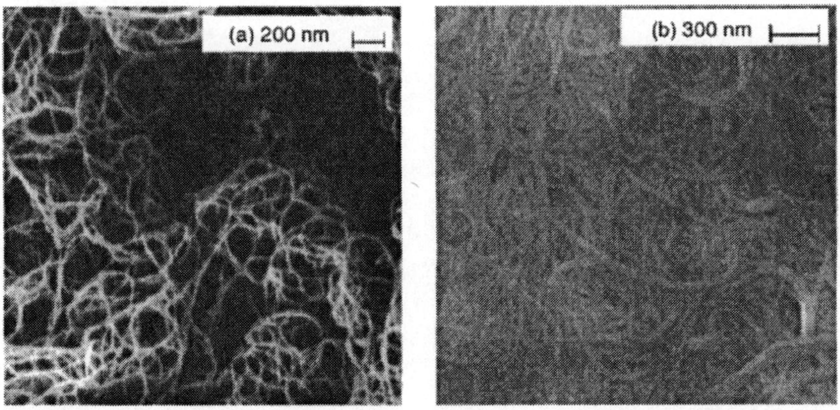

FIGURE 12.20 Scanning electron micrographs of (a) as-produced HiPco SWNT powder and (b) as-produced SWNT-PAN composite film.[36]

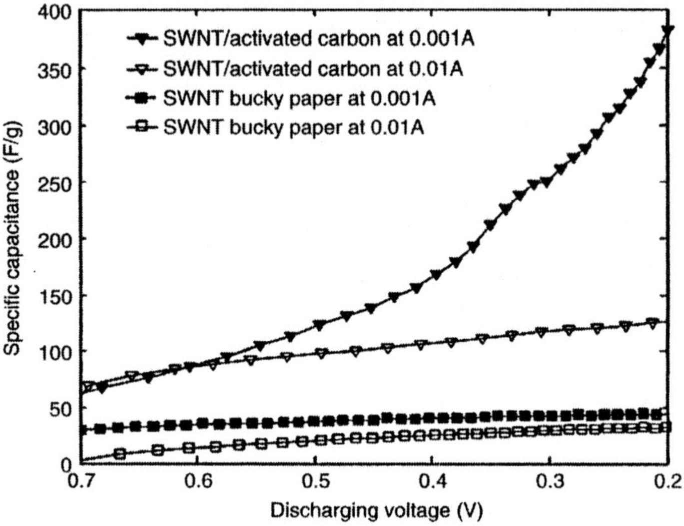

FIGURE 12.21 Specific capacitance as a function of discharging voltage.[36]

charging voltage is reduced from 0.7 volts to 0.2 volts, the specific capacitance increased from 60 F/g to 380 F/g. The Ragone plot (Fig. 12.22) shows that both the power and energy densities of SWNT-activated carbon composite film are significantly higher than that of SWNT bucky paper.

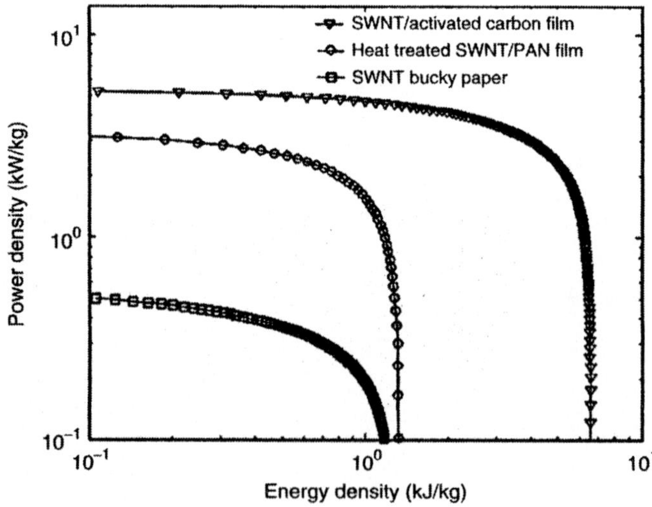

FIGURE 12.22 Ragone plot for various electrodes evaluated at discharging current of 0.01A.[36]

12.6 CONCLUDING REMARKS

This chapter summarizes studies of a number of polymeric systems containing carbon nanofibers as well as carbon nanotubes. Limited studies have also been reported on carbon nanotube materials containing no polymers. Composites have been processed using melt processing, solution processing, and *in situ* polymerization. Carbon nanotubes containing polymeric materials exhibit improved tensile strength, tensile modulus, strain to failure, torsional modulus, compressive strength, glass transition temperature, solvent resistance, and reduced shrinkage. In polyacrylonitrile containing 10 wt% SWNT, the modulus above glass transition temperature is more than an order of magnitude higher than the modulus of the control PAN fiber at that temperature. Theoretical studies suggest that with SWNT exfoliation and with improved orientation, similar property improvements can be expected at much lower SWNT content, making these materials affordable. It is noted that all the property improvements are not observed in every system. In addition, carbon nanotube-based materials will also have electrical and thermal properties. These materials will also find applications for hydrogen storage and various types of sensors such as gas, strain or stress, and heat, among others. Developments in production technology for carbon nanofibers, multiwall nanotubes (MWNT), double-wall nanotubes (DWNT), and single-wall nanotubes continue. Between 0.4 and 4 nm, there can be nearly 400 different types of SWNTs. A given production run today may contain about 20 different types of nanotubes, typically in the 0.7 to 1.4-nm diameter range. Synthesis of a specific diameter and chirality SWNT will represent a major breakthrough in the field and will unfold materials development that will be considered revolutionary, even by the standards of the material developments that have occurred in the twentieth century. This situation perhaps can be compared to the development of stereo-specific polymers that instantly made simple polymers such as polypropylene much more useful. Professor Smalley[76] has pointed out that there are only three types of polypropylene (atactic, syndiotactic, and isotactic), but there are nearly 400 possibilities for the SWNTs. This points to both the enormous potential of nanotubes and the complexity of the problem.

As we think about the future of potential materials development in the field, it will help to keep in mind that even visionaries and highly successful scientists and business people sometimes fail to comprehend and predict what is just around the corner, as can be seen from the following quotations:

Radio has no future; x-rays are a hoax.—Lord Kelvin (~1885).

I have not the smallest molecule of faith in aerial navigation other than ballooning.—Lord Rayleigh (1889).

I think there is a world market for about five computers.—Thomas J. Watson (IBM founder, 1943).

Having said this, the question to ponder is, what type of fiber development can we expect in the coming century? Will we develop fibrous materials with the following characteristics?

Ability to harvest and store the energy of the sun

Functionality of a computer and cell phone

Heating and cooling capability as needed
Production of light on demand
Ability to monitor health
Color and feel of a typical textile

ACKNOWLEDGMENTS

Funding from the Office of Naval Research (N00014-01-1-0657), Air Force Office of Scientific Research (F49620-03-1-0124), National Science Foundation, and Carbon Nanotechnologies, Inc. is gratefully acknowledged.

REFERENCES

1. Hounshell, D.A. and Smith, J.K., Jr., *Science and Corporate Strategy: Du Pont R&D, 1902–1980,* Cambridge University Press, 1988, p. 231.
2. U.S. Patent 4,565,684 (January, 21, 1986), G. G. Tibbetts and M. G. Devour (to General Motors Corporation).
3. Lake, M.L. and Ting, J.-M., in Burchell, T.D., Ed., *Carbon Materials for Advanced Technologies,* Pergamon, Oxford, U.K., 1999, pp. 139–167.
4. de Jong, D.P. and Geus, J.W., "Carbon Nanofibers: Catalytic Synthesis and Applications," *Catal. Rev. Sci. Eng., 42,* 4, 481, 2000.
5. Jacobson, R.L., Tritt, T.M., Guth, J.R., Ehrlich, A.C., and Gillespie, D.J., "Mechanical Properties of Vapor-grown Carbon Fiber," *Carbon, 33,* 9, 1217, 1995.
6. Ting, J., "Tensile Properties of VGCF Reinforced Carbon Composites," *J. Mater. Sci., 34,* 229, 1999.
7. Ishioka, M., Okada, T., and Matsubara, K., "Mechanical Properties of Vapor-grown Carbon fibers Prepared from Benzene in Linz-Donawitz Converter Gas by Floating Catalyst Method," *J. Mater. Res., 7,* 11, 3019, 1992.
8. Applied Sciences, Inc. (http://www.apsci.com)
9. Uchida, T., Anderson, D.P., Minus, M., and Kumar, S., "Morphology and Modulus of Vapor Grown Carbon Nano Fiber," *J. Mater. Sci.,* in press.
10. Tibbetts, G.G. and McHugh, J.J., "Mechanical Properties of Vapor-grown Carbon Fiber Composites with Thermoplastic Matrices," *J. Mater. Res., 14,* 7, 2871, 1999.
11. Lozano, K. and Barrera, E.V., "Nanofiber-reinforced Thermoplastic Composites. I. Thermoanalytical and Mechanical Analyses," *J. Appl. Polym. Sci., 79,* 1, 125, 2001.
12. Kuriger, R.J., Alam, M.K., Anderson, D.P., and Jacobsen, R.L., "Processing and Characterization of Aligned Vapor Grown Carbon Fiber Reinforced Polypropylene," *Composites A, 33,* 53, 2002.
13. Caldeira, G., Maia, J.M., Carneiro, O.S., Covas, J.A., and Bernardo, C.A., "Production and Characterization of Innovative Carbon Fiber Polycarbonate Composites," *Polym. Composites, 19,* 2, 147, 1998.
14. Carneiro, O.S. and Maia, J.M., "Rheological Behavior of (Short) Carbon Fiber/Thermoplastic Composites. Part II: The Influence of Matrix Type," *Polym. Composites, 21,* 6, 970, 2000.
15. Pogue, R.T., Ye, J., Klosterman, D.A., Glass, A.S., and Chartoff, R.P., "Evaluating Fiber-matrix Interaction in Polymer-matrix Composites by Inverse Gas Chromatography," *Composites A, 29,* 1273, 1998.

16. Cooper, C.A., Ravich, D., Lips, D., Mayer, J., and Wagner, H.D., "Distribution and Alignment of Carbon Nanotubes and Nanofibrils in a Polymer Matrix," *Composites Sci. Technol., 62,* 1105, 2002.
17. Sandler, J., Windle, A.H., Werner, P., Altstadt, V., Es, M.V., and Shaffer, M.S.P., "Carbon Nanofibre-reinforced Poly(ether ether ketone) Fibres," *J. Mater. Sci., 38,* 10, 2135, 2003.
18. Cadek, M., Le Foulgoc, B., Coleman, J.N., Barron, V., Sandler, J., Shaffer, M.S.P., Fonseca, A., van Es., M., Schulte, K., and Blau, W.J., *AIP Conference Proceedings: Structural and Electronic Properties of Molecular Nanostructures, 633,* 562, 2002.
19. Patton, R.D., Pittman, C.U., Wang, L., and Hill, J.R., "Vapor Grown Carbon Fiber Composites with Epoxy and Poly(phenylene sulfide) Matrices," *Composites A, 30,* 1081, 1999.
20. Uchida, T., Dang, T., Min, B.G., Zhang, X., and Kumar, S., "Processing, Structure, and Mechanical Properties of PBZT/carbon Nano Fiber Composite Fiber," *Composites B,* in press.
21. Kumar, S., Doshi, H., Srinivasarao, M., Park, J.O., and Schiraldi, D.A., "Fibers from Polypropylene/Nano Carbon Fiber Composites," *Polymer, 43,* 1701, 2002.
22. Ma, H., Zeng, J., Realff, M.L., Kumar, S., and Schiraldi, D.A., "Processing, Structure, and Properties of Fibers from Polyester/Carbon Nanofiber Composites," *Composites Sci. Technol., 63,* 1617, 2003.
23. Kozey, V.V., Jiang, H., Mehta, V.R., and Kumar, S., "Compressive Behavior of Materials. 2. High-performance Fibers," *J. Mater. Res., 10,* 4, 1044, 1995.
24. Zeng, J., Saltysiak, B., Johnson, W.S., Schiraldi, D.A., and Kumar, S., "Processing and Properties of Poly(methyl methacrylate)/Carbon Nano Fiber Composites," *Composites B, 35,* 173, 2004.
25. Iijima, S., "Helical Microtubules of Graphitic Carbon," *Nature, 354,* 56, 1991.
26. Baughman, R.H., Zakhidov, A.A., and de Heer, W.A., "Carbon Nanotubes—the Route toward Applications," *Science, 297,* 787, 2002.
27. Ajayan, P.M., "Nanotubes from Carbon," *Chem. Rev., 99,* 1787, 1999.
28. Ajayan, P.M. and Zhou, O.Z., in Dresselhaus, M.S., Dresselhaus, G., and Avouris, Ph., Eds., *Carbon Nanotubes: Synthesis, Structure, Properties and Applications,* Springer-Verlag, Heidelberg, 2001, pp. 391–425.
29. Thostenson, E.T., Ren, Z., and Chou, T.-W., "Advances in the Science and Technology of Carbon Nanotubes and their Composites: a Review," *Composites Sci. Technol., 61,* 1899, 2001.
30. Lau, K.-T. and Hui, D., "The Revolutionary Creation of New Advanced Materials—Carbon Nanotube Composites," *Composites B, 33,* 263, 2002.
31. Kumar, S., Dang, T.D., Arnold, F.E., Bhattacharyya, A.R., Min, B.G., Zhang, X., Vaia, R.A., Park, C., Adams, A.A., Hauge, R.H., Smalley, R.E., Ramesh, S., and Willis, P.A., "Synthesis, Structure, and Properties of PBO/SWNT Composites," *Macromolecules, 35,* 9039, 2002.
32. Sreekumar, T.V., Liu, T., Min, B.G., Guo, H., Kumar, S., Hauge, R.H., and Smalley, R.E., "Polyacrylonitrile Single-walled Carbon Nanotube Composite Fibers," *Advanced Mater., 16,* 1, 58, 2004.
33. Dunn, M.L., Ledbetter, H.L., Heyliger, P.R., and Choi. C.S., "Elastic Constants of Textured Short-fiber Composite," *J. Mech. Phys. Solids, 44,* 1509, 1996.
34. Salvetat, J.-P., Briggs, G.A.D., Bonard, J.-M., Bacsa, R.R., Kulik, A.J., Stockli, T., Burnham, N.A., and Forro, L., "Elastic and Shear Moduli of Single-walled Carbon Nanotube Ropes," *Phys. Rev. Letts., 82,* 944, 1999.

35. Liu, T. and Kumar, S., "Effect of Orientation on the Modulus of SWNT Films and Fibers," *Nano Lett., 3,* 647, 2003.
36. Liu, T., Sreekumar, T.V., Kumar, S., Hauge, R.H., and Smalley, R.E., "SWNT/PAN Composite Film-based Supercapacitors," *Carbon, 41,* 2440, 2003.
37. Uchida, T. and Kumar, S., "Single Wall Carbon Nanotube Dispersion and Exfoliation in Polymers," *J. Appl. Polym. Sci.,* in press.
38. Min, B.G., Sreekumar, T.V., Uchida, T., and Kumar, S., "Oxidative Stabilization of PAN/SWNT Composite Fiber," *Carbon, 43,* 599, 2005.
39. Sen, K., Bahrami, S.H., and Bajaj, P., "High-performance Acrylic Fibers," *J. Macromol. Sci. Rev. Macromol. Chem. Phys.,* C36, 1, 1996.
40. Martin, S.C., Liggat, J.J., and Snape, C.E., "*In Situ* NMR Investigation into the Thermal Degradation and Stabilisation of PAN," *Polym. Degrad. Stab., 74,* 407, 2001.
41. Donnet, J.B., Wang, T.K., Rebouillat, S., and Peng, J.C.M., *Carbon Fibers,* Marcel Dekker, New York, 1998.
42. Watt, W., Chemistry and Physics of the Conversion of Polyacrylonitrile Fibers into High-Modulus Carbon Fibers, in Watt, W. and Perov, B.V., Eds., *Strong Fibers,* Vol. 1, North-Holland, Amsterdam, 1985, pp. 327–387.
43. Johnson, W., The Structure of PAN Based Carbon Fibers and Relationship to Physical Properties, in Watt, W. and Perov, B.V., Eds., *Strong Fibers,* Vol. 1, North-Holland, Amsterdam, 1985, pp. 389–443.
44. Kumar, S., Anderson, D.P., and Crasto, A.S., "Carbon-fiber Compressive Strength and its Dependence on Structure and Morphology," *J. Mater. Sci., 28,* 423, 1993.
45. Kumar, S., Adams, W.W., and Helminiak, T.E., "Uniaxial Compressive Strength of High Modulus Fibers for Composites," *J. Reinforced Plastics Composites, 7,* 108, 1988.
46. Guo, H., et al., unpublished data.
47. Zhang, X., Sreekumar, T.V., Liu, T., and Kumar, S., "Gel Spinning of PVA/SWNT Composite Fiber," *Polymer, 45,* 8801, 2004.
48. Zhang, X., Liu, T., Sreekumar, T.V., Kumar, S., Moore, V.C., Hauge, R.H., and Smalley, R.E., "Poly(vinyl alcohol)/SWNT Composite Film," *Nano Lett., 3,* 9, 1285, 2003.
49. Liu, J., Wang, T., Uchida, T., and Kumar, S., "SWNT Core-polymer Shell Nano Fiber," *J. Appl. Polym. Sci.,* in press.
50. Dzenis, Y., "Spinning Continuous Fibers for Nanotechnology" *Science, 304,* 1917, 2004.
51. Yu, M., Files, B.S., Arepalli, S., and Ruoff, R.S., "Tensile Loading of Ropes of Single Wall Carbon Nanotubes and Their Mechanical Properties," *Phys. Rev. Lett., 84,* 24, 5552–5555, 2000.
52. Sreekumar, T.V., Liu, T., Kumar, S., Ericson, L.M., Hauge, R.H., and Smalley, R.E., "Single-wall Carbon Nanotube Films," *Chem. Mater., 15,* 175, 2003.
53. Zhang, X., Sreekumar, T.V., Liu, T., and Kumar, S., "Properties and Structure of Nitric Acid Oxidized Single Wall Carbon Nanotube Films," *J. Phys. Chem. B.* 108, 16435, 2004.
54. Calvert, P., "Nanotube Composites: A Recipe for Strength," *Nature, 399,* 210, 1999.
55. Holzinger, M., Steinmetz, J., Samaille, D., Glerup, M., Paillet, M., Bernier, P., Ley, L., and Graupner, R., "[2+1] Cycloaddition for Cross-linking SWCNTs," *Carbon,* 42, 944, 2004.
56. Kis, A., Csanyi, G., Salvetat, J.P., Lee, T.N., Couteau, E., Kulik, A.J., Benoit, W., Brugger, J., and Forro, L., "Reinforcement of Single-walled Carbon Nanotube Bundles by Intertube Bridging," *Nature Mater., 3,* 153, 2004.
57. Kuznetsova, A., Mawhinney, D.B., Naumenko, V., Yates, J.T., Jr., Liu, J., and Smalley, R.E., "Enhancement of Adsorption Inside of Single-walled Nanotubes: Opening the Entry Ports" *Chem. Phys. Lett., 321,* 292, 2000.

58. Yu, R., Chen, L., Liu, Q., Lin, J., Tan, K., Ng, S.C., Chan, H., Xu, G., and Andy Hor, S.T., "Platinum Deposition on Carbon Nanotubes via Chemical Modification." *Chem. Mater., 10,* 718, 1998.
59. Sun, Y., Huang, W., Lin, Y., Fu, K., Kitaygorodskiy, A., Riddle, L.A., Yu, Y.J., and Carroll, D.L., "Soluble Dendron-functionalized Carbon Nanotubes: Preparation, Characterization, and Properties," *Chem. Mater., 13,* 2864, 2001.
60. Hamon, M.A., Chen, J., Hu, H., Chen, Y., Itkis, M.E., Rao, A.M., Eklund, P.C., and Haddon, R.C., "Dissolution of Single-Walled Carbon Nanotubes," *Adv. Mater., 11,* 834, 1999.
61. Chen, J., Rao, A.M., Lyuksyutov, S., Itkis, M.E., Hamon, M.A., Hu, H., Cohn, R.W., Eklund, P.C., Colbert, D.T., Smalley, R.E., and Haddon, R.C., "Dissolution of Full-length Single-walled Carbon Nanotubes," *J. Phys. Chem. B, 105,* 2525, 2001.
62. Jia, Z., Wang, Z., Liang, J., Wei, B., and Wu, D., "Production of Short Multi-walled Carbon Nanotubes," *Carbon, 37,* 903, 1999.
63. Mawhinney, D.B., Naumenko, V., Kuznetsova, A., and Yates, J.T., "Infrared Spectral Evidence for the Etching of Carbon Nanotubes: Ozone Oxidation at 298 K," *J. Am. Chem. Soc., 122,* 2383, 2000.
64. Zhang, Y., Shi, Z., Gu, Z., and Iijima, S., "Structure Modification of Single-wall Carbon Nanotubes," *Carbon, 38,* 2055, 2000.
65. Zhang, M., Yudasaka, M., and Iijima, S., "Diameter Enlargement of Single-wall Carbon Nanotubes by Oxidation," *J. Phys. Chem. B, 108,* 149, 2004.
66. Hu, H., Zhao, B., Itkis, M.E., and Haddon, R.C., "Nitric Acid Purification of Single-walled Carbon Nanotubes," *J. Phys. Chem. B, 107,* 13838, 2003.
67. Bhattacharyya, A.R., Sreekumar, T.V., Liu, T., Kumar, S., Ericson, L.M., Hauge, R.H., and Smalley, R.E., "Crystallization and Orientation Studies in Polypropylene/single Wall Carbon Nanotube Composite," *Polymer, 44,* 2373, 2003.
68. Dresselhaus, M.S. and Eklund, P.C., "Phonons in Carbon Nanotubes," *Advan. Phys., 49,* 705, 2000.
69. Pimenta, M.A., Marucci, A., Empedocles, S.A., Bawendi, M.B., Hanlon, E.B., Rao, A.M., Eklund, P.C., Smalley, R.E., Dresselhaus, G., and Dresselhaus, M.S., "Raman Modes of Metallic Carbon Nanotube," *Phys. Rev. B, 58,* R16016, 1998.
70. Menna, E., Negra, F.D., Fontana, M.D., and Meneghetti, M., "Selectivity of Chemical Oxidation Attack of Single-wall Carbon Nanotubes in Solution," *Phys. Rev. B, 68,* 193412, 2003.
71. Brown, S.D.M., Jorio, A., Corio, P., Dresselhaus, M.S., Dresselhaus, G., Satio, R., and Kneipp, K., "Origin of the Breit-Wigner-Fano Lineshape of the Tangential G-band Feature of Metallic Carbon Nanotubes," *Phys. Rev. B, 63,* 155414–1, 2001.
72. Gommans, H.H., Alldredge, J.W., Tashiro, H., Park, J., Magnuson, J., and Rinzler, A.G., "Fibers of Aligned Single-walled Carbon Nanotubes: Polarized Raman Spectroscopy," *J. Appl. Phys., 88,* 2509, 2000.
73. Anglaret, E., Righi, A., Sauvajol, J.L., Bernier, P., Vigolo, B., and Poulin, P., "Raman Resonance and Orientational Order in Fibers of Single-wall Carbon Nanotubes," *Phys. Rev. B, 65,* 165426–1, 2002.
74. Hwang, J., Gommans, H.H., Ugawa, A., Tashiro, H., Haggenmueller, R., Winey, K.I., Fischer, J.E., Tanner, D.B., and Rinzler, A., "Polarized Spectroscopy of Aligned Single-wall Carbon Nanotubes," *Phys. Rev. B, 62,* R13310, 2000.
75. Haggenmueller, R., Gommans, H.H., Rinzler, A.G., Fischer, J.E., and Winey, K.I., "Aligned Single-wall Carbon Nanotubes in Composites by Melt Processing Methods," *Chem. Phys. Lett., 330,* 219, 2000.

76. Fischer, J.E., Zhou, W., Vavro, J., Llaguno, M.C., Guthy, C., Haggenmueller, R., Casavant, M.J., Walters, D.E., and Smalley, R.E., "Magnetically Aligned Single Wall Carbon Nanotube Films: Preferred Orientation and Anisotropic Transport Properties," *J. Appl. Phys., 93,* 2157, 2003.
77. Liu, T. and Kumar, S., "Quantitative Characterization of SWNT Orientation by Polarized Raman Spectroscopy," *Chem. Phys. Lett., 378,* 257, 2003.

13 Surface Patterning Using Self-Assembled Monolayers: A Bottom-Up Approach to the Fabrication of Microdevices

Lakshmi Supriya and Richard O. Claus

CONTENTS

13.1 INTRODUCTION

The bottom-up approach to the fabrication of microscale devices requires the ability to successfully pattern surfaces. This patterning can be achieved by the use of self-assembled monolayers. In this chapter, surface patterning results with respect to making photonic crystal devices and strain gauges are presented. Patterning is achieved using two approaches. For the photonic crystal devices, ultraviolet (UV) irradiation is used to pattern self-assembled monolayers, and for the strain gauge a photoresist-based process is used. The electrostatic self-assembly method is used to make patterned multilayers on silicon for the photonic crystal devices. Conducting films of gold were deposited on flexible substrates for making strain gauges. The films were deposited using a completely solution-based process. These approaches are very cost-effective compared to the present microdevice fabricating technologies discussed next, and the entire procedure can bè done on a laboratory benchtop.

One of the most important fabrication technologies in the microelectronics industry is the top-down approach to making microdevices, whether microelectromechanical systems (MEMS), microfluidics, optical systems such as micromirrors, or communication and information storage devices. In this approach, material is first laid down, and it is removed gradually from top to bottom in patterns depending on the application. The conventional microfabrication technologies have limited the choice of materials; silicon is still the most widely used material of choice. This in turn limits the applications of the devices. For example, it would not be possible to make a biocompatible device for use *in vivo* with conventional materials used for microelectronics. The bottom-up approach to fabricating micro- and nanodevices has been an area of considerable recent research.[1–2] The advantages of this approach over the conventional top-down approach used in the microelectronics industry are many. It offers considerable versatility in the materials used. Different types of organic materials that are more easily tailorable than ceramics, such as polymers, nanoparticles, and self-assembled monolayers, are commonly studied. This approach also minimizes waste of material and does not require sophisticated equipment, leading to much lower costs and to molecular-scale control in device fabrication.

The key to such an approach is to successfully pattern the substrates and selectively deposit materials forming three-dimensional structures. Patterning of substrates using self-assembled monolayers (SAMs) is an area that has generated much interest. Self-assembled monolayers[3] are long-chain alkanes that spontaneously chemisorb onto the surfaces of appropriate materials. These layers offer control of structure at the molecular level and can act as ultra-thin resists in lithographic processes. Many problems associated with current patterning techniques, such as the limit on the smallest feature size, optical diffraction effects, depth of focus, shadowing, and undercutting can be minimized. Self-assembly leads to equilibrium structures that are at or near the thermodynamic minimum, and as a result these systems tend to be self-healing and defect-rejecting.

In order to fabricate functional microdevices, it is essential to form three-dimensional structures on the self-assembled monolayer. These structures can be built if multilayers can be grown in patterns of the desired structure. There are a number of patterning techniques for SAMs that have been studied: soft lithography techniques such as microcontact printing,[4] replica molding,[5] microtransfer molding,[6] selective transformation of terminal functionalities by either photochemical[7–10] or chemical[11–13] methods, selective removal of monolayer film by irradiation[14–17] and micromachining,[18] and selective deposition of polyelectrolytes[19] and nanoparticles. A more recent method of patterning is the use of scanning probe microscopy tips, which has drastically reduced the sizes of the patterns formed and is so effective that it can be used to manipulate single molecules.[20] Techniques such as nanografting[21] and dip pen nanolithography[22] use the tip of an AFM probe to selectively remove and deposit molecules onto the surface. Scanning tunneling microscopy probes have also been used to pattern SAMs using electrochemical methods by applying a bias between the conductive probe and the substrate.[23]

In this chapter, results with regard to the fabrication of different types of devices using patterned SAMs using the bottom-up approach are presented. A brief description of the type of prototype devices investigated follows. A first set of representative experiments on silicon substrates was aimed at fabricating spatially periodic photonic crystals. Photonic crystals are a class of periodic composite structures consisting of spatially varying low and high refractive index materials that exhibit a forbidden band, or photonic band gap, of frequencies wherein incident electromagnetic waves cannot propagate.[24] Introducing defects into the ordinarily symmetrical structure disrupts the periodicity of the crystal-defining microcavities, and in some cases waveguides, where incident electromagnetic energy can become localized or allowed to propagate. The creation of patterns on silicon and formation of multilayers using the electrostatic self-assembly process[25] is a possible method of fabricating these devices.[26] Microstrain gauges are other spatially patterned devices that may be fabricated *in situ* using self-assembly. These can be achieved by patterning conducting films on flexible substrates. The basic theory behind these devices is that a change in the length and geometry of an electrical conductor causes a change in its resistance, and this change can be quantified and used to infer elongation. Using self-assembly, patterned gauges may be formed by depositing conducting gold on substrates such as polyethylene and measuring the change in resistance of the conductor versus strain.

13.2 EXPERIMENTAL PROCEDURE

13.2.1 MATERIALS

Three types of substrates for patterning were studied: silicon (111); Kapton (DuPont), a polyimide; and polyethylene. The SAMs used were octadecyltrichlorosilane (OTS) (Aldrich), heptadecafluoro 1, 1, 2, 2, tetrahydrodecyl trichlorosilane $CF_3(CF_2)_7CH_2CH_2SiCl_3$ (CF) (Gelest), 3-aminopropyltrimethoxy silane (APS) (Gelest) and 3-mercaptopropyltrimethoxy silane (MPS) (Gelest). The polyelectrolyte used was sodium polystyrene sulfonate (SPS), and the nanoparticles used were silicon dioxide (Nissan Chemical Industries), titanium dioxide, and gold, which were synthesized using a slight modification of previously described methods.[27]

13.2.2 SUBSTRATE PREPARATION

Silicon wafers, cut into small 2-cm squares, were rinsed thoroughly with dichloromethane and methanol and dried under a stream of nitrogen. The samples were treated with a 7 : 3 v/v mixture of H_2SO_4/H_2O_2 for 20 min at 80°C (piranha treatment) and rinsed thoroughly with nanopure water (18 MΩ) from a Barnstead nanopure water system. The Kapton and polyethylene were plasma treated using a March Plasmod plasma etcher in an argon atmosphere for ~2 min at 0.2–0.4 torr and 50 W power.

13.2.3 Deposition of SAM

OTS SAM was deposited in a glove bag under nitrogen atmosphere. The piranha-treated silicon substrates were immersed in a 1% (v/v) solution of OTS in dicyclohexyl (Aldrich) for 1 h with stirring. After the reaction was over, the substrates were rinsed thoroughly with dichloromethane. CF SAM was also deposited in a similar manner, with Isopar-G (Exxon) acting as the solvent in this case. The entire deposition process was carried out in a nitrogen atmosphere and dry conditions. The deposition of APS and MPS was carried out in ambient conditions. The substrates were immersed in a 1% (v/v) solution of APS/MPS for 15 min with stirring. After that they were rinsed thoroughly in methanol and heated at 110°C for $1^1/_2$ h and then dried under vacuum at 50°C. All the SAM depositions were carried out in less than an hour after the substrate preparation.

13.2.4 Patterning

Patterning of the SAMs was carried out by irradiation with an Hg(Ar) lamp (Model 6035, Oriel Instruments) at a wavelength of 184 nm. The OTS SAMs were photolyzed for 1 h, while all the others were photolyzed for 3 h. The masks consisted of meshes used for transmission electron microscopy, and masks were also made on aluminum sheets when larger scale patterns were required. Patterning on the flexible substrates was carried out by using a positive photoresist S1813 (Shipley). The photoresist was spin-coated onto the silane-coated substrates, and then irradiated with broadband UV (350–450 nm) (Oriel Instruments Model 92530) through a mask, developed using the Microposit 351 developer. Then gold was coated onto the entire substrate and the photoresist was removed using acetone. At that time the gold coated over the photoresist was also removed, leaving a pattern of conducting lines.

13.2.5 Deposition of Multilayers

Multilayers were deposited on silicon using electrostatic self-assembly. TiO_2 and SiO_2 were alternately deposited on patterned OTS, and TiO_2 and SPS on patterned CF. The patterned samples were first dipped in the positively charged solution for 3 min, sonicated in nanopure water for 2 min, dipped in the oppositely charged solution for 3 min and sonicated again, and this cycle was repeated to obtain the desired number of multilayers. On the polymer substrates, gold was deposited using previously described methods.[28]

13.3 RESULTS AND DISCUSSION

13.3.1 Patterning on Silicon

The piranha treatment cleans the surface exposing the hydroxyl groups on the surface of the silicon. The deposition of silanes on these surfaces occurs by the hydrolysis of the alkoxy or trichloro groups on the silane and subsequent reaction with the surface hydroxyl groups.[3] The deposition of OTS and CF was characterized

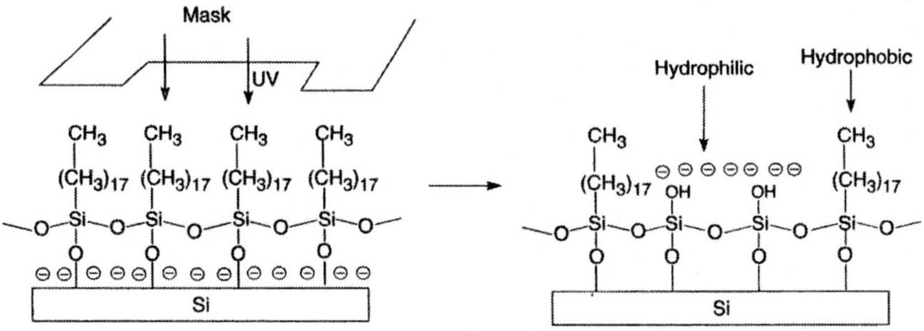

FIGURE 13.1 Conceptual illustration of the patterning process for OTS.

by contact angle measurements. After the piranha treatment the silicon substrate was completely wetted with contact angles as low as ~10°. The deposition of OTS or CF raised the contact angles. For OTS the angles were between 90° and 115°, and for CF the contact angles were slightly higher, around 120°. This was expected because CF has a much lower surface energy than OTS due to the presence of fluorine groups. After irradiation with UV light at 184 nm, both the SAMs were degraded and removed from the substrate.[17] This is evidenced by the drop in contact angles to ~10° upon photolysis. The CF took a longer time than OTS to photolyze completely and this can be attributed to the strong C–F bonds. Figure 13.1 illustrates the photolysis process on OTS.

On OTS and CF, the multilayers were formed by electrostatic self-assembly. Electrostatic self-assembly involves dipping a charged substrate alternately into solutions of an anionic charged species and a cationic charged species; the electrostatic force between oppositely charged molecules acts as the binding force holding the molecules together. Patterning on OTS and CF creates regions of charge and regions where there is no charge and the charged molecules become selectively adsorbed on the charged regions. TiO_2 was the positively charged nanoparticle for OTS and CF; SiO_2 was the negatively charged nanoparticle for the patterned OTS, and SPS was the negatively charged polyelectrolyte used for CF. After photolysis, the negatively charged substrate is exposed and the charged particles selectively adsorb on the charged regions and are bound by electrostatic interactions. Figure 13.2 shows SEM images of the patterns formed on OTS- and CF-coated substrates. The pattern contained 7.5 × 7.5 μm square holes with a center-to-center spacing of 12.5 μm. The lengths of all the scale bars in the figure were 10 μm. The dark regions are the regions of deposition in all the images except in Fig. 13.2(a), where the opposite applies. This is probably due to some differential charging effect when imaging the samples.

The deposition of the polyelectrolyte, SPS, was successful for the CF, while good patterns were not obtained for the OTS. This can be explained by the fact that the negatively charged groups on the polyelectrolyte are repelled by the electronegative fluorine atom on the CF and are thus selectively deposited on the regions where

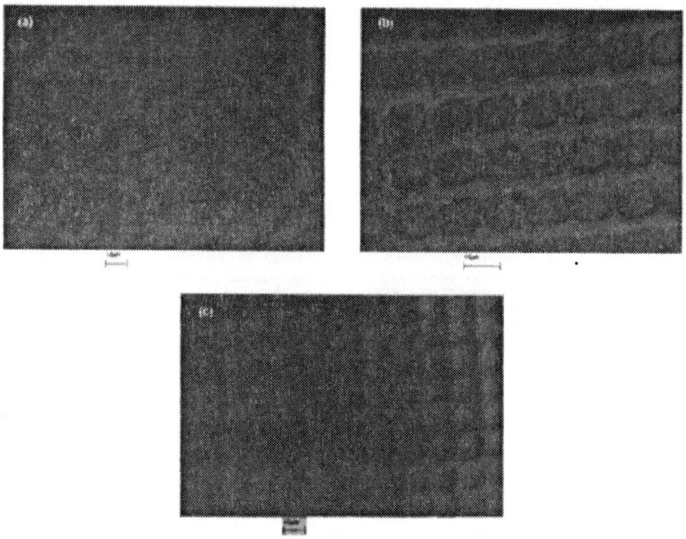

FIGURE 13.2 SEM images of patterned multilayers on silicon (a) 20, (b) 30 TiO_2-SiO_2 bilayers on OTS-treated silicon, (c) 50 TiO_2–SPS bilayers on CF-treated silicon.

CF is absent. On the OTS SAM there was no such interaction and SPS, being a large molecule, may be attached on the OTS regions also due to some hydrophobic interactions. These selectively grown ESA nanocomposites represent two-dimensional photonic crystal structures composed of high refractive index pillars suspended in air. Measurements on a Rudolph Auto EL ellipsometer revealed that the TiO_2-SiO_2 corresponded to high index pillars with an average refractive index of 1.6.

13.3.2 Patterning on Polyethylene and Kapton

Polyethylene and Kapton are flexible polymers suitable for making patterned electrodes to fabricate microstrain gauges *in situ*. This requires selective deposition of conductive layers, and gold was the material of choice because of the ease of deposition of a conductive film using a completely solution-based process. The resistance of a conductor is dependent on its length and cross-sectional geometry, and a change in the geometry would lead to a change in the resistance and this can be measured quantitatively. This principle is used in strain gauges also, where a change in the length will cause a change in the resistance and the strain can be measured. In order to increase the device sensitivity, the length of the conductor should be increased and this can be done using geometry as shown in Fig. 13.3.

The surface of the polymers is plasma treated to deposit the silanes. Plasma treatment of the polymers creates free radicals on the surface, which when exposed to air form oxygenated species.[29] These can then react with the silanes to form a monolayer on the surface. Table 13.1 shows the contact angles for substrates after

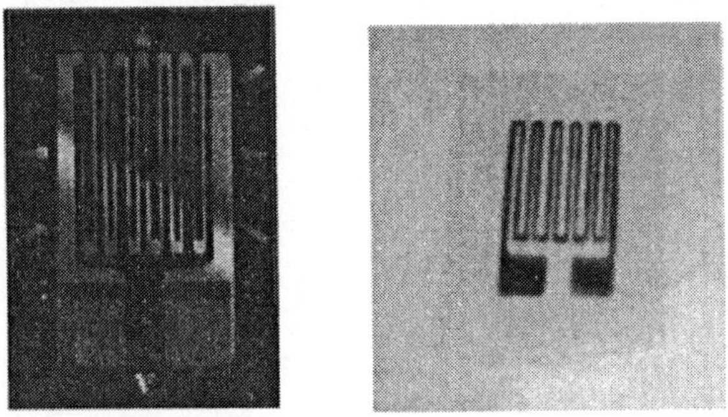

(a) (b)

FIGURE 13.3 (a) An illustration of the patterned geometry of a strain gauge (From http://www.vishay.com/brands/measurements_group/strain_gages/mm.htm). (b) Patterned and printed strain gauge.

plasma treatment and SAM deposition. Figure 13.4 shows the x-ray photoelectron spectroscopy (XPS) data corresponding to the different treatments on Kapton. A comparison of the untreated and the plasma-treated surfaces shows an increase in the oxygen concentration, as expected. The deposition of the SAM on the substrate gives rise to a polysiloxane backbone. Figure 13.4(c) shows the siloxane peak at 102.4 eV, which confirms the formation of the SAM. Similar results were also obtained for polyethylene.

The deposition of gold on the SAMs takes advantage of the fact that gold has a large affinity for groups such as amino, mercapto, and cyano compounds.[27] The choice of the SAMs was based on this, and amino- and mercapto-terminated SAMs were used in our experiments. The formation of conductive layers of gold nanoparticles is done by a seeding method as described in a previous study.[28] In brief, the SAM-coated substrates are dipped in colloidal gold solution for 2–3 h and gold particles are attached to the substrate. Then more gold is reductively deposited on the gold already attached, leading to complete coverage. Figure 13.5 shows the SEM images of the formation of complete coverage gold on Kapton. All scale bars are 100 nm.

TABLE 13.1
Contact Angles for Polymers after Plasma Treatment and SAM Deposition

	Kapton	Polyethylene
Untreated	71°	94°
Plasma treated	18°	41°
After SAM deposition	84°	95°

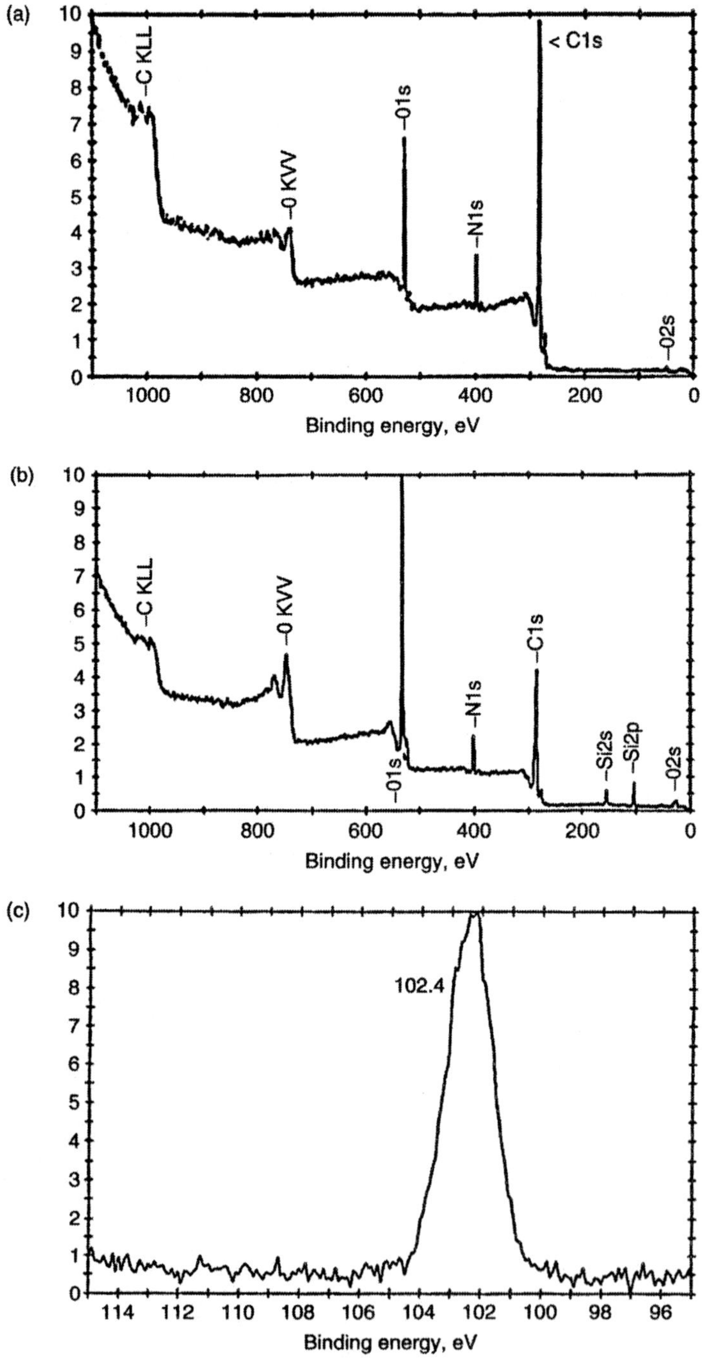

FIGURE 13.4 XPS spectra of (a) untreated (b) plasma treated, and (c) Si peak of APS-deposited Kapton. (Reprinted with permission from *Langmuir*, 2004, 20, 8870–8876. Copyright 2004, American Chemical Society.)

TABLE 13.2
Strain-Resistivity Data for Gold on Kapton

Load (N)	Resistivity × 10^4 (Ωcm)	Strain
0	0.95	0
10	1.28	0.078
20	1.26	0.17
70	3.2	1.875

The resistance of the samples was measured using a four-point probe. The resistivities on both polyethylene and Kapton were about 1 ohm/sq. The thickness of a layer of gold was found to be ~300 nm using cross-section SEM. Table 13.2 depicts the strain resistivity data for gold on Kapton, and as expected the resistance is found to increase as a function of length.

The patterning on these substrates was done to generate conductive lines, as shown in Fig. 13.3. Patterning was done using a method developed by Hua et al.[30] Figure 13.6 depicts the patterns generated on Kapton, which consist of alternate

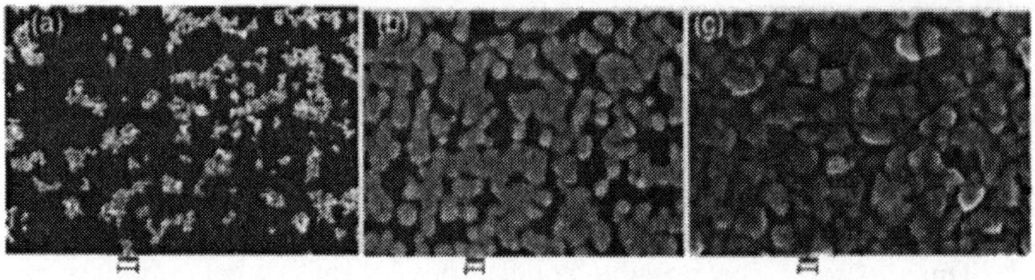

FIGURE 13.5 SEM images of (a) gold monolayer, (b) one-layer seeded gold, and (c) two-layer seeded gold on Kapton. (Reprinted with permission from *Langmuir*, 2004, 20, 8870–8876. Copyright 2004, American Chemical Society.)

FIGURE 13.6 Patterned gold on Kapton.

conducting lines. The size of the lines is about 1 mm. Although the patterns shown here are large, with masks of smaller size, patterns of 1 μm or smaller can also be made. The conductivity of the film was not affected by the patterning process and was the same as for the unpatterned substrates.

13.4 CONCLUSIONS AND APPLICATIONS

It is possible to selectively deposit and create three-dimensional structures on silicon using a combination of patterned self-assembled monolayers and electrostatic self-assembly. The size of the smallest pattern created was 7.5 μm, which was the smallest mask utilized. However, with appropriate masks the pattern size can be reduced even further. It was also possible to deposit conducting layers of gold on polymer substrates such as polyethylene and Kapton for the fabrication of strain gauges. Patterning was also successfully achieved to fabricate conducting lines. These results show that fabrication of microdevices can be achieved by methods that are more economical and easier than conventional microfabrication technologies that use sophisticated equipment and involve increased costs. This method is so simple that it can be performed on a laboratory benchtop. Future efforts are being directed toward decreasing the sizes of patterns and using a maskless approach to patterning.

PROBLEMS

1. List the important trade-offs when considering bottom-up and top-down fabrication technologies.
2. Perform a literature search on the subject of patterning of substrates using self-assembled monolayers and list any new advances in techniques that have been made relative to the techniques cited in this chapter.
3. Think of some new possible applications for surface patterning using the SAM technique and list potential benefits.

ACKNOWLEDGMENTS

The authors gratefully acknowledge financial support from NASA and the U.S. Army Research Laboratory and U.S. Army Research Office under contract/grant number DAAD19-02-1-0275, Macromolecular Architecture for Performance (MAP) MURI. We also thank Steve McCartney and the Materials Research Institute, Virginia Tech, for the SEM images, and Frank Cromer and the Department of Chemistry, Virginia Tech, for the XPS data.

REFERENCES

1. Faul Charl, F.J. and Antonietti, M., Ionic self assembly: facile synthesis of supramolecular materials *Adv. Mat., 15,* 9, 673–683, 2003.
2. Zhao, X.-M., Xia, Y., and Whitesides, G.M., Soft lithography methods for nanofabrication *J. Mater. Chem., 7,* 7, 1069–1074, 1997.

3. Ulman, A., Formation and structure of self-assembled monolayers, *Chem. Rev., 96,* 1533–1554, 1996.
4. Kumar, A. and Whitesides, G.M., Features of gold having micrometer to centimeter dimensions can be formed through a combination of stamping with an elastomeric stamp and an alkanethiol "ink" followed by chemical etching, *Appl. Phys. Lett., 63,* 2002–2004, 1993.
5. Xia, Y., Kim, E., Zhao, X.-M., Rogers, J.A., Prentiss, M., and Whitesides, G.M., Complex optical surfaces formed by replica molding against elastomeric masters, *Science, 273,* 347–349, 1996.
6. Zhao, X.-M., Xia, Y., and Whitesides, G.M., Fabrication of three-dimensional microstructures: microtransfer molding, *Adv. Mat., 8,* 837–840, 1996.
7. Bhatia, S.K., Hickman, J.J., and Ligler, F.S., New approach to producing patterned biomolecular assemblies, *J. Am. Chem. Soc., 114,* 4432–4433, 1992.
8. Collins, R.J., Shin, H., DeGuire, M.R., Heuer, A.H., and Sukenik, C.H., Low temperature deposition of patterned TiO_2 thin films using photopatterned self-assembled monolayers, *Appl. Phys. Lett., 69,* 6, 860–862, 1996.
9. Brandow, S.L., Chen, M.-S., Aggarwal, R., Dulcey, C.S., Calvert, J.M., and Dressick, W.J., Fabrication of patterned amine reactivity templates using 4-chloromethylphenylsiloxane self-assembled monolayer films, *Langmuir, 15,* 5429–5432, 1999.
10. Rozsnvai, L.F. and Wrighton, M.S., Selective electrochemical deposition of polyaniline via photopatterning of a monolayer-modified substrate, *J. Am. Chem. Soc., 116,* 5993–5994, 1994.
11. Wasserman, S.R., Tao, Y.T., and Whitesides, G.M., Structure and reactivity of alkylsiloxane monolayers formed by reaction of alkyltrichlorosilanes on silicon substrates, *Langmuir, 5,* 1074–1087, 1989.
12. Balachander, N. and Sukenik, C.N., Monolayer transformation by nucleophilic transformation: applications to the creation of new monolayer assemblies, *Langmuir, 6,* 1621–1627, 1990.
13. Netzer, L. and Sagiv, J., A new approach to construction of artificial monolayer assemblies, *J. Am. Chem. Soc., 105,* 674–676, 1983.
14. Tarlov, M.J., Burgess, D.R.F., and Gillen, G., UV photopatterning of alkanethiolate monolayers self-assembled on gold and silver, *J. Am. Chem. Soc., 115,* 5305–5306, 1993.
15. Dulcey, C.S., Georger, J.H., Jr., Krauthamer, V., Stenger, D.A., Fare, T.L., and Calvert, J.M., Deep UV photochemistry of chemisorbed monolayers: patterned coplanar molecular assemblies, *Science, 252,* 551–554, 1991.
16. Dressick, W.J. and Calvert, J.M., Patterning of self-assembled films using lithographic exposure tools, *Jpn. J. Appl. Phys., 32,* 1, 12B, 5829–5839, 1993.
17. Masuda, Y., Seo, W.S., and Koumoto, K., Selective deposition and micropatterning of titanium dioxide on self-assembled monolayers from a gas phase, *Langmuir, 17,* 4876–4880, 2001.
18. Abbott, N.L., Folkers, J.P., and Whitesides, G.M., Manipulation of the wettability of surfaces on the 0.1- to 1-micrometer scale through micromachining and molecular self-assembly, *Science, 257,* 1380–1382, 1992.
19. Clark, S.L., Montague, M., Hammond, P.T., Selective deposition in multilayer assembly: SAMs as molecular templates, *Supramol. Sci., 4,* 1–2, 141–146, 1997.
20. Grandbois, M., Beyer, M., Rief, M., Clausen-Schaumann, H., and Gaubi, H.E., How strong is a covalent bond? *Science, 283,* 1727–1730, 1999.
21. Xu, S. and Liu, G., Nanometer-scale fabrication by simultaneous nanoshaving and molecular self-assembly, *Langmuir, 13,* 127–129, 1997.

22. Piner, R.D., Zhu, J., Xu, F., Hong, S., and Mirkin, C.A., "Dip-pen" nanolithography, *Science, 283,* 661–663, 1999.
23. Schoer, J.K., Zamborini, F.P., and Crooks, R.M., Scanning probe lithography. 3. Nanometer-scale electrochemical patterning of Au and organic resists in the absence of intentionally added solvents or electrolytes, *J. Phys. Chem., 100,* 11086–11091, 1996.
24. Joannopoulos, J.D., Meade, R.M., and Winn, J.N., *Photonic Crystals: Molding the Flow of Light,* Princeton University Press, Princeton, NJ, 1995.
25. Decher, G. and Hong, J.D., Buildup of ultrathin multilayer films by a self-assembly process. 1. Consecutive adsorption of anionic and cationic bipolar ampiphiles on charged surfaces, *Makromol. Chem. Macromol. Symp., 46,* 321–327, 1991.
26. Huie, K.C., Chandran, A., Claus, R.O., Nelson, R., and Supriya, L., Photonic crystal fabrication using electrostatic self-assembly photonics, paper presented at 6th Annual Conference on Optoelectronics, Fiber-Optics, and Photonics, December 16–18, 2002, TIFR, Mumbai.
27. Grabar, K.A., Freeman, R.G., Hommer, M.B., and Natan, M.J., Preparation and characterization of Au colloid monolayers, *Anal. Chem., 67,* 735–743, 1995.
28. Brown, K.R., Lyon, L.A., Fox, A.P., Reiss, B.D., and Natan, M.J., Hydroxylamine seeding of colloidal Au nanoparticles. 3. Controlled formation of conductive Au films, *Chem. Mater., 12,* 314–323, 2000.
29. Inagaki, N., Tasaka, S., and Baba, T., Surface modification of polyimide film surface by silane coupling reactions for copper metallization, *J. Adhesion Sc. Tech., 15,* 7, 749–762, 2001.
30. Hua, F., Cui, T., and Lvov, Y., Lithographic approach to pattern self-assembled multilayers, *Langmuir, 18,* 6712–6715, 2002.

14 Enhancement of the Mechanical Strength of Polymer-Based Composites Using Carbon Nanotubes

Kin-Tak Lau, Jagannathan Sankar, and David Hui

CONTENTS

14.1 INTRODUCTION

Since the discovery of carbon nanotubes a decade ago,[1] many important studies and results related to these nanostructural materials in different scientific and engineering fields have emerged. The extraordinary mechanical, electrical, and thermal properties of the nanotubes are governed by their atomic architecture, commonly called their chiral arrangement. Ideally all carbon atoms in the nanotubes are covalently bonded and form repeated close-packed hexagonal structures in each layer or shell. Due to these chemically formed atomic arrangements, the carbon nanotubes possess superior mechanical properties, and the nanotubes are stronger than any known metallic materials. Many critical results have been reported recently by using nanotubes as atomic force microscope (AFM) probes, conductive devices in artificial muscles, and nanothermometers, and to store hydrogen for fuel cells.[2–5] In the United States, the investment in the development of fuel cells by storing hydrogen atoms inside the cavities of nanotubes to supply electricity to microelectromechanical (MEM) or even nanoelectromechanical (NEM) devices has been increasing. Although all this work is still at the research stage, it demonstrates a high potential for the development of nanotube-related products and components for real-world applications.

In the past few years, many researchers and engineers from the advanced composites community have attempted to use these tiny structural materials to enhance the properties of conventional advanced composite structures by altering their mechanical, thermal, and electrostatic behaviors for space and infrastructure applications. However, in order to achieve these goals, more work is needed for (1) understanding the mechanical properties of both single-walled and multi-walled nanotubes, (2) investigating appropriate fabricating processes of nanotube/polymer composites, (3) clarifying the interfacial bonding properties between the nanotubes and surrounding matrices, and (4) justifying the benefit based on the strength improvement of composites after mixing with the nanotubes. It is understandable that research related to the aforementioned issues still has a long way to go due to many uncertainties related to nanotube properties and their structural integrity in nanotube/polymer composites.

In this chapter, a critical review of the aforementioned aspects is based on recent research by the authors and other researchers. All these aspects cannot be considered individually in the development of nanotube/polymer composites. Therefore, a discussion of how these aspects affect each other, as well as a detailed discussion of each one, is given. Since this chapter is mainly focused on the mechanical properties of nanotube/polymer composites, the fundamental physics of nanotubes is briefly explained here.

14.2 PROPERTIES OF CARBON NANOTUBES

14.2.1 Experimental Measurements

A carbon nanotube is similar to a flat graphene layer rolled up to form a tube where both ends of the tube are sealed with semihemisphere caps. All the carbon atoms are tightly bonded to each other to form a close-packed hexagonal structure in the form of a circular tube, as shown in Fig. 14.1. If more than one layer of graphene layers are rolled together to form a coaxial tube, this nanotube is called a multi-walled nanotube (MWNT). The mechanical and electrical properties of nanotubes are governed by their atomic structures. In the past few years, many researchers have attempted to measure the mechanical properties of nanotubes. However, due to their small size, it is difficult to directly measure the strength of the nanotubes by using traditional testing methods. Yu et al.[6] mounted MWNTs on two AFM tips and conducted a tensile test. The whole stretching process was captured inside a scanning electron microscope (SEM) as shown in Fig. 14.2. Since different sizes and types

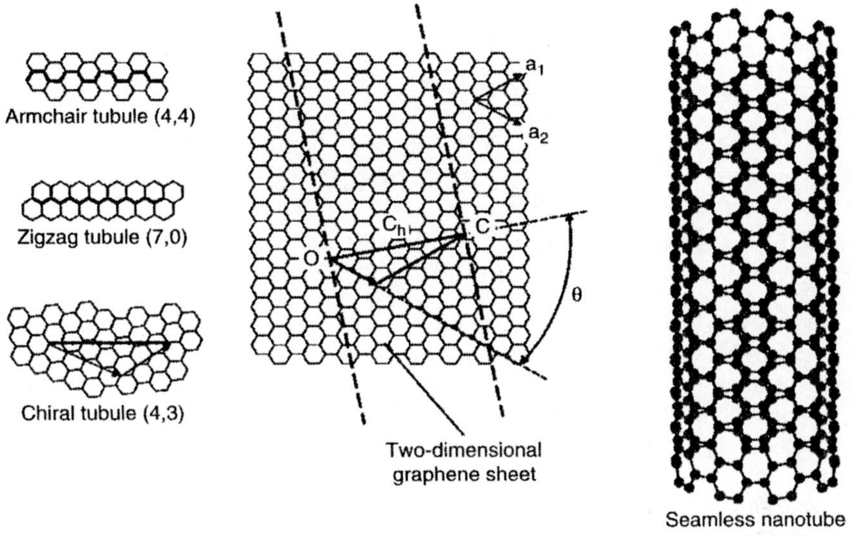

FIGURE 14.1 Examples of carbon nanotubes.

of multi-walled nanotubes were tested, the results were quite diverse. The Young's moduli of the nanotubes ranged from 0.32 TPa to 1.47 TPa. In strength testing, almost all the outer layers of the nanotubes broke first, and then pulling out of inner layers occurred. According to the results in the literature, the maximum stress applied to the nanotubes occurred in the outer layers. The larger diameter nanotubes could withstand a larger tensile force. This conclusion may not coincide with some estimates of strength from continuum models and molecular dynamics (MD) simulations.

An indirect tensile test on nanotubes by Demczyk et al.[7] demonstrated that the tensile strength of MWNTs was about 0.15 TPa, which is far below theoretical estimates and the known properties of a graphene layer. A telescoping action was seen in some fractured nanotubes. This finding agreed with the work done in

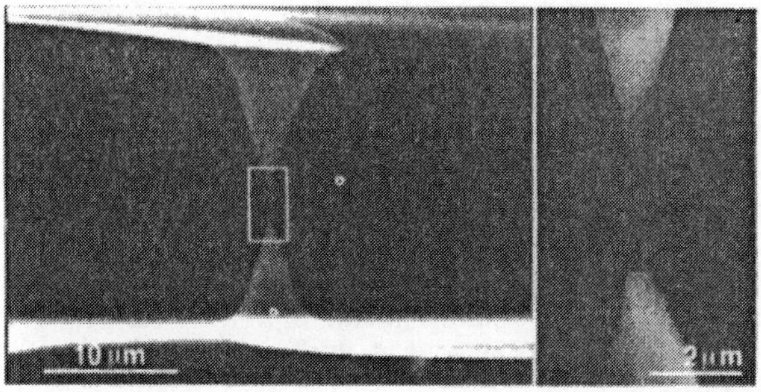

FIGURE 14.2 SEM images of a nanotube linked between two opposing AFM tips before tensile loading.[6]

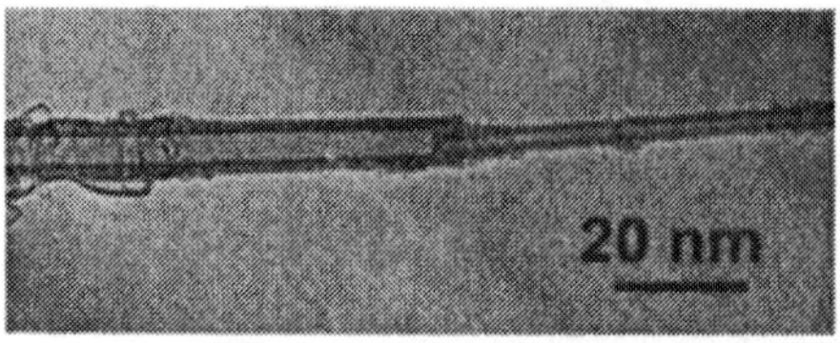

FIGURE 14.3 A telescoping action in a multi-walled nanotube.[7]

Reference 8, wherein a weak van der Waals attractive force between individual shells of the MWNTs resulted in generating an extremely low frictional force among the shells, and all the inner shells slid freely in their longitudinal direction (Fig. 14.3). The principal result is that all the inner shells cannot contribute any strength to the nanotubes while they are subjected to tension. Qi et al.[9] also measured the Young's moduli of MWNTs using a nanoindentation method. An indentor was pressed onto vertical aligned nanotubes and the bending stiffness of the nanotubes was measured and also calculated using classical bending theory. The effective bending and axial elastic moduli ranged 0.91 ~ 1.24 TPa and 0.90 ~ 1.23 TPa, respectively. Since the ineffectiveness of the stress transfer among different shells of the nanotubes was not considered in the calculation, particularly in the bending case, the moduli estimated by this method do not represent the true tensile moduli of the nanotubes.

14.2.2 Theoretical Study and Molecular Dynamics Simulation

Because the sizes of nanotubes are extremely small, direct measurement of the mechanical properties of nanotubes is extremely difficult. In the past few years, much effort has been spent on using MD simulations to predict the properties of nanotubes. Since the accuracy of the simulated results is highly dependent on the size of the models as well as the capacity (memory size) and running speed of computers, the results from different papers and approaches are therefore different. Tu and Ou-Yang[10] have used the local density approximation method associated with elastic shell theory to estimate the mechanical properties of single-walled nanotubes (SWNTs) and MWNTs. They found that the Young's modulus of the multi-walled nanotubes decreased with an increase in the number of shell layers. The Young's modulus of the nanotubes can be determined by the following equation:

$$E_m = \frac{n}{n-1+t/d} \cdot \frac{t}{d} \cdot E_s \tag{14.1}$$

where E_m and E_s represent the Young's moduli of MWNTs and SWNTs, respectively, and n, t, and d denote the number of shell layers, average thickness of the shell (≈0.75 Å), and spacing between the shells (≈0.34 Å), respectively. They have also reduced Equation 14.1 to the continuum limit form, and the classic shell theory can be used to describe the deformation of the nanotubes. Lau et al.[11] used the Tersoff-Brenner bond order potential to represent the interaction between carbon atoms to

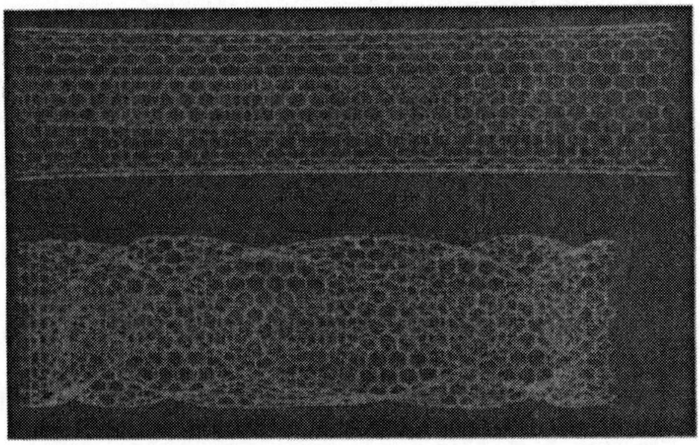

FIGURE 14.4 (Color figure follows p. 12.) MD-simulated results of three-layer MWNTs. The inner two layers would not be affected by the surface layer when the MWNTs are subjected to tensile (top) and torsional (bottom) loads.

study the stretching motion of MWNTs. In their study, it was found that the outer shells of the nanotubes took all the applied loads when the nanotubes were subjected to tensile and torsion motions (Fig. 14.4). Since only a weak van der Waals interaction exists between the shells, external motions of the outer shell cannot be easily transferred to inner shells. Eventually fracture at the outer shell was initiated (Fig. 14.5) and this phenomena was similar to that observed in Reference 7, where

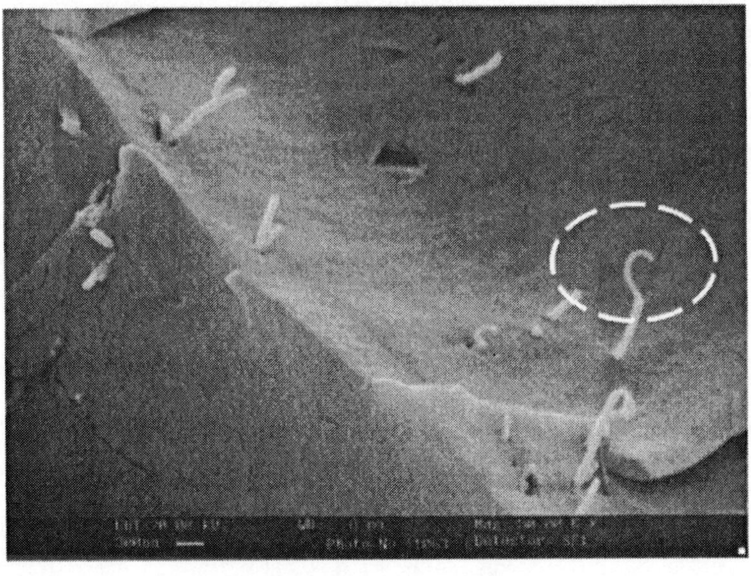

FIGURE 14.5 Outer shell of MWNT pulling out of a matrix.

pullout of nanotubes was observed. Lau[12] has also pointed out that the size and Young's moduli of nanotubes are highly dependent on their chiral arrangement. The diameter of a zigzag nanotube is generally smaller than that of an armchair type. The radius of the first layer (an inner layer) of nanotubes can be determined by using the rolled graphene sheet model:

$$\rho_o = \frac{\sqrt{m^2 + n^2 + mn}}{\rho} \cdot \sqrt{3}a_o \tag{14.2}$$

where ρ_o, m, n, and a_o are the nonrelaxed radius and indices of the SWNTs and C–C bond distance (1.42 Å), respectively.

The MD simulation has been popularly adopted in recent years to predict the properties of nanostructural materials at the atomic scale with high accuracy. At the early stage, several studies have used empirical force potential molecular dynamic simulations to estimate the Young's moduli of nanotubes and found that their value was four times that of the diamond. However, those calculations were based on the SWNTs of several angstroms in radius. To determine the mechanical properties of nanotubes, the details of their atomic arrangements have to be clearly understood. In general, there are three types of nanotube structures: zigzag (n, 0), armchair (n, n), and chiral (n, m where $n \neq m$). Lau and Hui[13] wrote a comprehensive review on the structures of nanotubes. To investigate the mechanical properties of materials at the atomic scale by using MD simulations, the interactions between neighboring atoms must be accurately calculated.

Two common approaches based on quantum mechanics and molecular mechanics are used to simulate these interactions. Both approaches attempt to capture the variation of system energy associated with the change in atomic positions by following Newton's second law, $F = ma$. In carbon nanotubes, the mutual interactions are described by force potentials from both bonding and nonbonding interactions. The nonbonding interactions are either due to the van der Waals force (which can be attractive or repulsive, depending on the distance between atoms), or to electrostatic interactions. The van der Waals force F_{VDW} is most often modeled using the Lennard-Jones potential function,[14] originally derived for inert gases. The general form of this potential is

$$\Phi(r) = \frac{\lambda_n}{r^n} - \frac{\lambda_m}{r^m} \tag{14.3}$$

For van der Waals forces arising from dipole-dipole interactions, the attractive part corresponds to $m = 6$. The most common form of this potential is the so-called (6–12) form:

$$\Phi(r) = 4\varepsilon\left[\left(\frac{\sigma}{r}\right)^{12}\left(\frac{\sigma}{r}\right)^{6}\right] \tag{14.4}$$

The minimum of $\Phi(r)$ is determined by equating to zero the first-order derivative of $\Phi(r)$ versus r. The van der Waals force between two carbon atoms can be estimated from

$$F_{VDW} = -\frac{d\phi}{dr} = \frac{24\varepsilon}{r}\left[2\left(\frac{\sigma}{r}\right)^{12} - \left(\frac{\sigma}{r}\right)^{6}\right] \tag{14.5}$$

The two parameters σ and ε can be estimated from experimental data such as the equilibrium bond length (lattice parameters at equilibrium), equilibrium bond energy (cohesive energy), and bulk modulus at equilibrium. The bonding energy (E_{bond}) is the sum of four different interactions among atoms, namely bond stretching (U_ρ), angle variation (U_θ), inversion (U_ω), and torsion (U_τ),[15] written as

$$E_{\text{bond}} = U_\rho + U_\theta + U_\omega + U_\tau \tag{14.6}$$

A schematic illustration of each energy term and corresponding bond structure for a graphene cell is shown in Fig. 14.6. The most commonly used functional forms are

$$U_\rho = \frac{1}{2}\sum_i K_i\left(dR_i\right)^2$$

$$U_\theta = \frac{1}{2}\sum_i C_j\left(d\theta_j\right)^2$$

$$U_\omega = \frac{1}{2}\sum_k B_k\left(d\omega_k\right)^2$$

$$U_\tau = \frac{1}{2}\sum_i A_i\left[1+\cos\left(n_i\tau_i\right)-\phi_i\right] \tag{14.7a–d}$$

where dR_i is the elongation of the bond identified by the label i, K_i is the force constant associated with the stretching of the i bond, and $d\theta_j$ and $d\omega_k$ are the variance of bond angle j and inversion angle k, respectively. C_j and B_k are force constants associated with angle variance and inversion, respectively. A_i is the barrier height to rotation of the bond i; n_i is the multiplicity that gives the number of minimums as the bond is rotated through 2π.[15]

To determine the tensile modulus of a SWNT subjected to uniaxial loading, it is useful to observe that at small strains the torsion, the inversion, the van der Waals, and the electrostatic interaction energy terms are small compared with the bond

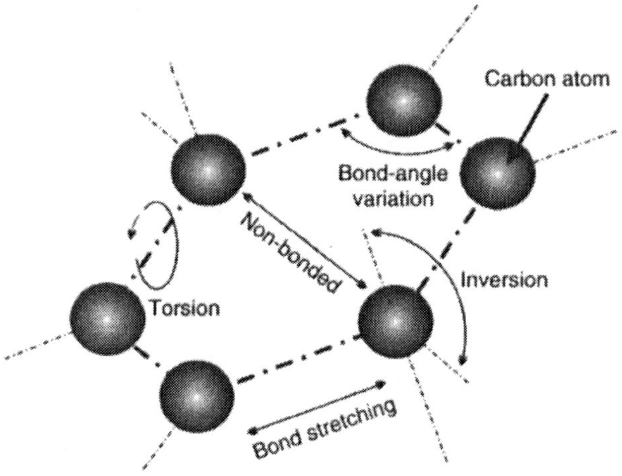

FIGURE 14.6 Bond structures and corresponding energy terms of a graphene cell.

stretching and the angle variation terms. Thus the total energy of the single-wall nanotube can be reduced to

$$E_{Total} = \frac{1}{2}\sum_i K_i (dR_i)^2 + \frac{1}{2}\sum_j C_j (d\theta_j)^2 \tag{14.8}$$

The force constants K_i and C_i can be obtained from quantum mechanics (*ab initio*). The average macroscopic elastic modulus and Poisson's ratio were estimated to be about 1.347 TPa and 0.261, respectively.[15] Such calculations may be performed using either the force or the energy approach, by measuring the mechanical forces developed between carbon atoms in nanotubes with different chiral arrangements. Lu[16] has used the empirical force potential molecular dynamic simulation to investigate the properties of nanotubes. The structure of the nanotubes was obtained by the conformational mapping of a graphene sheet onto a cylindrical surface. The nanotube radius was estimated by Equation 14.2. The average estimated tensile modulus of SWNTs and MWNTs was about 1 TPa. The elastic properties were the same for all nanotubes with a radius larger than 1 nm. Zhou et al.[17] have used the first principles cluster model within the local density approximation to evaluate the mechanical properties of a SWNT. The estimated values for tensile modulus, tensile strength, and Poisson's ratio were 0.764 TPa, 6.248 GPa, and 0.32,[17] respectively. The binding energy of the nanotube was less than that of graphite due to the curvature effect. Lier et al.[18] calculated the tensile modulus of zigzag and armchair SWNTs using the *ab initio* multiplicative integral approach (MIA), which is based on the energy of elongation of nanotubes in a simple tension (which is not constrained laterally or in any other way). They found that the moduli of SWNTs or MWNTs were larger than that of a graphene sheet. The MD simulations showed that the fracture behavior of zigzag nanotubes was more brittle than the fracture behavior of armchair nanotubes.[19] The formation of a local Stone-Wales defect (5–7–7–5) in the deformed armchair nanotube induced ductile deformation.

14.2.3 Finite Element Modeling

In the past few years, the demand for the development of faster methods to compute the mechanical properties of nanostructures has been increasing. The classical shell theory has been judged as too simple and less than accurate because it is limited by some unrealistic boundary conditions. The finite element modeling (FEM) method associated with the molecular dynamics (MD) or equivalent continuum (EC) model has been recently adopted to calculate the mechanical properties of nanotubes. Odegard et al.[20] have developed an equivalent continuum tube model to determine the effective geometry and effective bending rigidity of a graphene structure. Molecular mechanics considerations (see Equation 14.6 and Equation 14.7) were first used to determine linking forces between individual carbon atoms. This molecular force field was simulated by using a pin-joint truss model; i.e., each truss member represents the force between two atoms, as shown in Fig. 14.7. Therefore, the truss model allows the accurate simulation of the mechanical behavior of nanotubes in terms of atom displacements. As the nanotube was subjected only to a uniaxial load,[20] the bond stretching and bond-angle variation energies (see Equation 14.8) were considered. The strain energy of the whole system was used in the FEM computation to estimate the effective thickness of the nanotube layer. It was found that the effective thickness of the nanotubes (0.69 Å and 0.57 Å) was significantly larger than the interlayer spacing of graphite, estimated to be about ~0.34 Å.

Li and Chou[21–22] have worked out the contributions of van der Waals interactions between individual carbon atoms within nanotubes using the FEM truss model. The relationships of the structural mechanics parameters *EA*, *EI,* and *GJ* and the molec-

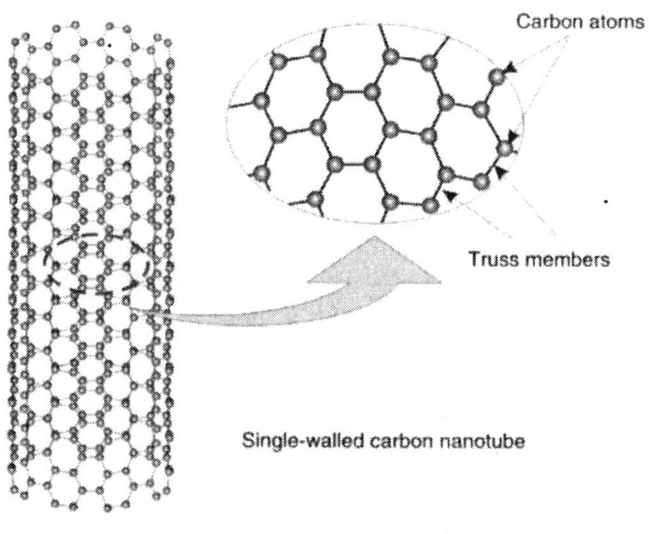

FIGURE 14.7 Truss model of carbon nanotube structure.

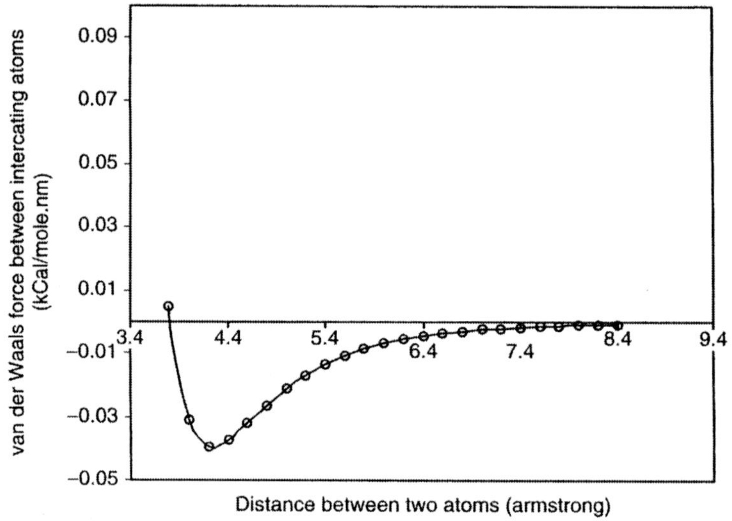

FIGURE 14.8 Van der Waals force versus the distance between two carbon atoms.

ular mechanics parameters K_ρ, C_θ, and A_τ are shown in Equations 14.7a, b, and d for each truss member:

$$\frac{EA}{L} = K_\rho; \quad \frac{EI}{L} = C_\theta \quad \text{and} \quad \frac{GJ}{L} = A_\tau \qquad (14.9\text{a–c})$$

In Fig. 14.8, the dependence of the van der Waals force versus the distance between two carbon atoms is plotted. Nonlinear truss elements were used in simulations, since the force between two carbon atoms was also nonlinear. A uniaxial load was applied uniformly at the ends of the nanotubes;[21–22] the effects due to the end caps were neglected. It was found that the Young's modulus of the nanotubes increased as the nanotubes' diameter was increased. The Young's and shear moduli of MWNTs were in the range of 1.05 ± 0.05 and 0.4 ± 0.05 TPa, respectively. For SWNTs, the Young's modulus was almost constant when the nanotubes' diameters were larger than 1.0 nm. The average Young's modulus of the zigzag nanotubes was slightly higher than the armchair type. Also the Young's moduli of MWNTs were generally higher (~7%) than those of SWNTs.

14.3 FABRICATION PROCESSES OF NANOTUBE/POLYMER COMPOSITES

To achieve desirable properties of nanotube polymer-based composites, control of the fabrication process in order to produce well-dispersed composites is an essential issue. Several parameters such as the selection of dispersion solvents, sonication time, stirring speed, and ambient temperature are the keys to disperse the nanotubes uniformly in the resin. Another crucial factor that influences the strength of the composites is the alignment of the nanotubes. Sonication is one of the most popular ways to disperse nanotubes into polymer-based resin. Park et al.[23] first demonstrated the use of *in situ* polymerization to disperse SWNT bundles in the polymer matrix. A dilute SWNT solution, typically around 0.05 wt% of nanotubes in dimethylformamide (DMF) subjected to sonication for $1^1/_2$ hours in an ultrasonic bath (40 kHz) was followed by mixing with the hardener. The long sonication time may cause the nanotubes to entangle and form bundles, which may result in losing the benefit of their high-strength properties in composite materials. Mukhopadhyay et al.[24] have reported that if the sonication time is greater than 4 h, it may destroy most of the graphene layers of the nanotubes and possibly cause the formation of junctions in the nanotubes.

Lu et al.[25] have found that the use of different chemical solvents as dilute solutions would greatly influence the integrity of composites. Since DMF has a high boiling temperature compared with ethanol and acetone, the solvent remains inside the resin and lowers the chemical reaction process between the resin and hardener. The boiling points of acetone, ethanol, and DMF are 56°C, 78°C, and 130°C, respectively. Eventually the overall mechanical and chemical properties of the composite decrease. In Table 14.1, the Vickers' hardness values of different nanotube/epoxy composites mixed by using different types of solvent are shown. Obviously the mechanical properties of the composites are directly affected by the melting temperatures of the solvent.

In Fig. 14.9, FTIR spectra of different types of composites are shown. Spectra *b*, *c*, and *d* represent acetone, ethanol, and DMF solvents, respectively, which were used to disperse the nanotubes in an epoxy matrix. Spectrum *a* is a pure epoxy. The most prominent feature in the spectra *b*, *c*, and *d* is the appearance of a new absorption band located at ~1650 cm^{-1}. Considering epoxide and hydroxyl groups are the only two reaction groups in the epoxy molecule, the ~1650 cm^{-1} band can be assigned to an amino group formed by the intermolecular nucleophilic substitution of hydroxyl at the amide functionality, which can therefore be used to estimate the relative amount of the product of the cure reaction. The band locations are different in the spectra *b*, *c*, and *d*, with values of 1645 cm^{-1}, 1649 cm^{-1}, and 1664 cm^{-1}, respectively. This indicates that there are functional and curing differences among epoxy resins with different solvent treatments that may be related to the variation of the mechanical

TABLE 14.1
Vickers' Hardness Values of Different Samples Determined from a Load Force of 100 g and a Dwell Time of 15 s

Composition	Solvent	Vickers' Hardness No.
CNTs/epoxy	Acetone	18.0 ± 0.11
CNTs/epoxy	Ethanol	14.4 ± 0.08
CNTs/epoxy	DMF	7.7 ± 0.10
Pure epoxy	—	17.8 ± 0.06

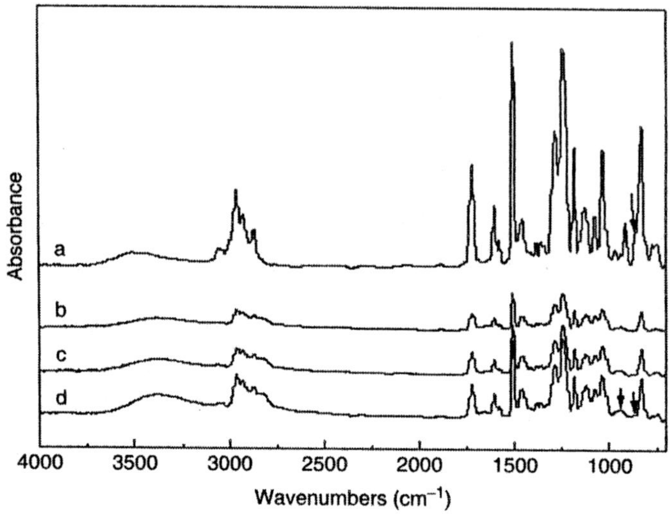

FIGURE 14.9 The representative FTIR spectra measured from the samples a to d.

properties discussed previously. Blanchet et al.[26] directly dispersed nanotubes into polyaniline (PANI) using a sonication method. The nanotubes were first sonicated in xylene and that dispersion afterward was sonicated in the DNNSA-PANI solution. It was found that all the nanotubes were well dispersed into the PANI and enhancement of the electrical conductivity was achieved. Tang et al.[27] have also introduced the use of a melt processing technique for making nanotube/high-density polyethylene (HDPE) composites. The nanotubes and HDPE were pre-melted together to form pellets, followed by feeding them into a twin-screw extruder to make composites (Fig. 14.10). In their work, one difficulty was controlling the uniformity of the nanotubes in the HDPE. Once the pellets were melted, some of the nanotubes were trapped inside the injection head of the extruder and entangled, forming bundles. Although research related to nanotube/polymer composites has been done for more than a half decade, the detailed study on how to fabricate well-dispersed and -aligned nanotube/polymer composites is still a critical issue.

14.4 INTERFACIAL BONDING PROPERTIES OF NANOTUBE/POLYMER COMPOSITES

14.4.1 Experimental Investigation

It has been recognized that high mechanical strength of nanotube/polymer composites can be achieved when all nanotubes are aligned parallel to the load direction, and an external stress should be effectively transferred to the nanotubes. Jin et al.[28] and Wood et al.[29] have studied the control of the alignment of nanotubes in polymer matrices for making high-strength advanced nanocomposite structures. However, the

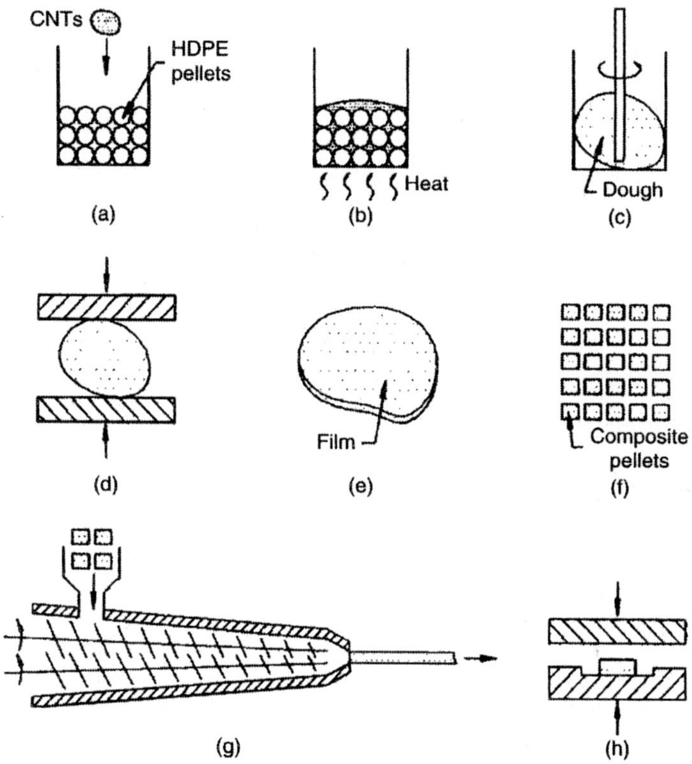

FIGURE 14.10 Process of preparing MWNT/HDPE nanocomposite films.[27]

investigation on the mass production and control of the alignment of such nanotubes toward the predetermined direction is still under investigation. The stress transfer mechanisms of different types of nanotubes in nanotube/polymer composites have not been comprehensively discussed elsewhere. Wagner et al.[30] and Qian et al.[31] have studied the load transfer properties of the nanotube/polymer composites by conducting many experimental investigations. They found the interfacial shear strength between the nanotube and polymer to be as high as 43.3 MPa when measured in a fragmentation test. However, in several experimental investigations, it was obvious that a poor adhesion property in the nanotube/polymer composites was found because of the pulling out of the nanotubes.[32] In addition, a weak attraction between different shells inside the multiwall nanotube also caused failures initiated only at the outer shells of the nanotubes. The latest experimental study conducted by Cooper et al.[33] demonstrated that the shear strength between a nanotube and epoxy matrix varied with the diameter, the embedding length, and the number of shells of the nanotubes. Two sections of an epoxy matrix were formed with a nanotube bridging across the sections. The two sections were pulled apart using the tip of a scanning probe microscope. The maximum shear strength ranging from 38

MPa to 376 MPa was recorded, and a high shear strength was reported for SWNT ropes. A pull-out of nanotubes from a solid polyethylene matrix was demonstrated using an AFM,[34] and the average interfacial stress required to remove a SWNT from the matrix was about 47 MPa. This experiment was closer to a realistic situation since all the nanotubes were physically bonded to the matrix and a typical pull-out test was actually conducted. However, as described in the previous section, the mechanical properties of different kinds of nanotubes may influence their bonding strength to the matrix. To better understand the interfacial bonding behavior of the nanotube and matrix, an improved experimental setup to perform pull-out tests for different types and geometries of nanotubes is needed.

14.4.2 Theoretical Study and Molecular Dynamics Simulation

The molecular dynamics (MD) simulation is a powerful tool to predict the interfacial bonding properties among different shells in nanotubes, as mentioned in Section 14.2.2, and also between the nanotubes and surrounding polymer molecules. Liao and Li[35] simulated a pull-out action of nanotubes in polystyrene (PS) matrices using the commercial software Hyerchem®. The adhesion strength between PS molecules and a graphene sheet was studied using a molecular mechanics model employing an empirical MM+ force field. A random coil of PS $[(\text{-}CH_2CHC_6H_6\text{-})_n]$ molecules with $n = 2, 4, 10, 20, 40$, and 80 was constructed and located near the surface layer (Fig. 14.11). The nanotubes were then subjected to a pull-out force to measure the interfacial shear strength. In their study, the shear stress between the nanotubes and polymer estimated from the simulation was about 160 MPa. An experiment was then subsequently carried out by one of the coauthors of Reference 35 and a strong interfacial adhesion between nanotubes and matrix was observed from electron microscopy.[36] Lordi and Yao[37] used a force field-based molecular mechanics calculation to determine binding energies and sliding frictional stresses between pristine nanotubes and a range of polymer substrates. They found that all of the substrates studied showed more sliding friction on a nanotube surface than friction between two graphene sheets and much more than that between MWNTs. The key factor in forming a strong bond between the polymer and nanotube lies in the polymer's morphology, specifically its ability to form large-diameter helices around individual nanotubes. Frankland et al.[38] studied the effect of chemical cross-links on the interfacial bonding strength between a SWNT and polymer matrix using MD simulations. Their models were composed of single-wall (10,10) armchair nanotubes embedded into either a crystalline or amorphous matrix. The nonbond interactions within the polyethylene matrix and between the matrix and nanotubes were modeled with Lennard-Jones 6-12 potentials. They found that even a relatively low density of cross-links can have a large influence on the properties of the nanotube-polymer interface. Ren and Liao[39] also found that pull-out energies are affected by the thermal condition of composites. However, in those simulated studies, due to the reduction of complexity, chemical interactions between the nanotubes and matrix were generally ignored. Only nonbond interactions, and electrostatic and van der Waals forces were assumed.

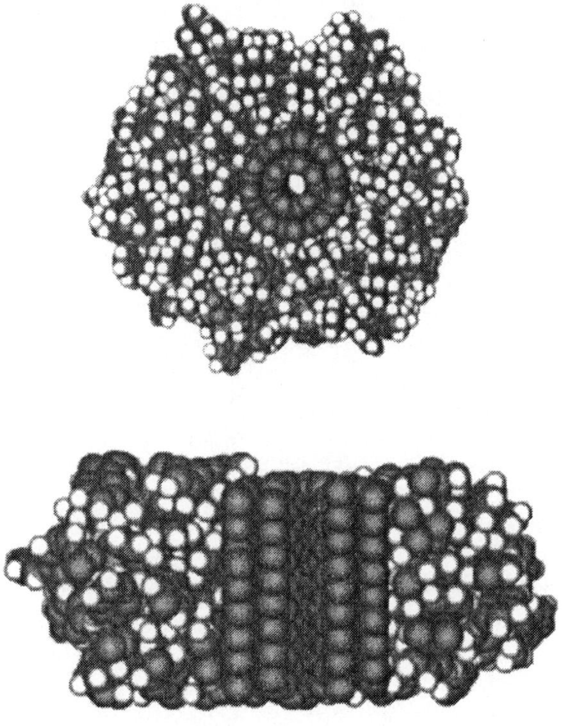

FIGURE 14.11 A molecular model of a double-walled carbon nanotube in a PS matrix.[35]

In the early stages of nanotube-related research, many diverse results on the interfacial-bonding characteristic between nanotubes and polymer-based matrices were found. Xu et al.[40] noted that high interfacial shear stress between a MWNT and an epoxy matrix was observed from a fractured sample. Wagner[41] first used the Kelly-Tyson model that has been widely used to study the matrix-fiber stress transfer mechanism in micron-size fiber composites to study the interfacial shear strength between the nanotube and polymer matrix. Since it was found that the binding force between inner layers is very low, and sliding failure always occurs, only a single-wall system was of interest in his work. In his study, it was assumed that an externally applied stress to a nanotube/polymer composite is fully transferred to the nanotubes via a nanotube-matrix interfacial shear mechanism at the molecular level; i.e., the length of the nanotubes (l) is larger than the critical length (lc). A single nanotube cylindrical model was used to study the stress transfer properties of the composite shown in Fig. 14.12. To consider the force balance in the composite system, the following equation is formed:

$$\tau_{NT} d_i dx = \left(\sigma_{NT} + d\sigma_{NT}\right)\left(\frac{d_o^2 - d_i^2}{4}\right) - \sigma_{NT}\left(\frac{d_o^2 - d_i^2}{4}\right) \qquad (14.10)$$

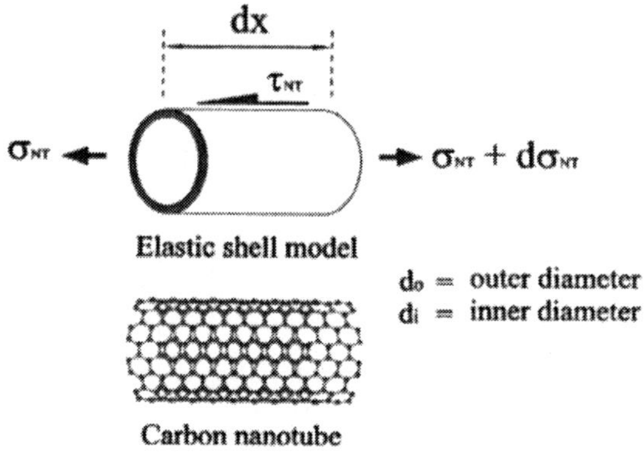

FIGURE 14.12 Stress transfer study using Kelly-Tyson model.

where τ_{NT} is the interfacial shear strength between the nanotube and matrix, σ_{NT} is the tensile strength of a nanotube segment of length dx, and d_o and d_i are outer and inner diameters of the nanotube, respectively.

After integrating Equation 14.10 and considering the critical length of a typical short fiber system in composite structures, the interfacial shear strength can be written as

$$\tau_{NT} = \sigma_{NT}\left(\frac{1}{2}\left(\frac{l_c}{d_o}\right)^{-1}\left(1-\frac{d_i^2}{d_o^2}\right)\right) \quad (14.11)$$

where l_c/d_o is the critical aspect ratio of the nanotube and d_i/d_o is the diameter ratio. As an external applied stress of 50 GPa was used in the literature, the interfacial strength was calculated for critical length values of 100, 200, and 500 nm. It was concluded that the interfacial shear stress was affected by several factors: (1) the critical length, and (2) the outer diameter of the nanotubes. Increasing the diameter of the nanotube resulted in increasing the interfacial shear strength at the bond interface.

Recently Lau[12] conducted an analytical study of the interfacial bonding properties of nanotube/polymer composites by using the well-developed local density approximation model described in Section 14.2.2,[10] and classical elastic shell theory and the conventional fiber pull-out model. In his study, several important parameters such as the nanotube wall thickness, Young's modulus, volume fraction, and chiral vectors of the nanotubes were considered. It was found that decrease of the maximum shear stress occurs by increasing the size of the nanotubes. Increasing the number of the walls of the nanotubes will cause (1) a decrease of Young's modulus of the nanotubes, (2) an increase of the effective cross-sectional area, and (3) an increase in the total contact surface area at the bond interface and the allowable pull-out force of the nanotube/polymer system. In Fig. 14.13 a plot of the interfacial shear stress of SWNTs with different chiralities is shown. The figure shows that the maximum shear stress of a zigzag nanotube (5,0) is comparatively higher than those of chiral (5,3) and armchair (5,5) nanotubes.

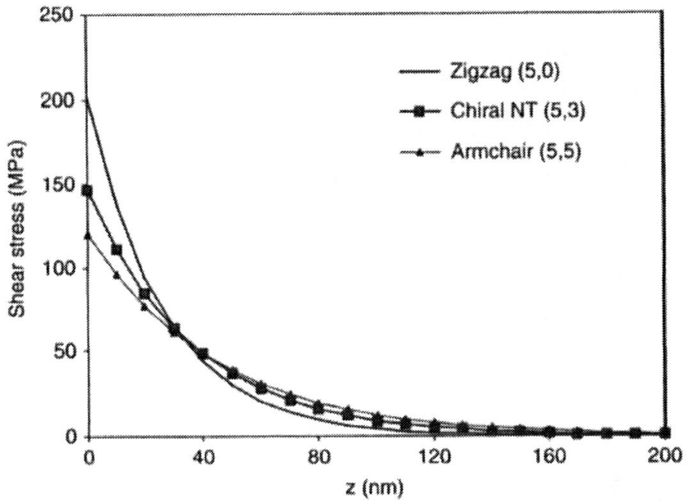

FIGURE 14.13 Plot of interfacial shear stress versus different types of nanotube.

14.5 CONCLUDING REMARKS

In this chapter, a critical review of the strength of nanotubes, fabrication processes, and interfacial bonding properties of nanotube/polymer composites is given. It was found that there are still many uncertainties in measuring the mechanical properties of nanotubes due to the difficulty of conducting nanoscale property tests. MD and theoretical analyses are mainly based on certain assumptions that may not be applicable to a real situation in nanotube/polymer composites. Dispersion properties and control of alignment of nanotubes are also important issues that govern the global properties of the composites. Presently no one has reported a practical large-scale manufacturing method to disperse and align nanotubes. This is the main factor limiting enhancement of the properties of nanotube/polymer composites. However, a good interfacial bonding between the nanotubes and matrix is also required. In the last section, we summarized some recent work on this particular issue. Two open problems are (1) whether the chemical bonding between nanotubes and matrix exists or not, and (2) whether nanotubes still maintain their extraordinary mechanical, electrical, and thermal properties in the presence of chemical bonding between the nanotube and matrix. Enhancement of the mechanical properties of advanced composite materials will require much further investigation, and this is definitely a challenging area for the composite community.

PROBLEMS

1. Which factors affect the mechanical, electrical, and thermal properties of nanotube/polymer composites?

2. What are the differences between the mechanical properties of single-walled and multi-walled carbon nanotubes? Do all inner layers of the nanotubes contribute to the strength of the multi-walled nanotubes?
3. Fabrication processes of nanotube/polymer composites are crucial factors that control the integrity of the composites. Which parameters have to be controlled or studied in making high-strength composites?
4. Based on your own understanding, which type of nanotube (MWNT or SWNT) is better to use to make nanotube/polymer composites?

ACKNOWLEDGMENTS

This work was supported by Hong Polytechnic University Grant G-T 861, Graduate Enhancement Fund from the University of New Orleans.

REFERENCES

1. Iijima, S., Helical microtubules of graphite carbon, *Nature, 354,* 56–58, 1991.
2. Snow, E.S., Campbell, P.M., and Novak, J.P., Single-wall carbon nanotube atomic force microscope probes, *Appl. Phys. Lett., 80,* 11, 2002–2004, 2002.
3. Kiernan, G., Barron, V., Blond, D., Drury, A., Coleman, J., Murphy, R., Cadek, M., and Blau, W., Characterization of nanotube based artificial muscles materials, *Proceedings of SPIE, 4876,* 775–781, 2003.
4. Gao, Y. and Bando, Y., Carbon nanothermometer containing gallium, *Nature, 415,* 599, 2002.
5. Liu, C., Yang, L., Tong, H.T., Cong, H.T., and Cheng, H.M., Volumetric hydrogen storage in single-walled carbon nanotubes, *Appl. Phys. Lett., 80,* 13, 2389–2391, 2002.
6. Yu, M.F., Files, B.S., Arepalli, S., and Ruoff, R.S., Tensile loading of ropes of single wall carbon nanotubes and their mechanical properties, *Phys. Rev. Lett., 84,* 24, 5552–5555, 2000.
7. Demczyk, B.G., Wang, Y.M., Cumings, J., Hetman, M., Han, W., Zettl, A., and Ritchie, R.O., Direct mechanical measurement of the tensile strength and elastic modulus of multiwalled carbon nanotubes, *Mater. Sci. Eng. A, 334,* 173–178, 2002.
8. Cumings, J. and Zettl, A., Low-friction nanoscale linear bearing realized from multiwall carbon nanotubes, *Science, 289,* 602–604, 2000.
9. Qi, H.J., Teo, K.B.K., Lau, K.K.S., Boyce, M.C., Milne, W.I., Robertson, J., and Gleason, K.K., Determination of mechanical properties of carbon nanotubes and vertically aligned carbon nanotube forests using nano-indentation, *J. Mech. Phys. Solids,* in press.
10. Tu, Z.C. and Ou-Yang, Z.C., Single-walled and multiwalled carbon nanotubes viewed as elastic tubes with the effective Young's moduli dependent on layer number, *Phys. Rev. B, 65,* 233407, 2002.
11. Lau, K.T., Gu, C., Gao, G.H., Ling, H.Y., and Reid, S.R., Stretching process of single- and multiwalled carbon nanotubes for nanocomposite applications, *Carbon, 42,* 423–460, 2004.
12. Lau, K.T., Interfacial bonding characteristics of nanotube/polymer composites, *Chem. Phys. Lett., 370,* 399–405, 2003.
13. Lau, K.T. and Hui, D., The revolutionary creation of new advanced materials: carbon nanotube composites, *Comp. Pt. B Engineering, 33,* 263–277, 2002.

14. Lennard-Jones, J.E., The determination of molecular fields: from the variation of the viscosity of a gas with temperature, *Proc. R. Soc., A106,* 441, 1924.
15. Chang, T.C. and Gao, H.J., Size-dependent elastic properties of a single-walled carbon nanotube via a molecular mechanics model, *J. Mech. Phys. Solid, 51,* 1059–1074, 2003.
16. Lu, J.P., Elastic properties of single and multi-layered nanotubes, *J. Phys. Chem. Solids, 58,* 11, 1649–1652, 1997.
17. Zhou, G., Duan, W.H., and Gu, B.L., First-principles study on morphology and mechanical properties of single-walled carbon nanotubes and graphene, *Chem. Phys. Lett., 326,* 181–185, 2000.
18. Lier, G.V., Alsenoy, C.V., Doren, V.V., and Geerlings, P., *Ab initio* study of the elastic properties of single-walled nanotubes, *Phys. Rev. B, 58,* 20, 14013–14019, 1998.
19. Nardelli, M.B., Yakobson, B.I., and Bernholc, J., Brittle and ductile behaviour in carbon nanotubes, *Phys. Rev. Lett.,* 81, 21, 4656–4659, 1998.
20. Odegard, G.M., Gates, T.S., Nicholson, L.M., and Wise, K.E., Equivalent-continuum modelling of nano-structured materials, *Comp. Sci. Tech., 62,* 1869–1880, 2002.
21. Li, C.Y. and Chou, T.W., A structural mechanics approach for the analysis of carbon nanotubes, *Inter. J. Solid Struct., 40,* 2487–2499, 2003.
22. Li, C.Y. and Chou, T.W., Elastic moduli of multi-walled carbon nanotubes and the effect of van der Waals forces, *Comp. Sci. Tech.,* in press.
23. Park, C., Ounaies, Z., Watson, K.A., Crooks, R.E., Smith, J., Jr., Lowther, S.E., Connell, J.W., Siochi, E.J., Harrison, J.S., and Clair, T.L., Dispersion of single wall carbon nanotubes by *in situ* polymerization under sonication, *Chem. Phys. Lett., 364,* 303–308, 2002.
24. Mukhopadhyay, K., Dwivedi, C.D., and Mathur, G.N., Conversion of carbon nanotubes to carbon nanofibers by sonication, *Carbon, 40,* 1369–1383, 2002.
25. Lu, M., Lau, K.T., Ling, H.Y., Zhou, L.M., and Li, H.L., Effect of solvents selection for carbon nanotubes dispersion on the mechanical properties of epoxy-based nanocomposites, *Comp. Sci. Tech.,* in press.
26. Blanchet, G.B., Fincher, C.R., and Gao, F., Polyaniline nanotube composites: a high-resolution printable conductor, *Appl. Phys. Lett., 82,* 8, 1290–1292, 2003.
27. Tang, W.Z., Santare, M.H., and Advani, S.G., Melt processing and mechanical property characterization of multiwalled carbon nanotube/high density polyethylene (MWNT/HDPE) composite films, *Carbon, 41,* 2779–2785, 2003.
28. Jin, L., Bower, C., and Zhou, O., Alignment of carbon nanotubes in a polymer matrix by mechanical stretching, *Appl. Phys. Lett., 73,* 9, 1197–1199, 1998.
29. Wood, J.R., Zhao, Q., and Wagner, H.D., Orientation of carbon nanotubes in polymers and its detection by Raman spectroscopy, *Comp. Pt. A, 32,* 391–399, 2001.
30. Wagner, H.D., Lourie, O., Feldman, Y., and Tenne, R., Stress-induced fragmentation of multiwall carbon nanotubes in a polymer matrix, *Appl. Phys. Lett., 72,* 2, 188–190, 1998.
31. Qian, D., Dickey, E.C., Andrews, R., and Rantell, T., Load transfer and deformation mechanisms in carbon nanotube-polystyrene composites, *Appl. Phys. Lett., 76,* 20, 2868–2870, 2000.
32. Lau, K.T. and Hui, D., Effectiveness of using carbon nanotubes as nano-reinforcement for advanced composite structures, *Carbon, 40,* 1597–1617, 2002.
33. Cooper, C.A., Cohen, S.R., Barber, A.H., and Wagner, H.D., Detachment of nanotubes from a polymer matrix, *Appl. Phys. Lett., 81,* 20, 3873–3875, 2002.
34. Barber, A.H., Cohen, S.R., and Wagner, H.D., Measurement of carbon nanotube-polymer interfacial strength, *Appl. Phys. Lett., 82,* 23, 4140–4142, 2003.

35. Laio, K. and Li, S., Interfacial characteristics of a carbon nanotube-polystyrene composite system, *Appl. Phys. Lett.*, *79,* 25, 4225–4227, 2001.
36. Wong, M., Paramsothy, M., Xu, X.J., Ren, Y., Li, S., and Liao, K., Physical interaction at carbon nanotube-polymer interface, *Polymer, 44,* 7757–7764, 2003.
37. Lordi, V. and Yao, N., Molecular mechanics of binding in carbon-nanotube-polymer composites, *J. Mater. Res.*, *15,* 12, 2770–2779, 2000.
38. Frankland, S.J.V., Caglar, A., Brenner, D.W., and Griebel, M., Molecular simulation of the influence of chemical cross-links on the shear strength of carbon nanotube-polymer interfaces, *J. Phys. Chem. B.*, *106,* 3046–3048, 2002.
39. Ren, Y., Fu, Y.Q., Li, F., Cheng, H.M., and Liao, K., Fatigue failure mechanisms of single-walled carbon nanotube ropes embedded in epoxy, *Appl. Phys. Lett.*, in press.
40. Xu, X.J., Thwe, M.M., Shearwood, C., and Liao, K., Mechanical properties and interfacial characteristics of carbon-nanotube-reinforced epoxy thin film, *Appl. Phys. Lett.*, *81,* 2833, 2002.
41. Wagner, H.D., Nanotubes-polymer adhesion: a mechanics approach, *Chem. Phys. Lett.*, *361,* 57, 2003.

15 Nanoscale Intelligent Materials and Structures

Yun-Yeo Heung, Inpil Kang, Sachin Jain, Atul Miskin, Suhasini Narasimhadevara, Goutham Kirkeria, Vishal Shinde, Sri Laxmi Pammi, Saurabh Datta, Peng He, Douglas Hurd, Mark J. Schulz, Vesselin N. Shanov, Donglu Shi, F. James Boerio, and Mannur J. Sundaresan

CONTENTS

15.1 INTRODUCTION

Since the time of the design of the lever and wheel, it has been a goal of mankind to improve the performance of engineered systems. To achieve large advances in the performance of current systems — starting from the simplest machine elements to automobiles, bridges, buildings, aircraft, helicopters, wind turbines, environmental monitoring systems, national security systems, and surgical and medical devices — will probably require that these systems become intelligent. Intelligent systems and machines will be autonomous, self-sensing, able to resist vibration, redistribute loads, and change shape in response to their environmental conditions, and therefore preserve their integrity, increase their lifetimes, optimize their performance, reduce the cost of ownership, and provide continuous safety during use. The critical ingredient needed to make structures and machines intelligent is intelligent materials. The problem is that current intelligent materials have limited capabilities and new classes of practical and enabling intelligent materials with vastly improved properties are needed. This chapter therefore tackles the problem of designing new intelligent materials and takes the approach of using nanoscale components and developing processes to build intelligent materials from the bottom up.

The terms *intelligent* and *smart* are often used interchangeably, and they can describe both materials and structures. Smart materials are defined as having sensing or actuation properties such as a piezoelectric response to dynamic strain that causes electrons to move. Smart structures use smart materials for sensing and actuation in analog or digital closed-loop feedback systems. *Intelligent* usually designates a more complex level of material behavior or control, possibly by using miniaturized electronics built into the material itself or by using a more advanced feedback system such as a digitally controlled structure programmed with a rudimentary ability to learn and reason. Smart structures can also be passive such as a wing designed with bend-twisting coupling to protect against flutter, or semiactive such as a magnetorheologic damper that uses a feedback signal to control the voltage field and viscosity of a fluid. Smart materials and structures that are fully active use feedback signals, processors, amplifiers, and actuators to apply forces to control the vibration or shape of a structure.

Present smart/intelligent materials and structures have fundamental limitations in their sensitivity, size, cost, ruggedness, and weight that have restricted their widespread application. To address the need for new intelligent materials and systems, researchers are investigating nanotechnology and also intersecting nanotechnology with biomimetics. Nanotechnology can provide the route to build perfect materials starting at the atomic scale, as in nature, while biomimetics can provide the inspiration and architecture for material system design. Multifunctionality is a universal trait of biological materials and systems and is an important attribute when developing intelligent materials and structures. Biological systems have been designed by nature from the smallest components upward over a long period of time, and they have material capabilities unmatched by man-made systems. The concept of biologically inspired nanotechnology, or bionanotechnology, can be described as the process of mimicking the chemical and evolutionary processes found in nature to synthesize unique almost defect-free multifunctional material systems starting from the nanoscale up. Bionanotechnology is thus a new frontier that can lead to new generations of smart materials and intelligent systems that can sense and respond to their environment.

Although we are at an early stage of research, nanoscale intelligent materials including carbon nanotubes, inorganic nanotubes, compound nanotubes, nanobelts, and other nanoscale materials have produced great excitement in the research community. This is because of their remarkable and varied electrical and mechanical properties including electrical conductance, high mechanical stiffness, light weight, electron-spin resonance, electrochemical actuation, piezoelectricity, piezoresistance, contact resistance, Coulomb drag power generation, thermal conductivity, luminescence, and the possibilities for functionalizing these materials to change their intrinsic properties. These attractive properties also have the potential to allow development of self-contained nanoscale intelligent materials based on analog or digital control of the material using built-in nanoscale electronics. The broad area of nanoengineering of intelligent materials is discussed in this chapter, starting with material synthesis, to the processing of nanostructured hybrid materials and then fabrication of devices such as strain sensors, biosensors, and wet and dry actuators. The focus is on nanotube-based materials that have novel strain sensing and force actuation properties and the associated material processing steps needed to build macroscale materials from nanoscale components, including digital control of the material behavior. Nanoscale smart materials are important because they often possess improved or different properties from macroscale materials due to their near-perfect construction, large surface area per volume, and quantum effects.[1–12]

While nanoscale materials, and in particular carbon nanotubes (CNTs), have extraordinary properties, utilizing these properties at the nanoscale and at the macroscale is presenting major challenges for scientists and engineers. Thus far, nanoscale intelligent materials have mostly been based on carbon nanotubes. A single-wall carbon nanotube (SWCNT) is a superelastic crystalline molecule that can have a length-to-diameter ratio of 1000 or more, and an enormous interfacial area of 500 m^2/gm or more. Custom designing of multifunctional materials by mixing nanotubes with bulk carrier materials is now producing a range of new properties such as piezoresistivity and electrochemical transduction, as well as enhanced strength, modulus, toughness, hardness, and thermal and electrical conductivity, while simultaneously decreasing thermal expansion and permeability. Controlling and using these properties for smart materials is dependent on understanding the composition, topography, and processing relationships that define nanomaterial behavior. Major difficulties in forming nanocomposite materials occur because CNTs do not easily transfer load to matrix materials because their small size can affect the molecular structure of the host material, and because van der Waals bonding is weak in the axial direction of the nanotubes and cannot prevent intertube longitudinal slipping. Also the van der Waals bonding is strong in the transverse direction, which makes it difficult to transversely separate and disperse the CNT. A consequence of this behavior is that bundles of parallel nanotubes are difficult to pull apart transversely, but they can axially slip and therefore can act as defects in polymer smart materials. Mechanically and chemically altering the nanotubes to disperse and to bond with the host material is helpful, but this can adversely affect the electrical and transduction properties of the material. In addition, nanotube electrochemical actuators must allow ion conductivity over the surface of the nanotube while simultaneously transferring actuation strain to and from the nanotube. While many materials have the possibility for electrochemical actuation, the effect is negligible in bulk materials because most of their atoms are in the interior. Conversely, in SWCNTs all the atoms on are the surface and the electrochemical actuation effect is large. When the nanotubes collect in bundles or ropes, the inner nanotubes are shielded from the ion exchange and the nanotubes also slip on one another. Therefore, the critical factors in the design of nanotube intelligent materials, and especially electrochemical actuators, are that the nanotubes must be separated and aligned for electrical conductivity and ion exchange, and the strain must be transferred from the nanotube to the host material. Otherwise macroscale nanotube intelligent materials and actuators will be weak and inefficient. There is also the possibility of using nanotube actuators at the nanoscale, but there is not much published research in this new area.

In the broadest sense, the goal of nanoscale materials research is to develop multifunctional intelligent materials. These materials would possess high strain-to-failure value and high stiffness, be lightweight, and have a high bandwidth for actuation or sensing. They would serve as strong structural materials that simultaneously have sensing, actuation, and other functional capabilities. There is no other smart material available today that is also a structural material. Therefore nanotube hybrid actuator and sensor materials could become an enabling technology for the improvement of all kinds of dynamic systems including helicopters, reconfigurable aerodynamic surfaces, launch vehicles, ultra-high energy density wireless motors, and nanodevices such as biomedical nanosensors, surgical robots, active catheters, and biomedical implants. Nanotechnology is attractive and exciting because it may produce the most efficient smart materials ever made. On the other hand, nanoscience is presently difficult to implement in practical applications. Various techniques for processing nanoscale materials are presented in the literature, but many of these are difficult to extrapolate to the macroscale or to use for large volume production due to the impracticality of the processing method or cost. Moreover some statements in the literature seem to inflate the potential of nanotechnology, which may lead to a backlash that will impede nanotechnology research. Stating that carbon nanotubes are 100 times stronger at one-sixth the weight of steel is misleading, because nanotubes are nanoscale orthotropic molecules and steel is an isotropic bulk material, and the two are not directly interchangeable. It would be more realistic to compare the properties of a SWCNT with those of a typical commercial reinforcing fiber used today. In Table 15.1, the elastic properties of the Hexcel IM8 carbon fiber[12] are compared to the elastic properties of a SWCNT. The elastic modulus of the SWCNT is 3.3 times greater than the elastic modulus of the carbon fiber, and the specific modulus of the SWCNT is 4.4 times greater than the specific modulus of the carbon fiber. The strength of the SWCNT is 8.9 times greater than the strength of the carbon fiber, and the specific strength of the SWCNT is 12 times greater than the specific strength of the carbon fiber. Also the elastic modulus of a unidirectional SWCNT-epoxy composite is 3.2 times greater than the elastic modulus of a unidirectional carbon fiber-epoxy composite, and the specific modulus of a unidirectional SWCNT-epoxy composite is 4.2 times greater than the specific modulus of a unidirectional carbon fiber-epoxy composite. Since a typical carbon fiber is a bulk material that is 3643 times larger in diameter than the SWCNT, there is a significant problem of processing the nanocomposite material to bring its nanoscale properties to the macroscale. Another seeming overstatement appeared in a New York Times article: "Today's Visions of the Science of Tomorrow," January 4, 2003. In the article, it was predicted that nanotechnology will lead to aerospace vehicles with 98% less structural mass. It is interesting to speculate how long it will take the aerospace companies to meet this prediction.

On the positive side, large advances in nanoscale materials development have occurred. Small percentages of CNT are being incorporated into polymers, carbon fibers, and metals for reinforcement, and these hybrid materials have shown significant improvements in elastic modulus or toughness, as compared to the host polymer material. Spinning of fibers from carbon nanotubes has produced strong fibers that can be incorporated into polymers to produce super-strong materials when the cost of the fiber decreases. Electrochemical actuation in carbon nanofiber composites has also been demonstrated, and this material could provide actuator capability to large structures. An important contribution of this chapter is to present useful approaches to tackle the problems of manufacturing nanoscale intelligent materials. Also presented are some ideas a little on the "wild side" for building futuristic intelligent structures, biodevices, and structural health monitoring systems. Readers should also pay attention to nonstructural and smaller material volume applications for nanoscale intelligent materials including biosensing, surgical manipulators, and smart machines, which are also discussed in the chapter. One such idea for applications is a fluid form of an intelligent material that may have properties that can be activated by contact with another material. There are potentially many types of nanotube-based fluids that take advantage of the small size and ability to functionalize nanotubes.

TABLE 15.1
Properties of Reinforcing Fibers and Their Composites with 62% Fiber Volume Fraction

Material	Diameter D [nm]	Density d [kg/m^3]	Strain to Fail [%]	Modulus Y [GPa]	Specific Modulus [Y/d]	Ult. Strength S [GPa]	Specific Strength [S/d]
SWCNT Fiber[1]	1.4	1340	>10	~1000	0.746	~50	37.3 e^{-3}
Hexcel IM8 carbon fiber[12]	5100	1790	2	304	0.17	5.589	3.12 e^{-3}
Ratio SWCNT/IM8 epoxy	1/3643	0.75	>5	**3.3**	**4.4**	8.9	12
	—	1121	7	3.2	0.0029	0.083	7.4 e^{-5}
SWCNT-epoxy	—	1252	—	~562 (calculated)	0.48	—	—
IM8 fiber-epoxy	—	1537	—	176 (measured)	0.115	2.76	1.8 e^{-3}
Ratio SWCNT-epoxy/IM8 epoxy	—	0.815	—	**3.2**	**4.2**	—	—

Note: Bolding indicates .

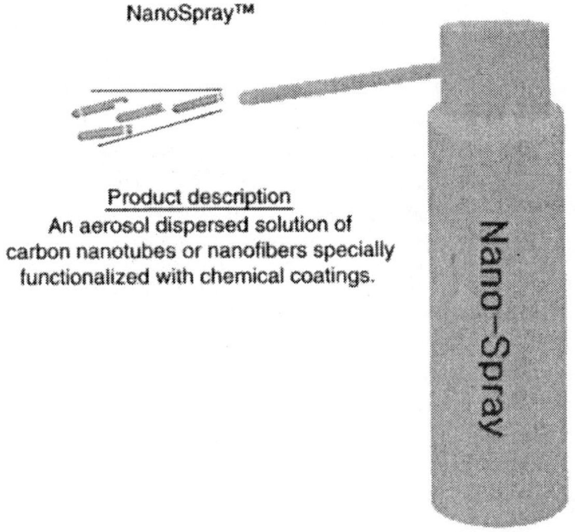

FIGURE 15.1 NanoSpray advertisement.

Figure 15.1 depicts a NanoSpray™ advertisement. An intelligent material is sprayed onto another material and a chemical reaction takes place. Possible applications of the spray are left to the reader's imagination. Some nanosprays are undergoing testing at University of Cincinnati. We have found that the nanotubes can be dispersed and sprayed as a liquid, or dispersed, dried, and sprayed as a nearly invisible gas. Ultrathin uniform smart coatings can be quickly sprayed on materials using this approach.

The remainder of this chapter is organized as follows. Smart materials and the geometric structures and properties of nanotubes are reviewed. Then a review of new types of sensors and actuators built using nanoscale materials is given. Next methods of processing nanoscale smart materials to manufacture macroscale materials are discussed. In nanotechnology research, we talk about the properties of nanoscale materials and the properties of macroscale materials built using the nanoscale component materials, and these two sets of properties can be very different. With this background, we cover the design of nanotube sensors and hybrid actuators and show some exciting results. Then we propose how to build future intelligent machines and devices using nanotubes. Finally conclusions based on research are noted.

15.2 REVIEW OF SMART MATERIALS

Smart materials are solid-state transducers that can have piezoelectric, pyroelectric, electrostrictive, magnetostrictive, shape memory alloy/ceramic, piezoresistive, electroactive, and other sensing and actuating properties and lack moving parts. Piezoelectric materials are the most important smart materials today.[13–29] In the mid-1980s piezoceramic PZT was developed and has a high strain energy density for actuating and sensing of structures. Active fiber composite (AFC) materials using PZT fibers were developed in 1997[21–23] by Hagood and Bent at MIT. In 2000, a variation of this material called the macrofiber composite (MFC) was developed by NASA.[24] The AFC/MFC material utilizes parallel piezoceramic fibers or ribbons in an epoxy matrix with an interdigitated electrode pattern on the top and bottom faces of the composite layer, but not touching the fibers. At the University of Cincinnati (UC), the construction of the AFC material was modified to develop piezoelectric artificial nerves for sensing of structures.[25] While AFCs represent a large advance in the development of high-impedance smart materials, AFCs are still relatively heavy and only partially load bearing; there is a strength mismatch between the AFC and host material, and their strain is ~0.2%. These are significant limitations for developing smart structures for advanced applications. There has been progress in developing higher strain single crystal piezoelectrics, relaxor ferroelectrics, and alkaline-based piezoelectric materials, but the processing methods for these materials are not yet scalable to commercial quantities, and the brittle heavy nature of the materials remains.

In 1991, multiwall carbon nanotubes (MWCNTs) were discovered by Sumio Iijima,[30] and since then carbon nanotube synthesis has been intensely studied.[31–34] See Part 1 of this book for current research in nanotube growth. In 1996 the C60 fullerene and SWCNT were synthesized by Richard Smalley et al.[32] Coupling CNTs to electronic circuits was then studied as cited in References 35–43. In 1999, electrochemical actuation of SWCNTs was developed by Ray Baughman et al.[44–68]; in 2000–2001 CNT ropes and aligned membranes were developed;[69–72] in 2001 composite materials strengthened with CNTs were studied by methods including Raman scattering[73–90]; and in 2001 the piezoelectric effects in carbon nanotubes (CNTs), boron nitride nanotubes (BNTs)[91–95] (also see Chapter 21 of this book), and silicon nitride nanotubes were studied.[96–100] Silicon oxide and vanadium oxide nanotubes were also studied. In 2002 electrochemical actuation of a single carbon nanotube was studied[49,54] and improvements in CNT actuation[58] were made. In 2003 vanadium oxide nanotubes (VNTs) were used for actuation.[98] Zinc oxide nanobelts that have piezoelectric properties are discussed in Reference 100 and in Chapter 4 of this book. Noncarbon nanotubes are discussed in References 101–114, and piezoelectric and other nanotubes are discussed in References 115–121. Nanotube electronic properties are given in References 122–135. These research efforts have verified that CNT-, VNT-, and BNT-based materials have extraordinarily high strength, super-elasticity, and sensing capability, all at the same time. These properties make nanotubes possibly the materials of choice for designing future smart materials and strain actuators. However, BNTs are not readily available yet and they have piezoelectric strain coefficients (d_{ij}) that are perhaps 20 times lower than existing piezoelectric ceramic materials, and it is not possible to separate the chiralities of BNT. VNTs have lower actuation properties than CNTs, but their cost is projected to be lower. Zinc oxide nanobelts may have good piezoelectric properties and strength, but they are not commonly available or used in composites yet. Because the BNT, VNT, and zinc oxide materials may have lower strain than electrochemical CNTs, and the materials are not as well developed, their potential as a bulk smart materials seems not yet as promising as CNTs, depending on the application. Hopefully piezoelectric nanoscale materials will become more available soon.

Based on theory, CNTs have large electrochemical actuation properties. In practice, bringing these properties to the macroscale is difficult. Transferring charge to the tubes in a SWCNT actuator and transferring the shear load through the actuator have been limitations to realizing the theoretically predicted ~2% strain and high performance of SWCNT actuators and sensors. Ray Baughman and his group at the University of Texas at Dallas and Matteo Pasquali and his group at Rice University in Houston (Chapter 11) have described great progress in spinning CNT fibers. The spun fibers are microns in diameter and the strength is greatly improved compared to previous CNT fibers. A trade-off of strong fibers is that the nanoscale actuation properties may be reduced compared to actuation of individual nanotubes because it is difficult to intercalate ions into the interiors of the dense fibers.

To overcome the limitations of present smart materials and actuators, carbon nanotube hybrid materials are being developed at UC with a focus on minimizing ion blockage and shear lag, and increasing the electrochemical process velocity and bandwidth of the material. Such CNT hybrid smart materials and the fiber-based approaches mentioned are anticipated to enable new material systems with applications in mechatronic devices and large structures for defense and civilian systems. Unlike other smart materials, these CNT hybrid materials and fibers can be unique because they are simultaneously structural, functional, and smart. The CNT hybrid material is predicted to have a large load-carrying capability, and high thermal and electrical conductivity, actuation, and sensing properties. For design use, mathematical models of the CNT material must be developed in terms of the electrochemical constitutive equations for a composite material using a solid polymer electrolyte (SPE) matrix. In preliminary work at UC, a small CNT electrochemical actuator material, a strain sensing material, and a power generation material have been built. Moreover a multiwall carbon nanofiber (CNF) PMMA electrochemical actuation material tested exhibited good actuation performance and is about 200 times lower in cost than a SWCNT actuator material. Development of a SWCNT-MWCNT-CNF-epoxy-SPE material that is a high-performance structural load-bearing material that actuates and senses is the "holy grail" of the research described in this chapter. If successful, it would provide actuation and sensing capabilities (i.e., intelligence) to large structures.

15.3 NANOTUBE GEOMETRIC STRUCTURES

The geometries of different types of nanotubes are briefly described in this section. The geometry plays an important role beyond the elastic properties — the geometry also defines the electronic properties of the nanotube and affects the sensing and actuation properties of the smart material. Detailed nanotube mechanics are discussed in References 136–142, and in Part 3 of this book.

15.3.1 Structures of Carbon Nanotubes

The crystal structure of graphite is layered and the carbon atoms within a layer are strongly bound to their neighbors at room temperature, but the layers are only loosely coupled by van der Waals forces, which makes graphite quite easily deformable in a particular direction. This is why graphite is used as the cores of pencils and for dry lubricants. Although nanotubes grow axially, a SWCNT can be most easily described as a rolled-up tubular shell of graphite sheet with the carbon atoms covalently bound to their neighbors. The bonding mechanism in a carbon nanotube system is similar to that of graphite, which is sp^2 hybridization. One of the properties in sp^2 hybridization is the occurrence of σ-bonds and π-bonds. The σ-bonds are strong covalent bonds that bind the atoms in the plane and result in the high stiffness and high strength of a CNT. On the other hand, the π-bond is the interlayer interaction of atom pairs and is much weaker than the σ-bonds. One of the exciting properties of nanotubes relates to their electronic band structure, which depends on the helicity and diameter of the nanotube. The armchair nanotubes are metallic, whereas the zigzag and chiral tubes can be either metallic or semiconducting. These results are major driving forces in the evolution of nanoelectronics including piezoresistance of nanotubes,[143–147] electrophoresis and dielectrophoresis of nanotubes,[148–156] and carbon nanotube flow sensors.[157–158] Functionalization is the chemical alteration of the surface properties of nanotubes for the purpose of improving adhesion to a matrix and improving the sensing[165–177] and actuation properties of nanotubes.[44–68] Apart from unique electronic properties, mechanical characteristics have provided excitement because they include high strength, high stiffness, low density, superelasticity, and structural perfection. These remarkable multifunctional properties of CNTs provide the potential to be used as smart materials for reinforcing, sensing, and actuating polymer composites.

15.3.2 Structures of Noncarbon Nanotubes

Nanotube materials can be categorized as organic (e.g., carbon and various peptides) and inorganic (e.g., silicon nitride, boron nitride) nanotubes. Inspired by the remarkable functions of tubular structures in biology, much research on organic nanotubes has focused on the biological functions *in vivo* and synthesis of various organic nanotubes for *in vitro* environments.[178–191] Special attention is needed to understand noncovalent processes of tubular materials such as lipic, peptidic, or steroidic systems in organic nanotubes. Since organic nanotubes have remarkable functions in biology, researchers have been trying to understand the role of organic nanotubes and synthesize them *in vitro*. Organic nanotubes were studied for biosensor applications. In addition, remarkable functions of organic nanotubes (e.g., motor proteins) can inspire researchers and may produce new types of biomimetic nanotube structures for sensors and actuators in the near future.[191]

The small size, high strength, and remarkable physical properties of CNTs are attractive, but nanotube properties depend on their diameter, thickness, helicity, and defects, and the random helicity and high synthesis cost for purified samples are barriers to applications. Therefore, researchers are increasingly investigating non-CNT materials or inorganic nanotubes and nanotubes of different inorganic materials. Using our understanding of CNT phenomenona as a background, the extraordinary properties of inorganic nanotubes can be predicted in such examples as $B_xC_yZ_z$, and dichalcogenides of transition metals including M_eS_2 (M_e = Mo,W, Nb,

S = S, Se, Te), MoO_3, SiO_2, Al_2O_3,V_2O_5, TiO_2, $NiCl_2$, InS, and Bi.[147–154] Also, lead titanate ($PbTiO_3$, PT), lead zirconate ($PbZrO_3$, PZ), and lead zirconium titanate ($PbZrO_3$-$PbTiO_3$, PZT) solid solution nanotubes were synthesized by a chelate sol-gel method.[110] These inorganic nanotubes and compound nanotubes can compensate for the lack of control of chirality and properties of smart carbon nanotube structures, making it possible to modify the electrical, optical, mechanical, charging pattern, and other properties of nanotubes over a wide range.[115–121] Therefore we cannot exempt the likelihood of using inorganic nanotubes for developing smart materials. In particular, the piezoelectric properties of boron nitride nanotubes (BNTs) were evaluated by calculating the polarization of a nanotube by the MNDO (neglect of diatomic differential overlap method) which is a semiempirical method for the calculation of molecular polarizabilities and hyperpolarizability. All types of dichalcogenide nanotubes remain semiconducting even at the smallest diameters, and BNTs are dielectric or semiconducting (n- or p-doped). Generally the width of the band gap increases as the diameter of the NT increases, but in the case of BNsT and MeS_2, the band gap remains constant. Another interesting feature of heteropolar non-CNT nanotubes is the charge transfer resulting in ionic interatomic bonding. Compared to CNT, these non-CNT nanotubes are of great interest and are enlarging the research field. More broad research not only in synthesis but also in determination of inorganic nanotubes and their properties in bundles and in polymers is needed to develop useful smart materials.

15.3.3 Designations of Nanotubes and Nanostructured Materials

Many types of nanotubes are being developed, and the terminology used to describe them is not universal. Table 15.2 lists some of the different types of nanotubes and and relevant references. The acronym for a carbon nanotube is CNT and they are described as single wall (SW), double wall (DW), and multiwall (MW). The nanotube may be amorphous or crystalline. Bamboo nanotubes have internal closeouts along their length and this occurs in boron nitride nanotubes (BNTs) and in some MWCNTs due to the growth mechanism. There are also carbon nanofibers (CNFs), which are different from MWCNTs because in the nanofiber the graphene walls are not parallel to the fiber axis. They are at an angle of 20° in a "Dixie cup" arrangement and terminate in groups at the outer wall of the fiber. This produces a rough surface and also exposes the ends of the tubes to the surface, which may affect their electrical conduction and electrochemical properties. SWCNTs have one wall and diameters of about 1.4 nm; MWCNTs have multiple concentric parallel walls and diameters of 10 nm and larger; CNFs have multiple concentric parallel walls at an angle to the fiber axis and have diameters of 70 nm or larger. For comparison, graphite fibers used in traditional composite materials have diameters on the order of 6000 nm.

Some of the important differences in the nanotubes/nanostructures are discussed next. There are also other nanostructured materials including nanobelts, nanohorns, nanoribbons, nanowires, nanoplates, nanonails, etc. that are not common and are mostly under development. The type of nanotube affects the processing method for

TABLE 15.2
Designations of Nanotubes

Descriptions and Sources
SWCNT: 0.3–2 nm *D*, 200–1000 nm *L*; grown by catalyzed CVD and other methods; cost of purified form is ~$500/gram[1,10,13]
DWCNT: 2–4 nm *D*, 1–50 micron *L*; grown by CVD of methane over cobalt nanoparticles supported on porous MgO nanoparticles[2,12]
MWCNT: 10–50 nm *D*, 1–50 micron *L*; grown by CVD; also bamboo MWCNT 20–40 nm *D*, 1–20 micron *L*, internal closeouts[2,10]
CNF (Pyrograf III®): 70–200 nm *D*, 50–100 micron *L*; nonconcentric nested tubes with walls angled 20° to the longitudinal axis, carbon crystal structure, produced as a vapor by decomposing hydrocarbons with a catalyst, hydrogen sulfide, and ammonia, different grades; cost ~$100/lb[3]
BNT: 3–40 nm *D*, 40–80 nm *D* bamboo/cone nanotubes and nanosize polyhedral particles in clusters[15]
SiCNT: 200 nm *D*, 60 micron *L*, amorphous nanocrystalline[11,141–142]
SiOx Nanostructure: 20–30 nm *D* tubes, bundles, amorphous, thermal evaporation[2]
Vanadium Oxide Nanotube (VNT): 15–100 nm *D*, MW[144]

smart materials; SWCNTs are the smallest in diameter, the most difficult to process into smart materials, and the most expensive. MWCNTs do not have as high or varied properties but are easier to process because of their larger diameter (~20 nm dia.). CNFs are similar to MWCNTs but there is a critical difference: The nanofibers are not continuous tubes and their surfaces show steps at the termination of the walls of groups of fibers. The nanofibers include PR 24 HT (~70 nm diameter) and the PR 19 (~130 nm diameter). The PR 19 has a CVD layer with a graphitic nature and these fibers may be more robust to breakage, but the electrical properties of the nanofiber are changed by the carbon coating. The PR 24 does not have a CVD coating. There are low- and high-density variations of these two nanofiber types. More information can be obtained from the manufacturer.[3] There are three main types of carbon nanotube raw materials that have potential for use as electrochemical actuators and smart materials: (1) SWCNTs, which have excellent electrochemical properties. The cost, however, is very high, ~$500/gm, and incorporating the nanotubes into polymers at high loadings is difficult. (2) MWCNTs, which have good electrochemical properties. The cost is high, ~$150/gm bulk or $1000 for a 1 cm^2 array. Incorporating the nanotubes into polymers might be done by growing arrays of nanotubes and casting the polymer around the arrays. (3) The PR 24 CNF. The electrochemical properties are good, the cost is low, ~$0.25/gm, and incorporating the nanofibers into polymers is easier because the fibers are large. All three materials separately and in combination are being used at the UC to develop intelligent materials. SWCNTs can be purchased commercially[1–2] or can be grown using a commercial nanofurnace[5]; MWCNTs can be purchased commercially[2] and also grown in arrays[5]; and CNFs can be purchased commercially.[3] A nanotube synthesis laboratory using chemical vapor deposition (CVD) is a feasible way to make nanotube materials for experimental purposes, (See Chapter 5). The differences in nanotubes affect the processing and performance of the smart material. The dispersion and functionalization processes depend on the type of nanotube. Dispersion of the nanotubes, preventing reagglomeration during curing, and the effect the nanotubes have on the polymer structures may be the most critical part of integrating any of the types of nanotubes into polymers.

15.4 MECHANICAL AND PHYSICAL PROPERTIES OF NANOTUBES

Because of the great properties of nanotubes, smart materials developed using nanotubes have the potential to improve the way we generate and measure motion in devices from the nano- to the macroscale in size. The advantages of using SWCNTs, MWCNTs, and CNFs are discussed here for building smart materials. One important property that is ideal for all CNT materials is the ability to functionalize (chemically modify the surface of) the CNT and alter its properties. SWCNT properties are discussed in the next section relative to forming CNT hybrid actuators and sensors. In general, the elastic and transduction properties are the greatest for SWCNTs. The properties decrease somewhat for MWCNTs and decrease further for PR 24 CNF. Finally graphite fibers have almost no actuation properties, although they can be used as electrical resistance-based sensors.

15.4.1 Elastic Properties

The Young's modulus of SWCNT is as high as 1 terapascal. Compared with aluminum, steel, or titanium, the strength-to-weight ratio of nanotubes in the axial direction is much greater. Compared to a composite with carbon fibers, the theoretical specific modulus of a SWCNT composite may be about four times greater, as shown in Table 15.1. The maximum strain of a SWCNT is 10% or more, which is higher than most high-strength structural materials. All of these strong mechanical properties are due to the C–C covalent bonding and the seamless hexagonal network. Thermal conductivity is also very high in the direction of the nanotube axis. The mechanics of nanotubes is discussed in detail in Part 3 of this book. The incorporation of nanotubes into polymers can initiate multifunctional improvements in the following areas with industrial applications noted in parentheses: (1) increased physical strength (in aircraft and engine structural components); (2) increased vibration damping (in aircraft structures, brake disks); (3) resistance to surface wear and scratching (in brake disks, leading edge components in aircraft, paint); (4) higher stiffness and dimensional stability (space structures and structural components); (5) increased thermal conductivity (in brakes, engines, spacecraft, electronics heat sinks, PWBs); (6) EMI shielding (defense electronics); (7) energy harvesting from active nanocomposites (future application for spacecraft, biomedical materials, communications); (8) increased fracture resistance (brake disks, due to FOD in engines); (9) increased actuation performance (future application for morphing of wings); (10) sensing for health monitoring (future application with carbon-based structural neural system at the UC); (11) corrosion and erosion resistance (aircraft edges); (12) clear nanocoatings and materials (on anodized, aluminum, and polished surfaces, windows); (13) increased electrical conductivity and photoluminescence (in flight vehicles); and (14) actuation of structures (shape changing).

15.4.2 Electrical Conductivity

Electronically, a carbon nanotube can be either metallic or semiconducting. Carbon nanotubes also have been shown to conduct current ballistically without dissipating heat. In theory, there are two propagating eigenmodes for a SWCNT. Mismatch of the eigenmodes between the nanotube and a mechanical conductor makes electrical contact difficult and high contact resistance may occur. The conductance (the inverse of resistance) of SWCNT is predicted to be $2G_0$ independent of the diameter and length, where $G_0 = 2e^2/h = 1/12.9$ kΩ, which is one unit of the conductance quantum, and e and h are the charge on one electron and Planck's constant, respectively. This means that SWCNTs are predicted to have a minimum resistance of about 6500 ohms, independent of their length. Temperature and magnetic fields affect the resistance of the nanotubes. Metallic SWCNTs behave as long ballistic quantum conductors with the charge carriers exhibiting a large phase-coherence length. The semiconducting nanotubes are predicted to have the strongest electrochemical properties, but this is difficult to verify in macroscale actuators. The properties of the nanotubes in a structural polymer electrolytes (SPEs) define the electrical and ion exchange and actuation performance. The metallic or semiconducting properties of SWCNTs actually depend on their helicity or chirality, which is denoted by a pair of integers (n, m). In particular, they are predicted to be metallic if $n - m = 3q$, where q is an integer. While armchair CNTs are metallic, electrical properties of chiral and zigzag CNTs vary with the diameter. Therefore rational design of nanotube smart structures requires a fundamental understanding of the conductive properties of nanotube materials and how conductivity depends on temperature and dimensionality.[122–123] Based on recent observations of electron transport and both theoretical and experimental results, SWCNTs are ballistic in nature, implying the absence of inelastic scattering.[125] Experimental results from metallic SWCNTs exhibited resistance below 100 kΩ at room temperature, approaching the theoretical lower limit of 6.5 kΩ.[126]

Transport in MWCNTs is controversial regarding ballistic and diffusive conductivity.[172] In experiments, Bachtold et al. showed that resistance per unit length of MWCNTs was above 5 kΩ/μm, resulting in diffusive conductivity.[122] Frank et al. demonstrated that conductivity of MWCNTs was ballistic and transport was confined to the outer layer of the tube.[128] C. Gerger et al. reported that resistance of MWCNTs was at most 200 Ω/μm. This implies that MWCNT conductivity is ballistic with almost a 30-μm mean free path. The most interesting result is that only the surface of MWCNT transport current, and high current densities can be applied.[129–130] Since our interest is in macroscopic smart nanotube structures, the conductivity properties of SWCNTs and MWCNTs must be brought to the microscale or macroscale in the form of composites. Then the transport features of CNT composites can be considered for optimal design of nanotube smart materials.[132–133] In ropes or bundles of SWCNT or in mats consisting of randomly oriented ropes, the resistance increasing with temperature has been observed. M. Radosavljevic et al. found that the coupling between nanotubes in the bundle is weak, and the current is carried predominantly by nanotubes on the surface of the bundle at low fields.[134] At high bias voltage just before complete breakdown, they observed an additional current increase caused by the coupling to the nanotubes in the interior of the bundle. These results indicate that nanotubes that are bundled may have different transduction properties compared to individual nanotubes.

15.4.3 Magnetoresistance

The CNTs also have spin-dependent transport properties or magnetoresistance. The direction of magnetization of the ferromagnetic electrodes used to contact the nanotube defines the spin direction of the charge carriers into and out of the nanotube and a change in the resistivity of the nanotube. Spintronic nanoscale devices in theory can be built using the superconductivity and magnetoresistance effects, where the nanotube–metal junction appears to have a strong effect on the spin-dependent transport. When a magnetic field is applied perpendicularly to the nanotube at low temperature, the resistance shows a change of 15%. Also a strongly suppressed conductance gap at zero bias voltage on a substrate can open at low temperature (4.2 K). The magnetoresistance effect would be difficult to measure for sensing/actuation of strain of the nanotube and for use in a smart material.

15.4.4 Piezoresistance

Piezoresistance is the change in electrical resistance with strain and is a useful property for self-sensing of intelligent structures. Piezoresistance is discussed in References 143–147. A pioneering experiment showed that the conductance of a metallic CNT could decrease by orders of magnitude when strained by an atomic force microscope tip.[147] It was shown that the band structure of a CNT can be dramatically altered by mechanical strain and the conductance of the CNT can increase or decrease based on the chirality of the nanotube. This happens because the strain changes the structure of the quantum states available to the electrons. Metals conduct electricity easily because their electrons have easy access to the quantum states that carry the electrons long distances. These states are in the conduction band of the electronic structure. In semiconducting nanotubes, there is a band gap that serves as an energy barrier that electrons must overcome to reach the conduction band. The extra energy push to overcome the band gap can come from heat or an electric field or strain. Actually strain changes the band structure, which changes the electrical properties, making the nanotube or nanocomposite material more or less conductive (piezoresistive), depending on the chirality of the nanotube, the polymer, and the percent loading of the nanotubes.[80,82]

Raman spectroscopy is a very useful tool to characterize carbon materials because it gives information about the size of graphite crystallites and the degree of ordering of the material.[166] Of particular importance, Raman shifts can represent all types, shapes, and orientations of nanotubes. Therefore, Raman spectroscopy was used to study the vibrational modes of carbon nanotubes and nanoropes and to determine the characteristic properties of armchair, zigzag, and chiral tubes. It is also reported that a polymer embedded with low-weight fractions of carbon nanotubes becomes Raman-active under an applied mechanical strain. Small amounts of SWCNTs embedded in a polymer matrix were used to sense the mechanical response of the polymer using Raman spectral data. The polarized Raman methods were very useful tools for mapping the stress in a polymer plate under uniaxial tension. This could be one way to sense in a smart structure, but there is still a limitation since a large and expensive Raman spectroscope instrument is required.

The electronic and mechanical degrees of freedom are tightly coupled in the CNT system. Many researchers tried to prove how bond stretching and twisting in nanotubes affects the electrical properties of nanotubes, since the electromechanical properties of SWNTs depend on nanotube chirality (m, n). These properties have been exploited to sense mechanical deformations by a change in conductance or resistance of the nanotube due to the potential application for nanoscale smart structures such as NEMS devices. Several groups calculated conductance variations due to mechanical deformation theoretically. Even for small bending angles, the conductance changed, and the conductance of metallic SWCNTs was lowered up to tenfold at the ~45° bending angle.[147] Experimental investigation of nanotubes was carried out with suspended nanotubes that were manipulated with an AFM tip while the resistance or conductance was monitored.[147] Conductance of the SWCNT decreased each time the AFM tip pushed the SWCNT down but recovered as the tip retracted. These observed changes are entirely due to the mechanical elastic deformation of the SWCNT. This phenomenon can be explained by the changes in local bonding configuration from sp^2 to nearly sp^3. As the AFM tip pushes the SWCNT down, local π-electron density is decreased significantly. Drastic reduction in π-electron density causes a significant decrease in conductance, because local π-electron density, which is delocalized, is responsible for the electrical conduction. Reversibility in the process and the dramatic variation in conductance are good properties of reversible electromechanical transducers for smart structures. In particular, Z.L. Wang et al.[120] reported that the conductance is quantized and it is independent of the length and width of the carbon nanotube. This property could be used to build smart sensors for health monitoring systems. Jien Cao et al.[171] showed that small band-gap semiconducting nanotubes exhibited the largest resistance changes and have good piezoresistive gauge factors under axial strains. This idea can be used for nanotube-based smart structures.

15.4.5 Electrokinetics of Nanotubes

In fluids, ponderomotive responses of particles can be produced by externally applied time-dependent electrical fields. The electrical properties (conductivity and dielectric constant) of a nanotube are usually different from those of a fluid. Therefore when a nanotube is in an electrolyte, it will attract ions of opposite electrical polarity, forming an electrical double layer. If a uniform DC electric field is applied to nanotubes suspended in an electrolyte, the electrical double layer surrounding the nanotube is distorted, and electrical charges that define the nanotube's structure are induced to appear at the interfaces.[148–156] Lateral movement may occur if the nanotube has a net electrical charge. This movement is called electrophoresis, which is the attraction and movement of charged particles due to an electric field in a fluid. If an AC field is used, the electrophoresis effect is predicted to become small at frequencies above about 1 KHz.

The distortion of the electrical double layer and the creation of interfacial charges cause the electric dipole moment and this allows the nanotube to be moved in an electric field. These distortions of the electrical double layer have a finite relaxation time and this effect becomes small at a much higher frequency exceeding about 50 KHz. If the nanotube is polar, the dipole moments due to polarization have an effect up to MHz frequencies. The total lateral electric force acting on a nanotube of net charge Q in a nonuniform electric field vector E is $F = QE + (m \cdot \nabla) \cdot E$ where ∇ is the Del operator that defines the field gradient and m is the dipole moment vector. If the electric charge is zero or the frequency exceeds about 1 KHz, the Coulombic interaction effect is small and the dipole moment and field gradient will dominate the motion of the nanotube. The time-averaged force acting on the nanotube is given by the real part of the dipole moment and is called dielectrophoresis, which is the attraction and movement of uncharged particles due to polarization induced by nonuniform electric fields in a fluid. The magnitude of the dipole moment caused by induced electrical polarization depends on the size of the nanotube and the complex conductivities of the nanotube and the suspending fluid medium. The dielectrophoretic force itself depends on the size and shape of the nanotube. The magnitude, polarity, and time response of the dipole moment induced in a nanotube in an imposed electric field provide a means to understand how electric fields can be used to manipulate nanotubes using electrophoresis, dielectrophoresis, electrorotation, and traveling fields. Electrorotation is the rotation of a nonuniform electric field to induce a torque on a particle causing the particle to rotate. Traveling wave dielectrophoresis is the introduction of a traveling electric field of specified wavelength for the selective manipulation, trapping, and separation of particles using a frequency variation of the applied dipole moments. Based on these properties, dielectrophoresis may have use in separation of zigzag nanotubes from semiconducting CNTs to build actuators and sensors and for aligning or dispersing CNT in solvents or resin systems. However, the forces involved in the movement of CNTs using electrophoresis and dielectrophoresis are very small and these effects are not considered suitable for high-performance actuation applications as in smart structures.

15.4.6 Piezoelectric Properties

Piezo-, pyro-, and ferroelectric materials have been in use for two decades, and many applications of piezoelectrics for sensors and actuators in smart structures have been developed. Some of the highlights include electrostrictive materials for nonlinear actuators, ferroelectric single crystals with very high electromechanical couplings for medical transducers, thin and thick PZT films for MEMS, and multilayer-type actuators and sensors for smart structures and structural health monitoring.[13–29] Piezoelectric materials have electromechanical coupling and are used for the health monitoring of structures since they can be embedded or attached to flexible structures as sensors or actuators. Recently it has been predicted that CNTs and BNTs exert piezoelectric effect, and there is some possibility for using these to develop smart structures. (See Chapter 21.)

N.G. Lebedev et al.[115–121] modeled the piezoelectric effect in SWCNTs and BNTs based on quantum chemical calculations of nanotube polarization in accordance with the stretch of nanotubes along their axis. They showed that the piezoelectric values are very small for CNTs and increase in accordance with the increasing diameter for BNTs. The physics group of North Carolina State University recently presented polarization and piezoelectricity studies of BNTs. They reported the existence of nonzero spontaneous polarization fields since most BNT are noncentrosymmetric and polar. However contrary to their predictions, the combined Berry phase and Wanier function analysis demonstrated that electronic and ionic spontaneous polarization cancel exactly in systems. However after the intrinsic helical symmetry is broken, their results suggested that the piezoelectric constants of zigzag BNTs are larger than those in the PVDF polymer family. Piezoelectric properties of zigzag BNT bundles and piezoelectric materials are also listed in Chapter 21.

Oxide-based pyroelectric and ferroelectric materials have the general formula ABO_3 and are employed for their dielectric, piezoelectric, electrorestrictive, pyroelectric, and electro-optic properties and their functional perovskite structure. Chang et al. reported the synthesis and characterization of perovskite nanotubes, which are lead titanate ($PbTiO_3$ [PT]), lead zirconate ($PbZrO_3$ [PZ]), and zirconium tianate ($PbZrO_3$–$PbTiO_3$ [PZT]).[110] These nanotubes have the potential to be used as intelligent materials for smart structures, active catheters, strain sensors, force actuators, accelerometers, and other applications. Certain nanotubes such as boron nitride nanotubes, as discussed, and zinc oxide nanowires and nanobelts are naturally polar in bundles, and the phoretic effects and piezoelectric effects will be present. However, the piezoelectric effect of BNTs is smaller than for piezoelectric ceramic materials, and the piezoelectric effect of ZnO is not well known. Also the BNTs and ZnO are not readily available. For the CNTs, the van der Waals, piezoresistive, and electrochemical effects are present while the piezoelectric effect is very small. Therefore using piezoelectric nanotubes, wires, and ribbons is promising but currently is in an early stage of developing high-strain actuator materials, as compared to using electrochemical nanotubes.

15.4.7 Electrochemical Effects

Introducing excess charge into CNTs produces mechanical deformations that do mechanical work. The charge injected into the valence or conduction band causes the electronic structure to shift. The electrochemical effect should produce up to 2% strain[44–68] based on the basal plane intercalation strain of graphite. The electrochemical property can generate large strains and forces using low voltages. Therefore, the electrochemical and piezoresistive properties of CNTs are considered promising for actuation and sensing. The CNT electrochemical actuation mechanism is a volume change of the nanotube induced by ion exchange when the nanotube is immersed in an electrolyte. If a voltage source is connected to the CNT, anions and cations in an electrolyte attach to the surface of a cathode CNT or anode CNT, respectively, because of the electrical charging effect. Consequently the interface surface of the CNT is expanded to produce an elongation of the actuator. This effect has been under study using SWCNTs by Baughman et al. since 1999[45] and recently at the UC.[66–68] Research is also being done at UC to investigate the use of different electrode geometries and applied voltage schemes for the remote actuation of SWCNTs in electrolytes. Electrochemical actuation using CNFs is also being studied at UC. The CNFs cannot produce macroscopic sheets because the van der Waals forces are smaller than for SWCNTs. Even in the SWCNTs, the van der Waals forces do not provide efficient shear transfer. Therefore, research is focusing on developing a polymer host for nanotubes that can provide ion exchange for the electrochemical effect.

Little research has been done using MWCNTs for electrochemical actuation. A forest-like aligned MWCNT actuator in an aqueous electrolyte has been tested,[46] but the actuation mechanism is an electrostatic repulsion among nanotubes. This electrical charging repulsion actuation mechanism produces a low actuation force. A comparison of the electrochemical actuation effect of CNTs with other actuator types is given in Table 15.3. The theoretical predicted energy density of CNT actuators is two orders of magnitude greater than for piezoceramic materials. The analogy of the piezoresistance effect in an electrolyte is called electrochemical impedance, and this effect is being studied related to self-sensing nanotube actuators.[66]

15.4.8 Nanotube Power Generation

Power generation is discussed in the literature.[157–158] The flow of an ionic fluid over SWCNT paper in the direction of the flow generates a charge. The voltage produced fits a logarithmic velocity dependence, and the magnitude of the voltage also depends on the ionic conductivity and the polar nature of the liquid. The dominant mechanism thought to be responsible for this highly nonlinear response involves a direct forcing of the free charge carriers in the nanotubes by the fluctuating Coulombic field of the liquid flowing past the nanotubes. This sensor can be scaled down to length.dimensions on the order of micrometers, and the sensor has high sensitivity at low velocities and a response time better than 1 ms. Therefore nanotube paper could also be used to generate charge in a flowing liquid environment, which may be useful in biomedical applications.

15.4.9 Nanotube Contact Phenomena

Contact phenomena between nanotubes or a nanotube and an electrode must be considered, including conductivity phenomena of the nanotube when an optimal configuration to design smart nanotube structures is needed. Compared to the contact resistance of CNTs, the intrinsic resistance of CNTs is small. Therefore the contact resistance was not significantly affected by the CNT's own properties such as defects or diameter. Alper Buldum reported that conductance values with simple end-end contacts are high with a negative differential resistance.[35] Experiments show that end contact

TABLE 15.3
Approximate Properties of Piezoelectric Ceramic and Nanotube Materials

Material	Temperature Limit [°C]	Elastic Modulus Y [GPa]	Band-width [Hz]	Volt Range	Density [Kg/m³]	Pk Strain ε_m	Strain/Volt ε_m/volts	Strain Energy $Y\varepsilon_m^2/2$ [J/m³]	Energy Density $Y\varepsilon_m^2/2\rho$ [J/Kg]	Piezo-Induced Stress e_{33} [C/m²]
PZT monolithic wafer[1]	175	55–69	10^6	+/–200	7700	.00026	1.3×10^{-6}	2332	0.3	5–10
Active fiber composite[2]	~110	35	10^5	+2.8Kv	4250	.0018	6.4×10^{-6}	56,700	13.34	6.2
Piezoelectric (5,0) BNT[3]	~1000	1220	—	—	1370	—	—	—	—	0.4 max
Piezoelectric (5,5) CNT[4]	600	1000	—	—	1330	—	—	—	—	.0003 max
Electrochemical CNT[5]	—	5.5	10^2	+/–2	1330	.002	1×10^{-3}	11,000	8.25	—
Electrochemical CNT[6]	—	640	10^3	+/–1	1330	.01	1×10^{-2}	29×10^6	2400	—
Aluminum nitride (ALN)[7]	~1000	—	—	—	—	—	—	—	—	0.58
SiCNT nanotube[8]	—	—	—	—	—	—	—	—	—	—
Zinc oxide nanobelt[9]	—	~39–65	—	—	—	—	—	—	—	—
V_2O_5 nanotube[10]	—	—	—	—	—	—	—	—	—	—

[1] PZT 5A.
[2] Testing.
[3] Nardelli, M. (theory).
[4] Lebedev; N. (theory).
[5] Zhang, Q.M. and Baughman, R. (bundled test).
[6] Nonbundled theory.
[7] Testing.
[8] Properties to be determined.
[9] Wang, Z.L.
[10] Baughman, R. and Spahr, M.

resistance of armchair tubes is low due to a strong coupling between metal and carbon atoms compared to weak coupling of the side contact. Reference 35 cites a two-terminal nanotube junction formed by bringing two tubes' ends together in parallel. (l is the contact length.) The transmission coefficient T of the two-armchair tube [(10-10)–(10-10)] junction for l = 564 Å was reported. The current-voltage characteristics of the [(10-10)–(10-10)] junction for l = 546 Å and a model of a four-terminal junction formed by crossing two nanotubes were also reported along with contact resistance of the [(18,0)–(10,10)] junction as a function of rotation angle θ. It was determined that contact resistance changes with the diameter of the nanotube, the chirality of the nanotube, the Fermi wave vector of the metal, and the area of contact. The contact resistance between nanotubes and various metals is important and low ohmic contacts are needed to develop smart material actuator and sensor systems. Based on the various experimental and theoretical results, metals with few *d* vacancies in atomic structure such as Fe, Ni, and Co reveal more finite solubility for carbon than Al, Au, and Pd, which have no *d* vacancies. This phenomenon is due to curvature-induced rehybridization of carbon sp^2 orbitals with the Ni *d* orbital. Also Ti and Nb, which are 3*d* and 4*d* metals with many *d* vacancies, form strong bonds with CNTs.[40–42] Although the Fe, Ni, and Co are better conductors, they are corroded in an electrolyte. Palladium would not be corroded in an electrolyte. Silver-filled conductive epoxy was shown to lower the contact resistance when contacting copper with a carbon nanotube epoxy material in a dry environment, and this process is used for making piezoresistive sensors at UC.

15.5 REVIEW OF NANOSCALE SENSORS AND ACTUATORS

This section reviews the literature discussing the modeling and design of nanoscale sensors and actuators.

15.5.1 Simulation of Nanotube Structures

In order to simulate smart nanostructured materials, the electromechanical coupling effect must be determined for the constitutive modeling of intelligent materials. The *ab initio* method, tight-binding method, and classical molecular dynamics (MD) are major categories of nanotube molecular simulation. The *ab initio* method could provide the exact solutions of the Schordinger equation, but assumptions and approximations are needed, and advanced algorithms and the duration time for calculation are weak points for realistic modeling of nanotube smart materials. Also only limited solutions are available for a specific class of problem. But the *ab initio* method can give the most accurate results among all methods. Many researchers have studied the mechanical, electrical, and optical properties of nanotubes based on the *ab initio* method.[136–142] Nanotube mechanics is presented in Part 3 of this book and provides more detail. D. Sánchez-Portal et al. used the *ab initio* method for the study of structural, elastic, and vibrational properties of SWCNTs with different radii and chiralities. M. Verissimo-Alves et al. obtained electromechanical behavior based on *ab initio* calculations of charged graphene and single-wall carbon nanotubes. Properties of non-CNT materials also were calculated.

The tight binding method initially attributed to Slater and Koster in 1954 is a semiempirical method for electronic structure calculations. The key role is that the Hamiltonian is parameterized and simplified before the calculation, rather than constructed it from first principles, and the total energy and electronic eigenvalues are deduced from the Hamiltonian matrix. This is a very convenient method for molecular simulation because it can consider larger systems than the *ab initio* method. S. Reich et al. derived an analytic expression for the tight-binding dispersion including up to the third-nearest neighbors, while V. Meunier presented the theoretical modeling of scanning tunneling microscopy of nanotubes for computation of STM images and current-voltage characteristics. The structures and electronic states of capped carbon nanotubes were investigated using a tight-binding method by Yusuke Kasahara. Also, A.A. Maarouf et al. presented a tight-binding theory to analyze the motion of electrons between carbon nanotubes bundled into a carbon nanotube rope.[142] In the case of non-CNTs, properties of charged borocarbide (BC) nanotubes were calculated by the density functional tight binding method. MD is a particle motion analysis based on Newton's second law. This method can consider larger systems than previous methods. Nan Yao presented a detailed investigation of the spring behavior of carbon nanotube caps based on MD simulation, and Jie Han et al. developed an algorithm using MD simulation techniques to model CNT bend junction structures formed by topological defects, i.e., pentagon–heptagon pairs. These three methods could give accurate results, but intensive computations for the Schordinger equation and the particle dynamics of Newton's second law are needed to predict the motion of nanotube materials. Thus it is difficult to make quantum mechanics simulations.

On the other hand, simplified continuum and multiscale approaches seem more practical for nanostructured smart materials. The modeling method must have an ability to consider not only exceptional DC electrical transport but also AC behavior. AC characteristics of nanotube materials are very important features for modeling smart materials. P.M. Ajayan et al. reported that electrical transport in carbon nanotubes depend on frequency. The results explained an intrinsic resonance at a fixed ultrasonic frequency of 37.6 kHz in the AC impedance spectra of nanotubes that did not depend on the types or lengths of nanotubes. And until reaching an 8-MHz frequency, the overall impedance showed a negative capacitance due to metal-tube connections. P.J. Burke used Luttinger liquid theory for modeling the gigahertz electrical properties and designed an RF circuit model of carbon nanotubes.[134] In addition to the conductance of SWCNTs and MWCNTs, the conductance of bundles (ropes) and networks (mats) of ropes in varying directions have to be determined both for metallic and nonmetallic behavior. A.B. Kaiser et al. reported that the peak in the frequency-dependent conductivity of single-wall carbon nanotube networks is consistent with metallic conduction interrupted by nonmetallic defects that act as barriers.[135] They further explored this model of metallic conduction interrupted by barriers and two additional properties of SWCNT networks: frequency dependent conductivity and nonohmic conduction. D. Qian and Y. Liu are developing multiscale methods for the analysis of nanocomposite materials[136–138] that may be applied to nanoscale intelligent materials. Overall there has been little modeling of the electromechanical actuation properties of carbon nanotubes. The elastic properties and the electrode kinetics at the nanotube–SPE interface are critical and have not been modeled explicitly.

15.5.2 Nanotube Strain Sensors

CNTs have been used as better alternatives to the conventional silicon tips in atomic force microscopes because CNTs have flexible and mechanically robust properties, and also the same repeated properties in air, vacuum, and water. Another enlarging research field for CNT sensors is chemical force microscopy (CFM), which can investigate the biological structure, for example, of a DNA sequence. Applying electrostatic voltages can deform nanotubes. P. Poncharal reported that the charge is located at the tip of a nanotube, and deflection is proportional to V_s^2, where V_s^2 is the static potential. It was also found that the resonance frequency excited by alternating the applied potential changes if masses from the picogram to femtogram level are attached to a CNT. This property can be used for a nanoscale smart mass sensor.[173]

As sensor systems are becoming more complex with many sensor nodes, a wiring network is another issue to be resolved in using nanotube sensors in structural health monitoring systems. Keat Ghee Ong et al. first discussed the possibility of passive wireless communication of nanotube sensors.[175–176] Such sensors are based on inductor-capacitor resonant-circuits (LCs), which can be remotely monitored through loop antenna. This passive wireless sensor enables the long-time monitoring of sensors without batteries. Other new technologies such as wireless-LAN or Bluetooth have the potential to be used as active wireless sensors for structural health monitoring systems.

15.5.3 Actuators Based on Nanoscale Materials

Researchers in the field of smart structures have been trying to overcome the limitation of small strains or small forces produced by smart materials. Dynamic systems are becoming more complex, and higher actuation efficiency is needed. In order to resolve this situation, locomotion morphology and the physiology of animals[51] and nanotechnology are topics of study for scientists and engineers. With the help of bionanotechnology, biomimetic materials such as nanotubes for smart actuators have considerable potential in nanoelectromechanical systems (NEMS). These actuators provide greater work per cycle than previous actuator technologies and may have better mechanical strength. In addition, CNTs offer high thermal stability, and the actuators can be used in high-temperature environments.

15.5.3.1 Carbon Nanotube Electrochemical Actuators

The first actuator made of CNT was a macroscale sheet of nanotubes termed *buckypaper* by Ray Baughman et al.[57–59] This actuator used the change in dimension of the nanotube in the covalently bonded direction caused by an applied electric potential. The charge injection led to strain of the nanotube paper causing the assembly to bend. This excess charge was compensated at the nanotube–electrolyte interface by electrolyte ions forming the double layer. Soon after the buckypaper actuator, their team demonstrated actuation of a single SWCNT in which the SWCNT was suspended over a trench, and its edges fixed at the surface with a metal layer.[54] The device was placed into an electrolyte, and a potential was applied to the SWCNT. The change in SWCNT length was monitored by an AFM tip. These experiments demonstrate that it is possible to use SWCNTs as actuators for driving pumps or controlling nanofluidic valves.

Compared to the best known ferroelectric, electrostrictive, and magnetostrictive materials, the very low driving voltage for CNT actuation is a major advantage for various applications such as smart structures, multilink catheters, micropumps, flaps for a microflying object (MFO), molecular motors, or nanorobots. Another advantage is direct conversion of electrical energy to mechanical energy and high actuation strain, strength, and elastic modulus. Since Faradaic actuators basically come from the charge–discharge–charge method such as a battery, it is possible to design a self-powered actuator that can actuate motion based on stored charge in the CNT structural material without an external power source. However, since these CNT actuators work in electrolyte, the bandwidth of an actuator is low and applications thus far have been focused on artificial muscle.[74–81] On the other hand, Von Klitzing's group in Germany used a solid electrolyte such as a PVA/Nafion/ H_3PO_4 mixture for a harder actuator and a wide bandwidth.[61] NASA's Jet Propulsion Laboratory in California tried to develop bimorph actuators and force sensors based on carbon nanotubes. This idea also came from double-layer charge injection that causes bond expansion. These devices could generate and control displacements and forces on a molecular scale, and can be embedded in circuits as NEMS devices.

15.5.3.2 Thermally Activated Actuators

The Nanoscale Science Research Group of the University of North Carolina demonstrated that a double-layered cantilever on a MWCNT changed its curvature under a temperature change. This is called a thermally actuated mobile system (TAMS).[63] The bilayer cantilever TAMS is an actuator/sensor that has two layers with different thermal stress or intrinsic stress between the two materials. They reported that greatest bending occurred at a thickness ratio of 1:7 (thickness of aluminum/ thickness of MWNT = tAL/tMWNT), and the TAMS actuator was controlled by laser heating.

15.5.3.3 Piezoelectric and Nanotweezer Actuators

The piezoelectric effect and high strength of BNT have the potential to be used in developing high bandwidth structural actuators. Most of the papers investigating the piezoelectricity of nanotubes are based on quantum mechanics, and there is little experimentation. Controlling the chirality and producing aligned BNTs for practical applications is still a problem. Eventually this synthesis problem will be solved and BNTs may be widely used in smart structures.[183] Different nanotweezers were developed using optical trapping. The optical system is complex, large, and expensive, but if SWCNTs are used for nanotweezers, they will provide an advantage in the development of various sizes and forms of nanotweezers.[183–184] This CNT nanotweezer has the ability to grab and manipulate nanosize objects in three dimensions and may find applications related to nanomechanical structures.

15.5.3.4 Shape Memory Alloy and Platinum Nanoscale Actuators

Shape-memory alloys (SMA) are already widely used in smart structures and artificial muscles.[203–207] Notably nickel–titanium and copper-based alloys are available and used to develop devices that can perform a wide variety of functions: as actuators, components for smart structures, medical and orthodontic devices, safety valves, eye glass frames, etc. But the actuation is indirect and requires high power to convert electrical energy to thermal energy to cause actuation. The SMA actuators are being formed as thin films to increase the cooling rate. Recently Weissmuller et al. reported that an electrolyte-filled assembly of pressure-compacted platinum nanoparticles could be used as a smart actuator.[185–186] They observed actuation strain up to 0.15% for the platinum electrode, which is the same as for commercial ferroelectric ceramics, but the applied voltage is small, similar to that of carbon nanotube electro-chemical actuator cells. They also said that the strain depends linearly on the potential, with little or no hysteresis.

15.5.3.5 Biological Molecular Actuators

Biological molecules are other possible types of nanoscale actuator. Three types of cytoplasmic motors — myosins that move on actin filaments, dyneins, and kinesins that use microtubules as tracks — are amazing biological machines that can be used with nanotube materials as sensors and actuators. These examples of extreme nanoengineering are fueled by ATP and convert its chemical energy into mechanical energy. For example, a carbon nanotube can serve as a casing material for an active motor or functionalized nanotube to be used as a molecular shuttle on a track. These kinds of structures of nanotube sensors and actuators are gaining the spotlight as *in vivo* bionanotube structures.[187–188] Compared with protein structures, DNA is another new and up-and-coming smart actuator since it is small, simple, and its function is well understood. The NASA Institute for Advanced Concepts has already started research to develop a DNA actuator with SWCNTs. Several viral protein linear (VPL) actuators were placed in parallel and series to multiply force and displacement, respectively, and a SWCNT-outfitted VPL actuator was shown[190] to open and close depending on the pH level of the environment. Basically this change in viral protein is required for the process of membrane fusion; i.e., the fusion of viral and cellular membranes is essential for infection of the cell.

Other biological molecules being developed for actuation uses inlude proteins. Changing the pH level causes the protein molecules to unfold, producing an actuation force. The force may be used for morphing structures and to form active membranes.

15.6 MANUFACTURING OF CARBON NANOTUBE AND NANOFIBER INTELLIGENT MATERIALS

Manufacturing of carbon nanotube and nanofiber intelligent materials involves the following steps: (1) nanotube synthesis and purification, (2) functionalization of the nanotubes, (3) dispersion and casting of the nanotubes or nanofibers in a structural polymer electrolyte (SPE), which includes forming the structural material, electroding, and encapsulating the material, and (4) testing and characterization of the material performance. These steps are discussed in the following sections.

15.6.1 Synthesis of Nanotubes

The SWCNT, MWCNT, and CNF materials can be used separately and in combination to make force actuators, strain sensors, and power generators. The actuation capability and cost can be tailored for a particular application. The synthesis step is important to improve the size and actuation properties of nanotube materials. The complete processing of smart materials starting from nanotube synthesis to final device fabrication is performed at UC. Nanotube synthesis is accomplished using a commercial EasyTube™ nanofurnace, as discussed in Chapter 5.[31] Objectives of nanotube synthesis experiments include growing long CNTs on or between substrates to improve the load transfer from the nanotube to the matrix and the transduction properties of CNT hybrid materials. During synthesis, MWCNTs have sufficient rigidity to overcome the gravity, buoyancy, and van der Waals forces to keep growing in the same direction up to a point, and then they bend. The MWCNT arrays grow at about 5 μm/min and millimeter lengths may be possible. The length of the nanotube is also controlled by the covering or depletion of the catalyst.

An array actuator using MWCNTs grown on substrates is being developed to overcome the problems of alignment, ion exchange, and load transfer for electrochemical actuation. MWCNT arrays that are 250 μ square by 130 μ tall blocks were grown on a silicon substrate by FirstNano, Inc. using the EasyTube nanofurnace.[31] Also, arrays of 2.5 mm long SWCNTs were grown Japan's Agency of Industrial Science and Technology[208] using the nanofurnace with the bubbler attachment. Water vapor was used to react with amorphous carbon to prevent the catalyst from being covered and becoming inactive, e.g., $H_2O + C_{am} \rightarrow CO + H_2$. This is a large advance in the development of smart materials because longer nanotubes will allow easier processing and provide greater actuation and material strength.

A key component of synthesis research is to couple the synthesis of the nanotubes with detailed atomic scale characterization of the structure, composition, and morphology of the tubes and to increase the lengths of the nanotubes. The structure, diameter, uniformity, and chirality of the MWCNT and the structure of crystallites formed by self-organization of the MWCNT can be determined by environmental scanning electron microscopy (ESEM), high-resolution transmission electron microscopy (HRTEM), atomic force microscopy (AFM), micro-Raman, and electron and diffraction techniques. ESEM is very useful for characterizing nanotube growth and nanocomposite failure surfaces without a need for much preparation of the material. HRTEM is useful to study the wall structure of

SWCNTs and MWCNTs. Raman is used to determine the nanotube diameter. The diameter is related to the (m, n) indices, and the chirality of SWCNTs can be determined by knowing the diameter based on the Raman results.

15.6.2 Functionalization of Nanotubes

Purification and functionalization are essential in defining active surface area, charge transfer rate, and adsorption/desorption at the CNT surface. Creating a strong bond (interface) between the nanotubes themselves and between nanotubes and the matrix materials in the case of composite materials is crucial for design of sensors and actuators. UC has developed a plasma polymerization process to coat or functionalize CNTs with different materials.[159–164] In a plasma gas environment, the surface atoms of the CNT become activated. At these active centers, a polymer film will form on the CNT surface. The polymerization conditions can be controlled so that the structure of the coating film has the desired number of functional groups. The plasma coating process is well described in Chapter 8.

Functionalization research related to smart materials is focusing on increasing the electrochemical ion exchange at the nanotube surface when the nanotubes are cast into a solid polymer electrolyte (SPE). The individual nanotubes are coated with an ultrathin coating of acrylic acid or polypyrrole using plasma polymerization before dispersing and mixing in the polymer electrolyte using simultaneous sonication and shear mixing. The different functionalization methods can be tailored to change the mechanical, electrical, and magnetic properties of the CNT due to a combination of dimensional changes and interface properties. In combination with plasma coating, surfactants used with ultrasonication appear to be a good approach to disperse nanotubes. Using 1% by weight of random SWCNTs or nano-fibers and the UC functionalization and dispersion methods, 30 to 40 wt% increases in elastic modulus of epoxy have been achieved. Up to 15% by weight of nanotubes was incorporated in epoxy, but the dispersion techniques for this large loading are still being developed. Nafion has also been used to coat CNTs to make electrochemical actuators. The experience in reinforcing composites is useful for developing CNT actuators. The challenge is to improve the speed of the ion exchange and increase the strain transfer of the nanotube actuator in the SPE. Sometimes the nanotubes may be filled with the SPE. A nanotube filled with the SPE is called a composite nanotube, because the nanotube is the host material, while integrating nanotubes in the SPE produces a nanotube composite because the SPE matrix is the host material.

15.6.3 Casting Nanotubes and Nanofibers in Structural Polymer Electrolyte

There are particular challenges that must be met before nanotube properties can be fully exploited. The main problem is to attain good dispersion and interface bonding between the CNT and the polymer matrix and achieve good load transfer from the matrix to the CNT during loading, since colloidal materials such as CNTs do not spontaneously suspend in polymers.[192–199] Dispersion may be the most critical part of integrating any of the types of nanotubes into polymers or other host materials. Sonication is often used to disperse nanotubes in solvents or resins. Inadequate sonication power level will not disperse the nanotubes satisfactorily. On the other hand, an excessive power level will break the nanotubes apart and shorten the nanotubes. During sonication, resonant bubbles are formed and collapse, generating very high local pressures and temperatures. Different sizes of nanotubes require different levels of sonication to achieve dispersion without shortening the nanotubes. To resolve these issues, many groups are trying to disperse CNTs homogeneously and uniformly in matrix materials. Solvents including amides, NN-dimethylformamide (DMF), and N-methylpyrrolidone (NMP) were reported to provide solubility of CNTs for different applications. Using appropriate surfactants, CNTs can also be solubilized in water.[90] Also, SWCNTs have been solubilized by functionalizing the end caps with long aliphatic amines or the side walls with fluorine and alkanes. By using an ultrasound mixer or using strong acids, good CNT dispersion can be achieved[68] for lower percentages of nanotubes from about 1 to 3 wt% of SWCNT. Dispersing larger percentages of nanotubes in polymers remains a problem. Recently NaDDBS has been reported to be an excellent surfactant for dispersing nanotubes in water.[90] A filtering and drying procedure is needed to prepare the surfactant-dispersed nanotubes for incorporation into epoxy and PMMA polymers. High shear mixing is also critical to separate nanotubes.

For the smart structures applications requiring the high conductivity of CNTs, coating of metals and organic conductive polymers (CPs) on the surface of CNTs to produce CNT-reinforced composites is another method.[192–199] CNTs coated with polypyrrole, Ni, Co, Ti, W, Pd, Au, Al, and Fe to form composites are not uniform and homogeneous, resulting in tangling and random distributions.[125] CPs such as polypyrrole, polyaniline, and polythiothene can be used as the conducting matrices, but these are mechanically weak and have to be oxidized and doped by a counter anion to achieve significant conductivity.[126,128,129] With large enough percentages of CNT in the insulating matrix, the nanotube-reinforced composite may be an ionically conductive structure that has satisfactory mechanical strength. Careful consideration to choosing the appropriate solid polymer matrix and coating materials is a key issue in synthesizing smart materials using nanotubes. Loading polymers with nanotubes can be done using SWCNTs, MWCNTs, or CNFs. SWCNTs that have perfect atomic structures are electrically the best fillers, but the cost is high and the chiralities cannot be easily separated. MWCNTs do not have good shear stress transfer between the individual shells, and the number of walls usually increases the number of defects. It appears that the inner layers provide mechanical support, unless electrical contact is made with the inner layers. CNFs have lower electrochemical properties than MWCNTs. At UC, the ultrasonication and shear mixing steps are combined in the procedure for the dispersion of nanotubes in epoxy. This shortens the time of sample preparation. The combined setup is shown in Fig. 15.2. A procedure for processing smart nanocomposite materials now in development includes (1) nanotube synthesis, (2) plasma polymerization or acid oxidization functionalization, (3) dispersion in a solvent, annealing, and then dispersing in the resin, (4) casting using

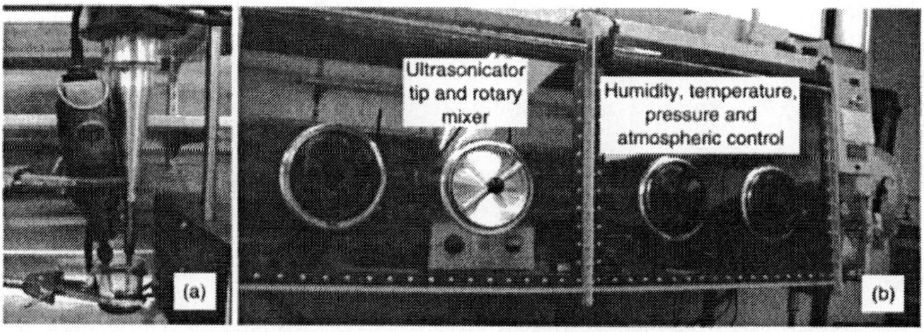

FIGURE 15.2 Simultaneous ultrasonication and shear mixing (a) of a nanotube polymer in a glove box (b).

high-pressure consolidation or vacuum degassing, (5) curing, and (6) cutting, electroding, and encapsulating the smart material.

Precise control and optimization of all processing steps are extremely important to fabricate nanostructured smart materials. Therefore we are developing a vacuum processing system for the fabrication of CNT smart materials. The vacuum processing system shown in Fig. 15.2 allows sonication and mechanical mixing of the nanotubes and SPE at elevated temperature in a partial vacuum or inert gas. High-pressure consolidation of the liquid nanocomposite material to fill the larger open nanotubes with the resin is also being done. This fabrication procedure along with using longer nanotubes is expected to significantly improve the performance of nanocomposite materials. The actuation and strength properties of the nanocomposites produced are determined using electrochemical impedance spectroscopy, load and displacement testing, mechanical strength testing, and microscopy. The following section discusses recent research and presents ideas to develop nanotube sensors and actuators for smart structures and nanotube-based health monitoring systems.[45,66–68,200–202]

15.6.4 Carbon Nanotube Composite Strain Sensors

Carbon nanotube piezoresistive sensors are being considered for self-monitoring of composite structures. The CNT and CNT-epoxy have a number of different sensing properties. There has been investigation of these properties and what properties are optimal for different applications. Figure 15.3(a) shows a ~2-mm thick CNT-epoxy sensor with 5% randomly oriented CNT. The piezoresistive effect (resistance change with pressure) is also shown. The contact resistance[35–43] of the CNT-epoxy/metal contacts causes the large slope in Fig. 15.3(a). Silver-filled epoxy can eliminate most of the high-contact resistance. Resistance that decreases with compressive loading may also be due to pushing the CNTs closer together improving hopping/tunneling and direct contact in the epoxy matrix. The electrostrictive effect (quadratic charge due to strain) of the matrix material is shown in

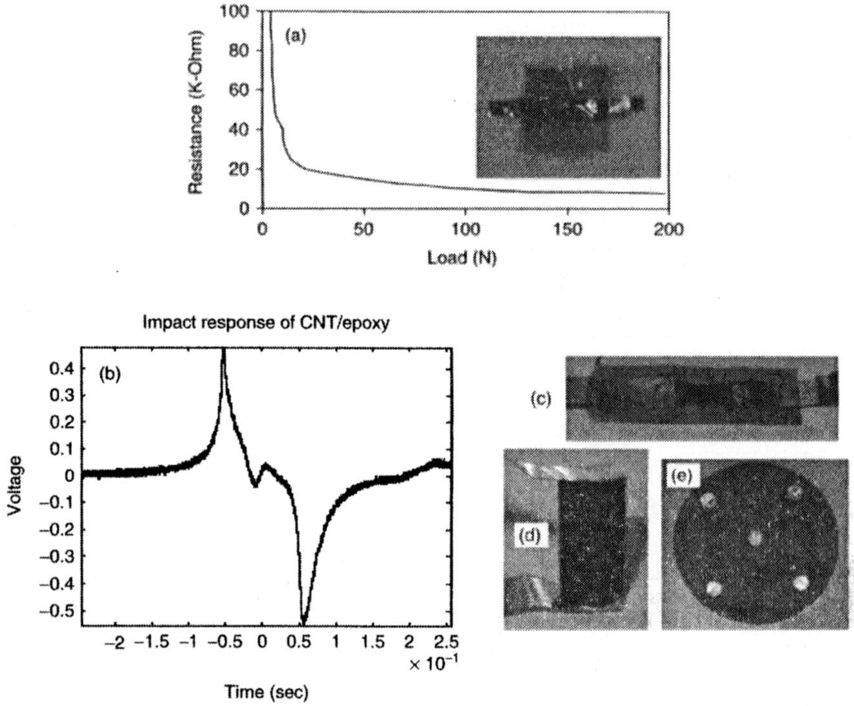

FIGURE 15.3 SWCNT sensors built at University of Cincinnati: (a) 5% CNT load sensor and resistance versus compressive load, (b) voltage (0.5 V peak) versus time due to impact on an electrostrictive polymer, (c) 100% SWCNT film sensor, $R = 12$ ohms, (d) pressure-cast CNT–epoxy barrel sensor, $R = 400$ ohms, (e) CNT–epoxy sensor plate with electrodes on top.

Fig. 15.3(b). A 0.5-volt peak signal is produced due to an impact; Fig. 15.3(c) shows a 100% CNT film sensor, $R = 12$ ohms; Fig. 15.3(d) shows a pressure-cast CNT-epoxy barrel sensor, $R = 400$ ohms; Fig. 15.3(e) shows a CNT-epoxy sensor plate with electrodes on top and bottom. These are early examples of self-sensing structural materials.

Since macroscale strain sensors are needed, the nanotube resistance properties must be brought to the microscale. Different nanotube systems including individual SWCNTs, MWCNTs, CNFs and ropes and mats can be used for making piezoresistive sensors. It may also be possible to build nanotube switches sensitive to very small pressure changes. A carbon nanotube neuron crack sensor for SHM is shown in Fig. 15.4. The crack sensor can be a highly reliable safe-life sensor that will ensure that no crack length larger than the spacing between two parallel carbon nanotube sensor nerves exists. Modeling and different polymers are being used to improve the sensitivity. NASA Langley researchers[200] are also developing a carbon nanotube strain sensor. Rice University researchers[177] have developed a buckypaper piezoresistive sensor.

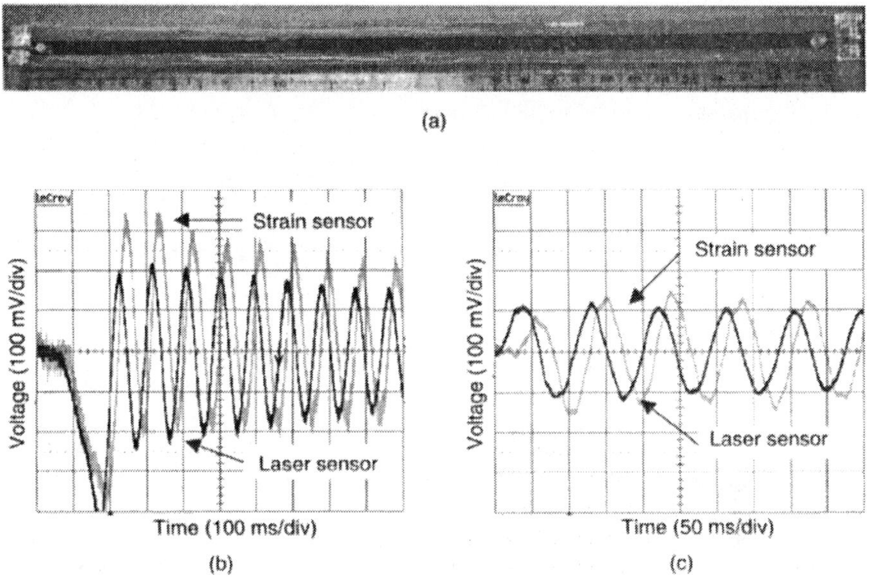

FIGURE 15.4 A CNT neuron continuous strain sensor and its dynamic strain response after low-pass filtering (cutoff freq 30 Hz, 60dB gain): (a) MWNT/PMM 10 wt% neuron (300 mm × 5 mm, 10 Kø resistance) on a glass fiber plate; (b) response due to an initial tip end displacement of 1 cm; and (c) response due to an impact at the tip of the beam. A laser sensor is used to measure the displacement of the beam for reference. (a) Composite Actuator (b) Linear Actuator (c) Tripod Actuator.

15.6.5 Carbon Nanotube-Based Biosensors

Biomedical applications of nanoscale intelligent materials have an especially large potential. Three applications can be categorized: (1) biomedical diagnostic techniques, (2) drugs, and (3) prostheses and implants. Biomedical applications for use outside the body, including diagnostic sensors such as DNA, RNA chips, lab-on-a-chip, and micro-total analysis systems (μTAS), are suitable for analyzing biomolecular samples. For internal use, several researchers are developing anticancer drugs, bio-MEMS devices, and gene therapy. Other researchers are working on prostheses and implants that include nanostructured materials.

Current biological sensors commonly rely on optical detection principles that are inherently complex, requiring multiple steps, multiple reagents, preparative steps, signal amplification, and relatively large sample sizes. Therefore, these techniques are highly sensitive and specific, but difficult to miniaturize. In order to overcome these limitations, a carbon nanotube array-based sensor is considered using the electrochemical properties of nanotubes for biosensors. Based on this idea, different possible uses of nanotubes for sensors were studied. Types include strain sensors, electrochemical biosensors, gas chemical sensors, temperature, and light sensors. A limiting factor for developing artificial organs such as eyes or ears is the interface between the biological component and an artificial silicon chip. A remarkable solution for this problem is a CNT array that can directly contact a cell *in vivo*, since CNTs provide highly biocompatible surfaces. One example is a vision chip consisting of an array

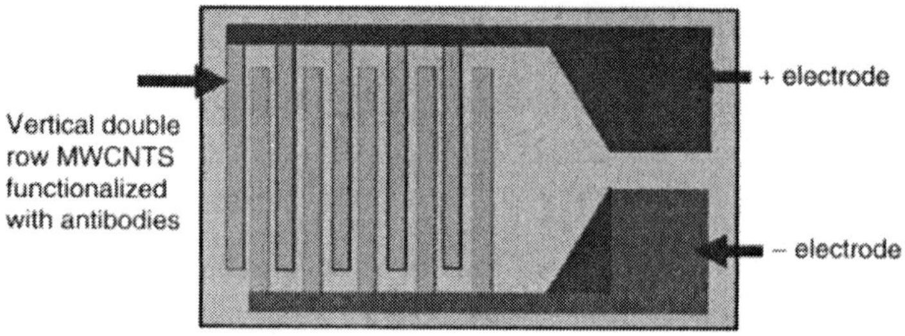

FIGURE 15.5 Concept of a nanotube interdigitated array biosensor.

of electrically conductive CNT towers grown directly on the surface of a silicon chip. A group has already tested the implant materials with retinal ganglion cells *in vitro* and proved the excellent biocompatibility.[201]

The nanotube-based electrochemical sensor can reduce the drawbacks of other electrode materials that include short life, low conductivity, and low reliability. For biosensing applications, carbon nanotubes have advantages of small size, large surface area, high sensitivity, fast response, and good reversibility at room temperature. At the same time, carbon nanotube biosensors can be integrated with microelectronics and microfluids systems to gain advantages in miniaturization, multiplexing, and automation. Carbon nanotubes functionalized with special chemical groups can form electrochemical biosensors for enzyme immobilization and stabilizers and mediators used in cancer research. UC is focusing on the development of a carbon nanotube array-based cancer sensor. The nanotube array becomes a miniature bioelectronic sensor with a large number of functionalized antibody sites for high sensitivity. The design of the nanotube array determines the sensitivity of the sensor. The research involves synthesizing MWCNTs in an interdigitated pattern, as shown in Fig. 15.5. Based on this array pattern design, the fabrication and processing can be done using semiconductor processing techniques. The complex CNT array pattern is expected to improve the detectability and fast response of the sensor and provide reliable data with minimal false positive and false negative readings. This array could also be put on a chip in a portable device to create a lab-on-a-chip system with a microflow channel.

The semiconductor processing steps that form the array are shown in Fig. 15.6. These steps are preliminary and are under early development and can be used to grow large planar arrays of MWCNTs that can be sectioned to provide many sensor elements. The critical step to develop the CNT array-based sensor is the fabrication process. Figure 15.6 illustrates the proposed detailed sequential steps in the fabrication process of the sensor segment. The starting substrate in Step 1 is either a highly resistive Si single crystal wafer or sapphire that provides an insulating base, which is required for the wiring of the segment. Step 2 involves deposition of highly conductive (doped) poly Si for planting the CNT catalyst on it. Doped Si provides good electrical contact between the CNT and the substrate. The CVD technique is used for depositing poly Si doped either with P or As. As an alternative solution, one can use an insulating Si wafer that is ion-implanted on the CNT growth side for increasing the electrical conductivity. Steps 3 to 7 involve patterning and require sputtering of a thin metal film

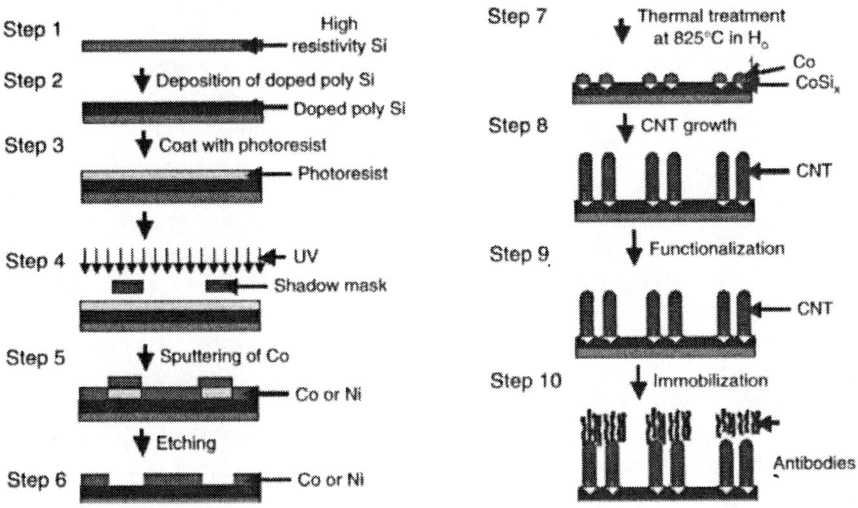

FIGURE 15.6 Fabrication steps for the double-row interdigitated nanotube array sensor shown in Fig. 15.5.

(20–40 nm) of catalyst such as Co and thermal annealing of the film in hydrogen at 825°C. During the annealing, catalytic semispheres of Co islands are formed, driven by surface tension to lower the total energy. Simultaneously a chemical reaction between the metal and the Si substrate takes place, producing highly conductive $CoSi_x$. This compound is expected to provide a strong mechanical bond between the CNT and the substrate and also serve as a low resistivity electrical junction. The patterned substrate with catalyst provides aligned growth of MWCNTs during Step 8. Functionalization of the array by solution methods or plasma coating is shown in Step 9. The patterning and functionalization will be optimized to match the specific antibodies to be immobilized and the growth factors and cytokine to be sensed. Medical doctors at UC are involved in the design. Immobilization of the antibodies is shown in Step 10. In this design, the electrical signal can be provided to each individual column of aligned carbon nanotubes or in almost any arrangement desired to tune the sensitivity.

The nanotube sensor array can be patterned to have a desired connectivity of the individual pairs of nanotube sensors shown in Step 10 of Fig. 15.6. The highest sensitivity of the array can be examined by modeling the impedance of the whole array when the antibody-growth factor coupling occurs. The nanotube spacing in the array and the nanotube diameter, length, and functionalization must be optimized to allow the greatest sensitivity while rejecting small changes in sensitivity that could lead to false positive and false negative indications. The functionalization improves the electrochemical response of the array to specific proteins, enzymes, and growth factors on the surface of CNTs. At UC, various chemical groups such as –COOH and $-NH_2$ are coated on the nanotubes to optimize the electrochemical response from the CNT array-based electrode acting as a host for different biospecies. The functionalization can be done in two ways for comparison. One is a reagent bath in which acids are used to attach specific function groups to the nanotubes. This is the current conventional method that is widely used to functionalize nanotubes. Plasma surface modification is the other very promising method for nanotube functionalization. In the plasma technique, for example, water and oxygen may be used to functionalize the surface of the MWCNT array instead of using a monomer such as styrene or pyrrole that is used for polymer coating. In the plasma phase, the carbon atoms on the surface are active and react with water or oxygen molecules to form –COOH and –OH groups. If $-NH_2$ is needed, ammonium hydroxide and NH_3 gas can be used as the monomer as well.

The carboxyl function groups functionalized on the nanotube surface will react with biospecies such as enzymes, antibodies, and oligonucleotides and form strong covalent bonds to attach biospecies to the surface of the nanotube. Depending on the different biospecies attached on the surface, the aligned CNT can be used to detect DNA, hydrogen peroxide, glucose, and other biomolecules. As an example, a carboxyl function group was attached to a MWCNT by surface modification. An amino function group on the oligonucleotide probe can react with the carboxyl function group to form covalent bonds and attach the probe on the surface of the nanotube. When this kind of biosensor is put into the test solution, the species in solution can attach on the probe. When immobilization occurs, a different electrochemical response occurs; the change in the impedance spectra or current is used to detect the presence and the amount of the complementary sequence. Figure 15.7 shows the experimental setup for the electrochemical bionanosensor for on-site analysis. The electrochemical system is based on amperometric or electrochemical impedance measurements.

Once the nanotubes are functionalized using the plasma system, immobilization of antibodies on the nanotubes must be done. Antibody immobilization formats can be broadly categorized into two classes: direct labeling experiments and dual antibody sandwich assays. These methods are possible for use with the nanotubes. However, based on our experience functionalizing carbon nanotubes, the carboxyl function group is very active and it can react with the $-NH_2$ function group via carbodiimide activation. A possible approach is to use the plasma coating method to functionalize the nanotubes based on our existing technique, use solution methods to immobilize the antibodies, and use techniques such as FTIR, and TOFSIMS to characterize the immobilization.

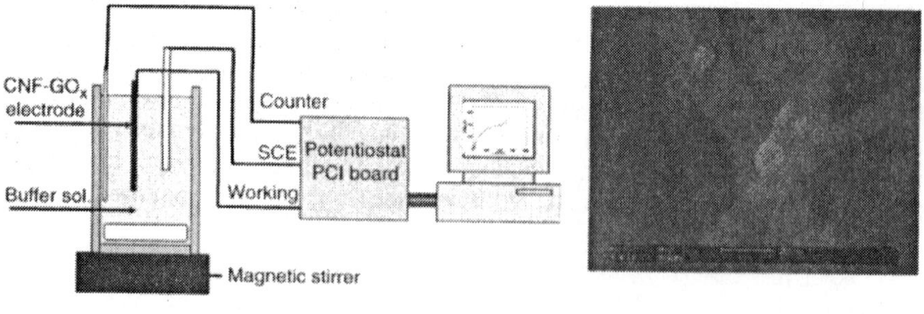

FIGURE 15.7 Electrochemical analysis setup using (a) a Gamry Potentiostat, and (b) functionalized CNF (ESEM image by Srinivas Subramaniam, University of Cincinnati Materials Characterization Center.)

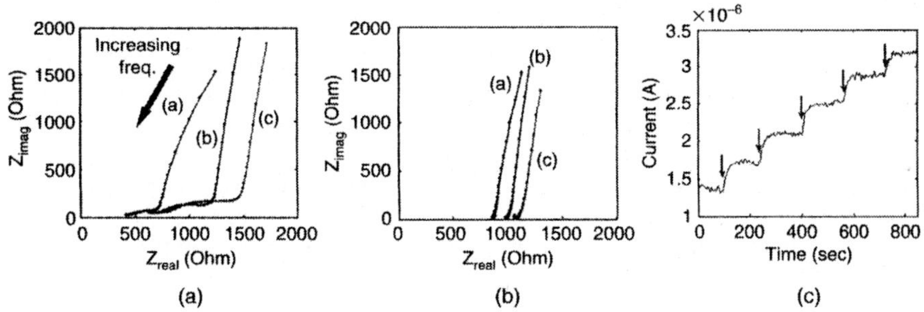

FIGURE 15.8 Biosensor characterization using glucose. (a) Electrochemical impedance spectra for the biosensor at DC potentials 0.2 V (a), 0.4 V (b), and 0.6 V (c) before immobilization of GO_x; (b) after immobilization of GO_x on the CNF; (c) amperometric response at an applied potential of 5 V versus the SCE reference electrode from the CNF-based sensor due to the successive addition of 1 mM glucose in a 0.05 M phosphate buffer solution (pH 7).

The amperometric method, especially cyclic voltammetry, is used because it can measure the concentrate-dependent current through the CNT array-based electrode coated with a specific chemical group. Each human has on the order of 50,000 genes that may play a critical health-related rôle. Biosensors that can perform genetic analysis, therefore, have tremendous potential. At UC, initial experiments to develop a nanotube biosensor were performed.

Figure 15.8 shows electrochemical impedance spectroscopy (EIS) data (Nyquist plot) with imaginary (capacitive) versus real (resistive) components of the complex impedance. Three different potentials were tested: 0.2 V, 0.4 V, and 0.6 V. Figure 15.8(a) shows the EIS results before the immobilization of GO_x on the surface of the CNF film and Figure 15.8(b) shows the EIS results after GO_x was immobilized on the surface of the CNF electrode. With the immobilization of the GO_x on the nanotubes, the impedance of the sensor changed as shown by comparison of the responses (a) to (c) in Fig. 15.8(a) and Fig. 15.8(b). The change in impedance is due to the change in the electrical resistance and capacitance of the electrode due to the immobilization of the GO_x. This result indicates the potential to use the bioelectronic sensor for label-free sensing. Figure 15.8(c) shows the amperometric response of the biosensor in which 1 mM glucose is added at each step at a potential of 0.5 V in PBS (pH 7). Even though there is slight noise in the signal caused by the 1-μA resolution level of the Gamry system, the current response increased immediately after each addition of 1 mM of glucose in the PBS solution. This change in response was repeated five times and retained the same value for 200 s. This high sensitivity and stability are advantages of CNF-Go_x biosensors for commercial applications. The structure of the CNF provides a large surface area to immobilize the GO_x and maintain this high activity. At the same time, CNFs could promote the electron transfer in the bioelectrochemical reaction.

A biosensor architecture based on carbon nanotubes may detect cancer and other pathogens at the earliest stages using electrochemical sensing of molecular signatures that can identify diseases. The nanotubes are good biosensors because their small size can allow detection of small numbers of biomolecules and the nanotube resistance and capacitance are sensitive to the attachment of biomolecules on the walls of the nanotube. The smaller size and higher sensitivity of the sensor may provide early detection and better outcomes for many diseases including cancer. The approach using interdigitated array architecture and a plasma coating method is expected to improve sensitivity and speed of detection as compared to existing types of sensors, and as compared to using a multiplexing approach with statistical analysis of separated individual arrays.

15.6.6 Carbon Nanotube and Nanofiber Hybrid Actuators

In order to develop actuators for practical structural applications, there are many challenges. Therefore at UC we have been developing a polymer host for nanotubes that can provide ion exchange for the electrochemical effect.[192–199] Little or no previous research has been done using CNF for electrochemical actuation. Even though the electrochemical properties of CNFs are fewer, we are focusing on CNFs for new actuators because the cost is low at $0.25/gm (~$100/lb), and incorporation of the nanofibers into polymers is easier because the fibers are large (~70–150 nm). The CNF material is also readily available in commercial quantities and large amounts can be made. The Bionanotechnology Lab at UC is developing for the first time a carbon nanofiber (CNF)–polymethylmethacrylate (PMMA) composite material that has electrochemical actuation properties. A combination of solvent casting and melt mixing was used to disperse CNF in PMMA and cast thin films of the material. Solution casting of the actuator material is shown in Fig. 15.9(a) for use as a wet actuator. Since PMMA–CNF composites have electrical resistance, sputtering of metal particles on the surface of PMMA/CNF composite is one approach being investigated to develop the activator and provide a uniform voltage field.

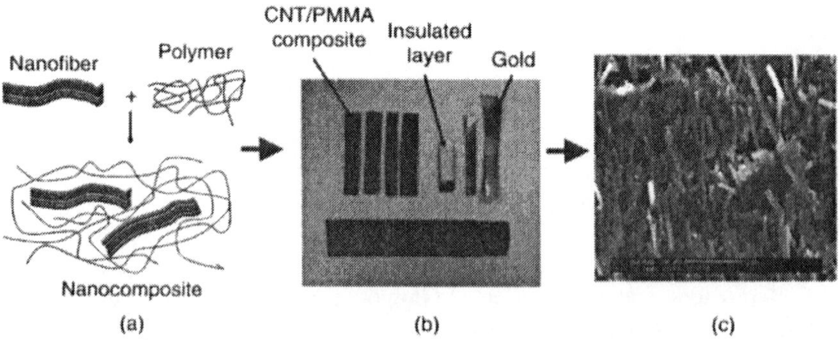

FIGURE 15.9 Manufacturing steps for the CNF–PMMA hybrid actuator. (a) CNF in a polymer matrix; (b) actuator strips as produced and with sputtered gold electrodes; (c) SEM image of fracture surface with fiber pull-out. (SEM image by Robert Gilliland).

The electrodes of the actuator should be insulated from the electrolyte because we have found that only the top part of the actuator provides most of the actuation. This is due to shunting through the electrolyte and a nonuniform voltage along the actuator. Tailoring of the voltage along the actuator and encapsulation of a partially hydrated actuator are approaches that are under investigation to overcome this problem and to further improve the actuation performance. The CNF–PMMA material produced using the solution casting procedure is shown in Fig. 15.9(b). The actuators shown are from 1 to 6 in. in length. Larger actuators and components can be made because of the low cost of the CNFs and chemicals. Fig. 15.9(c) shows the failure surface. The nanofibers are shown pulling out of the matrix material. The interfacial bonding of the CNF to the polymer is currently weak and will be improved using the plasma coating and dispersion methods.

Solution casting of the CNF–PMMA actuator for use in wet electrolytes is shown in Fig. 15.10(a). Figure 15.10(b) shows the experimental setup for the CNF–PMMA actuator. These aqueous-based actuators are immersed in a 2 M NaCl solution. Square

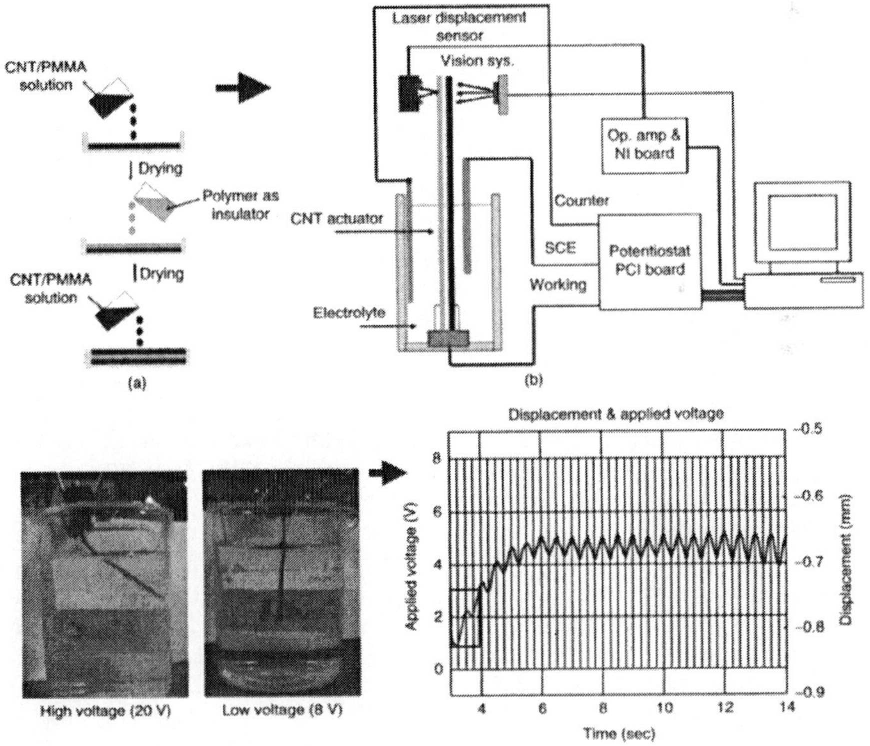

FIGURE 15.10 Solution casting and testing of the wet CNF–PMMA hybrid actuator: (a) solution casting; (b) typical experimental setup; (c) actuator response; (d) measured response near fixed end (out of water) of actuator.

wave potentials are applied between the platinum counter electrode and the working electrode. The displacement of CNF–PMMA actuator is shown in Fig. 15.10(c) and is measured by a laser displacement sensor (Keyence, LC-2400 Series) and monitored using a vision system (Intel Pro PC Camera). Note the very large actuation at 20 volts. This is due to surface actuation because of poor dispersion of the CNF. Also notice that the bending occurs mostly near the top of the actuator. This is because of shunting in the electrolyte and because the resistance of the actuator reduces the voltage in the lower part of the actuator. Improved dispersion, processing, an insulated electrode, and array methods will greatly improve the stiffness and bandwidth of the actuator. This is a very exciting material because of the low cost and good actuation.

A square wave with amplitude of +8 V was applied at frequencies ranging from 0.2 to 10 Hz. The response was measured near the fixed end of the actuator out of the water at 0.5 Hz and is shown in Fig. 15.10(d). The characterization of the actuators is done using a specially built test system for electrochemical actuators. This PC-based system can be built using readily available components for a total cost of about $30,000. The displacement of the actuators is measured using a laser displacement sensor (Keyence, LC-2400 Series). Square wave potentials are applied between two electrodes using a National Instruments PCI board and an operational amplifier designed and built to drive electrochemical actuators. The square wave excitation is an effective way to drive the amplifier and actuator with various amplitudes and frequencies using the PCI D/A board. In order to supply enough power to drive the actuator, a voltage follower and noninverting amplification were designed using the OPA2544 (TI) with a gain of 20:1.

Solution casting of a dry actuator material is similar to that for the wet actuator, Fig. 15.11(a). However, in the dry actuator procedure, the SPE contains the ion exchange material. A dry actuator was constructed according to Fig. 15.11(a) using individual CNF cast in a PMMA +$LiBF_4$ SPE. A loading of ~10% CNF is used in the actuator in this example. The actuator is formed using two CNF actuators, one on each side of tape. One actuator is used as ground and the other is switched between +2 and −2 volts. The test setup is shown in Fig. 15.11(b), and the actuation is shown in Fig. 15.11(c). Because the actuator is not hydrated, the actuation is slow and the amplitude is small compared to the wet actuator. Some creep also occurs, partly because the actuator is very flexible and the two actuators are not exactly the same size. Individually coating the CNF, uniformly dispersing the CNF, pressure casting, and aligning the CNF will improve the actuation. Allowing the material to absorb a small amount of moisture would also improve actuation. There is an unusual characteristic of this actuator. Note in the time response of the actuator shown in Fig. 15.11(d) that each time the voltage switches, the actuator returns to equilibrium quickly and then more slowly begins to actuate in the opposite direction. This is the interplay between the elasticity and ion exchange in the solid polymer electrolyte.

A summary of the main properties of carbon nanotube and carbon nanofiber hybrid actuators is given in Table 15.4. This is the first time some of these materials have been characterized, and much more detailed characterization is planned.

We are developing the electrical equations of a CNT actuator in an electrolyte. A 2 M NaCl electrolyte solution is used for the experiments. Details of the modeling are given in References 66–68. Electrochemical impedance spectroscopy (EIS) was

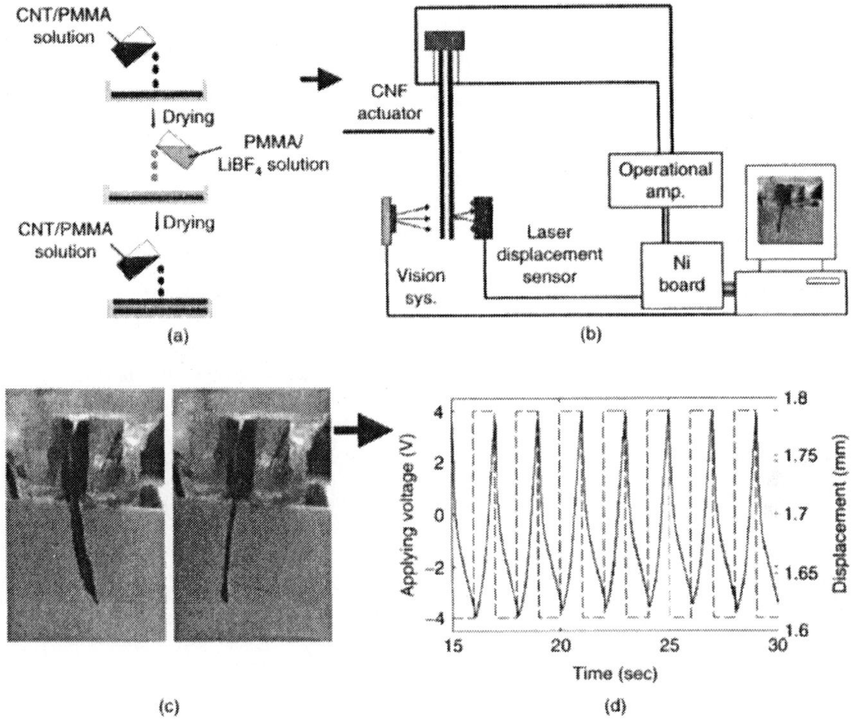

FIGURE 15.11 Solution casting and testing of the dry CNF–PMMA hybrid actuator: (a) solution casting; (b) test setup; (c) dry actuator with two-sided actuation; and (d) measured response due to 0.5 Hz ± 4 V excitation.

performed on three-electrode cells, with a dry CNF actuator as the working electrode, a saturated Calomel reference electrode (Gamry, Part No. 930-03), and using platinum as the counter electrode. EIS measurements were performed using a Gamry Potentiostat (Model: PCI4/750) coupled with the EIS software (Gamry, EIS300). The cell was equilibrated for several hours after each step. The electrochemical impedance spectroscopy (EIS) results are shown in Fig. 15.12.

The EIS results and modeling using Randle's circuit are discussed in References 67 and 68. Randle's circuit is an equivalent representing each component at the electrode–electrolyte interface and in the solution during an electrochemical reaction. The electrode kinetics model the change from electronic conductivity to ionic conductivity, in the simplest way, in parallel as a double-layer capacitor and a polarization resistance that models the electrode kinetics, and this parallel capacitor and resistor combination is in series with a resistor that models the electrolyte solution resistance.

The CNT actuator must now be modeled as it charges and discharges under an AC environment wherein the actuation is proportional to the basic charge transfer rate. Although the electrolysis mechanism with an AC voltage input has very complex characteristics, we can model this effect as an impedance element. The relationship

TABLE 15.4
Properties of Carbon Nanotube and Carbon Nanofiber Hybrid Actuators

Type of Actuator	Wet Environment (NaCl Solution)	Dry Environment (with SPE)
SWCNT buckypaper	Fast, large strain, expensive, good cohesion property to make buckypaper, low strength	Slower response and smaller actuation than wet
MWCNT buckypaper	Good actuation strain and lower cost than SWCNT	Slower response and smaller actuation than wet
CNF buckypaper	Low cost, difficult to make buckypaper, low mechanical strength, needs higher voltage and current to actuate	Slower response and requires larger voltage, good mechanical loading with SPE
CNF coated with polypyrrole to form bucky paper	Improved mechanical loading, similar actuation properties as CNF buckypaper	Slower response and requires larger voltage, good mechanical loading with SPE
SWCNT composite made with PMMA polymer	Fast, large strain, best among nanotube composites, expensive	TBD
MWCNT composite made with epoxy polymer	Good strength and actuation, need to laminate	TBD
CNF composite made with epoxy polymer	Needs higher voltage and current, good mechanical properties, dispersion needs to be improved, has smart structures applications	TBD

TBA = To be determined.

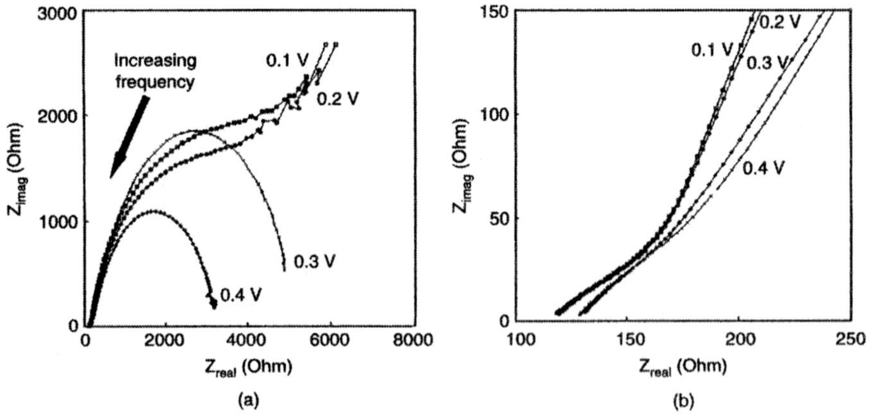

FIGURE 15.12 The EIS responses at cell potentials of 0.1 to 0.4 V: (a) 0.05 to 100 KHz; and (b) enlarged view of high frequencies between 100 Hz and 100 KHz.

between mechanical properties and electrical properties can be expressed approximately by the linear relation:

$$\begin{bmatrix} q(t) \\ \varepsilon(t) \end{bmatrix} = \begin{bmatrix} x & y \\ \alpha C_v & 1/E \end{bmatrix} \begin{bmatrix} V(t) \\ \sigma(t) \end{bmatrix} \quad (15.1)$$

Here, ε is the actuator strain, q is the charge per unit volume, σ is stress, V is input voltage, t is time, α is the strain-to-charge ratio, E is the elastic modulus, the double layer capacitance is C_v, and x and y are constants that depend on the construction of the nanotube smart material and the type of electrolyte used. In this method, the SPE will be designed to increase the capacitance and strain-to-charge ratio, while maintaining the highest elastic modulus of the material as possible.

Based on a concept developed at UC, by monitoring the voltage source, a self-sensing actuator can be formed as shown in Fig. 15.13. To implement this practical self-sensing actuator, an electrical bridge circuit was used to measure strain, as shown in Fig. 15.13(a). The electrolyte and CNT contact act as a resistance; R, the interface between the polymer and the electrolyte is modeled as the double layer capacitance, C_v, and the diffusion element is Z_D. The self-sensing actuator will be used in the feedback control system shown in Fig. 15.13(b) that allows the scheme to take advantage of the guaranteed stable collocated output feedback control. Compared to the best known ferroelectric, electrostrictive, and magnetostrictive materials, the very low driving voltage of CNT is a major advantage for developing various actuator applications. Other predicted future advantages are the direct conversion of electrical to mechanical energy, high actuation strain, strength, and modulus.

The reference model of Fig. 15.13(a) can determine the strain in the actuator due to the piezoresistive effect of the nanotube composite. We have demonstrated this effect in which the electrical resistance of the material changes with strain: Tension increases the resistance, and compression decreases the resistance. This sensing idea can allow the same material to act as an actuator and as a strain sensor. Actually there are three interesting effects acting together in the nanotube actuator: the electrochemical actuation, the piezoresistive effect, and the electrochemical

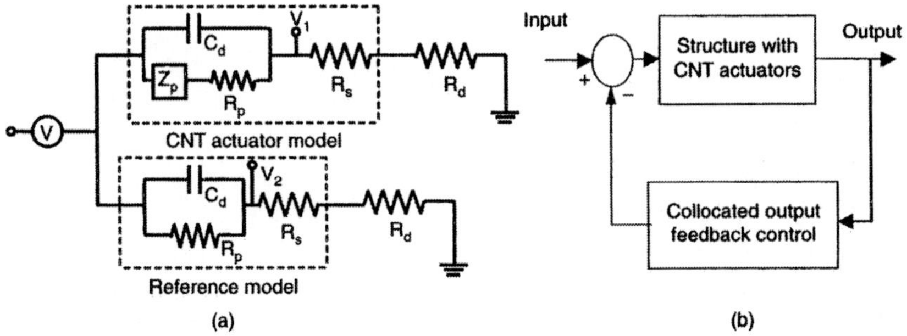

FIGURE 15.13 Models of CNT actuators developed at University of Cincinnati: (a) self-sensing CNT circuit; (b) controller design.

charge generation effect in which charge is generated by electrolyte flow over the nanotube material without strain. These effects will be studied separately and then a combined model will be discussed.

The density of the nanotubes in the SPE has an effect on the electrochemical actuation. Well-aligned but separated carbon nanotubes produce a large surface area supercapacitor with high power density. The capacitance of the CNT actuator SPE material is $C = Q/V = \varepsilon_o, \varepsilon_r(A_s/t)$, where Q and V are the charge and voltage, respectively, and ε_o, ε_r, A_s, and t are the permittivity of free space, the relative permittivity, the surface area of the electrode, and the separation of the plates, respectively. In electrochemical double-layer capacitors (ECDLCs) the inter-plate spacing t is reduced to the Helmholz double layer, which is 0.34 nm. The active area of the actuator depends on the density of nanotubes in the composite and the length and diameter of the nanotubes. A density of at least 50 million CNFs/cm^2 is expected, and the density of SWCNTs will be much greater. The density achievable will depend on how well the nanotubes can be functionalized and dispersed to provide effective ion exchange and load transfer. The relative permittivity is dependent on the type of nanotube including the chirality, number of walls, spacing between the walls, and diameter of the nanotube. There have not been many studies of the electrochemical properties of carbon nanotubes in a conductive polymer composite. The transduction properties of the CNT material must be studied experimentally to determine parameters to model the strain and to find the optimum material and design to maximize the capacitance and the ion exchange.

Synthesis of arrays of parallel nanotubes is another approach to making nanotube smart materials. Embedding planar arrays in layers will allow build-up of composite materials, as shown in Fig. 15.14(a). Another actuator type can be made by attaching

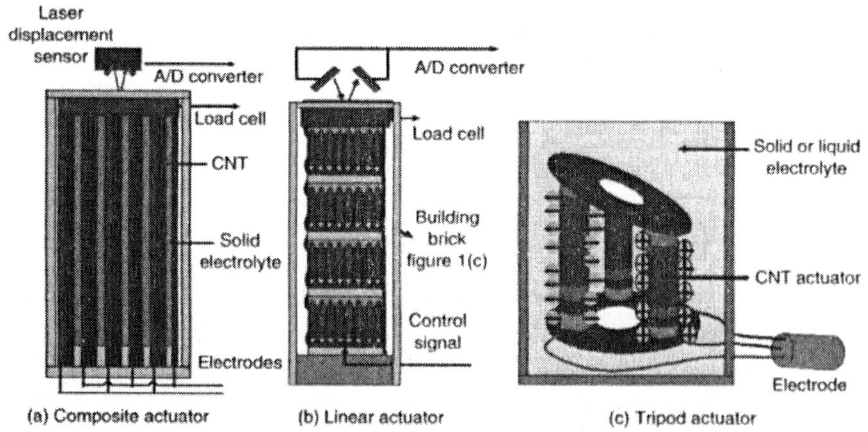

FIGURE 15.14 Concepts for CNT actuators; (a) SPE hybrid actuator; (b) linear actuator built with 1-mm bricks; and (c) tripod actuator 1 mm high where the rotation is exaggerated for illustration. Background shows a 1-mm-high MWCNT array grown by Yi Tu of First Nano Inc. The array is separated to show the nanotube density.

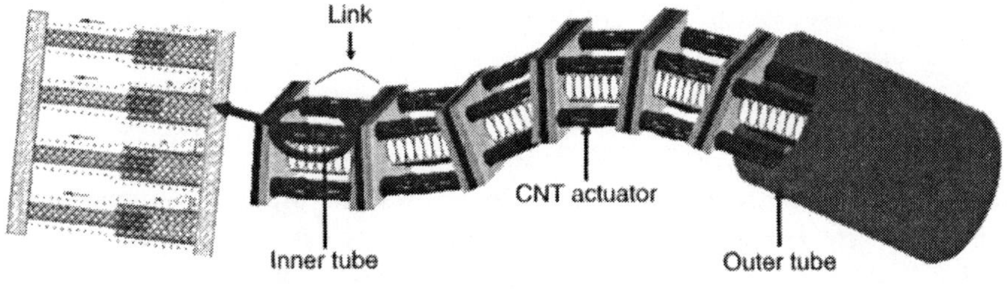

FIGURE 15.15 Concept for an active catheter using CNT electrochemical actuation and self-sensing using the principle of electrochemical actuation or telescoping actuation.

"building bricks" of the array in series by bonding or direct growth in a nanofurnace, as shown in Fig. 15.14(b). A single segment or building brick can be used to form a multisegment tilting actuator. A concept for this actuator is shown in Fig. 15.14(c), and the deflection is exaggerated. About 1% strain is expected and the angle of deflection will depend on how closely the nanotube "posts" are spaced. The linear or tilting segments may be connected to design active devices from the nanoscale to the macroscale in size. Inch-size planar arrays can be synthesized and current efforts are focused on increasing the length of MWCNTs. The tilting motion may be beneficial for alignment and pointing applications. A computer control system will be used to program the voltage for the actuation.

Based on these experiments, a concept for an electrochemical active catheter has been devised, as shown in Fig. 15.15. Benjamin Franklin invented the first catheter. If smart materials could be used to build the active catheter proposed and other miniature manipulators, they would have many medical applications. The catheter has potential for use in angioplasty, stent insertion, plaque removal, hydrocephaly treatment, remote surgery, and other applications. Medical device and surgical innovation is a growing area where nanoengineering can make contributions. In particular, methods to increase actuation stroke are needed. One concept is telescoping MWCNTs illustrated in Fig. 15.15.

15.7 FUTURE DIRECTIONS FOR INTELLIGENT MATERIALS

This section presents ideas for (near) future applications of nanoscale intelligent materials.

15.7.1 Carbon Structural Neural Systems

Structural health monitoring (SHM) is a subarea of intelligent structures wherein the integrity of a structure is monitored in real-time by an *in situ* sensor system. Several methods of SHM using smart materials have been proposed in the literature and include damage identification using the vibration response of a structure, fiber-optic strain measurement, stress wave propagation techniques, and wireless MEMS.

These techniques for *in situ* real-time SHM may not be cost-effective for applications to large structures that have complex geometries and varying thicknesses and curvatures such as joints, ribs, fasteners, hybrid materials, and highly damped devices such as honeycomb sandwich structures. Although there are several types of SHM systems being developed, practical on-board SHM systems should add the lowest complexity with the highest benefit. This section describes a concept for a new sensor system architecture[9] that can make practical the use of large numbers of miniature sensors to monitor mechanical, civil, and environmental systems that can benefit from highly distributed sensors and massively parallel processing using a small number of channels of data acquisition. In biological systems, millions of identical parallel nerves are used for sensing. This architecture allows billions of bits of sensory information to be processed in the neural, auditory, and visual systems in an efficient hierarchical order and in harmony and researchers are attempting to replicate the mechanisms of this biological system. This intersection of biological function with nanoscale perfection may lead to an efficient and intelligent structural neural system (SNS).[202]

One of many possible architectures for the SNS is in the form of an array of orthogonal intersecting nerves built with carbon nanotubes. This approach is based on a design in which piezoceramic receptors are used to detect acoustic emissions or high strains caused by damage propagation.[9,20,25–27] The carbon nanotube SNS works broadly on the principle of the biological nervous system.[27] The sensor fibers act as neurons and each neuron is connected to other neurons. In the event of high strain or damage to the material, the piezoresistive sensor/neuron changes resistance and changes the voltage of the neuron. In a biological system, the signal from one neuron is transmitted by special molecules called neurotransmitters that cross the synaptic gaps to receptor sites on the target neurons. The signal from a biological neuron is inhibitory or excitatory, not both. Excitatory signals tend to fire the target neurons, and inhibitory signals tend to prevent firing. In our SNS, the firing of the neuron depends on the threshold voltage of the sensor. If the voltage is above a certain level, the sensor/neuron sends an excitatory signal that in turn combines with other firing neurons, and thus the signal is transmitted. Additional layers of hierarchy are possible. If the sensor/neuron is far away from the damage event, it will generate a voltage that is below the threshold, so the neuron will not be allowed to fire and it will act as an inhibitory signal to subsequent neurons. The threshold can be determined by having a capacitor/resistor in the circuit. This is analogous to synaptic transmission of a signal in a biological nervous system. The signal from the closest sensor/neuron can be processed to extract the information about the damage. The other neurons can be prevented from firing (occluded) to eliminate unwanted signals. The threshold voltage can be determined by experimentation and simulation, depending on the size and type of SNS. This approach uses receptors that cross but electrically independent. This allows individual row and column neurons to be formed in horizontal rows and vertical columns, as shown in Fig. 15.16. A single neuron is shown in Fig.15.4(a).

A unit cell of the SNS consists of two row receptors and two column receptors. When damage occurs near one receptor, the row and columns of that receptor fire

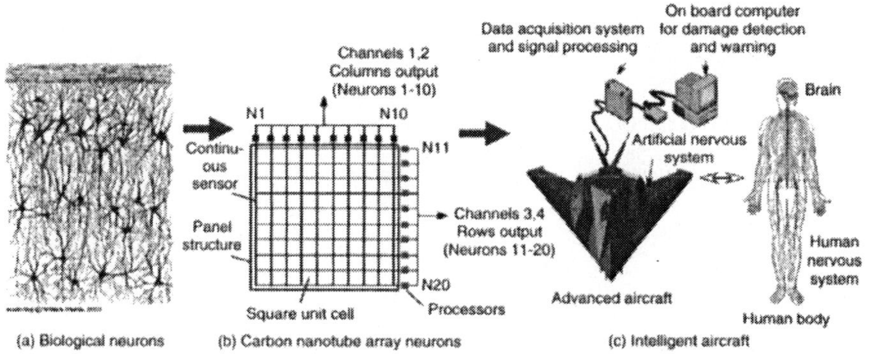

FIGURE 15.16 The biological neural system and an array carbon structural neural system.

and the other rows and columns are inhibited or occluded, depending on the design and training of the neurons. The intersection of the rows and columns that are firing will show the location of the damage. Also any change in the resistance of the neuron can be related to the strain and damage in the structure. This system is redundant and minimizes false signals. If damage causes one neuron to become inoperable, a problem would be indicated by the failed neuron, and the damage would also be detected when it propagates to the neighboring neuron. Since the neurons can be scaled to the micron or smaller level, an arbitrary degree of sensitivity can be obtained.

This SNS can use low frequency data acquisition, but requires a very high sensor density to detect initiating damage. For example, crack propagation around rivets requires a small-scale unit cell around each rivet. This is practical due to the small size and high strain sensitivity of CNT strands. No diagnostic signal or amplifiers or large sets of historical data are needed in this simple approach. In addition, the size of the carbon SNS can be miniaturized such that the nerves can be put inside composite materials or at the bonding interfaces of joints without affecting the integrity of the structural material. This approach overcomes many of the problems of embedding sensors inside composites. The piezoresistive sensors are highly distributed, simple, less expensive, less intrusive, and more sensitive than existing SHM sensor technology. Figure 15.16(a) shows biological neurons, and Fig. 15.16(b) shows an array SNS with ten row and ten column neurons (N1–N20), which are continuous sensors that sense along their entire length. The row and column neurons in the array are routed into four signal channels — two channels of time response signals and two channels for neuron firing. Using information from these four channels, it is possible to locate the damage source and measure the signal amplitude. The analog to digital conversion is only needed at the PC level, and this reduces the complexity of the instrumentation and the cost, size, and weight of the SHM system. It is anticipated that continued development of the array SNS using nanotube-based sensors will lead to ultra high-density high sensitivity sensor systems that can be built at either the macroscopic or the microscopic scale for use on safety-critical structures such as aircraft (Fig. 15.16(c)). The SNS architecture may also have applications in health monitoring of living systems and the environment.

15.7.2 High-Temperature Nanoscale Materials

Since piezoelectric ceramic materials have been under development for a long time, new materials are sought that can provide increased performance at elevated temperatures. There are several nanoscale materials under development that may be suitable for sensing at high temperatures. These materials are in early stages of development and are briefly discussed next. References are given to allow readers to obtain more information.

Nanocomposites with ~0.5 to 1 wt% of SWCNTs can become self-sensing structures by the addition of electrodes to the composite surfaces, but a host material that can withstand high temperatures is needed. The piezoelectric effect is small in SWCNTs, and it does not seem practical for use in sensing. The electrochemical effect of the SWCNT is based on double-layer charge injection that causes bond expansion. A limitation is that the nanotubes must have an ion exchange for the double-wall charge injection to work. For high temperature applications, a molten electrolyte might be used as suggested by Baughman et al.[53] The SWCNTs are formed at temperatures from 700 to 1000°C. They are expected to operate at temperatures up to 500°C, beyond which they can burn in an oxygen environment. Use of a molten electrolyte would pose problems in some applications.

Noncarbon nanotubes have received much less attention in recent years and remain relatively unexplored. The BNTs are predicted to be piezoelectric in bundles and the maximum piezoelectric-induced stress constant occurs for the specific (5,0) chirality, but the piezoelectric d coefficient is about 20 times smaller than for piezoceramics. (See Chapter 21.) BNT materials are the subjects of different investigations that started around 2000 and their characteristics offer the possibility of piezoelectric actuation tailored over a wide bandwidth, and the possibility of replacing piezoceramic materials with BNTs. However, the synthesis of BNTs is at an early stage (see Chapter 21) and the information on this issue in the scientific literature is limited. There appears to be no experimental verification of the properties of the BNTs, and it is not possible to control the chirality of the nanotubes produced. BNTs are formed at temperatures above 1000°C. They are expected to operate at temperatures up to 1000°C. BNTs represent one possible material for high temperature applications.

Silicon carbide nanotubes (SiCNTs) are under early stages of synthesis at NASA Glenn, as described on the web site.[10] Macroscopic SiC is a very hard and heat-resistant ceramic material. The objective of this approach is to take advantage of the properties of SiC to synthesize and characterize SiCNTs for high temperature and high radiation conditions. Multiple synthetic approaches are being considered that parallel the direct CNT formation, as well as an indirect approach involving derivatization of a CNT to a SiCNT. Recently SiCNTs were produced in-house at NASA Glenn. A CVD-template synthesis method, in which the template functions as a nanomold, was employed to produce highly ordered arrays of SiCNTs. The arrays are composed of individual SiCNTs (»200 nm in diameter, 60 μm in length) aligned vertically, with a density of approximately 109 tubes/cm^2. The electrical and mechanical properties of the SiCNTs are being characterized and compared with theoretical SiCNT modeling results. Electrical properties studies of nanotubes at high temperatures could determine the suitability of nanotube high-temperature sensors. The electrical activity of SiCNTs could also be studied as a function of adsorbates, which could ultimately lead to applications such as nanogas sensors for harsh environments and sensors for the working areas of jet engines.

Bismuth nanowires (see Chapter 3), ZnO nanobelts (see Chapter 4) and other inorganic materials such as WS_2–MoS_2, NbS_2, and TiS_2 may have piezoelectric or other interesting properties that might be useful as sensors.

In summary, existing piezoelectric materials do not have satisfactory properties at high temperatures. Some of the different forms of nanotubes that are naturally polar in bundles or otherwise fullerene-like may be useful for high temperature applications. Further research is needed to develop and test the piezoelectric nanotubes.

15.7.3 Power Harvesting Using Carbon Nanotubes

In preliminary experiments we tested an induced charge generator in an electrolyte for power harvesting in mechanical and structural systems. The CNT material is shown in Fig. 15.17(a). The voltage due to vertical motion of the film in the electrolyte was a 7-mV peak in the electrolyte. The film was taken out of the electrolyte and placed in deionized water and the initial peak voltage was 17 mV, as shown in Fig. 15.17(b), but upon continued cycling, the voltage reduced to near zero. This result showed that a micron thin film with a volume of 5 mm by 20 mm by 0.018 mm = 1.8 mm^3 can generate a small amount of power (~10 nW) due to electrolyte flow over the nanotubes, not by strain of the nanotubes. It was also determined that the flow direction should be in the long direction of the nanotube actuator to generate the greatest voltage. A bending type motion produced a 2-mV signal. These are interesting results that are partly discussed in previous literature.[157–158] This experiment showed that charge may be partially shunted due to the ionic conductivity of the electrolyte, and that increased power generation might be obtained by partially encapsulating a hydrated material to prevent charge loss through the electrolyte. This is currently under investigation; see Reference 66. It is also interesting that the charge produced is flow dependent and greater when the flow is in the long direction of the nanotube paper. The flow in the transverse direction yields only 1–2 mV. The width of the film, electrode location, and some alignment of the nanotubes in the long direction may explain this result. One possible application of the nanotube film is to have a self-powered biosensor made of the patterned nanotube array described earlier. This self-contained sensor could be inside the body to detect diseases and monitor critical signals from the body and transmit signals wirelessly from inside the body using power harvested due to blood flow through a CNT array. The MWCNT array might also simultaneously act as a cancer sensor and generate power to send signals wirelessly. Figure 15.17(c) shows an approach planned to demonstrate the power generation of the smart material. A flow-induced charge generator MWCNT array used for power harvesting in ionic fluidic systems is shown. Flow or horizontal vibration will be applied

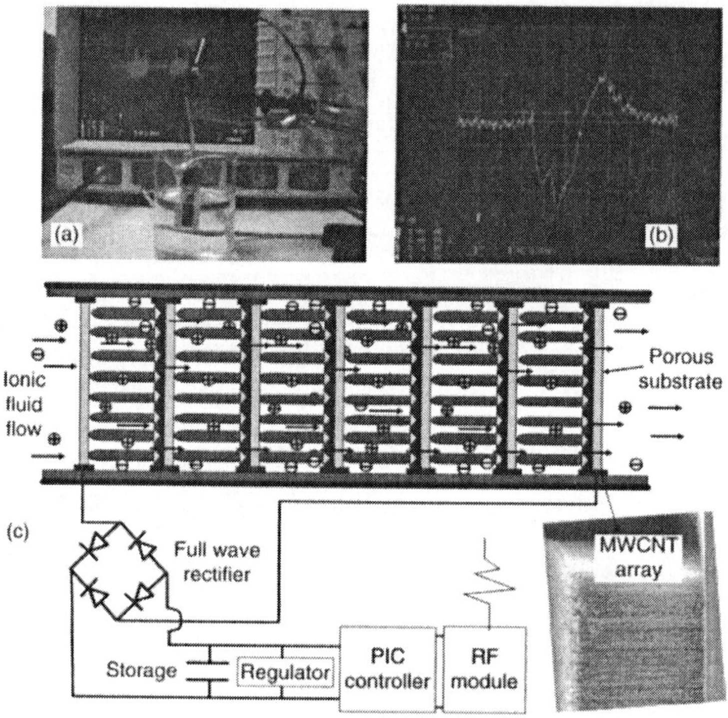

FIGURE 15.17 Power generation by vertical fluid flow over a SWCNT Nafion film: (a) film on two sides of a stiff backing; (b) initial 17 mV generated in deionized water after 7 mV generated in 5 M NaCl; and (c) concept for flow-induced power harvesting with six stacked MWCNT arrays.

to the CNT material, and a circuit will be used to measure the power developed. An RF module may be used to test wireless transmission for later use in a SHM system. The CNT power generator can power a batteryless RF module that transmits a short-range wireless ID code to a nearby server. Using the RF module, structural states, vibration, strain, or corrosion can be monitored wirelessly and in real time. While monitoring the structural state, we can identify sensors sending an abnormal signal, since every sensor has its own identification. A LABVIEW system will be used to control the power generation experiment.

15.7.4 Intelligent Machines

The intelligent machine is a new concept for manufacturing, self-repair, and demanufacturing based on the use of nanoscale materials. One concept is to develop a carbon nanocomposite material for assembly and disassembly of machines and to fabricate parts used to manufacture machines. The nanocomposite material is produced by incorporating carbon nanotubes and carbon nanofibers into a polymer matrix. The nanocomposite material has piezoresistive sensing to measure large strains and detect damage in parts. The material also has electrical conductivity that can be used to heat and recure the material in the appropriate area to partially self-repair damage, and also to heat the material to disassemble a product when it reaches the end of its useful life, either due to wear or obsolescence. A machine with the capabilities for sensing damage, partial self-repair of damage, and controlled disassembly is described here as an intelligent machine.

Carbon nanotubes added to polymers can be viewed as new smart materials that have satisfactory strength, sensing, and actuation capabilities. A novel application of this nanocomposite smart material is to form machine parts or for use as an adhesive to assemble parts. This nanocomposite material can be turned on to repair parts by electrical heating to recure and heal the material in the damaged area, or by heating to debond and partially disassemble the product to replace or repair parts. Also, at the end of the product life, the smart material can be heated to debond and allow complete disassembly of the product efficiently and without damaging reusable or hazardous components. The demanufactured components will be sorted and either recycled or remanufactured into new products. Carbon nanotube polymers have recently been used at UC to form different structural smart materials with electrical conductivity, sensing, and actuation capabilities. The properties of these materials can be characterized and optimized for applications related to intelligent machines.

The benefit of utilizing nanotechnology in manufacturing is that manufacturing and consumer use of products would change from traditional use of fasteners for assembly to quick assembly using adhesives. Repair of damage would be done by reforming and curing the material, and large structural parts could be made of a nanocomposite material that is self-sensing and self-repairing. In addition, discarding old products would cease and be replaced by a new paradigm in which the purchaser would receive a credit to return obsolete or worn-out products to the manufacturer. A reverse manufacturing process, defined here as demanufacturing, would use a disassembly line to apply the appropriate stimulation to the part to disassemble itself into useful smaller components that would subsequently be sorted and reused, disposed of, or shipped to other remanufacturing facilities. If this new manufacturing ideology is successfully developed, it will have a large impact on manufacturing in the following areas: (1) reducing the costs of manufacturing by efficiently reusing parts or materials from old products; (2) complex machines could be designed to be assembled faster with less need for critical tolerances; (3) parts count could be reduced using adhesive joining; (4) components could be disassembled easily for repair by electrical heating in some cases combined with ultrasound, which would eliminate fasteners for many applications; (5) machines assembled using smart materials and parts from old machines that could be efficiently disassembled using smart materials could be remanufactured at a cost lower than the cost of a new machine; (6) demanufacturing would promote conservation by reusing materials and reduce overall product costs; and (7) the new paradigm of intelligent machines could revitalize manufacturing industries because new types of products would be produced and recycled. In addition, manufacturing plants are located almost everywhere and the new idealogy would provide a good opportunity to educate students and bring nanotechnology to industry. As an example, students in cooperative education programs could work in industries and government laboratories to investigate the use of smart materials for manufacturing and demanufacturing of different products.

15.7.5 Telescoping Carbon Nanotubes

Carbon nanotube polymer materials being developed have electrical and ionic conductivity properties wherein the polymer is actually a semiconducting structural material that can function as a sensor and actuator. There is another concept for a potential actuation and sensing mechanism that can be used with a conductive polymer electrolyte material. This concept mechanism is telescoping of MWCNTs. Telescoping of the inner tubes in a MWCNT that has the outer tube opened has been shown using an atomic force microscope.[212] It has also been suggested that telescoping and oscillating actuation is possible and the MWCNTs have been proposed for use as rotational and translational bearings, and as nuts and screws for building nanomachines by taking advantage of the spiral chirality of nanotubes.[209-217] Because of the potentially huge displacement of this type of actuator, telescoping nanotube arrays (TNAs) are of interest for linear actuation applications based on electrostatic repulsion actuation. The telescoping nanotube has also been shown to change electrical resistance with telescoping length.[209, 211] Thus, the TNA can also become a displacement sensor. Moreover, based on our testing of MWCNT power cells to generate electricity, we believe the TNA can also be used for power generation due to telescoping of the material when it is encapsulated in an electrolyte. The electrically telescoping nanotube has not been demonstrated and is an area of current investigation and future work at the University of Cincinnati and elsewhere.

15.8 CONCLUSIONS

The attempt in this chapter was to provide an up-to-date review of nanotube intelligent materials, structures, and applications. Special nanotube properties were explained, including electrical conductivity, contact resistance, piezoelectricity, piezoresistance, and electrochemical actuation by charge injection that might be useful to generate ideas for new intelligent materials and structures. Various types of nanostructured smart sensors and actuators were also proposed. Although significant progress has been made in understanding nanoscale smart materials, there is much still to be learned, especially about bringing the nanoscale properties to the macroscale. It can, however, be concluded that nanoscale intelligent materials will engender fundamental changes in the field of smart materials and structures in the near future, and there are some emarkable possibilities for building intelligent structures using nanotechnology.

PROBLEMS

1. In the chapters in this book, different types of nanoscale materials are discussed. These include carbon nanotubes, boron nitride nanotubes, zinc oxide nanobelts, and bismuth nanowires. What might be the advantages and disadvantages of building smart materials using these different types of nanoscale materials?

2. Describe the principle of the Potentiostat/Galvanostat. Describe the saturated Calomel reference electrode (SCE). You may go to the manufactures' Websites for information on electrochemistry.
3. What is a motor protein? Why are DNA and motor proteins possibly good actuators *in vivo*? Can a biological smart machine be made?
4. Define solid polymer electrolyte. What is the limitation of a solid polymer electrolyte in a structural application? Propose ways to improve the mechanical properties of solid polymer electrolytes.
5. Describe the sp^2 hybridization in the CNT structure and the σ-bonds and π-bonds.
6. A machine with the ability to prevent failure is a smart machine. A manufacturer wants to develop a way to provide short-time emergency lubrication for the bearings in an aircraft engine. This would allow the aircraft to operate long enough to safely land if the normal bearing lubrication system fails. Think of a nanoscale material that could be a lubricant. Design a system that automatically delivers this lubricant to the bearing for a short time if the standard lubrication system fails.
7. Search the literature for the properties of nanoscale materials and try to compute and fill in the missing properties in Table 15.3.

ACKNOWLEDGMENTS

This work was sponsored in part by the University of Cincinnati (UC) Summer Student Fellowship, the university's Research Council, FirstNano, Inc., the National Renewable Energy Laboratory, and the Ohio Aerospace Institute. This funding support is gratefully acknowledged. In addition, the following student researchers assisted in the development of nanoscale and smart materials at UC: Courtney Brown, Nicole Reinart, Jacob Hause, Jennifer Chase, Payal Kaul, Robert Gilliland, and Subrahmin Srivinas. Other help was provided by Chris Sloan, Dennis Adderton, and Yi Tu of FirstNano, and Bo Westheider, Dale Weber, George Kreishman, Larry Schantman and Dave Breiheim of UC. Assistance in the catheter concept was provided by doctors Randall Wolf and Charles Dorn of the Department of Surgery, Center for Surgical Innovation at UC. Drs. Zhongyun Dong and Abdul Rahman Jazieh of the Department of Internal Medicine at UC provided the ideas for using antibodies for cancer detection. Hong-Chao Zhang of the Center for Applied Research in Advanced Manufacturing, Lubbock, TX, Jay Lee of the University of Wisconsin, and John Sutherland provided ideas for the smart machine. All their help and cooperative work are gratefully acknowledged and made this chapter possible.

REFERENCES

1. Carbon Nanotechnologies, Inc., Houston, TX, www.cnanotech.com; contact person, Tom Pittstick.
2. NanoLab, Inc., info@nano-lab.com.
3. Applied Sciences, Inc. and Pyrograf Products, Inc.; contact person, David Burton.

4. Dresselhaus, M.S., Dressehaus, G., and Eklund, P.C., *Science of Fullerenes and Carbon Nanotubes*, Academic Press, San Diego, 1996.
5. FirstNano, Inc., Santa Barbara, California, 93111.
6. Dresselhaus, M.S., Dresselhaus, G., and Avouris, Ph., Carbon nanotubes: synthesis, structure, properties, and applications, in *Springer, 80th Topics in Applied Physics*, 2000, Springer-Verlag, New York.
7. Goddard, W.A., Brenner, D.W., Lyshevski, S.E., and Lafrate, G.J., *Handbook of Nanoscience, Engineering, and Technology*, CRC Press, Boca Raton, FL, 2003.
8. Nalwa, H.S., *Nanostructured Materials and Nanotechnology*, Academic Press, San Diego, 2000.
9. Smart Structures Bio-Nanotechnology Laboratory, University of Cincinnati, OH, http://www.min.uc.edu/~mschulz/smartlab/smartlab.html.
10. NASA Glenn Research Center. http://www.grc.nasa.gov/WWW/SiC/redhot.html
11. Department of Electronic Materials Engineering, The Australian National University, Canberra. http://www.anutech.com.au/TD/Info%20Sheets/ Nanotubes %20brochure.pdf
12. Mallick, P.K., *Fiber Reinforced Composites*, Marcel Dekker, New York, 1988. http://www.hexcelfibers.com/Markets/Products/default.htm
13. Cady, W.G., *Piezoelectricity: An Introduction to the Theory and Applications of Electromechanical Phenomenon in Crystals*, Dover, New York, 1964.
14. Jaffe, B. and Cook, W.R., Jr., *Piezoelectric Ceramics*, Academic Press, San Diego, 1971.
15. Uchino, K., *Piezoelectric Actuators and Ultrasonic Motors*, Kluwer Academic Publishers, Boston, 1966.
16. APC International, Piezoelectric Ceramics: Principles and Applications, http://www.americanpiezo.com/materials/index.html
17. Dubois, M.-A. and Muralt, P., Measurement of the effective transverse piezoelectric coefficient $e_{31,f}$ of AlN and $Pb(Zr_x,Ti_{1-x})O_3$ thin films, *Sensors Actuators A, 77*, 106–112, 1999.
18. Eitel, R.E., Randall, C.A., Shrout, T.R., Rehrig, P.W., Hackenberger, W., and Park, S., New high temperature morphotropic phase boundary piezoelectrics based on $Bi(Me)O_3$-$PbTiO_3$ ceramics, *Jpn. J. Appl. Phys., 40*, 5999–6002, 2001.
19. Schulz, M.J., Sundaresan, M., McMichael, J., Clayton, D., Sadler, R., and Nagel, W., 2003, Piezoelectric materials at elevated temperature, *J. Intelligent Mater. Syst. Struct., 14*, 11, 693–705.
20. Martin, W.N., Ghoshal, A., Schulz, M.J., and Sundaresan, M.J., Structural health monitoring using an artificial neural system, *Recent Res. Dev. Sound Vibration*, 2003.
21. Bent, A.A., *Active Fiber Composites for Structural Actuation*, Ph.D. dissertation, Massachusetts Institute of Technology, Cambridge, MA, 1997.
22. Bent, A.A and Hagood, N.W., Piezoelectric fiber composites with interdigitated electrodes, *J. Intell. Mater. Syst. Struct.*, 8, 1998.
23. CeraNova Corporation, Franklin, MA.
24. Wilkie, K., NASA LaRC Macro-Fiber Composite Actuator, *NASA News Release*, September 19, 2000.
25. Datta, S., *An Active Fiber Continuous Sensor for Structural Health Monitoring*, M.S. thesis, University of Cincinnati, 2003.
26. Sundaresan, M.J., Ghoshal, A., and Schulz, M.J., Sensor Array System, U.S. patent 6,399,939 B1, 2002.

27. Sundaresan, M.J., Schulz, M.J., Ghoshal, A., and Pratap, P., A Neural System for Structural Health Monitoring, paper presented at SPIE 8th Symposium on Smart Materials and Structures, March, 4–8, 2001.
28. Piezo-Systems, Inc. www.piezo.com, and, http://www.piezo.com/psi5a4.html.
29. Measurement Specialties. http://www.msiusa.com/default/index.asp.
30. Iijima, S., Helical microtubules of graphitic carbon, *Nature, 354,* 56–58, 1991.
31. EasyTube CVD Nanofurnace, FirstNano, Inc., Santa Barbara, CA. http://www.frist-nano.com.
32. Hafner, J.H., Bronikowski, M.J., Azamian, B.R., Nikolaev, P., Rinzler, A., Colbert, D.T., Smith, K.A., and Smalley, R.E., Catalytic growth of single-wall carbon nanotubes from metal particles, *Chem. Phys. Lett., 296,* 195–202, 1998.
33. Huang, S., Cai, X., and Liu, J., Growth of millimeter long and horizontally aligned single-walled carbon nanotubes on flat substrates, *JACS, 125,* 5636–5637, 2003.
34. Wei, B.Q., Vajtai, R., Zhang, Z.J., Ramanath, G., and Ajayan, P.M., Carbon nanotube-magnesium oxide cube networks, *J. Nanosci. Nanotechnol., 1,* 35–38, 2001.
35. Buldum, A. and Lu, J.P., Contact resistance between carbon nanotubes, *Phys. Rev. B, 63,* 161403, 2001.
36. Tian, W. and Datta, S., Aharonov-Bohm-type effect in graphene tubules: a Landauer approach, *Phys. Rev. B, 49,* 5097–5100, 1994.
37. Ananntram, M.P., Datta, S., and Xue, Y., Coupling of carbon nanotubes to metallic contact, *Phys. Rev. B, 61,* 20, 14219, 1999.
38. Mortensen, N.A., Johnsen, K., Jauho, A.-P., and Flensberg, K., Contact resistance of quantum tubes, *Superlattices Microstructures, 26,* 6, 351–361, 1999.
39. Bachtold, A., Henny, M., Terrier, C., Strunk, C., Schonenberger, C., Salvetat, J.P., Bonard, J.M., and Forro, L., Contacting carbon nanotubes selectively with low ohmic contacts for four-probe electric measurements, *Appl. Phys. Lett., 73,* 2, 274–276, 1998.
40. Menon, M., Andriotis, A., and Froudakis, G., Curvature dependence of metal catalyst atom interaction with carbon nanotubes wall, *Chem. Phys. Lett., 320,* 42–434, 2000.
41. Tachibana, T., Williams, B., and Glass, J., Correlation of the electrical properties of metal contacts on diamond films with the chemical nature of the metal-diamond surface 2, titanium contacts: a carbide forming metal, *Phys. Rev. B, 45,* 11975–11981, 1992.
42. Zhang, Y., Franklin, N., Chen, R., and Dai, H., A metal coating study of suspended carbon nanotubes and its implications to metal-tube interactions, *Chem. Phys. Lett., 331,* 35–41, 2000.
43. Benoit, J.M., Buisson, J.P., Chauvet, O., Godon, C., and Lefrant, S., Low-frequency Raman studies of multiwalled carbon nanotubes: experiments and theory, *Phys. Rev. B, 66,* 073417, 2002.
44. A. Ahuwalia, R., Baughman, D., De Rossi, A., Mazzoldi, M., Tesconi, A., Tongnetti, G., Vozzi, Microfabricated electroactive carbon nanotube actuators, *Proc. SPIE Conf. Smart Struct. Mater., 4329,* 2001.
45. Jain, S., Kang, P., Yeo-Heung, Y., He, T., Pammi, S.L., Muskin, A., Narsimhadevara, S., Hurd, D., Schulz, M.J., Chase, J., Subramaniam, S., Shanov, V., Boerio, F.J., Shi, D., Gilliland, R., Mast, D., and Sloan, C., Developing Smart Materials Based on Carbon Nanotubes, paper presented at SPIE 11th International Symposium on Smart Structures, San Diego, 2004.
46. Gao, M., Dai, L., Baughman, R.H., Spinks, G.M., et al., Electrochemical properties of aligned nanotube arrays: basis of new electromechanical actuators, electroactive polymer actuators and devices, *SPIE Proceedings,* 18–24, 2000.

47. Mazzoldi, D.D. and Baughman, R.H., Electromechanical behavior of carbon nanotube sheets in electrochemical actuators: electroactive polymer actuators and devices, *SPIE Proceedings*, 25–32, 2000.
48. El-hami, K. and Matsushige, K., Covering single walled carbon nanotubes by the PVDF copolymer, *Chem. Phys. Lett., 368,* 168–171, 2003.
49. Knez, M., Sumser, M., Bittner, A.M., Wege, C., Jeske, H., Kooi, S., Burghard, B., and Kern, K., Electrochemical modification of individual nano-objects, *J. Electroanaly Chem., 522,* 70–74, 2002.
50. Baughman, R.H., Conducting polymer artificial muscles, *Synth. Met., 78,* 339–353, 1996.
51. Dickinson, M.H., Farley, C.T., Full, R.J., Koehl, M.A.R., Kram, R., and Lehman, S., How animals move: an integrative view, *Science, 288,* 100–106, 2003.
52. Inganas, O. and Lundstrum, I., Carbon nanotube muscles, *Science, 284,* 31–32, 1999.
53. Baughman, R.H., Cui, C., Zakhidov, A.A., Iqbal, Z., Barisci, J.N., Spinks G.M., Wallace, G.G., Mazzoldi, A., De Rossi, D., Rinzler, A.G., Jaschinski, O., Roth, S., and Kertesz, M., Carbon nanotube actuators, *Science, 284,* 1340–1344, 1999.
54. Fraysse, J., Minett, A.I., Gu, G., Roth, S., Rinzler, A.G., and Baughman, R.H., Towards the demonstration of actuator properties of a single carbon nanotube, *Curr. Appl. Phys., 1,* 407–411, 2001.
55. Minett, A., Fraysse, J., Gang, G., Kim, G.-T., and Roth, S., Nanotube actuators for nanomechanics, *Curr. Appl. Phys., 2,* 61–64, 2002.
56. Fraysse, J., Minett, A.I., Jaschinski, O., Dueshberg, G.S., and Roth, S., Carbon nanotubes acting like actuators, *Carbon, 40,* 1735–1739, 2002.
57. Spinks, G.M., Wallace, G.G., Lewis, T.W., Fifield, L.S., Dai, L., and Baughman, R.H., Electrochemically driven actuators from conducting polymers, hydrogels, and carbon nanotubes, *Proc. SPIE Int. Soc. Opt. Eng. 4234 (Smart Materials),* 223–231, 2001.
58. Barisci, J.N., Spinks, G.M., Wallace, G.G., Madden, J.D., and Baughman, R.H., Increased actuation rate of electromechanical carbon nanotube actuators using potential pulse with resistance compensation, *Smart Mater. Struct., 12,* 549–555, 2003.
59. Baughman, R.H., Conducting polymer artificial muscles, *Synth. Met., 78,* 339–353, 1996.
60. Tahhan, M., Truong, V.-T., Spinks, G.M., and Wallace, G.G., Carbon nanotube and polyaniline composite actuators, *Smart Mater. Struct., 12,* 626–632, 2003.
61. http://www.mpi-stuttgart.mpg.de/klitzing/research/nano/nanoactuators.html
62. http://www.nasatech.com/Briefs/Sep01/NPO21153.html
63. http://www.cs.unc.edu/Research/nano/tams.html
64. Roth, S. and Baughman, R.H., Actuators of individual carbon nanotubes, *Curr. Appl. Phys., 2,* 311–314, 2002.
65. Livage, J., Actuator materials: towards smart artificial muscle, *Nature Mater., 2,* 297–299, 2003.
66. Inpil Kang, Yun Yeo Heung, Jay H. Kim, Jong Won Lee, Ramanand Gollapudi, Srinivas Subramaniam, Suhasini Narasimhadevara, Douglas Hurd, Goutham R. Kirikera, Vesselin Shanov, Mark J. Schulz, Donglu Shi, Jim Boerio, Carbon nanotube smart materials, submitted, *Comp. B J.*
67. Yun Yeo-Heung, Atul Miskin, Phil Kang, Sachin Jain, Suhasini Narasimhadevara, Douglas Hurd, Mark J. Schulz; Vesselin Shanov, Tony He, F. James Boerio, Donglu Shi, Subrahmin Srivinas; Carbon nanofiber hybrid actuators. Part I: Liquid electrolyte-Based, *J. Int. Mater. Smart Struct.,* submitted, special issue.

68. Yun Yeo-Heung, Atul Miskin, Phil Kang, Sachin Jain, Suhasini Narasimhadevara, Douglas Hurd, Mark J. Schulz, Vesselin Shanov, Tony He, F. James Boerio, Donglu Shi, Subrahmin Srivinas; Carbon nanofiber hybrid actuators. Part II: solid electrolyte-based, *J. Int. Mater. Smart Struct.*, submitted, special issue.
69. Davis, V.A., Ericson, L.M., Saini, R., Sivarajan, R., Hauge, R.H., Smalley, R.E., and Pasquali, M., Rheology, phase behavior, and fiber spinning of carbon nanotube dispersions, *AIChe Annual Meeting,* 2001.
70. Girifalco, L.A., Hodak, M., and Lee, R.S., Carbon nanotubes, buckyballs, ropes, and a universal graphitic potential, *Phys. Rev. B, 62,* 13104–13110, 2000.
71. Liu, H.T., Cheng, F.L., Cong, G.S., et al., Synthesis of macroscopically long ropes of well-aligned single-walled carbon nanotubes, *Advan. Mater., 12,* 1190–1192, 2000.
72. Walters, M.J., Casavant, X.C., Qin, C.B.H., Boul, P.J., et al., In-plane-aligned membranes of carbon nanotubes, *Chem. Phys. Lett., 338,* 14–20, 2001.
73. Shenton, W., Douglas, T., Young, M., Stubbs, G., and Mann, S., Inorganic-organic nanotube composites from template mineralization of tobacco mosaic virus, *Advan. Mater., 11,* 3, 253–256, 1999.
74. Li, D., Zhang, X., Sui, G., Wu, D., and Liang, J., Toughness improvement of epoxy by incorporating carbon nanotubes in the resin, *J. Mater. Sci. Lett., 22,* 791–793, 2003.
75. Rege, K., Raravikar, N.R., Kim, D.-Y., Schadler, L.S., Ajayan, P.M., and Dordick, J.S., Enzyme-polymer-single walled carbon nanotube composites as biocatalytic films, *Nano Lett., 3,* 6, 829–832, 2003.
76. Schaefer, D.W., Zhao, J., Brown, J.M., Anderson, D.P., and Tomlin, D.W., Morphology of dispersed carbon single-walled nanotubes, *Chem. Phys. Lett., 375,* 369–375, 2003.
77. Watts, P.C.P., Hsu, W.K., Kotzeva, V., and Chen, G.Z., Fe-filled carbon nanotube-polystyrene:RCL composites, *Chem. Phys. Lett., 366,* 42–50, 2002.
78. Dujardin, E., Ebbesen, T.W., Krishnan, A., and Treacy, M.M.J., Wetting of single shell carbon nanotubes, *Advan. Mater., 10,* 17, 1472–1475, 1998.
79. Lozano, K., Bonilla-Rios, J., and Barrera, E.V., A study on nanofiber-reinforced thermoplastic composites (2): Investigation of the mixing rheology and conduction properties, *J. Appl. Polym. Sci., 80,* 1162–1172, 2001.
80. Hughes, M., Shaffer, M.S.P., Renouf, A.C., Singh, C., Chen, G.Z., Fray, D.J., and Windle, A.H., Electrochemical capacitance of nanocomposite films formed by coating aligned arrays of carbon nanotubes with polypyrrole, *Advan. Mater., 14,* 5, 382–385, 2002.
81. Chen, G.Z., Shaffer, M.S.P., Coleby, D., Dixon, G., Zhou, W., Fray, D.J., and Windle, A.H., Carbon nanotube and polypyrrole composites: coating and doping, *Advan. Mater., 12,* 7, 522–526, 2000.
82. Ounaies, Z., Park, C., Wise, K.E., Siochi, E.J., and Harrison, J.S., Electrical properties of single wall carbon nanotube reinforced polyimide composites, *Composites Sci. Technol., 63,* 2003.
83. Vander Wal, R.L. and Hall, L.J., Nanotube coated metals: New reinforcement materials for polymer matrix composites, *Advan. Mater., 14,* 18, 1034–1308, 2002.
84. Tahhan, M., Truong, V.-T., Spinks, G.M., and Wallace, G.G., Carbon nanotube and polyaniline composite actuators, *Smart Mater. Struct., 12,* 626–632, 2003.
85. Stephan, C., Nguyen, T.P., Lamy de la Chapelle, M., Lefrant, S., Journet, C., and Bernier, P., Characterization of single walled carbon nanotubes–PMMA composites, *Synth. Met., 108,* 139–149, 2003.
86. Fan, J., Wan, M., Zhu, D., Chang, B., Pan, W., and Xie, S., Synthesis and properties of carbon nanotube-polypyrrole composites, *Synth. Met., 102,* 1266–1267, 1999.

87. Byron Piper, R., Frankland, S.J.V., Hubert, P., and Saether, C., Self-consistent properties of carbon nanotubes and hexagonal arrays as composite reinforcements, *Composites Sci. Technol., 63,* 1349–1358, 2003.
88. Shinoda, H., Oh, S.J., Geng, H.Z., Walker, R.J., Zhang, Z.B., McNeil, L.E., and Zhou, O., Self-assembly of carbon nanotubes, *Advan. Mater., 14,* 12, 899–901, 2002.
89. Chen, J.H., Huang, Z.P., Wang, D.Z., Yang, S.X., Wen, J.G., and Ren, Z.F., Electrochemical synthesis of polypyrrole/carbon nanotube nanoscale composites using well-aligned carbon nanotube array, *Appl. Phys. A, 73,* 129–131, 2001.
90. Islam, M.F., Rojas, E., Bergey, D.M., Johnson, A.T., and Yodh, A.G., High weight fraction surfactant solubilization of single-wall carbon nanotubes in water, *Nano Lett., 3,* 2, 2003.
91. Nardelli, M.B., Polarization Effects in Nanotube Structures, presentation notes, North Carolina State University. http://nemo.physics.ncsu.edu/~nardelli/index.html
92. Lebedev, N.G., Zaporotskova, I.V., and Chernozatonskii, L.A., On the Estimation of Piezoelectric Modules of Carbon and Boron Nitride Nanotubes, Volograd State University and Institute of Biochemical Physics, Moscow, Russia, 2001.
93. Rubio, A., Corkill, J., and Cohen, M., Theory of graphitic boron nitride nanotubes, *Phys. Rev. B, 49,* 5081, 1994.
94. Bando, Y., Ogawa, K., and Golberg, D., Insulating "nanocables": Invar Fe-Ni alloy nanorods inside BN nanotubes, *Chem. Phys. Lett., 347,* 349, 2001.
95. Erkoc, S., Structural and electronic properties of single-wall BN nanotube, *J. Molec. Struct., 542,* 89–93, 2001.
96. Lao, J.Y., Wen, J.G., Wang, D.Z., and Ren, Z.F., Synthesis of Amorpohous SiOx Nanostructures, Department of Physics, Boston College. http://www.physics.bc.edu/faculty/Ren.html
97. Fagan, S.B., Baierle, R.J., Mota, R., da Silva, A.J.R., and Fazzio, A., *Ab initio* calculations for a hypothetical materials: silicon nanotubes, *Phys. Rev. B, 61,* 9994–9996, 2000.
98. Spahr, M.E., Stoschitzki-Bitterli, P., Nesper, R., Haas, O., and Novak, P., Vanadium oxide nanotubes, a new nanostructured redox-active material for the electrochemical insertion of lithium, *J. Electrochem. Soc., 146,* 8, 2780–2783, 1999.
99. Gu, G., Schmid, M., Chiu, P.-W., Minett, A., Fraysse, J., Kim, G.-T., Roth, S., Kozlov, M., Muñoz, E., and Baughman, R.H., V_2O_5 nanofibre sheet actuators, *Nature, 2,* 316–319, 2003.
100. Hughes, W.L. and Wang, Z.L., Nanobelts as nanocantilevers, *Appl. Phys. Lett., 82,* 17, 2004.
101. Bong, D.T., Clark, T.D., Granja, J.R., and Ghadiri, M.R., Self-assembling organic nanotubes, *Angew. Chem., 40,* 988–1011, 2001.
102. Pokropivny, V.V., Non-carbon nanotube (review) 1, synthesis methods, *Powder Metallurgy Met. Ceram., 40,* 485–496, 2001.
103. Pokropivny, V.V., Non-carbon nanotube (review) 2, types and structure, *Powder Metallurgy Met. Ceram., 40,* 582–594, 2001.
104. Pokropivny, V.V., Non-carbon nanotube (review) 3, properties and applications, *Powder Metallurgy Met. Ceram., 40,* 123–135, 2001.
105. Zambov, L., Zambova, A., Cabassi, M., and Mayer, T.S., Template-directed CVD of dielectric nanotube, *Chem. Vapor Deposition, 9,* 1, 26–33, 2003.
106. Pu, L., Bao, Z., Zou, J., and Feng, D., Individual alumina nanotubes, *Angew. Chem., 113,* 8, 1538–1541, 2001.
107. Sun, X. and Li, Y., Synthesis and characterization of ion-exchangeable titanate nanotubes, *Chem. Eur. J., 9,* 2229–2238, 2003.
108. Tremel, W., Inorganic nanotubes, *Angew. Chem. Int. Ed., 38,* 15, 2175–2179, 1999.

109. Chen, Q., Xhou, W., Du, G., and Peng, L.-M., Tritianate nanotubes made via a single alkali treatment, *Advan. Mater., 14,* 17, 1208–1211, 2002.
110. Chang, K.-S., Hernandez, B.A., Fisher, E.R., and Dorhout, P.K., Sol-gel template synthesis and characterization of PT, PZ and PZT nanotube, *J. Korean Chem. Soc., 46,* 3, 242–251, 2002.
111. Sha, j., Niu, J., Ma, X., Xu, J., Zhang, X., Yang, Q., and Yang, D., Silicon nanotubes, *Advan. Mater., 14,* 17, 1219–1221, 2002.
112. Mele, E.J. and Kral, P., Electric polarization of heteropolar nanotubes as a geometric phase, *Phys. Rev. Lett., 88,* 5, 056803, 2002.
113. Stephan, P., Ajayan, M., Colliex, C., Redlich, Ph., Lambert, J.M., Bernier, P., and Lefin, P., Doping graphitic and carbon nanotube structures with boron and nitrogen, *Science, 266,* 1683–1685, 1994.
114. Redlich, Ph., Loeffler, J., Ajayan, P.M., Gill, J., Aldinger, F., and Ruhle, M., B-C-N nanotubes and boron doping of carbon nanotubes, *Chem. Phys. Lett., 260,* 465–470, 1996.
115. Zaporotskova, I.V. and Lebedev, N.G., Some Features of Hydrogenization of Single-Walled Carbon Nanotubes, Volgograd State University and Institute of Biochemical Physics, Moscow, Russia.
116. Zaporotskova, I.V. and Lebedev, N.G., The Elastic Modules of Carbon and Boron Nitride Nanotubes in a Molecular Cluster Model, Volgograd State University and Institute of Biochemical Physics, Moscow, Russia.
117. Nakhmanson, S.M., Designing Novel Polar Materials through Computer Simulations, paper presented at Mardi Gras Physics Conference, Louisiana State University, Baton Rouge LA, February, 2003.
118. Nakhmanson, S.M., Spontaneous Polarization and Piezoelectric Properties of Boron-Nitride Nanotubes, in *Nanotube 2002,* Boston College, Boston MA, 2002.
119. Sai, N. and Mele, E.J., Nanotube piezoelectricity, *Cond. Mat., 1,* 2003.
120. Wang, Z.L., Gao, R.P., Poncharal, P., de Heer, W.A., Dai, Z.R., and Pan, Z.W., Mechanical and electrostatic properties of carbon nanotubes and nanowires, *Mater. Sci. Engin. C, 16,* 3–10, 2001.
121. Spinks, G.M., Wallace, G.G., Fifield, L.S., Dalton, L.R., Mazzoldi, A., De Rossi, D., Kharyrullin, I.I., and Baughman, R.H., Pneumatic carbon nanotube actuators, *Advan. Mater., 14,* 23, 1728–1732, 2002.
122. Bachtold, A., Hadley, P., Nakanishi, T., and Dekker, C., Logic circuits with carbon nanotube transistor, *Science, 294,* 1317–1320, 2001.
123. White, C.T. and Todorov, T.N., Carbon nanotubes as long ballistic conductors, *Nature, 393,* 21, 240–242, 2001.
124. Liang, W., Bockrath, M., Bozovic, D., Hafner, J.H., Tinkham, M., and Park, H., Fabry-Perot interference in a nanotube electron waveguide, *Nature, 411,* 7, 665–669, 2001.
125. Andoa, T., Matsumuraa, H., and Nakanishib, T., Theory of ballistic transport in carbon nanotubes, *Physica B, 323,* 4450, 2002.
126. Ebbesen, T.W., Lezec, H.J., Hiura, H., Bennett, J.W., Ghaemi, H.F., and Thio, T., Electrical conductivity of individual carbon nanotubes, *Nature, 382,* 4, 54–56, 1996.
127. C. Schonenberger, A., Bachtold, C., Strunk, et al., Interference and interaction in multi-wall carbon nanotubes, Appl. Phys. A, 69, 283–295, 1999.
128. Frank, S., Poncharal, P., Wang, Z.L., and de Heer, W.A., Nanocapillarity and chemistry in carbon nanotubes, *Science, 280,* 1774, 1998.

129. Roche, S., Triozon, F., Rubio, A., and Mayou, D., Electronic conduction in multiwalled carbon nanotubes: role of intershell coupling and incommensurability, *Phys. Lett. A, 285,* 94–100, 2001.
130. Berger, Y., Yi, Z., Wang, L., and de Heer, W.A., Multiwalled carbon nanotubes are ballistic conductors at room temperature, *Appl. Phys. B, 74,* 363–365, 2002.
131. Radosavljevic′, M., Lefebvre, J., and Johnson, A.T., High-field electrical transport and breakdown in bundles of single-wall carbon nanotubes, *Phys. Rev. B, 64,* 241307, 2003.
132. Appenxeller, J., Martel, R., and Aavouris, Ph., Phase-coherent transport in ropes of single-wall carbon nanotubes, *Phys. Rev. B, 64,* 121404, 2001.
133. Zhao, Y.-P., Wei, B.Q., Ajayan, P.M., Ramanath, G., Lu, T.-M., and Wang, G.-C., Frequency-dependent electrical transport in carbon nanotubes, *Phys. Rev. B, 64,* 201402, 2002.
134. Burke, P.J., An RF circuit model for carbon nanotubes, *IEEE Trans. Nanotechnol., 2,* 1, 55–58, 2003.
135. Kaiser, A.B., Challis, K.J., McIntosh, G.C., Kim, G.T., Yu, H.Y., Park, J.G., Jhang, S.H., and Park, Y.W., Frequency and field dependent conductivity of carbon nanotube networks, *Curr. Appl. Phys., 2,* 163–166, 2002.
136. Qian, D., Wagner, G.J., Liu, W.K., Yu, M.F., and Ruoff, R.S., Mechanics of carbon nanotubes, *ASME, Appl. Mech. Rev., 55,* 6, 2002.
137. Qian, D., Liu, W.K., and Ruoff, R.S., Effect of interlayer interaction on the buckling of multi-walled carbon nanotube, *J. Nanosci. Nanotechnol.,* under review, 2002.
138. Liu, Y.J. and Chen, X.L., Evaluations of the effective materials properties of carbon nanotube-based composites using a nanoscale representative volume element, *Mechan. Mater., 35,* 69–81, 2002.
139. Qian, D., Wagner, G.J., and Liu, W.K., Mechanics of carbon nanotubes, *ASME, 55,* 6, 495–533, 2002.
140. Yu, M.-F., Files, B.S., Arepalli, S., and Fuoff, R.S., Tensile loading of ropes of single wall carbon nanotubes and their mechanical properties, *Phys. Rev. Lett., 84,* 24, 5552–5555, 2002.
141. Kahn, D. and Lu, J.P., Vibration modes of carbon nanotubes and nanoropes, *Phys. Rev. B, 60,* 9, 6535–6540, 1999.
142. Maarouf, A.A., Kane, C.L., and Mele, E.J., Electronic structure of carbon nanotube ropes, *Phys. Rev. B, 61,* 16, 61, 2000.
143. Paulson, S., Helser, A., Nardelli, M.B., Taylor, R.M., II, Falvo, M., Superfine, R., and Washburn, S., Tunable resistance of a carbon nanotube-graphite interface, *Science, 290,* 1742–1744, 2000.
144. Oomman, K., Varghese, D.G., Paulose, M., Ong, K.G., Dickey, E.C., and Grimes, C.A., Extreme changes in the electrical resistance of titania nanotubes with hydrogen exposure, *Advan. Mater., 15,* 7–8, 624–627, 2003.
145. Fischer, J.E., Dai, H., Thess, A., Lee, R., Hanjani, N.M., Dehaas, D.L., and Smalley, R.E., Metallic resistivity in crystalline ropes of single-wall carbon nanotubes, *Phys. Rev. B,* 55, R4921–R4924, 1997.
146. Nath, M., Teredesai, P.V., Muthu, D.V.S., Sood, A.K., and Rao, C.N.R., Single-walled carbon nanotube bundles intercalated with semiconductor nanoparticles, *Curr. Sci., 85,* 7, 2003.
147. Tombler, T.W., Zhou, C., Alexseyev, L., Kong, J., Dal, H., Liu, L., Jayanthl, C.S., Tang, M., and Wu, S.-Y., Reversible electromechanical characteristics of carbon nanotubes under local-probe manipulation, *Nature, 405,* 769–772, 2000.
148. Dielectrophoresis tutorial, http://www.ibmm-microtech.co.uk/pages/science/basic.htm.

149. Vykoukal, J., Sharma, S., Vykoukal, D.M., and Gascoyne, P.R.C., Engineered dielectric microspheres for use in microsystems, University of Texas, M.D. Anderson Cancer Center, Department of Molecular Pathology.
150. Gascoyne, P.R.C. and Vykoukal, J., Particle separation by dielectrophoresis, *Electrophoresis,* 1973–1983, 2002.
151. Fuller, C.K., Hamilton, J., Ackler, H., Krulevitch, P., Boser, B., Eldredge, A., Becker, F., Yang, J., and Gascoyne, P., Microfabricated Multi-Frequency Particle Impedance Characterization System, fuller14@llne.gov
152. Huang, Y., Yang, J., Wang, X., Becker, F.F., and Gascoyne, P.R.C., The removal of human breast cancer cells from hematopoietic CDD34+ stem cells by dielectrophoretic field-flow fractionalization, *J. Hematother. Stem Cell Res., 8,* 481–490, 1999.
153. Cummings, E.B. and Singh, A.K., Dielectrophoretic Trapping without Embedded Electrodes, Sandia National Laboratories, Livermore, CA. ebcummi@sandia.gov
154. Cummings, E.B., A comparison of theoretical and experimental electrokinetic and dielectrophoretic flow fields, *AIAA,* 2002–3193.
155. Three Dimensional Nano Manipulation by Using Dielectrophoresis. http://www.mein.nagoya-u-ac.jp/activity/1999-e/NANO_99E.html
156. Dielectrophoresis, Cell Analysis Ltd. http://www.cell-analysis.com/SciTech/Dielectro/DielectroBody.html
157. Kral, P. and Shapiro, M., Nanotube electron drag in flowing liquids, *Phys. Rev. Lett., 86,* 1, 2001.
158. Ghosh, S., Sood, A.K., and Kumar, N., Carbon nanotube flow sensors, *Science, 299,* 2003.
159. Shi, D., Lian, J., He, P., Wang, L.M., Xiao, F., Yang, L., Schulz, M.J., and Mastt, D.B., Plasma coating of carbon nanotubes for enhanced dispersion and interfacial bonding in polymer composites, *Appl. Phys. Lett., 83,* 5301, 2003.
160. Bahr, J.L. and Tour, J.M., Covalent chemistry of single-wall carbon nanotubes, *J. Mater. Chem., 12,* 1952–1958, 2002.
161. Zhao, J., Buldum, A., Han, J., and Lu, J.P., Gas molecule adsorption in carbon nanotubes and nanoparticle bundles, *Nanotechnology, 13,* 195–200, 2002.
162. Bahr, J.L. and Tour, J.M., Covalent chemistry of single-wall carbon nanotubes, *J. Mater. Chem., 12,* 1952–1958, 2002.
163. Appetecchi, G.B., Croce, F., and Scrosati, B., *Electrochim. Acta, 40,* 991, 1995.
164. Sebhon, S.S., Arora, N., Singh, B., Chandra, A., Protonic polymer gel electrolytes based on carboxylic acids: ortho and inductive effects, *J. Mater. Sci. Lett.,* 37 10, 2002.
165. NASA Glenn Research. http://www.lerc.nasa.gov/WWW/ictd/content/hilight20020607.html
166. Ahao, Q., Frogley, M.D., and Wagner, H.D., Direction-sensitive strain-mapping with carbon nanotube sensors, *Composites Sci. Technol., 62,* 147–150, 2002.
167. Chiarello, G., Maccallini, E., Agostino, R.G., Formoso, V., Cupolillo, A., Pacile, D., Colavita, E., Papagno, L., Petaccia, L., Larciprete, R., Lizzit, S., and Goldoni, A., Electronic and vibrational excitations in carbon nanotubes, *Carbon, 41,* 985–992, 2003.
168. Shahinpoor, M. and Kim, K.J., Ionic polymer-metal composites: I. Fundamentals, *Smart Mater. Struct., 10,* 819–833, 2001.
169. Liu, B., Sundqvist, B., Andersson, O., Wagberg, T., Nyeanchi, E.B., Zhu, X.-M., and Zou, G., Electric resistance of single-walled carbon nanotubes under hydrostatic pressure, *Solid State Commn., 118,* 31–36, 2001.
170. Peng, S., O'Keeffe, J., Wei, C., and Cho, K., Carbon Nanotube Chemical and Mechanical Sensors, paper presented at 3rd International Workshop on Structural Health Monitoring, September 12–14, 2001.

171. Cao, J., Wang, Q., and Dai, H., Electromechanical properties of metallic, quasimetallic and semiconducting carbon nanotubes under stretching, *Phys. Rev. Lett., 90,* 15, 157601, 2003.
172. Rochefort, A., Salahub, D., and Avouris, P., The effect of structural distortions on the electronic structure of carbon nanotubes, *Chem. Phys. Lett., 297,* 45–50, 1998, 2000.
173. Poncharal, P., Wang, Z.L., Ugarte, D., and de Heer, W.A., Electrostatic deflections and electromechanical resonances of carbon nanotube, *Science, 283,* 1513, 1999.
174. Varghese, O.K. Kichambre, P.D., Gong, D., Ong, K.G., Dickey, C.C., and Grimes, C.A., Gas sensing characteristics of multi-wall carbon nanotube, *Sensors Actuators B, 81,* 32–41, 2001.
175. Ong, K.G. and Grimes, C.A., A resonant printed-circuit sensor for remote query monitoring of environmental parameters, *Smart Mater. Struct., 9,* 421–428, 2000.
176. Ong, K.G., Zeng, K., and Grimes, C.A., Wireless, passive carbon nanotube-based gas sensor, *IEEE Sensors J., 2,* 2, 82–88, 2002.
177. Dharap, P., Li, Z., Nagarajaiah, S., and Barrera, E.V., Nanotube film based on single-wall carbon nanotubes for strain sensing, *Nanotechnology, 15,* 2004.
178. Schnur, J.N., Lipid tubules: A paradigm for molecularly engineered structures, *Science, 262,* 1669, 1993.
179. Lord, E.A., Helical structures: The geometry of protein helices and nanotubes, *Struct. Chem., 13,* 305–314, 2002.
180. Terech, P., de Geyer, A., Struth, B., and Talmon, Y., Self-assembled monodisperse steroid nanotubes in water, *Advan. Mater., 14,* 7, 495–498, 2002.
181. Chung, Y., Kong, M., Modeling of polynucleotide translocation through protein pores and nanotubes, *Electrophoresis, 23,* 2697–2703, 2002.
182. Frusawa, H., Fukagawa, A., Ikeda, Y., Araki, J., Ito, K., John, G., and Shimizu, T., Aligning a single-lipid nanotube with moderate stiffness, *Angew. Chem., 115,* 76–78, 2003.
183. Kim, P. and Liever, C.M., Nanotube nanotweezers, *Science, 286,* 2148–2150, 1999.
184. Nakayama, Y., Scanning probe microscopy installed with nanotube probes and nanotube tweezers, *Ultramicroscopy, 91,* 49–56, 2002.
185. Weissmuller, J., Viswanath, R.N., Kramer, D., Zimmer, P., Wurschum, R., and Gleither, H., Charge-induced reversible strain in a metal, *Science, 300,* 312–315, 2003.
186. Baughman, R.H. et al., Muscles made from metal, *Science, 300,* 268–269, 2003.
187. Hess, H. and Vogel, V., Molecular shuttles based on motor proteins: active transport in synthetic environments, *Rev. Molecular Biotechnol., 82,* 67–85, 2001.
188. Schliwa, M. and Woehlke, G., Molecular motor, *Nature, 422,* 759–765, 2003.
189. Ait-Haddou, R. and Herzog, W., Force and motion generation of myosin motors: muscle contraction, *J. Electromyography and Kinesiology, 12,* 435–445, 2002.
190. VPL actuators. http://npg.nature.com/npg/servlet/Content?data=xml/01_reprints.xml &style=xml/01_reprints
191. Mavroidis, C., Yarmush, M., Dubey, A., Thronton, A., Nikitszuk, K., Tomassone, S., Papadimitrakopoulos, F., and Yurke, B., Protein based nano-machines for space applications. NiAC Grant Final Report, http://bionano.rutgers.edu/Mavroidis_Final_Report.pdf
192. Mustarelli, P., Quartarone, E., Tomasi, C., and Magistris, A., New materials for polymer electrolytes, *Solid State Inonics, 135,* 81–86, 2000.
193. Stephan, A.M., Saito, Y., Muniyandi, N., Renganathan, N.G., Kalyanasundaram, S., and Nimma, E.R., Preparation and characterization of PVC/PMMA blend polymer electrolytes complexed with $LiN(CF_3SO_2)_2$, *Solid State Inonics, 148,* 467– 473, 2002.
194. Rajendran, S. and Uma, T., Lithium ion conduction in PVC–LiBF electrolytes gelled with PMMA, *J. Power Sources, 88,* 282–285, 2000.

195. Sukeshini, A.M., Nishimoto, A., and Watanabe, M., Transport and electrochemical characterization of plasticized poly(vinyl chloride) solid electrolytes, *Solid State Inonics, 86–88,* 385–393, 1996.
196. Gupta, P.N. and Singh, K.P., Characterization of H_3PO_4 based PVA complex system, *Solid State Inonics, 86–88,* 319–323, 1996.
197. Lewis, T.W., Kim, B.C., Spinks, G.M., and Wallace, G.G., Evaluation of solid polymer electrolytes for use in conducting polymer/nanotube actuators, *Proc. SPIE, 3987,* 351–357, 2000.
198. Lehtinen, T., Sundholm, G., Holmberg, S., Sundholm, F., Bjornbomd, P., and Burselld, M., Electrochemical characterization of PVDF-based proton conducting membranes for fuel cells, *Electrochimica Acta, 43,* 12–13, 1881–1890, 1998.
199. Kaiser, A.B., Flanagan, G.U., Stewart, M., and Beaglehole, D., Heterogeneous model for conduction in conduction polymers and carbon nanotubes, *Synth. Met., 117,* 67–73, 2001.
200. Smits, J., Wincheski, B., Ingram, J., Watkins, N., and Jordan, J., Controlled deposition and applied field alignment of single walled carbon nanotubes for CNT device, NASA Langley, 2003.
201. NASA Ames. http//ettc.usc.edu/ames/nano/visionchip.html
202. Kirikera, G.R., Datta, S., Schulz, M.J., Ghoshal, A., and Sundaresan, M.J., Mimicking the biological neural system using active fiber continuous sensors and electronic logic circuits, paper presented at SPIE 11th International Symposium on Smart Structures, San Diego, March, 2004.
203. Mineta, T., Mitsui, T., Watanabe, Y., Kobayashi, S., Haga, Y., and Esashi, M., Batch fabricated flat meandering shape memory alloy actuator for active catheter, *Sensors Actuators A, 88,* 112–120, 2001.
204. Lim, G., Park, K., Sugihara, M., Minami, K., and Esashi, M., Future of active catheters, *Sensors Actuators A, 56,* 113–121, 1999.
205. Dae-Kyu, K., Yeo-Heung, Y., Kee-Ho, Y., Seong-Cheol, L., Simulation of an Active Catheter Actuator Using Shape Memory Alloy, Korea Society of Precision Engineering 2000 Fall Conference, 72–75, 2000.
206. Krishna, G.M. and Rajanna, K., Tactile sensor based on piezoelectric resonance, *IEEE Sensors J. A*, No, 5, 2004.
207. Sherrit, S., Bao, X., Sigel, D.A., Gradziel, M.J., Askins, S.A., Dolgin, B.P., Bar Cohen, Y., Characterization of transducers and resonators urder high drive levels, IEEE International Ultra Sciences Symposium, Atlanta, 2001.
208. Hata, K. Futabal, D.N., Mizuno, K., Namai, T., Yumura, M., and Iijima, S., Water-assisted highly efficient synthesis of impurity-free single-walled carbon nanotubes, *Science,* 306, Novmeber 19, 2004.
209. Cumings, J. and Zettl, A., Localization and nonlinear resistance in telescopically extended nanotubes, *Phys. Rev. Lett. 93,* 086801, 1–4, 2004.
210. Cumings, J., Collins, P.G., and Zettl, A., Peeling and sharpening multiwall nanotubes, *Nature, 406*, 586, 2000.
211. Cumings, J. and Zettl, A., Resistance of Telescoping Nanotubes, in *Structural and Electronic Properties of Molecular Nanostructures*, Kuzmanyh, H. et a1., Eds., American Institure of Physics, Washington, D.C., 2002.
212. Cumings, J. and Zettl, A., Low-friction nanoscale linear bearing realized from multiwall carbon nanotubes, *Science,* 28, Vol 289, July 2000.
213. Lozovikl, Yu.E., Minogin, A.V., and Popov, A.M., *Nanomachines Based on Carbon Nanotubes,* Institute of Spectroscopy, Russian Academy of Science, Moscow, Russia.

214. Forro, L., *Nanotechnology: Beyond Gedanken Experiments,* Department of Physics, Ecole Polytechnique Federale de Lausanne, Lausanne, Switzerland.
215. Kim, D.H., Chang, K.I., Electron transport in telescoping carbon nanotubes, *Physical Review B* 66, 2002.
216. Lozovik. Y.E., Nikolaev, A.G., Popv, A.V., *Atomic Scale Design of Carbon Nanotubes: the Way to Produce Bolt-and-Nut Pairs,* Institute of Spectroscopy of Russia Academy of Science, Moscow, Russia.
217. Legoas, S.B., Coluci, V.R., Braga, S.F., Coura, P.Z., Dantas, S.O., Galvao, D.S., *Gigahertz Nanomechanical Oscillators Based on Carbon Nanotubes,* TNT, Sept. 15–19, 2003, Salamanca Spain.

16 Thermal Properties and Microstructures of Polymer Nanostructured Materials

Joseph H. Koo and Louis A. Pilato

CONTENTS

16.1 INTRODUCTION

The utility of introducing inorganic nanomaterials as additives into polymer systems has resulted in polymer nanostructured materials exhibiting multifunctional, high-performance polymer characteristics beyond what traditional polymer composites possess. Multifunctional features attributable to polymer nanocomposites consist of improved thermal resistance, flame resistance, moisture resistance, decreased permeability, charge dissipation, and chemical resistance. Through control and/or alteration of the additive at the nanoscale level, one can maximize property enhancement of selected polymer systems to meet or exceed the requirements of current military, aerospace, and commercial applications. The technical approach involves the introduction of nanoparticles into selected polymer matrix systems whereby nanoparticles may be surface-treated to provide hydrophobic characteristics and enhanced inclusion into the hydrophobic polymer matrix.

The objectives of this chapter are to summarize our research activities in the areas of (1) developing processes to disperse nanoparticles uniformly in the different types of polymers, (2) using wide-angle x-ray diffraction (WAXD), transmission electron microscopy (TEM), and scanning electron microscopy (SEM) techniques to characterize polymer nanocomposite structures, (3) studying the structure–property relationships of these types of new materials, and (4) evaluating the thermal performances of these materials for different applications using established laboratory devices.

Four different processing methods were used to disperse nanoparticles in a variety of polymers in this study: (1) high shear mixing for liquid resins, (2) three-roll milling for liquid resins, (3) Brabender-type mixing for high viscosity resins, and (4) twin screw extrusion for solid polymers. The degree of nanodispersion was characterized by WAXD, TEM, and SEM analyses. These imaging techniques allowed us to screen formulations and distinguish compositions that exhibited both favorable and unfavorable nanodispersed nanoparticle/polymer blends. Furthermore these analytical techniques facilitated and provided guidelines in the scale-up of favorable compositions.

Different types of polymers including (1) thermosets, (2) thermoplastics, and (3) thermoplastic elastomers are presented as examples in this chapter. Three types of nanoparticles including (1) montmorillonite (MMT) organoclays, (2) carbon nanofibers (CNFs), and (3) polyhedral oligomeric silsesquioxanes (POSS®) were used to disperse into the preceding polymers. Selected laboratory devices such as a cone calorimeter, simulated solid rocket motor (SSRM), and subscale solid rocket motor were used to evaluate material performance for fire and rocket propulsion applications.

16.2 SELECTION OF NANOPARTICLES

16.2.1 Montmorillonite Nanoclays

Exfoliation of montmorillonite (MMT) clays into polymers has been shown to increase mechanical properties, barrier performance, and application processing.[1] Achieving exfoliation of organomontmorillonite in various polymer continuous phases is a function of the surface treatment of the MMT clays and the mixing efficiency of the dispersing apparatus. Surface treatment of MMT is classically accomplished with the exchange of inorganic counterions, e.g., sodium, with quaternary ammonium ions. The choice of the quaternary ammonium ion is a function of the hydrophilic and hydrophobic nature of the continuous phase; the hydrophilic and hydrophobic nature of the interface of MMT is balanced with the hydrophobic and hydrophilic nature of the continuous phase. Cloisite® 30B, a MMT clay from Southern Clay Products, had been used in several of our application studies and will be described in detail as an example for nanoclay. It is a surface modified

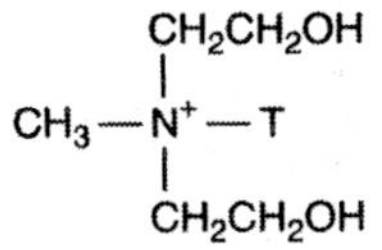

Where T is tallow (~65% C18; ~30% C16; ~5% C14)
Anion: Chloride

(1) MT2EtOH: methyl, tallow, bis-2-hydroxyethyl, quaternary ammonium

FIGURE 16.1 Chemical structure of Cloisite 30B. T = tallow (~65% C18; ~30% C16; ~5% C14). Anion: Chloride. MT2EtOH = methyl, tallow, bis-2-hydroxyethyl, quaternary ammonium.

montmorillonite clay [tallow bishydroxyethyl methyl, $T(EOH)_2M$] manufactured by Southern Clay Products (SCP).[2] The chemical surface structure of Cloisite 30B is shown in Fig. 16.1. Cloisite 30B is used as an additive for plastics to improve various physical properties, such as reinforcement, heat distortion temperature, and barrier properties. It has a ternary ammonium salt modifier at 90 meq/100 g clay and is off-white in color. Its typical dry-particle size and physical properties are shown in Table 16.1 and Table 16.2, respectively.

Cloisite 30B in loadings of 5, 10, and 15 wt% was dispersed in a liquid phenolic resole resin (SC-1008, Borden Chemicals).[3–6] SC-1008 is the matrix used in the manufacture of MX 4926, an ablative material for solid rocket nozzles. Mixing efficiency in the phenolic resin can be achieved with standard paint mixing or mixing equipment (IKA Works, Inc., Wilmington, NC) that is designed to provide high-shear mixing if the phenolic resin is in solution. The TEM analyses allowed us to determine the degree of dispersion/exfoliation of the nanoclay before committing to a 20-lb production run of these nanoparticle/resin mixtures to make prepregs at Cytec Engineered Materials (CEM). This proved to be a very cost-effective and efficient technique for screening different formulations. This nanoclay has also been used with cyanate ester,[7–9] thermoplastic elastomer,[10,11] epoxy,[12] and nylon 11[13] in our research.

16.2.2 Carbon Nanofibers (CNFs)

CNFs are forms of vapor-grown carbon fiber, which is a discontinuous graphitic filament produced in the gas phase from the pyrolysis of hydrocarbons.[14–17] In regard to physical size, performance improvement, and product cost, CNFs complete a

TABLE 16.1
Typical Dry-Particle Sizes for Cloisite 30B (Microns, by Volume)

10% Less Than	50% Less Than	90% Less Than
2 μ	6 μ	13 μ

TABLE 16.2
Physical Properties of Cloisite 30B (Microns, by Volume)

Property	Cloisite 30B
Loose bulk density (lb/ft^3)	14.25
Packed bulk density (lb/ft^3)	22.71
Specific gravity	1.98
d_{001} (Å)	18.5

continuum bounded by carbon black, fullerenes, and single- and multi-wall carbon nanotubes on one end and continuous carbon fibers on the other.[16] CNFs are able to combine many of the advantages of these other forms of carbon for reinforcement in engineered polymers. They have transport and mechanical properties that approach the theoretical values of single-crystal graphite, similar to the fullerenes, but can be made in high volumes at low cost — ultimately lower than conventional carbon fibers. In equivalent production volumes, a CNF is projected to have a cost comparable to E-glass on a per-pound basis, yet possess properties that far exceed those of glass and are equal to or exceed those of much more costly commercial carbon fiber. Maruyama and Alam published an excellent review of carbon nanotubes and nanofibers in composite materials.[17]

CNFs are manufactured by Applied Sciences, Inc./Pyrograf® Products by pyrolytic decomposition of methane in the presence of iron-based catalyst particles at temperatures above 900°C. Pyrograf-III is a patented, very fine, highly graphitic CNF. Pyrograf-III is available in diameters ranging from 50 to 200 nm and a length of 50 to 100 μm. Therefore CNFs are much smaller than conventional continuous or milled carbon fibers (5 to 10 μm) but significantly larger than carbon nanotubes (1 to 10 nm). Compared to PAN and pitch-based carbon fiber, the morphology of CNFs is unique in that there are far fewer impurities in the filament, providing for graphitic and turbostatic graphite structures, and the graphene planes are more preferentially oriented around the axis of the fiber. Consequences of the circumferential orientation of high-purity graphene planes are a lack of cross-linking between the graphene layers, and a relative lack of active sites on the fiber surface, making it more resistant to oxidation, and less reactive for bonding to matrix materials. Also in contrast to carbon fiber derived from PAN or pitch precursors, CNFs are produced only in a discontinuous form, where the length of the fiber can be varied from about 100 μm to several cm, and the diameter is of the order of 100 nm. The two commonly used CNFs in our application studies are PR-19-PS and PR-24-PS.[3–12]

CNFs exhibit exceptional mechanical and transport properties, allowing them excellent potential as components for engineering materials. Table 16.3 lists the properties of vapor-grown carbon fibers, both as-grown and after a graphitizing heat treatment to 3000°C. Note that due to the difficulty of direct measurements on the nanofibers, the values shown are measured on vapor-grown fibers that have been thickened to several microns in diameter.[16] Such fibers consist almost

TABLE 16.3
Properties of CNFs

Property (Units)	As Grown	Heat Treated
Tensile strength (GPa)	2.7	7.0
Tensile modulus (GPa)	400	600
Ultimate strain (%)	1.5	0.5
Density (g/cc)	1.8	2.1
Electrical resistivity (μΩ-cm)	1000	55
Thermal conductivity (W/m-K)	20	1950

exclusively of chemical vapor deposition (CVD) carbon, which is less graphitic and more defective than the catalytically grown carbon core that constitutes the CNF. Thus the properties listed in the table represent an estimate of the properties of CNFs.

One of the goals for CNF broad utility is to provide mechanical reinforcement comparable to that achieved with continuous carbon fiber at a price that approaches that of glass fiber reinforcement and with low-cost composite fabrication methods such as injection molding. Theoretical models[16] suggest that reinforcement by discontinuous fibers such as CNFs can closely approach that of continuous fibers as long as the aspect ratio of the fibers is high and the alignment is good. Work is ongoing to improve the mechanical benefits of CNFs through fiber surface modification to provide physical or chemical bonding to the matrix. Such modifications have resulted in strength and modulus improvements of four to six times the values of the neat resin; however, these values are still a modest fraction of what may be anticipated from idealized fiber–matrix interphase and alignment of the fibers within the matrix. The more immediate opportunities for use in structural composites lie in the prospect for modifying the properties of the matrix material. For example, use of small volume loadings of CNFs in epoxy may allow for improvement of interlaminar shear strength of PAN or pitch-based composites. The CNF additives to fiberglass composites could provide benefits to a suite of properties, including thermal and electrical conductivity, coefficient of thermal expansion, and mechanical properties, as suggested by the data in Table 16.4.[16] Figure 16.2 is an SEM

TABLE 16.4
Thermoset Polyester/Pyrograf-III Composite Properties

Fiber Content (Wt%)	Tensile Strength (MPa)	Tensile Modulus (GPa)	Electrical Resistivity (ohm-cm)
17% PR-19	51.5	4.55	3.2
17% PR-19, OX	47.4	4.55	7.1
5% PR-19 in 10% ¼" glass	44.1	11.52	5.0
5% PR-19, OX in 10% ¼" glass	33.8	8.92	7.0

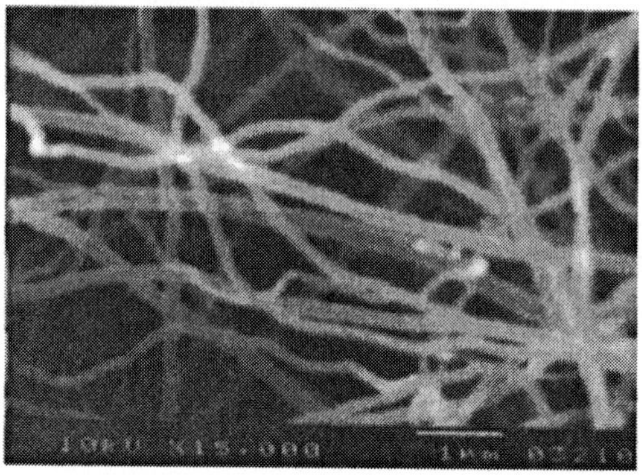

FIGURE 16.2 SEM micrograph of CNF bundles.

micrograph of a CNF bundle.[16] Figure 16.3 is a TEM micrograph of an individual CNF with a hollow core in two distinct regions: catalytic and deposited.[16]

Recently Glasgow et al.[18] demonstrated that the achievement of significant mechanical reinforcement in CNF composites requires high fiber loadings and is somewhat dependent on generating an appropriate interphase between the CNF and the matrix. Novel surface treatments under development have yielded good improvements in the tensile modulus and strength of CNF-reinforced polypropylene. Adding surface functional groups, particularly oxygen groups, also demonstrated benefits for interphase development. Carboxyl and phenolic groups contributing to a total surface oxygen concentration in the range from 5 to 20 atom percent have been added to CNF used to fabricate epoxy polymer matrix composites, providing

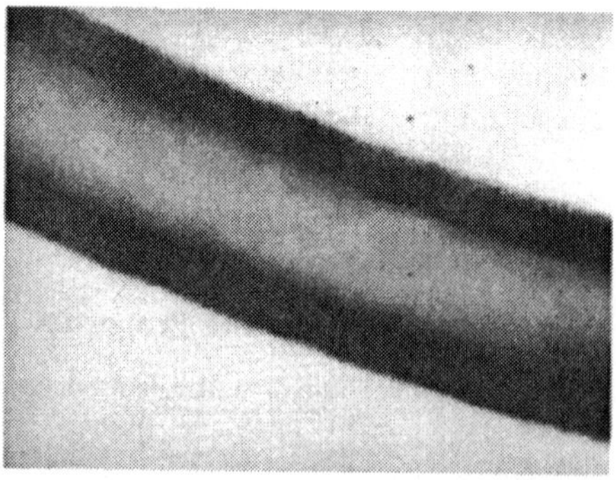

FIGURE 16.3 TEM micrograph of individual CNF showing hollow core and two distinct regions: catalytic and deposited.

improved flexural strength and modulus. The effect of similarly functionalized CNF in bismaleimide (BMI) polymer matrix composites also shows promise. Data for polypropylene, epoxy, and BMI/CNF-reinforced composites indicate that higher fiber volume loadings will find a role in structural composite markets as price and availability improve.

16.2.3 Polyhedral Oligomeric Silsesquioxane (POSS)

Representing a merger between chemical and filler technologies, POSS nanostructured materials can be used as multifunctional polymer additives, acting simultaneously as molecular level reinforcements, processing aids, and flame retardants. POSS nanostructured materials have two unique structural features: (1) the chemical composition is a hybrid, intermediate ($RSiO_{1.5}$) between those of silica (SiO_2) and silicones (R_2SiO); (2) the POSS molecules are nanoscopic in size, ranging from approximately 1 to 3 nm. POSS molecules have multifunctional properties and an inorganic framework of silicone oxide. It has been proven that this is ceramic-like in nature, especially in severe environments. The other key factor for POSS property enhancement is the length scale effect.

Lee[19] followed the approach in using POSS as an additive alloyed into epoxy systems and as a copolymer compatibilizer. A POSS-alloyed epoxy sample preparation procedure was developed. The epoxy was preheated for an hour, and then POSS in different weight percentages was added. The two epoxies used were the DER 332 system, which is the difunctional epoxy of the DGEBA variety and the tetrafunctional epoxy of 4,4′-methylene bisaniline. The trisilanolphenyl-POSS was selected for his work. A very low amount of loading (0.20 to 1.00 wt%) of the POSS compound was used. At 0.20 wt% of POSS, he reported a significant increase in T_g from 84.2 to 92.1 as shown in Table 16.5. For the tetrafunctional epoxy also mixed with Jeffamine, the same increase in the glass transition temperature was observed.[19] When a Dytek A curing agent was used on the DER-332 and 4,4

TABLE 16.5
Summary of DMA and DSC Thermal Data for Jeffamine D-230 Cured DER-332 Epoxy System

POSS (wt%)	Tg DMTA (ΔT)*	Tg DSC (ΔT)**
0%	84.2 (12)	86.6 (6)
0.20%	92.1 (10)	90.1 (5)
0.40%	91.2 (9)	91.4 (5)
0.60%	90.6 (9)	94.4 (6)
0.80%	91.6 (10)	91.1 (5)
1.00%	91.7 (12)	91.7 (7)

* Tg is determined by using tan δ curve; ΔT is the full width at half the tan δ peak.
** ΔT is endset temperature—onset temperature.

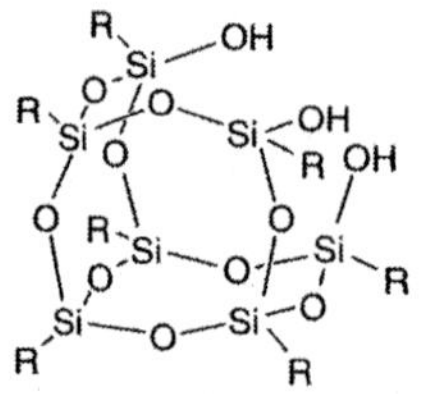

R = phenyl

SO1458
TriSilanolPhenyl-POSS

$C_{42}H_{38}O_{12}Si_7$ MW: 931.34 g/mole

Soluble: tetrahydrofuran, chloroform, ethanol

Insoluble: hexane

Appearance: white powder

FIGURE16.4 Chemical structure of trisilanolphenyl-POSS.

methylenebis (N-N-diglycidylaniline) epoxy system, he did not observe any change in DSC. The particular diamine used to cure the system is critical.

Hybrid Plastics' SO1458 trisilanolphenyl-POSS ($C_{42}H_{38}O_{12}Si_7$) has been used in several of our studies. It is the material of choice for use with phenolic,[3–6] epoxy,[12] cyanate ester,[7–9] and thermoplastic elastomer. [10,11] This was also used by Lee.[19] Figure 16.4 shows the chemical structure of the trisilanolphenyl-POSS. POSS is currently very expensive at the research and development material level because it is a new nanoparticle with low-production volume.

16.3 DISCUSSION OF RESULTS

The challenges to form polymer nanocomposites are (1) choice of nanoparticles involving an interfacial interaction or compatibility with the polymer matrix, (2) proper processing techniques to uniformly disperse and distribute the nanoparticles or aggregates within the polymer matrix, and (3) selective analytical tools to characterize the morphology of the nanocomposites. We selected four application areas as examples for this chapter: (1) fire retardant nanocomposite coatings, (2) nanostructured materials for rocket propulsion systems with improved thermal insulation, (3) nanocomposite rocket ablative materials, and (4) nanomodified carbon–carbon composites. These examples are used to illustrate the processing techniques to form the resulting nanocomposite system, the analytical and imaging techniques used to characterize the nanocomposites, and the thermal property and performance characteristics of these nanocomposites for their specific applications.

16.3.1 TEM Analyses of Nanoparticles

Three types of nanoparticles were dissolved in 0.5% isopropanol (IPA) solvent overnight to create specimens for TEM analyses. Figure 16.5 shows that the Cloisite 30B nanoclay layers are in an intercalated state and the scale bar is 500 nm. Figure 16.6 shows the PR-24-PS CNFs are entangled and the scale bar is 1 μm. Figure 16.7 shows trisilanolphenyl-POSS-SO1458 ($C_{48}H_{38}O_{12}Si_7$) showing micron and nanometer particles in the system and the scale bar is 1 μm. Neat nanoparticles dissolved in isopropanol solvent exhibit particle intercalation, entanglement, and a wide assortment of micron and nanometer particles.

FIGURE 16.5 TEM micrograph of Cloisite 30B nanoclays dissolved in IPA solvent. Scale bar = 500 nm.

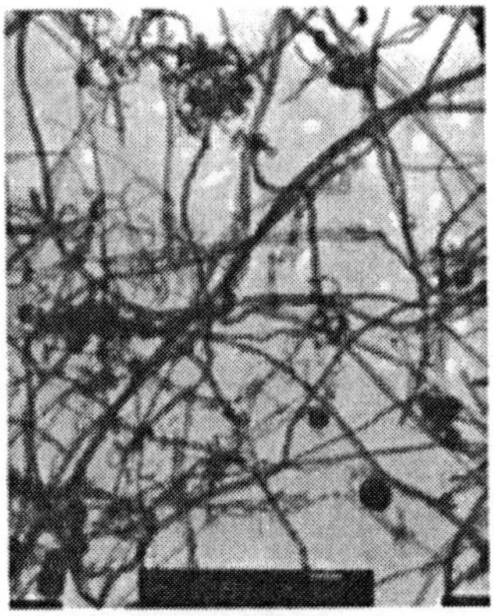

FIGURE 16.6 TEM image of PR-24-PS CNF dissolved in IPA solvent. Scale bar = 1 μm.

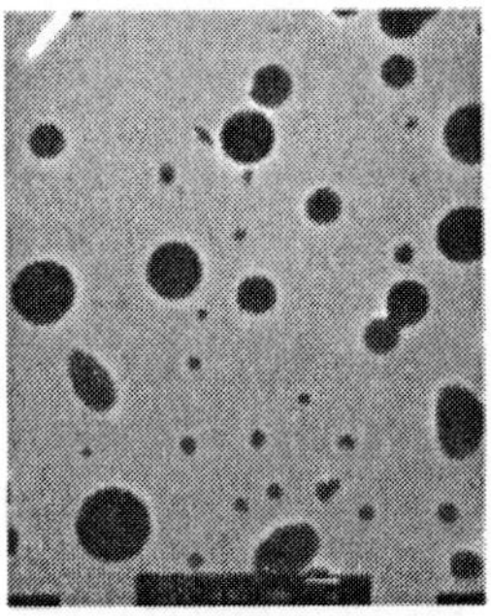

FIGURE 16.7 TEM image of trisilanolphenyl-POSS-SO1458 ($C_{48}H_{38}O_{12}Si_7$) dissolved in IPA solvent. Scale bar = 1 μm.

16.3.2 Fire-Retardant Nanocomposite Coatings

Vinyl acetate–acrylic copolymer (PVA 5174), ammonium polyphosphate (APP), melamine and dipentaerythritol supplied by StanChem (Hartford, CT), and nanoclay (Cloisite Na^+) supplied by Southern Clay Products were used as received. The nanoclay was dispersed in deionized water with Dispermat (high-shear mixer) at high speed for about 1 h. The slurry was allowed to stay at room temperature overnight. The slurry was then mixed with PVA 5174 and/or intumescent components (APP, dipentaerythritol, and melamine) for about 2 h for subsequent coating. Douglas fir plywood sized 10 by 10 by 1.8 cm was used as the substrate for all test specimens. A 5-ml plastic syringe was used to measure and place the amount of coating on the sample. It was evenly spread (visual monitoring only) with a brush presaturated with the coating. The coating thickness was monitored with a digital caliper for a dry thickness of 0.28 to 0.30 mm (0.011 to 0.012 in.). The edges were then fully coated and the coating was allowed to dry at room temperature for about 24 h before testing. Samples were exposed to a mass loss calorimeter under a heat flux of 50 kW/m^2 for 15 min. The calorimetry data reported here represent the average of three replicates.[20]

Figure 16.8 shows the heat release rates of PVA–clay fire-retardant (FR) coating systems with different clay loading at a heat flux of 50 kW/m^2. It is very clear that nanoclay plays a key role in reducing the flammability of these coating systems. With only 10% clay, the peak high release rate (PHRR) was dramatically reduced by almost 50%. With more clay applied, the PHRR continued to drop. Actually with

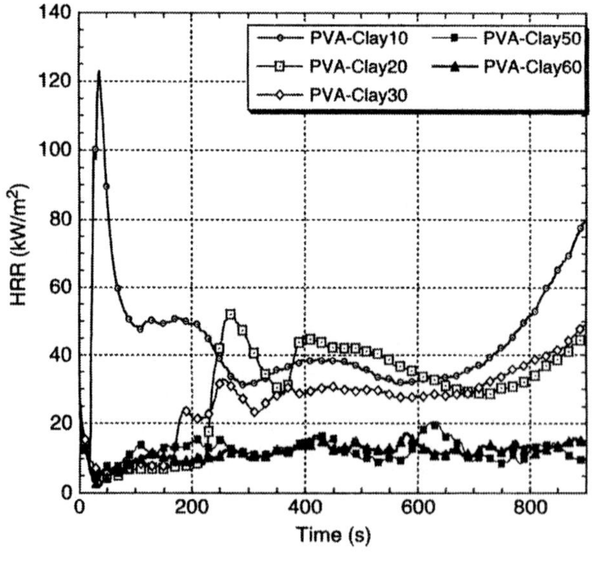

FIGURE 16.8 Comparison of HRR values for PVA–clay fire-retardant coatings at 50 kW/m^2.

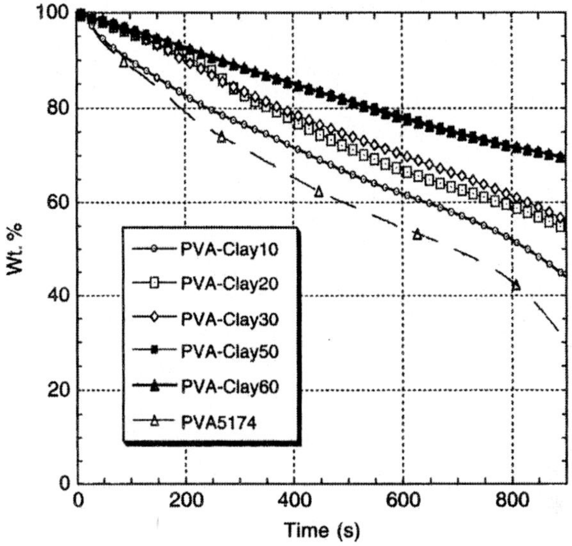

FIGURE 16.9 Comparison of residual mass for PVA–clay fire-retardant coatings at 50 kW/m^2.

50% and 60% clay loading, the testing specimen could not be ignited throughout the 15-min test. Figure 16.9 shows the mass loss of the PVA–clay fire-retardant coating systems. It can be seen that with the addition of clay, the residual mass has been raised from 32% to about 70%.

A discernable effect of the MMT clay on the HRR data is supported by Fig. 16.10. Without any MMT clay, the wood was severely burnt. As a result, a higher HRR was observed throughout the test. However, with the MMT clay loading up to 30%, it is clear that a uniform ceramic layer has been formed on the top of the wood block, which protects the wood from heat throughout the test. Therefore a lower HRR has been obtained.

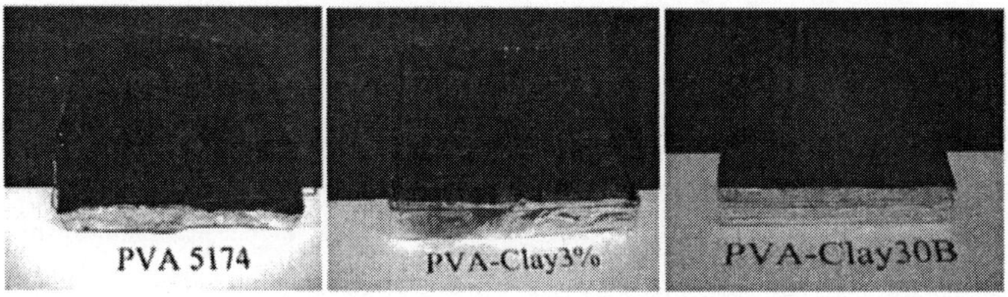

FIGURE 16.10 Digital photos of combustion residues of wood blocks coated with PVA 5174 (left), PVA–3% MMT clay (center), and PVA–30% MMT clay (right).

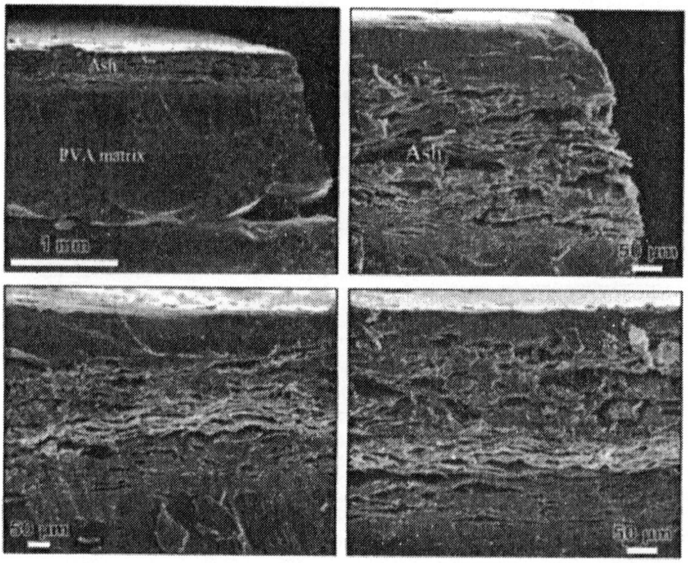

FIGURE 16.11 SEM micrographs of post-test PVA–30% clay specimen showing ash and unburned PVA matrix (top left), layers of ash/MMT clay (top right), and different views of ash layers (bottom left and bottom right).

It has been shown that incorporation of MMT clay in the PVA coating systems significantly improved its fire retardant properties at clay loading of 30% or higher. A thin section of the post-test specimen of PVA–clay nanocomposite fire-retardant coating system with 30% clay loading was studied by scanning electron microscopy (SEM) and transmission electron microscopy (TEM). The SEM micrographs in Fig. 16.11 show a post-test specimen of PVA–30% clay loading at low and high magnifications. It is evident that the MMT clay formed protective layers underneath the charred top surface of the coating. The TEM micrographs of PVA–30% clay post-test residue are shown in Fig. 16.12. Distinct nanoclay layers in the post-test ash residue of PVA–30% clay loading specimen were observed.

16.3.3 Nanostructured Materials for Propulsion Systems

Polymer chopped fiber-filled systems are used as insulative materials in rocket propulsion systems. This research program is aimed to develop new classes of nanostructured materials that are lighter and have better erosion and insulation characteristics than current insulative materials (Kevlar®-filled EPDM rubber). The TPSiV™ X1180 thermoplastic elastomer (TPE) is a polyamide-based vulcanized silicone thermoplastic resin manufactured by Dow Corning and was selected as a potential replacement material.[10,11] Its typical uses include profiles for automotive fuel and vapor line covers, brake hose covers, and industrial applications involving extruded profiles exposed to harsh environments. Table 16.6 shows the chemical compositions of three thermoplastic elastomer/nanoparticle blends produced by twin screw extrusion.

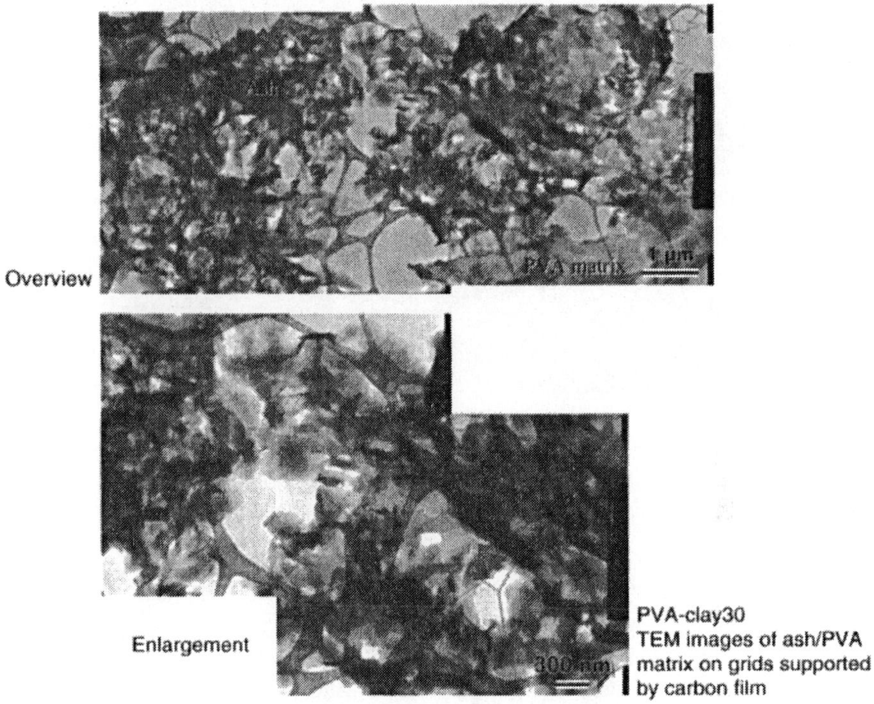

FIGURE 16.12 TEM micrographs of PVA–clay 30% post-test ash showing layers of clay residues at low (top) and high magnification (bottom) on grids supported by carbon film.

TEM analyses were conducted on the aforementioned three blends. Figure 16.13 shows the TEM images of the Dow Corning polyamide silicone TPSiV X1180 in two separate phases; the dark color is the silicone phase and the light color is the polyamide phase. Figure 16.14 shows the TEM images of the 7.5 wt% Cloisite 30B–92.5 wt% TPSiV X1180 blend. We speculate that the Cloisite 30B nanoclays are dispersed only in the polyamide phase, since the silicone phase is already cross-linked. TEM images of the 15 wt% PR-24-PS CNF85 wt% TPSiV X1180 blend (Fig. 16.15) and 10% $Ph_{12}T_{12}$–POSS 90 wt% TPSiV X1180 blend (Fig. 16.16)

TABLE 16.6
Thermoplastic Elastomer/Nanoparticle Blends

Material	TPE (wt%)	Nanoparticle (wt%)
1	TPSiV™ X1180 (90%)	$Ph_{12}T_{12}$-POSS (10%)
2	TPSiV™ X1180 (92.5%)	Cloisite 30B (7.5%)
3	TPSiV™ X1180 (85%)	PR-24-PS CNF (15%)

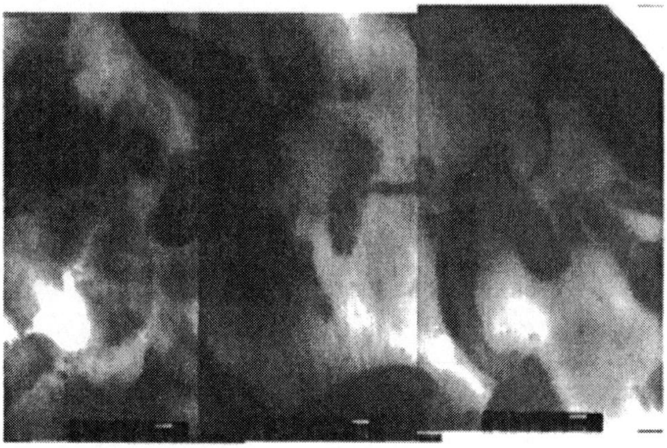

FIGURE 16.13 TEM micrographs of polyamide silicone TPSiV X1180 (scale bar = 500 nm).

showed similar results. It was determined that the TPSiV X1180 TPE is difficult to process with any of the three selected nanoparticles. None was compatible with the resin system.

The preceding three blends with the baseline Kevlar-filled EPDM rubber were tested for ablation resistance using a subscale solid rocket motor. Figure 16.17 shows the ablation rates of the materials at low, medium, and high Mach number regions inside the rocket motor.[10] All three TPEs were out-performed by the baseline material. These observations clearly demonstrate that if the nanoparticles are poorly dispersed in the polymer matrix, no thermal performance improvement can be expected.

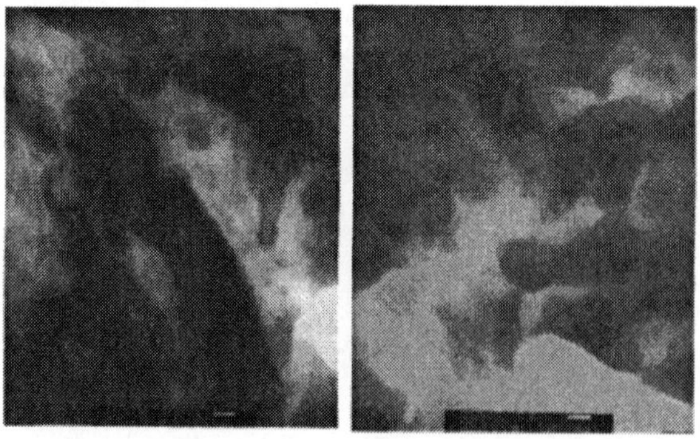

FIGURE 16.14 TEM micrographs of 92.5% polyamide silicone TPSiV X1180 with 7.5% Cloisite 30B (scale bar = 500 nm).

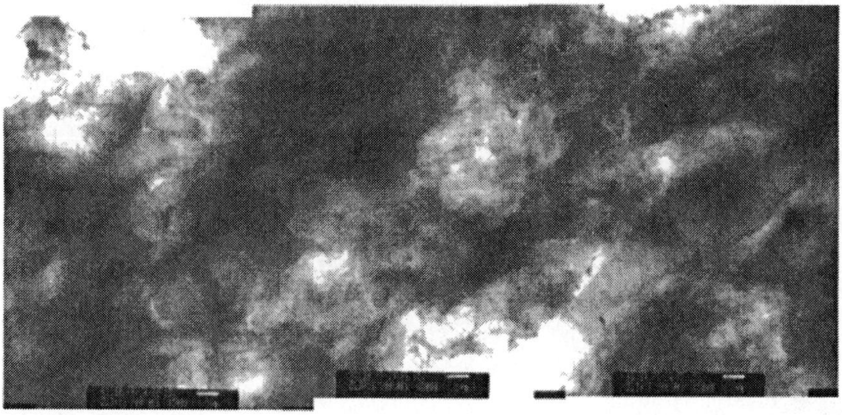

FIGURE 16.15 TEM micrographs of 85% polyamide silicone TPSiV X1180 with 15% PR-24-PS CNF (scale bar = 1 μm).

16.3.4 Nanocomposite Rocket Ablative Materials

This program is aimed at developing nanocomposite rocket ablative materials for solid rocket nozzles.[3–6] The candidate materials were evaluated for dispersion uniformity using WAXD and TEM prior to full ablation testing. Borden SC-1008 resole phenolic in isopropanol (IPA) is the baseline resin system. Our first attempts were to modify this resin by the incorporation of MMT organoclays. The surface treatment on these clays (usually with quaternary ammonium ions) is critical for blend compatibilization and ease of dispersion of the surface-treated clay into the hydrophobic phenolic resin matrix. This treatment is a quaternary ammonium ion (quat). These blends were dispersed using high-shear, nonsparking paint mixing equipment. Neat resin castings (without fiber reinforcement) were made for WAXD and TEM analyses.

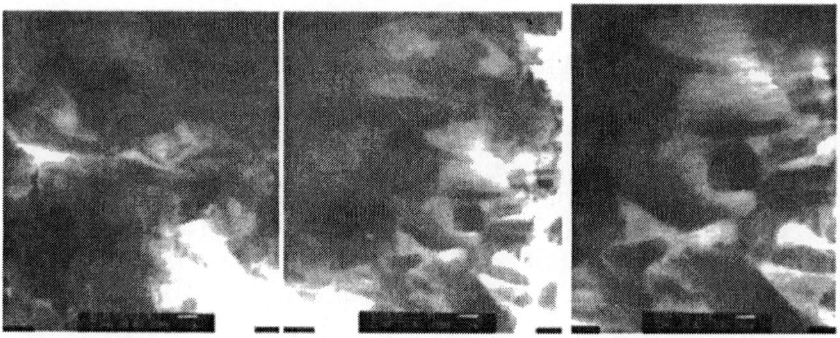

FIGURE 16.16 TEM micrographs of 90% polyamide silicon TPSiV X1180 with 10% $Ph_{12}T_{12}$-POSS (scale bars = 1 μm and 500 nm).

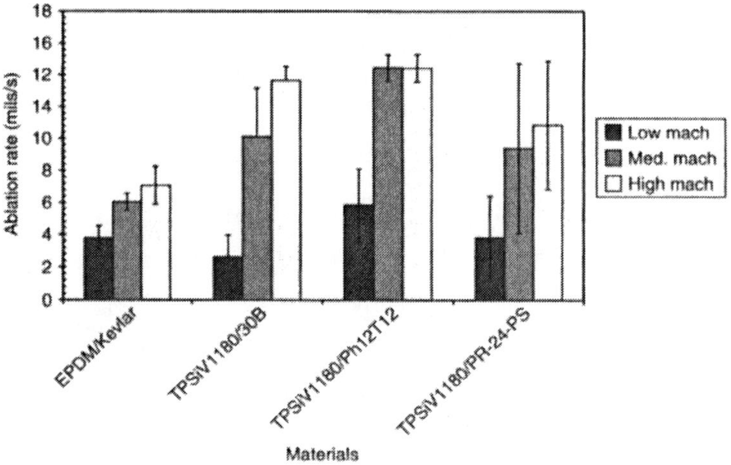

FIGURE 16.17 Ablation rates of polyamide silicone TPSiV X1180 with Cloisite 30B, $Ph_{12}T_{12}$-POSS, and PR-24-PS CNF as compared to EPDM/Kevlar in low, medium, and high Mach number regions inside a rocket motor.

MX-4926 is a rayon carbon fabric impregnated with SC-1008 containing carbon black (CB) particles. Based on TEM analyses, we concluded that CB particles in the baseline MX-4926 caused interference of the dispersions of all nanoparticles in the SC-1008. As a result, we eliminated CB in our subsequent blending experiments. Cloisite 30B in loadings of 5, 10, and 15 wt% was dispersed in the SC-1008. Figure 16.18

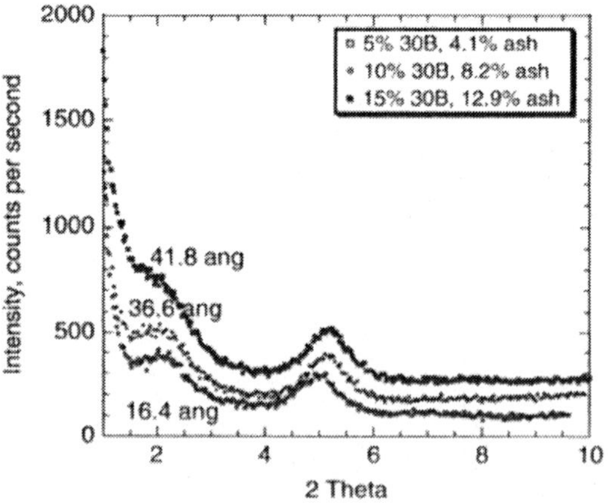

FIGURE 16.18 WAXD of 5, 10, and 15 wt% Cloisite 30B in 95, 90, and 85 wt% SC-1008 phenolic resin, respectively.

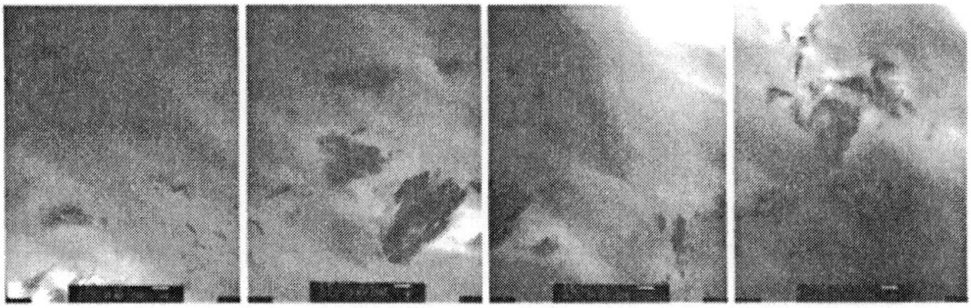

FIGURE 16.19 TEM images of 5 wt% Cloisite 30B in 95 wt% SC-1008 (scale bar = 1 μm).

shows the WAXDs of the three nanoclay loadings in SC-1008 and indicates good dispersibility of the nanoclay into SC-1008. Figure 16.19, Fig. 16.20, and Fig. 16.21 are TEM images of the 5 wt% Cloisite 30B in 95 wt% SC-1008 (Fig. 16.19), 10 wt% Cloisite 30B in 90 wt% SC-1008 (Fig. 16.20), and 15 wt% Cloisite 30B in 85 wt% SC-1008 (Fig. 16.21). The TEM images indicate intercalation and not

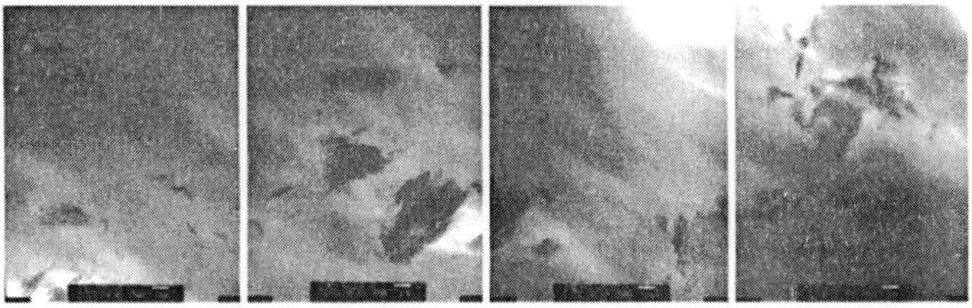

FIGURE 16.20 TEM images of 10 wt% Cloisite 30B in 90 wt% SC-1008 (scale bar = 1 μm).

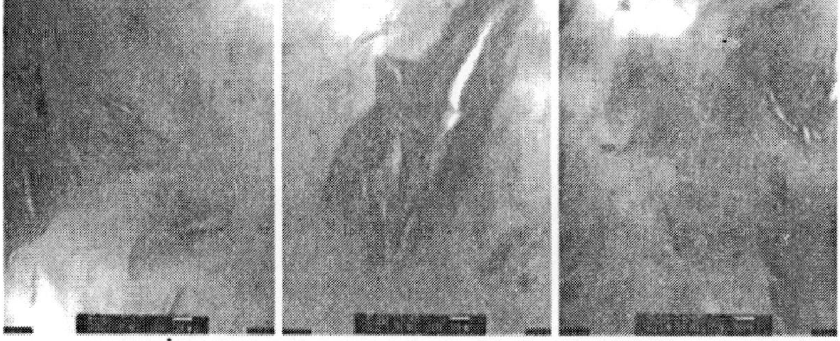

FIGURE 16.21 TEM images of 15 wt% Cloisite 30B in 85 wt% SC-1008 (scale bar = 500 nm).

exfoliation of the nanoclay in the resin system. The TEM analyses assisted in determining the degree of dispersion/exfoliation of the nanoclay before committing to a 20-lb run of these nanoparticle/resin mixtures to fabricate prepregs at Cytec Engineered Materials (CEM). This procedure is a very cost-effective and efficient technique for screening different formulations.

We selected SC-1008 with 5, 10, and 15 wt% Cloisite 30B to replace 15 wt% of carbon black in the original MX-4926 formulation. CEM prepared three versions of MX-4926 alternates designated MX-4926 ALT 5%, MX-4926 ALT 10%, and MX-4926 ALT 15%. Three loadings of PR-24-PS in 20, 24, and 28 wt% were dispersed in SC-1008 without the rayon carbon fiber reinforcement. Three loadings of trisilanolphenyl-POSS in 2, 6, and 10 wt% were also dispersed in SC-1008. The POSS–SC-1008 mixture was used with the rayon carbon fabric to produce prepregs. Table 16.7 shows the chemical compositions for the laminates used for ablation testing. Figure 16.22 compares the densities of three types of nanocomposite rocket ablative materials (NRAMs) with nanoclay, CNF, and POSS at various loading levels of nanoparticles. All CNF NRAMs and POSS NRAMs had densities lower than MX-4926.

A small-scale supersonic liquid-fueled rocket motor (SSRM) burning kerosene and oxygen was used to study the ablation and insulation characteristics of the ablatives. It was demonstrated as a cost-effective laboratory device to evaluate different

TABLE 16.7
Specimen Configuration for SSRM Laminates Fabrication

Material ID	Density (g/cc)	Rayon Carbon Fiber Reinforcement (wt%)	Resin SC-1008 Phenolic (wt%)	Filler (wt%)
MX-4926 (Control)	1.44	50	35	15 carbon black (CB)
MX-4926 ALT Clay 5%	1.42	50	47.5	2.5 Cloisite 30B
MX-4926 ALT Clay 10%	1.43	50	45	5 Cloisite 30B
MX-4926 ALT Clay 15%	1.43	50	42.5	7.5 Cloisite 30B
PR-24-PS 20%/SC-1008	1.35	None	80	20 PR-24-PS CNF
PR-24-PS 24%/SC-1008	1.38	None	76	24 PR-24-PS CNF
PR-24-PS 28%/SC-1008	1.41	None	72	28 PR-24-PS CNF
MX4926 ALT SO-1458 2%	1.41	50	49	1 Trisilanolphenyl-POSS
MX4926 ALT SO-1458 6%	1.38	50	47	3 Trisilanolphenyl-POSS
MX4926 ALT SO-1458 10%	1.40	50	45	5 Trisilanolphenyl-POSS

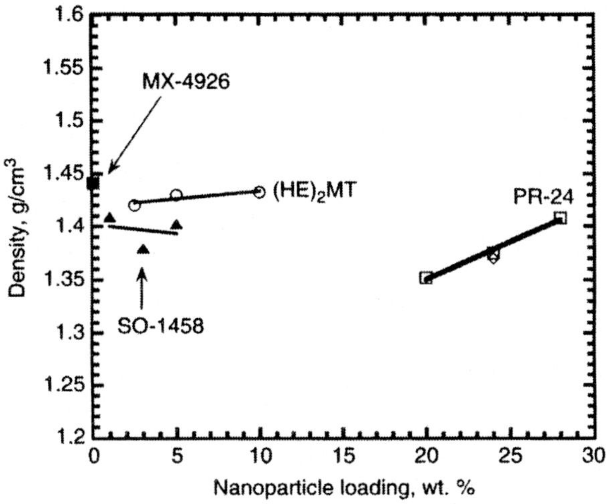

FIGURE 16.22 Densities of MX-4926 and NRAMs with different nanoparticle loadings.

ablatives under identical conditions for initial material screening and development.[3–6] Figure 16.23 shows the peak erosion levels of MX-4926 and MX-4926 ALTs at the three heat fluxes with Al_2O_3 particles. As expected, peak erosion decreased as heat flux decreased. At high heat flux (1000 Btu/ft^2-sec) the MX-4926 and MX-4926 ALTs showed more difference in peak erosion performance, while the medium- and low-heat flux conditions showed little difference in peak erosion. MX-4926 ALT 5 and

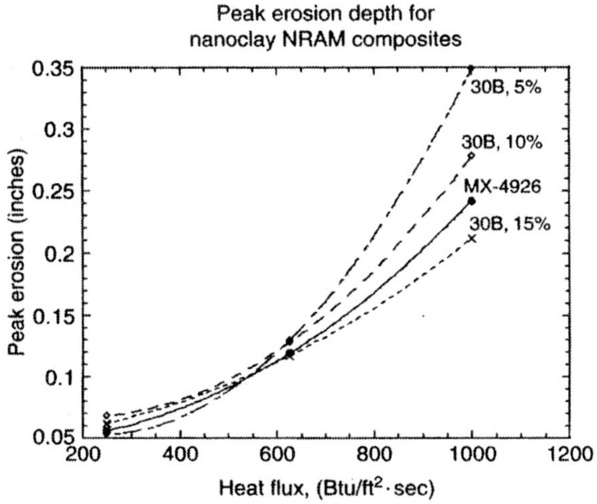

FIGURE 16.23 Peak erosions of nanoclay NRAMs at 250, 625, and 1000 Btu/ft^2-s with particles.

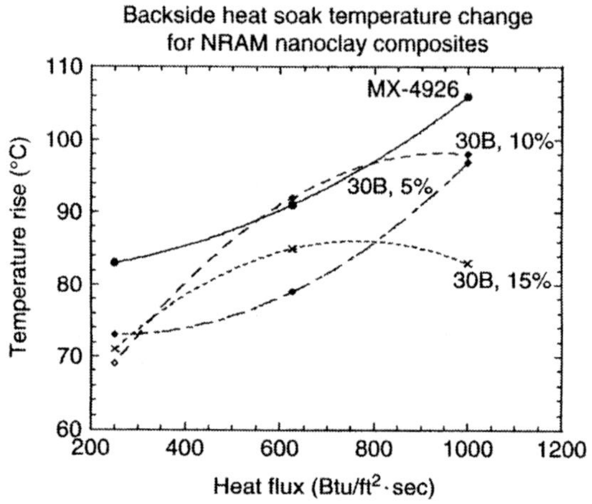

FIGURE 16.24 Maximum backside heat-soaked temperature rises for nanoclay NRAMs at 250, 625, and 1000 Btu/ft^2-s with particles.

10% Cloisite 30B samples were not as erosion resistant as the control MX-4926. The MX-4926 ALT 15% showed better erosion characteristics than MX-4926 by 14%. Figure 16.24 shows all nanoclay NRAM composites exhibited lower maximum backside heat-soaked temperature rises than MX-4926 by 28%. Surface temperature of MX-4926 ALTs were lower than MX-4926.

Figure 16.25 shows both PR-19-PS and PR-24-PS CNF NRAMs (without rayon carbon fabric) have significantly less peak erosion than MX-4926 at 1000 Btu/ft^2-s

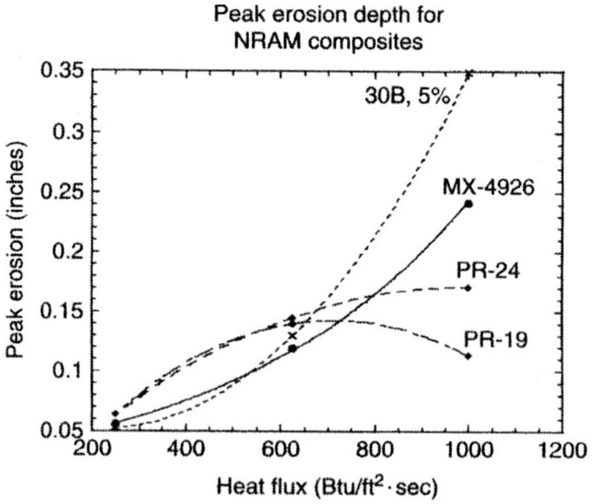

FIGURE 16.25 Peak erosion depths of nanoclay and CNF NRAMs at 250, 625, and 1000 Btu/ ft^2-s with particles.

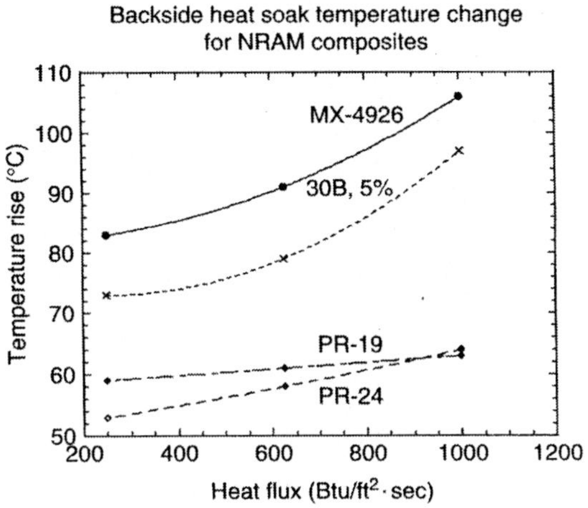

FIGURE 16.26 Maximum backside heat-soaked temperature rises at 250, 625, and 1000 Btu/ft^2-s with particles.

by about 42%. Figure 16.26 shows both PR-19-PS and PR-24-PS CNFs also have substantial lower maximum backside heat-soaked temperature rises than MX-4926 by about 68% at all three levels of heat fluxes. Surface temperatures of the CNF samples were hotter than the MX-4926 and MX-4926 ALTs. This suggests we may have produced better radial heat transfer than axial heat transfer supported by the glowing heat of the surface observed during materials testing.

Figure 16.27 shows the ablation rates of MX-4926 and three groups of NRAMs: clay NRAM [$(HE)_2MT$ nanoclay], CNF NRAM (PR-24), and POSS NRAM (SO-1458). The

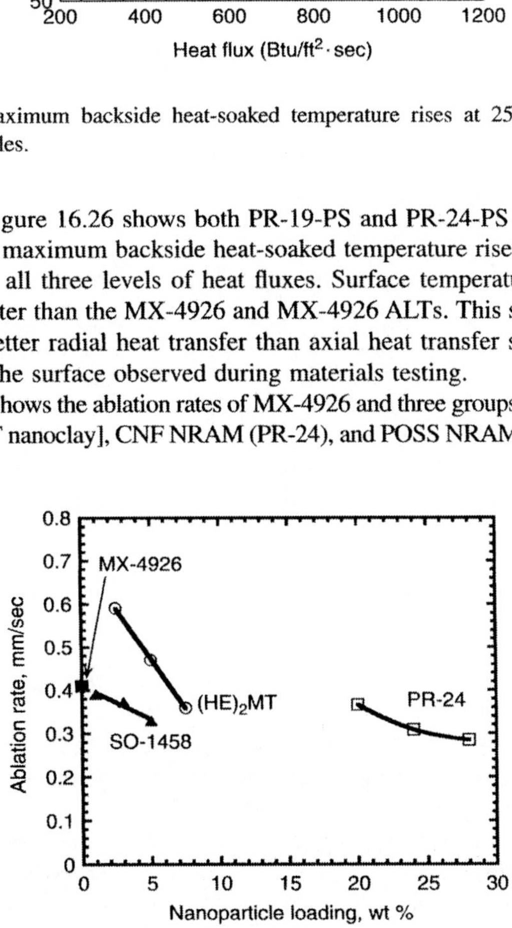

FIGURE 16.27 Ablation rates of MX-4926 and NRAMs with different types of nanoparticles at various loading levels.

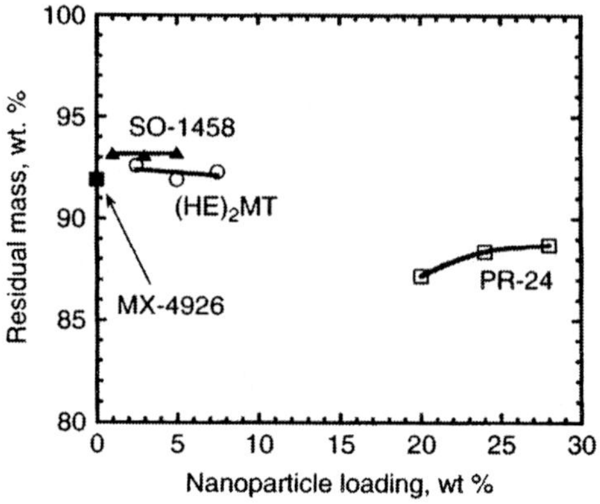

FIGURE 16.28 Residual masses of MX-4926 and NRAMs with different types of nanoparticles at various loading levels.

ablation rate of MX-4926 is about 0.4 mm/s. For the clay NRAM group, only the 7.5 wt% clay NRAM has a lower ablation rate than MX-4926. For the CNF NRAM group, all three loadings have lower ablation rates than MX-4926 with 28% CNF NRAM being the lowest. For the POSS NRAM group, all three loadings have lower ablation rates than MX-4926 with 5% POSS NRAM being the lowest. The loadings of POSS are 1, 3, and 5 wt%, the lowest of the three NRAM groups. The loadings of clay are 2.5, 5, and 7.5 wt%, the medium of the three NRAM groups. The loadings of CNF are 20, 24, and 28 wt% without the rayon carbon reinforcements, the highest of the three NRAM groups. Figure 16.28 shows the residual masses of MX-4926 and all the NRAMs. The residual mass of MX-4926 is about 92%. The POSS NRAM group has the most residual mass, about 93% for all three loadings. The clay NRAM group has about the same residual mass as the MX-4926. The CNF NRAM group has about 86 to 88 wt% residual mass, the lowest of all.

Figure 16.29 shows the maximum backside heat-soaked temperature rises of MX-4926 and the NRAMs. All NRAMs exhibited lower maximum backside heat-soaked temperature rises than MX-4926. The backside heat-soaked temperature rise of MX-4926 was about 106°C. It is obvious that the CNF-NRAM group had substantial lower maximum backside heat-soaked temperature rises than MX-4926, from 54° to 72°C. The POSS-NRAM group has the second lowest than MX-4926, from 75° to 86°C. The clay NRAM group had the third lowest from MX-4926, from 82° to 98°C.

An IR pyrometer was used to measure the surface temperatures of all materials during SSRM firings. Figure 16.30 shows the surface temperatures of MX-4926 and the NRAMs. Surface temperatures of MX-4926 were about 1700°C. Surface temperatures of the CNF NRAM samples were hotter than the MX-4926, clay NRAMs, and POSS NRAMs. This suggests better radial heat transfer than axial heat transfer, supported by the glowing heat of the surface observed during material testing. This phenomenon was observed by other researchers[21] and needs further study. The

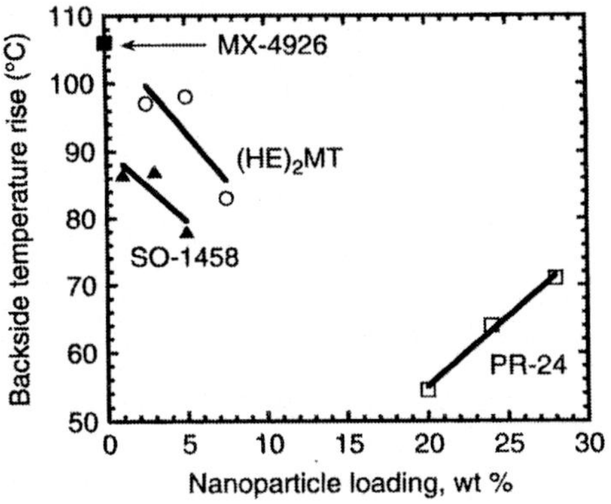

FIGURE 16.29 Backside temperature rises of MX-4926 and NRAMs with different types of nanoparticles at various loading levels.

surface temperatures of clay NRAMs and POSS NRAMs are lower than MX-4926. There is essentially no effect of the amount of nanoclay in the clay NRAMs. The amount of POSS in the POSS NRAMs has a significant effect on the surface temperature of the POSS NRAMs.

16.3.5 Nanomodified Carbon/Carbon Composites

The objective of this materials program is to develop an improved carbon/carbon (C/C) composite with enhanced thermo-oxidative resistance at intermediate temperatures (700° to 1200°F).[7–9] We proposed that a nanophase be introduced into the C/C

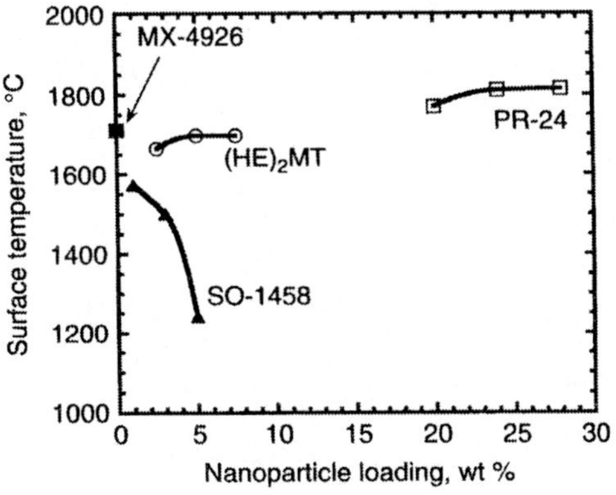

FIGURE 16.30 Surface temperatures of MX-4926 and NRAMs with different types of nanoparticles at various loading levels.

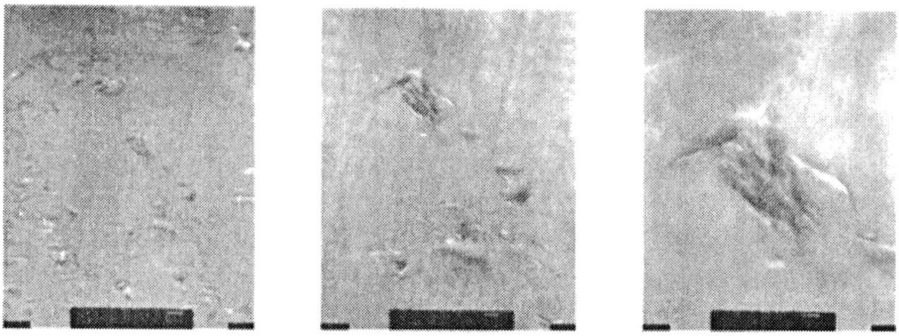

FIGURE 16.31 Higher magnification of PT-30/Cloisite 30B (97.5/2.5) THF showing nanoclay in an intercalated state in the PT-30 cyanate ester resin where scale bars are 500 nm (left), 200 nm (center), and 100 nm (right).

composites (CCCs), prior to cure, to provide improved and maintained mechanical strength by preventing oxidation of the composites. The candidate materials were evaluated for dispersion using WAXD and TEM prior to full scale-up. Baseline phenolic resin 134A and cyanate ester resins PT-15 and PT-30 were used as matrix resins. Lab scale dispersion of (1) PT-30 cyanate ester (CE) resin with different wt% POSS, nanoclays, and CNF, (2) PT-15 CE resin with different wt% of POSS and nanoclays, and (3) 134A phenolic resole resin with different wt% POSS and nanoclays were conducted. The morphology of selective resin/nanoparticle systems was characterized using TEM and SEM analyses. Detailed processing and characterization of PT-30/nanoparticle, PT-15/nanoparticle, 134A/nanoparticle systems are reported elsewhere.[9] A brief discussion of PT-30/nanoparticle and PT-15/nanoparticle systems is included in this section.

Cloisite 30B was first dispersed in THF before blending with the PT-30 resin. Cloisite 30B was uniformly dispersed in PT-30/30B THF system. At higher magnification, Cloisite 30B was in an intercalated state in the PT-30 resin, and larger clay tactoids were observed than when PT-15 was mixed with Cloisite 30B and cured in the same manner (Fig. 16.31).

Different weight percents (wt%) of the seven POSS chemicals were blended with PT-30 for a total of 15 blends using a lab scale high-shear mixer. Appearance in terms of transparency, translucency, or opaqueness was examined during mixing, before and after curing, and was recorded for all blends and densities. Based on visual observation, trisilanolphenyl-POSS is the only potential POSS compound that works well with PT-30 using a direct melt blending process. All other blends were either opaque or translucent and large, poorly dispersed particles were observed. Phase separation had clearly taken place.

Selective candidates were cured for TEM analyses. PT-15 and PT-30 were cured thermally. The SO1458 POSS particles, when directly blended into PT-30 resin, showed very few undissolved POSS particles in the resin matrix, as shown in Fig. 16.32. Some molecular dispersion in the PT-30/SO1458 POSS (95/5) was achieved, as shown in Fig. 16.32. Significant Si was detected in the resin matrix where no phase separation could be detected. SO1458 is the preferred POSS system for

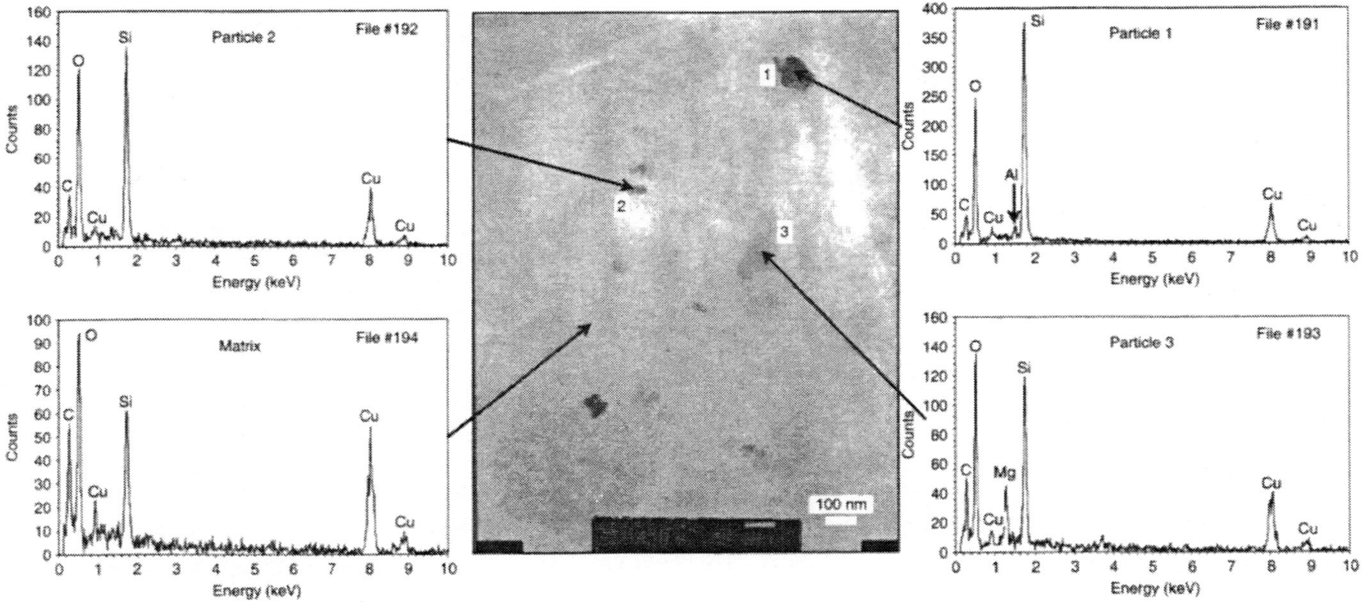

FIGURE 16.32 TEM micrographs of PT-30/SO1458 POSS (95/5) at high magnification showing molecular dispersion of SO1458 POSS with some POSS particles in the PT-30 cyanate ester. Significant Si was detected in the resin matrix.

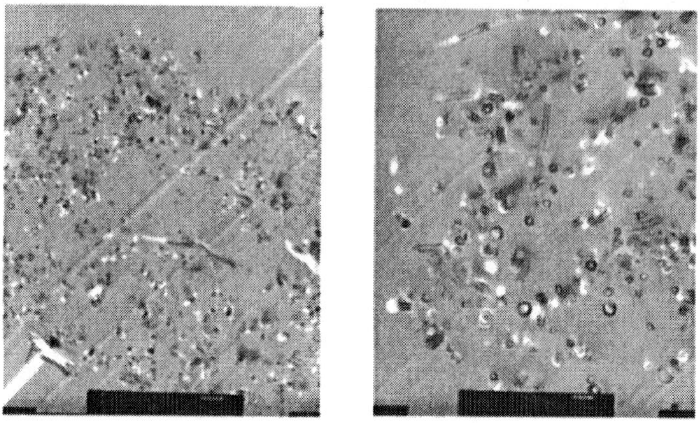

FIGURE 16.33 TEM micrographs of PT-30/PR-24-PS (99/1) at high magnification showing good dispersion of PR-24-PS CNF in the PT-30 cyanate ester. Scale bars = 500 nm (left) and 200 nm (right).

PT-30 resin. PT-30/PR-19-PS and PT-30/PR-24-PS were blended in 99.5/0.5 and 99/1 wt% (Fig. 16.33), and a total of four CNF blends were prepared.

PT-15 was blended with several clays (Nanomer® I.28E, Nanomer I.30E, PWG, Cloisite 10A, and Cloisite 30B) using different mixing methods, and a total of 34 blends were recorded. Resulting samples were either opaque or translucent. PT-15/30B (97.5/2.5) THF, PT-15/10A (97.5/2.5) THF, and PT-15/I.30E (97.5/2.5) THF blends were prepared using THF as a carrier medium in these samples to facilitate clay dispersion. When Cloisite 30B, Cloisite 10A, and Nanomer I.30E were compared in higher magnification TEM micrographs, Cloisite 30B was dispersed slightly more uniformly than the other two clays. Cloisite 30B was in a partial exfoliated state in PT-15, as shown in Fig. 16.34. Small tactoids are present in this nanodispersion. Figure 16.35 shows higher magnification TEM micrographs of PT-15/Cloisite 30B

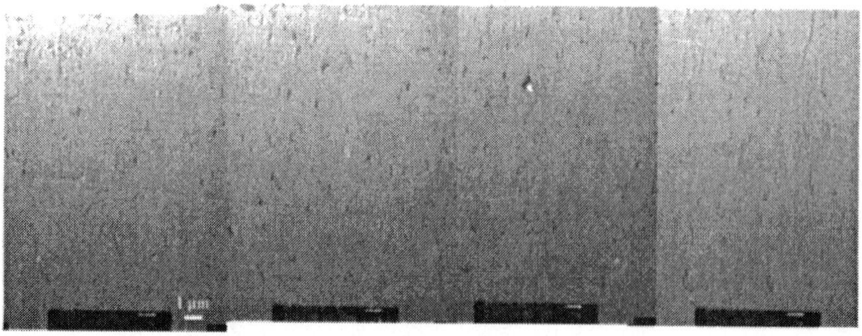

FIGURE 16.34 TEM micrographs of PT-15/Cloisite 30B (97.5/2.5) THF showing uniform dispersion of Cloisite 30B in the PT-15 cyanate ester. Scale bar = 1 μm.

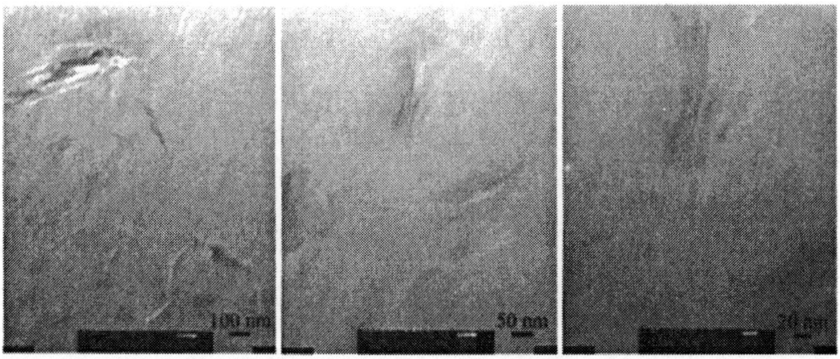

FIGURE 16.35 Higher magnification TEM micrographs of PT-15/Cloisite 30B (97.5/2.5) THF showing Cloisite 30B clays are exfoliated in PT-15 cyanate ester.

(97.5/2.5) THF showing that Cloisite 30B clays are exfoliated in PT-15 cyanate ester. Cloisite 30B is our preferred clay for the PT-15 resin.

PT-15 was blended with poly(phenyl silsesquioxane), PPSQ, in THF and trisilanol-phenyl POSS in different wt%, and a total of 13 blends were recorded. Only PT-15/trisilanol phenyl POSS in 99/1 and 97/3 were completely transparent. TEMs of PT-15/SO1458 (97/3) and PT-15/SO1458 (95/5) blends were prepared. Figure 16.36 shows molecular dispersion of SO1458 POSS in the PT-15/SO1458 (97/3) system. Where particles are noted, one finds their Si content is higher than in the open resin matrix. Significant Si is detected in the resin matrix in regions where no particles are observed. Clearly the POSS is partially dissolved in the matrix and partially nano-dispersed, resulting in molecular dispersion of SO1458 in PT-15. This is a good candidate system for a conversion to CCC. SO1458 POSS is the preferred POSS system for the PT-15 resin.

A summary of all the nanomodified carbon/carbon composite (NCCC) candidates and CC139 (baseline commercial carbon/carbon composite prepared with 134A phenolic resin) with their weight loss percentage, density, and ranking based on weight loss data is shown in Table 16.8. Samples were designated of –1 through –4 to indicate room temperature, 700°C, 1200°, and retained conditions, respectively. An air leakage was detected in our first set of experiments, and as a result the 134A/CLO/3-1A, 134A/CLO/3-2A, 134A/CLO/3-3A, 134A/POSS/3-1A, 134A/POSS/3-2A, 134A/POSS/3-3A, PT15/ POSS/5/-1B, PT15/POSS/5/-2A, PT15/POSS/5/-3A, PT30/PR24PS/1-1A, and PT30/PR24PS/1-2A specimens were overexposed to unrealistic thermo-oxidative conditions. They are denoted with asterisks in Table 16.8. The 134A/CLO/3-4B and 134A/POSS/3-4D specimens were retested at 1200°F using the fourth spared panels of the 134A/CLO/3 and 134A/POSS/5 candidates designated. The PT15/POSS/5-4 spared specimen was lost, and as a result the PT15/POSS/5 candidate was unable to be retested for the 1200°F conditions of this project. Regrettably the behavior of molecularly dispersing POSS into PT-15 (Table 16.8, entries 9 to 12) could not be assessed for improved thermo-oxidative stability due to overexposure or loss.

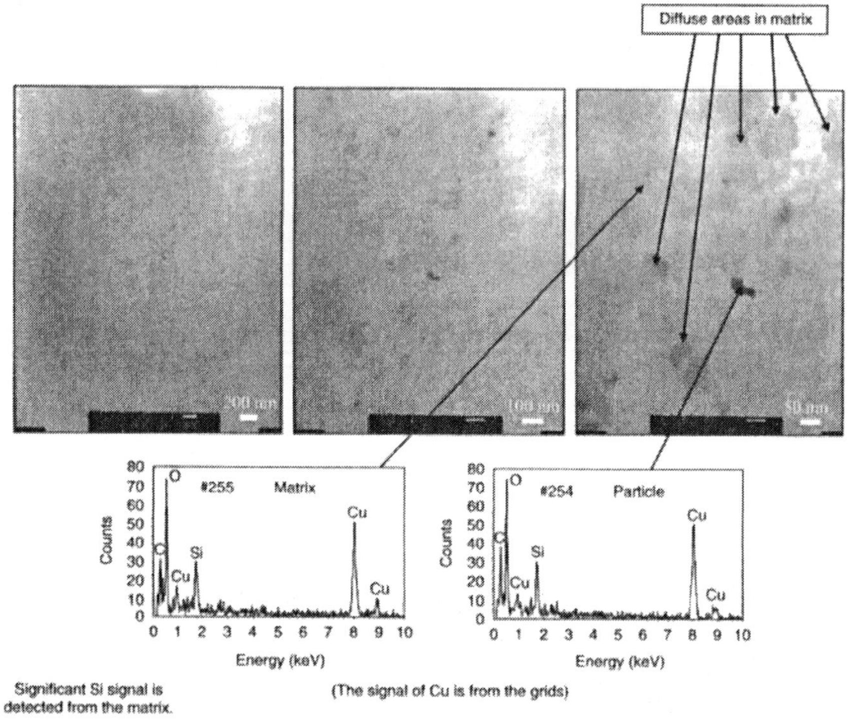

FIGURE 16.36 Progressive magnification TEM micrographs of PT-15/SO1458 POSS (97/3) showing molecular dispersion of SO1458 POSS in PT-15 cyanate ester. Si from POSS molecules is dispersed in the resin matrix. The dark particle has the same Si/O/C composition as the resin matrix. Significant Si signal is detected from the matrix. The signal of Cu is from the grids.

Based on the thermo-oxidative analyses, the PT30/CLO/5 candidate is the most thermo-oxidative resistant material; weight losses were 5.2%, 5.1%, and 3.7% at room temperature (RT), 700°F, and 1200°F, respectively. PT30 ranked as first or best. The PT15/CLO/5 candidate was the second best thermo-oxidative resistant material and weight losses were 6.5%, 5.1%, and 3.7% at RT, 700°F, and 1200°F conditions, respectively. The PT30/PR24PS/1 NCCC showed a low 6.8% weight loss and was the third best thermo-oxidative resistant material NCCC candidate. The standard CC139 was fourth with weight losses of 8.9%, 9.0%, and 13.6% for RT, 700°F, and 1200°F, respectively. The 134A/CLO/3-4B and 134A/POSS/3 had high weight losses of 30.5% and 23.3%, respectively.

Weight losses of different NCCCs were compared, as shown in Fig. 16.37. It is evident that the cyanate ester resin-based PT30/PR24PS/1, PT30/CLO/5, and PT15CLO/5 specimens had better thermo-oxidative resistance than the phenolic resin-based CC139 standard CCCs. The phenolic resin-based 134A/CLO/3 and 134A/POSS/3 specimens exhibited worse thermo-oxidative resistant characteristics

TABLE 16.8
Summary of Thermo-Oxidative Data and Ranking

P/N	Initial wt. (g)	TGA recorded wt. (g)	Weight Loss (%)	Sample Density (g/cc)	Rank
134A/CLO/3-1A*	2.053	1.728	−15.8	1.619	
134A/CLO/3-2A*	2.145	1.772	−17.4	1.581	
134A/CLO/3-3A*	2.056	1.152	−44.0	1.567	
134A/CLO/3-4B	2.036	1.414	−30.5	1.585	6
134A/POSS/3-1A*	1.958	1.441	−26.4	1.616	
134A/POSS/3-2A*	1.989	1.216	−38.9	1.575	
134A/POSS/3-3A*	1.933	1.666	−13.8	1.611	
134A/POSS/3-4D	2.055	1.573	−23.2	1.603	5
PT15/POSS/5-1B*	2.469	2.178	−11.8	1.650	
PT15/POSS/5-2A*	2.462	2.146	−12.9	1.595	
PT15/POSS/5-3A*	2.577	1.782	−30.8	1.651	
PT15/POSS/5-4	Lost				
PT15/CLO/5-1	2.52	2.36	−6.5	1.630	2
PT15/CLO/5-2B	2.36	2.28	−3.4	1.643	2
PT15/CLO/5-2C	2.40	2.24	−6.8	1.643	2
PT15/CLO/5-3B	2.34	2.26	−3.5	1.684	2
PT15/CLO/5-3C	2.38	2.29	−3.8	1.684	2
PT30/PR24PS/1-1A*	2.586	1.244	−51.9	1.653	
PT30/PR24PS/1-2A*	2.574	1.406	−45.4	1.618	
PT30/PR24PS/1-3B,C	2.573	2.399	−6.8	1.656	3
PT30/CLO/5-1B,C	2.522	2.358	−5.2	1.653	1
PT30/CLO/5-2B,C	2.383	2.261	−5.1	1.656	1
PT30/CLO/5-3B,C	2.363	2.265	−3.7	1.618	1
CC139-1A	2.673	2.436	−8.9	1.627	4
CC139-2A	2.506	2.881	−9.0	1.638	4
CC139-3A	2.536	2.191	−13.6	1.630	4

*Air leakages were detected in these experiments. room temperature, 700°F, and 1200°F specimens were designated as −1, −2, and −3, respectively. Asterisks indicate overexposed specimens.

than the CC139 standard. From our TEM analyses we concluded that the nanoparticles were dispersed very well in the PT-30 and PT-15 cyanate resin systems and poorly in the 134A phenolic resin system. Figure 16.38 shows a comparison of the densities of the different NCCCs. It is also evident that PT15/CLO/5-3B,C had the highest density (1.684 g/cc), followed by PT30/CLO/5-2B,C and PT30/PR24PS/1-3B,C (1.656 g/cc), compared with the CC139 specimens (1.627 to 1.638 g/cc), while the densities of 134A/POSS/3-4D and 134A/CLO/3-4B were 1.603 g/cc and 1.585 g/cc, respectively. Figure 16.39 compares densities versus weight losses of the different NCCCs. A trend of higher density NCCCs exhibiting

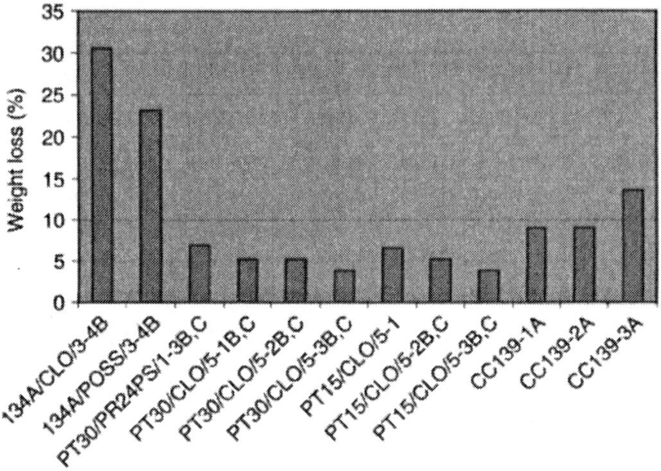

FIGURE 16.37 Comparison of weight losses of different NCCCs.

lower weight loss while lower density NCCCs show higher weight loss is proposed. Lower weight loss for NCCCs from cyanate ester resin systems is expected due to few or no volatiles from cured cyanate esters as well as proposed improved thermo-oxidative stability attributable to the presence of nanophases in the NCCC cyanate ester materials.

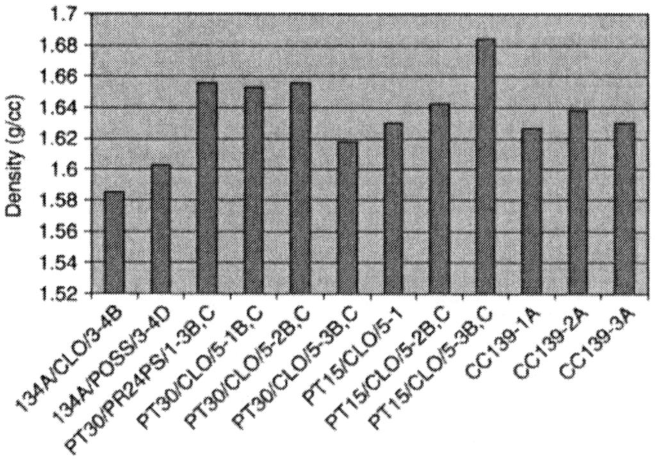

FIGURE 16.38 Comparison of densities of different NCCCs.

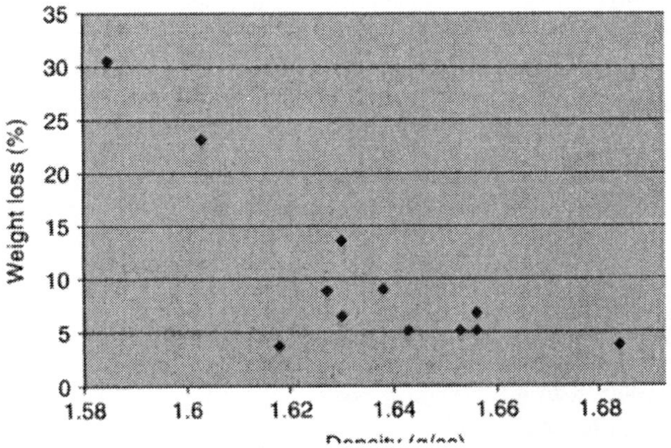

FIGURE 16.39 Comparison of densities and weight losses of different NCCCs.

16.4 SUMMARY AND CONCLUSIONS

We have been developing polymer nanocomposites to enhance materials properties for high-temperature applications. The following conclusions are presented:

1. The feasibility of using polymer nanocomposite for high-temperature applications was clearly demonstrated.
2. MMT organoclays, CNFs, and POSS can be easily incorporated into various polymers using conventional processing and manufacturing techniques to form nanocomposites.
3. The degree of dispersion of nanoparticles in the polymer matrix is essential to achieve the desired properties enhancement.
4. SEM and TEM analyses techniques have been demonstrated to be effective and efficient screening tools to determine nanodispersion in the polymer matrix.

PROBLEMS

1. What are the shortcomings of using wide-angle x-ray diffraction (WAXD) to determine the degree of dispersion in polymer–clay nanocomposites? How can one compensate for the shortcomings of this technique?
2. What is the disadvantage of using transmission electron microscopy (TEM) to determine the degree of nanoparticle dispersion in polymer nanocomposites? How can one compensate for this technique?
3. Discuss other types of techniques that one can use to study the morphology of these polymer nanocomposites.

4. What other types of nanoparticles can be used to enhance the thermal and mechanical properties of polymer nanostructured materials? Explain and give examples.
5. Discuss the similarities and differences of different types of nanoparticles and give examples.
6. Discuss other thermal applications of polymer nanostructured materials in the literature.
7. How can one enhance the interfacial interaction and compatibility with different nanoparticles in a specific polymer matrix?

Readers should read the references in this chapter and current nanocomposite literature to answer the preceding questions.

ACKNOWLEDGMENTS

The authors would like to thank Dr. Charles Y-C. Lee of the Air Force Office of Scientific Research for sponsoring several of our research activities through the AFOSR STTR programs. Support from Dr. Shawn Phillips of the Air Force Research Laboratory and Edwards Air Force Base and W. Casey West of StanChem is appreciated. The authors also would like to express their appreciation to numerous colleagues who have contributed to our nanomaterials research.

REFERENCES

1. Krishnamoorti, R. and Vaia, RA., Ed., *Polymer Nanocomposites: Synthesis, Characterization, and Modeling,* Symposium Series 804, American Chemical Society, Washington, D.C., 2001.
2. Cloisite 30B technical data sheet, Southern Clay Products, Gonzales, TX.
3. Koo, J.H., Stretz, H. et al., Phenolic-clay nanocomposite ablatives for rocket propulsion systems, *Proceedings of International Symposium,* SAMPE, Covina, CA, 1085, 2002.
4. Koo, J.H., Stretz, H. et al., Nanocomposite rocket ablative materials: processing, characterization, and performance, *Proceedings of International Symposium,* SAMPE, Covina, CA, 1156, 2003.
5. Koo, J.H., Chow, W.K. et al., Flammability properties of polymer nanostructured materials, *Proceedings of International Symposium,* SAMPE, Covina, CA, 954, 2003.
6. Koo, J.H., Stretz, H. et al., Nanocomposites rocket ablative materials: subscale ablation test, *Proceedings of International Symposium,* SAMPE, Covina, CA, 2004.
7. Koo, J.H., Pittman, C. et al., Nanomodified carbon/carbon composites for intermediate temperature: processing and characterization, *Proceedings of 35th International SAMPE Technical Conference,* SAMPE, Covina, CA, 2003.
8. Koo, J.H., Pilato, L. et al., Nanomodified carbon/carbon composites for intermediate temperature, AFOSR STTR Phase I Final Report, submitted to Air Force Office of Scientific Research, Arlington, VA, Jan. 2004.
9. Koo, J.H., Pilato, L. et al., Thermo-oxidative studies and mechanical properties of nanomodified carbon/carbon composites, *Proceedings of International Symposium,* SAMPE, Covina, CA, 2004.

10. Blanski, R., Koo, J.H. et al., Polymer nanostructured materials for solid rocket motor insulation: — ablation performance, *Proceedings of 52nd JANNAF Propulsion Meeting,* CPIA, Columbia, MD, 2004.
11. Koo, J.H., Marchant, D. et al., Polymer nanostructured materials for solid rocket motor insulation—Processing, microstructure, and mechanical properties, *Proceedings of 52nd JANNAF Propulsion Meeting,* CPIA, Columbia, MD, 2004.
12. Koo, J.H., Pilato, L.A. et al., Nanocomposites for carbon fiber-reinforced polymer matrix composites, AFOSR STTR Final Report, Arlington, VA, Oct. 2004.
13. Koo, J.H., Pilato, L.A., Wissler, G., Ervin, M., and Firestone, K., Innovative selective laser sintering rapid manufacturing technique using nanotechnology, unpublished data.
14. Tibbetts, G.G. and McHugh, J.J., Mechanical properties of vapor-grown carbon fiber composites with thermoplastic materials. *J. Mater. Res., 14,* 2871–2880, 1999.
15. Lake, M.L. and Ting, J.-M., Vapor-grown carbon fiber composites in carbon materials for advanced technologies., in *Carbon Materials for Advanced Technologies,* Burchell, T.D., Ed., Pergamon, Oxford, UK, 1999.
16. Lake, M.L., Low-Cost Carbon Fibers for Industrial Applications. Applied Sciences, Inc., Cedarville, OH, 11/2002.
17. Maruyama, B. and Alam, K., Carbon nanotubes and nanofibers in composite materials. *SAMPE J., 38,* 3, 59, 2002.
18. Glasgow, D.G., Tibbetts, G.G., Matuszewski, M.J., Walters, K.R., and Lake, M.L., Surface treatment of carbon nanofibers for improved composite mechanical properties, *Proceedings of International Symposium,* SAMPE, Covina, CA, 2004.
19. Lee, A., *Proceedings of POSS® Nanotechnology Conference,* Huntington Beach, CA, Sept. 25–27, 2002.
20. Hu, X., Koo, J.H. et al., Flammability studies of water-borne fire retardant nanocomposites coating on wood, *Proceedings of Polymer Nanocomposites Symposium, American Chemical Society* Southwest Regional Meeting, San Antonio, TX, Oct. 17–20, 2001.
21. Patton, R.D., Pittman, C.U., Wang, L., Hill, J.R., and Day, A., Ablation, mechanical and thermal conductivity properties of vapor growth carbon fiber/phenolic matrix composites, *Composites Part A, 33,* 243–251, 2002.

17 Manufacturing, Mechanical Characterization, and Modeling of a Pultruded Thermoplastic Nanocomposite

Samit Roy, Kalivarathan Vengadassalam, Farzana Hussain, and Hongbing Lu

CONTENTS

17.1 INTRODUCTION

The use of polymer matrix composites (PMCs) in aerospace and civil engineering as well as in sports and leisure applications is rapidly increasing. PMCs have found a wide range of applications in structural components in which the substitution of PMCs for metals has substantially improved performance and reliability. The high tensile strength of PMCs is mainly derived from the high strength of the carbon or glass fibers embedded in the matrix. Fibers typically have high strength in tension. However, their compressive strength is generally much lower due to the fact that under compression the fibers tend to fail through buckling well before compressive fracture occurs. Also, fiber misalignment and presence of voids during manufacturing processes contribute to a further reduction in compressive strength. In fact, the overall compressive strength of a PMC is only about 50% of the tensile strength. The strength of the surrounding polymer matrix plays a key role in characterizing the critical buckling loads of the fibers by constraining the fibers from buckling. Consequently high-strength polymer resins such as PEEK and PPS — although very expensive — are typically used for applications requiring high compressive strength. Low-cost commodity resins such as polypropylene (PP) suffer primarily due to low compressive strength. Because the critical buckling stress in the fiber is directly related to the stiffness and yield strength of the surrounding matrix material, any increase in these matrix properties would directly result in an improvement in the compressive strength of the composite. In this context, it has been observed that the addition of small amounts of nanostructured montmorillonite (MMT) clay (~5 wt%) significantly improves the stiffness, strength, gas barrier, and fire resistance properties of most thermoplastic resins, with only small effect on flexibility. These nanoclay fillers are only slightly more expensive than glass (a few dollars per pound), yet generally far less expensive than carbon fibers (~$100/lb) or carbon nanotubes (~$50,000/lb). Additionally the small amounts of nanofillers required to enhance properties enables these materials to compete more effectively with traditional glass-fiber reinforcements.

17.1.1 NANOCLAYS

A nanoclay is a mineral with a high aspect ratio and with at least one dimension of the particle — typically the thickness — in the nanometer range. Reinforcements in the nanometer size range closely approach the molecular size of the polymer, enabling an intimate encounter between the two materials. When properly modified, the filler particles and polymer interact to create constrained regions at the particle surface. This immobilizes a portion of the polymer chain, creating a reinforcement effect. Purity and cation exchange capacity are two characteristics critical to the success of nanoclays as polymer reinforcing agents.

Because a nanofiller contains so many individual particles in such a small amount of material, it requires very low loading to obtain a high concentration of constrained areas within the polymer. For example, 5% loading by weight of nanoclay leads to a reinforcing effect equal to about 12 to 15% glass fiber. The nanofiller also creates a tortuous path for the penetration of gaseous vapors and liquids into the polymer, which leads to better chemical and moisture resistance.

17.1.2 Montmorillonite

Montmorillonite (MMT) has the widest acceptability for use in polymers. It is a type of smectite clay that can absorb water, and it has a layered structure, with aluminum octahedron sandwiched between two layers of silicon tetrahedron. Each layered sheet is slightly less than 1 nm thin (10 Å), with surface dimensions extending to about 1 μm or 1000 nm. The aspect ratio is about 1000:1 and the surface area is in the range of 750 m^2/g.

Depending on the degree of polymer penetration into the silicate tactoid, two idealized nanocomposite structures are feasible: intercalated and exfoliated. The process of opening the spaces between the clay platelets, which are known as galleries, is called intercalation. Intercalation increases the separation distance between the nanoclay platelets within a clay tactoid and is brought about by surfactant molecules. Without this separation, the polymer molecules will not be able to penetrate into the galleries between the platelet layers. In the intercalated systems, the multilayer structure of the silicates is retained, with alternating polymer/silicate layers and a repeat spacing larger than that of the original clay.

In exfoliated nanocomposites, the primary particles of the organically modified clay are delaminated into individual nanometer-thick silicate platelets, which are dispersed uniformly in the polymer matrix due to extensive polymer penetration and intercalation. In the exfoliated form, nanofillers have very small flexible platelet-type structures. As mentioned earlier, the thickness of the platelet is in the nanometer range, while the width and length are between 0.1 and 2 μm. Because of this, a single gram of exfoliated nanoclay will contain over a million individual particles.

17.1.3 Dispersion of Montmorillonite Clay Platelets in Polypropylene

For nanoclay to improve mechanical strength of a polymer at a low clay loading level, the clay must be exfoliated into platelets having the thickness of one unit cell of the crystal structure, about 1 nm. Dispersion of montmorillonite (MMT) into polypropylene (PP) is especially challenging because the ionic surface of the clay repels the aliphatic hydrocarbon polymer. Dispersion of MMT into PP in an extruder has been accomplished with (1) surfactant modified clay, (2) PP modified with maleic anhydride (PP-MA), and (3) a combination of both modified clay and PP-MA. The most common surfactant for making the clay organophilic is octadecylamine (ODA). Other amines and ammonium ions such as trialkylimidazolium ions, polyamine dendrimers, and amine-terminated polyethylene and polypropylene are also known to aid dispersion of clays. Composites containing both glass fibers and clays modified with cationic surfactants also appear promising. Favorable enthalpic interaction between the clay and the PP, which is not achieved by use of ODA-modified clay and unmodified PP, would aid dispersion of exfoliated clay.

17.1.4 Compression Strength of Nanocomposites

Earlier work by Rosen,[1] which assumed extensional (transverse) and shear microbuckling of the fibers embedded in an elastic matrix, consistently overpredicted the compressive strength when compared with experimentally measured values. Argon[2] was among the first to recognize that fiber-reinforced composites made by normal manufacturing processes, including pultrusion, have regions of fiber misalignment. Argon,[2] and later Budiansky,[3,4] showed that fiber misalignments present in fiber-reinforced composites could lead to yielding of the polymer matrix in shear. The yielding of the matrix would, in turn, result in loss of matrix stiffness that could eventually trigger fiber microbuckling with resulting kink band formation leading to final failure.[4] Consequently, it was recognized that the dominant compressive failure mode in aligned-fiber polymer matrix composites is localized compressive buckling or kinking. Argon[2] proposed a simple yet elegant formula for composite compressive strength given by $\sigma_C = \sigma_Y/\phi_0$ where σ_Y is the matrix yield strength in shear, and ϕ_o is the initial fiber misalignment angle with respect to load direction, sometimes referred to as fiber misorientation. Further it is likely that any increase in composite longitudinal shear stiffness (G_{12c}) due to the presence of nanoclay particles would result in an increase in the compressive strength of the composite, as indicated by a modified form of the Argon formula proposed by Budiansky that includes both elastic and plastic buckling behavior as limiting cases,[4–6]

$$\sigma_c = \frac{G_{12c}}{\left(1+\dfrac{\phi_0}{\gamma_Y}\right)} \tag{17.1}$$

where $\gamma_Y = \sigma_Y/G_{12c}$ is the composite yield strain in longitudinal shear. The failure mechanism described in Equation 17.1 is still a shear mode of fiber buckling, but unlike Rosen's elastic analysis, the local nature of the imperfections, coupled with (perfectly) plastic yielding of the matrix in shear, results in local microbuckling and kink band formation, as shown in Fig. 17.1. However, it should be noted that Equation 17.1 assumes zero kink band angle and an elastic–perfectly plastic matrix material.

From Equation 17.1, it quickly becomes apparent that improvements in the compressive strength of the composite may be achieved by (1) improving the yield strength and stiffness of the surrounding matrix in shear and (2) reducing fiber misalignment in the composite through optimization of manufacturing process variables, such as pull speed, preformer temperature, nanoparticle alignment, and dispersion. It is therefore the aim to accomplish this objective by concurrently increasing matrix yield strength and matrix shear modulus, as well as by decreasing fiber misalignment, thereby making use of synergies that may exist between these effects. There exist very

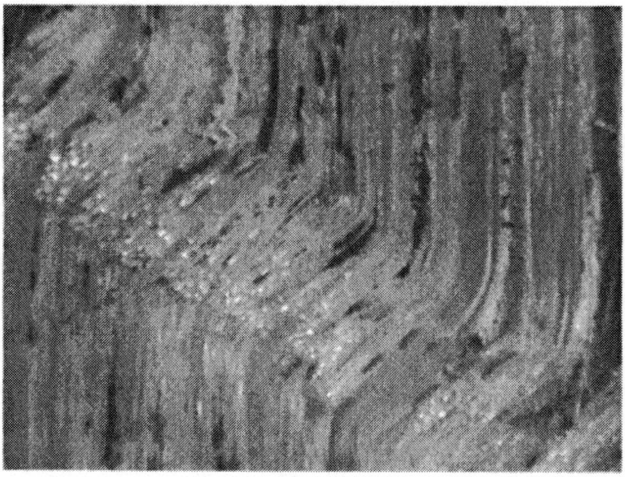

FIGURE 17.1 Compression failure of pultruded specimen showing kink band.

few published data on the compressive strength of pultruded thermoplastic composites. Fibers in pultruded material are in general well aligned. However, the composite compressive strength is very sensitive to even slight misalignment in fiber orientation.[5,6]

Typically a thermoplastic resin matrix offers better structural performance than thermosetting resin matrix. Thermoplastics exhibit superior fracture toughness[7] and impact resistance, and have better postprocess formability and improved reparability. With improved manufacturing techniques, such as ultrasonically activated consolidation dies, it is now feasible for thermoplastic pultrusion to yield higher production rates than thermoset composites. Recyclability of thermoplastics is another attractive feature. In this chapter, the effect of incorporating nanoclay particles on mechanical properties such as compressive strength and compressive modulus of pultruded E-glass/PP unidirectional composites will be presented and discussed.

17.2 EXPERIMENTAL PROCEDURE

17.2.1 Materials

Marlex HGL-350 polypropylene homopolymer (antistatic) was used as neat resin (density 902 kg/m^3 and melt flow rate 35 g/10 min). Material systems evaluated in the experimental studies were unidirectional E-glass fibers melt-impregnated with polypropylene with and without nanoclay particles using a single-screw extruder. To pultrude unidirectional composites shapes using either E glass/PP or nanoclay/E glass/PP, prepreg tapes were fed to the pultrusion machine. Subsequently 2.54-cm wide and 1.27-cm thick composite beams for different nanoclay content (0%, 1%, 2%, 3%, 4%, 5%, and 10% by weight) were pultruded. The pultruded samples were machined according to ASTM D695 specifications to obtain compression test specimens. The clay used was organically modified montmorillonite, 1.28 E supplied by Nanocor.

17.2.2 Processing of Fiber–Reinforced Nanocomposites

17.2.2.1 Thermoplastic Pultrusion Process

The use of thermoplastics as matrix materials for polymer composites has gained popularity recently due to their high-performance properties compared with thermoset resins.[8] Pultrusion is among the most rapid and efficient processes to fabricate composites when low cost or high volume is required.[9,10] The versatility of this process enables the production of profiles with many different sectional shapes, in what has largely contributed to the rapid increase of pultruded composites in the construction industry. It remains one of the most efficient and inexpensive methods for producing continuous fiber composites. In its simplest form, continuous reinforcing fibers (or tapes) are coated with a resin, and then shaped and cured by pulling the mass through a heated die. The process consists of the following general steps: (1) preimpregnated reinforcement feeding (the preimpregnation of fibers is usually done in a separate step using an extruder); (2) die shaping/curing; (3) pulling; and (4) cutting. The preimpregnated reinforcing materials (continuous fibers, mats, etc.) are guided through carding plates and other style guides to assemble them in the right place and order. Next the fiber and resin mass flows through a heated preformer die that causes the thermoplastic resin to melt before it enters the consolidation die. The operating parameters (temperature and pull speed) are adjusted so that the composite is fully consolidated as it leaves the die. Finally a mechanical means for pulling is provided, e.g., a belt puller. A system for cutting the pultruded composite as it is continuously pulled is also provided. In the case of the novel process utilized by the pultrusion machine at Oklahoma State University (OSU), the prepreg is guided through a heated preformer and then through an ultrasonically activated final consolidation die. A schematic of the pultrusion machine with its ultrasonic die assembly is shown in Fig. 17.2. The ultrasonic vibrations directed into the preheated prepreg by means of a waveguide help stimulate the flow of resin in the prepreg. Pressure is also applied to the prepreg within the die to consolidate the prepreg to the chosen profile. The ultrasonic vibration at low power input (maximum power 50 watts) does result in some incidental heating in the prepreg, but the increased flow rates of the resin indicate a reduction in viscosity that is far greater than expected in relation to

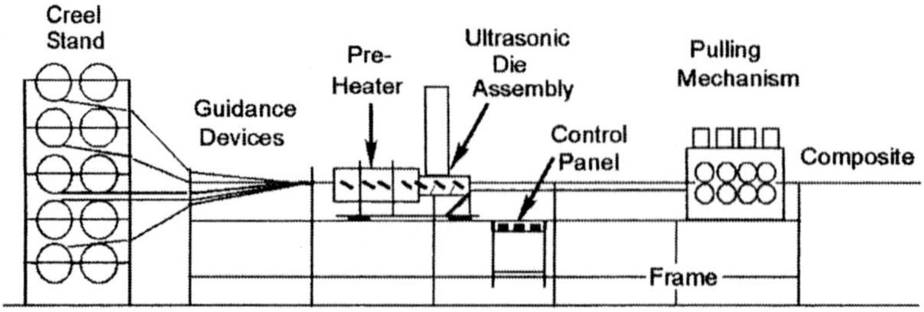

FIGURE 17.2 Schematic of pultrusion machine with ultrasonic consolidation die.

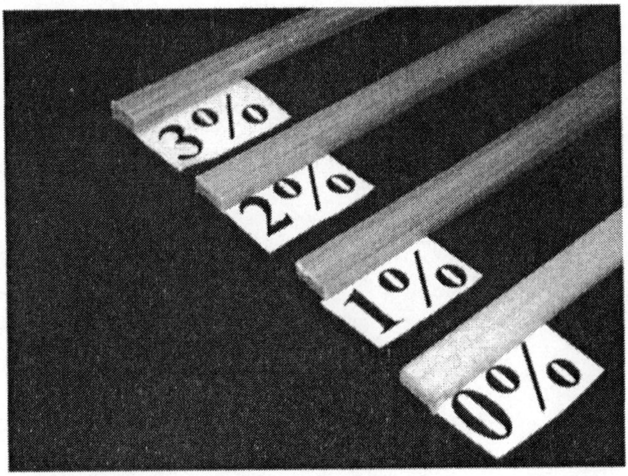

FIGURE 17.3 Pultruded E glass/PP samples with different nanoclay loadings.

the temperature increase. The net effect of the ultrasonic die is to reduce effective resin viscosity as well as friction at the composite–die interface, thereby significantly reducing void content and pull-force requirement. Pultruded E glass/PP samples that are 2.54 cm wide and 1.27 cm thick are shown in Fig. 17.3 for 0%, 1%, 2%, and 3% nanoclay loading by weight. The change in color of the samples due to the presence of clay is noticeable.

17.2.2.2 Process Parameter Control

The most critical processing parameters to be controlled are speed and preformer temperature profile. For any fixed die length the process speed will determine the residence time of material. Speed also depends on the sectional thickness of the part. In this research, E glass/PP composites were pultruded successfully at speeds up to 5 mm/s. Initially pulling speed was maintained as low as 0.5 mm/s to ensure correct consolidation of the thermoplastic composite. The preformer die consisted of three temperature zones. Temperature set points as measured by the thermocouples at these three zones in the preformer die were 220°C, 300°C, and 325°C, respectively. The preformer die temperature profile determined the rate of heating of material and influenced the resin viscosity and degree of cure attained during residence in the preformer. Subsequently the composite passed through the final consolidation die, where ultrasonic vibrations directed into the preheated prepreg by means of a waveguide helped stimulate the flow of resin in the prepreg. Pressure was applied to the prepreg within the die to consolidate the prepreg to the chosen profile. Finally the pultruded laminates were sectioned using a precision cutting saw that traveled at the same speed as the part being pultruded and examined in an optical microscope in order to determine the size and distribution of voids and unwetted fibers. Table 17.1 summarizes the pultrusion processing condition.

TABLE 17.1
Pultrusion Processing Variables

Prepreg Precursor	No. of Tapes	Preformer Temperature Profile (°C)			Die Pressure (kPa)	Pulling Speed (mm/s)	Ultrasound Energy (Watts)
		Zone 1	Zone 2	Zone 3			
PP/E glass	38						
PP/E glass–1% clay	37						
PP/E glass–2% clay	37	220	300	325	35–70	0.5–5.0	22–25
PP/E glass–3% clay	37						
PP/E glass–4% clay	36						
PP/E glass–5% clay	36						
PP/E glass–10% clay	36						

17.2.3 Mechanical Test Methodology

Two separate sets of compression tests were performed. In the first set, ASTM's D695 procedure was used to determine the compressive properties of neat PP and nanoclay-modified PP resin samples using an MTS (22.24 kN range) testing machine with two parallel platens. Load was measured by load cells attached to the tops of

FIGURE 17.4 Cylindrical polypropylene neat resin specimen after compression testing.

FIGURE 17.5 Pultruded compression test specimen (10% nanoclay loading) after failure.

the platens. The tests were carried out using compression-molded cylindrical test specimens that were 12.7 mm in diameter and 25.4 mm in length, as shown in Fig. 17.4. From applied load and displacement data, stress versus strain curves were plotted using Instron fast track data acquisition software.

In the second set of tests, compressive strength and compressive modulus of pultruded E glass/PP with different loadings of nanoclay particles were determined by performing compressive tests using a standard 245-kN Servo hydraulic MTS machine in displacement mode. Again the ASTM D-695 test method was used to determine the compressive strength and modulus of elasticity of the E glass/PP nanocomposites. A test specimen after failure is shown in Fig. 17.5, with a close-up view of the resulting kink band shown in Fig. 17.1. At least four replicate test specimens were used for each nanoclay loading.

17.3 NANOCOMPOSITE MORPHOLOGY

17.3.1 Transmission Electron Microscopy (TEM)

For thin foil preparation, cylindrical specimens of 3-mm diameter were cut by an ultrasonic disc cutter. This procedure was performed for compression-molded resin specimens as well as for pultruded E glass/PP nanocomposite samples. These cylinders were then sliced in a diamond saw to 0.5 mm or smaller and then ground gently with a 600-grade silicon carbide paper to a thickness of about 200 μm and dimpled by mechanical polishing. These dimpled discs were then ion milled and thinned to a thickness of less than 100 nm to allow electron beam transmission. The thin foils so prepared were examined with a 200-KeV TEM (Hitachi H 8000) equipped with EDS facilities to determine elemental composition of the specimens.

17.3.2 Scanning Electron Microscopy (SEM)

Microstructure and fracture surfaces of the composites were evaluated by scanning electron microscopy (JOEL) JXM 6400, at an accelerating voltage of 20 kV. A Balzer MED 010 sputtering system was used with an Au-Pd sputter coater.

17.4 RESULTS AND DISCUSSION OF TEST DATA

17.4.1 Transmission Electron Microscopy

Transmission electron microscopy (TEM) was applied to characterize the hierarchical morphology present in the nanocomposites. Figure 17.6(a) shows a bright-field TEM micrograph of polypropylene with 3% clay loading in a dogbone sample. Figure 17.6(b) shows the selected area diffraction pattern for 3% polypropylene-based nanocomposites. The diffraction pattern shows several reflections with varying degrees of intensity. Figure 17.6(a) and Figure (b) reveal the presence of aggregates, but there are regions where completely delaminated sheets are dispersed individually, showing the dark lines of 10-Å thickness. It appears that the clay exists as expanded aggregates made up of 2 to 10 platelets. The sizes of these

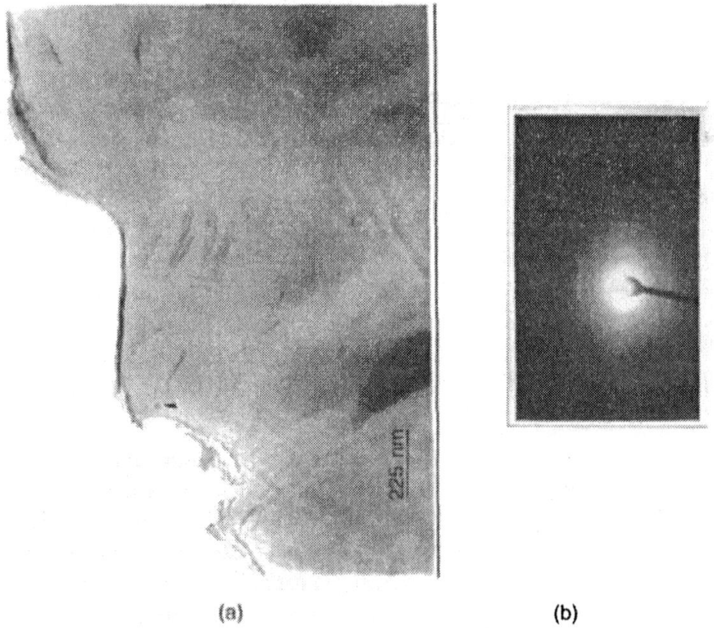

FIGURE 17.6 (a) Bright-field TEM micrograph of polypropylene with 3% clay loading in a dogbone sample. Note the irregular lamellar morphology of the polymer microstructure. A small nanosize clay particle is present in the picture. (b) Selected area diffraction pattern. Scale bar = 225 nm.

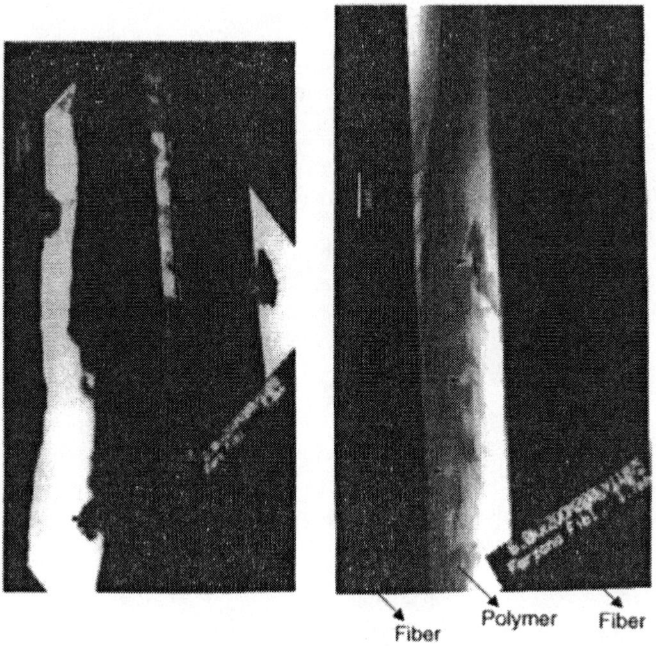

(a) (b)

FIGURE 17.7 (a) Bright-field TEM micrograph of compression molded E glass/polypropylene with 3% clay loading sample. The micrograph shows a few relatively large particles on the surface of the fiber, positions designated by arrows. (b) TEM bright-field micrograph showing how polymer is entrapped between two fibers. Positions of nanoclay particles (3% nanoclay) within polymer are indicated by arrows.

aggregates and the numbers of platelets in them increase with the percentage of clay within the PP matrix. Indeed the TEM pictures show all possible platelet morphologies, namely exfoliated, intercalated, and stacked structures within the samples. There appears to be reasonably good compatibility of the surface-treated clay with the polypropylene resin.

Figure 17.7(a) shows a typical arrangement of a group of such glass fibers present in a thin film of E glass/polypropylene with 3% nanoclay loading. Nanosized clay particles present in the composites were also found randomly distributed in the polypropylene within the individual spacing of E glass. Figure 17.7(b) a bright-field micrograph indicating the existence of a nanosized clay particle embedded in the matrix of polypropylene and in contact with the E glass fibers. This provides evidence of particle–fiber interaction in which the clay particles are actually in contact with the fiber wall. The orientation of the clay with respect to the fiber surface is unclear. A close-up view in Fig. 17.8 reveals the existence of a nanoclay agglomerate (A), as well as intercalated particles (B) and an exfoliated particle (C).

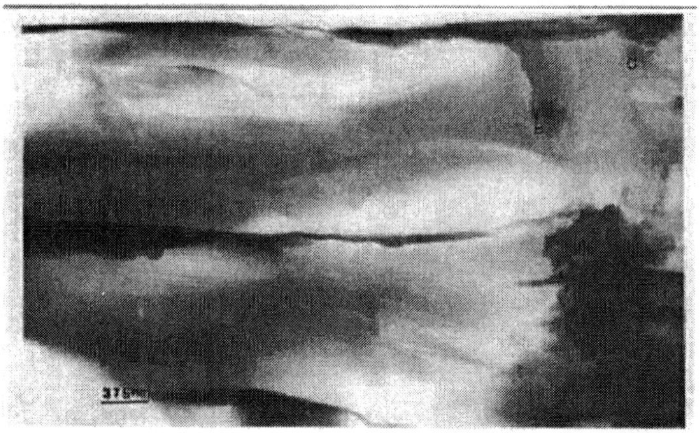

FIGURE 17.8 Bright-field TEM micrograph of polypropylene with 3% clay loading fiber-reinforced composite sample. Fine structure of polymer is evident in the micrograph. Positions denoted by A, B, and C with arrows indicate groups of particles, a couple particles, and single particle, respectively. Scale bar = 375 nm.

17.5 MECHANICAL PROPERTIES CHARACTERIZATION

17.5.1 Uniaxial Compression Tests

Compression strength data for neat polypropylene (PP) resin for different nanoclay loading are tabulated in Table 17.2. Figure 17.9 is a bar chart showing the progressive improvement in the normalized compressive strength of nanoclay-reinforced PP for different nanoclay content. A 134% increase in compressive strength was observed for PP resin with 10% clay loading.

TABLE 17.2
Compression Test Data for Neat Polypropylene Resin and Nanoclay-Modified Resin

Sample	Clay (wt%)	Avg. Compressive Strength (MPa)	% Increase Compressive Strength	Avg. Standard Deviation
PP	0.0	6.51	—	0.045
PP	1.0	8.29	26	0.29
PP	2.0	9.42	42	0.22
PP	3.0	10.33	57	0.23
PP	4.0	11.01	67	0.32
PP	5.0	12.77	94	0.48
PP	10.0	15.37	134	0.89

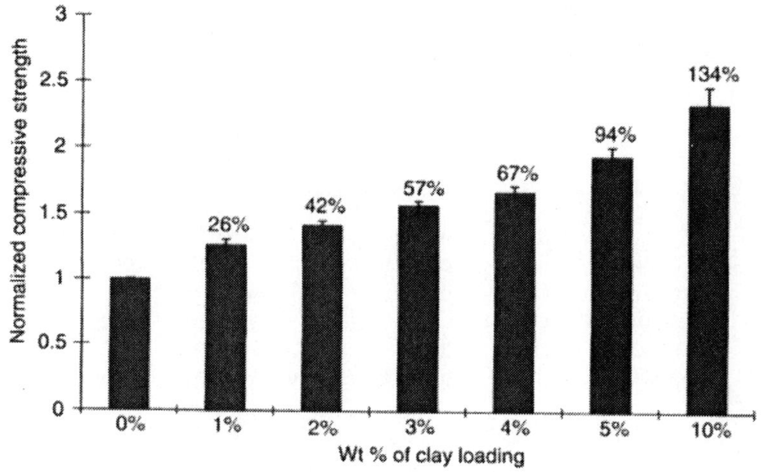

FIGURE 17.9 Normalized compressive strength versus clay content for neat polypropylene.

Stress versus strain curves for nanoclay-reinforced E glass/PP composite are shown in Fig. 17.10 for different nanoclay loadings, and the compressive strength data are summarized in column 3 of Table 17.3. Because our primary objective was to obtain strength data, the data shown in Fig. 17.10 were not corrected for machine compliance.

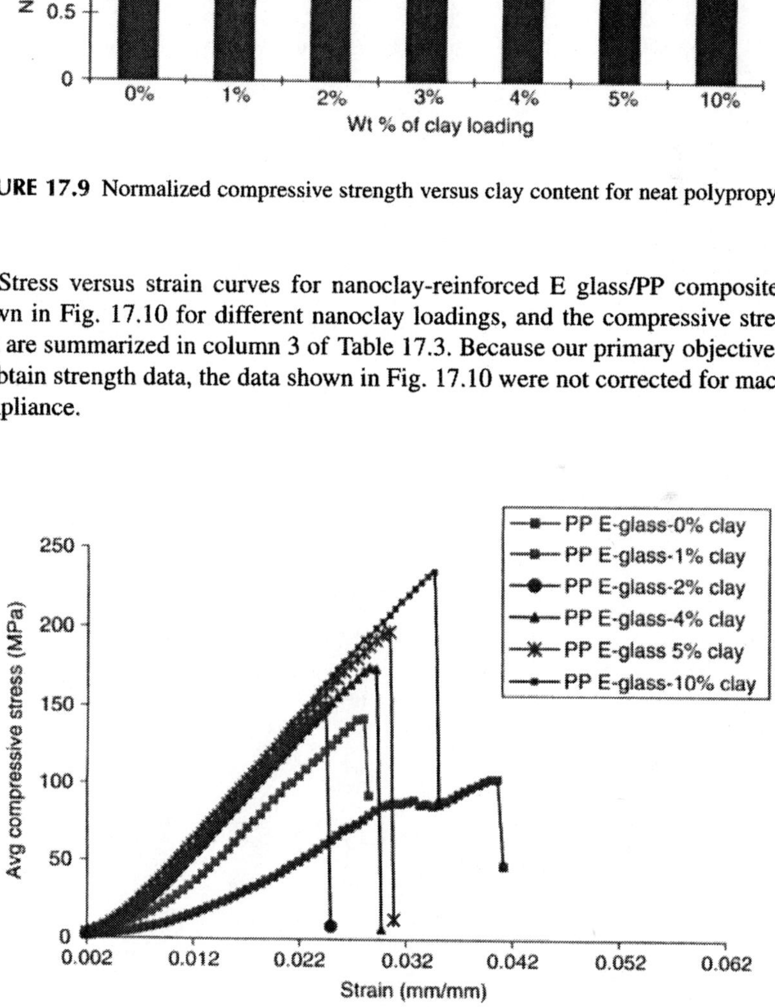

FIGURE 17.10 Comparison of stress versus strain curves from compression test for E glass/PP pultruded nanocomposites with different nanoclay loadings.

TABLE 17.3
Shear Modulus, Compressive Strength, Strain at Yield and Shear Strain in the Kink Band for Pultruded E Glass/PP/MMT

% Clay	$G_{Composite}$ (GPa)	$\sigma_{Critical}$ (GPa)	$\sigma_{Critical}/G_{Composite}$	$\frac{\tilde{\phi}}{\gamma_y}$	γ_y at $\tilde{\phi}$ = 0.2176	γ_c	γ_c/γ_y
0	0.5242	0.1053	0.2009	8.0887	0.0269	0.1581	5.8783
1	0.6845	0.1430	0.2089	7.5574	0.0288	0.1604	5.5716
2	0.7031	0.1449	0.2061	7.7387	0.0281	0.1596	5.6765
3	0.7418	0.1728	0.2330	6.2443	0.0348	0.1674	4.8045
4	0.7520	0.1763	0.2344	6.1747	0.0352	0.1679	4.7634
5	0.7688	0.1928	0.2508	5.4756	0.0397	0.1728	4.3481
10	0.8499	0.2435	0.2865	4.2973	0.0506	0.1840	3.6340

Figure 17.11 shows the progressive improvement in the normalized compressive strength of pultruded E glass/PP composites with different nanoclay content. Each bar represents the average of four test specimens, with a standard deviation of roughly 5%. A 122% improvement in strength for 10% clay content was observed compared with baseline specimens with zero nanoclay content. A similar dramatic enhancement was measured for normalized compressive modulus (stiffness), as shown in Fig. 17.12, where 10% clay reinforcement led to a 110% improvement in

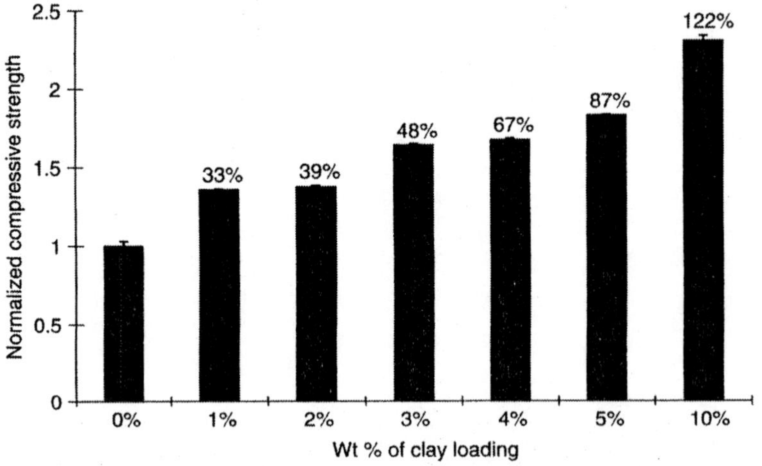

FIGURE 17.11 Normalized compressive strength versus clay content for pultruded E glass/MMT/PP.

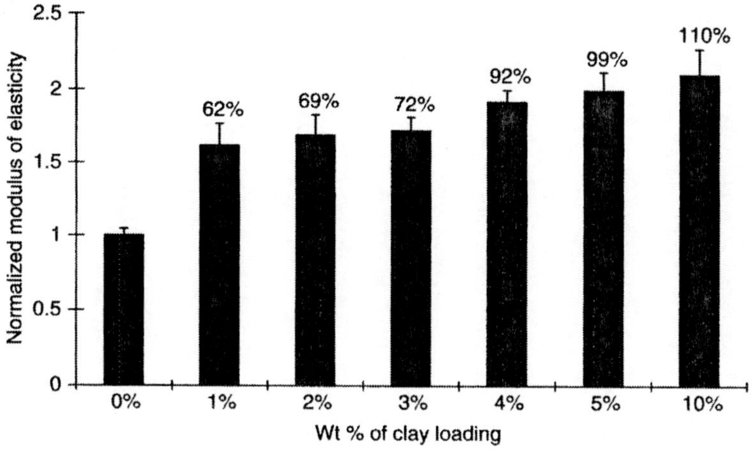

FIGURE 17.12 Normalized compressive modulus versus clay content for pultruded E glass/MMT/PP composite.

modulus. From these charts it is apparent that a significant improvement in compressive strength (87%) and modulus (99%) is attainable for E glass/PP systems with only 5% nanoclay content while using purely mechanical (unoptimized) means of nanoclay dispersion, namely, a single-screw extruder and sonication at the die. Examination of TEM micrographs of pultruded samples indicates the presence of unexfoliated clay agglomerates, which when fully exfoliated would presumably yield even better mechanical properties. As evidenced in Fig. 17.11, any negative effect of the ultrasonic treatment on fiber alignment is small compared with the large increase in matrix strength due to the presence of nanoclay particles. Further the test data indicate that it is feasible to pultrude nanoclay-reinforced E glass/PP laminates with up to 27% fiber volume fraction and up to 10% nanoclay content without a dramatic increase in resin viscosity that could clog the consolidation die or result in poor wetting of the fibers.

17.5.2 Inelastic Kinking Analysis for Pure Compression Loading

Microbuckling under axial compression of unidirectional composites is a failure mechanism in which the fibers undergo concurrent kinking in a narrow band, as depicted in Fig. 17.13. β is the angle of the kink band relative to the transverse direction, ϕ is the initial fiber misalignment angle, and ϕ is the additional rotation of the fibers in the kink band at critical stress. As mentioned in Section 17.1, Equation 17.1 is valid for elastic–perfectly plastic matrix with kink band angle equal to zero, which is an unrealistic assumption for the E glass/PP test specimens under consideration. Thus Budiansky and Fleck[4] modified Equation 17.1 to include (1) matrix

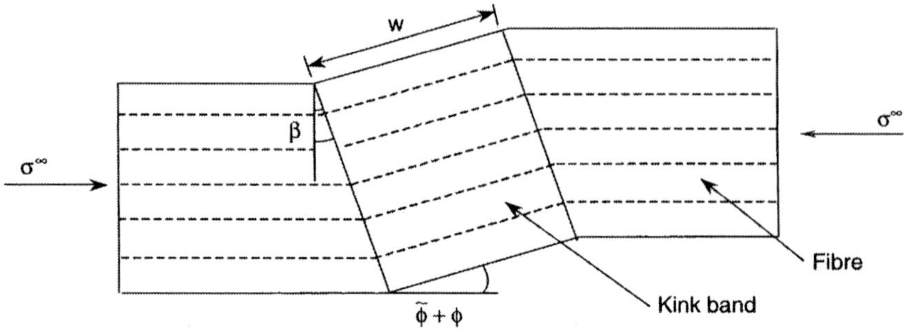

FIGURE 17.13 Kink band geometry and parameters.

strain-hardening assuming a Ramberg-Osgood relation and (2) kink band angle $\beta >$ 0. The modified kinking equation takes the form

$$\frac{\sigma_c}{G} = \frac{1 + R^2 \tan^2 \beta}{1 + n(3/7)^{1/n} \left[\frac{\phi / \gamma_Y}{n-1} \sqrt{1 + R^2 \tan^2 \beta} \right]^{(n-1)/n}} \quad (17.2)$$

where R is a constant defining the eccentricity of the yield ellipse, i.e., $R = \sigma_{TY}/\tau_Y$, and n is the strain hardening parameter. Based on the work presented in Reference 4, a step-by-step procedure for obtaining the critical shear strain (γ) within the kink band is developed and presented in the following section. The objective of this analysis is to be able to predict the change in the critical shearing strain in the kink band (assuming steady state) with increasing nanoclay loading in the matrix.

For kink band angle $\beta > 0$, from equilibrium of the kink band in steady state, the applied longitudinal compressive stress (Fig. 17.13) on the composite can be expressed in terms of transverse stress (σ_T) and shear stress (τ) in the kink band as

$$\sigma^\infty = \frac{\tau + \sigma_T \tan\beta}{\phi + \tilde{\phi}} \quad (17.3)$$

With the assumption that far field shearing strain $\gamma^\infty = 0$, and that the additional rotation of fibers ϕ is positive, Equation 17.3 yields[4]

$$\gamma = \phi \quad (17.4)$$

The transverse normal strain in the kink band is given by

$$e_T = \gamma \tan\beta \quad (17.5)$$

Defining an effective shear strain,

$$\gamma_e = \sqrt{\gamma^2 + R^2 e_T^2} \tag{17.6}$$

Substituting Equation 17.5 in Equation 17.6 yields

$$\gamma_e = \gamma\sqrt{1 + R^2 \tan^2\beta} \tag{17.7}$$

Also, it can be shown that kink band shear stress

$$\tau = \left(\frac{\tau_e}{\gamma_e}\right)\gamma \tag{17.8}$$

And kink band transverse stress

$$\sigma_T = \gamma \tan\beta R^2 \left(\frac{\tau_e}{\gamma_e}\right) \tag{17.9}$$

Hence the numerator of Equation 17.3 can be expressed as

$$\tau + \sigma_T \tan\beta = \left(\frac{\tau_e}{\gamma_e}\right)\gamma + \gamma R^2 \tan^2\beta \left(\frac{\tau_e}{\gamma_e}\right) = \tau_e \sqrt{1 + R^2 \tan^2\beta} \tag{17.10}$$

Defining a constant parameter $\alpha = \sqrt{1 + R^2 \tan^2\beta}$, Equation 17.3 can then be written as

$$\sigma^\infty = \frac{\alpha\tau_e}{\gamma + \tilde{\phi}} = \frac{\alpha\tau_e}{\dfrac{\gamma_e}{\alpha} + \tilde{\phi}}$$

$$\frac{\sigma^\infty}{G} = \frac{\alpha\, \tau_e/\tau_y}{\dfrac{1}{\alpha}\left(\dfrac{\gamma_e}{\gamma_y} + \dfrac{\tilde{\phi}}{\gamma_y/\alpha}\right)}$$

or

$$\frac{\sigma^\infty}{\alpha^2 G} = \frac{\tau_e/\tau_y}{\dfrac{\gamma_e}{\gamma_y} + \dfrac{\tilde{\phi}}{\gamma_y/\alpha}}$$

giving

$$\frac{\sigma^{\infty}}{G^{*}} = \frac{\tau_e/\tau_y}{\dfrac{\gamma_e}{\gamma_y} + \dfrac{\tilde{\phi}}{\gamma_y^{*}}} \tag{17.11}$$

where $G^{*} = \alpha^2 G$, and $\gamma_y^{*} = \gamma_y/\alpha$. As mentioned earlier, the elastic–plastic strain hardening behavior of the polypropylene matrix is modeled using a Ramberg-Osgood relation given by Reference 4:

$$\frac{\gamma_e}{\gamma_y} = \frac{\tau_e}{\tau_y} + \frac{3}{7}\left(\frac{\tau_e}{\tau_y}\right)^{n} \tag{17.12}$$

where n is the hardening parameter. Substituting Equation 17.12 in Equation 17.11 gives

$$\frac{\sigma^{\infty}}{G^{*}} = \frac{\tau_e/\tau_y}{\dfrac{\tau_e}{\tau_y} + \dfrac{3}{7}\left(\dfrac{\tau_e}{\tau_y}\right)^{n} + \tilde{\phi}/\gamma_y^{*}} \tag{17.13}$$

In order to find the value of τ_e/τ_y for which Equation 17.13 reaches an extremum, the derivative of Equation 17.13 with respect to τ_e/τ_y was obtained and set equal to zero:

$$\frac{\partial\left(\dfrac{\sigma^{\infty}}{G^{*}}\right)}{\partial\left(\dfrac{\tau_e}{\tau_y}\right)} = -\frac{\dfrac{\tau_e}{\tau_y}\left(1 + \dfrac{3}{7}n\left(\dfrac{\tau_e}{\tau_y}\right)^{-1+n}\right)}{\left(\dfrac{\tau_e}{\tau_y} + \dfrac{3}{7}\left(\dfrac{\tau_e}{\tau_y}\right)^{n} + \dfrac{\tilde{\phi}}{\gamma_y^{*}}\right)^{2}} + \frac{1}{\dfrac{\tau_e}{\tau_y} + \dfrac{3}{7}\left(\dfrac{\tau_e}{\tau_y}\right)^{n} + \dfrac{\tilde{\phi}}{\gamma_y^{*}}} = 0$$

yielding the condition in the kink band

$$\frac{\tau_e}{\tau_y} = \left[\frac{7}{3}\frac{(\tilde{\phi}/\gamma_y^{*})}{(n-1)}\right]^{1/n} \tag{17.14}$$

Substituting Equation 17.14 in Equation 17.12, an analytical expression for the effective shearing strain at kinking (γ_e) is obtained:

$$\frac{\gamma_e}{\gamma_y} = \left[\frac{7}{3}\frac{(\tilde{\phi}/\gamma_y^{*})}{(n-1)}\right]^{1/n} + \frac{\tilde{\phi}/\gamma_y^{*}}{(n-1)} \tag{17.15}$$

Finally combining Equation 17.5 and Equation 17.6 with Equation 17.15, the critical steady-state shear strain in the kink band (γ) can be obtained:

$$\gamma = \frac{\gamma_Y}{\sqrt{1+R^2\tan^2\beta}}\left(\left[\frac{7}{3}\frac{(\tilde{\phi}/\gamma_y^*)}{(n-1)}\right]^{1/n} + \frac{\tilde{\phi}/\gamma_y^*}{(n-1)}\right) \qquad (17.16)$$

Based on data reported by Budiansky et al.,[4] the values assumed for the constant parameters are $R = 2$, $\beta = 20°$, and $n = 3$. The kink band angle (β) measured from failed composite specimens was roughly around 25°. It should be noted that the compressive strength of the composite is not very sensitive to the strain hardening parameter.[4] The predicted values for the yield strain in shear (γ_y) and critical shear strain in kink band (γ) as functions of nanoclay loading are tabulated in Table 17.3.

17.5.3 Calculation of Fiber Volume Fraction, Composite Shear Modulus, and ϕ/γ_y^* Ratio

From stoichiometric analysis, the fiber volume fraction (V_f) in the pultruded E glass/PP composite was found to be 27.16%. It should be noted that Budiansky and Fleck's formula (Equation 17.2) was originally applied to composites with high fiber volume fraction (~60%). The following stepwise procedure was used in calculating the composite shear modulus and the ϕ/γ_y^* ratio:

1. Modulus of elasticity of the resin, E_m, was calculated using the stress–strain data collected from resin compression tests for different percentages of clay loadings.
2. The Poisson's ratio, υ, of neat PP resin (0% clay loading) was 0.47.[11] It is assumed that Poisson's ratio does not change with increasing clay loading.
3. The shear modulus (G_m) of the resin was calculated using the isotropic formula

$$G_m = \frac{E_m}{2(1+\upsilon)}$$

4. From Rosen's formula, the shear modulus of the composite is related to the matrix shear modulus through

$$G = \frac{G_m}{1-v_f}$$

where v_f = volume fraction of the fiber (~27%).

5. σ_c was obtained from composite compression test data reported in Table 17.3.

6. Knowing G_m and V_f, G was calculated, as reported in Table 17.3.
7. Knowing σ_c/G for a given nanoclay loading, the ϕ/γ_y^* ratio can be calculated using Equation 17.2, from which critical shear strain in kink band (γ) can be calculated using Equation 17.16, as reported in Table 17.3.
8. For neat PP matrix, shear strain at yield is $\gamma_Y = 0.0269$.[11] An initial fiber misalignment of $\bar{\phi} = 0.2176$ radians (~12.5°) was obtained using Equation 17.2 for the 0% clay loading case. The somewhat high fiber misalignment angle is likely due to the fact that prepreg tapes were used instead of yarns in the pultruder. It was assumed in all subsequent analysis that initial misalignment $\bar{\phi}$ remains unchanged with increased clay loading.

17.5.4 Discussion of Results

To be able to better visualize the effect of nanoclay on compressive strength and shear modulus of E glass/PP composite, experimental data for σ_c versus G from Table 17.3 are plotted in Fig. 17.14 for different nanoclay loadings. It is observed that both compressive strength and composite shear modulus exhibited monotonic increases with increasing nanoclay loading, as indicated by the symbols. In addition, considering that the line $\bar{\phi}/\gamma_y = 0$ corresponds to the Rosen solution for elastic buckling, the fact most of the data points are located between $\bar{\phi}/\gamma_y = 4$ and $\bar{\phi}/\gamma_y = 8$ indicates that the compressive failure for the nanoclay-reinforced pultruded E glass/PP is definitely inelastic and that the formation of the kink band entails elastic–plastic deformations for this material system. Figure 17.15 shows values of matrix yield

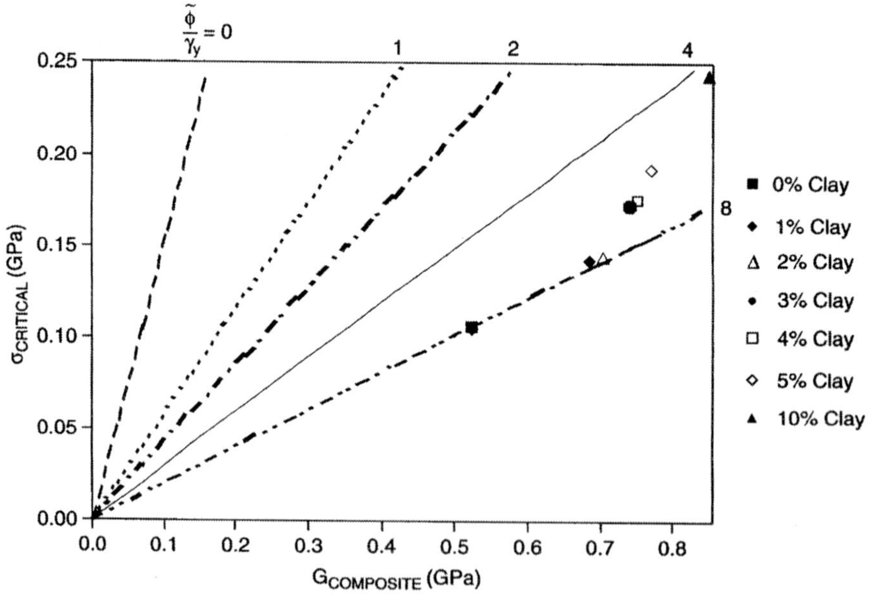

FIGURE 17.14 Compressive strength versus composite shear modulus plot showing test data and elastic-strain hardening–plasticity predictions for $\beta = 20°$.

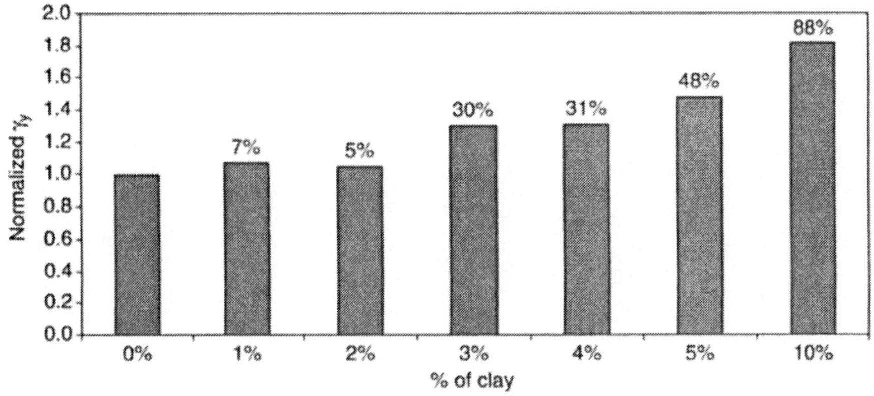

FIGURE 17.15 Change in matrix shear strain at yield with nanoclay loading.

strain in shear (γ_Y) plotted as a function of clay loading calculated using Equation 17.2, following the stepwise procedure described in the previous section. Again a significant increase (88%) in matrix yield strain in shear with nanoclay loading was observed. Figure 17.16 depicts values of the critical shear strain in the kink band (γ) plotted as a function of clay loading that was predicted using Equation 17.16. An approximately 17% increase in critical shear strain in the kink band was predicted at 10% nanoclay loading, presumably due to the presence of nanoclay in the resin matrix. Finally Fig.17.17 depicts values of the ratio of critical shear strain in the kink band (γ) to matrix yield strain in shear (γ_Y) plotted as a function of clay loading. The fact that this ratio remains well above 1 for all nanoclay loadings corroborates our earlier observation that the compressive failure process for pultruded E glass/PP is inelastic, even though the degree of inelasticity appears to decrease with increasing nanoclay loading.

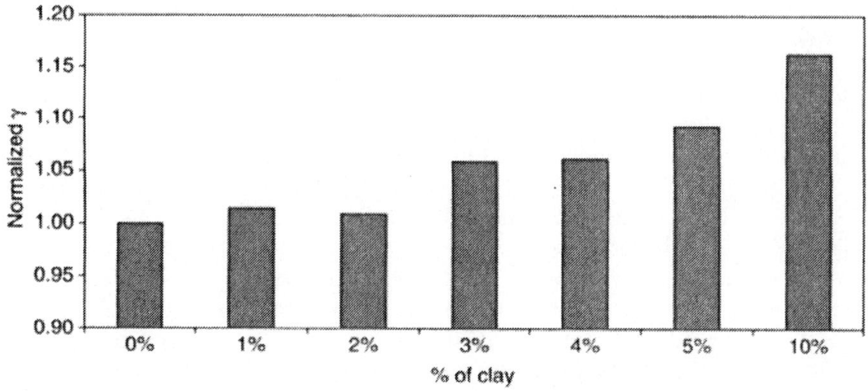

FIGURE 17.16 Change in shear strain in kink band with nanoclay loading.

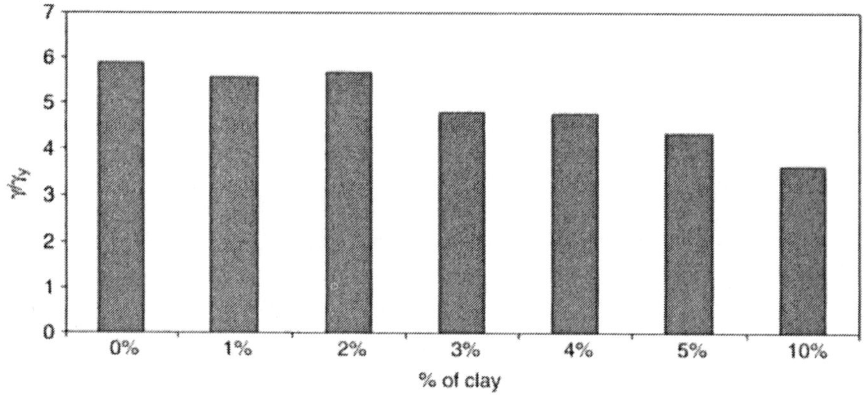

FIGURE 17.17 Variation in critical shear strain:yield strain ratio in kink band with nanoclay loading.

17.6 SUMMARY AND CONCLUSIONS

The determination of compressive strength of pultruded composites is of considerable importance, as these materials are generally weak in compression. This area is one of the least understood fields in composites, especially nanocomposites, because of the various parameters that affect the compressive behavior as well as difficulty in determining the strength experimentally. Unidirectional E glass fiber-reinforced pultruded composites were prepared in which the matrix was polypropylene with layered silicate nanoclay reinforcement. Thermoplastic pultrusion was used to prepare the composites. The compressive properties exhibited significant increases with increasing clay loadings (up to 10 wt%). Continuum-based analytical modeling was employed to predict the increase in critical shear strain in the kink band assuming elastic strain hardening–plastic matrix behavior and nonzero kink band angle. The model was able to establish that the kinking mechanism in E glass/PP nanocomposite is definitely inelastic. However, the current model does not explain the exact strengthening mechanism of the nanoclay reinforcement. Multiscale simulations of the nanoclay–polymer–glass fiber interactions would be necessary to adequately address this issue. TEM studies indicate a favorable interaction between the silicate clays and the glass fibers, which are also silicon based, indicative of enhanced matrix–fiber adhesion. The clays are also presumed to decrease the CTE mismatch, significantly reducing residual stresses and leading to higher quality laminates.

Additional characterization experiments to determine resin and composite shear properties are currently underway for additional verification of compressive failure data. Experimental verification of predicted shear strains within the kink band will be attempted using digital image correlation techniques. Another important aspect that needs to be studied is the effect of processing variables, such as pull speed, preformer temperature profile, and ultrasound on the orientation and dispersion of nanoclay platelets in the pultruded specimens.

PROBLEMS

1. Describe in detail the difference between intercalation and exfoliation of nanoclay platelets. Why are these stages important in the processing of high-quality nanoclay composites?
2. Explain why unidirectional fiber-reinforced composites are much weaker in compression than in tension. What is the most frequently encountered failure mode in compression of unidirectional composites?
3. What are some of the limitations of using the modified Argon formula (Equation 17.1) to predict compressive failure in a unidirectional fiber composite?
4. What were the changes made to the modified Argon formula (Equation 17.1) by Budiansky and Fleck (Equation 17.2)? Explain why.
5. Make a list of all of the processing variables used in the thermoplastic pultrusion process. Discuss the effects of ultrasonic vibrations at the consolidation die on the pultruded composite.
6. Discuss in detail the effect of addition of nanoclay particles on the following composite properties:
   a. Compressive strength
   b. Compressive modulus
   c. Matrix yield strain in shear
   d. Critical kinking strain

ACKNOWLEDGMENTS

This work was financially supported through a NASA-EPSCoR Research Initiation Grant and through a seed grant from the Oklahoma NanoNet. Special thanks are due to Dr. Shamsuzzoha at University of Alabama for his kind cooperation with TEM characterization and Dr. Scott Taylor of Vybron Composites, Inc. for fabricating prepreg tapes. We also acknowledge Nanocor, Inc. for supplying us with the surfactant-modified clay particles.

REFERENCES

1. Rosen, B.W., Mechanisms of composite strengthening, in *Fiber Composite Materials*, American Society of Metals, Cleveland, OH, pp. 37–75, 1964.
2. Argon, A.S., Fracture of composites, in *Treatise on Material Science and Technology*, Vol. 1, Academic Press, New York, pp. 79–114, 1972.
3. Budiansky, B., Micromechanics, *Computers Struct.*, *16*, 3–12, 1983.
4. Budiansky, B. and Fleck, N.A., Compressive failure of fiber composites, *J. Mechan. Phys. Sol.*, *41*, 183–211, 1993.
5. Creighton, C.J. and Clyne, T.W., The compressive strength of highly aligned carbon fibre/epoxy composites produced by pultrusion. *Composites Sci. Technol.*, *60*, 525–533, 2000.
6. Creighton, C.J., Sutcliffe, M.P.F., and Clyne, T.W., A multiple field image analysis procedure for characterization of fibre alignment in composites, *Composites Part A*, *32*, 221–229, 2001.

7. Lu, H., Roy, S., Sampathkumar, P., and Ma, J., Characterization of the Fracture Behavior of Epoxy Nanocomposites, *Proceedings of 17th Annual Technical Conference*, American Society for Composites, West Lafayette, IN, October 21–23, 2002 (CD-ROM).
8. Taylor, S. and Thomas, W., High Speed Pultrusion of Thermoplastic Composites, *Proceedings of 22nd International SAMPE Technical Conference*, November 6–8, 1990, Boston, MA, 1990.
9. Carsson, A. and Astrom, B.T., Experimental investigation of pultrusion of glass fiber reinforced polypropylene composites, *Composites Part A*, *29A*, 585–593, 1998.
10. Liskey, A.K., Pultrusion on a fast track, *Advan. Mater. Proc.*, *135*, 2, 31–35, 1989.
11. Maier, C. and Calafut, T., *Polypropylene: The Definitive User's Guide and Databook*, Plastics Design Library, p. 268, 1998.

Part 3

Modeling of Nanoscale and Nanostructured Materials

18 Nanomechanics

Young W. Kwon

CONTENTS

18.1 INTRODUCTION

As nanoscale devices and structures such as nanowires and nanotubes are being designed and developed, it is important to analyze such devices and structures using nanomechanics. There are multiple techniques for analyses of nanoscale structures. *Ab initio* simulation methods and classical molecular dynamics are examples of those techniques. The *ab initio* or first-principles method is a numerical scheme to solve rigorous quantum many-body Schrödinger equations and can deal with a very small number of atoms. On the other hand, classical molecular dynamics are based on Hamilton's classical equation of motion and can model a relatively large number of atoms. Tight-binding molecular dynamics fall between the *ab initio* technique and classical molecular dynamics. The tight-binding technique is based on simplification of the full quantum many-body problem so that its accuracy and the number of atoms to be analyzed practically lie between those of the *ab initio* and classical molecular dynamics techniques. Those simulation techniques are useful for design and development of nanostructures.

Four different topics at the nanoscale are presented here: the static atomic model, coupling between the discrete atomic model and the continuous finite element model, fatigue analysis at the atomic level, and carbon nanotubes. The static atomic model is beneficial for predicting static properties of atomic structures such as Young's modulus. Furthermore coupling between the atomic model and a continuous medium will be useful for multiscale analysis ranging from nanoscale to macroscale. The fatigue analysis at the atomic level is to provide insight of fatigue damage processes at the nanolevel, leading to eventual prediction of fatigue, life cycle, and design of tougher materials. Finally heterogeneous carbon nanotubes are discussed, which may be useful in future nanotechnology development.

18.2 STATIC ATOMIC MODEL

Classical molecular dynamics are based on Newton's second law of atomic motions. The forces applied to atoms are derived from the energy potential among the atoms. As a result, the theory of classical molecular dynamics solves second-order differential equations of motion of atoms using a numerical algorithm, i.e., using a time integration scheme. There are many different numerical time integration techniques. However, the two most popular techniques have been Varlet's algorithm and Gear's predictor-corrector algorithm.

The classical molecular dynamics model computes positions, velocities, and accelerations of atoms as a function of time. Then static and dynamic properties are computed from the time history of the solutions. This is a time-consuming process. Some properties such as elastic modulus can be determined more quickly through static analysis without time integration of equations of motion. As a result, the discrete static atomic model[1] is presented here for static analysis.

The discrete atomic model is based on static equilibrium of interatomic forces among atoms. Interatomic forces are computed from the interatomic potential energy expression. There are different mathematical forms of potential energies depending on the bonding types of atoms. Simple potential expressions consider two-body interaction, i.e., interaction between two atoms, without considering effects of other atoms. More complex potentials include multibody interactions. Some of the potentials are provided next.

One of the old and commonly used potential expressions is the Lennard–Jones potential:[2]

$$\Phi(r_{ij}) = \kappa\mu\left[\left(\frac{\alpha}{r_{ij}}\right)^{n} - \left(\frac{\alpha}{r_{ij}}\right)^{m}\right] \tag{18.1}$$

where

$$\kappa = \frac{n}{n-m}\left(\frac{n}{m}\right)^{m/(n-m)} \tag{18.2}$$

Here $\Phi(r_{ij})$ is the pair potential between two atoms i and j with distance r_{ij}. In addition, α is the distance to the zero potential and μ is the energy at the minimum of the potential energy. For London's theory, which is one of van der Waals forces, $n = 12$ and $m = 6$. Then the first term models the short-range repulsion, while the second term indicates the attractive potential. The Lennard–Jones potential has been applied to amorphous solids or fluids.

The Morse potential is as simple as the Lennard–Jones potential, and it is expressed in terms of exponential functions:

$$\Phi(r_{ij}) = D\left[e^{-2\alpha(r_{ij}-r_o)} - 2e^{-\alpha(r_{ij}-r_o)}\right] \quad (18.3)$$

in which D and α are constants with dimensions of energy and reciprocal distance, respectively. In addition, r_o is the equilibrium distance of the two atoms with $\Phi(r_o) = -D$, as shown in Equation 18.3.

The Gaussian-type potential is also a pair potential between two atoms such as the following:

$$\Phi(r_{ij}) = Ae^{-\alpha r_{ij}^2} - Be^{-\beta r_{ij}^2} \quad (18.4)$$

in which A, B, α, and β are material constants.

In the embedded atoms method,[3–5] which has been used for metallic atoms, the energy of a collection of atoms is determined by the electron density, i.e., the distribution of electrons in the system. As a result, the energy consists of two parts: the energy to embed each atom into the electron density of the neighboring atoms and the energy by a short-range pair interaction for the core–core repulsion. The energy of a system can be represented by

$$\Phi = \sum_i E_{ee}(\rho_{h,i}) + \frac{1}{2}\sum_{i,j} \phi_{ij}(r_{ij}) \quad (18.5)$$

where $\rho_{h,i}$ is the total electron density seen by atom i due to the surrounding atoms in the system, E_{ee} is the embedding energy for placing an atom into that electron density, and ϕ_{ij} is the short-range pair interaction for the core–core repulsion between atoms i and j separated by distance r_{ij}.

In Equation 18.5, the total electron density seen by an atom is computed by summation of atomic densities in surrounding atoms. That is,

$$\rho_{h,i} = \sum_{j \neq i} \rho_j^a(r_{ij}) \quad (18.6)$$

in which ρ_j^a is the electron density contributed by atom j with distance r_{ij} from atom i. The short-range, repulsive pair potential used in References 3–5 is expressed as

$$\phi_{ij}(r) = Z_i(r)Z_j(r)/r \quad (18.7)$$

The function $Z(r)$ is determined by fitting to the universal binding function.

The Abell–Tersoff–Brenner potential[6] for carbons and hydrogens is expressed as

$$\Phi(r_{ij}) = V_R(r_{ij}) - \bar{B}_{ij} V_A(r_{ij}) \tag{18.8}$$

where

$$V_R(r_{ij}) = f_{ij}(r_{ij}) D_{ij}^{(e)} \Big/ (S_{ij} - 1) e^{-\sqrt{2S_{ij}}\beta_{ij}(r_{ij} - R_{ij}^{(e)})} \tag{18.9}$$

$$V_A(r_{ij}) = f_{ij}(r_{ij}) D_{ij}^{(e)} S_{ij} \Big/ (S_{ij} - 1) e^{-\sqrt{2/S_{ij}}\beta_{ij}(r_{ij} - R_{ij}^{(e)})} \tag{18.10}$$

$$f_{ij} = \begin{cases} 1 & r_{ij} \le R_{ij}^{(1)} \\ \left[1 + \cos\left(\dfrac{\pi(r_{ij} - R_{ij}^{(1)})}{R_{ij}^{(2)} - R_{ij}^{(1)}}\right)\right] \Big/ 2 & R_{ij}^{(1)} < r_{ij} < R_{ij}^{(2)} \\ 0 & r_{ij} \ge R_{ij}^{(2)} \end{cases} \tag{18.11}$$

$$\bar{B}_{ij} = (B_{ij} + B_{ji})/2 + F_{ij}\left(N_i^{(t)}, N_j^{(t)}, N_{ij}^{conj}\right) \tag{18.12}$$

$$B_{ij} = \left[1 + \sum_{k(\neq i,j)} G_i(\theta_{ijk}) f_{ik}(r_{ik}) e^{\alpha_{ijk}\left\{(r_{ij} - R_{ij}^{(e)}) - (r_{ij} - R_{ij}^{(e)})\right\}} + H_{ij}\left(N_i^{(H)}, N_i^{(C)}\right)\right]^{-\delta_i} \tag{18.13}$$

$$G(\theta) = a_o\left[1 + \frac{c_o^2}{d_o^2} - \frac{c_o^2}{\left\{d_o^2 + (1 + \cos\theta)^2\right\}}\right] \tag{18.14}$$

Here θ is the angle between the lines connecting atoms i–j and i–k. The constant values in these equations are provided for carbon, hydrogen, and hydrocarbons in Reference 6.

The force acting on atoms is computed from

$$F(r_{ij}) = -\frac{\partial \Phi}{\partial r_{ij}} n_r \tag{18.15}$$

in which $\vec{n}_r$ is the unit position vector between the two atoms. The forces among N atoms along with any external load in a given system must be in equilibrium.

Consider any two atoms located at positions i and j, called atom i and atom j, as shown in Fig. 18.1. In that figure, solid circles indicate present positions of the two atoms under equilibrium before applying external loads. As external loads are applied to the system, the two atoms move to the positions denoted by open circles, as seen in Fig. 18.1. Then displacement vectors from initial positions to final positions of the atoms are expressed as u_i and u_j, respectively. The position vectors of the two atoms at the initial and final positions, respectively, are denoted by r_{ij} and R_{ij}, as

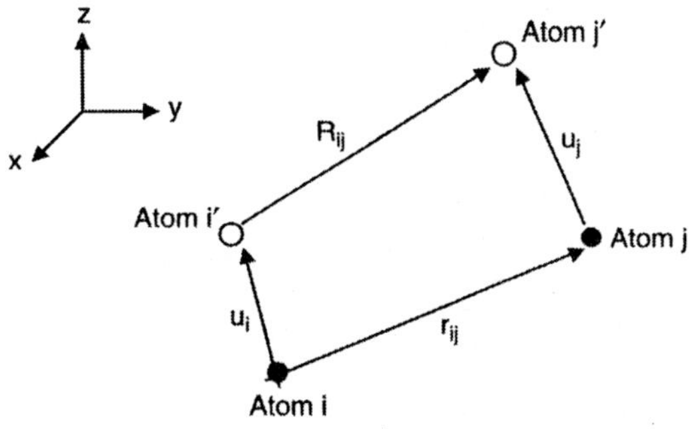

FIGURE 18.1 Positions of two atoms before and after displacements.

shown in Fig. 18.1. The displacement vectors and position vectors are related as

$$R_{ij} = r_{ij} + u_j - u_i = r_{ij} + \Delta u_{ij} \tag{18.16}$$

where $\Delta u_{ij} = u_j - u_i$ is the relative displacement vector of the two atoms.

The force between the two atoms i and j at their new equilibrium positions is expressed as Equation 18.17,

$$F_{ij}(R_{ij}) = F_{ij}(R_{ij})n_R \tag{18.17}$$

in which R_{ij} is the distance between the two displaced atoms. n_R is the directional unit vector along the vector R_{ij} and it is expressed as

$$n_R = \frac{r_{ij} + \Delta u_{ij}}{R_{ij}} \tag{18.18}$$

Equation 18.18 is substituted into Equation 18.17 and the resultant expression is

$$F_{ij}(R_{ij}) = F_{ij}(R_{ij})\frac{\Delta u_{ij}}{R_{ij}} + F_{ij}(R_{ij})\frac{r_{ij}}{R_{ij}} \tag{18.19}$$

Applying Equation 18.19 to atoms i and j yields a matrix expression

$$\begin{bmatrix} k & 0 & 0 & -k & 0 & 0 \\ 0 & k & 0 & 0 & -k & 0 \\ 0 & 0 & k & 0 & 0 & -k \\ -k & 0 & 0 & k & 0 & 0 \\ 0 & -k & 0 & 0 & k & 0 \\ 0 & 0 & -k & 0 & 0 & k \end{bmatrix} \begin{pmatrix} u_{ix} \\ u_{iy} \\ u_{iz} \\ u_{jx} \\ u_{jy} \\ u_{jz} \end{pmatrix} = \begin{pmatrix} F_{ix} \\ F_{iy} \\ F_{iz} \\ F_{jx} \\ F_{jy} \\ F_{jz} \end{pmatrix} \tag{18.20}$$

where

$$k = \frac{F_{ij}(R_{ij})}{R_{ij}} \tag{18.21}$$

$$F_{ix} = -F_{jx} = \frac{F_{ij}(R_{ij})}{R_{ij}}(x_i - x_j) \tag{18.22}$$

$$F_{iy} = -F_{jy} = \frac{F_{ij}(R_{ij})}{R_{ij}}(y_i - y_j) \tag{18.23}$$

$$F_{iz} = -F_{jz} = \frac{F_{ij}(R_{ij})}{R_{ij}}(z_i - z_j) \tag{18.24}$$

Here (x_i, y_i, z_i) and (x_j, y_j, z_j) are the coordinates of the atoms, as located in Fig. 18.1.

The matrix expression in Equation 18.20 is computed for all atoms that interact with one another and is assembled into the system matrix consisting of all atoms' displacements. This resultant system matrix equation is nonlinear and it is solved after applying constraints of atoms. Because R_{ij} and $F_{ij}(R_{ij})$ are not known *a priori*, u_i , u_j , and R_{ij} are assumed initially or obtained from the previous iteration solutions with a proper relaxation technique.

Figure 18.2 shows initial equilibriums of atoms in square arrays with a dislocation oriented at 27°, 45°, and 63°, respectively, to the horizontal axis. The atoms at the bottom row are constrained from any movement, while the atoms at the top row are subjected to uniform tensile loading. As the tensile load is applied, the displacements of atoms into the new equilibrium in each case are depicted in Fig. 18.3. This figure illustrates how a slanted dislocation affects the movement of atoms under a uniform load.

Figure 18.4 shows two different types of carbon nanotubes: armchair and zigzag shape. The armchair model is a single-wall nanotube of the chiral vector of (6,6), while the zigzag single-wall nanotube has the chiral vector of (12,0). Using the Abell–Tersoff–Brenner potential function,[6] the equilibrium state of the initial armchair model is computed. After the atomic equilibrium state, the axial tensile force was applied to some atoms in the longitudinal direction of the tubes. As a result of the force, the nanotubes had elongations. Then the effective stiffness of the nanotubes was calculated as

$$E = \frac{\sigma}{\varepsilon} \tag{18.25}$$

where σ is the applied stress and $\varepsilon = \Delta l/l$ where Δl is the elongated length of tube and l is initial length of the tube. The computed elastic modulus of zigzag nanotube was 1.4 TPa, which compared well with other available data.

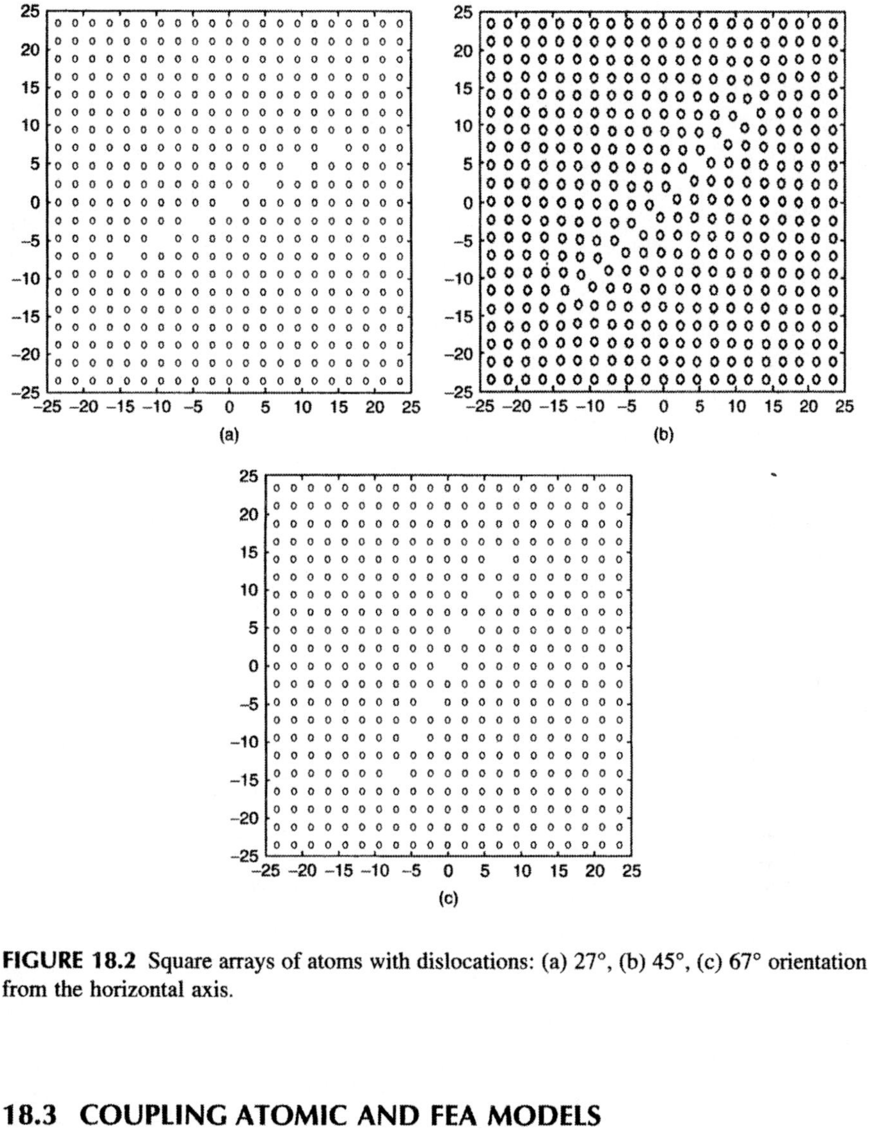

FIGURE 18.2 Square arrays of atoms with dislocations: (a) 27°, (b) 45°, (c) 67° orientation from the horizontal axis.

18.3 COUPLING ATOMIC AND FEA MODELS

The atomic model is a discrete system and it is useful to couple the atomic model to a continuum model. A practical size of the discrete system for computational modeling is very small because atoms are separated by distances one-tenth on the order of a nanometer. On the other hand, a continuum can be modeled in any size. Therefore coupling of the discrete atomic and continuous models provides a unique benefit. In other words, a relatively large size of domain can be analyzed in conjunction of accurate representation of atomic behavior, which is not obtainable in a continuous model.

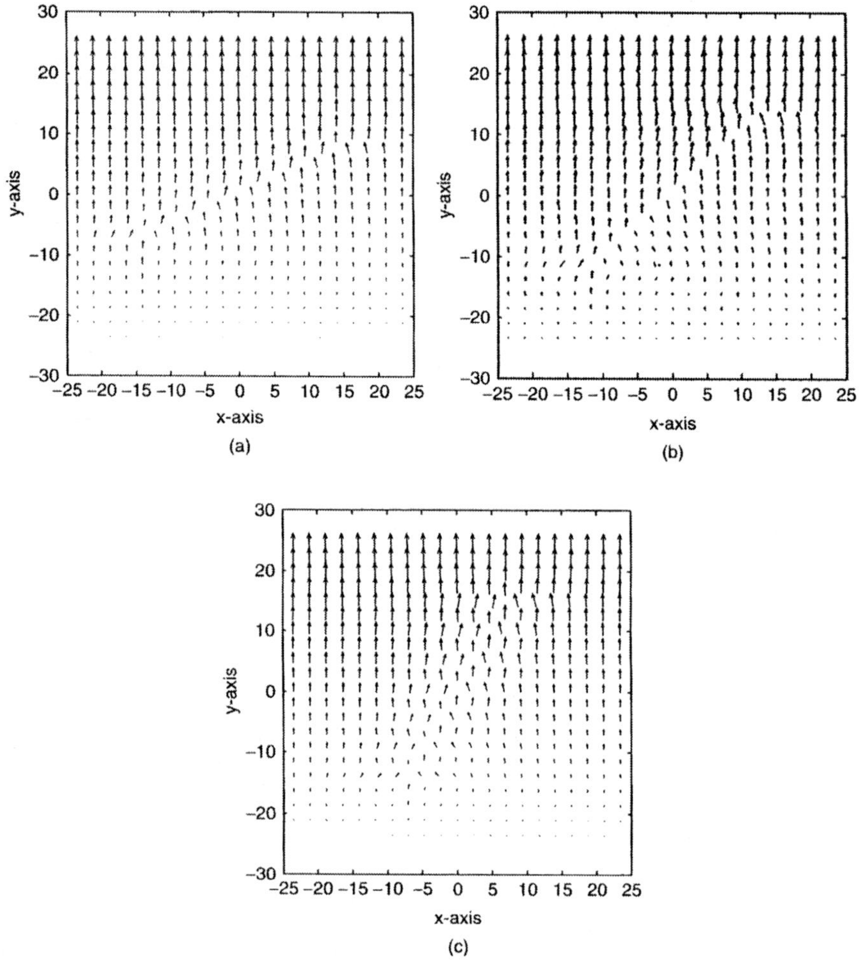

FIGURE 18.3 Movements of atoms with dislocations subjected to vertical tensile loads: (a) 27°, (b) 45°, (c) 67° orientation from the horizontal axis.

There are different approximation techniques to analyze continuous systems. One of the most common techniques is the finite element method. As a result, the coupling technique is described here between the finite element model and the atomic model. If another method such as the boundary element, finite difference, or meshless technique is used, the overall procedure remains the same except for some modifications in details, as necessary.

For simplicity, consider a two-dimensional coupling between the two models even though the same algorithm can be applied to three-dimensional bodies. Figure 18.5 shows discrete atoms surrounded by a continuous medium discretized

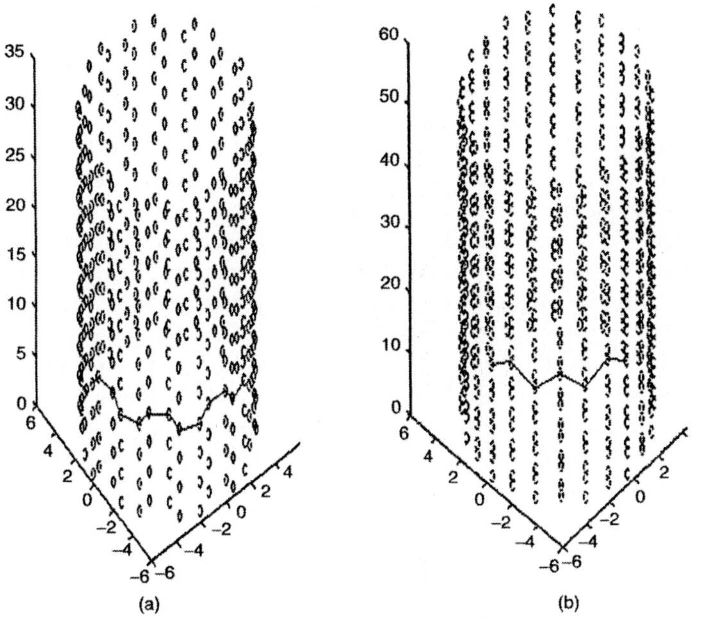

FIGURE 18.4 (a) Armchair and (b) zigzag carbon nanotubes.

for a finite element mesh. In the figure, the inner domain is the atomic domain, while the outer domain is the finite element domain of the continuous medium. The intermediate domain bounded by bold lines where both atoms and finite element-meshes overlap each other is called the interface domain for coupling. The atoms

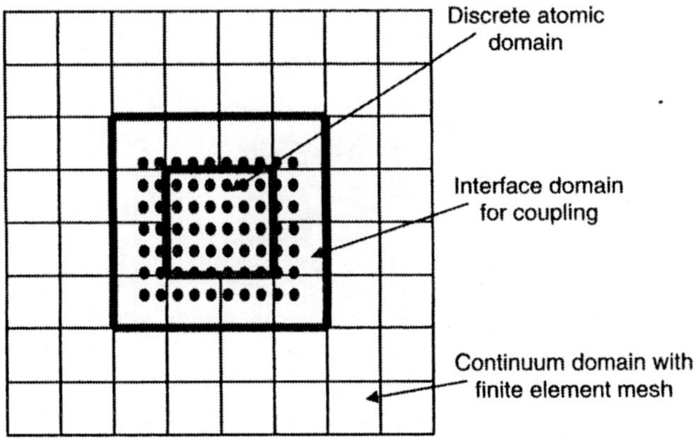

FIGURE 18.5 Coupling between discrete atomic model and continuous FEA model.

in the interface domain are called embedded atoms. The solution procedure with staggered solutions between the atomic domain and the finite element domain is as follows:

1. Solve the finite element matrix equation $[K_f]\{u_f\} = \{F_f\}$ of the continuous domain. From now on, subscripts *f* and *a* are used to denote finite element and atomic domains, respectively.
2. Compute the embedded atoms' displacements in the interface domain from the finite element nodal displacements using the finite element shape functions such as $\{u_a^e\} = [N]\{u_f^e\}$, where superscript *e* indicates the embedded atoms and finite elements in the interface domain, and [*N*] is the shape function matrix of finite elements in the interface domain.
3. Compute the new positions $\{x_a^e\}$ of the embedded atoms by adding the displacements computed in step 2 to the previous positions.
4. Using the atomic model, solve for the new positions of the atoms in the atomic domain with fixed embedded atoms' positions.
5. Compute the interactive forces $\{F_a^e\}$ on the embedded atoms exerted by all atoms.
6. Compute the equivalent nodal forces of the finite elements at the interface using $\{F_f^e\} = \int_{V_e} [N]^T \{F_a^e\} dV$ where V^e is the finite element volume.
7. With the nodal forces computed in step 6, the new finite element solution is obtained. Then continue the process from Step 2.

Figure 18.6 shows a coupling between the atomic domain and a finite element domain with a crack. The close-up view of the atomic domain of Fig. 18.6 is shown in Fig. 18.7. In this case, boundary conditions such as external loading and constraints can be applied to the finite element model. The problem domain was subjected to a tensile load perpendicular to the crack orientation at the top boundary, while the bottom

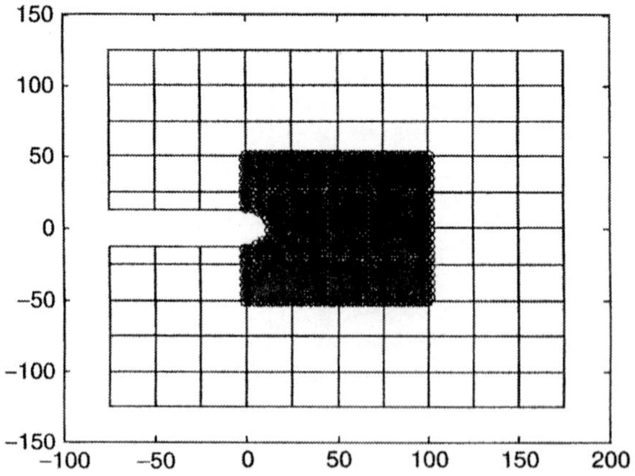

FIGURE 18.6 Coupling between atomic model and finite element model with a crack.

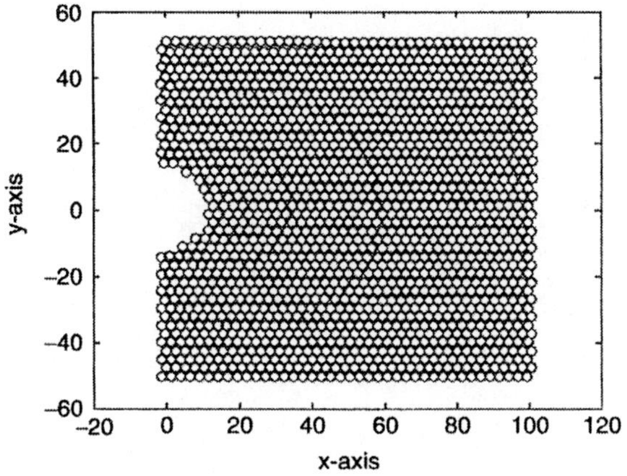

FIGURE 18.7 Close-up view of the atomic domain of Figure 18.6.

boundary was fixed. The deformed finite element domain is plotted in Fig. 18.8. The associated atoms' movements are plotted in Fig. 18.9 with the atoms' final positions shown in Fig. 18.10.

18.4 FATIGUE ANALYSIS AT ATOMIC LEVEL

Most structural materials, especially metals, fail due to repeated loads. This is called fatigue failure. Fatigue is a damage accumulation process in materials under repeating loads. Material defects at the atomic level, such as impurities, inclusions, vacancies, and dislocations, play an important role in fatigue process. When a material is

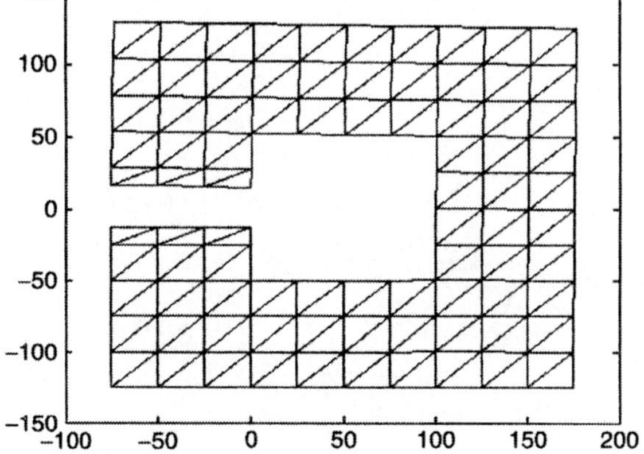

FIGURE 18.8 Deformed finite element domain with vertical tensile loading of Figure 18.6.

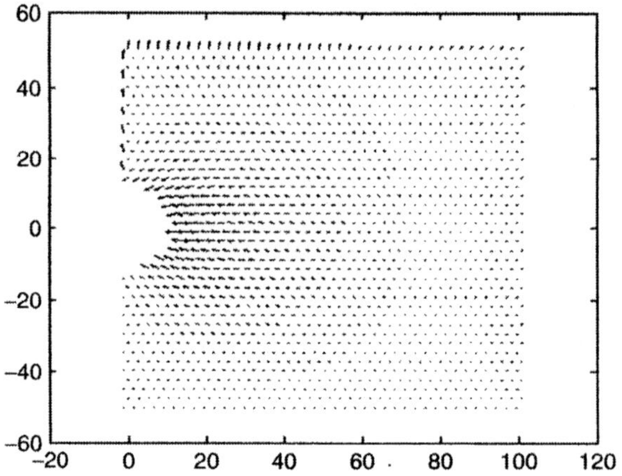

FIGURE 18.9 Atomic displacements associated with Figure 18.7.

stressed even in the elastic range in macroscale, there are irreversible changes within the material that are so small that they are not measurable in the stress-strain curve. However, as the number of cycles in the loads increases significantly, the accumulated damage leads to ultimate failure of the materials.[7]

In the past, prediction of fatigue failure mainly relied on experimental observation and data, which were time-consuming processes. Fatigue failure is influenced by many different parameters. Considering all the parameters affecting fatigue failure in experiments is impossible and practically prohibitive. Furthermore, because nanotechnology

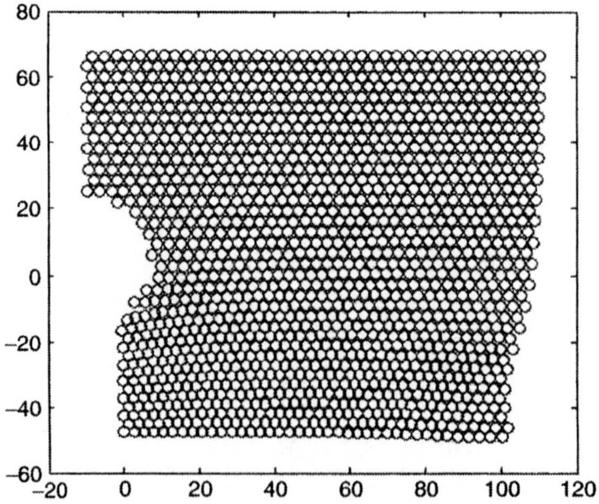

FIGURE 18.10 Atomic positions associated with Figure 18.7.

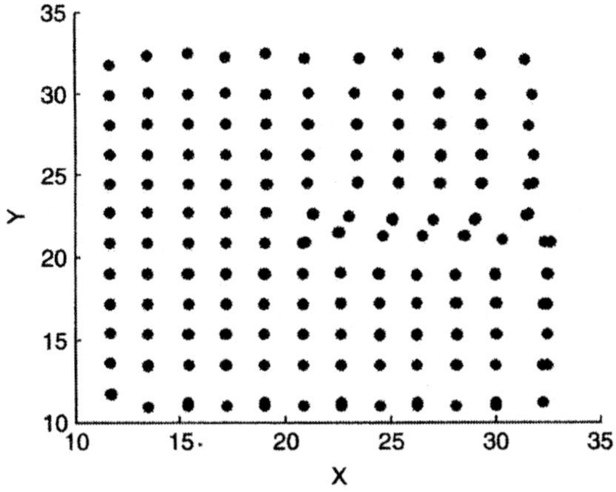

FIGURE 18.11 Plane view of copper atomic arrangement with dislocation.

has been developed recently, many devices have been devised at nano- or at most microscale. Even those scale specimens were shown to fail as a result of fatigue.[8] Experimental testing for fatigue behavior at nano- or microscale is quite a challenge.

Analytical modeling and simulation of fatigue processes are very beneficial to alleviate the heavy burden of experimental work and to enhance the understanding of fatigue processes. The present task is to utilize the classical molecular dynamics to understand the fatigue damage mechanisms in metallic materials.

A pure metal such as copper was studied using the embedded atom method.[3,4] The atomic structure with dislocation is shown in Fig. 18.11. The atomic arrangement was determined after applying a number of cyclic loads, as described later. For visual clarity, dislocation was placed through the z-axis and the xy-plane view is depicted in the figure. The atoms on the left face were constrained from any movement and a cyclic axial loading was applied uniformly to atoms on the right face under an isothermal condition. The applied cyclic loading (force versus. time) is shown in Fig. 18.12 at a constant room temperature. The applied peak stress was 100 or 200 MPa and the period of load cycle was 200 times the time step size.

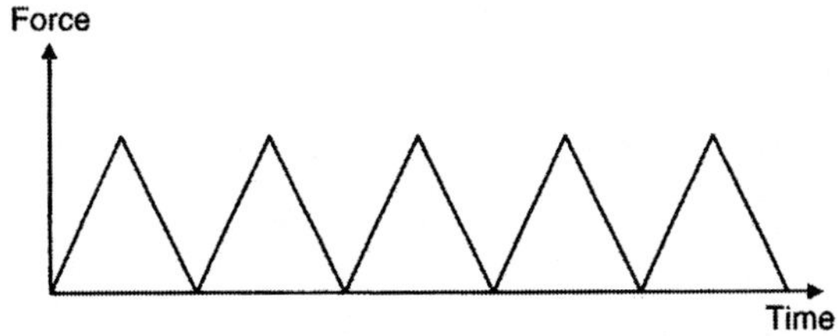

FIGURE 18.12 Cyclic loading.

Because of the cyclic load, there is a change of potential energy in the copper atoms. Figure 18.13 and Fig. 18.14 plot the average potential energy per copper atom as a function of time as the cyclic loading continues. There are fluctuations of the average potential energy, as seen in the figures. However, linear curve fitting of the fluctuations indicates the increase of the average potential energy along with cyclic loading. The increase of the average potential energy is larger in Fig. 18.14 than in Fig. 18.13 because a larger stress is applied to the system in Fig. 18.14. With the linear curve fitting, the slope of the curve may predict when the separation of atoms occurs, i.e., the fatigue failure of the material. In other words, the life cycle (number of cycles to failure) can be estimated from the rate of potential energy increase per cycle and the potential energy change necessary for atomic separation.

In order to explain why there is an increase in potential energy, a simple atomic model was presented. Let us consider an interaction between two atoms whose interaction is described in terms of a nonlinear potential function, for example, Morse potential for mathematical simplicity. One of the two atoms is fixed, while the other vibrates. Assuming one-dimensional motion, the dynamic equation of the vibrating atom can be written as

$$m\ddot{x} + f(x) = 0 \tag{18.26}$$

where m is the atomic mass, x is the displacement of the vibrating atom from its equilibrium position, and $f(x)$ is the interaction force between the atoms, which is

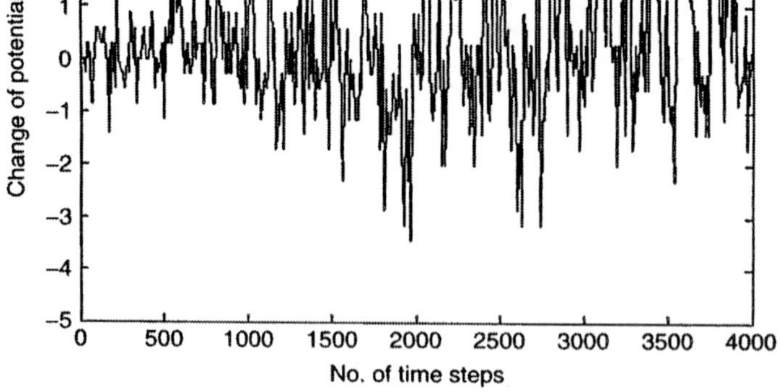

FIGURE 18.13 Change of potential energy per atom resulting from a cyclic load between 0 and 100 MPa versus the number of time steps.

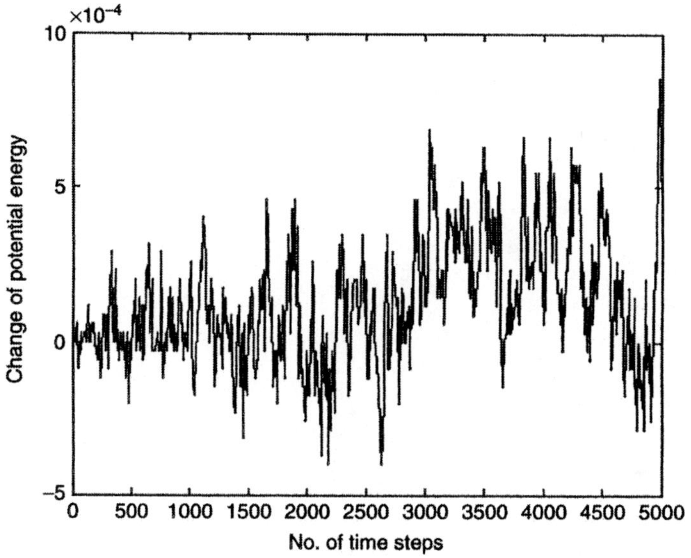

FIGURE 18.14 Change of potential energy per atom resulting from a cyclic load between 0 and 200 MPa versus the number of time steps.

derived from the interaction potential. Rewriting the equation in terms of state variables, displacement $x_1 = x$ and velocity $x_2 = \dot{x}$ yields

$$\begin{aligned} \dot{x}_1 &= x_2 \\ \dot{x}_2 &= -f(x_1)/m \end{aligned} \tag{18.27}$$

The plot of phase trajectory of Equation 18.6 indicates the stability of the system and it depends on the initial conditions of the displacement and velocity.

Velocities of atoms under equilibrium obey the Maxwell–Boltzmann law.[9] Considering a one-dimensional problem, the fraction of atoms having velocities between v and $v + dv$ can be expressed as

$$g(v)dv = \sqrt{\frac{m}{2\pi kT}} \exp\left(-\frac{mv^2}{2kT}\right) dv \tag{18.28}$$

where k is Boltzmann's constant and T is the temperature. Therefore, some of the atoms have much higher velocities than the average velocity of the system. Those atoms with high velocities show unstable phase trajectory, as shown in Fig. 18.15, while atoms with low velocities have stable phase trajectory, as shown in Fig. 18.16. The unstable phase trajectory means an increase of position or increase of potential energy. In other

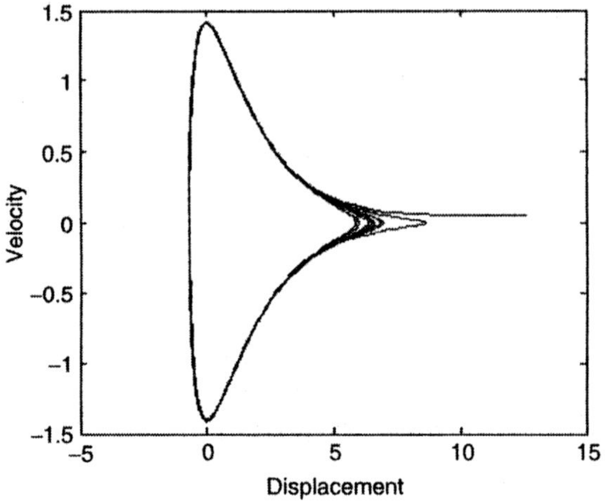

FIGURE 18.15 Unstable phase trajectory for an atom with an initial high velocity.

words, when a system of atoms is subjected to an external loading, some atoms at unstable equilibrium will increase potential energy, thus contributing to the increase of the total potential energy in the system.

Furthermore because not all atoms vibrate at the same frequency and amplitude, the vibrating frequencies of some atoms can be near or at the excitation frequency or frequencies of the applied repeated loading. Then the displacements of the atoms increase, resulting in the increase of potential energy.

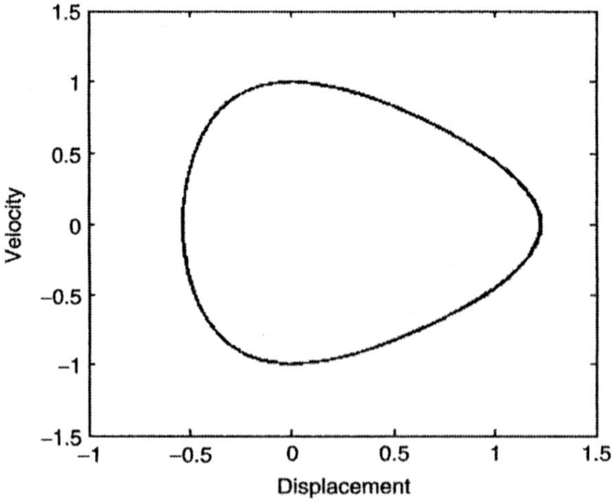

FIGURE 18.16 Stable phase trajectory for an atom with an initial normal velocity.

18.5 HETEROGENEOUS CARBON NANOTUBES

Conventional carbon nanotubes (CNTs) have uniform atomic structures along their lengths as shown in Fig. 18.17. In order to have a nonhomogeneous arrangement of atoms along the tube wall, a bamboo structural nanotube (BSNT) was devised by combining two single-wall nanotubes (SWNTs), as seen in Fig. 18.18.

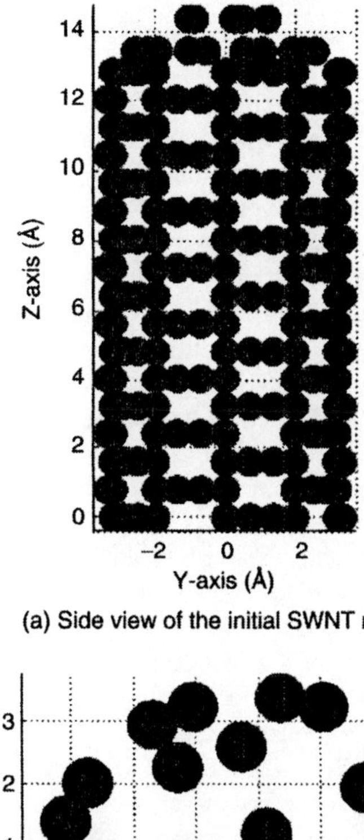

(a) Side view of the initial SWNT model

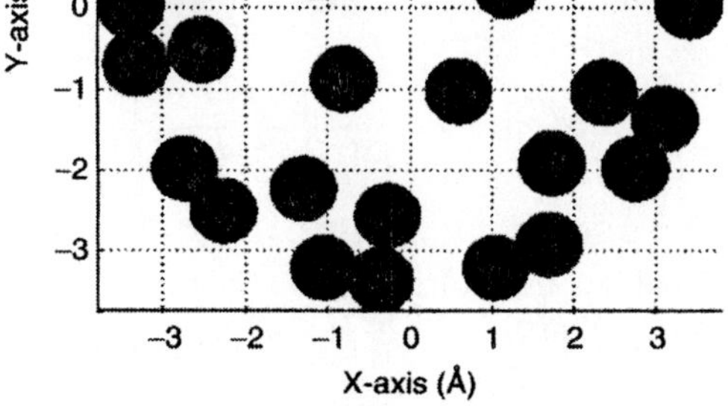

(b) Top view of the initial SWNT model

FIGURE 18.17 Initial (5,5) armchair SWNT model from geometrical consideration.

Heterogeneous carbon nanotubes with specific purposes may be controlled and shaped in the near future.

For the BSNT model, let us assume that there are two SWNTs as CNT *A* and CNT *B* with the same diameter: CNT *A* has capped ends at both top and bottom, while CNT *B* has the capped top and the open bottom. When the bottom of CNT *B* is physically contacted with the capped top of CNT *A* along the tube's *z*-axis with a revolved horizontal angle of 0.5 radian, CNT *A* and *B* can be jointed as a heterojunction region to lower the total system energy. When two nanotubes were coupled with a horizontal joint angle of 0 radian, the heterojunctions could not be formed during equilibrium simulation due to the chemical and geometrical bonding limits.

Figure 18.18 shows the side and top views of a BSNT model. The two (5,5) armchair SWNTs, upper and lower SWNTs, were prepared for modeling BSNT with the average C–C distance 1.42 Å and diameter 6.78 Å. The resultant tube length of the initial BSNT model was 15.14 Å measured from the origin coordinate (0,0,0) at the bottom of the lower tube to the tip of the upper tube end along the *z*-axis. The number of carbon atoms for the BSNT model is 210.

Since additional pentagon defect rings are introduced in the junction region, the mechanical properties of the BSNT such as Young's modulus will be different from the previous SWNT model. This section describes the method to calculate Young's modulus of SWNT and BSNT by introducing the freestanding room temperature vibration method presented in Reference 10. According to the kinetic molecular theory, carbon atoms are vibrating about their equilibrium positions with a mean vibrational kinetic energy that increases with the temperature as 3/2 kT, where k is the Boltzmann constant and T is temperature. From this fundamental principle, Krishnan et al. [10] estimated Young's modulus of single-wall carbon nanotubes as $\langle Y \rangle = 1.25 - 0.35/ + 0.45$ TPa by observing a SWNT's free-standing room temperature vibrations in a transmission electron microscope (TEM).

Their formula to calculate Young's modulus was derived from the relationship between the motion of a vibrating clamped cylindrical cantilever rod governed by the classical fourth-order wave equation and the quantum mechanics statistical probability theory given by the Boltzmann factor. The resultant relationship between Young's modulus Y, length L, inner and outer tube radii b and a, which form a cylindrical CNT wall, and the root mean square (*rms*) of displacement δ, which is represented by the vibration amplitude at the tip of a carbon nanotube at a temperature T, can be expressed by

$$\delta^2 = 0.8486 \frac{L^3 kT}{YWG(W^2 + G^2)} \tag{18.29}$$

where W is the SWNT's width (diameter), $G(= a - b)$ is the graphite interlayer spacing of 0.34 nm, and k is the Boltzmann constant.

For the present study, Equation 18.29 is used to predict Young's modulus:

$$Y = 0.8486 \frac{L^3 kT}{\delta^2 WG(W^2 + G^2)} \tag{18.30}$$

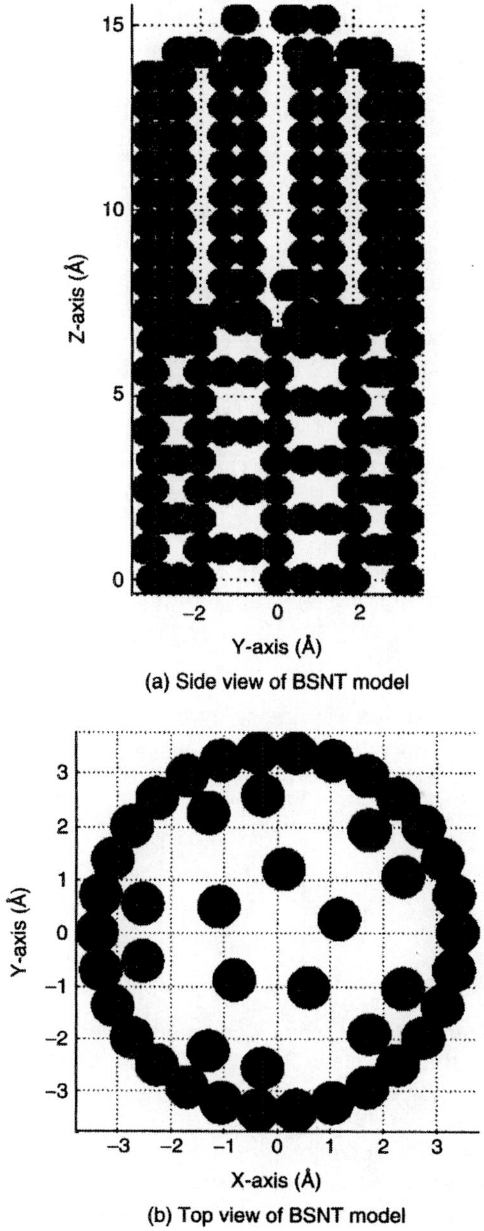

FIGURE 18.18 Side and top views of BSNT model with 210 atoms.

From this equation, it should be noticed that Young's modulus depends strongly on the *rms* displacement δ at the tip of the SWNT for a given length L, width W of the CNT, and the given ambient temperature T.

How far will a carbon atom at the tip of SWNT or BSWNT vibrate in a given parameter such as time interval, diameter, and temperature? In the three-dimensional Cartesian coordinate $\{x, y, z\}$, a position vector r_i that locates carbon atom i in the CNT is defined by three linearly independent vectors as

$$r_i = Xx + Yy + Zz \tag{18.31}$$

If the atoms are allowed to move from their equilibrium positions by an amount of displacement vector $u_i = (u_{ix}, u_{iy}, u_{iz})$ at a time t_0, then the actual position at the time of any carbon atom under the influence of a vibration (thermal fluctuation) is given by

$$R_i(t_o) = r_i(t_o) + u_i(t_0) \tag{18.32}$$

After a time interval t, the changed position of atom can be written as

$$R_i(t) = r_i(t_o) + u_i(t) \tag{18.33}$$

From Equation 18.32 and Equation 18.33, the mean square displacement (*msd*) of a carbon atom i at the tip of SWNT at given time t can be determined by

$$msd_i(t) = \langle (R_i(t) - R_i(t_0))^2 \rangle \tag{18.34}$$

where the notation $\langle \ldots \rangle$ denotes averaging over all the atoms.

When this quantity is averaged over all N atoms consisting of the end tip of the SWNT or BSNT, the *rms* displacement δ of the carbon nanotube-tip can be written as

$$\delta = \left(\frac{\sum_{i=1}^{N} msd_i(t)}{N} \right)^{1/2} \tag{18.35}$$

In a molecular dynamics simulation to find Young's moduli for SWNT and BSNT models under the equilibrium state, Equation 18.31 through Equation 18.35, played an important role to calculate the numerical *rms* displacement values from the simulated trajectory data of the carbon atoms, which were then applied in Equation 18.30 to compute Young's moduli.

The two models generated previously were tested under the molecular dynamics simulation at room temperature, 300 K, and then Young's moduli of SWNT and BSNT were evaluated based on Equation 18.30 and Equation 18.35. After that the numerical values were compared with the theoretical and experimental results presented from other studies.

Figure 18.19 histograms demonstrate the spread in the evaluated Young's moduli for SWNT and BSNT by running 100 simulations for each model. The measured

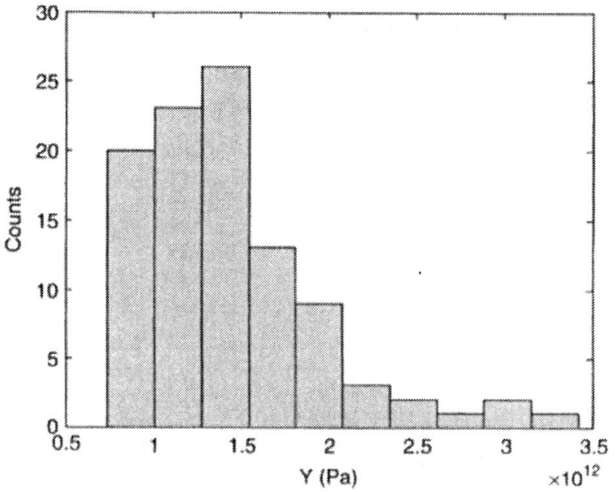

(a) Histogram of SWNT Young's modulus

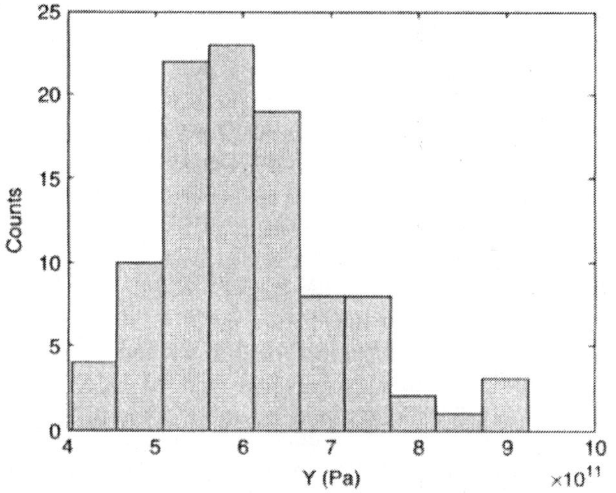

(b) Histogram of BSNT Young's modulus

FIGURE 18.19 Histograms of SWNT and BSNT Young's modulus.

objects were the carbon atoms of the pentagon ring poles at the tips in both models. The simulation code was designed to calculate the *rms* displacements of the tip carbon atoms using Equation 18.31 through Equation 18.35 when the instantaneous kinetic energy per atom of both models reached 0.039 eV (~3/2 kT) at given room temperature and at 300 K. Then Young's modulus was calculated using Equation 18.30. In both cases, the use of 100 simulations was considered sufficient to reduce statistical errors due to the random initial velocities for both models. The simulation results are tabulated in Table 18.1.

TABLE 18.1
Statistical Data Measured from Molecular Dynamics Simulations of SWNT/BSNT Model

Parameter	SWNT	BSNT
Number of simulations	100	100
Mean Young's modulus (TPa)	1.4243	0.6041
Standard deviation (TPa)	0.342	0.1002
Maximum Y value (TPa)	3.4202	0.9241
Minimum Y value (TPa)	0.7402	0.4046

The mean value of Young's modulus for the SWNT model was 1.42 TPa with the standard deviation of 0.34 TPa. Based on this data, the expected mean Young's modulus for the real population of SWNT can be predicted as 1.42 ± 0.23 TPa with 50% confidence intervals. These values are in good agreement with the theoretical and experimental Young's modulus values reported in Reference 10.

In particular, considering about 48% of Young's moduli for the SWNT model, which were distributed within the range 1.0 ~ 1.5 TPa, and comparing with the previously mentioned mean Young's modulus of SWNT, ⟨1.25 TPa⟩ as evaluated by the same thermal vibration method presented by Krishnan et al.,[10] the consistency of both results is comparable. In their method,[10] the tube's length and tip vibration amplitudes were estimated directly from the digital micrographs (TEM images). From this study, the molecular dynamics simulation was proved to be reasonable to evaluate Young's modulus of the SWNT model, and it can be also used to simulate the BSNT model.

The molecular dynamics simulation results of the SWNT and BSNT models show two important features. One is that though present SWNT results and prior studies are consistent with each other, the Young's modulus value (mean 1.42 TPa) of the present SWNT model calculated from the molecular dynamics simulation shows a somewhat higher value than the Young's modulus value (mean 1.25 TPa) used with the same thermal vibration method reported by Krishnan et al.[10] The second is that the calculated Young's modulus value of the BSNT model shows an obvious difference with respect to the SWNT model. The mean Young's modulus of the BSNT model was calculated as 0.604 TPa with a standard deviation 0.1 TPa. Consequently the expected mean Young's modulus for a BSNT population would be predicted to be 0.604 ± 0.07 TPa with a 50% confidence level. The results show a significantly lower Young's modulus of BSNT than that of SWNT. Unfortunately there are thus far no available data for the heterogeneous CNT's Young's modulus to be compared with the value of the BSNT model. Nevertheless it might be expected that this study will play a starting role to speed up the progress of heterogeneous-type nantotube synthesis in further studies.

Why is the Young's modulus of the SWNT model higher than that of the BSNT model by roughly twice, as shown in Fig. 18.19? Earlier studies reported that the Young's modulus of CNTs depends on the presence of structural imperfections such as the nesting of tubular cylinders, which can create a joint or knuckle, thereby weakening the tube, and structural defects of the pentagon (5) and heptagon (7) rings.

In the BSNT model shown in Fig. 18.17(b), the defects along the circumference can prove the arguments presented by Srivastava et al.,[11] who stated that heterogeneous CNTs induced 5–7 defect rings, which absorbed partial energy from the total energy of the system because it required defect formation energy.

Thus the defects at the junction regions of the BSNT model may absorb energy. Then the energy needed for the atomic vibration in the BSNT tips would increase, and Young's modulus of the BSNT model would be lower than that of the SWNT model based on the relationship of Equation 18.30. This analytical interpretation is also consistent with the results of Fig. 18.19(b), as mentioned previously.

The deformation behaviors of SWNTs and BSNTs under an external tension load were also simulated with the molecular dynamics code to compare Young's moduli of both models. When a system is subjected to an external load, the system is forced out of its initial equilibrium into a new equilibrium with the given load. Series of the molecular dynamics simulation plots at the nanoscale are presented to investigate the mechanical behaviors and reactions of both models.

Figure 18.20 and Fig. 18.21 are molecular dynamics simulation plots of the tensile loading test that were performed on both models at various forces of 30 ~ 60 (nN/atom), respectively. Figure 18.20(a) and Fig. 18.21(a) show the initial set-up of the SWNT and BSNT after the equilibrium relaxation. The bottom atoms of both models are constrained as if they are anchored at a substrate material. Then, z-directional tensile forces were applied to the carbon atoms in the cap regions of both SWNT and BSNT such as a conventional tensile specimen.

Figure 18.20(b) and Fig. 18.21(b) show the configurations of both models when the applied tensile force is 30 (nN/atom), which means that both models are within the small strain region. Although separation gaps have appeared in both models, their average separation distances are less than the maximum potential influence distance of 2 Å between carbon atoms. Beyond the scope of this tensile force, both carbon nanotubes show characteristic buckling and kicking features at the side tubular wall. These features are also shown in Fig. 18.20(c) and Fig. 18.21(c) where the applied force is 50 (nN/atom).

Figure 18.20(d) and Fig. 18.21(d) demonstrate that both SWNT and BSNT models separate into two halves (top and bottom) beyond the maximum tensile force (~60 nN/atom), leading to a fracture, which means that the average separation gaps are larger than 2 Å. This separation and fracture are not rapid, but occur gradually due to local interactions or rearrangement between the carbon atoms. However, slip taking place at about 45° to the loading direction, which is a general phenomenon in the case of ductile metals, was not observed explicitly. Rather, under the strong tensile force, they exhibited a brittle feature similar to a diamond, which is known as the hardest material consisting of carbon atoms with an sp^3 bonding structure.

Figures 18.21(e) and (f) show the changes in the cross-sectional area and distribution of atoms seen from the top view of the BSNT model from the case of initial configuration (e), to the case of large applied tensile force of 75 nN/atom (f). A precise value for the changed cross-sectional area is difficult to obtain, because the number of distortions of circumferential carbon atoms in the side wall of the BSNT model is not obvious. This unclear cross-sectional area that appears during the molecular dynamics simulation will prevent computing the applied stress needed for calculating Young's modulus from molecular dynamics results. Furthermore

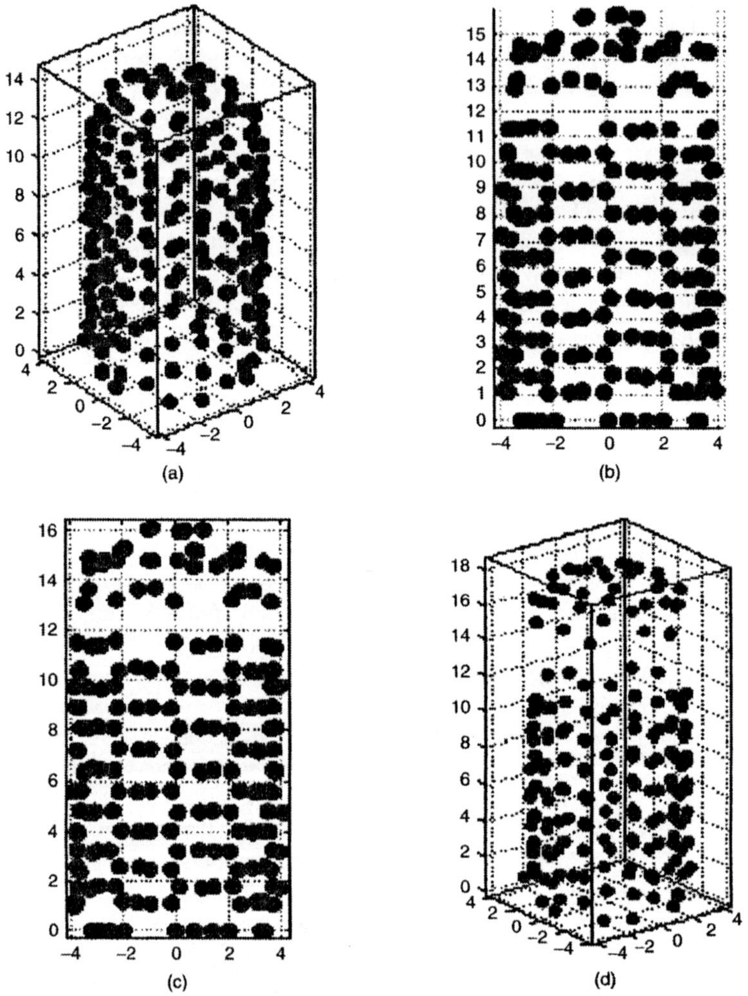

FIGURE 18.20 Configurations of the SWNT model under various magnitudes of tensile loads: (a) no loading, (b) 30 nN/atom, (c) 50 nN/atom, (d) separation of atoms.

carbon atoms in the dome-shaped cap region of the BSNT model show the rotational motions on the axis as the applied force increases. Both observations demonstrate that carbon atoms under the potential and an external force look for the appropriate position to minimize the system energy. Thus Fig. 18.21(f) directly proves that the binding between atoms can be formed or broken due to the induced defects on the BSNT model during the simulation.

Next a relative Young's modulus ratio concept for the SWNT and BSNT will be used with the force-strain diagram extracted from the molecular dynamics simulation, which will then be compared with the Young's modulus values calculated from the previous simulation.

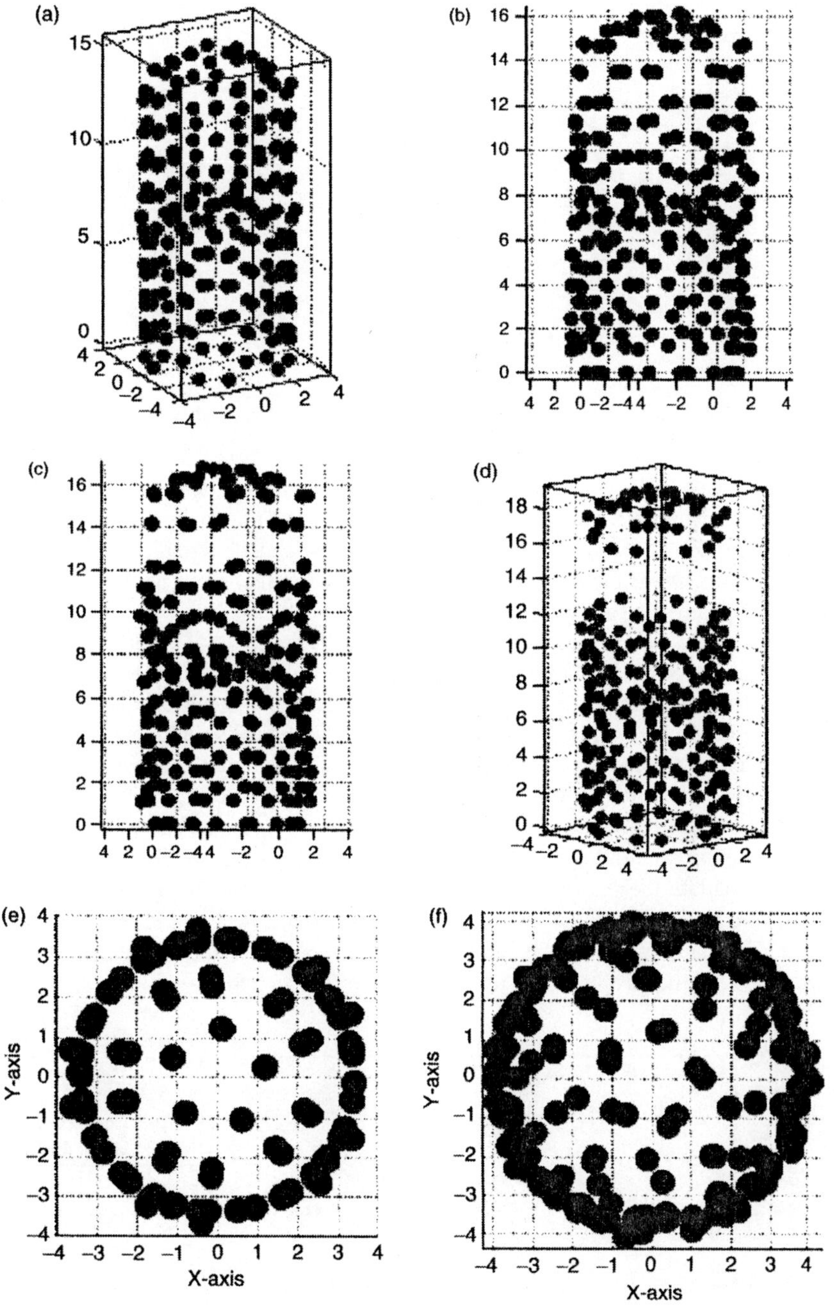

FIGURE 18.21 Configurations of BSNT model under various magnitudes of tensile loads: (a) no loading, (b) 30 nN/atom, (c) 50 nN/atom, (d) separation of atoms, (e) top view of initial atomic configuration, (f) top view of atoms under a large tensile load.

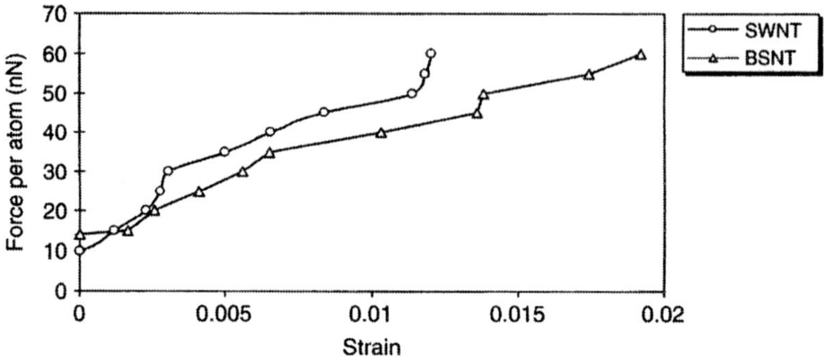

FIGURE 18.22 Tensile force-strain per atom for the SWNT and BSNT models.

By conventional continuum mechanics, the ultimate tensile strength and Young's modulus for a bulk material can be determined from the force-displacement data under the external tensile loading test. The ultimate tensile strength is measured as the maximum stress prior to fracture. The modulus is measured as the slope within the small strain limit range.[12]

The tensile force-strain diagram for the SWNT and BSNT obtained from the molecular dynamics simulation is plotted in Fig. 18.22. The applied tensile forces per atom of the SWNT and BSNT were recorded on the y-axis, and the average strains of carbon atoms in the location right below the forced atoms were recorded on the x-axis. Overall the observed force-strain diagrams show a significant nonlinear relationship. In particular, it is worthwhile observing that the strain of BSNT is more linearly increased than that of SWNT as the applied force increases. The plot shows that the Young's modulus of BSNT is less than the modulus of SWNT.

To calculate the relative Young's modulus between two materials, SWNT and BSNT, the linear regression analysis was applied as shown Fig. 18.23. Based on the linear regression analysis, the ratio of force to strain of a specific material can be represented as the slope of the regression line. In Fig. 18.23, while the data set of

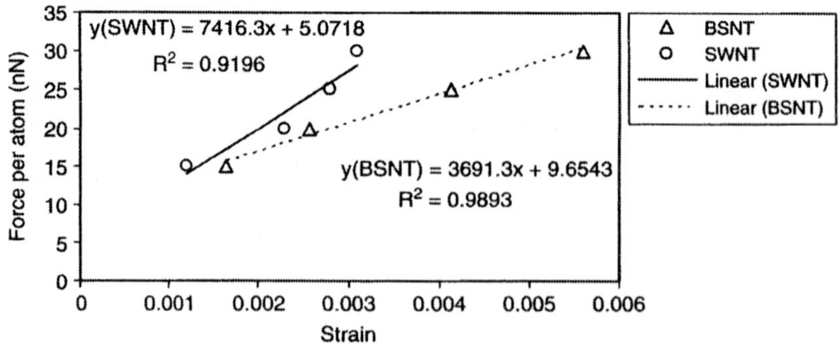

FIGURE 18.23 Linear regression fitting for SWNT and BSNT models within elastic limits.

SWNT shows a more nonlinear characteristic than those of BSNT within the same applied force range, the linear regression analysis can be used for both data sets due to a good fitting statistical coefficient R^2, which quantifies a goodness of fit and is a fraction between 0.0 and 1.0, with no unit. Higher R^2 values indicate that the fitting line or curve comes closer to the data. In this case, the computed R^2 using the statistical software package was 0.91 for SWNT and 0.99 for BSNT. These numerical values indicate that the linear assumption used to calculate the slope of a regression line for SWNT and BSNT is appropriate.

The resultant slope of the regression line is 7416.3 for SWNT and 3691.3 for BSNT, as seen in Fig. 18.23. Consequently the relative Young's modulus ratio of the BSNT with respect to the SWNT was calculated as 0.498. This means that the average Young's modulus of the BSNT model is 49.8% that of SWNT.

From the results of the previous equilibrium simulations, the evaluated mean Young's moduli of SWNT and BSNT were 1.424 TPa and 0.604 TPa, respectively. Based on these two values, the ratio of Young's modulus is 0.424 (42.4%). This is consistent with the aforementioned value, 0.498 (49.8%), within an error of ±15%. Consequently Young's modulus of the BSNT model was observed to be lower than that of the SWNT model, indicating that heterogeneous CNTs have lower Young's moduli than pure CNTs. However, the heterogeneous CNT still has a larger modulus than conventional metals. The heterogeneous CNTs may be great potential candidates promising high-strength and ultra-lightweight composite materials and nanoscale devices in the near future.

The last example considered natural vibrations of SWNT and BSNT. The natural frequencies and mode shapes were computed from the equation of motion:

$$[M]\{\ddot{u}\}+[K]\{u\}=0 \tag{18.36}$$

where the mass matrix of the carbon nanotube, $[M]$, was determined from the distributed masses of carbon atoms and the stiffness matrix, $[K]$, was computed from Equation 18.20. The stiffness matrix was linearized assuming a small displacement. The first several mode shapes of both BSNT and SWNT are plotted in Fig. 18.24 as both ends of the tubes were constrained. The figure shows both axial and bending vibrations. The first five natural frequencies are listed in Table 18.2. Both BSNT and SWNT have almost the same natural frequencies because the inner cap of BSNT has little effect on the axial and bending mode shapes. The natural frequencies of BSNT were generally higher than those of SWNT.

TABLE 18.2
Comparison of First Five Lowest Natural Frequencies of SWCNT and BSWCNT

	1st	2nd	3rd	4th	5th
SWNT (THz)	4.34	5.66	8.21	8.66	9.38
BSNT (THz)	3.99	6.57	8.71	9.72	9.75

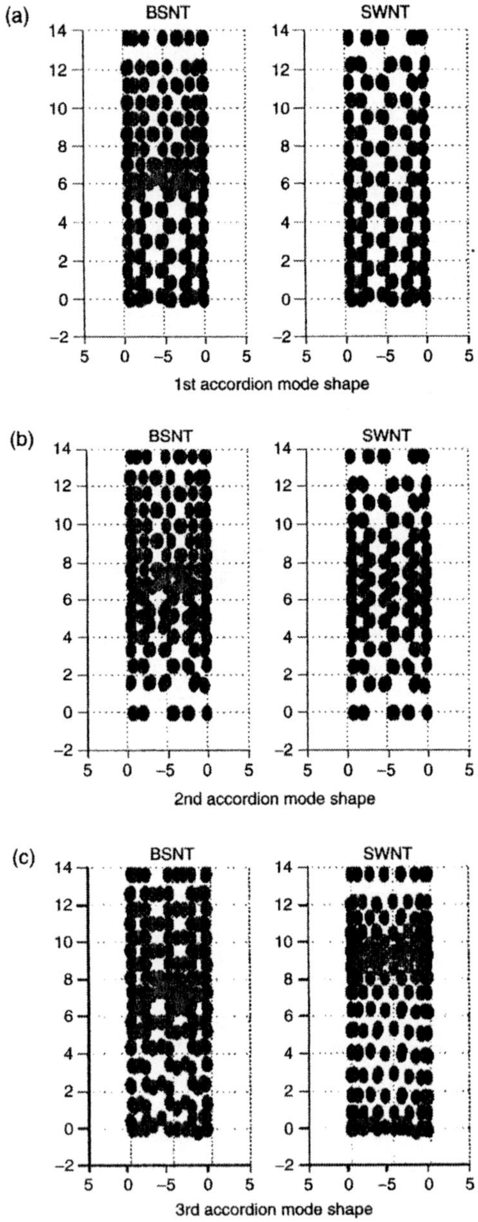

FIGURE 18.24 Comparison of vibrational mode shapes of BSNT and SWNT models.

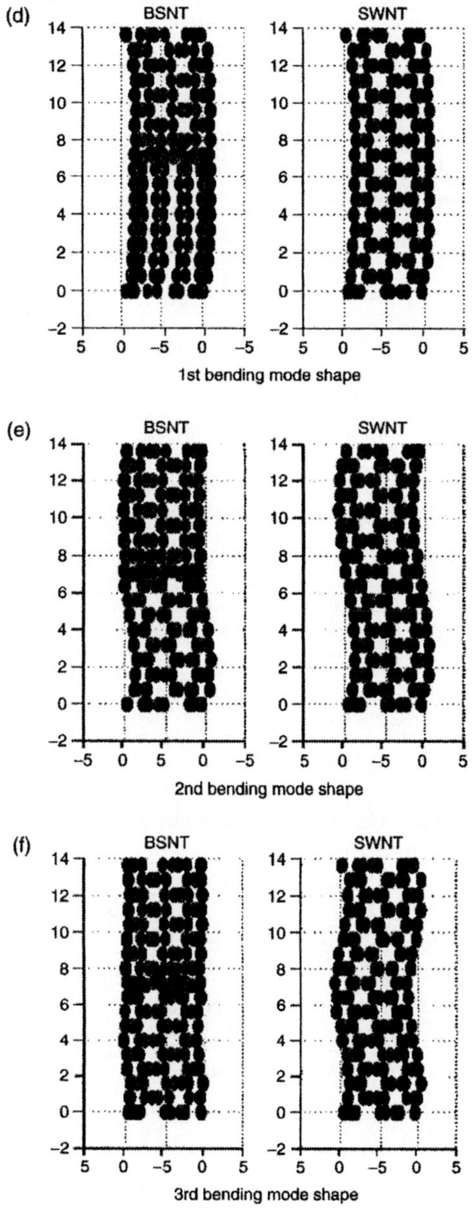

FIGURE 18.24 (*Continued*)

18.6 SUMMARY

In this chapter the concepts pertaining to the static atomic model, coupling between the discrete atomic model and the continuous finite element model, and fatigue analysis at the atomic level, and carbon nanotubes were explored. The static atomic model addressed methodology to predict the static properties of atomic structures such as Young's modulus. Then coupling between the atomic model and a continuous medium was presented for multiscale analysis ranging from the nanoscale to the macroscale. The fatigue analysis at the atomic level provides understanding of the fatigue damage process at the nanolevel, leading to eventual prediction of the fatigue life cycle of the material and to the design of tougher materials.

PROBLEMS

1. What is the difference between the *ab initio* simulation method and the classical molecular dynamics method as applied to nanomechanics?
2. Discuss briefly the embedded atom method.
3. Explain clearly how the atomic and finite element models are coupled together.
4. How is fatigue analysis at the atomic level performed?
5. What are the advantages of performing fatigue analysis at the atomic level as opposed to conventional fatigue analysis?
6. Compare different atomic energy potentials in terms of their characteristics and applications.
7. Construct the matrix equation of motion for a carbon nanotube with chiral vector (n, n) or $(2n, 0)$ where n is an integer.
8. Apply the central difference technique for time integration to the matrix equation of motion derived in Problem 7.
9. Apply Newmark's technique for time integration to the matrix equation of motion derived in Problem 7.
10. Determine the natural frequencies and mode shapes of the nanotubes discussed in Problem 7 with appropriate assumptions of the length of the tube, end closures, and constrained conditions.
11. Model paired dislocation pileups at a free surface and apply a cyclic load to reveal the movements of the dislocations.

ACKNOWLEDGMENTS

The author greatly appreciates contributions of his former graduate students, R.A. Duff, S.H. Jung, W.S. Lunt, J.J. Oh., and C. Manthena, in preparing the manuscript.

REFERENCES

1. Kwon, Y.W., Discrete atomic and sheared continuum modeling for static analysis, *Engin. Computations, 20,* 8, 964–978, 2003.

2. Lennard-Jones, J.E., The determination of molecular fields. I. From the variation of the viscosity of a gas with temperature, *Proc. Roy. Soc. (London), 106A,* 441–462, 1924.
3. Daw, M.S. and Baskes, M.I., Embedded atom method: derivation and application to impurities, surfaces, and other defects in metals. *Phys. Rev. B, 29,* 12, 6443–6453, 1984.
4. Daw, M.S., Model of metallic cohesion: the embedded-atom method. *Phys. Rev. B, 39,* 11, 7441–7452, 1989.
5. Hoagland, R.G., Daw, M.S., Foiles, S.M., and Baskes, M.I., An atomic model of crack tip deformation in aluminium using an embedded atom potential. *J. Mater. Res., 5,* 313–324, 1990.
6. Brenner, D.W., Empirical potential for hydrocarbons for use in simulating the chemical vapor deposition of diamond films, *Phys. Rev. B, 42,* 15, 9458–9471, 1990.
7. Felbeck, D.K. and Atkins, A.G., *Strength and Fracture of Engineering Solids,* Prentice-Hall, Englewood Cliffs, NJ, 1984.
8. Kahn, H., Ballarini, R., Bellante, J., and Heuer, A.H., Fatigue failure in polysilicon not due to simple stress corrosion cracking, *Science, 298,* 8, 1215–1218, 2002.
9. Chapman, S. and Cowling, T.G., *The Mathematical Theory of Non-uniform Gases,* Cambridge University Press, London, 1961.
10. Krishnan, A., Dujardin, E., Ebbesen, T.W., Yianilos, P.N., and Treacy, M.M.J., Young's modulus of single-walled nanotubes, *Phys. Rev. B, 58,* 14013–14019, 1998.
11. Srivastava, D., Wei, C., and Cho, K., Nanomechanics of carbon nanotubes and composites, *Appl. Mech. Rev., 56,* 2, 215–230, 2003.
12. Lu, J.P., Elastic properties of carbon nanotubes and nanoropes, *Phys. Rev. Lett., 79,* 1297–1300, 1997.

19 Continuum and Atomistic Modeling of Thin Films Subjected to Nanoindentation

J. David Schall, Donald W. Brenner, Ajit D. Kelkar, and Rahul Gupta

CONTENTS

19.1 INTRODUCTION

Nanocomposite thin films (Fig. 19.1) usually involve ceramic or polymeric matrices. The large surface area-to-volume ratio has found large-scale applications in diverse fields. These materials exhibit improved hardness and remarkable magnetoresistance. They also exhibit improved properties compared to the parent materials.[1,2] There is

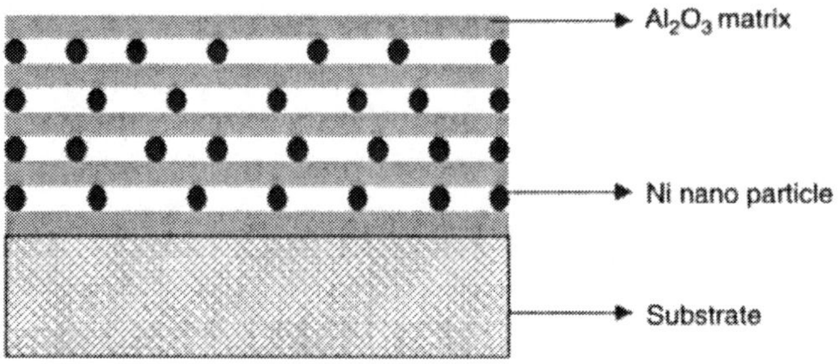

FIGURE 19.1 Schematic of a thin-film nanocomposite.

also the possibility of new properties not found in the parent constituent materials. Some of the characteristics of nanocomposite thin films are as follows:

Superior mechanical properties such as strength, hardness, modulus, and fracture toughness
Important applications of the large surface-to-volume ratio in heterogeneous catalysis, heat exchangers, and magnetic devices
Thermal stability and heat distortion temperature
Flame retardancy and reduced smoke emissions
Improved chemical and wear resistance
High levels of magnetoresistance and electrical conductivity

To understand the deformation and damage mechanics of these thin films, several techniques are being developed and applied. The properties of thin films can be different from those of bulk materials even when the chemical composition of the thin film and the bulk material are identical. There are several methods to characterize the mechanical behavior of thin films. One of the popular methods is predicting mechanical properties using depth sensing indentation. Although it is possible to model depth sensing indentation of thin films using the conventional finite element method based on a continuum approach, for very thin films, substrate effects can be significant and the conventional continuum principles may not be applicable. This chapter will address both continuum and atomistic modeling of thin films subjected to nanoindentation.

The first major contribution in indentation experiments was made by Sneddon.[3] Sneddon studied the load, displacement, and contact area relationships for many simple punch geometries and presented an equation, $P = \alpha h^m$, where P is the indenter load, h is the elastic displacement of the indenter, and α and m are constants.[4] Values of the exponent m for some common punch geometries are $m = 1$ for flat cylinders, $m = 2$ for cones, $m = 1.5$ for spheres in the limit of small displacements, and $m = 1.5$ for paraboloids of revolution.[4]

When plasticity is considered while modeling indentation, the constituent equations become nonlinear and additional material parameters such as yield strength and work hardening coefficient must be included. Analytical solutions for these nonlinear equations are difficult to obtain. Therefore the problems involving plasticity are studied through experimentation and finite element simulation.[4]

Early indentation studies were done on metals using hardened spherical indenters by Tabor.[5] Later Stillwell and Tabor examined the behavior of conical indenters. They observed that at least in metals, the impression formed by a spherical indenter is still spherical with a slightly larger radius than the indenter, and the impression formed by a conical indenter is still conical with a large included tip angle. Tabor showed that the shape of the entire unloading curve and the total amount of recovered displacement could be accurately related to the elastic modulus and the size of the impression for both spherical and conical indenters.[5] Another important observation made during these studies was that the diameter of the contact impression in the surface formed by conical indenters does not recover during unloading and only the depth recovers.

The indenter must be loaded and unloaded a few times before the load displacement behavior becomes perfectly reversible. The plasticity can be dealt with by taking into account the shape of the perturbed surface in the analysis of the elastic unloading curve. Therefore the shape of the unloading curve and the recovered displacement characterizes the elastic modulus of the material.

Typical load-displacement data obtained from a nanoindenter XP is shown in Fig. 19.2. This load-displacement data is further analyzed for hardness and modulus measurements. Stiffness, S, is calculated from the unloading portion of the load displacement data as follows:

$$S = \frac{dP}{dh} = \frac{2}{\sqrt{\pi}} E_r \sqrt{A} \tag{19.1}$$

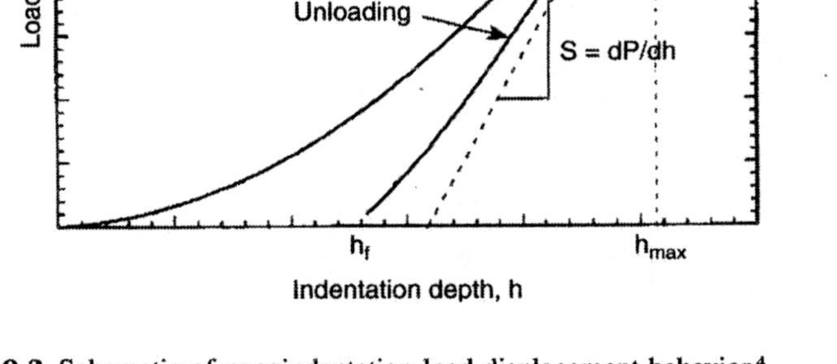

FIGURE 19.2 Schematic of nanoindentation load-displacement behavior.[4]

$S = dP/dh$ is the experimentally measured stiffness of the upper portion of the unloading data, A is the projected area of the elastic contact, and E_r is the reduced modulus (accounting for the effect of nonrigid indenters) defined by the equation,

$$\frac{1}{E_r} = \frac{(1-\upsilon^2)}{E} + \frac{(1-\upsilon_i^2)}{E_i} \tag{19.2}$$

where E and υ are Young's modulus and Poisson's ratio for the specimen and E_i and υ_i are the same parameters for the indenter. By measuring initial unloading stiffness, the modulus is calculated using Equation 19.1 and Equation 19.2. Equation 19.1 originated from elastic contact theory and was used initially only for conical indenters. Later it was shown that the equation holds well for spherical and cylindrical indenters also. Oliver and Pharr[4] proved that Equation 19.1 applies to any indenter that can be described as a body of revolution of a smooth function. Finite element simulations by King[6] proved that Equation 19.1 works well for flat-ended punches and triangular cross-sections that cannot be described as bodies of revolution, with deviations of 1.2% and 3.4%, respectively. King found out that for all the three geometries, the unloading stiffness can be rewritten as

$$S = \frac{dP}{dh} = \beta \frac{2}{\sqrt{\pi}} E_r \sqrt{A} \tag{19.3}$$

where β is a constant and the values of β for different indenter geometries are given in Table 19.1.

Keeping in mind the difficulties and time required for imaging very small indentations, the need for some means other than direct observation of the hardness impressions was felt. Oliver, Hutchings, and Pethica[7] suggested a method based on measured indentation load-displacement data and an area function (or shape function) i.e., the cross-sectional area of the indenter as a function of the distance from its tip. This notion assumes that at the peak load, the material conforms to the shape of the indenter to some depth.

Doerner and Nix[8] devised a comprehensive method for determining hardness and modulus from indentation load-displacement data based on the observation that during the initial stages of unloading, the elastic behavior of the indentation contact

TABLE 19.1
Constant β Value for Different Indenter Geometries

Indenter Geometry	Value of β
Circular	1.000
Triangular	1.034
Square	1.012
Berkovich	1.034

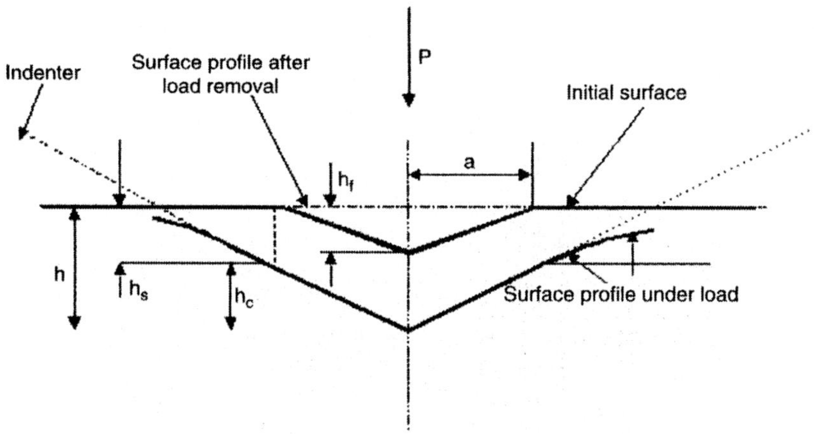

FIGURE 19.3 Sectional view of indentation with various quantities used in the analysis.

is similar to that of a flat cylindrical punch. Thus the area of contact remains constant during unloading; i.e., the initial portions of the unloading curves are linear. The unloading stiffness dP/dh is related to the modulus and contact area through Equation 19.1. Doerner and Nix proposed a simple empirical method based on extrapolation of the initial linear portion of the unloading curve to zero load (Fig. 19.2) and using the extrapolated depth with the indenter shape function to determine the contact area.[8]

Doerner and Nix assumed that the unloading stiffness could be computed from a linear fit of the upper one-third of the unloading curve. They suggested that the plastic depth h_c, also termed contact depth (Fig. 19.3), is the distance the indenter is in contact with the plastically deformed material along the indenter axis during most of the unloading cycle.[9]

The contact area remains constant during the unloading. Once the contact area is determined for a known indenter shape, using Equation 19.6, the modulus is calculated from Equation 19.1 and the hardness is calculated according to its definition:

$$H = \frac{P_{max}}{A} \tag{19.4}$$

where P_{max} is the peak indentation load and A is the projected area of the impression. The Doerner and Nix method is based on their experimental observations of metals, i.e., that the unloading portion of load-displacement curve is linear.

Oliver and Pharr conducted depth-sensing indentation tests on a large number of materials and suggested that the unloading portion of the load-displacement curve is not always linear, even in the initial portions of unloading.[4,10] This is particularly evident while working with hard materials such as ceramics and glass that exhibit a continuous decrease in indentation depth during unloading due to elastic recovery within the indentation.[9] Oliver and Pharr reasoned that unloading data are well described by a simple power law relation given by

$$P = \lambda(h - h_f)^m \quad (19.5)$$

where λ, m and h_f are constants and are determined by a least square fitting procedure.

Oliver and Pharr modified the Doerner and Nix analysis by taking the elastic recovery into account. The exponent, m, ranges from 1.2 to 1.6.[4] The depth, h_f, is the final depth at which the indenter is in contact with the material (Fig. 19.3). The Oliver and Pharr method accounts for the nonlinear portion of the unloading curve. This method provides a procedure for determining the contact depth that should be used in the area function to determine the contact area at peak load and thus the hardness. The initial unloading slope is found by differentiating Equation 19.4 at the peak load and displacement. The contact area A is a function of the contact depth h_c given by

$$A(h_c) = 24.5h_c^2 + C_1h_c^1 + C_2h_c^{\frac{1}{2}} + C_3h_c^{\frac{1}{4}} + C_4h_c^{\frac{1}{8}} + \quad + C_8h_c^{\frac{1}{128}} \quad (19.6)$$

The first term in Equation 19.6 describes a perfect Berkovich indenter and the other terms describe deviations from Berkovich geometry due to blunting at the tip. Since the trailing coefficients, $h_c^{1/128}$, begin to approach unity, convergence is achieved by appropriate decay of the C_n for large n.

The contact depth h_c at maximum load, P_{max}, is given by

$$h_c = h_{max} - h_s = h_{max} - \varepsilon\frac{P_{max}}{S} \quad (19.7)$$

where $\varepsilon = 1$ for flat punch and $\varepsilon = 0.72$ for conical indenter (Fig. 19.2). For the paraboloid of revolution, $\varepsilon = 0.75$.[3,10] Thus, similar to the Doerner and Nix method, the modulus is obtained from Equation 19.2 and Equation 19.3, and hardness from Equation 19.4.

The Doerner and Nix method, which considers the flat punch assumption, though simple, underestimates the distance the indenter deflects the plastically deformed material along the indenter axis.[9] Unless the tip is a flat punch, the contact area decreases substantially upon unloading due to elastic recovery within the indentation, which leads to $\varepsilon < 1$ in Equation 19.7. The Oliver and Pharr method accounts for elastic recovery within the indentation and is therefore more accurate than the Doerner and Nix method. The Doerner and Nix method is a special case of the Oliver and Pharr method when $m = 1$. The Oliver and Pharr method initially assumed the unloading to be elastic. It may be affected by time-dependent deformations such as plasticity and thermal drift. Plasticity has not been observed in hard materials and performing the tests in a thermally controlled environment can control thermal drift.

In the present investigation, the elastic properties of multiple-layer aluminum nitride (AlN) and titanium nitride (TiN) thin films of various periodicities deposited on silicon (111) substrates are considered. These films were grown using the PLD method as abrupt immiscible multiple layers of AlN and TiN, as well as single layers of the AlN and TiN components.

Conventional finite element analysis (FEA), which is based upon a continuum description of solids, has been used by several investigators for analyzing indentation. FEA provides insight into the contact area, plasticity, and effects of a differentiated material substrate on the load-displacement behavior of a thin film. FEA requires that the indented body act as a continuum. In the present investigation, an initial three-dimensional FEA model based on the experiment was 20 mm × 20 mm × 2 mm, with initial element edge size set at 0.1 mm. The model consisted of 44,000 eight-node three-dimensional (3-D) brick elements and had 48,492 nodes. The relevant details of the finite element model are discussed in the following section.

19.1.1 Finite Element Model

A number of nanoindentation experiments were performed on thin film (AlN/TiN) with silicon substrate. Since experimental results were available for this thin-film system, a finite element model was developed for the same system. The mechanical properties of the thin film were not known, and it was decided to use the properties as obtained from the nanoindentation experiments. In the present study, finite element modeling following substrate and thin film properties was used.

1. Substrate (silicon (111)): 20 mm × 20 mm × 2 mm
   E = 185 GPA, ν = .266
2. Thin film thickness = 250 nm (multiple layers of AlN/TiN)
   E = 215* GPA, ν = .25

*Thin film modulus was assumed to be 215 GPa.

Thin film thickness usually is not uniform and can vary with a slope of 10 to 60 nm/mm as measured along the width of the sample. A load of 100 mN was applied through the nanoindentor tip. Multiple cases were studied with mesh edge dimension H descending continuously from the 0.1 mm (1E–4M) to the 1.0 micron (1E-6M) level.

Measured thin film thicknesses ranged from about 250 to 350 nm. Thin film hardness and elastic modulus were measured by a frequency-specific depth-sensing nanoindentation method, which provided continuous stiffness data throughout the indentation process from the film surface up to a depth of 1000 nm. In addition, the thin film material was assumed to be initially stress free, which may not have been the actual case. Both film and substrate were assumed to be homogeneous and isotropic with perfect elastic-plastic behavior.

In the present finite element simulations, several mesh sizes were utilized. It was observed that the mesh size has a significant effect on the prediction of the through-the-film thickness displacements. The mesh size was varied from 1E–4M to 1E–6M. Figure 19.4 shows the three-dimensional finite element model of the thin film with substrate. Figure 19.5 shows the simulated through-the-film thickness indentation, whereas Fig. 19.6 shows the experimental indentation for the load of

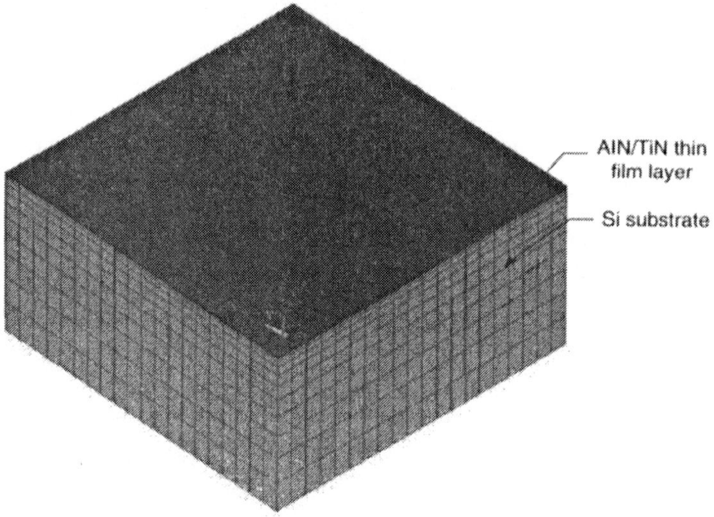

FIGURE 19.4 Three-dimensional finite element model of thin film showing 2-mm substrate with 250-nM thin-film coating.

100 mN. The present FEA included both material and geometric nonlinearity. The finite element results agreed well with the experimental results. Although the film thickness in the present simulations and actual experiments was only 250 nm, the loading on the thin film was continued until the through-the-film thickness displacements far exceeded 250 nm. (Simulations were performed until the through-the-film thickness displacements were about 750 nm.)

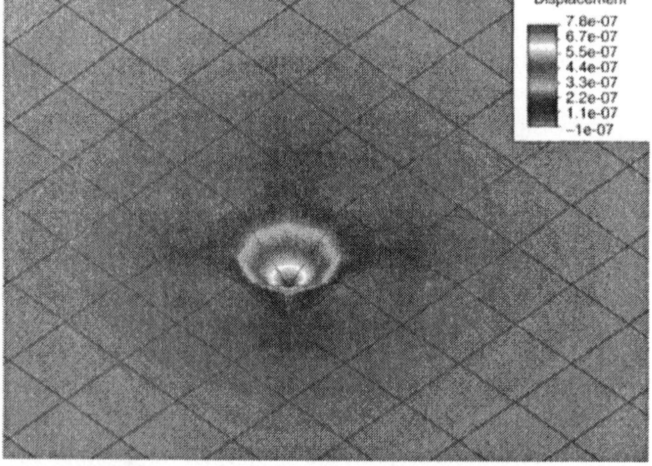

FIGURE 19.5 Finite element simulation of nanoindentation process.

FIGURE 19.6 Atomic force microscope snapshot of nanoindentation.

Figure 19.7 and Fig. 19.8 clearly show that the experimental hardness and modulus values within the thin film thickness region are about 16 GPa and 225 GPa. The modulus value is the same as the one used in the finite element simulations. Through-the-film thickness indentations beyond the film thickness result into simulating nanoindentation experiments on the substrate, and hence the modulus has to be far lower than the thin film. Finite element simulations exactly predicted similar behavior.

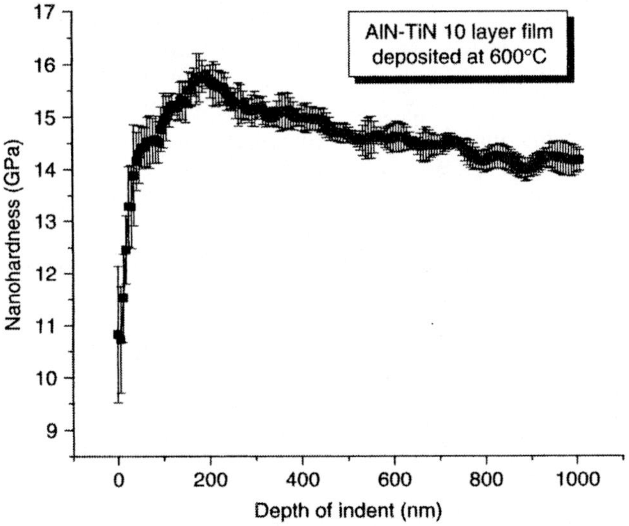

FIGURE 19.7 Nanohardness (GPa) versus nanoindentation depth (nM).

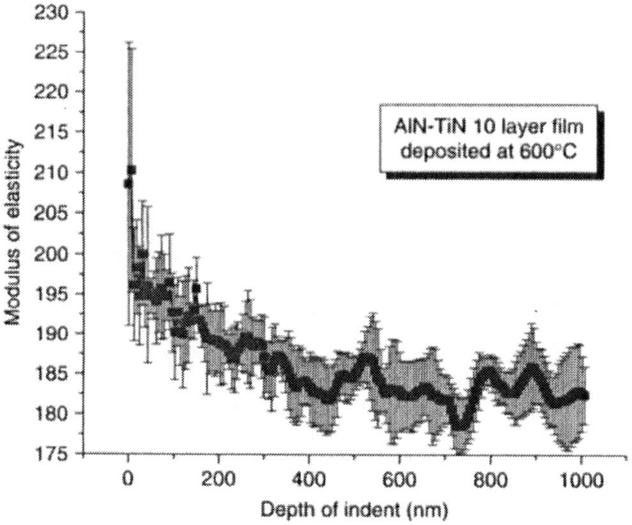

FIGURE 19.8 Modulus of elasticity (GPa) versus nanoindentation depth (nM).

19.1.2 Need for Atomistic Model

The FEA of the nanoidentation process is highly nonlinear because (1) the large nodal displacement forces the stiffness matrix to reformulate throughout the deformation analysis, (2) the nonlinear behavior of the material transitions from the elastic regime to the plastic, and (3) the boundary conditions at the sliding indenter contact with indented material are continually changing throughout the deformation. Furthermore as the thin film thickness gets smaller than 100 nm, substrate effects start coming into the picture. In addition, the finite element analysis also ignores the residual stresses in the thin films, which can significantly affect the prediction of load displacement behavior in ultrathin films. An alternative approach is to perform a full three-dimensional nonlinear analysis using LSDYNA-3D that would render superior characterization for nanoindentation process as well as material behavior and properties since LSDYNA-3D is an explicit solver with automatic surface-to-surface contact capability. It should also characterize material behavior and properties better since LSDYNA-3D has the ability to dynamically adjust the contact elements and geometry to account for changing surface contact under load.

The present investigation also revealed that although a continuum mechanics approach works reasonably well when the film thickness is 250 nm or thicker, smaller film thicknesses give rise to convergence problems for nanoindentation modeling, which represents a large deformation problem. An alternative approach is to model indentation using atomistic simulation. With today's supercomputers, quantum mechanics-based analysis of a micron cube of matter for a microsecond would require over 10 million years of computer time.[11] Therefore large-scale atomistic simulations typically use some analytic approximation to interatomic forces. One such bonding model typically used for metals, the embedded atom method (EAM), is discussed in the next section in the context of nanoindentation modeling.

19.2 MODELING OF NANOINDENTATION

Many applications of thin films and coatings require knowledge of their mechanical properties, which may be significantly different from those of bulk materials. Based on analysis of indentation data discussed earlier, by using the continuum finite element model or by using Hertzian mechanics, it is possible to obtain both local elastic moduli and hardness of thin films. However, thin films often possess residual stresses, and it is well established experimentally that load-displacement curves measured by nanoindentation can be altered by stresses in the sample. Atomistic simulation is a powerful tool for addressing these challenges.

A number of atomistic studies of nanoindentation have been conducted in the past. Tadmor and coworkers developed a quasicontinuum method capable of representing very large system sizes and used this model to predict the resolved shear stress required for emission of a dislocation dipole.[12] Their work also led to important insights into the onset of plastic deformation, dislocation nucleation, and dislocation image forces due to nanoindentation. However, these calculations were based on an idealized two-dimensional model, which limited its ability to be compared experimentally. A number of workers have applied the embedded atom method to model nanoindentations. In 1990, Landman and coworkers studied the influences of repeated contact.[13] While limited in size, their work was useful in explaining the jump-to-contact phenomena observed in experiments. More recently indentation simulations by workers such as Kelchner,[14] Zimmerman,[15] Rodriguez,[16] and Lilleodden[17,18] have achieved length scales approaching that of earlier experiments. These simulations used the embedded atom method along with an energy minimization scheme to explore the issues of onset of plasticity, dislocation nucleation, and various structural effects such as surface steps and grain boundaries. Zimmerman and Kelchner[15] also developed a parameterization that allows dislocations and other defects in the simulated material to be easily visualized and characterized.

The remainder of this chapter will be dedicated to a general discussion of molecular dynamics simulation, the embedded atom method along with a brief description of the computational method and parallel algorithm implemented in one commonly used embedded atom method code, and a description of the typical temperature control method used in many molecular dynamics simulations. Details of one commonly used set-up for molecular dynamics simulation of nanoidentation are also presented and include descriptions of the typical system configuration, two types of virtual indenters, and results of tests of these indenters.

19.2.1 MOLECULAR DYNAMICS SIMULATION

Molecular dynamics are used to move atoms using classical equations of motion according to a model for the interatomic forces. Simulations of this type can be used to determine equilibrium (and minimum energy) structures or explore nonequilibrium dynamics.

Atomic-scale dynamics are simulated by numerically integrating the classical equation of motion, i.e., Newton's equation, for each atom:

$$F = ma \tag{19.8}$$

$$-\frac{dU}{dr} = m\frac{d^2r}{dt^2} \tag{19.9}$$

where U is the potential energy of the system, r and m are the atomic positions and masses of each atom, respectively, and t is time.[19] The potential energy U is given by an analytic expression that yields energy as a function of the relative positions of the atoms. We present one such energy function known as the embedded atom method; however, a wide range other of potential functions exist for a wide range of materials.

19.2.2 Embedded Atom Method (EAM)

The selection of the EAM for the energy function in molecular dynamics simulations is a popular choice for close-packed metals.[20–22] It combines the computational simplicity required to model large systems with a physical picture that includes many-body effects and the moderation of bond strengths by local coordination. The research community has investigated many problems of interest using the EAM, including point defects, melting, alloying, grain boundary structure and energy, dislocations, segregation, fracture, nanoindentation, surface structure, and epitaxial growth.[22] In general, most of the EAM calculations have been carried out in close connection with experimental work.

The EAM is based on effective medium theory. In this theory, each atom in a solid is viewed as an impurity embedded in a host material, and the energy of the atom is a function of the electron density. In the EAM, this density is taken as the contribution of electron density from neighboring atoms at the site at which the atom resides. More rigorous theories take this electron density as a sum of the density in some region surrounding an atom that is determined such that the total electron density of the solid is included in the calculation. In EAM, however, the relationship between the site electron density and the potential energy of the atom, together with a pair term that mimics core–core repulsion, is empirically fit to a range of solid properties. This makes the EAM computationally attractive, as it is relatively short-ranged and can be computed by summing over pair-wise interactions, while at the same time it is able to reasonably reproduce a wide range of bulk and defect properties, including properties not used in fitting the analytic form of the potential.

In the EAM formalism, the total potential energy U is taken as a sum of energies associated with each atom i given by

$$U_i = \sum_j \phi(r_{ij}) + F_{embed}\left(\sum_j \rho(r_{ij})\right) \tag{19.10}$$

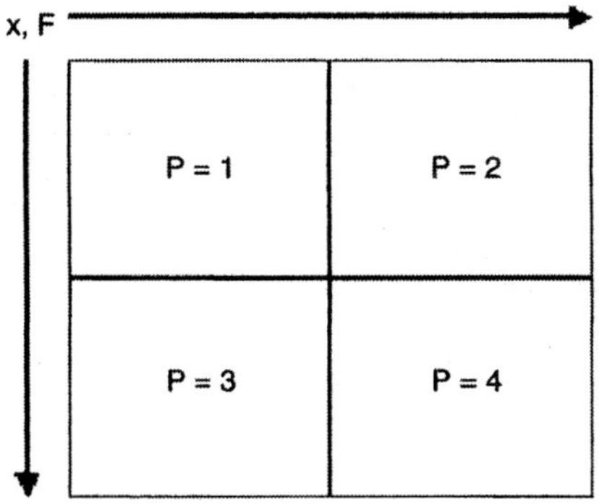

FIGURE 19.9 Processor assignment of the $N \times N$ force matrix. Each processor need only store two pieces of the position and embedding function vectors, each of length $N/\sqrt{P}$ to compute the matrix elements. Processors then communicate this information along rows and columns.

where $\phi(r_{ij})$ is a core–core pair repulsion of atoms i and j at a radial distance of r_{ij}, and F is an embedding function defined as the energy required to embed atom i in the electron density $\rho(r_{ij})$. This density is determined through a linear superposition of the surrounding j atoms. The ρ and F embedding functions are fit to various experimentally determined quantities such as equilibrium density, sublimation energy, elastic constants, and vacancy formation energy. The core–core repulsion term $\phi(r_{ij})$ is based on a Coulombic potential with a parameterization derived from s and p electron densities, as determined by Hartee–Fock wave functions.

19.2.3 PARADYN

PARADYN, a parallel implementation of Daw's EAM code DYNAMO, is used for the MD simulations. PARADYN uses an algorithm developed by Plimpton[23] that utilizes force-domain decomposition to achieve $N/\sqrt{P}$ scaling of computational time where N and P are the numbers of particles and processors. Each processor is assigned N/P atoms. The force for all atoms can be described as an $N \times N$ force matrix, illustrated in Fig. 19.9, where element ij is the ϕ, ρ, or force interaction between atoms i and j. The total force on atom i is the sum of all elements in row i of the matrix. The matrix is sparse due to the short range nature of the interatomic forces and symmetric due to Newton's Third Law. The algorithm for the parallel implementation of the EAM energy and forces is outlined as follows for the system illustrated in Fig.19.9

1. Get $N/\sqrt{P}$ positions of atoms in their rows. (Processor 4 communicates with Processor 3.)
2. Transpose its N/P positions with appropriate processor. (Processor 3 trades with Processor 2.)
3. Acquire $N/\sqrt{P}$ positions of atoms in their columns. (Processor 4 communicates with Processor 2.)
4. Calculate electron density matrix elements in each block.

5. Sum values across row. (Processor 4 sums with Processor 3.)
6. Compute embedding functions (F and F') for each block of N/P atoms using summed density values.
7. Repeat the first three steps with F' instead of coordinates to get all the F' values for the $N/\sqrt{P}$ atoms in rows and columns of the matrix.
8. Calculate the force vectors in each block using F'
9. Sum forces across rows and update positions.

These steps are easy to perform on a parallel machine. Communication takes place between small groups of $\sqrt{P}$ processors and requires information exchange of length $N/\sqrt{P}$ or shorter (versus length N for atom decomposition). Newton's Third Law can be used to halve the required communication. This is achieved by checkerboarding the force matrix. This way each pair interaction is calculated only once in the matrix.

19.2.4 Nordsieck–Gear Predictor–Corrector Algorithm

Given the molecular positions and their successive derivatives (velocity, acceleration, etc. at time t), molecular dynamics attempts to determine the positions, velocities, etc. at time $t + \delta t$ to some degree of accuracy.[29] By assuming that the classical trajectory is continuous and that no forces act on the particles, the positions, velocities, accelerations, etc. at $t + \delta t$ can be estimated by Taylor series expansion.

$$\begin{aligned}
\mathbf{r}^p(t+\delta t) &= \mathbf{r}(t) + \delta t\mathbf{v}(t) + 1/2\delta t^2\mathbf{a}(t) + 1/6\delta t^3\mathbf{b}(t) + \\
\mathbf{v}^p(t+\delta t) &= \mathbf{v}(t) + \delta t\mathbf{a}(t) + 1/2\delta t^2\mathbf{b}(t) + \\
\mathbf{a}^p(t+\delta t) &= \mathbf{a}(t) + \delta t\mathbf{b}(t) + \\
\mathbf{b}^p(t+\delta t) &=
\end{aligned} \tag{19.11}$$

The superscript p indicates predicted values of position ($\mathbf{r}$), velocity ($\mathbf{v}$), acceleration ($\mathbf{a}$), and higher derivatives. Unfortunately Equation 19.11 will not predict the correct trajectories because the forces at each time step have not been taken into account. To get the corrected positions, etc., the new positions, $\mathbf{r}^p$, are used to calculate the force using the first derivative of the potential energy (i.e., Equation 19.10) and hence the correct acceleration ($\mathbf{a}^c(t + \delta t) = \mathbf{F}/m$) at time $t + \delta t$ using the embedded atom formalism (or in principle any other formalism relating force to positional information). These values are compared to the predicted accelerations to estimate the error of the prediction step:

$$\Delta\mathbf{a}(t + \delta t) = \mathbf{a}^c(t + \delta t) - \mathbf{a}^p(t + \delta t) \tag{19.12}$$

TABLE 19.2
Coefficient Values for a Second-Order Nordsieck–Gear[24,25] Predictor–Corrector Using Four Values

$c_0 = 1/6$	$c_1 = 5/6$	$c_2 = 1$	$c_3 = 1/3$

The error and results from the predictor step, Equation 19.11 and Equation 19.12, enter into the corrector step as follows

$$\begin{aligned}
\mathbf{r}^c(t+\delta t) &= \mathbf{r}^p(t+\delta t) + c_0\Delta\mathbf{a}(t+\delta t) \\
\mathbf{v}^c(t+\delta t) &= \mathbf{v}^p(t+\delta t) + c_1\Delta\mathbf{a}(t+\delta t) \\
\mathbf{a}^c(t+\delta t) &= \mathbf{a}^p(t+\delta t) + c_2\Delta\mathbf{a}(t+\delta t) \\
\mathbf{b}^c(t+\delta t) &=
\end{aligned} \tag{19.13}$$

The values of c_0, c_1, $c_2, \ldots$ are chosen to optimize numerical stability and accuracy. These values for the Nordsieck–Gear [24,25] second-order predictor–corrector using the first three derivatives of position are given in Table 19.2. While it is possible to iterate the predictor–corrector algorithm to achieve higher accuracy, the force calculation is the most time-consuming part of simulation. Thus a large number of predictor–corrector steps would be inefficient.

19.2.5 Temperature Control

In the majority of the work presented, temperature control is achieved via the generalized Langevin equation (GLE).[26] Based on the theory of Brownian dynamics,[27] this technique assumes that the energy of single particles is exchanged with the remainder of the system in two ways. The first is friction, which models the dissipation of energy into the bulk. The second is by a fluctuating force, which mimics the thermal vibrations of the surrounding atoms. The two balance to produce the desired average. The equation of motion, which describes these processes, is given by

$$m\frac{d\mathbf{v}}{dt} = -m\beta\mathbf{v} + \mathbf{F} \tag{19.14}$$

where m and $\mathbf{v}$ are the mass and velocity of the particle, β is a generalized friction coefficient, and $\mathbf{F}$ is a random force. The random forces used in the generalized Langevin equation have a Gaussian distribution centered about zero given by

$$\mathbf{F} = \sqrt{\frac{2\beta k_B T}{dt}}\mathbf{R}_{Gauss} \tag{19.15}$$

where β is a constant derived from the Debye frequency, k_B is Boltzmann's constant, T is the desired temperature, dt is the time step size, and $\mathbf{R}_{\text{Gauss}}$ is a random number with a Gaussian distribution given by

$$R_{Gauss} = \sqrt{-2\log(R_1)\cos(2\pi R_2)} \tag{19.16}$$

where R_1 and R_2 are randomly generated numbers between 0 and 1. In PARADYN the opposing frictional term βv and the random force F are added to the total force before the correction step in the integrator.

19.3 MOLECULAR DYNAMICS SIMULATION OF NANOINDENTATION

The following sections give an overview of the techniques commonly used to conduct molecular dynamics simulation of nanoindentation. The first section describes an example simulation set-up and lists parameters for the simulated material presented in this example, in this case, EAM gold. The next two sections discuss a repulsive potential and rigid indenter. The last section presents a series of tests of the indenters in EAM gold.

19.3.1 Simulation Set-Up

As described previously, the molecular dynamics code PARADYN developed by Plimpton[26] was used to model the indentation of gold surfaces. PARADYN utilizes embedded-atom method potentials such as those developed by Foiles, Daw, and Baskes[21] to model atomistic behavior of close-packed metals. For such materials, the embedded atom method has proven to be particularly good at modeling bulk and defect properties such as energies and structures and has been used extensively to model the onset of plasticity during the nanoindentation of gold in quasistatic (0 temperature) simulations and to a smaller extent with molecular dynamics simulations at nonzero temperatures. The materials constants derived from the Au potential used in this study are given in Table 19.3. A crystal lattice was generated by tesselating a unit cell with the desired crystallographic orientation by a desired number

TABLE 19.3
Materials Properties of Gold Derived from the Embedded Atom Potential: Equilibrium Lattice Constant, Bulk Modulus, Cohesive Energy, Vacancy Formation Energy, and Elastic Constants

a_o	4.08 Å	C_{11}	1.83 nN/Å^2
B	1.67 nN/Å^2	C_{12}	1.59 nN/Å^2
E_{coh}	3.93 eV	C_{44}	0.45 nN/Å^2
E_v^f	1.03 eV		

Note: As reported by Foiles, Daw, and Baskes.[21]

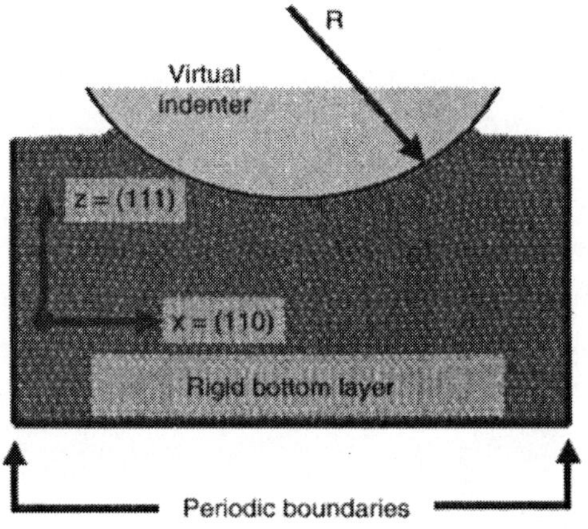

FIGURE 19.10 Schematic diagram of the set-up for molecular dynamics simulation. The x-axis is along a $\langle 110 \rangle$ direction, the z-axis is oriented in the $\langle 111 \rangle$ direction, and the y-axis forms a right-hand coordinate system along a $\langle 211 \rangle$ direction (normal to the plane of the illustration). Periodic boundaries are held in the x and y directions. The bottom layer of atoms is held rigid. R is the radius of the virtual indenter.

of repeat units in the x, y, and z directions. The majority of the simulations conducted in this work were oriented such that there was a $\langle 111 \rangle$ direction normal to the free surface. A schematic diagram of the model used in the simulated indentation tests is shown in Fig. 19.10. In the figure, the z direction is taken to be the normal to a $\langle 111 \rangle$ plane. The x direction is taken to be along a $\langle 110 \rangle$ close-packed direction and y is along a $\langle 112 \rangle$ direction. The bottom layer of atoms is held rigid in the z-direction and the sides are maintained through periodic boundaries in the x and y directions.

19.3.2 Repulsive Potential Indentation

In the next two sections, two types of so-called virtual indenters will be discussed. The first relies on a repulsive potential to generate atomic displacements by the indenter. The second simply displaces the atoms rigidly in a manner that describes the geometry of the indenter. In the repulsive potential formulation for the virtual indenter, the indenter acts as if it were a giant atom with a repulsive force between the indenter and atoms the indenter contacts. The force from the indenter I on substrate atom i is given as

$$\begin{aligned} F_i &= 0 \quad \text{for } r_{il} \geq R \\ F_i &= -k(R - r_{il})^n \quad \text{for } r_{il} < R \end{aligned} \tag{19.17}$$

where R defines the radius of a virtual spherical indenter, r_{il} is the radial distance from the center of the indenter to atom i, and n and k are constants related to the stiffness of the indenter (Fig. 19.11). A strictly repulsive interaction, which is similar

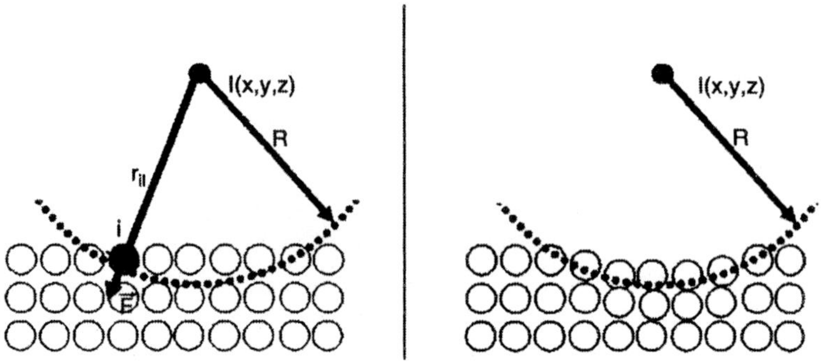

FIGURE 19.11 Schematic diagram of the repulsive potential indenter. A normal force is applied normal to the surface of the indenter (left panel). The actual displacement of atoms in contact with the indenter depends on the constants in Equation 19.17.

to other recent simulations, is intended to mimic nonadhesive contact.[24–28] The total force on atom i is then

$$F_i = F_{EAM} + F_{il} \tag{19.18}$$

where F_{EAM} is the total force determined by the embedded atom method and F_{il} is the force on atom i from the indenter. To simulate loading or unloading of the indenter, the center of the indenter is placed above the substrate in preparation for the indention. Typically it is centered in x and y with respect to the periodic boundaries of the system. The z component for the center of the indenter is then given as the maximum z coordinate of the substrate atoms plus the radius of the indenter plus some small increment of displacement. This initial placement of the indenter ensures that the indenter is not in contact with the surface at the start of the simulation. To load or unload the substrate, the z coordinate of the indenter is decreased or increased by $I_{\text{depth}}/n_{\text{steps}}$ at each molecular dynamics time step where n_{steps} is the number of time steps required to indent to a depth of I_{depth}. The force on the indenter is then calculated by summing the forces normal to the surface for all atoms in contact with the indenter ($r_{il} < R$). The force data and the displacement of the indenter are used to generate load displacement curves.

19.3.3 Rigid Indentation

Placement of the rigid indenter is identical to that of the repulsive potential indenter. However, instead of adding a force to each atom, a rigid displacement of contacted atoms is made. At each step the magnitude of the radial vector $\mathbf{V}_j$ between the coordinate of atom $\mathbf{A}_j$ and the indenter position $\mathbf{I}(x, y, z)$ is calculated $\mathbf{V}_j = \mathbf{I} - \mathbf{A}_j$. Contact between atoms and the indenter is defined as follows. For $|\mathbf{V}_j| < R$ where R is the radius of the indenter, with the atom in contact, the atom is moved along the radial vector $\mathbf{V}_j$ to the point where $|\mathbf{V}_j^{new}| = R$. This generates a rigid displacement of atoms by the indenter that exactly matches the geometry of the indenter; see Fig. 19.12. The atom is made rigid by zeroing its velocity, acceleration, and higher order

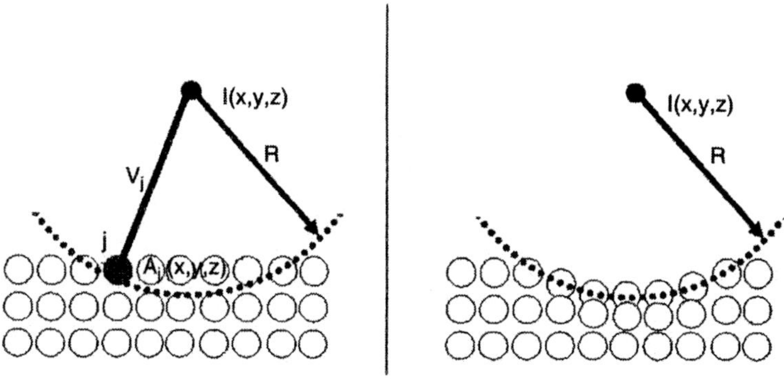

FIGURE 19.12 Schematic diagram of the rigid indenter. Atoms are rigidly displaced normal to the surface of the indenter such that $|\mathbf{V}_j| = R$.

derivatives so that the shape of the contact is maintained, i.e., atoms cannot move back inside the indenter. Once an atom loses contact with the indenter, for instance, during unloading, it is no longer constrained. For $|\mathbf{V}_j| > R$, there is no contact. This type of indenter gives a more accurate estimate of the elastic properties of the material being indented; however, it works only for loading. During unloading, atoms lose contact with the indenter and no force is registered.

19.3.4 Tests and Comparisons of Indenter Functions

Lilleodden et al.[17,18] have spent considerable effort to examine indenter stiffness on the virtual indenter function expressed by Equation 19.17. They settled on values of n and k to be 2 and 10 eV/A^3, respectively, and utilized a conjugate gradient energy minimization scheme to determine the equilibrium structure at each indentation displacement before moving to the next. This is equivalent to indenting at a very slow indentation rate but at zero temperature. Molecular dynamics simulations were performed only at specific points along the load displacement curve where dislocation emission occurred as indicated by discontinuities in the load displacement curves. We conducted several small test cases to qualify the indenter function. The results are significantly different from those reported by Lilleodden and others where a conjugate gradient minimization scheme was implemented. This led us to the selection of different values for the indenter stiffness parameters n and k. In our simulations we have chosen to simulate the entire indentation cycle using molecular dynamics. Due to the computational limitations inherent to molecular dynamics simulations, namely, the limitation of time-scale, the trade-off of this choice is a high indentation rate. The particles in each simulation are held at the thermal equivalent of 300 K via the generalized Langevin equation thermostat.[26] For a typical indentation simulation of gold, a molecular dynamics time step of 0.005 ps was used. At each time step the indenter is moved 0.01 Å down or up for loading or unloading. This results in an indentation rate of 2 Å/ps. Although much faster than experimental indentation rates, which are on the order of angstroms per second, the indentation rate in our simulations is still approximately eight times slower than the theoretical dislocation velocity in

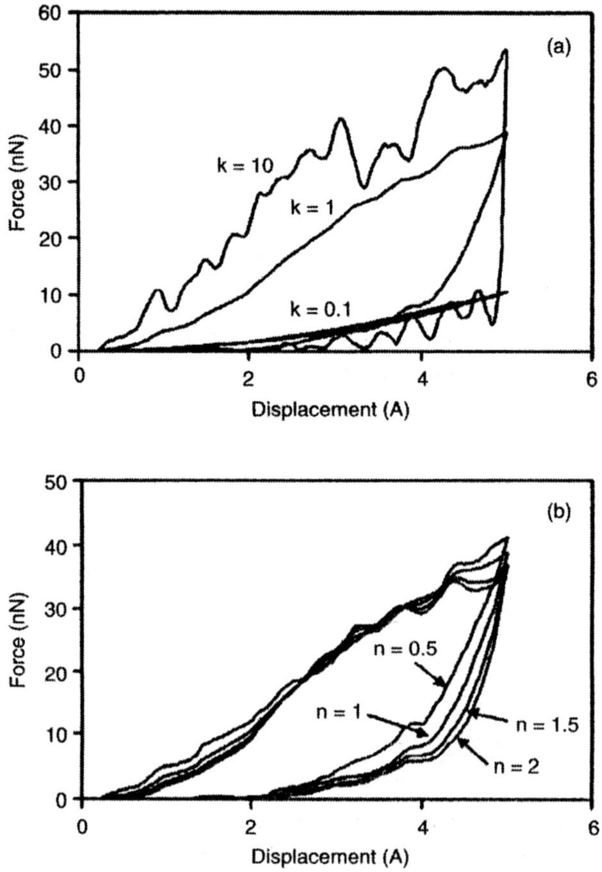

FIGURE 19.13 Load displacement curves for (a) varying values of the constant k with $n = 1$, and (b) for varying values of the constant n with $k = 1$.

bulk gold. As a result the underlying plastic deformation mechanisms are largely independent of the indentation rate with the benefit that any thermally activated processes present in a typical indentation experiment may be accurately represented. Illustrated in Fig. 19.13(a) are the load displacement curves of several small-scale molecular dynamics simulations of indentation with varying values of the constant k at a constant value $n = 1$. A 10 Å radius indenter was used to indent a ⟨111⟩-oriented substrate with dimensions 70 Å × 70 Å × 70 Å to a depth of 5 Å at a constant displacement rate of 2 Å/ps. Small values of k (>0.5 eV/Å^2) suggest an overly compliant indenter that leads to little or no plastic deformation of the substrate. Large values of k (>5 eV/Å^2) cause bouncing of the atoms as they make and lose contact with the tip. This leads to noisy data that makes analysis of the load-displacement curves difficult. Shown in Fig. 19.13(b) are load displacement curves of the same substrate with varying values for the exponent n with a constant $k = 1$ eV/Å^{n+1}. Over the range of n evaluated no particular value appears to be better than another.

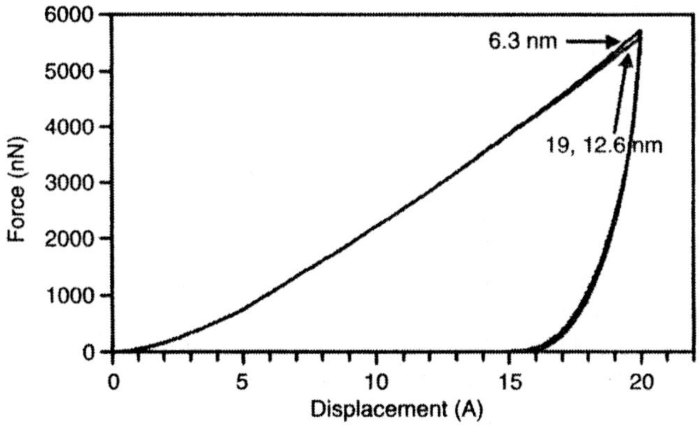

FIGURE 19.14 Load displacement curves for substrate thicknesses of 6.3, 12.6, and 19 nm indented to a depth of 2 nm with a 10-nm radius indenter. The curves overlap completely for the 12.6- and 19-nm thick substrates with only a slight increase in the force with displacement for the 6.3-nm substrate with respect to the other substrates.

Based on these tests, we have found suitable values of n and k in Equation 19.17 to be 1.0 and 1.0 eV/Å^2, respectively. For the range of indenter stiffness parameters tested it does not appear possible to back-calculate a value for the modulus or Poisson's ratio of the indenter, a problem also noted by Lilleodden.[18] Therefore all modulus values reported are for reduced modulus instead of the substrate modulus.

Finite element simulations by several other groups have reported observable periodic boundary effects for substrate thicknesses less than 50 times the indention depth.[28,29] To determine the effect of the rigid boundary layer on indentation results in our simulations, three substrates with thicknesses of 6.3, 12.6, and 19 nm were indented to a maximum depth of 2 nm with a 10 nm radius indenter. The x and y dimensions were held constant at 30 nm and 30 nm for each substrate indented. The simulated load displacement curves are shown in Fig. 19.14.

Only minor differences were observed. The load displacement curves for the 19- and 12.6-nm thick substrates overlap completely, while the 6.3-nm thick substrate shows only a slight increase in the force with displacement over the 19- and 12.6-nm substrates. This translates into an increase in hardness of only 2% and a negligible change in the reduced modulus. These results are consistent with the finite element simulations of soft thin films on hard substrates conducted by Chen and Vlassak.[30] They observed that film thickness effects have significant importance only for indent depths greater than 50% of the film thickness. Our current computational resources limit us to simulations consisting of roughly 1,000,000 atoms, which corresponds to the substrate size of approximately 30 nm × 30 nm × 19 nm, and we limit the indentation depth to 2 nm, roughly 10% of the total thickness. Additional analysis of atomic stresses during indentation showed that the in-plane periodic boundary conditions x and y do not influence the results significantly.

Consider the equation for the reduced modulus:

$$\frac{1}{E_r} = \frac{\left(1-\upsilon_s^2\right)}{E_s} + \frac{\left(1-\upsilon_i^2\right)}{E_i} \quad (19.19)$$

The repulsive potential indenter does not allow accurate calculation of the substrate modulus, because the indenter modulus E_i and Poisson's ratio υ_i are indeterminate. For the rigid indenter, E_i goes to infinity, and therefore

$$\frac{1}{E_r} = \frac{\left(1-\upsilon_s^2\right)}{E_s} \quad (19.20)$$

giving an expression for the reduced modulus that does not include the modulus of the indenter. The rigid indenter does not allow for unloading, in effect making the stiffness in the equation

$$E_r = \frac{S}{2}\sqrt{\frac{\pi}{A}} \quad (19.21)$$

go to infinity. The solution lies in the Hertzian analysis. This relation assumes a linear elastic relation between the indenter and the surface, i.e., no plasticity and no adhesion. If this requirement is met, the relationship between load and deformation closely follows the Hertzian model for a rigid, noninteracting parabolic punch deforming an elastic half-space that can be expressed as a power law expression:

$$P = \frac{4}{3}E_R\sqrt{R}d^{3/2} \quad (19.22)$$

where P is the applied load, R is the radius of the indenter, and d is the depth of the indentation. If the initial portion of the load displacement curve is fit to this equation, a direct measure of the elastic response of the material is possible. The simplest way to make the fit is by taking the natural log of Equation 19.22.

$$\ln P = 3/2 \ln d + \ln\left(\frac{4}{3}E_r\sqrt{R}\right) \quad (19.23)$$

which is simply the equation for a straight line ($y = mx + b$). Given the radius R of the indenter, solving for the reduced modulus E_r is straightforward. An example load displacement curve for a ⟨001⟩-oriented substrate is shown in Fig. 19.15(a). The sample was indented to a depth of 2.5 Å at a rate of 0.5 Å/ps with a 40-Å radius indenter. No plastic damage was observed. The rigid indenter creates singularities in the force as individual atoms come in contact with the indenter. It also incorporates the random frictional forces that are imposed by the GLE thermostat. As a result,

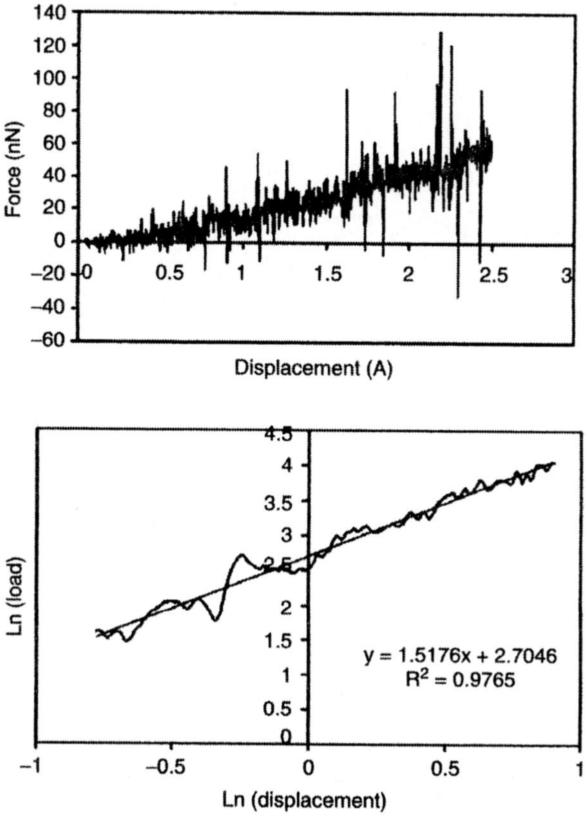

FIGURE 19.15 Load displacement curves for a ⟨001⟩-oriented substrate indented to 2.5 Å at a rate of 0.5 Å/ps with a 40-Å radius indenter. Top: raw load displacement data, Bottom: smoothed Ln–Ln data and linear fit.

the data appear quite noisy compared to the load displacement curves shown in Fig. 19.13 and Fig. 19.14 for the repulsive potential indenter. To generate good data from the noise, a running average with a 21-point centered spread is calculated. The results of this averaging and fitting to a Ln–Ln plot are shown in Fig. 19.15(bottom).

The linear fit to the data shown in Fig. 19.15 yields a value for the exponential of 1.52, which is in very good agreement with the Hertzian exponent value of 1.5, as seen in Equation 19.22. Calculating the modulus with the intercept value of 2.70 and the indenter radius of 40 Å gives an indentation modulus of 1.77 nN/Å^2. This value is in reasonably good agreement (order of magnitude) with the value of 0.55 nN/Å^2 for the indentation modulus of bulk EAM gold obtained using the elastic constants given in Table 19.3. When considering the 12 orders-of-magnitude difference in the indentation rates between simulation and experiment, this order of magnitude agreement is remarkable. The bulk elastic constants are determined in a static energy calculation. The use of these constants in the calculation of the indentation

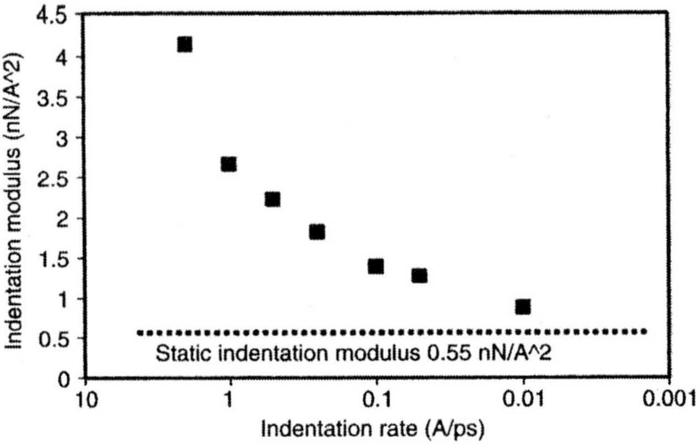

FIGURE 19.16 Indentation modulus from simulated indentation of a ⟨001⟩-oriented substrate as a function of the indentation rate. As the rate decreases, the indentation modulus appears to approach the static value for the indentation modulus (as shown by the dotted line) determined using the bulk elastic constants.

modulus should in principle only be valid in the case of static loading. This is clearly not the case in the simulations of nanoindentations presented here. A series of simulated nanoindentation experiments with indentation rates varying between 2 and 0.01 Å/ps was conducted on a ⟨001⟩-oriented gold substrate. Even slower indentation rates (<0.01 Å/ps) become very computationally expensive. The indentation modulus from these simulations has been plotted as a function of indentation rate, shown in Fig. 19.16. At slower rates, the indentation modulus appears to converge toward the static value determined from the bulk elastic constants supporting this hypothesis. It is important to note that these approximations are valid only for shallow indentations where no plasticity is present.

19.4 CONCLUSIONS

Two approaches to modeling indentation have been discussed in this chapter: FEA, a continuum-based approach, and molecular dynamics simulation, in which the motion of individual atoms is explicitly followed. For a more comprehensive review of the literature associated with atomic-level indentation simulations, the reader is referred to three recent review articles.[31,33] Which method is more appropriate depends on the conditions to be studied. FEA has the advantage that large structures and time scales that are comparable can be modeled. A potential disadvantage to this approach is that the continuum approximation begins to break down as thin film sizes become comparable to the spatial extent of defects created during indentation. On the other hand, fully atomistic simulations are able to describe a wide variety of defect structures provided that an appropriate interatomic force model is used, enough atoms can be modeled, and there are enough time steps to allow nucleation processes (e.g., defect nucleation) to proceed. Enough progress has been made over the last two decades in the development of quantum-based analytic potential functions and in the availability of parallel computing platforms that the first two points can be satisfied for most systems. On the other hand, numerically integrating classical equations of motion is done in a stepwise fashion, and generally speaking, faster (rather than more) processors are needed to address the third point. While progress in this area is being made, it has been much slower than developments in parallel processing.

Not discussed in the previous paragraph are hybrid continuum atomistic modeling approaches that are now being explored by a large number of research groups. The most mature of these is probably the quasicontinuum method developed by Phillips and coworkers.[34] In this approach, atomic forces are used to govern the deformation of finite elements in response to external deformations, including indentation. In related work, Liu[35] has developed a method in which groups of atoms are assigned to a given finite element region. Where uniform stress is applied, the finite elements respond, and the atoms in that region follow the continuum deformation. In regions where the stress distribution has subelement deviations, individual atoms are allowed to move according to the interatomic forces and any external stresses, and the finite element containing those atoms distorts in response to the local atomic motion. This approach allows one to use continuum and atomic dynamics simultaneously in a manner that can respond to the overall deformation of the system. These and related multiscale approaches hold great promise for indentation simulations, and with the rapid advances being made in computing power and numerical algorithms we soon will be able to routinely model details of indentation across disparate length scales.

PROBLEMS

1. Perform linear finite element simulation of the nanoindentation problem described in Section 19.1.1 using the same material properties and Berkovich indentor loading. Generate the force deflection curve. Evaluate the effect of mesh size and perform convergence study.
2. Perform nonlinear finite element simulation of the nanoindentation problem described in Section 19.1.1 using the same material properties and Berkovich indentor loading. (Include the effects of material and geometry nonlinearity.) Generate the force deflection curve. Evaluate the effect of mesh size and perform convergence study. Compare the linear and nonlinear analysis results.
3. Explore the effect of thin film thickness on the nanoindentation results for thin film thickness greater than 250 nM as well as lower than 250 nM.
4. Describe the basic differences between the continuum-based finite element approach and the EAM-based molecular dynamics modeling approach for the nanoindentation problem.

REFERENCES

1. Derby, B., Alumina/Silicon carbide nanocomposites by hybrid polymer/powder processing: microstructures and mechanical properties, *Solid State Mater. Sci.*, *3*, 490–495, 1998.

2. Sternitzke, M., Review: structural ceramic nanocomposites, *J. Eur. Ceramics Soc., 17,* 1061–1082, 1997.
3. Harding, J.W. and Sneddon, I.N., The elastic stresses produced by the indentation of the plane of a semi-in finite elastic solid, *Proc. Cambridge Philos. Soc., 41,* 12, 1945.
4. Oliver, W.C. and Pharr, G.M., An improved technique for determining hardness and elastic modulus using load and displacement sensing indentation experiments, *J. Mater. Res., 7,* 6, 1564–1583, 1992.
5. Tabor, D., Adams, M.J., Biswas, S.K., and Briscoe, B.J., Indentation hardness and material properties, *Solid-Solid Interactions,* Imperial College Press, London, 1996.
6. King, R.B., Elastic analysis of some punch problems for a layered medium, *Int. J. Solids Structures, 23,* 1657, 1987.
7. Pethica, J.B., Hutchings, R., Oliver, W.C., Hardness measurement at penetration depths as small as 20 nm, *Philos. Mag. A, 48,* 593, 1983.
8. Doerner, M.F. and Nix, W.D., A method for interpreting the data from depth-sensing indentation instruments, *J. Mater. Res., 1,* 601, 1986.
9. Baker, S.P., The analysis of depth-sensing indentation data, *Mater. Res. Soc. Symp. Proc., 308,* 209, 1993.
10. Harding, D.S., Oliver, W.C., and Pharr, G.M., Thin films: stresses and Mechanical Properties V, *Mater. Res. Soc. Symp. Proc., 356,* 663–668, 1995.
11. Chung, P.W. and Namburu, R.R., *Nanotechnology: Novel Computational Mechanics Methods for Nanoscale and Nanostructured Solids, Computational and Information Sciences Directorate,* U.S. Army Research Laboratory, Aberdeen Proving Ground, Maryland.
12. Tadmor, E.B., Miller, R., and Phillips, R., Nanoindentation and incipient plasticity, *J. Mater. Res., 14,* 2233, 1999.
13. Landman, U., Atomistic mechanisms and dynamics of adhesion, nanoindentation and fracture, *Science, 248,* 454, 1990.
14. Kelchner, C.L. and Plimpton, S.J., Dislocation nucleation and defect structure during surface indentation, *Phys Rev B., 58,* 11085, 1998.
15. Zimmerman, J.A., Kelchner, C.L., Klein, P.A., Hamiltion, J.C., and Foiles, S.M., Surface step effects on nanoindentation, *Phys. Rev. Lett., 87,* 165507, 2001.
16. Rodriguez de Fuente, O., Zimmerman, J.A., Gonzalez, M.A., de la Figuera, J., Hamilton, J.C., Pai, W.W., and Rojo, J.M., Dislocation emission around nanoindentations on ⟨001⟩ fcc metal surface studied by scanning tunneling microscopy and atomistic simulations, *Phys. Rev. Lett., 88,* 36101, 2002.
17. Lilleodden, E.T., Indentation-Induced Plasticity of Thin Metal Films, PhD thesis, Stanford University, 2001.
18. Lilleodden, E.T., Zimmerman, J.A., Foiles, S.M., and Nix, W.D., Atomistic simulations of elastic deformation and dislocation nucleation during nanoindentation, *J. Mech. Solids, 51,* 901, 2003.
19. Allen, M.P. and Tildesley, D.J., *Computer Simulations of Liquids.* Clarendon Press, Oxford, 1987.
20. Daw, M.S. and Baskes, M.I., Embedded-atom method: derivation and application to impurities, surfaces, and other defects in metals, *Phys Rev B., 29,* 6443, 1984.
21. Foiles, S.M., Baskes, M.I., and Daw, M.S., Embedded atom method functions for the FCC metals Cu, Ag, Au, Ni, Pd, Pt, and their alloys, *Phys. Rev. B., 33,* 7983, 1986.
22. Daw, M.S., Foiles, S.M., and Baskes, M.I., The embedded atom method: a review of theory and applications. *Mater. Sci. Reports, 9,* 251, 1993.

23. Plimpton, S.J. and Hendrickson, B.A., Parallel molecular dynamics with the embedded atom method, *Mater. Res. Soc. Symp. Proc., 291, 37,* 1993.
24. Gear, C.W., *Numerical Initial Value Problems in Ordinary Differential Equations,* Prentice-Hall, Englewood Cliffs, NJ, 1971.
25. Nordsieck, A., On numerical integration of ordinary differential equations, *Math Comp., 20,* 130, 1962.
26. Kubo, R., Fluctuation–dissipation theorem, *Rep. Prog. Theor. Phys., 33,* 425, 1965.
27. McQuarrie, D.A., *Statistical Mechanics,* Harper and Row, New York, 1976.
28. Bolshakov, A., Oliver, W.C., and Pharr, G.M., *J. Mater. Res., 11,* 760, 1996.
29. Pharr, G.M. and Bolshakov, A., *J. Mater. Res., 17,* 2660, 2002.
30. Chen, X. and Vlassak, J.J., *J. Mater. Res., 16,* 2974, 2001.
31. Harrison, A., Stuart, S.J., and Brenner, D.W., Atomic-scale simulation of tribological and related phenomena, in Bhushan, B., Ed., *CRC Handbook of Micro/Nanotribology,* 2nd ed., CRC Press, Boca Raton, FL, 1998, pp. 525–596.
32. Brenner, D.W., Shenderova, O.A., Schall, J.D., Areshkin, D.A., Adiga, S., Harrison, and Stuart, S.J., Contributions of molecular modeling to nanometer-scale science and technology, in Goddard, W., Brenner, D., Lyshevski, S., and Iafrate, G., Eds., *Nanoscience, Engineering and Technology Handbook,* CRC Press, Boca Raton, FL, 2002.
33. Schall, J.D., Brenner, D.W., Atomistic simulation of the influence of pre-existing stress on the interpretation of nanoindentation data, *J. Mater. Res.,* 19, 3172–3180, 2004.
34. Phillips, R., Dittrich, M., and Schulten, K., Quasi-continuum representation of atomic-scale mechanics, *Ann. Rev. Mater. Res., 32,* 219–233, 2002.
35. Zhang, L., Gerstenberger, A., Wang, X.D., and Liu, W.K., Immersed finite element method, *Comp. Meth. Appl. Mechan. Engin., 193,* 2051–2067, 2004.

20 Synthesis, Optimization, and Characterization of AlN/TiN Thin Film Heterostructures

Cindy K. Waters, Sergey Yarmolenko, Jagannathan Sankar, Sudhir Neralla, and Ajit D. Kelkar

CONTENTS

20.1 INTRODUCTION

Multilayer coatings and nanothin layers of films are being developed and investigated every day.[1–8] Previous studies have shown that multilayer coatings can lead to benefits in performance over comparable single-layer coatings and can combine the attractive properties of different materials in a single protective layer. There are many advanced coating systems currently under investigation and many methods are being researched to create these films, as shown in Fig. 20.1[8] (Elshabini-Riad). The reason for the enormous amount of research in this area is that these coatings are useful in a plethora of different fields from national defense, to biomedical applications, to manufacturing.[2,4,9–11] Each area of use mandates a different set of properties and these properties must be analyzed and quantified by researchers.

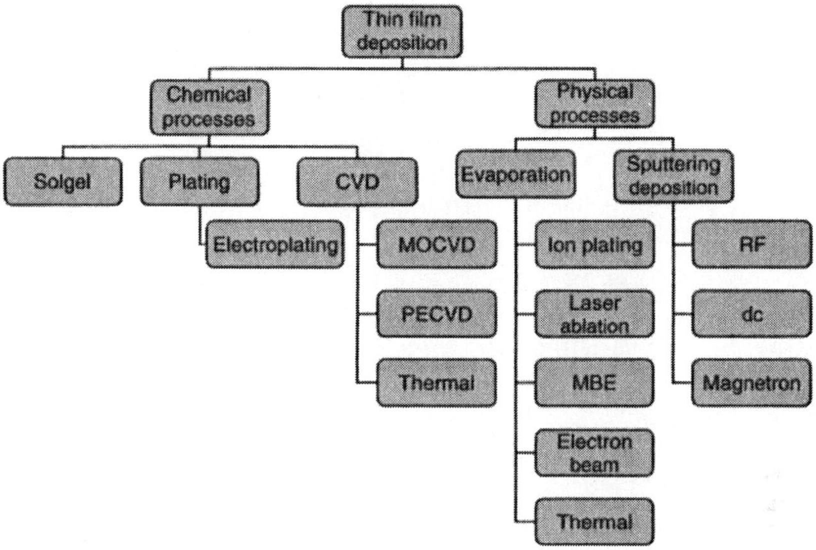

FIGURE 20.1 Classification of the most common deposition processes (Elshabini-Riad[8]).

20.1.1 Nitrides

Nitride single-layer coatings have been studied and long used on an assortment of substrates for their exemplary properties. An interesting note concerning their mechanical and physical properties is that they behave more as a ceramic than their constituent metal. A couple of the most popular group III nitrides include aluminum nitride (AlN) and titanium nitride (TiN). AlN is a wide-band-gap (energy gap ~6.2 eV) semi-conductor, while TiN is a transition metal nitride thin film. TiN has the highest melting point and hardness when compared to IV-B nitrides (Zr and Hf). AlN also possesses respectable hardness and melting temperature, as shown in Table 20.1

20.1.2 Multilayer Coatings

Researchers have worked on improving the toughness, hardness, and residual stress state of coated layers in order to create new and enhanced tribological coatings, often using nanostructural or multilayered coating designs.[12] Multilayer coatings have many advantages over single-layer films.[1–5,9–11,13,14] Multilayering can combine the attractive properties of single layers into a solitary protective coating.

In addition to a permutation of the best properties, dislocation theory reveals that the introduction of multiple interfaces parallel to the substrate and potential crack growth direction will deflect or slow any crack growth. This phenomenon will in turn increase the toughness and hardness of the coating. The variation in shear modulus among the materials in superlattice nitride coatings is the primary factor leading to increased hardness. Thin-film superlattices composed of metal layers, ceramic layers, and a combination of ceramic and metal layers have been reported

TABLE 20.1
Properties of Various Nitrides

Periodic Group	IV B			V B		VI B		III A		IVA
Elements	**Ti**	**Zr**	**Hf**	**V**	**Nb**	**Cr**	**Mo**	**B**	**Al**	**Si**
				Nitride Properties						
Melting point (K)	3220	3150	3130	2619	2846	2013	2223	3300	2523	2173
Hardness (bulk, GPa)	21	15	17	15	14	11	17	33	12	17
Crystal structure	fcc	fcc	Cubic	fcc	fcc		hcp/fcc	Cubic/ hcp	hcp	
Dominant bond type	Metallic							Covalent		

fcc = face-centered cubic.
hcp = hexagonal close-packed.

to exhibit strength and hardness increases, often >100%, over homogeneous materials and rule-of-mixtures values. The barriers to dislocation motion provided by superlattice layers with different shear moduli usually account for these results. Koehler was the first to apply the expressions describing image forces on a dislocation near a heterointerface to superlattices.[15] While variations of this proposal have been reported, the strength enhancement is normally predicted to be proportional to the quotient given by Equation 20.1:

$$Q = (G_A - G_B)/(G_A + G_B) \quad (20.1)$$

where G_A and G_B are the shear moduli of the superlattice layers, A and B, respectively. There are other mechanisms of superlattice strengthening proposed: inhibition of dislocation motion due to layer coherency strains, increases in total interface area due to glide across the layers, misfit dislocation arrays at the interfaces, the Hall–Petch effect, and the supermodulus effect.[16]

Figure 20.2 depicts possible reactions with a layered interface that will influence hardness and toughness according to the various methods mentioned. There are three main classifications of nanofilm-layered morphologies.[1–3] Coating with a limited number of single layers composes the majority of coatings in use today. These coatings are useful because they break up the column grain growth and combine different materials. The second class is composed of coatings with high numbers of nonisostructural single layers.

The last group is the coating classified as the superlattice. The superlattice is composed of isostructural single-layer materials that have similar periodic table properties (i.e., chemical bonding, atomic radii, and lattice dimensions are relatively close). The thickness of the single layer is similar to its lattice dimension. These superlattices exhibit increases in hardness and strength when the multilayer superlattice period is in the range of 5 to 10 nm. Prior work by Mirkarimi et al. with specific nitride

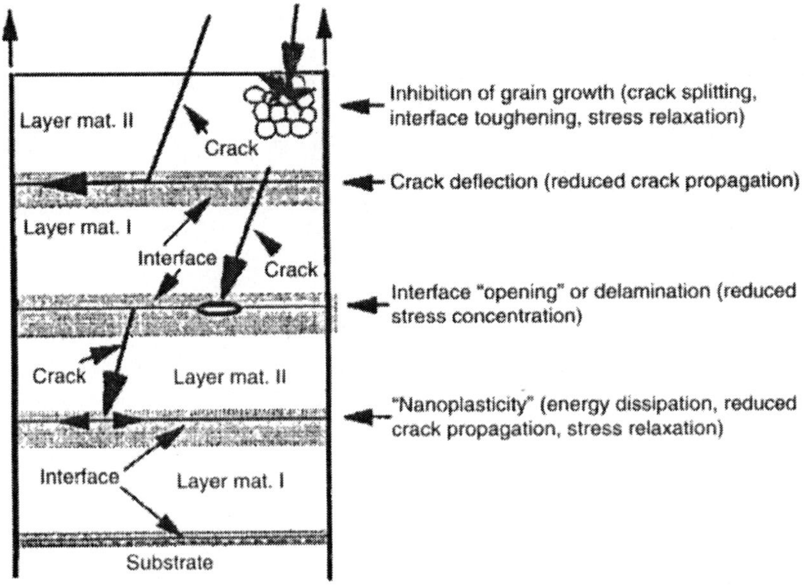

FIGURE 20.2 Periodic layered structure and toughening mechanisms (Holleck[3]).

superlattice coatings has achieved maximum hardness values for coatings of TiN/VN (~56 GPa), and of TiN/NbN (~51 GPa) that were much higher than the hardness of 17 to 23 GPa for homogeneous single-crystal coatings of TiN, VN, and NbN.[7,17,18] The use of TiN/AlN multilayered thin-film coatings has proven to be successful. Research has proved that 0.5-μm thin films achieve the same performance as mono-nitride TiN films with more than six times the thickness.[18]

Studies involving alternating layers of Ti and TiN multilayer coatings in the Ti–N system were performed by Bull et al. to demonstrate the advantages of compositionally and structurally modulated coatings over conventional single-layer titanium nitride coatings in tribological applications.[4] Bull classified two different multilayer coatings:

1. Structural multilayers — The applied substrate bias is cycled between two values to produce layers with different levels of residual stress. In this case, bias voltages of –30 V and –60 V were used, which produce compressive stresses in single-layer coatings of ~3 GPa and ~6 GPa, respectively. The multilayer coating should thus have a lower residual stress than a comparable single-layer coating deposited at –60 V bias.
2. Compositional multilayers — The flow of reactive gas to the chamber (nitrogen in this case) is stopped periodically to produce alternating layers of titanium and titanium nitride. Owing to the relatively high process pressures used in SIP and the restrictions in pumping required to maintain these pressures, it takes a few minutes to completely remove the nitrogen from the chamber. For this reason, the coatings are expected to have graded composition across the Ti/TiN interface.[4]

An important concept in the production of multilayer coatings is the multilayer period λ, which is the repetitive spacing in the structure. Bull assessed the properties of the multilayer coatings as a function of λ. For structural multilayers, at low λ the hardness was less than that of a comparable single-layer film, but increased up to a maximum at λ~10 nm. The hardness of very thick structural multilayers is indistinguishable from the average value of the single-layer coating hardness.[4] The period thickness in this investigation is varied to address different types of films. Interatomic distances for AlN (0001) and TiN (111) are used to calculate the lattice mismatch at the AlN/TiN interface, which is found to be 3.9%.

20.1.3 Aluminum Nitride

20.1.3.1 AlN Properties

Thin films of AlN are very promising dielectric materials because of their wide band gap, e.g., 6.2 eV. Generally speaking, a wide band gap semiconductor has an energy band gap wider than approximately 2 eV. Other examples of wide band gap semiconductors are gallium nitride, 3.4 eV, and silicon carbide, between 2.2 and 3.25 eV. AlN maintains a wurtzite crystal structure with lattice constants: a = 3.112 Å and c = 4.982 Å, as shown in Fig. 20.3. A is the typical wurtzite crystal structure, B is the view along the [0001] direction, C is the view along the [1120] direction, and D is the view along the [$0\bar{1}10$] direction. Dark atoms are the aluminum atoms and the lighter atoms are the nitrogen atoms.

AlN thin films grown on silicon are oriented with the c-axis of the hexagonal lattice perpendicular to the substrate.[19] AlN has been shown to form a nonequilibrium metastable cubic zinc blende. AlN has also been found to have high thermal conductivity with a thermal expansion coefficient close to that of silicon.[20] This is the

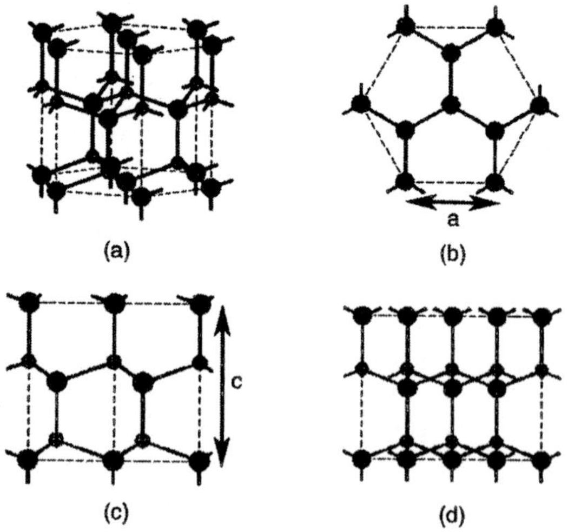

FIGURE 20.3 Various views of AlN wurtzite structure.

reason for its wide use as the initial layer on silicon substrates for many heterostructures. Surface tensions can be reduced with a greater match in thermal and crystallographic properties such as those maintained by AlN.

20.1.3.2 AlN Applications

AlN exhibits a remarkably high thermal conductivity, which is second only to beryllia. Its thermal conductivity exceeds that of copper at moderate temperatures (~200°C). This high conductivity coupled with high volume resistivity and dielectric strength leads to its use as a substrate and as packaging for high power or high density assembly of microelectronic components. Substrates made from AlN provide more efficient cooling than conventional or other ceramic substrates; hence, they are used as chip carriers and heat sinks. Other properties of AlN lead to its use in electro-optic applications. AlN, with its large energy gap, may also be the material of choice for passivation and the MISFET gate insulator due to its substantial resistivity and breakdown electric fields.[21] There is a heightened interest in wide-band-gap semiconductors for two important classes of applications, namely blue/green light emitters and high power/high temperature electronics.[22] Basic properties of both AlN and TiN thin films are listed in Table 20.2.

20.1.4 Titanium Nitride

20.1.4.1 TiN Properties

Titanium nitride thin crystalline films usually exist as δ-TiN with a NaCl-type lattice.[23] This is an fcc-type structure, as shown in Fig. 20.4. In this figure the dark atoms are the titanium atoms and the lighter atoms are the nitrogen atoms. The arrangement of the transition metal nitrides crystal structure is determined by the proportion of the atomic radius of the interstitial element and the transition metal,

TABLE 20.2
Basic Aluminum Nitride and Titanium Nitride Thin-Film Properties

	Aluminum Nitride	Titanium Nitride
Lattice constants (Å)	wurtzite $a = 3.112$; $c = 4.982$ zinc blende $a = 4.38$	FCC (NaCl) $a = 4.24$
Band gap (eV @ 300 K)	wurtzite 6.2 zinc blende 5.11 (theory)	N.A.
Index of refraction (3 eV)	2.15 ± 0.05	N.A.
Electrical resistivity (Ω-cm)	10^{13}	14×10^{-6}
Thermal expansion ($\times 10^{-6}$/°C)	$\Delta a/a = 4.2$ $\Delta c/c = 5.3$	$\Delta a/a = 9.4$
Thermal conductivity (W/mK)	200 typical 320 pure single crystal	11–67
Density	3.257 g/cm^3	

NA = not available.

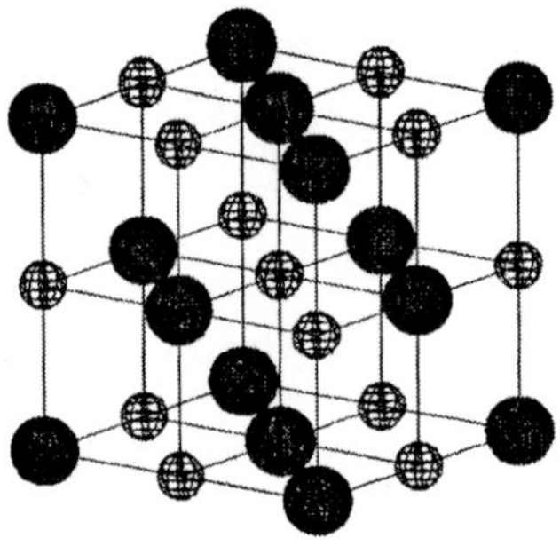

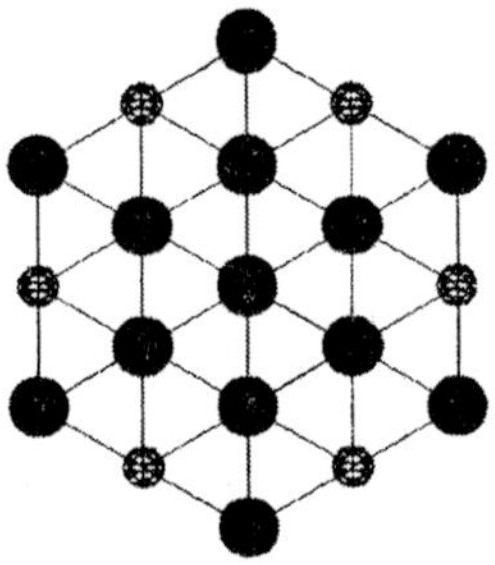

FIGURE 20.4 Left: crystal structure of titanium nitride, FCC; right: preferred growth orientation for TiN [111].

$r = r_x/r_{Me}$, according to the Hagg rule.[24] If r is less than 0.59 ($r_N/r_{TiN} = 0.56$), simple structures such as NaCl or a simple hexagonal will be formed, and if r is larger than 0.59, the transition metal and the interstitial atoms will form very complicated structures with unit cells that contain as many as 100 atoms.[10]

The elastic strain energy for TiN is minimum when the [111] direction lies normal to the film and substrate.[25] The dominant crystal orientations are those that exhibit high growth rates because they tend to take over the other directions so that the trend is growth with a [111] preferred orientation.

These films have the following properties: excellent corrosion and wear resistance, relative inertness, maintenance of a high sublimation temperature, high thermal and electrical conductivity, and high hardness. These properties are displayed in Tables 20.2 and 20.3. The mechanical and electrical properties of titanium nitride thin films have a significant dependence on the microstructure and therefore on the growth conditions. As in most known materials, their imperfections play a great role in their properties. Voids and impurities notably influence the mechanical properties such as hardness and the electrical properties such as resistivity.[9] In addition, kinetic restrictions imposed on the PLD growth process by nonequilibrium processes influence the microstructure.

TiN is utilized because of these properties and its low friction and admirable resistance to oxidation at temperatures below 500°C. These properties are valuable when choosing a material for an optical or electronic device.

TABLE 20.3
Mechanical Properties of Aluminum Nitride and Titanium Nitride Films

	Aluminum Nitride	Titanium Nitride
Modulus of elasticity (GPa)	260–350	350–450
Hardness (GPa)	18–25	20–40
Poisson's ratio	0.24	0.25

20.1.4.2 TiN Applications

The growth of ultrathin TiN heterostructures on various substrates is a subject of intense research because there is an increasing miniaturization drift in microelectronics. Smaller appliance sizes and increased densities require the development of reliable electronic devices at the nanometer scale.[26,27] Special attention is given to the morphological evolution of TiN heterostructures.

Known for its temperature stability and mechanical hardness, TiN is currently utilized in integrated circuits and has been applied in a wide range of industrial areas, such as hard coatings, coatings for corrosion resistance and decoration, and diffusion barriers in microelectronic devices.[1–3,7–8,28–30] To study the dielectric properties of laser-processed $BaTiO_3$ thin films on Si (100), researchers have used a conductive buffer layer of TiN because it has the following advantages: low diffusion coefficients of metal atoms in the TiN layer, low solubility, and chemical and thermal stability.[31] TiN is a good contact material for wide-bandgap nitrides, because its electrical conductivity is high and its contact resistance is low.

TiN coating is frequently selected as a protective film for a variety of metal surfaces in order to increase service life. The coating of stainless steel is a widely known application.[32] The corrosion resistance of stainless steel can be increased and its surface hardness can be improved by coating with a TiN film. This also adds a golden decorative color.[33]

Recently TiN research has focused on its applications in semiconductor device technology. The packing density of ultralarge integrated circuits is steadily increasing and this creates a need for corresponding decreases in the thickness of the metal interconnect layers and diffusion barriers.[34] The early stages of the growth of TiN thin layers have been studied extensively and there is a great demand for highly conductive, ultrathin, and ultrasmooth TiN layers.[27] Table 20.3 displays the standard mechanical thin films properties for AlN and TiN.

20.1.5 AlN/TiN Heterostructures

In many instances the properties of a single material are not sufficient for an application. This difficulty is overcome by the use of a multilayer coating that merges the attractive properties of several materials to solve problems in the application. Simple examples of this include the use of interfacial bonding layers to promote adhesion and thin inert coatings on top of wear-resistant layers to reduce the corrosion of cutting tools.[4] Research now shows that the multilayer structure produced by the deposition of many alternating layers of two materials improves performance more than the use of a mixed coating produced by the introduction of new interfaces. This occurs even when the two materials do not have definite functional requirements in the intended application.

In order to be used as a cutting tool material, the layers in a material must have the ability to be employed above 1000°C and the AlN/TiN multilayers are useful for this reason. They are not miscible at high temperatures (>800°C) and have a high oxidation resistance.[36]

In Japan, Setoyama et al. have investigated the TiN/AlN polycrystalline superlattice system. These researchers deposited the films by reactive cathodic arc deposition in an opposed cathode system and discovered a peak hardness of approximately 39 GPa at a superlattice period of 2.5 nm.[7] It has been discovered that the initial oxidation temperature for the TiN/AlN superlattice film is 930°C, which is significantly higher than the value of 790°C that has been reported for TIN.[7] TiN/AlN-coated tools maintain their exemplary wear properties for several reasons. When utilized as cutting tools at very high speeds, the tool tip temperature can reach 1000–1100°C and the tool must resist oxidation as well as interdiffusion of individual layers in the superlattice films. If oxidation occurs, the films lose their shielding properties because the oxides do not have the mechanical strength to withstand the loads caused by the cutting procedures.

Sproul et al. found a curious result in their research. When TiN/AlN films were manufactured as a superlattice, the TiN layers in the TiN/AlN superlattice manipulated the structure of the AlN layers. At a superlattice period of 2.5 nm, the AlN layers assume the same structure as TiN (i.e., the NaCl structure).[7,8] AlN forms a hexagonal structure during equilibrium, but in this circumstance TiN acted as a model and forced AlN into the NaCl structure.

In addition to cutting tool advantages, AlN/TiN structures are useful as capacitors and transistor gate elements of high-temperature electronic circuits. AlN is a covalent material and therefore does not adhere well to metallic substrates, so it performs poorly as a single layer. This issue can be resolved when AlN is adhered to covalent hard materials such as TiN. This creates bonded multilayer composites that can be used as wear-resistant materials, with significant improvements in the high-temperature properties.[2]

20.1.6 Motivation, Objective, and Organization

Thin films offer many advantages over bulk materials and researchers are continually searching for smaller and more efficient devices. Pulsed laser deposition (PLD) is a good technique for the growth of many new thin, stoichiometric films. By changing PLD parameters, the AlN and TiN thin-film properties can be changed and optimum values can be obtained. Material composition and microstructure will be adjusted in this research to analyze the optimization of strength and toughness in the coatings. This AlN/TiN multilayer study will report hardness gains and losses in the multilayer thin films. The overall mission of this research is to understand the synthesis, optimization, and characterization of AlN/TiN thin-film heterostructures.

20.2 PULSED LASER DEPOSITION

The thin films grown for this work were deposited using a vacuum deposition technique. Figure 20.1 lists other deposition methods, which vary as to how the deposition flux is generated, the means used to control the arrival rate of the film species, the energy of the deposition flux, and the pressure in the vacuum chamber. The PLD method was used for the deposition of films in this research. PLD has been used as a deposition technique since the 1960s. This process is discussed completely in Chapter 10.

20.2.1 Laser Energy Influences

The reach and contour of the plume can be controlled by varying the fluence. Laser fluence is the energy in the area where the laser spot hits the target. It is calculated in J/cm^2 and can be varied by changing the laser pulse energy or by regulating the laser spot size with a change in the aperture through which the laser beam exits the laser chamber. This influences the plume's appearance both in shape and size. The fluence has a minimum value that corresponds to the threshold fluence of the material. Below this threshold fluence, the energy is not great enough to create an explosion. A higher fluence will produce a larger explosion, which will make the plume more round shaped, but not necessarily longer.

The quick, strong heating of the target surface by the intense laser beam (typically up to temperatures of more than 5000 K within a few ns, corresponding to a heating rate of about 1012 K/s) guarantees that all target components, even those with different partial binding energies, evaporate simultaneously. When the ablation rate is high enough (as with laser fluences well above the ablation threshold), a Knudsen layer is formed and further heating creates a high temperature plasma that then expands adiabatically in a direction perpendicular to the target surface.

Researchers established that the $YBa_2Cu_3O_7$ thin film's greatest thickness increases with an increase in the pulse energy density.[37] A significant part of the irradiated volume is ejected as a plume consisting of molecular clusters of different sizes as well as individual molecules. At lower energies, larger clusters are ejected and a rougher surface remains after the ablation. Below the threshold value, the amount of ejected material drops sharply and single molecules are primarily desorbed.[38] A nonlinear increase in the maximum thickness of the deposit is also found. The maximum thickness increases very rapidly for low energy densities. However, Fahler found that for stainless steel, laser energy density increases with the deposition rate.[39] Another study demonstrated that high laser energy density can produce better quality film surfaces and this may be due to the higher fluence producing a higher particle density.[40] Studies with diamond-like-carbon (DLC) coatings established that hardness and modulus increase with an increase in laser intensity.[41] The energy of the laser will affect several properties including the film thickness and its uniformity, deposition rate, particulate density, and the mechanical properties of the film.[39,40] The pulse energy density plays a very important role in controlling the thickness and composition of PLD thin films. Singh and

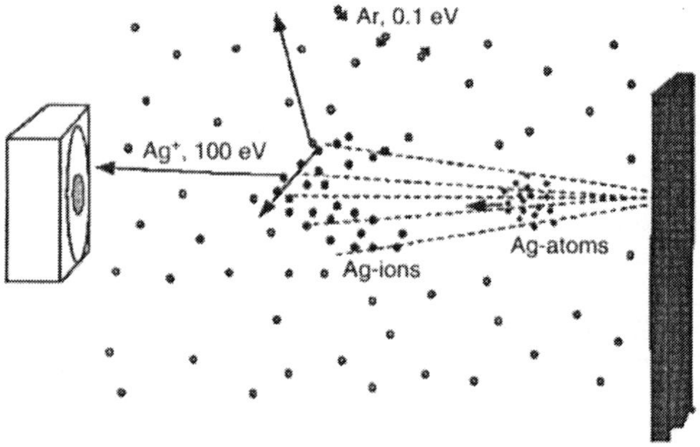

FIGURE 20.5 Model for dense cloud scattering of ablated material in an Ar gas (Sturm[43]).

and protective coatings were investigated, it was found that the film refractive index decreased with an increase in gas pressure.[42] Similar experimental results have been explained by the scattering of a dense cloud of ablated material in a diluted gas. Sturm studied scattering of Ag in argon gas. Since the mass of an Ar atom is sufficiently dense for a large scattering angle, approximately 90% of all scattering angles are larger than 10°, as estimated from ballistic collisions (Fig. 20.5).[43] Therefore, the scattered ions are not deposited onto the substrate. As the colliding Ar atoms are removed from the flight path, the ensuing slower ions and atoms fly through a significantly reduced Ar atom density and are more likely to reach the substrate. Sturm found that the faster ions are scattered and not deposited, and that they open a channel for the deposition of the slower particles. This model was independent of the exact values of the cross-section and the material.

Deposition in an inert gas changes the kinetic energy distribution of the deposited particles and notably lessens the resputtering at 0.04 mbar. The dependence of the deposition rate on pressure below 0.1 mbar can be explained by the scattering of a dense cloud of ablated material moving through a diluted gas.

Other studies confirmed this model. Geohegan states that "ablation into background gases results in scattering and attenuation of the laser plume".[44] Research with carbon films found that the ablated carbon species are decelerated to a degree that depends on the background pressure when PLD occurs in a gas ambient.

Alexandrou's research found that during CN_x deposition, the carbon species react effectively with N_2 at the beginning of their trajectories. However, when they reach the substrate, they usually do not have adequate kinetic energy to form a hard coating. Additional studies on the effects of ambient helium gas pressure and ambient argon gas pressure on the optical, structural, and physical properties of carbon (C) films deposited by pulsed laser ablation (PLD) using a camphoric carbon target have shown that the surface morphology, structural, and physical properties of deposited films are strongly dependent on the choice and pressure of the ambient inert gas.

Research performed specifically on nitrides has demonstrated that the nitrogen content and film quality of SiN_x films depend on the laser fluence and the N_2 gas pressure.

The film properties are determined by reactions between the plume and the N_2 gas, in particular, by reactions between Si species and nitrogen radicals in the plume and collisions between SiN_x clusters and nitrogen molecules.

20.2.3 Substrate Temperature Effects

The temperature of the substrate has a considerable effect on the early development and growth of the film. The homologous temperature, $T_h = T_s/T_m$, where T_s is the temperature of the substrate and T_m is the melting point of the film material in Kelvin, has been found to affect the overall microstructure. During this study the temperature was varied from 200°C up to 800°C, for example, when work was performed at $T_s = 600°C$, $T_h =$ 0.27 and 0.34 for TiN and AlN, respectively. A higher substrate temperature increases the atoms' energy, which increases their mobility. When the particles have more energy, the grain boundary has more mobility, and grains grow in size to lower their strain energy.

The details of film nucleation and growth demonstrate the dependence of film growth on material parameters and deposition conditions. The temperature at which the film is grown is a fundamental parameter governing the thin-film microstructure. The paramount temperature for optimal quality film growth is the regime where there is adequate surface diffusion to permit surface atoms to minimize their surface energy, and therefore become thermodynamically stable.[45] Research demonstrates that higher energy atomic fluxes, on the order of 5 eV to 15 eV, improve a variety of film properties.

Past studies on the ablation of TiN on various substrates have shown that TiN formation is controlled by the reactivity of the ablated and scattered species formed by the strong interaction between the ablation plume and gas pulse, which is maintained by subsequent adiabatic expansion to the substrate.[46] True epitaxial growth, however, is controlled by surface diffusion, which is also shown to be dependent on the substrate temperature.[46] The surface temperature of the substrate determines the atom's surface diffusion ability. High temperature favors rapid and defect-free crystal growth, whereas low temperature or large supersaturation crystal growth may be overwhelmed by energetic particle impingement, resulting in disordered or even amorphous structures. Studies by other researchers have shown that high temperature can reduce film surface roughness. CdO-doped SiO_2 films were found to have an optimum substrate temperature for quality crystalline films free from the CdO sites between 200°C and 300°C. Crystalline size has been shown to increase when the substrate temperature is increased for V_2O_5 films. Xiaohua Liu, on the other hand, found that TiO_2 thin-film surface roughness increased with an increase in substrate temperature.[47]

20.3 CHARACTERIZATION OF THIN FILMS

20.3.1 Thin-Film Thickness

Paramount to the measurement of any film's mechanical or electrical property is the measurement of the film's thickness t. The thickness of each film created was measured with an optical interferometric profilometer, the Wyko RST-500, as displayed in Fig. 20.6, and a Pacific Nanotechnology Nano-R AFM.

FIGURE 20.6 Wyko RST-500 optical interferometric profilometer.

The Wyko RST-500 is a phase-shift interferometry-based metrology instrument. An optical profilometer has advantages over the traditional stylus profilometer. Sensitive surfaces can be measured nondestructively and with the use of camera images. The results are not aggregated linear measurements but contiguous three-dimensional surface maps. An optical interferometer functions on the principle of interference light waves that divide a wavefront into two or more parts, a reference wavefront and a measurement wavefront, which then travel different paths before combining to produce an interference pattern, also known as an interferogram. A phase shifting method is used to extract the height information. The intensity is measured at each point in the interferogram as the phase between the measurement and reference waves is changed in discrete steps.

The steps were measured for each deposition condition (i.e., temperature, laser energy, and ambient gas pressure). This step measurement was performed by masking each sample with a clean-edged silicon strip. A large silicon sample equal to the size of the heater was mounted and a deposition was run in the initial stage of the research. These results suggested a much thicker film directly under the plume. The schematic of the masked sample is displayed in Fig. 20.7 and the resulting sample after deposition is shown in Fig. 20.8. Several traces from representative thickness profiles obtained with the Wyko profilometer are displayed in Fig. 20.9.

20.3.2 Hardness Testing

20.3.2.1 Testing Basics

Hardness implies the resistance to local deformation. Depending on the material, hardness involves different things. For example, hardness is a gauge of the indentation depth for a plastic material (i.e., a solid material that is indented beyond its yield point).

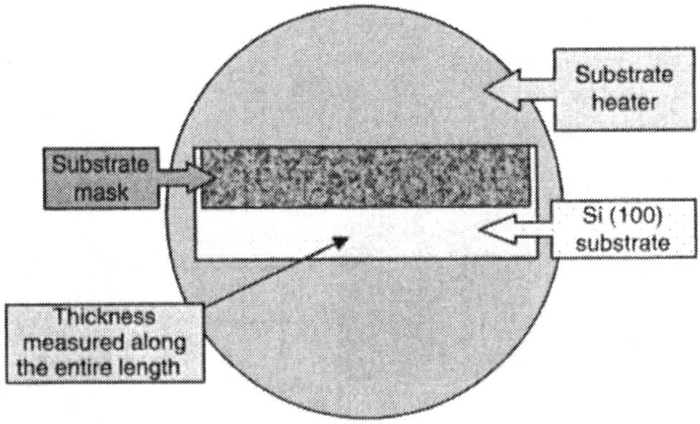

FIGURE 20.7 Model of sample used to mask the substrate and measure the film thickness.

For a brittle material, hardness is defined in the course of scratch experiments. Scratch tests reveal the oldest form of hardness measurements, typically referred to as Mohs' hardness.[48] The quasistatic indentation hardness method by Tabor of more familiar today.

20.3.2.2 Microhardness Testing

The Vickers hardness test is a standard form of microhardness testing that executes calculations from the size of an impression produced under load by a pyramid-shaped diamond indenter, as shown in Fig. 20.10. Created in the 1920s by engineers at Vickers, Ltd. in the United Kingdom, the diamond pyramid hardness test allows

FIGURE 20.8 Large masked silicon wafer and resulting deposition pattern.

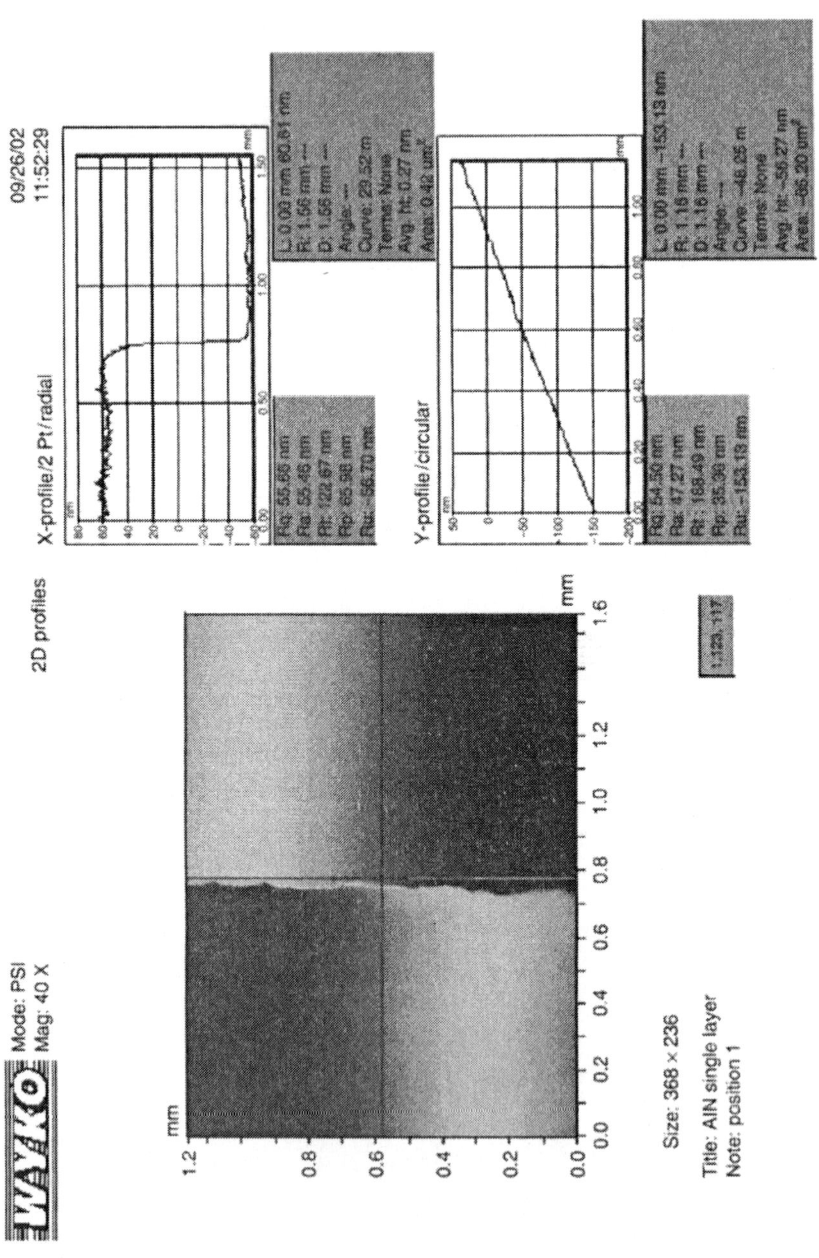

FIGURE 20.9 Wyko film thickness profile taken for an AlN thin film.

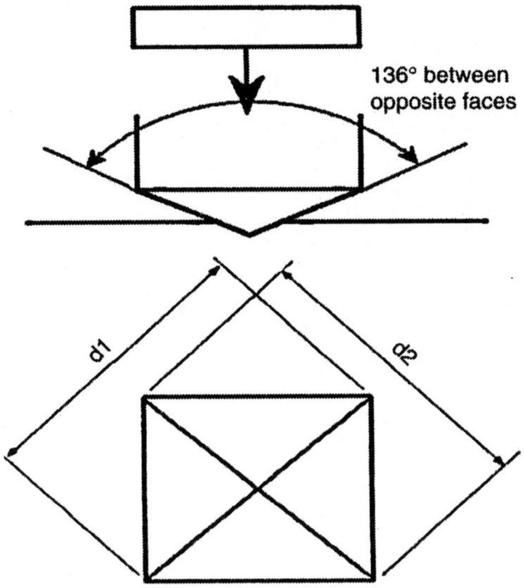

FIGURE 20.10 Schematic of a Vickers hardness indenter tip.

the establishment of a continuous scale of comparable numbers that accurately reflect the wide range of hardnesses found in steels. The indenter employed in the Vickers test is a square-based pyramid whose opposite sides meet at the apex at an angle of 136°. The diamond is pressed into the surface of the material at loads ranging up to approximately 120 kg force, and the size of the impression (usually no more than 0.5 mm) is measured with the aid of a calibrated microscope. The Vickers number (HV) is calculated using the formula, HV = $1.854(F/D^2)$, with F being the applied load (measured in kilograms force) and D^2 the area of the indentation (measured in square millimeters). The applied load is usually specified when HV is cited. The Leco M400-H1 hardness tester is displayed in Fig. 20.11.

20.3.2.3 Nanohardness Testing

Nanoindentation has been proven particularly useful in measuring the hardness and stiffness of thin films by reducing the hardness size range by one step. The MTS nanoindentation tester is shown in Fig. 20.12. The indentation response of the film on a substrate is a complicated function of the plastic and elastic properties of both the film and the substrate. As the indenter is driven into the material, both elastic and plastic deformation occurs, which results in the development of a hardness impression conforming to the contour of the indenter to some contact depth h_c.

It is possible to accurately observe both the load and the displacement of the indenter during indentation experiments in the nanometer range. By means of the established techniques of Oliver and Pharr[49] or those of Doerner and Nix,[50] the hardness and elastic modulus can be found from the peak load and the initial slope of

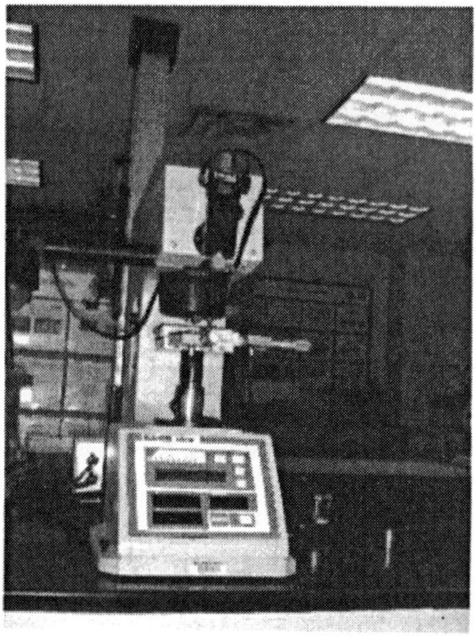

FIGURE 20.11 Leco microhardness tester.

the unloading curves. Both of these established methods depend on estimating the contact area under the load, which is difficult especially when pile-up occurs.[51]

Modeling the indentation to include plasticity is a more complex problem. The equations are nonlinear and additional material parameters must be included (e.g.,

FIGURE 20.12 MTS nanoindenter XP.

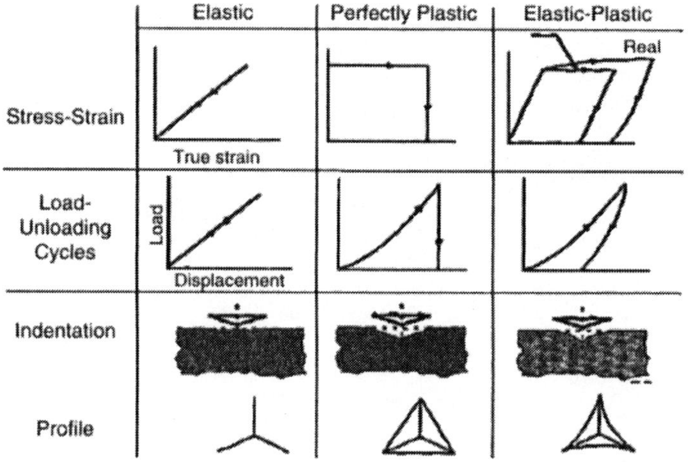

FIGURE 20.13 Theoretical examples of indentations and their resulting curves. (From Bhushan, B. Ed., *CRC Handbook of Micro/Nanotribology*, 2nd ed., CRC Press, Boca Raton, FL, 1998. With Permission.)

yield strength and work hardening). As a result, plasticity in indenter contact problems requires a great deal of experimentation and finite element simulation to be understood.[49]

Early experiments with metals by Tabor[48] revealed that the impression created by a spherical indenter remains spherical with a slightly larger radius than the indenter, and the impression formed by a conical indenter remains conical with a larger tip angle.[49] This confirmed that the diameter of the contact impression does not recover during unloading; only the depth recovers. Samples of various types of loading situations are displayed in Fig. 20.13. The significance of these experiments is that the plasticity can be dealt with by taking into account the shape of the perturbed surface in the analysis of the elastic unloading curve.[49] Therefore, the shape of the unloading curve and the recovered displacement can be related to the elastic modulus.

The basic theory of indentation testing is clear-cut. As illustrated in Fig. 20.14, during indentation, the force P and the penetration depth h are recorded as functions of time, and the load versus displacement relationship is obtained, as shown in Fig. 20.15.

The deformation pattern of an elastic-plastic sample indentation exhibits the following parameters:

h_f residual hardness impression (after elastic recovery)
h_c contact depth (i.e., depth the indenter is actually in contact with material)
h_{max} calculated (projected) indentation at the peak of the load
h_s depression of the sample around the indentation (i.e., $h_s = h - h_c$)
h_p slope extrapolated indentation depth reminder

$$h_p = h_{max} - \frac{L_{max}}{S_{max}} \tag{20.2}$$

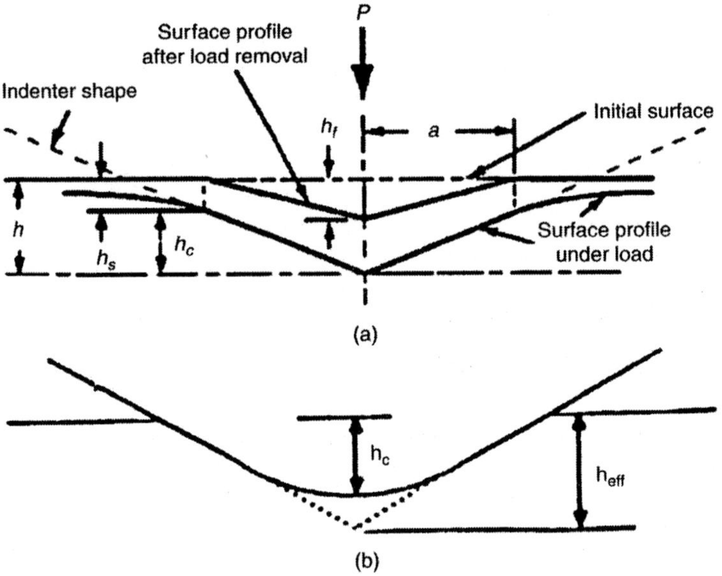

FIGURE 20.14 (a) Schematic of ideal tip indentation; (b) schematic of realistic indentation with "dull" tip. (From Bhushan, B. Ed., *CRC Handbook of Micro/Nanotribology*, 2nd ed., CRC Press, Boca Raton, FL, 1998. With Permission.)

S_{max} is the stiffness (inverse compliance) of the sample alone, which corresponds to the slope of the unloading curve at the maximum load. This process begins by fitting the unloading curve to the power law relation,

$$P = B(h - h_f)^m \tag{20.3}$$

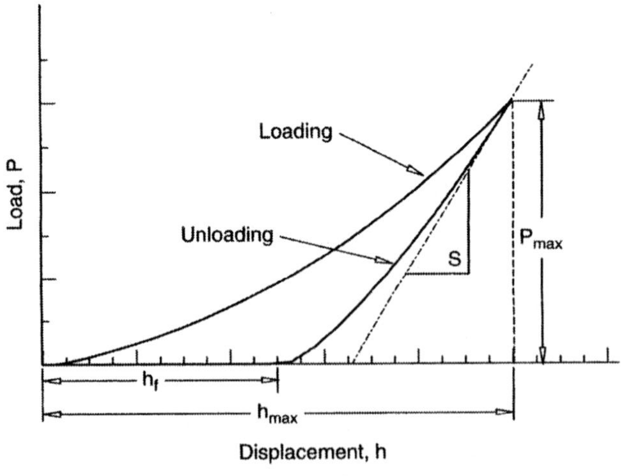

FIGURE 20.15 Schematic of nanoindentation load displacement behavior.

where P is the indentation load, h is the displacement, B and m are empirically determined fitting parameters, and h_f is the last displacement after complete unloading, for a Berkovich indenter $h_c > h_p$. When Equation 20.3 is differentiated at the point of maximum indentation, $h = h_{max}$, the stiffness is derived.

$$S = \frac{dP}{dH}(h = h_{max}) = mB(h_{max} - h_f)^{m-1} \tag{20.4}$$

The contact depth, h_c, is found by an approximation of the load displacement data such that

$$h_c = h_{max} - \varepsilon \frac{P_{max}}{S} \tag{20.5}$$

In this case, P_{max} is the peak indentation load and ε is a geometry-dependent coefficient so that ε~0.75 for a Berkovich indenter, as shown in Fig. 20.16.[47] Other typical tests may involve a spherical, conical, or pyramidal indenter, or some variety of standard geometry that is impressed into the sample with the area A of the generated deformation used to characterize the hardness.

The ratio between a corresponding load, P, and the characteristic indentation area defines the hardness number in GPa (or kg/mm^2) such that

$$H = P/A \tag{20.6}$$

The force that is utilized is calculated with certainty from the machine but the selection of an area to use for A is more difficult. The nanohardness tests can determine behavior on a nanoscale but the indentation area is not easily visible on the nanoscale without the aid of an SEM or an AFM.[51]

A Berkovich tip is a three-sided pyramid. Three sides are chosen because the tip mathematically intercepts at a single point. A sharply pointed tip ensures localized indentation, assuming no blunting has occurred. The broad opening angle is

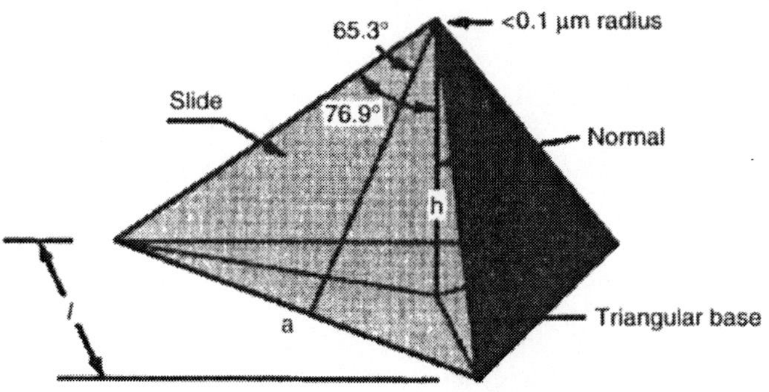

FIGURE 20.16 Sharp Berkovich indenter.

advantageous to circumvent cutting of the surface. The contact area for a Berkovich tip is approximated as $A = 0.433a^2 = 24.5h_c^2$, where$_c$ a is the tip triangular base side, and h is the tip length. This contact area is related to the actual indentation only if the material is fully plastic. The relationship between the area and the system compliance is called the area function. The presence of tip imperfections requires introduction of careful but uncertain calibrations that usually provide a function of the following form:

$$A(h_c) = 24.5h_c^2 + C_1h_c^1 + C_2h_c^{\frac{1}{2}} + C_3h_c^{\frac{1}{4}} + C_4h_c^{\frac{1}{8}} + \cdots + C_8h_c^{\frac{1}{128}} \tag{20.7}$$

The area function is usually calibrated from indentations on hard and plastic material surfaces, such as fused silica, in order to reduce the elastic and viscoelastic effects of the response. However, minor imperfections due to the machine compliance may be subtly included. The first term in Equation 20.7 describes a perfect Berkovich indenter and the other terms describe deviations from Berkovich geometry due to blunting at the tip.[52] Thurn and Cook have developed a simpler, two-parameter form of the area function:[53]

$$A_p = \frac{\pi h_c^2}{\cot^2 \alpha} + \sum_{i=0}^{\infty} C_i h_c^{-2i} \tag{20.8}$$

Their two-parameter area function acquired from the harmonic average of a spherical tip profile and a perfect conical profile describes the Berkovich indenter tip relatively well. They examined the function over the entire load range. The calculated projected contact areas were verified by SEM observation.[53]

The total compliance (compliance = deformation per unit force) during an indentation experiment is related to a two-spring system in series such that C_i and C_s are the compliances of the indenter and the sample and total compliance is given by $C = C_i + C_s$. The compliance for a Vickers and Berkovich indenter is

$$C_s = \frac{1}{S_{\max}} = \frac{dh}{dL} \approx \frac{1}{2E_r}\left(\frac{\pi}{A}\right)^{\frac{1}{2}} \tag{20.9}$$

derived by King (1987), where S is the slope of the unloading curve at maximum load and the residual elastic modulus, E_r (assumed to be a constant), is defined as

$$E_r = \frac{1-\upsilon_s^2}{E_s} + \frac{1-\upsilon_i^2}{E_i} \tag{20.10}$$

In the case of indentation with an infinitely sharp, ideal Berkovich or Vickers indenter, a relationship between the load, P, and the displacement, h, during the loading part can also be derived, yielding

$$P = \frac{E_r h^2}{\left[\left(\frac{1}{\sqrt{24.5}}\right)\sqrt{\frac{E_r}{H}} + \varepsilon\sqrt{\frac{\pi}{4}}\sqrt{\frac{H}{E_r}}\right]^2} \tag{20.11}$$

In general, indentations with contact depths of less than 10–20% of the film thickness are desired in order to attain the intrinsic film properties and to circumvent the substrate effect. This becomes complicated as time progresses and the theory evolves, because the thickness of films continues to decrease. For example, the thickness of barrier films currently used in semiconductor devices is well below 50 nm, and the film stack thickness used for magnetic data storage is less than 100 nm.[54]

20.3.2.4 Substrate Effect

Experimental equipment limitations pose problems in indentation. Equipment may not possess adequate sensitivity to test at shallow indentation depths. It is quite likely, in these circumstances, for the substrate to affect the final measurements. Chen and Vlassack have developed a substrate effect factor, which is essentially a correction factor that allows one to determine the yield stress of a film from hardness measurements.[34] Bull has also developed a method to modify values due to the substrate effect.[4] For a soft film on a hard substrate ($\sigma_f/\sigma_s \leq 1$), the influence of the substrate is not appreciable until $\delta > h/2$, δ being indentation depth. The yield stress of a film, σ_f, can be measured directly from the relation $H = P/A = c_b\sigma_f$, where the constant c_b is a constraint factor that depends on the indenter shape and material properties. This equation is valid as long as the indentation depth is less than 50% of the film thickness.

During nanoindentation measurements of coated systems, the influence of the substrate on hardness cannot be neglected. There are several models that have been developed to address the substrate effect.[55–59] For hard thin films, the substrate involvement is more important than other effects. In this case, hardness of the film alone can be calculated from measured composite values. Among several analytical models reported for this purpose, two were found to fit experimental data well. The Bhattacharya and Nix model[57] considers elastic-plastic deformation of a coating-substrate system and its contribution as proportional to the volume fraction involved in the indented area:

$$H_C = H_S + (H_f - H_S) \times \exp[-(H_f\sigma_S)/(H_S\sigma_f)(E_S/E_f)^{1/2}(t/t_f)] \tag{20.12}$$

where E_f and E_s are the Young's moduli, σ the yield strength, H_f and H_S the hardness of the film and substrate, respectively, and H_c the composite measured hardness.

The Tuck–Korsunsky model is based on a dimensional analysis and a separation of the work of indentation between the substrate and the coating. The model uses a work-of-indentation approach in terms of load work partition between the coating and the substrate and proposes the phenomenological expression for H_c

$$H_c = H_s + \frac{H_f - H_s}{1+\kappa\beta^2} \tag{20.13}$$

The model has been further modified to the form

$$H_c = H_s + \frac{H_f - H_s}{1+\left(\frac{\beta}{\beta_0}\right)^x} \tag{20.14}$$

where H_c is the composite hardness, H_s and H_f are the substrate and film hardness, β is the relative indentation depth, which is equal to the ratio of indentation depth over the film thickness, κ is a dimensionless transition parameter, x is a power exponent that describes how steeply the transition occurs, and β_0 is the value of the relative indentation depth at which the fractional hardness improvement is equal to exactly 50%. Durand-Drouhin et al. used this model in a study of TiAl(Si)N single- and multilayer thin films.[13] They concluded that crack morphology depends on indentation depth; lateral microcracks appear at shallow depths, while radial corner cracks occur at deeper indentation depths. Further results by Karimi showed that brittle films have lower values of x (typically $x <$ 3.0), while crack-resistant films exhibit higher values varying between 5 and 7.[14] In fact, x describes how steeply the transition is between the two hardnesses, H_f and H_s.

Plastic pile-up in the film is improved by the presence of the hard substrate.[34] However, the presence of the substrate in a hard film–soft substrate composite enhances the sink-in outcome. This is because the plastic region in the soft substrate is much larger than that in the hard film and the substrate is unable to support the large indentation load. These pile-up and substrate considerations will be discussed below. Experimental comparisons of all of the substrate methods are beyond the scope of this work. However, the substrate is known to play a significant role in thin-film hardness, fracture behavior, and the degree of indentation pile-up. Greater consideration is now given to the matter of indentation pile-up. There are many considerations that must be reviewed when pile-up is a significant factor.

20.3.2.5 Indentation Pile-Up

Chen and Vlassak proposed a new technique to determine the contact area in a nanoindentation experiment.[34] Their finite element results suggest that the indentation modulus of a thin film on a substrate was a function of the contact area between the indenter and the film. If the Young's moduli of both film and substrate are known, then the contact area of an indentation experiment can be found by solving the subsequent equation for the contact area A:

$$S = \beta \frac{2}{\sqrt{\pi}} E_R \sqrt{A} \tag{20.15}$$

where β is a constant and depends on the geometry of the tip. The values of β for different indenter geometries are known. If the following equation is subsequently used

$$H = \frac{P}{A} \Rightarrow P = H \times A \tag{20.16}$$

and the contact area is eliminated from Equation 20.15 and Equation 20.16,

$$\frac{P}{S^2} = \frac{1}{\beta^2}\left(\frac{\pi}{4}\right)\frac{H}{E_r^2} \tag{20.17}$$

Saha and Nix discussed this problem and stated that the "true contact pressure or hardness of the film can be determined by accounting for the pile-up effects on the contact area. This can be accomplished by adopting the method of Joslin and Oliver, i.e., using the P/S^2 parameter." This parameter can be used only when the material is elastically homogeneous and the indentation modulus is known.[35]

The advantage of the P/S^2 method is that, unlike the Oliver–Pharr technique, it is insensitive to plastic pile-up. The parameter involves only load and stiffness values that are directly measured during an indentation test. This is a great advantage because it cuts out the need to measure tip shape calibration and other ambiguous assumptions concerning material behavior.

It should be noted that this method requires knowledge of Young's modulus of the film. Usually the stiffness of a film can be equivalent to that of the bulk material when the film is not very porous and the film material can be made into bulk target material.[34] H and E are generally stable for homogeneous materials as the indentation depth changes and this results in a constant P/S^2 with indentation depth. When P/S^2 is plotted as a function of indentation depth for fused silica, the result is a constant curve. This result changes when a film and substrate are present, because P/S^2 will vary with depth depending on the hardness and elastic modulus of the film and substrate. Analyzing P/S^2 provides a more complete picture of the nanoindentation properties of a film/substrate system than can be obtained using the Oliver–Pharr method alone.

The nature of the resulting indentation provides clues to the deformation properties of the film or surface, as is shown in Fig. 20.14. The indents must be analyzed with an AFM to assess trends. Pile-up after indentation is a factor that lends uncertainty to the aforementioned nanohardness calculations. During indentation of a rigid plastic solid, the displaced material appears in the pile-up rim around the periphery of the indentation site, but with elastic-plastic materials, most, if not all, of the displaced material is accommodated by radial expansion of the elastic

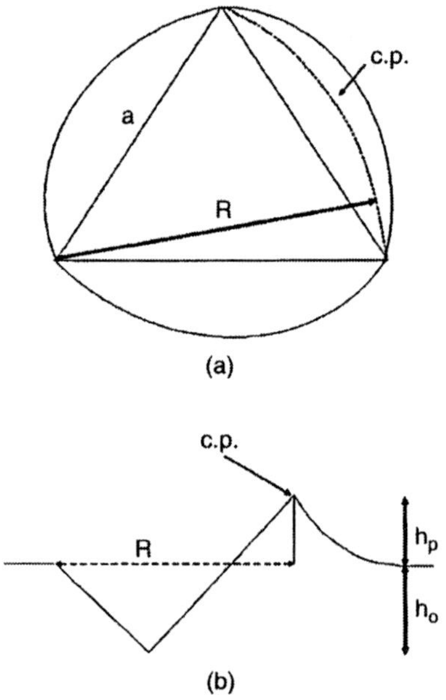

FIGURE 20.17 (a) Projected area determination and (b) cross-section showing contact point determination (Saha[35]).

surroundings, with an imperceptible change in the surface dimensions of the indented material.

The contact point (c.p.) is assumed to be the maximum point of the pile-up, as shown by the dotted line in Fig. 20.17. The area is then calculated by assuming that the load supporting area of the pile-up forms an arc of radius R at the indent edges. By measuring the indent edge a and the radius R, it is possible to calculate the total pile-up area:

$$A_{pile-up} = 3\left(\frac{\pi R^2}{6} - \frac{a}{2}\sqrt{R^2 - \frac{a^2}{4}}\right) \tag{20.18}$$

The triangular indent area is then calculated from

$$A_{triangle} = 0.433a^2 \tag{20.19}$$

The total projected contact area is now calculated and used in

$$H_c = \frac{P_{max}}{A_c f(h_c)} \tag{20.20}$$

to determine the film hardness.

The enhancement of pile-up at larger indentation depths for soft films on hard substrates has been reported.[54] It has been suggested that the plastic flow in the film is restricted by the relatively nondeformable substrate in a manner that causes soft film material to flow preferentially toward the surface. This enhances the pile-up of the material. The extent of pile-up and the depth of penetration of the indenter have been found to depend on sample hardness. Less pile-up has been observed for films with lower hardness and vice versa.[60] The pile-up of the material is associated with the deformation zone. The deformation zone in harder films is smaller than that in softer films. This may be due to the fact that harder film layers are relatively intact and dislocation theory suggests that there are no large slip bands.

Korsunsky's new approach to understanding the hardness of coated systems helps address the pile-up issue.[59] Composite hardness is known to vary depending on the applied load or indentation depth. Composite hardness is considered a simple function of the relative indentation depth β (which is the indentation depth normalized with respect to the coating thickness) and the substrate and coating hardnesses in the Korsunsky model.

Figure 20.18 outlines the plastic zone that develops when the relative indentation depth is increased and represents the primary deformation mechanisms in the indentation response of a coating-substrate system dominated by coating fracture.[59] In Fig. 20.18, Region 0 displays only an elastic response. Region I demonstrates the initial highly localized coating elastoplasticity and fully elastic substrate before the generation of an elastoplastic enclave in the substrate that leads to coating conformal stretching or deformation. Radial fracture is possible along the indentation diagonals. Region IIa is characterized by substrate deformation and conformal coating deformation sufficient

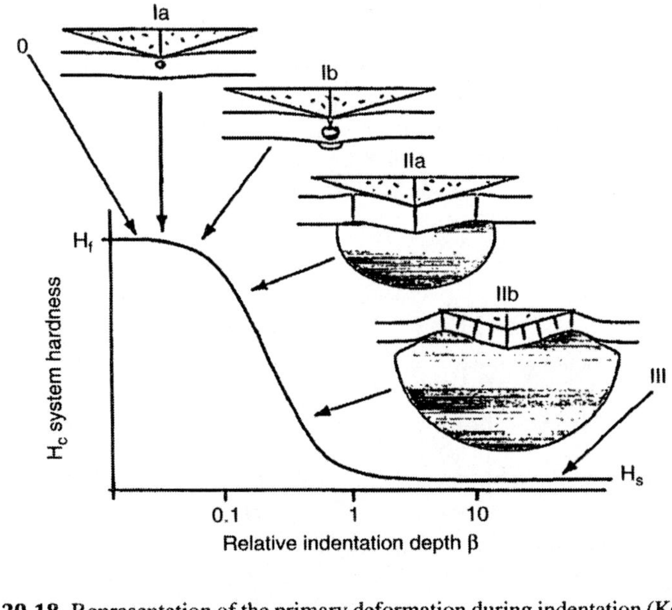

FIGURE 20.18 Representation of the primary deformation during indentation (Korsunsky[59]).

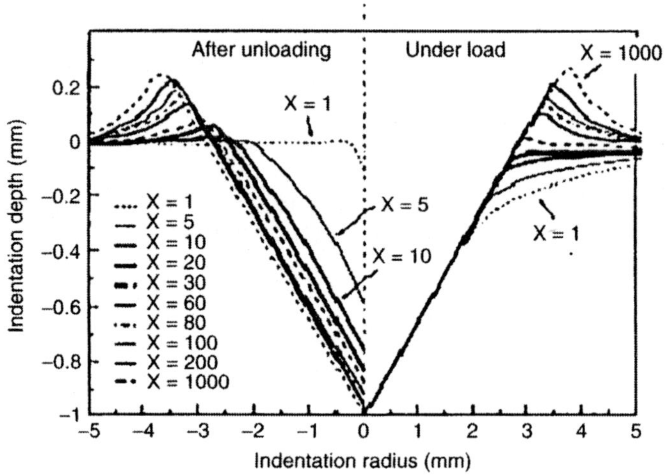

FIGURE 20.19 Cross-sectional view of elastoplastic indentation after unloading (Raymond-Angélélis[61]).

for circumferential through-thickness fracture that grows to completion, creating an island of dead coating. Region IIb represents the repeated circumferential or picture-frame fracture that occurs in the coating as it is bent to conform to the plastically deforming substrate. True substrate-only behavior occurs in Region III, where the energy absorbed by coating, stretching, flexure, and fracture is insignificant compared to the amount dissipated by substrate deformation.[59]

The function developed by Korsunsky and Tuck contains a single fitting parameter κ that describes a wide range of composite and indenter properties such as coating plasticity, interfacial strength, and indenter geometry.[59]

Additional work on pile-up is displayed in Fig. 20.19. The results were obtained by a 2-D finite element model, which demonstrated the nature of the pile-up for increasing values of indent load.[61] Much more work is required to quantify the relationship between material properties with the pile-up behavior.

20.3.2.6 Continuous Stiffness Measurement Method

Another technique, continuous stiffness measurement (CSM), offers a significant improvement in nanoindentation testing. CSM is accomplished by imposing a small, sinusoidally varying signal on top of a DC signal that drives the motion of the indenter. By analyzing the response of the system with a frequency-specific amplifier, data are obtained. This allows the measurement of contact stiffness at any point along the loading curve and not just at the point of unloading, as in conventional measurement. The CSM technique makes the continuous measurement of mechanical properties of materials possible in one sample experiment without the need for discrete unloading cycles, and with a time constant that is at least three orders of magnitude smaller than the time constant of the more conventional method of determining stiffness from the slope of an unloading curve. The measurements can be made at exceedingly small penetration depths. Thus this technique is ideal for mechanical property measurements of nanometer-thick films. Furthermore its small time constant makes it especially useful for measuring the properties of polymeric materials. In nonuniform materials, such as graded materials and multilayers, the microstructure and mechanical properties change with indentation depth. Continuous measurements of the mechanical properties of these materials during indentation are greatly needed. Utilizing the CSM technique, creep measurements on the nanoscale can be performed by monitoring changes in displacement and stress relaxation. Because CSM is performed at frequencies greater than 40 Hz, it is less sensitive to thermal drift. Load cycles of a sinusoidal shape at high frequencies allow the performance of fatigue tests at the nanoscale.

20.3.3 Atomic Force Microscope (AFM)

An atomic force microscope (AFM) senses interatomic forces that occur between a probe tip and a substrate. The AFM probe tip, which is similar in purpose to the stylus on a phonograph, is normally integrated into a microfabricated, thin-film cantilever. Once contact between the probe tip and the sample surface has been established, the sample is translated laterally relative to the probe tip, while the vertical position of the cantilever is monitored. Variations in sample height cause the cantilever to deflect up or down, which changes the position sensor output. This generates the error signal that the feedback circuit uses to maintain a constant cantilever deflection (constant force). Normal imaging forces are in the 1- to 50-nN range, and cantilever deflections of less than 0.1 nm can be detected.

Contact is normally considered a repulsive interaction between two surfaces, where there is pushing on the surface followed by pushing back. This occurs only when the distance between the surfaces is very small. Immediately outside the range at which repulsive interactions dominate, the force between the two surfaces is actually attractive, due to van der Waal's interactions. The AFM can sense both types of force. Repulsive forces are considered positive, and attractive forces are considered negative. These interactions are often described by use of the Lennard–Jones potential, which gives the potential energy of two atoms separated by a distance *r*.

A laser beam is trained on the back surface of the cantilever, and the reflected beam is sent to a photodiode that is divided into two sections, A and B, as shown in Fig. 20.20. Due to the macroscopic length of the reflected light path, any deflection of the cantilever causes a magnified lateral displacement of the reflected laser spot on the photodiode. The relative amplitudes of the signals from the two segments of the photodiode change in response to the motion of the spot-the-difference signal. The photodiode is very sensitive to cantilever deflection; detection of deflections of <0.1 nm is readily achieved. Optical lever detection is currently used in all but a few commercial AFM instruments. Compared to the scanning electron microscope, AFM provides extraordinary topographic contrast in direct height measurements and clear views of surface features with no coating necessary.

A new era in imaging began when scientists introduced a system for use of the noncontact mode. This mode is used in circumstances where tip contact would modify the sample in slight ways. In this mode, the tip hovers 50 to 150 Å above

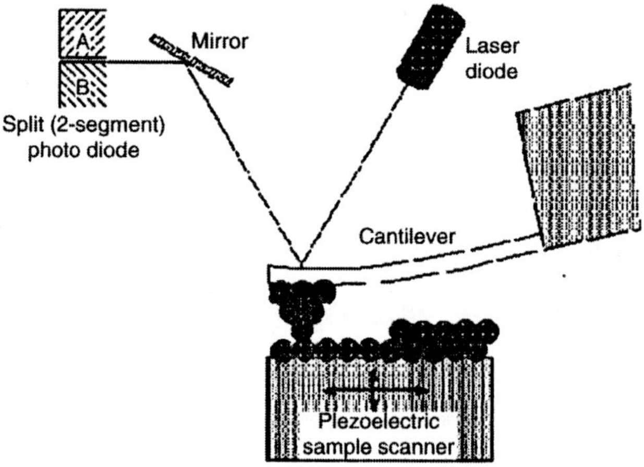

FIGURE 20.20 Schematic of an atomic force microscope.

the sample surface. Attractive van der Waal's forces between the tip and the sample are detected, and topographic images are created by scanning the tip above the surface. Unfortunately the attractive forces from the sample are considerably weaker than the forces used in the contact mode. Therefore the tip must be slightly oscillated so that AC detection methods can be used to sense the small forces between the tip and the sample by measuring the variation in amplitude, phase or frequency of the oscillating cantilever in response to force gradients from the sample. For highest resolution, it is necessary to measure force gradients from van der Waal's forces that may extend only a nanometer from the sample surface. Examples of AFM screens are displayed in Fig. 20.21 and Fig. 20.22. Figure 20.22 is a typical indent trace that would be

FIGURE 20.21 AFM scanned indent.

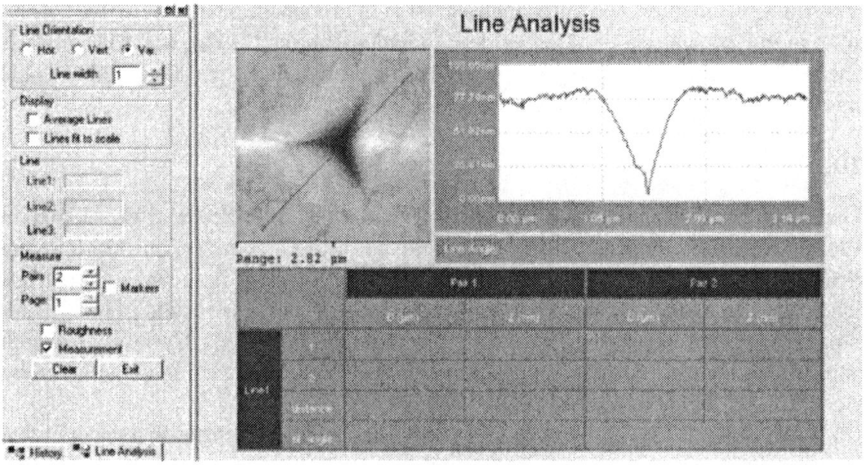

(a)

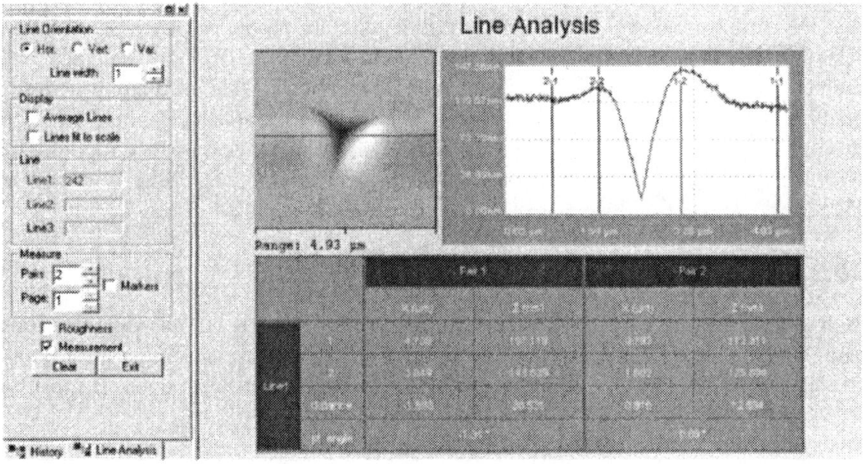

(b)

FIGURE 20.22 (a) Screen image from an indent displaying no significant pile-up; (b) screen image from an indent displaying significant pile-up.

taken to image analysis to measure the depth of the indent and the height of the pile-up, if any occurs. This figure displays two different screens: one demonstrating pile-up with the appropriate measurements and the other demonstrating no pile-up.

20.3.4 X-Ray Diffraction (XRD)

X-ray diffraction (XRD) was performed on the PLD-deposited thin films to determine their crystallinity and crystal structure orientation. A sample from several of the deposition processes was investigated. A Rigaku XRD double crystal diffractometer with Cu Kα_1 radiation and a symmetrical Si (111) monochromator were used, operating in the 2Θ scan mode. The incident radiation source (Cu Kα_1 = 0.154 nm) was operated at 30 kV and 20 mA. The Cu Kβ reflection of sapphire (0006) and Si (111) was used for calibrating the lattice constant for each substrate.

20.3.5 Scanning Electron Microscope (SEM)

The scanning electron microscope (SEM) was first described and developed in 1942.[62] Electron microscopy benefits from the wave nature of quickly moving electrons. Visible light has wavelengths from 4000 to 7000 Å; however, electrons accelerated to 10,000 KeV have a wavelength of 0.12 Å. This results in the greater capabilities of the SEM as compared to conventional microscopy. Currently research improvements in SEMs are reducing wavelength capabilities even more and thereby increasing magnification possibilities.

The scanning electron microscope produces a beam of electrons in a vacuum that is collimated by electromagnetic condenser lenses, focused by an objective lens, and scanned transversely on the surface of the sample by electromagnetic deflection coils. The chief imaging technique is accomplished by collecting the secondary electrons that the sample releases. The secondary electrons are monitored by a scintillation material that generates flashes of light from the electrons that are detected and enlarged by a photomultiplier tube. The sample position is then used along with the resulting signal to form an image that is analogous to that which would be seen through an optical microscope. The shadowed image displays a surface topography with a natural appearance. A schematic of the workings of the SEM is shown in Fig. 20.23 and the SEM used for these experiments is depicted in Fig. 20.24.

20.3.6 Transmission Electron Microscope (TEM)

A transmission electron microscope examines a structure by passing electrons through a specimen. An image is formed as a shadow of the specimen on a phosphorescent screen. In order for electrons to pass through the specimen, it must be

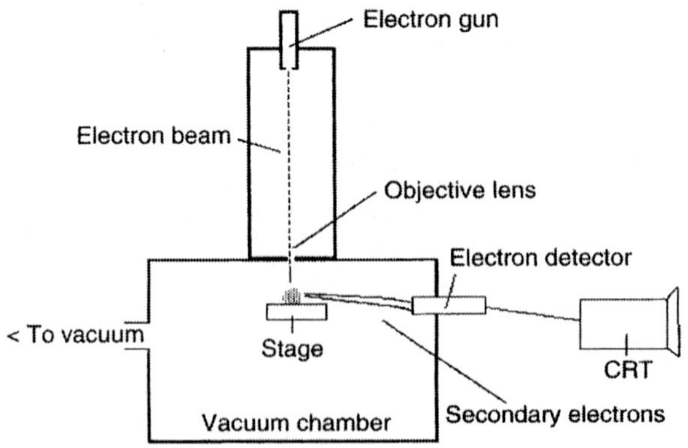

FIGURE 20.23 Schematic diagram of an SEM.

FIGURE 20.24 JEOL S-5000 SEM system.

very thin (usually less than 100 nm or approx. 1/25,000 in.). This requires considerable preparation, equipment, time, and skill. TEM is typically used to examine the internal structures of objects as varied as biological specimens, rocks, metals, and ceramics.

The transmitted beam is projected onto the viewing screen, forming an enlarged image of the slide. The image strikes the phosphor image screen and light is generated, allowing the user to see the image. The darker areas of the image represent those areas of the sample that fewer electrons are transmitted through (they are thicker or denser). The lighter areas of the image represent those areas of the sample that more electrons are transmitted through (they are thinner or less dense). The JEOL 2010 displayed in Fig. 20.25 is a high-resolution TEM with a spatial resolution of 0.194

FIGURE 20.25 JEOL 2010-F TEM.

nm. For the TEM investigations, the specimens were prepared using the conventional sandwich technique.

20.4 PERFORMANCE EVALUATION OF THIN FILMS

20.4.1 APPROACH

The research included depositions characterization, mechanical testing of the films, and finally analysis of the results. The procedure was completed in the following steps:

1. AlN monolayer film deposition and optimization
2. TiN monolayer film deposition
3. AlN/TiN heterostructure film deposition
4. Film characterization
5. Hardness and modulus property measurements and finite element work
6. XRD and TEM analysis

All depositions were performed on silicon (100) substrates. Before a silicon substrate was mounted to the heater, each silicon piece was ultrasonically cleaned in acetone for 15 min and then given an acetone vapor bath. Finally, immediately prior to the mounting of the samples onto the heater with a thin layer of silver conductive paste, the silicon pieces were soaked for 10 min in a weak HF etch to remove the SiO_2 layer. The Lambda Physik LPX300 PLD system used to create all the films is displayed in Fig. 20.26.

The constant laser parameters used for this experimentation are noted in Table 20.4. These parameters resulted in a thin-film growth rate of approximately 1 Å/s. The targets were also cleaned and polished before each deposition to ensure that the track created by the laser did not interfere with the plume and subsequently the

FIGURE 20.26 Lambda Physik LPX300 PLD system.

TABLE 20.4
Constant Excimer Laser Parameters Used for Thin-Film PLD of All Films Deposited

Parameter	Value
Wavelength	248 nm
Pulse repetition rate	10 pulses per second
Pulse duration	30 nsec

deposition rate. Results of the target analysis are provided in Section 20.5. The pulse repetition rate was set at the standard 10 Hz.

20.4.2 AlN Monolayer Film Deposition and Optimization

The laser energy was varied from 300 mJ to 750 mJ. When the gas for the laser was newly filled, it had the capacity to reach higher power values, but over time the laser energy capacity diminished. The first tests changed the laser energy, while holding constant the substrate temperature, the ambient gas pressure, and the target material. The aperture through which laser beam passed was 12 mm × 6 mm. The AlN target was mounted, and the vacuum was started. After a vacuum of 5 to 8×10^{-6} torr was achieved, the laser was turned on. After a warm-up period, pre-ablation of 500 pulses was run. The shield over the sample was then removed and 20,000 pulses at 10 Hz were performed. This test was similarly repeated at 300, 400, 500, 600, 650, and 700 mJ. The thickness of the resultant film was measured by the Wyko profileometer and the results are discussed in Section 20.5.

This process was repeated in a similar fashion with the laser energy constant and the ambient gas pressure and the substrate temperature varied. All the results are presented in Section 20.5.

20.4.3 TiN Monolayer Film Deposition

The TiN monolayer films were deposited at optimal process conditions with changes in only the substrate temperature. Tests were performed in the fashion previously described: 20,000 pulses at 10 Hz were used to deposit the films at 200°C, 500°C, 600°C, 700°C, and 800°C. The results of the tests are covered in Section 20.5. A value of 20,000 pulses was selected as the standard total.

20.4.4 AlN/TiN Heterostructure Deposition

The depositions of AlN/TiN were used to create multilayers whose periods varied from 2.4 nm to 140 nm. A constant film thickness of ~250 nm for all depositions was the goal. The samples created are discussed in Section 20.5. The total number of pulses was held constant at 20,000. The hardness and modulus were found via the CSM method on an MTS nanoindenter as discussed in Section 20.3.

20.4.4.1 Hardness and Modulus versus Deposition Temperature

The temperature was varied from 200°C to 800°C for the different multilayered films. The sample deposition parameters and properties are explained in Section 20.5. An annealing study was performed on a multilayered, low-temperature (200°C) deposited sample. The sample was hardness tested. It was then annealed at 500°C for 2 h and the hardness was retested. The sample was then annealed at 800°C for 2 h and the hardness was tested again. The hardness plots are displayed in Section 20.5.

20.4.4.2 Hardness versus Depth of Indentation

Hardness tests were performed at various different indentation loads, therefore creating different indentation depths. The purpose of this study was to use the Nix and Korsunsky models and determine their appropriateness for the AlN/TiN material system and the substrate effect. A TiN film was analyzed in greater detail to discern the correct model for describing the behavior. Further studies were performed to determine the constants needed for the Nix–Bhattacharya and Tuck–Korsunsky models discussed in Section 10.3. These models were then compared for their compatibility with the current data. The results are discussed in Section 10.5.

20.4.4.3 Hardness versus Layer Characteristics

The hardnesses of the various multilayered samples were compared in relation to their period thickness λ. The results are presented in Section 20.5.

20.4.4.4 Indentation Pile-Up Considerations and FEM

AFM imaging of the indents was performed to determine the extent and nature of the pile-up found around the indents. The line profiles of these images were determined and analyzed to discover the nature of the pile-up and assess the need for explanation of the pile-up and the substrate effect. Images and line scans are displayed in Section 20.5. The material parameters were used in an LS DANA finite-element program to model the indent plastic behavior. The results are presented in Section 20.5.

20.4.5 X-Ray Diffraction

X-ray diffraction (XRD) was performed on the pulsed laser-deposited thin films to determine their crystallinity and crystal structure orientation. A few samples were tested to ascertain the quality of the deposition and the effect the deposition temperature had on the crystallinity. A Rigaku XRD double-crystal diffractometer with Cu Kα_1 radiation and a symmetrical Si (111) monochromator was used, operating in the 2Θ scan mode. The incident radiation source (Cu Kα_1 = 0.154 nm) was operated at 30 kV and 20 mA. The Cu Kβ reflection of silicon (100) was used for calibrating the lattice constant for the substrate. Results are displayed in Section 20.5.

20.4.6 Transmission Electron Microscopy (TEM)

Samples were prepared according to the detailed specifications in Chapter 3. Sample B33, which was an AlN/TiN ten-layer film deposited at 500°C, was analyzed in the JEOL 2010-F TEM under different conditions. Results are presented in Section 20.5.

20.5 OPTIMIZATION OF RESULTS

20.5.1 Laser Energy Effect Results

Laser energy studies were performed and the results are shown in Fig. 20.27. The temperature (500°C) and the ambient gas pressure (15 mtorr) were held constant for each laser energy level. At each deposition condition, the quality of the resultant film was examined and the film thickness was quantified as a means of quality analysis. Notice the peak in film thickness at an energy range of approximately 600 mJ. This is the energy chosen for the remainder of the tests.

20.5.1.1 Ambient Gas Pressure Effects

When nitrides are deposited, the process should occur in a nitrogen-rich environment. However, as shown in Fig. 20.5, there is an interaction of the plume with the ambient gas. These collisions with the inert gas change the kinetic energy distribution of the deposited particles. The data from this work verify the models previously introduced. The ambient gas pressures used were 0, 15, 50, 135, and 255 mtorr of N_2. Laser energy was constant at 400 mJ, temperature was constant at 500°C, and the number of pulses was also constant at 20,000. As shown in Fig. 20.28, hardness tends to decrease as the internal gas pressure and, in turn, the number of plume–gas collisions increase. Samples had a constant temperature, 500°C, constant laser energy 400 mJ,

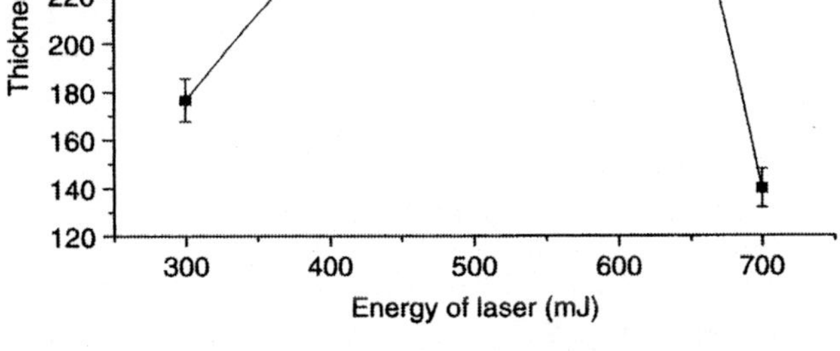

FIGURE 20.27 Graph of thickness of the AlN film versus laser energy.

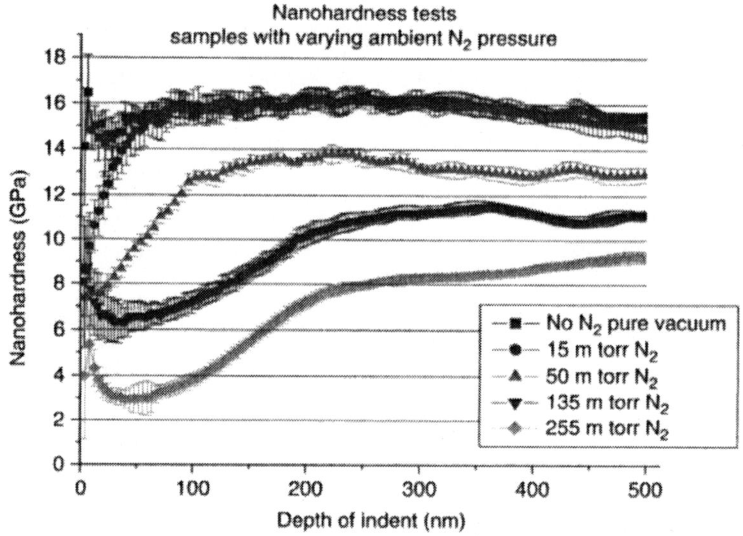

FIGURE 20.28 Nanohardness of AlN monolayer films at different nitrogen pressures.

and constant number of pulses. The chosen ambient gas pressure was 15 mtorr, which diminished the effect of the gas collisions.

20.5.1.2 Location of Deposition on Heater

The size and shape of the plume have been discussed. A study of the thickness along the entire width of the heater was performed. The results of step thickness measurements across the 2-in. width of the heater are displayed in Fig. 20.29. Each data point is an average of four step measurements. The thickness steps were measured across the sample shown in the smaller box with the start and end locations shown by the arrows. The results verify the work of other researchers who have achieved a 1-cm optimum deposition area.

20.5.2 AlN Monolayer Deposition and Properties

AlN films were deposited at 200°C, 500°C, and 800°C for this investigation. Initial hardness tests were run on the monolayer films of the AlN. Hardness and modulus of samples were determined by means of depth-sensing nanoindentation using the continuous stiffness measurement methods. An example of an indent from the AlN indentation is shown in Fig. 20.30. This film was deposited at 500°C with 15 mtorr ambient N_2 and on a Si (100) substrate.

The hardness (~16 GPa) and the modulus (190 GPa) of the AlN monolayer film were below the values published from earlier work (Rawdanowicz). The film is very thin; therefore the 10% rule for indentation testing is violated. The substrate affects the hardness to a great degree. Figure 20.31 displays the hardness and elastic modulus of the monolayer AlN film. The thickness is ~250 nm and is deposited on Si (100) at 500°C.

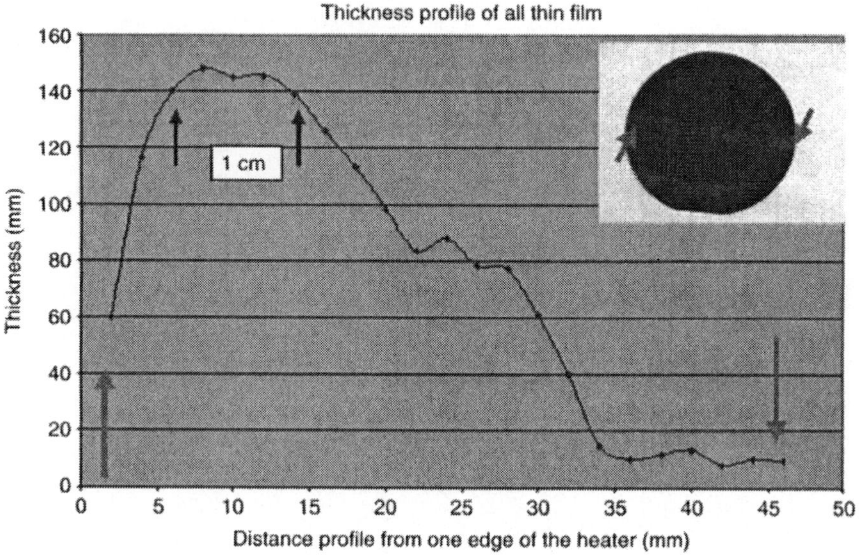

FIGURE 20.29 Thickness measurements across the face of the heater.

20.5.3 TiN Monolayer Deposition and Properties

TiN film samples were deposited at 200°C, 500°C, 700°C, and 800°C with 20,000 pulses. Initial hardness tests were run on the monolayer films of the AlN. The hardness and modulus of the samples were determined by means of depth-sensing

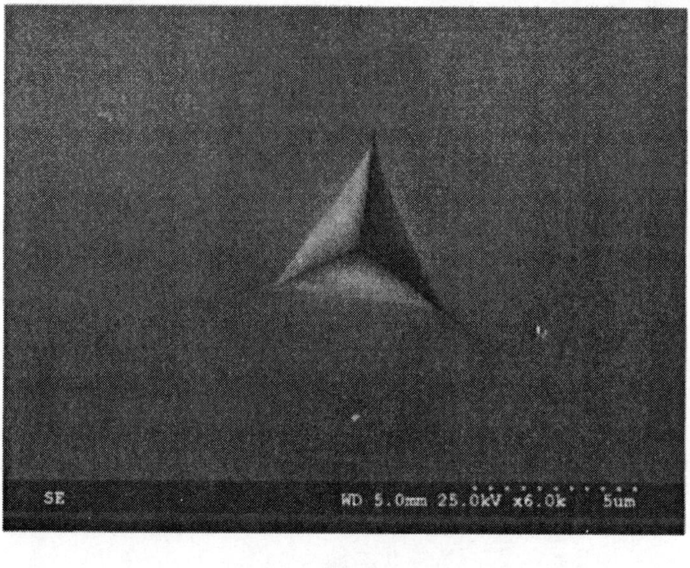

FIGURE 20.30 SEM micrograph of an indent on AlN monolayer film.

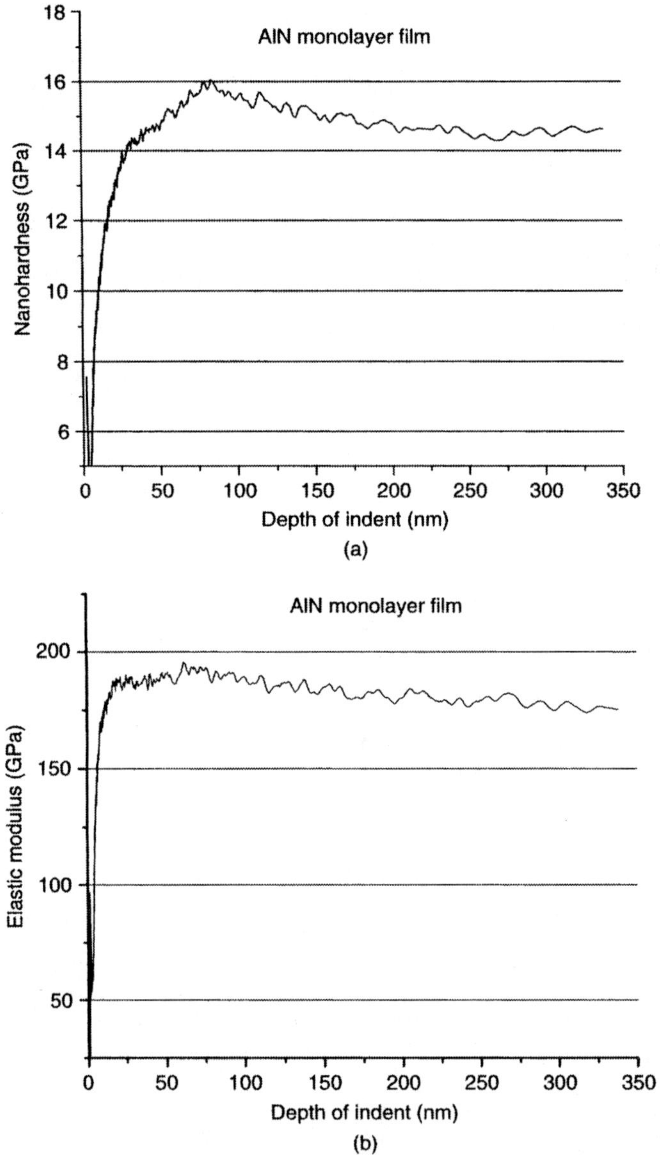

FIGURE 20.31 (a) Nanohardness and (b) elastic modulus results of monolayer AlN film.

nanoindentation using continuous stiffness measurement methods. An example of the indent from the TiN indentation is shown in Fig. 20.32. This film was deposited at 500°C.

The shapes of the load displacement curves for two representative TiN samples are displayed in Fig. 20.33. The solid line was deposited at 200°C, and the dashed

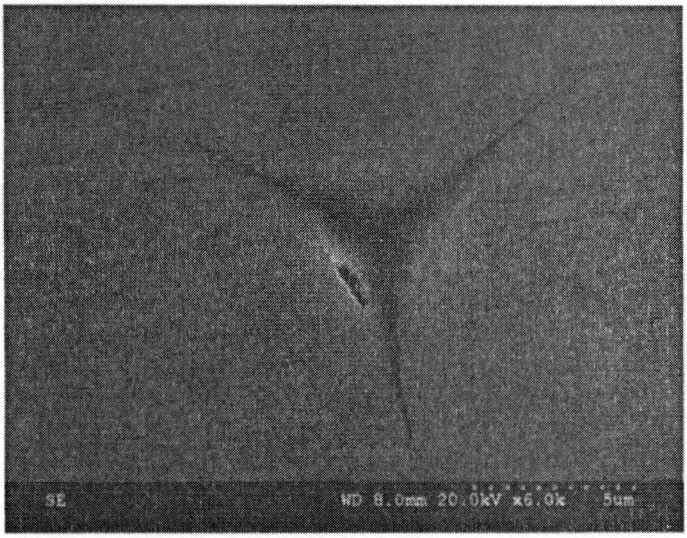

FIGURE 20.32 SEM micrograph of an indent on a TiN film.

line was deposited at 800°C. The shape of the unloading curve was examined in detail to verify the application of the power law relationship given in Equation 20.3. The initial portion of the unloading curve from Fig. 20.33 is expanded and plotted in Fig. 20.34. The unloading segment is dominated by elastic behavior. The displacements were shifted laterally by subtracting the final depth, h_f, for each data point by the total depth, h, to force all of the curves to pass through the origin. The

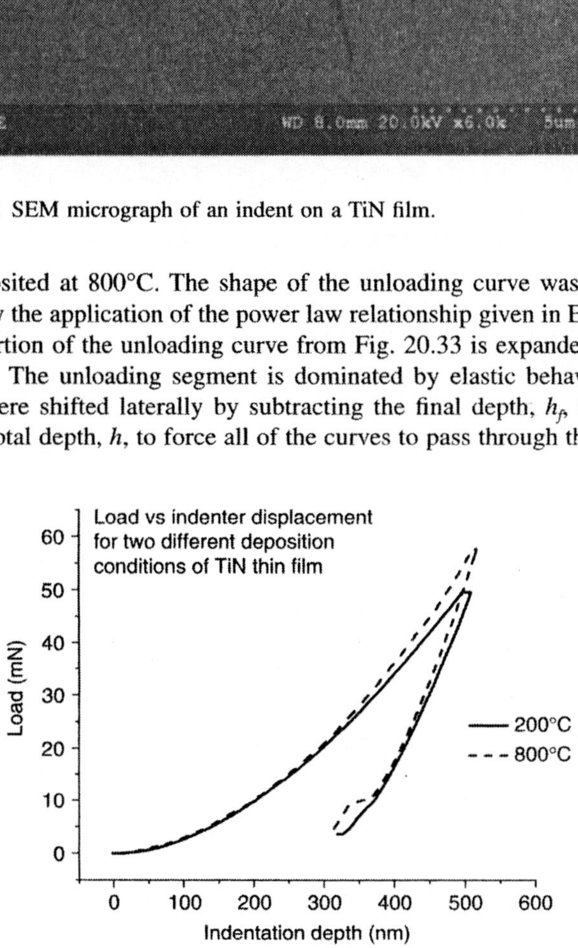

FIGURE 20.33 Load versus indentation depth plots for TiN thin films deposited at two different temperatures.

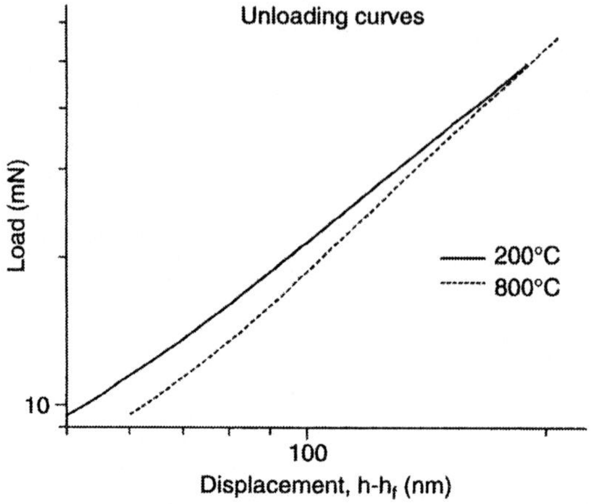

FIGURE 20.34 Data from Figure 20.33, adjusted for unloading, plotted on a log scale.

fact that the data is linear on logarithmic axes implies that the unloading curves are appropriately described by a power law relation.

The jog in the unloading portion of Fig. 20.33 is evidence of a large crack formed during unloading. When the load on the indent was removed, the stresses were relaxed, and the crack appeared as a discontinuity in the load displacement curve. The shallower slope of the 200°C sample displayed in Fig. 20.34 verifies its soft behavior. The resulting variation of the hardness and elastic modulus as a function of indentation depth is displayed in Fig. 20.35. This film was deposited at 500°C and indented with the Berkovich pyramid indenter. The primary influence of the Berkovich indenter tip with radius of approximately 60 nm is the rapid increase in the hardnesses and elastic moduli that occurs in the first 40 nm to 60 nm of the indentation. The indenter, in most cases, broke through the films and contacted the substrate. The thickness of the thin films located directly under the plume was greater than that in the area located at the plume's edge. Therefore exact thicknesses of the pulsed laser-deposited thin films were unknown. The films were too thin to restrict the indentation depth to approximately 10% of the film thickness. The TiN displayed hardness comparable to the hardnesses in Table 20.3, but the modulus was below what was expected. The substrate effect is the likely cause of the lower modulus because the film is thin and the plastic zone from the indenter would almost immediately extend into the substrate. This affects the elastic modulus to a degree.

20.5.4 AlN/TiN Multilayered Film Properties

Nanoindentation tests were performed on the multilayered films using an MTS Nanoindenter® XP by the process explained in previous chapters.

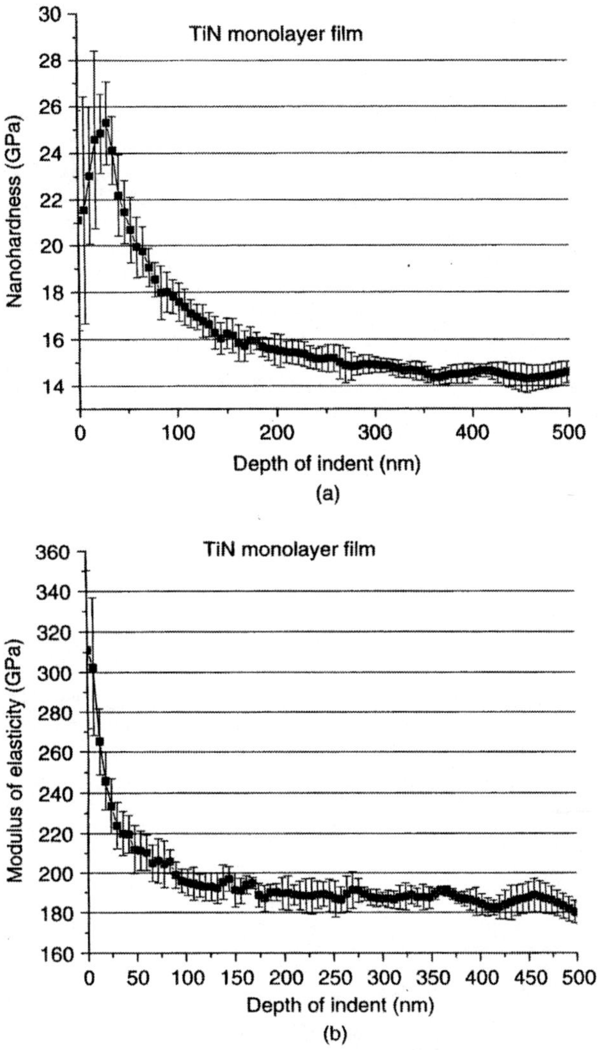

FIGURE 20.35 (a) Nanohardness and (b) elastic modulus plots for TiN monolayer film.

20.5.4.1 Hardness and Modulus versus Deposition Temperature

As the temperature of the deposition increased, there was a difference in the film characteristics. This difference was first noted in the values of nanohardness and subsequently in the SEM images and cracking pattern of indents studied during fracture mechanics analysis. The temperature of the deposition is known to affect the atomic behavior of the film (i.e., whether the films are crystalline or amorphous). If the film is crystalline, then the grain size is also affected by the deposition

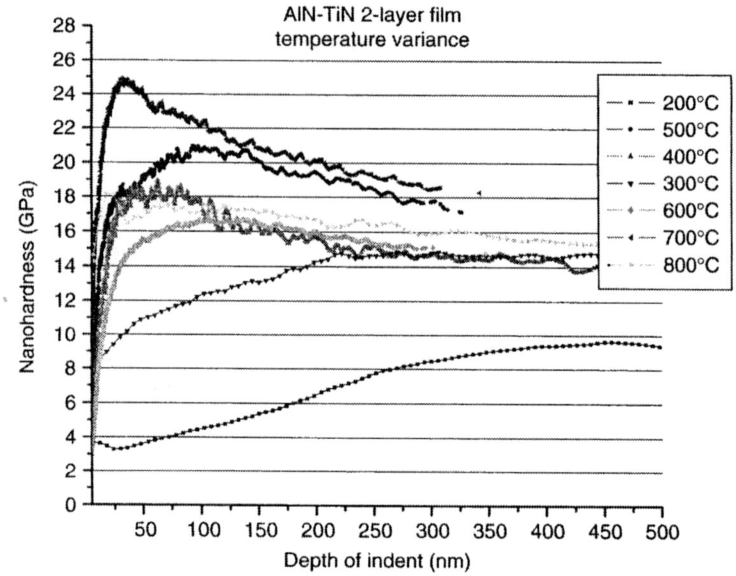

FIGURE 20.36 Nanohardness plots for bilayer AlN/TiN films.

temperature. The higher the temperature of the deposition, the larger the grains will be, because at higher temperatures the atoms have greater energy to move and rearrange once on the target. As seen in Fig. 20.36 through Fig. 20.39, the highest deposition temperature does not always produce the hardest film. The highly energetic

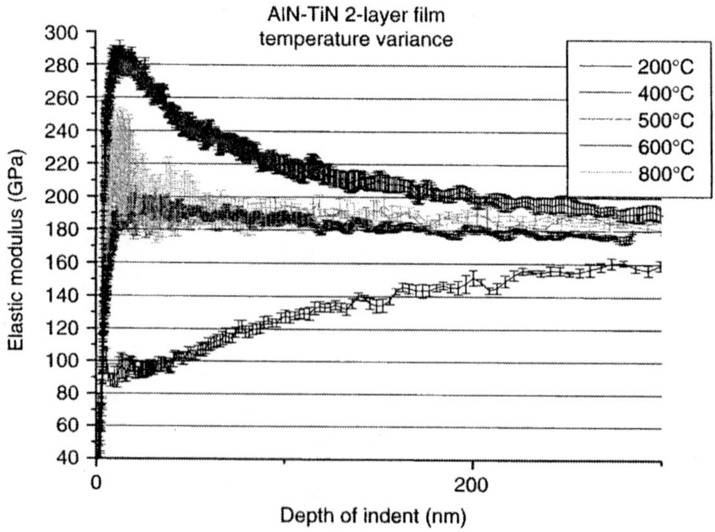

FIGURE 20.37 Elastic modulus plots for bilayer AlN/TiN films.

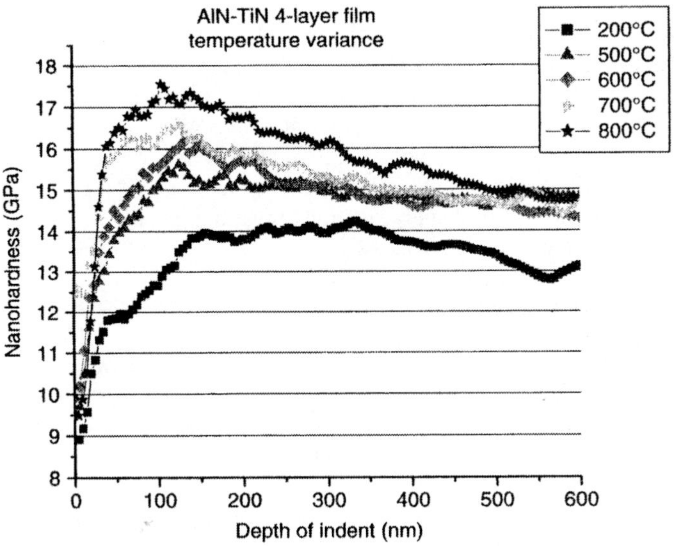

FIGURE 20.38 Nanohardness plots for four-layered AlN/TiN films.

nature of the laser deposition also leads to unique properties. The paramount temperature for optimal-quality film growth occurs when there is adequate surface diffusion to permit surface atoms to minimize their surface energy, and therefore become thermodynamically stable.[43] The current tests demonstrate that, for tests

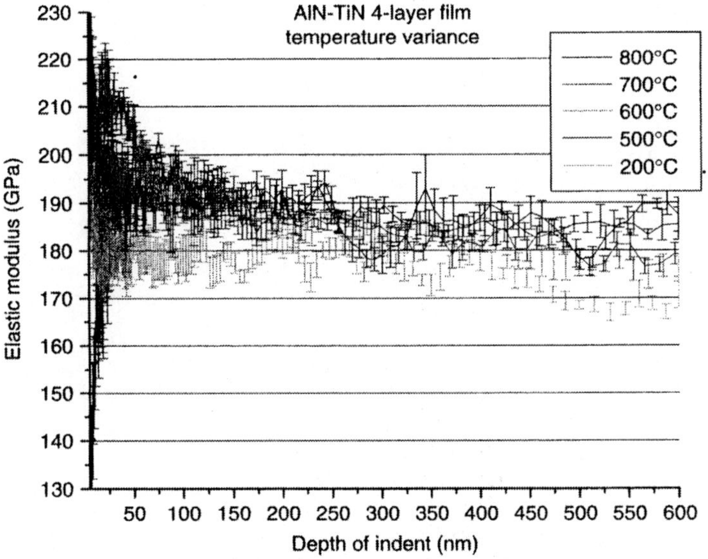

FIGURE 20.39 Elastic modulus plots for four-layered AlN/TiN films.

containing TiN as the outer layer, a deposition temperature between 700°C and 800°C produces the hardest film. Further XRD analysis, depicted in Fig. 20.45 through Fig. 20.48, shows that the higher temperature film was indeed of a crystalline nature and that the low temperature deposit was amorphous.

All of the hardness plots demonstrated a convergence to the Si (100) substrate hardness (12.75 GPa) or modulus (180 GPa). The increase in properties was not a linear increase but all curves displayed a trend toward greater hardness and modulus as the deposition temperature increased.

20.5.4.2 Hardness versus Layer Characteristics

The hardnesses of various samples were compared to their respective period thicknesses λ and plotted in Fig. 20.40. Prior research has shown a trend toward a harder film as the period decreases.[4] As stated in Chapter 1, the properties of multilayer coatings vary as a function of λ. Bull found the hardness of the structural multilayers was less than those of comparable single-layer films at low λ. However, it increased up to a maximum at $\lambda \sim 10$ nm. The hardness of very thick structural multilayers was indistinguishable from the average value of single-layer coating hardness.[4] Prior research has shown that the TiN layers in the TiN/AlN superlattice manipulate the structures of the AlN layers when TiN/AlN films are created as a superlattice. At a superlattice period of 2.5 nm, the AlN layers take on the same structure as the TiN (i.e., the NaCl structure[7,8]). AlN forms in a hexagonal structure during equilibrium, but the TiN acts in this circumstance as a model and forces AlN into the NaCl structure.

The graph in Fig. 20.40 was created after testing many films created by the PLD process. The data demonstrate that multilayering appears to only slightly affect

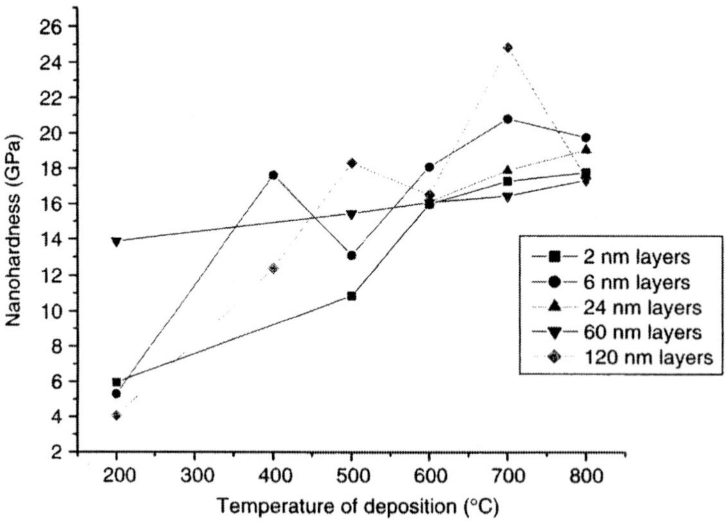

FIGURE 20.40 Nanohardness of films deposited at different conditions.

hardness, while it significantly delays the occurrence of indentation-induced large cracks. This can be deduced from the higher values of the fracture toughness. The smallest λ, the 2-nm curve, is not the hardest material. This result is consistent with Bull's findings.

20.5.4.3 Indentation Pile-Up Considerations and FEM

AFM imaging of the indents was used to determine the extent and nature of the pile-up around the indents. In Fig. 20.41, pile-up in the AlN/TiN film is evident, and the line scan of the indent is displayed. This was an AlN/TiN ten-layer film deposited at 700°C, with a 150-mN indent load. Figure 20.42 displays a representative image without significant pile-up deposited at 800°C, 30-mN indent load.

The true contact pressure or hardness of a film can be determined by accounting for the pile-up effects on the contact area. This can be accomplished by adopting the method of Joslin and Oliver (i.e., using the P/S^2 parameter). The P/S^2 parameter can be used only when the material is elastically homogeneous and the indentation modulus is known. It is necessary to choose a film–substrate combination that is elastically homogeneous for a film on a substrate,

$$\frac{P}{S^2} = \frac{1}{\beta^2}\left(\frac{\pi}{4}\right)\frac{H}{E_r^2} \tag{20.17}$$

The true contact pressure or hardness may be determined by measuring the load and contact stiffness, even when pile-up occurs. Joslin and Oliver were the first to use the P/S^2 parameter in a study of the mechanical properties of ion-implanted Ti alloys. They suggested that the P/S^2 parameter is a direct measure of true hardness

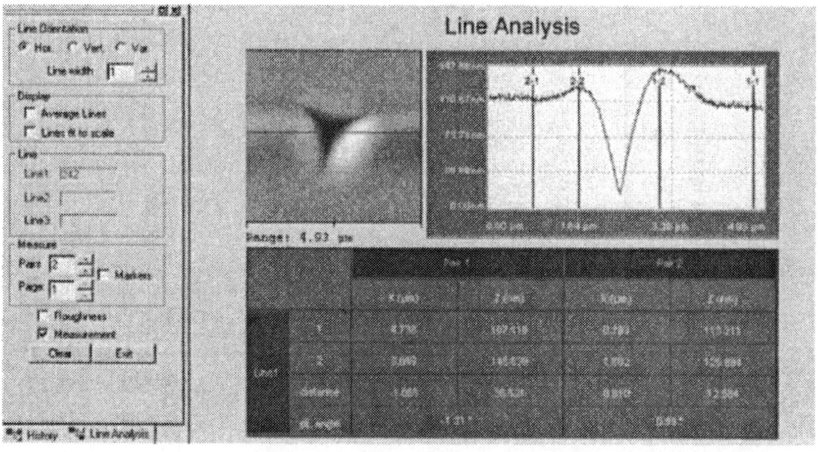

FIGURE 20.41 AFM trace showing pile-up height and indent depth.

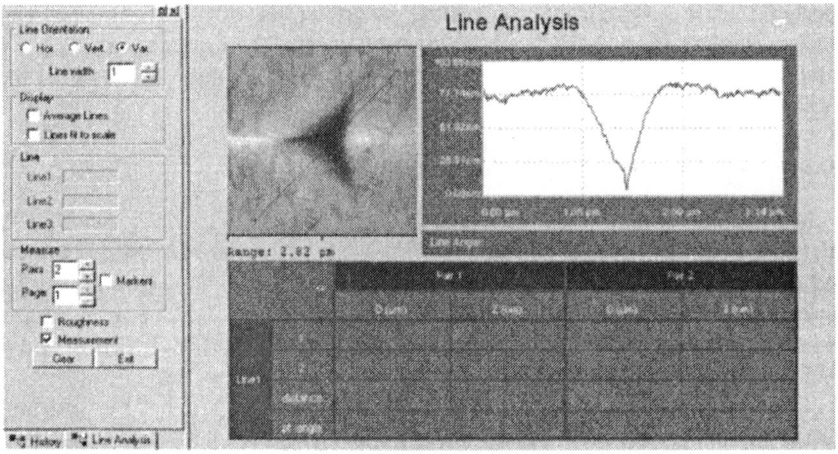

FIGURE 20.42 AFM trace showing no significant pile-up. AlN/TiN ten-layer film.

since it is proportional to *H*. This parameter is most accurate when the film acts as a part of a homogeneous system with the substrate. While not the most accurate, information is also gleaned from nonhomogeneous systems. These calculations will be applied to the current data in the future.

Figure 20.43 shows the pile-up area measured by AFM. The area of pile-up was found mathematically using an integral program from Origin Software. The area was found to increase with an increasing penetration load as similarly found by

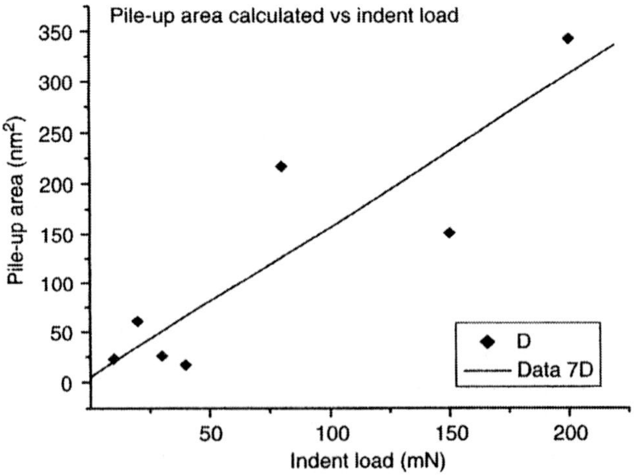

FIGURE 20.43 Pile-up area calculated from line profile of an AFM image.

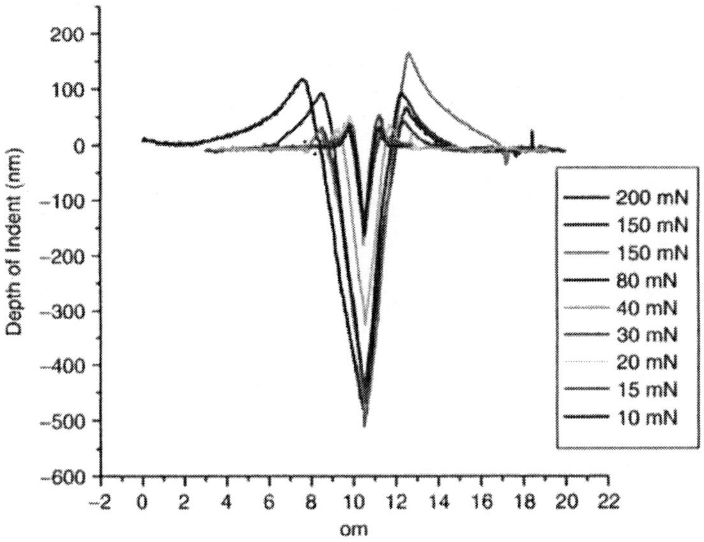

FIGURE 20.44 Graph of pile-up height versus indent load.

Chen and Beegan.[34,63] Figure 20.44 demonstrates the nature of the pile-up for increasing values of the indent load and is a direct verification of 2-D finite element modeling by Ramond-Angélélis, as shown in Fig. 20.21.

20.5.5 X-Ray Diffraction (XRD) Data

First, comparisons were made between two TiN samples deposited at different temperatures. In Fig. 20.45, the Bragg-Brentano 2Θ scans show that the TiN film grown on silicon (100) at 800°C is highly textured, as there are only (h00) peaks with $h = 2$ and 4. The smaller peak height of the Ti (400) line is in perfect accordance with the theoretical integrity (8%) of the (400) line with respect to a 100% peak of the (200) line. Figure 20.46, which displays a film deposited at 200°C, displays none of the crystalline peaks evident in Fig. 20.45. XRD data from the two figures verifies that the structure of Fig. 20.46 is a completely amorphous TiN film, whereas the film deposited at 800°C is crystalline. The film displayed in Fig. 20.47 is an AlN monolayer tested at Rigaku laboratories with the Rigaku Ultima III equipment. The peaks are similarly located as those in Fig. 20.48, which displays a multilayered AlN/TiN film with a layer thickness, λ~20 nm. The peaks at ~33° and 37° are consistent with values published for hexagonal AlN (JCPDS).

The film density and roughness measurements from the Rigaku Ultima III XRD machine are displayed in Table 20.5. These measurements also calculate the film density as 3.235 g/cm^3, which is consistent with values in the literature for a

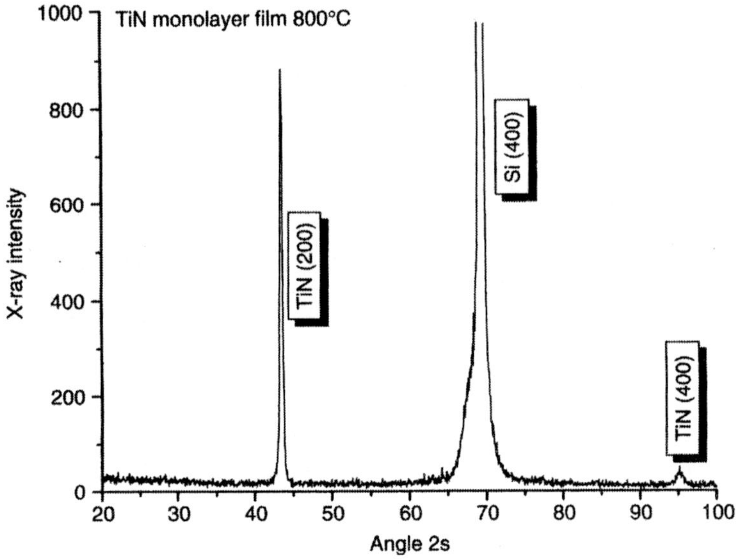

FIGURE 20.45 XRD analysis of TiN monolayer film deposited at 800°C on Si (100).

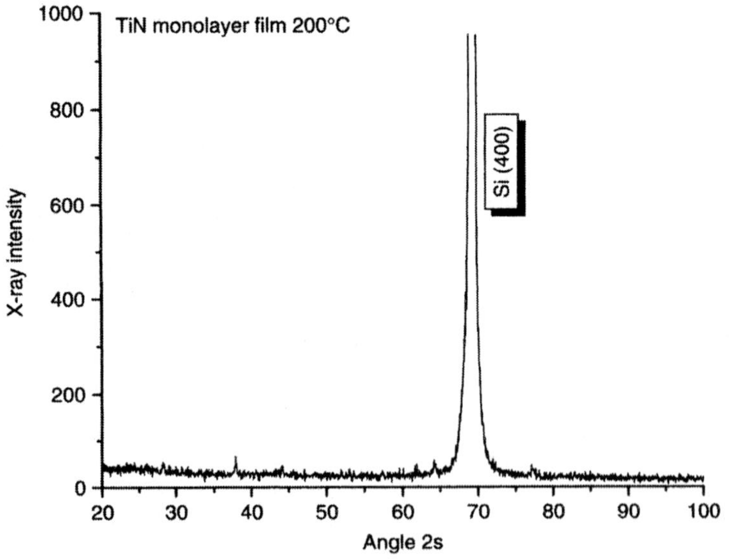

FIGURE 20.46 XRD analysis of TiN monolayer film deposited at 200°C on Si (100).

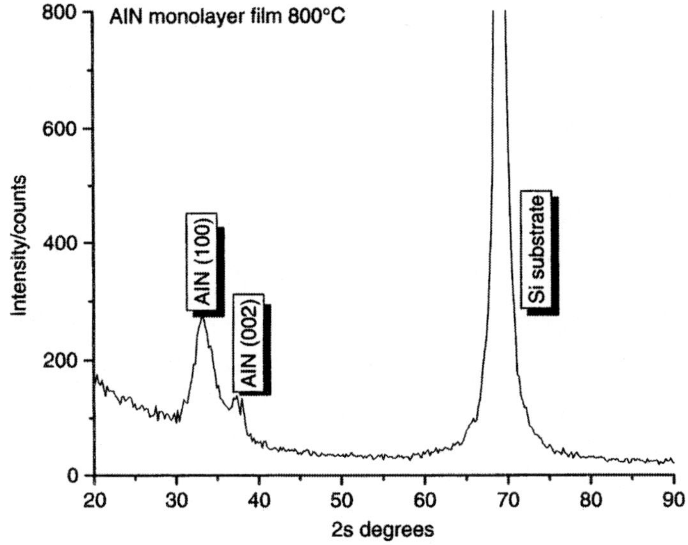

FIGURE 20.47 XRD analysis of AlN monolayer film deposited at 800°C on Si (100).

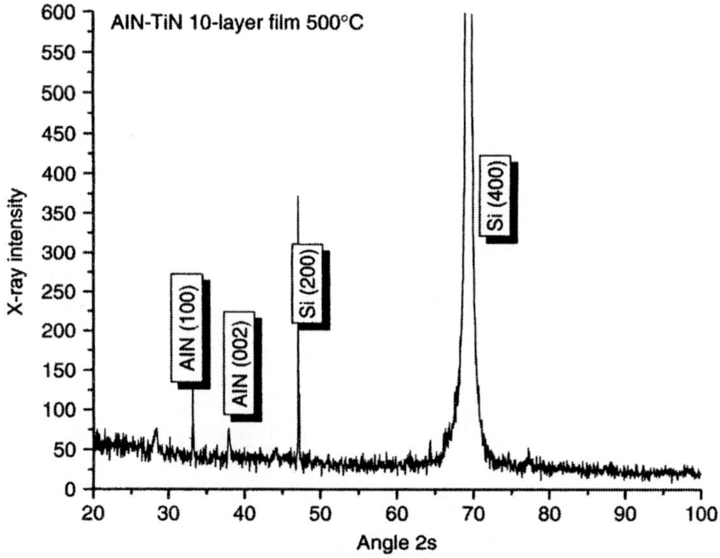

FIGURE 20.48 XRD analysis of AlN/TiN ten-layer film deposited at 500°C on Si (100).

TABLE 20.5
Reflectivity Analysis Results

Layer	Density (g/cm³)	Thickness (nm)	Roughness (nm)
AlN oxide	2.286	2.24	0.621
AlN	3.235	235	0.154

hexagonal AlN structure specified in Chapter 1. The film was deposited at 500°C, so strong crystalline peaks were not expected and in fact are not evident.

20.5.6 TEM Data

Figure 20.49 displays the results of the high-resolution STEM-Z contrast of the cross-section. This image exhibits the substrate–film interface along with the layer thickness. Notice the overall film structure and the relative layer thickness. The darker areas of the film are the TiN layers, which are narrower than the AlN layers. Figure 20.50 presents the cross-sectional STEM-Z micrograph at a greater magnification. The crystalline nature of the substrate and the amorphous AlN first layer is evident. The bright areas of the substrate are representative of the Si (100) orientation.

Figure 20.51 displays the film cros-section, with the darker area inhabited by the lighter elements. TiN, the heaviest element, emerges as the brighter layer. The first layer is AlN, but it is darker than the intermediate AlN layers. The TiN has likely diffused into the other AlN layers and they therefore appear lighter than the first layer. The first layer grows epitaxially on the Si substrate and is therefore more stable.

FIGURE 20.49 High-resolution STEM-Z contrast image of AlN/TiN ten-layer film deposited at 500°C.

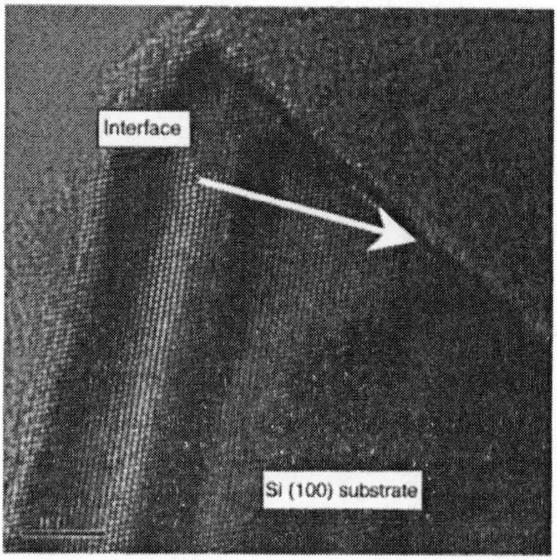

FIGURE 20.50 Cross-sectional TEM image of Sample B33 exhibiting crystalline Si (100) substrate adjacent to the mainly amorphous AlN first layer.

20.6 CONCLUSIONS

The PLD process was optimized for AlN/TiN thin films. The optimum laser energy was discovered to be approximately 600 mJ, and the optimum ambient N_2 pressure was found to be 15 mtorr. Pulsed laser-deposited films can be created in this system

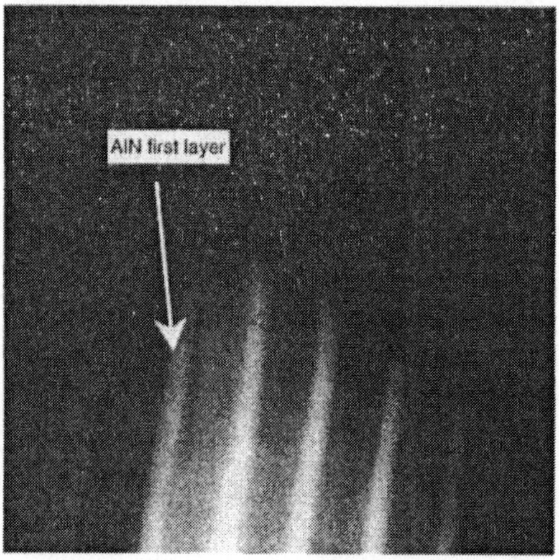

FIGURE 20.51 Cross-sectional STEM-Z micrograph of AlN/TiN layers deposited on an Si (100) substrate.

with maximum thickness in an area 1 cm^2 in size. The hardness values and fracture morphology were reported for these pulsed laser-deposited multilayer AlN/TiN films deposited on Si (100). The temperature of the substrate during the deposit was found to play an important role in the crystalline characteristics and mechanical properties. The hardness and elastic modulus values for pulsed laser-deposited multilayer AlN/TiN films deposited on Si (100) have been reported. The hardnesses for many 2.5- to 120-nm AlN/TiN bilayers deposited on silicon substrates concurred with recently reported results. However, these multilayer hardness values of 18 to 23 GPa were notably less than the typically expected rule-of-mixture values for AlN and TiN thin-film component values of 24 GPa and 35 GPa, respectively. No strong correlation between the individual layer thickness and the hardness was found. The data show that multilayering appears to slightly affect hardness but significantly delay the occurrence of indentation-induced large cracks. This can be deduced from the higher values of the fracture toughness.

When annealing was performed on the low temperature-deposited sample, the annealed hardness did not reach close to the hardness of the higher temperature-deposited sample. This was analyzed to mean that film quality is not just a result of temperature. Heat must be combined with the action of the highly energetic plume hitting the substrate surface. In other words, shortcuts during deposition cannot be easily addressed after deposition.

The continuous nanoindentation stiffness measurement method for measuring the hardness and elastic modulus provides sound hardness measurement results. This method is optimal for thin-film measurement. The substrate effect is well documented and evident in this body of work. The XRD analysis verified the crystal natures of the hardest films that were created at the higher temperature range 700 to 800°C. In order to create the hardest films, deposition must occur in the 700 to 800°C range. The AlN XRD data verified that the AlN was of a hexagonal structure with a density of 3.235 g/cm^3. Initial TEM work demonstrated the exact layer thickness, the lack of crystallinity at 500°C, and the possible diffusion of TiN into the AlN via the Z-contrast image.

PROBLEMS

1. Compare the mechanical properties of thin film and bulk of the same material system.
2. Describe the role of growth temperatures on the mechanical and the structural properties of AlN and TiN heterostructures.
3. Discuss the limitations of nanoindentation technique in the measurement of hardness and elastic moduli of thin films.

REFERENCES

1. Holleck, H. and Schulz, H., *Thin Solid Films, 153,* 11, 1987.
2. Holleck, H. and Schier, V., *Surf. Coat. Technol., 76–77,* 328, 1995.
3. Holleck, H., *J. Vac. Sci. Technol A, 4,* 6, 2661, 1986.

4. Bull, S.J. and Jones, A.M., *Surf. Coat. Technol., 78,* 173–184, 1996.
5. Helmersson, U., Todorova, S., Marbert, L., Barnett, S.A., Sundgren, J.E., and Geen, J.E., *J. Appl. Phys., 62,* 481, 1987.
6. Nakonechna, O., Cselle, T., Morstein, M., and Karimi, A., *Thin Solid Films,* 2003.
7. Sproul, W.D., *Surf. Coat. Technol, 81,* 1, 1996.
8. Elshabini-Riad, A., Barlow, F., Eds., *Thin Film Technology Handbook,* McGraw Hill, 1998.
9. Sundgren, J.-E., *Thin Solid. Films, 128,* 21, 1985.
10. Sundgren, J.-E. and Hentzell, H.T.G., *J. Vac. Sci. Technol.* A, 5, 2259, 1986.
11. Sundgren, J.E., Birch, J., Hakansson, G., Hultman, L., and Helmersson, U., *Thin Solid Films, 193–194,* 818, 1990.
12. Veprek, S. and Reiprich, S., *Thin Solid Films, 268,* 64, 1995.
13. Durand-Drouhin, O., Santana, A.E., Karimi, A., Derflinger, V.H., and Schutze, A., *Surf. Coat. Technol., 163–164,* 260–266, 2003.
14. Karimi, A., Wang, Y., Cselle, T., and Morstein, M., *Thin Solid Films, 420–421,* 275, 2002.
15. Koehler, J.S., *Phys. Rev. B, 2,* 547, 1970.
16. Yoshii, K., Takagi, H., Umeno, M., and Kawabe, H., *Metall. Trans. A, 15,* 1273, 1984.
17. Tsakalokos, T. and Hilliard, J.E., *J. Appl. Phys., 54,* 734, 1983.
18. Mirkarimi, P.B., Hultman, L., and Barnett, S.A., *Appl. Phys. Lett., 57,* 2654, 1990.
19. Baumvol, I.J.R., Stedile, F.C., Schreiner, W.H., Freire, Jr., F.L., and Schröer, A., *Surf. Coat. Technol., 59,* 187, 1993.
20. Godbole, V.P. and Narayan, J., *J. Mater. Res., 11,* 1810, 1996.
21. Bathe, R., Vispute, R.D., Habersat, D., Sharma, R.P., Venkatesan, T., Scozzie, C.J., Ervin, M., Geil, B.R., Lelis, A.J., Dikshit, S.J., and Bhattacharya, R., *Thin Solid Films, 398–399,* 575, 2001.
22. Talyansky, V., Vispute, R.D., Ramesh, R., Sharma, R.P., Venkatesan, T., Li, Y.X., Salamanca-Riba, L.G., Jones, K.A., and Iliadis, A.A., *Thin Solid Films, 323,* 37, 1998.
23. Pearton, S.J., *Wide Bandgap Semiconductors—Growth, Processing and Applications,* Noyes Corporation, 2003.
24. Manaila, R., Biro, D., Barna, P.B., Adamik, M., Zavaliche, F., Craciun, S., and Devenyi, A., *Appl. Surf. Sci., 91,* 295, 1995.
25. Toth, L.E., *Transition Metal Carbides and Nitrides,* Academic Press, New York, 1971.
26. McKenzie, D.R., Yin, Y., McFall, W.D., and Hoang, N.H., *J. Phys. Condens. Matter, 8,* 5883, 1996.
27. Patsalas, P. and Logothetidis, S., *J. Appl. Phys., 93,* 989, 2003.
28. Bae, Y.W., Lee, W.Y., and Blau, P.J., *Appl. Phys. Lett., 66,* 1895, 1995.
29. Bergman, E., Kaufmann, H., Schmid, R., and Vogel, J., *Surf. Coat. Technol., 42,* 237–251, 1990.
30. Munz., M.D., *J. Vac. Sci. Technol A, 4,* 2717, 1986.
31. Shu, N., Kumar, A., Alam, M.R., Chan, H.L., and You, Q., *Appl. Surf. Sci., 109,* 366, 1997.
32. Nose, M., Nagae, T., Yokota, M., Saji, S., Zhou, M., and Nakada, M., *Surf. Coat. Technol., 116–119,* 296, 1999.
33. Chou, W., Yu, G., and Huang, J., *Surf. Coat. Technol., 149,* 7, 2002.
34. Chen, Xi and Vlassak, J.J., *J. Mater. Res., 19,* 2974, 2001.
35. Saha, R. and Nix, W.O., *Mater. Sci. and Eng. A.,* 319–321, 898–901, 2001.
36. Auger, M.A., Sanchez, O., Ballesteros, C., Jergel, M., Aguilar-Frutis, M., and Falcony, C., *Thin Solid Films, 433,* 211, 2003.
37. Singh, R.K. and Narayan, J., *Phys. Rev. B, 41,* 8843, 1990.

38. Zhigilei, L.V., Kodali, P.B.S., and Garrison, B.J., *Chem. Phys. Lett., 276,* 269, 1997.
39. Fahler, S., Stormer, M., and Krebs, H.U., *Appl. Surf. Sci., 109–110,* 433, 1997.
40. Sharma, A.K., Thareja, R.K., Willer, U., and Schade, W., *Appl. Surf. Sci., 206,* 137, 2003.
41. Tabbal, M., Mérel, P., Chaker, M., El Khakani, M.A., Herbert, E.G., and Lucas, B.N., *Surf. Coat. Technol., 116–119,* 452, 1999.
42. Gottmann, J. and Kreutz, E.W., *Surf. Coat. Technol., 116,* 1189, 1999.
43. Sturm, K., Fahler, S., and Krebs, H.U., *Appl. Surf. Sci., 154–155,* 462–466, 2000.
44. Geohegan, D.B., *Thin Solid Films, 220,* 1–2, 138, 1992.
45. Hubler, G.K., Chrisey, D.B., and Hubler, G.K., Eds., *Pulsed Laser Deposition of Thin Films,* John Wiley, New York, 1994, p. 327.
46. Timm, R., Willmott, P.R., and Huber, J.R., *Appl. Phys. Lett., 71,* 1966, 1997.
47. Liu, X., Yin, J., Liu, Z.G., Yin, X.B., Chen, G.X., and Wang, M., *Appl. Surf. Sci., 174,* 35, 2001.
48. Tabor, D., *The Hardness of Metals,* Oxford University Press, London, 1951.
49. Oliver, W.C. and Pharr, G.M., *J. Mater. Res., 7,* 1564, 1992.
50. Doerner, M.F. and Nix, W.D., *J. Mater. Res., 1,* 601, 1986.
51. Cheng, Y.T. and Chen, C.M., *Appl. Phys. Lett., 73,* 5, 614, 1998.
52. Linchinchi, M., Lenardi, C., Haupt, J., and Vitali, R., *Thin Sol. Films, 312,* 240, 1998.
53. Thurn, J. and Cook, R.F., *J. Mater. Res., 17,* 1143, 2002.
54. Tsui, T.Y., Vlassak, J., and Nix, W.D., *J. Mater. Res., 14,* 2196, 1999.
55. Buckle, H., *Science of Hardness Testing and Its Research Applications,* Westbrook, J.H. and Conrad, H., Eds., ASM, Metals Park, OH, 1973, p. 453.
56. Fabes, B.D., Oliver, W.C., McKee, R.A., and Walker, F.J., *J. Mater. Res., 7,* 3056, 1992.
57. Bhattacharya, A.K. and Nix, W.D., *Int. J. Sol. Struct., 24,* 1287, 1998.
58. Korsunsky, A.R., McGurk, M.R., Bull, S.J., and Page, T.F., *Surf. Coat. Technol., 99,* 1–2, 171, 1998.
59. Tuck, J.R., Korsunsky, A.M., Bhat, D.G., and Bull, S.J., *Surf. Coat. Technol., 139,* 63, 2001.
60. Barshilia, H.C. and Rajam, K.S., *Surf. Coat. and Technol., 155,* 2–3, 195, 2002.
61. Raymond-Angélélis, C., Ecole Nationale Supérieure des Mines de Paris, 1998.
62. Zworykin, V.K., Hiller, J., and Snyder, R.L., *ASTM Bull., 15,* 117, 1942.
63. Alexandrou, I., Scheibe, H.J., Kiely, C.J., Popworth, A.J., Amaratunga, G.A.J., and Schultrich, B., Carbon Films with an sp^2 Network Structure, *Phy. Rev. B*, 60, 10903, 1999.
64. Joslin, D.L. and Oliver, W.C., A New Method for Analyzing Data from Continuous Depth-Sensing Microindentation Tests, *Journal of Materials Research,* 5(1): 123–126, Jan. 1990.

21 Polarization in Nanotubes and Nanotubular Structures

Marco Buongiorno Nardelli,
Serge M. Nakhmanson, and Vincent Meunier

CONTENTS

21.1 INTRODUCTION

A material is known as polar if, in one way or another, it displays macroscopic electric polarization. All polar materials exhibit piezoelectricity: they develop nonzero electric polarization when strained and change shape when placed into an external electric field. Structures that have nonzero intrinsic polarization — also called spontaneous polarization — even when no strain is applied to them are known as pyroelectrics. The prefix *pyro*—from the Greek word for fire—which is used to indicate the permanent electric moment associated to these materials, arises from the fact that pyroelectricity was first discovered in certain crystals when they were warmed. However, it was soon established that these materials also had permanent dipole moments at lower temperatures as well, only they were "hidden" by the adsorption of polar ions on the faces of the samples. Once these ions were desorbed at high temperatures, the permanent moment would appear. Usually the existence of pyroelectricity is attributed to the reduced symmetry in the geometrical structure of certain crystalline materials. In practice, if we model a crystal as a superposition of microscopic dipoles, these dipoles will cancel each other in most cases due to the symmetry of the solid. However, if the crystal has reduced symmetry, the lower number of symmetry operations that are allowed could lead to an only partial cancellation of the internal dipolar fields, leaving the system electrically polar. Ferroelectrics are pyroelectrics in which the direction of spontaneous polarization can be changed (for example, rotated by 180°) by applying a mechanical deformation or an external electric field. All these phenomena find a great number of applications in modern technology, from sensors and actuators, to memory cells and spintronics devices. In general, pyro- and piezoelectric materials to be used in modern technological applications should display an excellent piezoelectric response, combined with high mechanical stability and low environmental impact. Existing materials, which can be broadly divided into the families of perovskite crystals, ceramics, and polymers, can only partially fulfill the aforementioned requirements. Lead zirconate titanate (PZT) ceramics materials, for example, are strong piezo- and pyroelectrics[1,2] with spontaneous polarizations of 0.3 to 0.7 C/m^2 ($Coulomb/m^2$), but unfortunately they are also brittle, heavy, and toxic. On the other hand, polymers such as polyvinylidene fluoride (PVDF) are lightweight, flexible, and virtually inert, but display polar properties an order of magnitude weaker than those of PZT.[3] In Table 21.1 we summarize the main characteristics of modern piezo- and pyroelectric materials, compared with the predicted polar properties of BN nanotubes, which we will discuss at length in this chapter.

In this context, carbon nanotubes and related nanotubular structures can play an important role in advancing the field toward new and improved materials for polarization-based applications. After their discovery,[4] it was almost immediately shown that carbon nanotubes possess outstanding potential for technological applications due to their mechanical and electrical properties.[5] The advent of carbon nanotubes, together with recent progress in nanomaterials design and processing, has led to a quest for other novel graphene-based materials with technologically desirable properties. Furthermore the closely related boron nitride nanotubes (BNNTs) and mixed BN-C systems,[6–11] now produced in macroscopic quantities, have electronic properties that are complementary to pure carbon nanotubes and could therefore be useful in a variety of novel electronic devices. For instance, an early theoretical study predicted that BN-C junctions may well be a practical way to realize stable, nanoscale heterojunctions,[12] as exemplified in Fig. 21.1.

BNNTs, extensively investigated since their initial prediction[6] and succeeding synthesis,[7] are well known for their excellent mechanical properties.[13] However, unlike carbon nanotubes, the majority of BN structures are noncentrosymmetric and polar, a fact that suggests the existence of nonzero spontaneous polarization fields.

TABLE 21.1
Summary of Properties of Polar Materials and Boron Nitride Nanotubes

		Properties (C/m^2)			
Material Class	Representatives	Polarization	Piezoelectricity	Pros	Cons
Lead zirconate titanate (PZT) ceramics	$PbTiO_3$ $PbZrO_3$ $PbZr_xTi_{1-x}O_3$	Up to 0.9	5–10	Good pyro- and piezoelectric properties	Heavy, brittle, toxic
Polymers	Polyvinylidene fluoride (PVDF), PVDF copolymers	0.1–0.2	Up to 0.2	Light, flexible	Pyro- and piezoelectric properties weaker than in PZT ceramics
Boron nitride nanotubes	Zigzag BN nanotubes	Single NT: 0 Bundle: ~0.01	Single NT: 0.25–0.4 Bundle: ?	Light, flexible; good piezoelectric properties; could be used in nanodevices	Expensive?

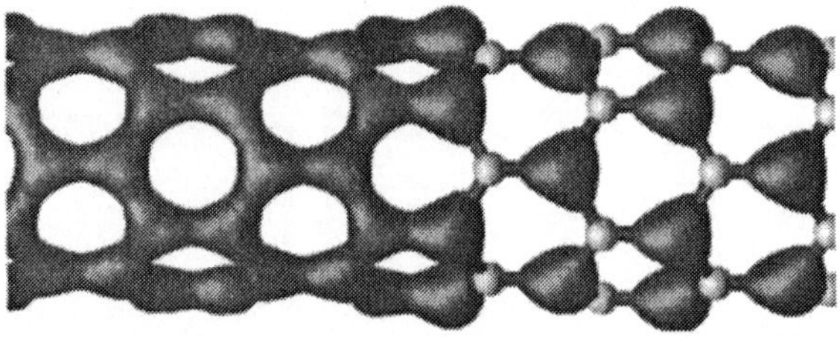

FIGURE 21.1 Electronic density in a nanotube containing a section of C (left) and a section of BN (right). Note the difference in bonding, from quasiplanar sp^2 in C to ionic in BN.

Recently these properties have been partially explored by Mele and Kràl, using a model electronic Hamiltonian,[14] which predicted that BNNTs are piezo- and pyroelectric materials whose electronic polarization orientation depends on the geometrical structures of the tubes.

In this chapter, we will review what we have learned in the past few years about spontaneous polarization and piezoelectricity in BNNTs and composite BN-C nanotube structures.[15,16] We will estimate their potential usefulness in various pyro- and piezoelectric device applications, and we will elucidate the interplay between symmetry and polarization in nanotubular systems. Finally we will discuss some recent experimental results and relate them to our theoretical understanding.

In Section 21.2 we briefly show how modern polarization theory is formulated employing the Berry phase or the Wannier functions approaches and present the details of the numerical techniques we have employed to compute polarizations in Section 21.3. (Subsections 21.2.1, 21.2.2, and 21.2.3 are quite technical and could be skipped during the first reading.) In Section 21.4 we will discuss the results of our computations of spontaneous polari-zation and piezoelectricity in BNNTs utilizing both of the prescriptions given in Section 21.2. Section 21.5 will be devoted to the analysis of the piezoelectric properties of these systems, while in Section 21.6 we will discuss a few instances where polarization effects can be indeed observed in nanotubular structures. Conclusions and further discussion will end the chapter.

21.2 MODERN THEORY OF POLARIZATION

Macroscopic polarization of a dielectric is a central concept of classical electrostatics, well known from elementary introductory physics classes. At a first glance, it might seem that it is a well-defined quantity easily obtainable via simple analysis or calculations. Most introductory condensed matter physics books contain a straightforward recipe to compute polarization in a solid that looks like this:

$$\mathbf{P} = \frac{1}{V}\left[\sum_l eZ_l\mathbf{b}_l + \int \mathbf{r}\rho(\mathbf{r})d\mathbf{r}\right] \tag{21.1}$$

Here the first term accounts for the polarization associated with ions of charge Z_l positioned at $\mathbf{b}_l$ and the second term represents the electronic contribution to polarization, where $\rho(\mathbf{r})$ is the density of electronic charge. If the summation and integration in Equation 21.1 are extended over the whole macroscopic sample, i.e., index l runs over all ions and $\rho(\mathbf{r})$ contains information about all electronic charges including the ones on the crystal boundary, then the sum of the two terms is absolutely convergent and the polarization $\mathbf{P}$ is well defined. However, in real calculations it is usually impossible to sum and integrate over the whole sample. Instead a unit cell representative of the characteristic features of the system under investigation is chosen and periodically repeated over the whole space. In other words, a finite piece of solid with a boundary is substituted with an infinite and periodic solid without a boundary. For the latter, most condensed matter textbooks introduce a dipole moment associated with the unit cell by limiting the summation and integration in Equation 21.1 to the atoms and electrons within the cell. In this case, index l runs over all ions in the cell and $\rho(\mathbf{r})$ is the cell-periodic density of electronic charge. The total polarization in the sample is then computed as a sum over the unit cell dipole moments. There is a serious problem with this definition since the aforementioned sum is now only conditionally convergent, i.e., it can converge to a number of different values based on the choice of a given order of summation. This means that with the exception of a very special case, when the solid can be clearly partitioned into a set of charge neutral, weakly interacting entities (the so-called Clausius-Mossotti limit), polarization becomes an ill-defined quantity. Of course, the Clausius-Mossotti limit cannot be applied to a covalent solid where electrons are delocalized.

One of the most relevant results in the modern theory of polarization is that macroscopic polarization of a material cannot be predicted on the sole basis of its microscopic charge distribution. However, it is obvious that it should also be a bulk property of the material, i.e., independent of the shape of the sample or the choice of the unit cell. The modern theory states that polarization in a solid can be defined analogously to actual measurement procedures used in experiments. In reality, such experiments never measure an absolute value of polarization, but rather a difference $\Delta\mathbf{P}$ between two states of the material. This difference is indeed an intrinsic property and does not depend on the truncation of the sample. The great achievement of modern polarization theory is the correct definition and an operative way to compute polarization differences in materials. In what follows, we will briefly outline two different, although equivalent, ways to compute polarization in materials and nanostructures: the Berry phase method and the Wannier functions method. For a comprehensive formal treatment of modern polarization theory, we refer the reader to the papers by D. Vanderbilt and R. Resta.[17,18]

21.2.1 Computing Polarization Using Berry Phase Method

The problem of computing polarization in materials is very subtle and can be approached by the so-called Berry phase method, introduced only a decade ago by Vanderbilt and King-Smith[17] and Resta.[18] Although a full description of the method is beyond the scope of this chapter, it is instructive to outline the main ideas behind this approach. In the modern theory of polarization, the polarization difference between two states of a material is computed as a geometrical quantum phase. In practice, this difference, $\Delta\mathbf{P} = \mathbf{P}^{(\lambda_1)} - \mathbf{P}^{(\lambda_0)}$, can be obtained if one can find an adiabatic transformation λ, which connects the states λ_1 and λ_0, while leaving system insulating. In the spirit of Reference 19, $\mathbf{P}^{(\lambda)}$ is conveniently split into two parts, $\mathbf{P}_{ion}^{(\lambda)}$ and $\mathbf{P}_{el}^{(\lambda)}$, corresponding to the ionic and electronic contributions, respectively. In the case of paired electron spins, the expression for the total polarization of the system can be written as follows:

$$\mathbf{P}^{(\lambda)} = \mathbf{P}_{ion}^{(\lambda)} + \mathbf{P}_{el}^{(\lambda)} = \frac{e}{V}\sum_{\tau} Z_{\tau}^{(\lambda)}\mathbf{b}_{\tau}^{(\lambda)} - \frac{2ie}{8\pi^3}\sum_{i\,occ}\int_{BZ} d\mathbf{k}\left\langle u_{i\mathbf{k}}^{(\lambda)}\middle|\nabla_{\mathbf{k}}\middle|u_{i\mathbf{k}}^{(\lambda)}\right\rangle \quad (21.2)$$

where V is the volume of the unit cell, $Z_{\tau}^{(\lambda)}$ and $\mathbf{b}_{\tau}^{(\lambda)}$ are the charge and position of the τth atom in the cell, and $u_{i\mathbf{k}}^{(\lambda)}$ is the occupied cell-periodic Bloch state of the system. For the electronic part, an electronic phase $\varphi_{\alpha}^{(\lambda)}$ (Berry phase) defined modulo 2π (that is, an uncertainty amounting to an integer number of 2π in the determination of $\varphi_{\alpha}^{(\lambda)}$) can be introduced as

$$\varphi_{\alpha}^{(\lambda)} = V\mathbf{G}_{\alpha} \cdot \mathbf{P}_{el}^{(\lambda)}/e \quad (21.3)$$

where $\mathbf{G}_{\alpha}$ is the reciprocal lattice vector in the direction α along which we would like to compute polarization (for example, along the axis of a nanotube). Of course, it is hard to determine from Equation 21.3 that $\varphi_{\alpha}^{(\lambda)}$ is truly an angular variable-defined modulo 2π. However, if we substitute $\mathbf{P}_{el}^{(\lambda)}$ from Equation 21.2 into 21.3 and do all the algebra (see Reference 17 for details), we can prove that it is indeed the case. In a similar fashion, one can construct an angular variable for the ionic part (which is just a sum over the point charges), called in what follows the ionic phase, so that the total geometrical phase is

$$\Phi_{\alpha}^{(\lambda)} = \sum_{\tau} Z_{\tau}^{(\lambda)}\mathbf{G}_{\alpha} \cdot \mathbf{b}_{\tau}^{(\lambda)} + \varphi_{\alpha}^{(\lambda)} \quad (21.4)$$

The total polarization in the direction α becomes

$$\mathbf{P}_{\alpha}^{(\lambda)} = e\Phi_{\alpha}^{(\lambda)}\mathbf{R}_{\alpha}/V$$

where $\mathbf{R}_{\alpha}$ is the real-space lattice vector corresponding to $\mathbf{G}_{\alpha}$, i.e., $\mathbf{R}_{\alpha} \cdot \mathbf{G}_{\alpha} = 1$.

21.2.2 Polarization from Wannier Functions

Alternatively the electronic polarization of a system can be expressed in terms of the centers of charge of the Wannier functions (WFs) of its occupied bands:[17,18]

$$\mathbf{P}_{el}^{(\lambda)} = -\frac{2e}{V}\sum_{i}\left\langle W_{i\mathbf{R}}^{(\lambda)}\middle|\mathbf{r}\middle|W_{i\mathbf{R}}^{(\lambda)}\right\rangle d\mathbf{r} = -\frac{2e}{V}\sum_{i}\mathbf{r}_{i}^{(\lambda)}$$

where $W_{i\mathbf{R}}^{(\lambda)}(\mathbf{r})$ is the WF for band i, and $\mathbf{r}_i^{(\lambda)}$ is the corresponding center. As a reminder for the readers, a Wannier function, labeled by the Bravais lattice vector **R** and the band index i, is usually defined via a unitary transformation of the Bloch functions $\Psi_{i\mathbf{k}}(\mathbf{r})$ of the ith band:

$$W_{i\mathbf{R}}(\mathbf{r}) = \frac{V}{(2\pi)^3} \int_{BZ} \psi_{i\mathbf{k}}(\mathbf{r}) e^{-i\mathbf{k}\cdot\mathbf{R}} d\mathbf{k}$$

where V is the volume of the unit cell and the integration is performed over the entire Brillouin zone (BZ). It is easy to show that the WFs, as defined previously, form an orthonormal basis set, and that any two of them, for a given index i and different **R** and $\mathbf{R}_0$, are only translated images of each other.‡

The advantage of this formula is that it allows the recasting of our problem from the summation over distributed charges to a summation over localized WF charge centers, which corresponds to the Clausius-Mossotti limit when the polarization is well defined. The localized WFs can be computed by a number of different techniques, for example, by the direct minimization of the spread functional as the sum of the second moments of the WFs corresponding to a particular choice of translational lattice vector:[20]

$$\Omega = \sum_j \left(\langle W_{i\mathbf{0}} | \mathbf{r}^2 | W_{i\mathbf{0}} \rangle_j - \langle W_{i\mathbf{0}} | \mathbf{r} | W_{i\mathbf{0}} \rangle_j^2 \right).$$

In either case (Berry phase or WFs), due to the arbitrariness in the choice of the phases of the Bloch functions, $\mathbf{P}_{el}^{(\lambda)}$ can be obtained only modulo $2e\mathbf{R}/V$. However, the difference in polarization $\Delta\mathbf{P}$ remains well defined provided that $|\Delta\mathbf{P}_{el}| \ll |2e\mathbf{R}/V|$. The same indetermination issues apply to $\mathbf{P}_{ion}^{(\lambda)}$.[19]

21.2.3 A Simple Way to Compute Berry Phases on a Computer

The main ingredient that has to be computed in order to obtain the Berry phase $\varphi_\alpha^{(\lambda)}$ is the following matrix element (see Equation 21.2): $\langle u_{i\mathbf{k}}^{(\lambda)} | \nabla_\mathbf{k} | u_{i\mathbf{k}}^{(\lambda)} \rangle$. If in the express-ion for $\mathbf{P}_{el}^{(\lambda)}$ in Equation 21.2 we split the 3-D integration over the Brillouin

‡ Any crystal can be viewed as a lattice, a periodic set of points in space that obey a number of symmetry operations. In particular, there exist only 14 different lattice types, the so-called Bravais lattices, each characterized by a set of translational vectors **R** that can be used to generate lattice sites. In the solid, atoms or groups of atoms are associated with each lattice site, resulting in a definite crystal structure. The periodicity of the solid also determines basic symmetry properties of the quantum mechanical behavior of the electrons. Electrons in solids are described by wave functions (Bloch states) that have the same periodicity as the crystalline lattice and an energy spectrum that reflects the symmetric properties of the crystal. These are usually characterized in terms of reciprocal lattice vectors **G**, the Fourier transforms of the translational vectors **R**. Reciprocal lattice vectors define the reciprocal of the crystal lattice, the so-called Brillouin zone. As the Bravais lattice reflects the symmetry of the structure of the solid, the Brillouin zone reflects the symmetry of the electronic states. For more on the geometrical and electronic structures of solids see, for instance, N.W. Ashcroft and N.D. Mermin, *Solid State Physics*, Saunders College, Philadelphia, 1976.

zone into the linear integral along $\mathbf{G}_\alpha$ and an area integral in the perpendicular direction, we can rewrite Equation 21.3 as

$$\varphi_\alpha^{(\lambda)} = \frac{2VG_\alpha}{(2\pi)^3} \int_A d\mathbf{k}_\perp \phi^{(\lambda)}(\mathbf{k}_\perp), \tag{21.5}$$

where

$$\phi^{(\lambda)}(\mathbf{k}_\perp) = -i \sum_{nocc} \int_0^{G_\alpha} dk_\alpha \left\langle u_{nk}^{(\lambda)} \middle| \frac{\partial}{\partial k_\alpha} \middle| u_{nk}^{(\lambda)} \right\rangle .$$

The matrix elements $\langle u_{nk}^{(\lambda)} | \frac{\partial}{\partial k_\alpha} | u_{nk}^{(\lambda)} \rangle$ can be computed by using finite differences on a fine 1-D mesh (or a string) of k-points $\{k_j\}$ along $\mathbf{G}_\alpha$, as shown in the upper panel of Fig. 21.2. An expression for such a string phase can be written in the following fashion:

$$\phi^{(\lambda)}(\mathbf{k}_\perp) = \lim_{J\to\infty} \phi_J^{(\lambda)}(\mathbf{k}_\perp) = \lim_{J\to\infty} \operatorname{Im} \ln \left(\prod_{j=0}^{J-1} \det \left\langle u_{nk_j}^{(\lambda)} \middle| u_{mk_{j+1}}^{(\lambda)} \right\rangle \right). \tag{21.6}$$

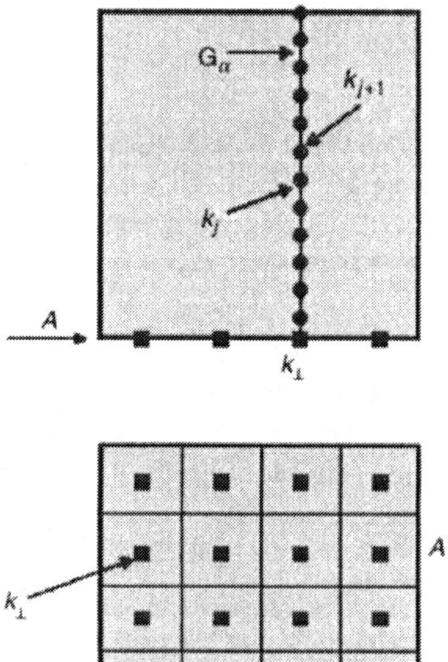

FIGURE 21.2 Partitioning of the Brillouin zone for the computation of the Berry phase (example for the cell of orthorhombic symmetry). G_α is the reciprocal lattice vector in the direction in which we compute polarization. A is a 2-D slice of the Brillouin zone, perpendicular to the direction of G_α. A 2-D mesh of k-points $\{k_\perp\}$ is chosen in A. Then a fine 1-D mesh of k-points $\{k_j\}$ in the direction of G_α is introduced for each point $k_\perp$.

The two sets of wave functions required to obtain each of the matrix elements in Equation 21.6 are computed in the neighboring k-points k_j and k_{j+1} along the string, also shown in the upper panel of Fig. 21.2. In the limiting case when $k_j = k_J$, i.e., we are positioned at the top of the string and the following condition should be satisfied for the Bloch-state sets at k-points k_0 and k_J:

$$u_{nk_J}^{(\lambda)}(\mathbf{r}) = e^{-i\mathbf{G}_\alpha \mathbf{r}} u_{nk_0}^{(\lambda)}(\mathbf{r}) .$$

Finally we turn the 2-D integral in Equation 21.5 into a sum over a set of k-points $\{k_\perp\}$, shown in the lower panel in Fig. 21.2, obtaining the operational expression for computing the Berry phase as an average of the string phases $\phi_J^{(\lambda)}(\mathbf{k}_\perp)$ over the area perpendicular to $\mathbf{G}_\alpha$:

$$\varphi_\alpha^{(\lambda)} = \frac{2}{N_{k_\perp}} \sum_{k_\perp} \phi_J^{(\lambda)}(\mathbf{k}_\perp) .$$

As we can see, in order to obtain the Berry phase $\varphi_\alpha^{(\lambda)}$ we need to know all the occupied Bloch states $u_{nk}^{(\lambda)}(\mathbf{r})$ in every point of the 3-D k-point mesh $\{k_j, k_\perp\}$ and also compute the overlap matrix elements $\langle u_{nk_j}^{(\lambda)} | u_{mk_{j+1}}^{(\lambda)} \rangle$ for the neighboring k-points within the same string. The electronic part of the polarization is then recovered from the Berry phase with the help of the following formula:

$$\left(\mathbf{P}_{el}^{(\lambda)}\right)_\alpha = e\varphi_\alpha^{(\lambda)} \mathbf{R}_\alpha / V.$$

21.3 COMPUTATIONAL DETAILS

The starting point for our procedure is the first principles calculations of the electronic structures of the systems. We adopt a standard electronic structure method based on self-consistent total energy and force minimization, which allows for optimizing simultaneously the atomic positions and the corresponding electronic wave functions. The electronic structure is described within the density functional theory (DFT).[21] We used a multigrid-based total energy method, employing a real-space grid as a basis,[22] as well as a more standard plane wave-based approach for all the polarization calculations.[23] The internal consistency of the results obtained using these two methods has been carefully checked. The results were obtained in the local density approximation (LDA),[24] and although more sophisticated exchange correlation functionals (e.g., generalized gradient approximations) can obviously be used, they do not alter the major conclusions of this investigation. The electron–ion interaction is described via norm-conserving pseudopotentials[25] in the form of Kleinman and Bylander.[26]

To isolate the contributions of individual nanotubes, we performed polarization calculations for periodic crystals of noninteracting (i.e., positioned sufficiently far apart) nanotubes (Fig. 21.3(a)). In order to assess the importance of close-packing, the calculations were repeated for nanotubes arranged in a hexagonal lattice as depicted in Fig. 21.3(b). The electronic structure calculations were carried out using

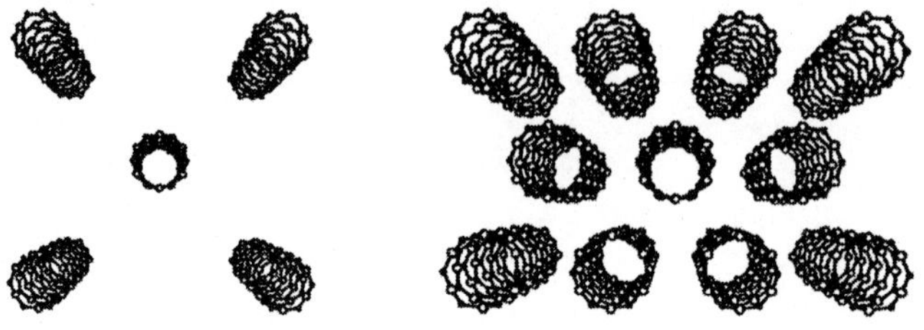

FIGURE 21.3 Examples of two nanotube arrangements used in the simulations: (a) array of nanotubes sufficiently far apart that can be considered noninteracting; (b) geometry of a unit cell containing interacting nanotubes (rope).

two special k-points along the Γ-A direction in the hexagonal or Γ-Z direction in the tetragonal Brillouin zone. The k-space integration to compute $\varphi_z^{(\lambda)}$ was done on a string of 20 k-points uniformly distributed along the same direction and shifted to avoid the Γ-point.

21.4 POLARIZATION IN NANOTUBES

Calculations of spontaneous polarization in BNNTs with the techniques described in the previous section require a reference state of the system that is intrinsically nonpolar. The polarization difference $\Delta\mathbf{P}$ between the polar and the nonpolar states of the system will then give the actual value of spontaneous polarization. An obvious choice for the nonpolar state of a BNNT is a nanotube of the same geometry, but with boron and nitrogen atoms substituted by carbon atoms. Such nanotubes are naturally nonpolar because of the purely covalent bonds connecting all the atoms. However, we might also want to compute the polarization difference in many points between the two limiting cases of BNNT and CNT to make sure that each such difference remains smaller than the polarization quantum $2e\mathbf{R}/V$. This can be done by employing a so-called virtual crystal procedure where carbon atoms in the CNT are substituted by pseudocarbon atoms, which are 50% boron and 50% nitrogen. The adiabatic transformation parameter λ corresponds to the content (in atomic %) of a site that is gradually transformed from pseudocarbon ($\lambda = 50\%$) to pure boron or nitrogen ($\lambda = 100\%$), as shown in Fig. 21.4. Because of the different alignments of the polar bond with respect to the nanotube axis, we anticipate that the symmetry of the nanotube will play an important role in determining the magnitude of the spontaneous polarization field. In particular, since the zigzag geometry maximizes the axial dipole moment, we expect to observe the strongest effects in (n,0) nanotubes.

21.4.1 Berry Phase Method

The ionic part of the polarization in zigzag BNNTs presented in Fig. 21.5 is large and directly proportional to the nanotube's index. This is in contrast, for instance, with the corresponding wurtzite III–V and II–VI systems,[27,28] where the spontaneous

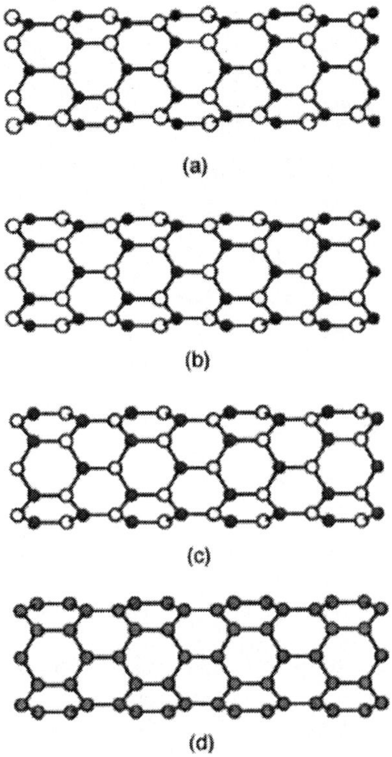

FIGURE 21.4 Virtual crystal procedure. Going from A to D, the B and N atoms are gradually transformed into C, according to the variation of the parameter λ.

polarization can be viewed as the difference between the polarizations of the wurtzite (polar) and zinc blende (nonpolar) geometries. Since these configurations become geometrically distinct only in the second shell of neighbors, their ionic phases are very close. The major contribution to the spontaneous polarization in wurtzite materials is then due to the difference between the electronic polarizations (which are 0.04 to 0.08 C/m^2), while in BNNTs *both* the ionic and the electronic contributions are essential.

The ionic phase differences $\Delta\varphi_{ion}$ between the polar and nonpolar configurations of zigzag nanotubes were evaluated via the virtual crystal approximation. The inset in Fig. 21.5 shows the results obtained by a simple lattice summation over the ionic charges (the first term in Equation 21.2), with the phases translated into the $[-\pi, \pi]$ interval. The phases plotted in the main graph were unfolded by eliminating all the 2π discontinuities and setting the phase of the nonpolar reference configuration to zero. For the unfolded phases, as the diameter of a nanotube increases, i.e., as another hexagon is added around the circumference of the tube, the ionic phase increases by $\pi/3$, so that the total ionic phase for a $(n,0)$ BN nanotube amounts to $n\pi/3$.

In Fig. 21.6 we show the electronic phase differences $\Delta\varphi_{el}$ between the polar and nonpolar configurations for zigzag nanotubes. These data suggest a natural

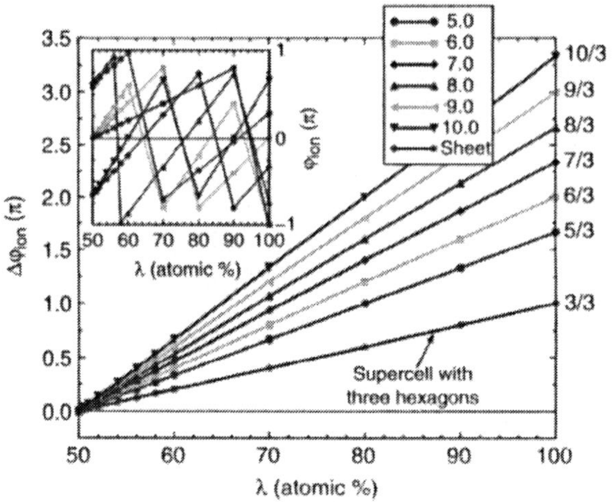

FIGURE 21.5 Ionic phase (see Equation 21.4) differences between the polar and nonpolar configurations for zigzag nanotubes; the ionic phase of the nonpolar configuration is set to zero. Inset: ionic phases wrapped into the $[-\pi, \pi]$ interval. Phases are given in units of π.

division of the nanotubes into three families with different $\Delta\varphi_{el}$: $\pi/3$ for $n = 3l - 1$, $-\pi/3$ for $n = 3l + 1$, and $-\pi$ for $n = 3l$, where l is an integer, which is similar to the result obtained by Mele and Král.[14] It is important to note that in the limiting case of flat C or BN sheets, the electronic polarization is zero for any value of λ, due to the existence of a three-fold symmetry axis perpendicular to the surface of the sheet. The existence of the three classes of behavior in BN nanotubes is surprising, given that the ionic character of the electronic charge density (associated with

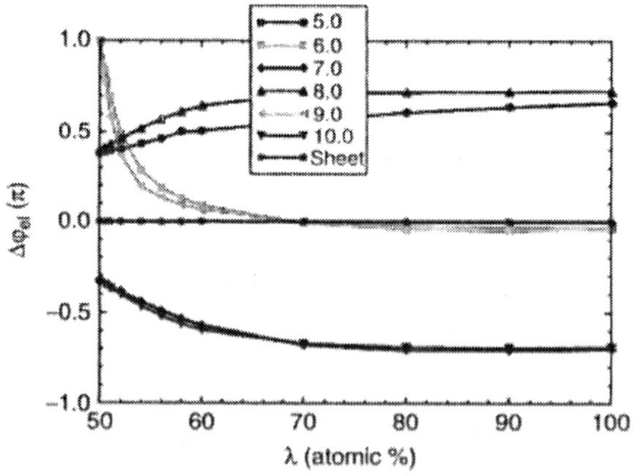

FIGURE 21.6 Electronic phase (see Equation 21.3) differences between the polar and non polar configurations for zigzag nanotubes.

the B–N bond) does not change with the nanotube index. Additionally, there is an important difference between our results and those of Reference 14, where electronic polarization of heteropolar nanotubes was studied within a simple π-orbital tight-binding (π-TB) approximation. In Reference 14 the $n = 3l$ family has a zero electronic phase instead of $-\pi$.

This discrepancy is due to the ambiguity of the definition of electronic polarization as a multivalued quantity,[19] which can assume a lattice of values corresponding to Berry phases that differ by arbitrary multiples of 2π. Unlike the ionic phase model, where discontinuities in $\varphi_{ion}^{(\lambda)}$ can be easily monitored, Berry phase calculations always produce phases that are smoothly folded into the $[-\pi, \pi]$ interval and cannot be extrapolated. To obtain an unambiguous determination of the spontaneous polarization of BNNTs of arbitrary diameters, one has to compute the polarization in a different way, using the centers of charge of the WFs of the occupied bands. Note that this approach does not solve the problem of branch indetermination, since as Berry phases are defined modulo 2π, Wannier centers are defined modulo a lattice vector **R**.[#] However, by shifting the indetermination from the phase to the lattice vector, we are able to map the electronic polarization problem onto a simple electrostatic model, where the unfolding of the electronic phase is straightforward.[19]

21.4.2 Maximally Localized Wannier Functions

The results of the maximally localized WF calculations for BNNTs are summarized in Fig. 21.7, where examples of the WFs for C and BN zigzag nanotubes of arbitrary diameter are shown, together with a schematic drawing that illustrates the shift of the Wannier centers in the adiabatic transformation from C to BN. Since

$$\mathbf{P}_{el}^{BN} = -\frac{2e}{V}\sum_i \left(\mathbf{r}_i^{BN} - \mathbf{r}_i^{C}\right)$$

the magnitude of the shift of the centers is directly proportional to the electronic polarization of the BNNT with respect to the nonpolar CNT.

The σ-band WFs are centered in the middle of the C–C bonds in carbon nanotubes while in BNNT's they are shifted toward the cations because of the different electronegativities of B and N atoms. Since these shifts have the same magnitude along each of the three bond directions, the vector sum of all shifts is zero (see bottom panel of Fig. 21.7), and the σ orbitals do not contribute to the total polarization of the system. The π-band WFs are centered on the cations in BNNTs, while in CNTs they have a peculiar V-shape, with centers somewhat outside of the C–C bond. The sum of the shifts of the π-band Wannier centers is nonzero only for the axial component, which means that the electronic polarization in BNNTs is purely axial.

The bottom panel of Fig. 21.7 shows the projection of the π WF centers onto the axis of the tube. The projections of the centers have an effective periodicity of half of the axial lattice constant c, which leads to the indetermination of the electronic

[#] Wannier functions and their centers have the periodicity of the lattice.

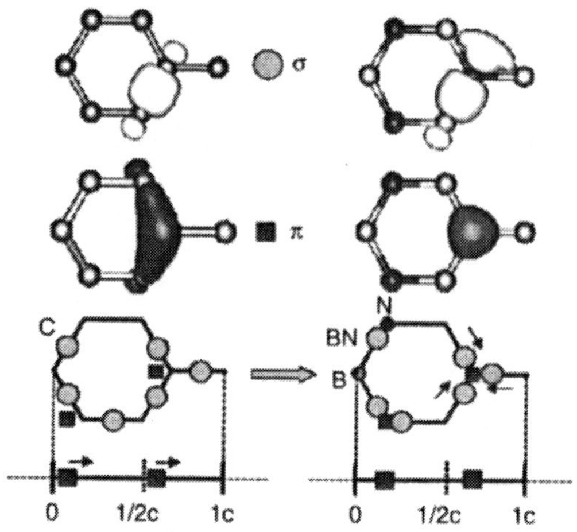

FIGURE 21.7 Upper panel: examples of Wannier functions (WFs) of the σ and π occupied bands of C (left panel) and BN (right panel) nanotubes. Lower panel: schematic positions of the centers of the Wannier functions in C and BN hexagons, and the projections of the π WFs onto the nanotube axes. The positions of the centers of σ WF are indicated by circles, and those of π by squares. The direction of the shifts of σ and π WFs in an adiabatic transformation from C to BN is indicated by arrows. The projections of shifts of the σ WFs cancel, so that the σ WFs do not contribute to polarization (see text).

phase by multiples of π. Moreover the WF description allows for an unambiguous unfolding of the electronic phase. In analogy to the ionic phase, we find that each individual hexagon carries a phase of $-\pi/3$, leading to a total electronic phase of $-n\pi/3$ for a $(n,0)$ nanotube. This result demonstrates that the direction of the electronic polarization in a BNNT is specified by the orientation of the B–N bond and does not oscillate in direction with the nanotube diameter, contrary to the model Hamiltonian predictions.[14] We should point out that the Wannier function results are completely consistent with the Berry phase calculations, since an electronic phase of $-n\pi/3$ for any n can be folded modulo π into the three families found previously.

When we combine the results for the ionic and electronic phases into a general formula for the phase of any (n,m) BNNT (n and m being any integer number), we find that the two contributions cancel exactly and that the total spontaneous polarization in any BNNT remains zero, i.e., the Wannier centers are positioned in such a way as to completely compensate the polarization due to ions. This result was confirmed by two-point (corresponding to $\lambda = 50$ and 100%) calculations of the Berry phase difference for a number of chiral nanotubes ((3,1), (3,2), (4,1), (4,2), (5,2), and (8,2)) and the exact cancellation of the polarization contribution was obtained for all BNNTs, except for those having diameters smaller than about 4 Å, where a residual polarization subsists as the consequence of the very high curvature. In such nanotubes, Wannier centers cannot fully compensate the ionic polarization because of the severe distortion of the atomic bonds, which makes these systems weakly pyroelectric. For example, we found that $P_3 = 0.11$ C/m^2 in (3,1), 0.008 C/m^2 in (7,0), and 0.002 C/m^2 in (12,0) nanotubes.

The exact cancellation is a result of the overall chiral symmetry of the nanotubes, which, although not centrosymmetric, are intrinsically nonpolar. However, cancellation of ionic and electronic polarizations is exact only in the limit of an isolated BNNT. The spontaneous polarization in a nanotube bundle, where the chiral symmetry is effectively broken, differs from zero. For example, in a bundle of (7,0) tubes positioned at a wall-to-wall equilibrium distance of 3.2 Å, the polarization amounts to $P_3 \approx 0.01$ C/m^2. It must be pointed out that in this example it is difficult to estimate the relative contributions to the polarization of the bundling, extreme curvature and elastic deformation of the nanotubes. Although smaller than in polymers or PZT, this polarization is comparable to some wurtzite pyroelectrics: e.g., $P_3 = 0.06$ C/m^2 in w-ZnO.[28]

21.5 PIEZOELECTRICITY IN NANOTUBES

The Berry phase method can also be employed to compute piezoelectric properties of BNNTs that are directly related to the polarization differences between strained and unstrained tube geometries. In the linear regime, the change in polarization due to strain can be decomposed into a sum of two terms: a uniform axial strain and a relative displacement of the two sublattices. It is therefore natural to describe the geometry of a BNNT of a given radius in terms of an axial lattice constant c and an internal parameter u, uc therefore being the length of the vector connecting the anion to the cation. With this choice, the axial piezoelectric polarization is

$$\Delta P_3 = e_{33}\varepsilon_3 = \frac{\partial P_3}{\partial c}(c - c_0) + \frac{\partial P_3}{\partial u}(u - u_0) \tag{21.7}$$

where the strain $\varepsilon_3 = (c - c_0)/c_0$, and c_0 and u_0 are the equilibrium values of c and u. Then the only surviving piezoelectric stress tensor component is

$$e_{33} = e_{33}^{(0)} + \frac{ec_0^2}{V} NZ^* \frac{du}{dc}$$

where N is the number of B–N pairs in the supercell. Here

$$e_{33}^{(0)} = c_0 \frac{\partial P_3}{\partial c}$$

is the so-called clamped-ion piezoelectric constant representing the effect of strain on the electronic structure, and

$$Z^* = \frac{V}{eNc_0} \frac{\partial P_3}{\partial u}$$

is the axial component of the Born dynamical charge tensor. Both polarization derivatives were computed as finite differences, changing c or u by ±1%. The parameter $\xi = c_0 du/dc$, which is a measure of the bond length change under axial strain, was obtained by rescaling c (together with the associated components of ionic coordinates) and then relaxing the geometry of the system. For all the systems considered subsequently, the value of ξ is approximately the same and equal to −0.085.

We have calculated the piezoelectric properties for various bundles composed of zigzag BNNTs with individual diameters ranging from 3.9 to 10.2 Å. These results are summarized in Table 21.2 and compared to a few well-known piezo- and pyroelectric materials. While the piezoelectric constants of zigzag BNNTs are modest when compared with inorganic compounds, they are still substantially larger than those in the PVDF polymer family. In addition, we should point out that piezoelectric properties of multiwall BNNTs, where all individual nanotube layers have the same orientation of the B–N bond, would be greatly magnified (the total piezoelectric constant will be equal to or larger than the sum of the constants of the individual tubes composing the multiwall BNNT) and could reach the same order of magnitude as in ferroelectric ceramics. In a recent paper by Sai and Mele [29] both uniaxial and shear piezoelectric distortions were considered in heteropolar nanotubes and a model connecting piezoelectric properties of nanotubes with the limiting case of the flat sheet was constructed, allowing for a simple derivation of piezoelectric constants in chiral nanotubes.

TABLE 21.2
Piezoelectric Properties of Zigzag BNNT Bundles

(n, m)	Diameter (Å)	Z^*	$\lvert e_{33} \rvert$ (C/m²)	Ref.
(5,0)	3.91	2.739	0.389	
(6,0)	4.69	2.696	0.332	
(7,0)	5.47	2.655	0.293	
(8,0)	6.24	2.639	0.263	
(9,0)	7.04	2.634	0.239	
(10,0)	7.83	2.626	0.224	
(11,0)	8.57	2.614	0.211	
(12,0)	9.38	2.609	0.198	
(13,0)	10.26	2.605	0.186	
w-AlN		2.653	1.50	[27]
w-ZnO		2.110	0.89	[28]
$PbTiO_3$			3.23	[2]
P(VDF/TrFE)			≈0.12	[3]

The corresponding values for a few piezoelectric materials are listed for comparison. The data reported in the table are for nanotube bundles where we assume a hexagonal close-packed geometry with intertube equilibrium distance of 3.2 Å.

In addition to the piezoelectric stress constants $e_{\alpha\beta}$ connecting polarization and strain (see Equation 21.7), it is informative to take a look at piezoelectric strain constants $d_{\alpha\beta}$ that couple polarization and stress or, conversely strain and electric field.[30] Knowledge of these constants (again, only the d_{33} constant is important due to the reduced dimensionality of the system) would provide some insight about possible application areas for BNNTs. Piezoelectric strain constant d_{33} can be obtained from e_{33} by dividing it by the tube's Young's modulus. By using Table 21.2 and a typical value of 0.9 TPa for the Young's modulus,[31] we can estimate d_{33} as approximately 0.3 pC/N. This allows us to evaluate actuating abilities of zigzag BN nanotubes. For example, to obtain a 3% elongation (strain) of the tube, an electric field in the order of 100 V/nm has to be applied. This leads to the conclusion that BN nanotubes will probably be much better sensors (high piezoelectric stress constants) than actuators (low piezoelectric strain constants). In ferroelectric polymers such as PVDFs the situation is inverse. Their piezoelectric strain constants are two orders of magnitude larger than in BN nanotubes (30 pC/N is a typical value) and piezoelectric stress constants are low (about 0.1 C/m^2).

21.6 POLARIZATION EFFECTS IN NANOTUBULAR STRUCTURES

Although they have zero macroscopic polarization when isolated, and they are only mildly pyroelectric when bundled, BN nanotubes, when properly used in complex systems, can still be viewed as intrinsically polar materials and their properties can therefore be exploited toward technological applications. They are only intrinsically nonpolar in the ideal infinitely long configuration, but nonvanishing polar properties should manifest themselves when they are part of more complex designs, such as nanotube superlattices and heterostructures or, in general, nanostructures with symmetry properties reduced with respect to the cylindrical geometry of the nanotubes. In general, superlattices alternating polar and nonpolar materials should electronically respond to the internal electric fields, with phenomena such as charge separation and screening, or as we will see, enhanced field emission properties in finite, tip-shaped samples. Incidentally our calculations on BN-C heterojunctions and BN-C coaxial systems show that the spontaneous polarization fields associated with B and N atoms do indeed dramatically enhance field emission properties and lead to attractive electronic devices.

Carbon nanotubes are generally considered to be good electron emitters, owing to their electronic properties and exceptional aspect ratios. The remarkable emitting properties may be further improved by exploiting the electronic properties of BN-doped carbon systems. The idea here is to make use of the intrinsic polarization field as means to reduce the difficulty to extract electrons, i.e., the work function, at the tip of the system, thereby enhancing the extraction of electrons from the system.

21.6.1 NANOTUBE SUPERLATTICES AND HETEROSTRUCTURES

The orderly introduction of BN into otherwise pure carbon nanotubes may be realized in two different manners: either by substituting one CC pair with one BN

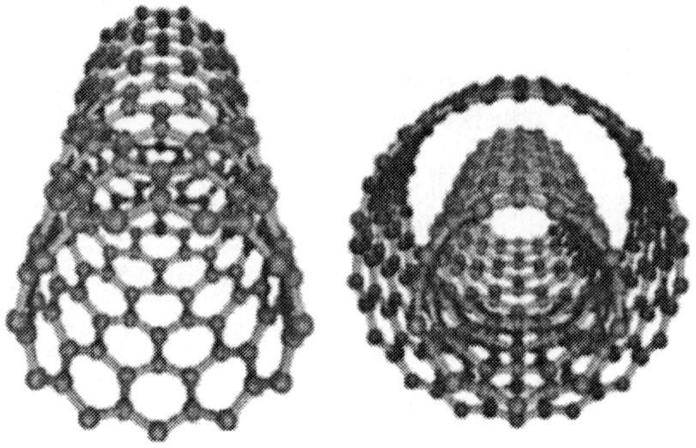

FIGURE 21.8 Prototypes of BN-C hybrids: (a) a BN-C heterojunction based on an (11,0) nanotube and (b) a (9,0) C and (17,0) BN biwalled system.

pair in the carbon nanotube lattice and therefore mixing the C and BN phases (Fig. 21.8(a)), or by creating a system that presents two coaxial homogeneous phases of C and BN (Fig. 21.8(b)). While early attempts to design BN-C heterojunctions by self-assembly have been so far unsuccessful, and actually led to the creating of biwalled BN-C systems, recent progress in the now well controlled substitution reaction (in which CC pairs are substituted by BN pairs) offers a potentially new route for the design of BN-C heterojunctions.

Since BN-C nanotube superlattices are quasi-one-dimensional (1-D) heterojunctions, their electronic properties are first and foremost characterized by their band offsets. In contrast to 3-D solids, the asymptotic band offsets of such heterojunctions are conveniently computed from the bulk properties of their isolated components, since the vacuum level may be used as unequivocal reference energy.[32] Pure BN nanotubes are wide band gap semiconductors with band gaps of 5.5 eV. Carbon nanotubes, on the other hand, may be either metallic or semiconducting, depending on their chiral indices. The $(l,0)$ zigzag tubes on which we will concentrate here are either metallic ($l = 3n$, where n is an integer) or semiconducting. In BN-C nanotube systems, the valence and conduction band offsets are spatially direct, with typical VLDA values given in Table 21.3. From this table, it is clear that simple matching of two straight nanotubes will give rise to a variety of band alignments with band offsets that are sensitive to the helicities and radii of the tubes. In particular, for

TABLE 21.3
Conduction (E_c) and Valence (E_v) Band Offsets for a Set of Zigzag BN-C Heterostructures Relative to Vacuum Level

	(7,0)	(8,0)	(9,0)	(10,0)
ΔE_c (eV)	−2.09	−2.50	−2.42	−2.52
ΔE_v (eV)	0.82	1.12	1.53	1.46

metallic carbon nanotubes, the heterojunction will behave as a Schottky diode. Thus combining the BN and carbon nanotubes of various symmetries offers a broad range of opportunities for band offset engineering and for the design of a number of different nanotubular heterojunctions.

The majority of BN-C nanotube superlattices are characterized by an axial dipolar electric field, which originates from the polar nature of the B–N bond. In a periodic superlattice of polar and nonpolar materials (such as BN and C), any polarization field will affect the behavior of the total electrostatic potential of the system. In fact, the latter displays a typical sawtooth behavior that is the signature of the presence of the field superimposed onto the periodic crystalline potential.[33] For convenience, we have computed the planar average of the electrostatic potential over the directions perpendicular to the tube axis, as shown in Fig. 21.9(a). The average potential displays strong oscillations due to the varying strengths of the ionic potentials. To eliminate this effect, we implemented the procedure of Reference 34, and thereby calculated the one-dimensional macroscopic average of the electrostatic potential of the system. The value of the dipole field was then obtained from the slope of the macroscopic

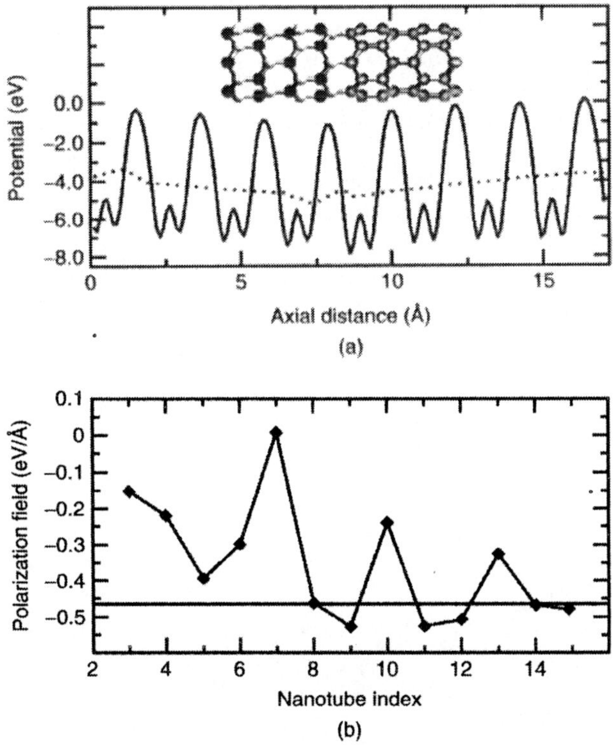

FIGURE 21.9 (a) Average electrostatic potential (solid line) and macroscopic average (dotted line) along a BN-C (6,0) superlattice (insert) and (b) macroscopic polarization field for zigzag (l,0) tubes as a function of the helicity index. The asymptotic value for planar BN is represented by the horizontal line.

average potential and is shown in Fig. 21.9(b). Since the geometry of the zigzag nanotubes maximizes the dipole moment of the B–N bond, the strongest effects will be observed in them. Hence we have primarily concentrated on zigzag nanotubes, while considering the flat sheet system as the limiting case for large diameter tubes (see Fig. 21.9(b)). In contrast, the (l,l) armchair nanotubes, in which one-third of the bonds are perpendicular to the tube axis, are not expected to display any dipole field. This is because any individual nanotube ring is charge neutral and therefore no field can develop. Chiral nanotubes have fields that are between the zigzag and armchair values and depend on the individual terminations. From Fig. 21.9(b), we see that the magnitude of the macroscopic electric fields for the various superlattices has a strong oscillatory character. Nanotubes with helicity indices 7, 10, . . . , $3n + 1$ are characterized by a substantially smaller field. This effect originates from the fact that the carbon sections effectively screen the macroscopic polarization field with an efficiency that depends on the nanotube helicity.

To understand that effect, one should analyze the electronic state at the valence band edge of the BN-C superlattice, which as the BN-C band offset calculations have shown, is always spatially localized in the carbon region.[16] This state plays an important role in determining the response to the macroscopic electric field induced by the BN section. For small-diameter nanotubes, the valence state displays either a longitudinal or a transverse symmetry as exemplified by the (7,0) and (8,0) tubes on Fig. 21.10.[35] Specifically for $(l = 3n + 1,0)$ nanotubes, such as the (7,0) tube, the valence state always assumes a longitudinal character. The axial distribution of valence electrons therefore induces a depolarization field that is opposite to the one induced by the BN section. This field is very effective in small-diameter nanotubes, so that for the (7,0) nanotube the screening is almost total and the observed macroscopic field is close to zero. For larger diameter nanotubes, the symmetry of the valence state gradually loses its strong axial or longitudinal character and the macroscopic field asymptotically approaches the value found for a flat BN sheet in the zigzag direction.

Despite the screening by the valence electrons, which are distributed over the carbon portion of the BN-C superlattices, a net polarization field is built up along any typical zigzag structure. The existence of such a macroscopic field clearly influences the extraction of electrons from BN-C systems. Qualitatively a good electron emitter is characterized by a large geometrical field enhancement factor β and a small work function ϕ, the latter being an intrinsic property of the emitter material.

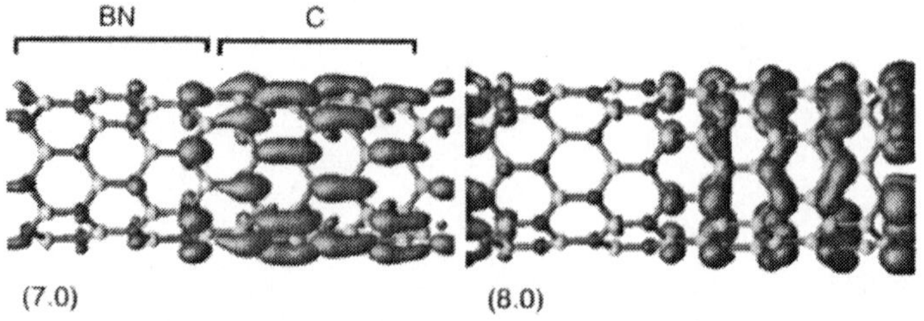

FIGURE 21.10 (Color figure follows p. 12.) Three-dimensional plots of the valence state for (7,0) and (8,0) BN-C superlattices.

Quantitatively the current at the emitter tip is calculated from the well-known Fowler–Nordheim relationship.[36]

To assess the efficacy of BN-C systems as emitters, we have computed the field enhancement factor and the work functions for tubes 15.62 Å in length in an applied field of 0.11 V/Å. As expected, the applied field was very well screened inside the system and a local field enhancement factor of 2.1 was obtained for the different BN-C systems. This value is very close to the one obtained on the corresponding pure carbon nanotube, which is not surprising since the field enhancement factor typically depends only on the emitter geometry. Since the field enhancement factor increases linearly with the sizes of the nanotubes, large enhancement factors are clearly possible for long nanotubes.[37] Turning to the effect on the work functions, we have built a set of finite-sized (6,0) zigzag structures using B, N, and C in various combinations. The work functions were then computed as the difference between the vacuum level and the Fermi energy of the system. The former was obtained using the previously discussed potential average procedure, while the latter was simply given by the highest occupied Kohn–Sham eigenstate of the system. In order to avoid any spurious effects arising from the periodic calculation, we have used the so-called dipole correction in a unit cell surrounded by a vacuum of at least 15 Å.[38] Due to the net polarization field experienced by the electrons, the work function in a NB-C heterostructure is reduced to 5.04 eV at the C tip and increased to 7.52 eV at the N tip. The same trend is observed for the BN-C system, for which the work function at the B tip is equal to 5.00 eV, while it takes a value of 6.45 eV at the C tip. The work function is therefore decreased by a significant 1.40 eV, as compared to the pure carbon system. This large difference leads to significant macroscopic effects. According to the Fowler–Nordheim relationship, the logarithm of the current density (J) depends upon the work function ϕ as in

$$J = \frac{C}{\phi}\beta E^2 e^{\frac{-B\phi^{3/2}}{\beta E}}$$

where E is the applied electric field, ϕ is the work function at the tube tip, and β is the field enhancement factor. It follows that the insertion of BN segments in C nanotubes may increase the current density by as much as two orders of magnitude as compared to pure carbon nanotube systems.

21.6.2 Multiwalled Hybrids

The previous discussion showed that the presence of BN atoms induces a significant improvement of field emission properties at the carbon tip in a NB-C heterojunction. However, the use of such devices as performant field emitters is limited by the major issue related to their synthesis. From a practical point of view, the related nested BN-C or C-BN systems offer a more direct way to take advantage of the polarity associated with the BN phase. To assess the field emission properties of BN-C devices, we have performed calculations of the field enhancement factor and of the work function. We have specifically investigated the effect of the presence of a finite BN nanotube shell surrounding a C nanotube: the C (5,5)-BN (18,0), as shown on Fig. 21.11. As expected

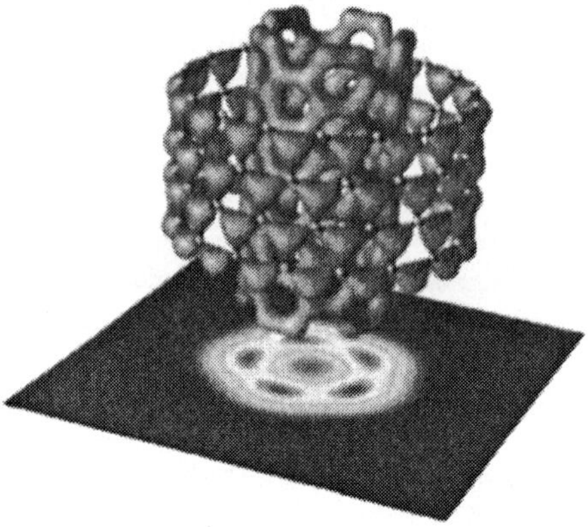

FIGURE 21.11 Electron density 5 Å from the tip of a C (5,5)-BN (18,0) biwalled system.

from classical electrostatics, the field enhancement factor at the carbon tip is not modified by the presence of the BN shell, since the insulating BN layer does not disturb the outstanding screening properties of the underlying C tube. On the contrary, the dipole field induced by the BN shell produces a net decrease of the work function (by about 0.5 eV) and a subsequent increase of the emitted current (by up to a factor of five), as compared to the current emitted by an isolated (5,5) carbon nanotube.

21.6.3 Experimental Results

Although field emission properties of carbon nanotubes are at the forefront of applied nanotechnology, a number of parameters for composite BNC and BN nanotubes remain not well established and a direct connection between theoretical predictions and experimental measurements cannot yet be made at the time of this writing. However, there are already indications that given a better control of the growth and the structures of these systems, their polarization properties might play a significant role in advanced applications. CN_x and BNC_x nanotubes have already shown remarkable, up to 4 μA, emission currents. This value exceeds by a factor of two the maximum stable field emission current obtained from pure carbon nanotube samples.[39] The same excellent field emission properties have been observed in aligned multiwalled nanotubes where the outer layer is a BN-rich nanotube that insulates semiconducting BNC layers.[40]

21.7 CONCLUSIONS AND FUTURE PROSPECTIVES

From the preceding discussion, it is clear that the engineering of nanostructures and nanocomposites that incorporate BN nanotubes could potentially lead to advanced device architectures for strain sensors and actuators, not to mention field effect devices. Three main technological challenges that have to be overcome before any useful devices with BN nanotubes can be produced are the following.

First, BN nanotubes have to be grown in quantity. Substantial progress has been achieved in this area in recent years with numerous techniques for BN nanotube synthesis — such as, for example, plasma arc discharge, laser ablation/heating, substitution reaction from carbon nanotubes or various chemical methods — becoming available. It is noteworthy that BN nanotubes *prefer* to grow in zigzag conformations,[41] which eliminates the problem of separating the tubes with different chiralities, an issue that has not yet been completely solved for carbon nanotubes. However, a different synthesis-related problem arises for BN nanotubes. In order for multiwall tubes to be piezoelectric, they must be grown in such a way that all the individual layers composing the multiwall tubes have the same direction as the polar B–N bond. A possible approach here could be to grow BN nanotubes in a strong electric field so that only a certain B–N bond direction would be preferred.

Second, efficient self-assembly procedures are required to make ropes or possibly molecular crystalline structures from individual tubes, again preserving the same direction of B–N bond among them.

Finally, advanced nanomanipulation techniques should be developed to convert pyro- and piezoelectric or field-emitting nanotube ropes and/or crystals into useful devices. This includes, for example, attaching contacts or covering them with protective coatings.

Of course, the aforementioned techniques for incorporation of BN- and carbon-based nanotubular structures into composite materials and their harnessing into useful geometries for sensing or actuating purposes are still in their infancy and require further investigation. However, we believe that the theoretical description of the microscopic properties that characterize the polarity and piezoelectric responses of these systems is a necessary and fundamental step for their possible industrial application.

PROBLEMS

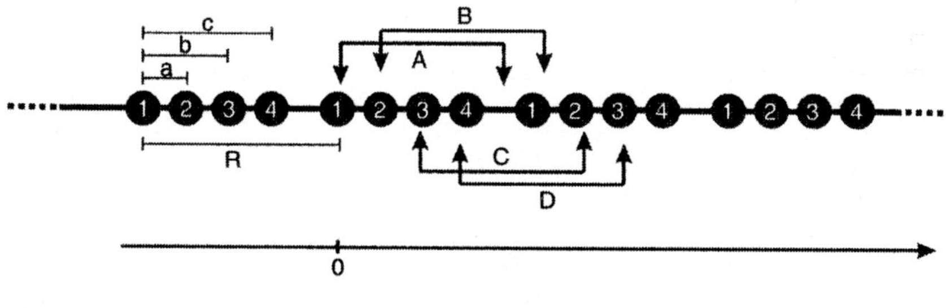

Consider a hypothetical one-dimensional periodic (infinite) system, with lattice constant R and with a unit cell composed of four atoms numbered 1 to 4 and positioned relative to one another, as depicted in the figure. At equilibrium the system is such that the net charge on atoms 1, 2, 3, and 4 is equal to δq, $-\delta q$, $-\delta q$, and δq, respectively.

1. Assuming that the ionic polarization is given by the dipole moment in a unit cell per unit volume, explain the apparent discrepancy between the values of the polarization computed using different choices of supposedly equivalent unit cells A, B, C, and D defined on the figure.
2. Supposing that the system is truncated into a finite system, find the optimal (in terms of maximum polarization) way to cut the system,

keeping the number of different atomic species equal. Show that the charge surface actually plays a crucial role in defining the magnitude of the polarization.

ACKNOWLEDGMENTS

The authors wish to thank Dr. Arrigo Calzolari for his help in computing polarization properties from Wannier functions.

This research was sponsored in part by the Mathematical, Information and Computational Sciences Division, Office of Advanced Scientific Computing Research of the U.S. Department of Energy under Contract DE-AC05-00OR22725 with UT-Battelle, LLC.

REFERENCES

1. Kumazawa, T., Kumagai, Y., Miura, H., Kitano, M., and Kushida, K., Effect of external stress on polarization in ferroelectric thin films, *Appl. Phys. Lett.*, **72**, 608, 1998.
2. Sághi-Szabó, G., Cohen, R.E., and Krakauer, H., First-principles study of piezoelectricity in PbTiO3, *Phys. Rev. Lett.*, **80**, 4321, 1998; First-principles study of piezoelectricity in tetragonal PbTiO3 and PbZr1/2Ti1/2O3, *Phys. Rev. B*, **59**, 12771, 1999.
3. Furukawa, T., Piezoelectricity and pyroelectricity in polymers, *IEEE Trans. Electr. Insul.*, **24**, 375, 1989; Eberle, G., Schmidt, H., and Eisenmenger, W., Piezoelectric polymer, *IEEE Trans. Diel. Electr. Insul.*, **3**, 624, 1996; Samara, G.A., Ferroelectricity revisited — advances in materials and physics, *Solid State Phys.*, **56**, 239, 2001.
4. Iijima, S., Helical microtubules of graphitic carbon, *Nature*, **354**, 56, 1991.
5. Bernholc, J., Brenner, D., Buongiorno Nardelli, M., Meunier, V., and Roland, C., Mechanical and electrical properties of nanotubes, *Annu. Rev. Mater. Res.*, **32**, 347, 2002.
6. Rubio, A., Corkill, J.L., and Cohen, M.L., Theory of graphitic boron nitride nanotubes, *Phys. Rev. B*, **49**, 5081, 1994.
7. Chopra, N.G., Luyken, R.J., Cherrey, K., Crespi, V.H., Cohen, M.L., Louie, S.G., and Zettl, A., Boron nitride nanotubes, *Science*, **269**, 966, 1995.
8. Stephan, O., Ajayan, P.M., Colliex, C., Redlich, P., Lambert, J.M., Bernier, P., and Lefin, P., Doping graphitic and carbon nanotube structures with boron and nitrogen, *Science*, **266**, 1683, 1994.
9. Loiseau, A., Willaime, F., Demoncy, N., Hug, G., and Pascard, H., Boron nitride nanotubes with reduced numbers of layers synthesized by arc discharge, *Phys. Rev. Lett.*, **76**, 4737, 1996.
10. Cumings, J. and Zettl, A., Mass production of boron nitride double-wall nanotubes and nanococoons, *Chem. Phys. Lett.*, **316**, 211, 2000.
11. Lee, R.S., Gavillet, J., de la Chapelle, M.L., Loiseau, A., Cochon, J.-L., Pigache, D., and Willaime, J.T.F., Catalyst-free synthesis of boron nitride single-wall nanotubes with a preferred zig-zag configuration, *Phys. Rev. B*, **64**, 121405, 2001.
12. Blase, X., Charlier, J.C., DeVita, A., and Car, R., Theory of composite BxCyNz nanotube heterojunctions, *Appl. Phys. Lett.*, **70**, 197, 1997.

13. Zhang, P. and Crespi, V.H., Plastic deformations of boron–nitride nanotubes: an unexpected weakness, *Phys. Rev. B*, **62**, 11050, 2000; Bettinger, H.F., Dumitricâ, T., Scuseria, G.E., and Yakobson, B.I., Mechanically induced defects and strength of BN nanotubes, *Phys. Rev. B*, **65**, 041406, 2002.
14. Mele, E.J. and Král, P., Electric polarization of heteropolar nanotubes as a geometric phase, *Phys. Rev. Lett.*, **88**, 056803, 2002.
15. Nakhmanson, S.M., Calzolari, A., Meunier, V., Bernholc, J., and Buongiorno-Nardelli, M., Spontaneous polarization and piezoelectricity in boron nitride nanotubes, *Phys. Rev. B*, **67**, 235406, 2003.
16. Meunier, V., Roland, C., Bernholc, J., and Buongiorno Nardelli, M., Electronic and field emission properties of boron nitride/carbon nanotube superlattices, *Appl. Phys. Lett.*, 81, **47**, 2002.
17. King-Smith, R.D. and Vanderbilt, D., Theory of polarization of crystalline solids, *Phys. Rev. B*, **47**, 1651, 1993; Vanderbilt, D. and King-Smith, R.D., Electric polarization as a bulk quantity and its relation to surface charge, *Phys. Rev. B*, **48**, 4442, 1993.
18. Resta, R., Macroscopic polarization in crystalline dielectrics: the geometric phase approach, *Rev. Mod. Phys.*, **66**, 899, 1994.
19. Vanderbilt, D., Berry phase theory of proper piezoelectric response, *J. Phys. Chem. Solids*, **61**, 147, 2000.
20. Marzari, N. and Vanderbilt, D., Maximally localized generalized Wannier functions for composite energy bands, *Phys. Rev. B*, **56**, 12847, 1997; Souza, I., Marzari, N., and Vanderbilt, D., Maximally localized Wannier functions for entangled energy bands, *Phys. Rev. B*, **65**, 035109, 2002.
21. Dreizler, R.M. and Gross, E.K.U., *Density Functional Theory: An Approach to the Quantum Many-Body Problem*, Springer-Verlag, Berlin, 1990.
22. Briggs, E.L., Sullivan, D.J., and Bernholc, J., Real-space multigrid-based approach to large-scale electronic structure calculations, *Phys. Rev. B*, **54**, 14362, 1996.
23. Baroni, S., Dal Conso, A., De Gironcoli, S., and Giannozzi, P., Plane wave-based calculations have been performed using the PWscf package, http://www.pwscf.org.
24. Ceperley, D.M. and Alder, B.J., Ground state of the electron gas by a stochastic method, *Phys. Rev. Lett.*, **45**, 566, 1980; Perdew, J.P. and Zunger, A., Self-interaction correction to density-functional approximations for many-electron systems, *Phys. Rev. B*, **23**, 5048, 1981.
25. Fuchs, M. and Scheffler, M., *Ab initio* pseudopotentials for electronic structure calculations of poly-atomic systems using density-functional theory, *Comput. Phys. Commun.*, **119**, 67, 1998; Troullier, N. and Martins, J.L., Efficient pseudopotentials for plane-wave calculations, *Phys. Rev. B*, **43**, 1993, 1991.
26. Kleinman, L. and Bylander, D.M., Efficacious form for model pseudopotentials, *Phys. Rev. Lett.*, **48**, 1425, 1982.
27. Zoroddu, A., Bernardini, F., Ruggerone, P., and Fiorentini, V., First principles prediction of structure, energetics, formation enthalpy, elastic constants, polarization, and piezoelectric constants of AlN, GaN, and InN: comparison of local and gradient-corrected density-functional theory, *Phys. Rev. B*, **64**, 045208, 2001.
28. Bernardini, F., Fiorentini, V., and Vanderbilt, D., Spontaneous polarization and piezoelectric constants of III-V nitrides, *Phys. Rev. B*, **56**, R10024, 1997; Accurate calculation of polarization-related quantities in semiconductors, *Phys. Rev. B*, 63, 193201, 2001.

29. Sai, N. and Mele, E.J., Microscopic theory for nanotube piezoelectricity, *Phys. Rev. B*, **68**, 241405(R), 2003.
30. Nye, J.F., Physical Properties of Crystals, Claredon Press, Oxford, 1985., Chap. 7.
31. Hernandez, E., Goze, C., Bernier, P., and Rubio, A., Elastic properties of C and $B_xC_yN_z$ composite nanotubes, *Phys. Rev. Lett.*, **80**, 4502, 1998.
32. Resta, R., private communication; Kleinman, L., Comment on the average potential of a Wigner solid, *Phys. Rev. B*, **24**, 7412, 1981.
33. Posternak, M., Baldereschi, A., Catellani, A., and Resta, R., *Ab initio* study of the spontaneous polarization of pyroelectric BeO, *Phys. Rev. Lett.*, **64**, 1777, 1990.
34. Baldereschi, A., Baroni, S., and Resta, R., Band offsets in lattice-matched heterojunctions: a model and first-principles calculations for GaAs/AlAs, *Phys. Rev. Lett.*, **61**, 734, 1988.
35. Kane, C.L. and Mele, E.J., Broken symmetries in scanning tunneling images of carbon nanotubes, *Phys. Rev. B*, **59**, R12759, 1999.
36. Fowler, R.H. and Nordheim, L.W., Electron emission in intense electric fields, *Proc. Roy. Soc. Lond.*, **A119**, 173, 1928.
37. Han, S. and Ihm, J., Role of the localized states in field emission of carbon nanotubes, *Phys. Rev. B*, **61**, 9986, 2000.
38. Bengtsson, L., Dipole correction for surface supercell calculations, *Phys. Rev. B*, **59**, 12301, 1999.
39. Golberg, D., Dorozhkin, P.S., Bando, Y., Dong, Z.-C., Tang, C.C., Uemura, Y., Grobert, N., Reyes-Reyes, M., Terrones, H., and Terrones, M., Structure, transport and field-emission properties of compound nanotubes: CN_x vs. BNC_x ($x < 0.1$), *Appl. Phys. A*, **76**, 499, 2003.
40. Golberg, D., Dorozhkin, P.S., Bando, Y., Dong, Z.-C., Grobert, N., Reyes-Reyes, M., Terrones, H., and Terrones, M., Cables of BN-insulated B-C-N nanotubes, *Appl. Phys. Lett.*, **82**, 1275, 2003.
41. Golberg, D., Bando, Y., Kurashima, K., and Sato, T., Ropes of BN multi-walled nanotubes, *Solid State Commun.*, **116**, 1, 2000; HRTEM and electron diffraction studies of B/N-doped C and BN nanotubes, *Diam. Relat. Mater.*, **10**, 63, 2001; Lee, R.S., Gavillet, J., Lamy de la Chapelle, M., Loiseau, A., Cochon, J.-L., Pigache, D., Thibault, J., and Willaime, F., Catalyst-free synthesis of boron nitride single-wall nanotubes with a preferred zig-zag configuration, *Phys. Rev. B*, **64**, 121405(R), 2001.

22 Multiscale Modeling of Stress Localization and Fracture in Nanocrystalline Metallic Materials

Vesselin Yamakov, Dawn R. Phillips, Erik Saether, and Edward H. Glaessgen

CONTENTS

22.1 INTRODUCTION

Uneven stress distribution and stress localization during deformation are key factors for fracture and failure in polycrystalline and nanocrystalline metals. The inhomogeneous polycrystalline microstructure that consists of grains of different sizes and shapes joined together at different angles and forming various types of grain boundaries (GBs) creates inhomogeneous deformation fields under homogeneous loading. There are a number of factors responsible for the appearance of inhomogeneous deformation inside the polycrystal. The coexistence of grains of different sizes and orientations with anisotropic elastic properties is one factor. The difference in the structures and properties of the GBs between grains of various misorientations is a second factor for uneven stress distribution. A third factor is grain-boundary sliding

(GBS), i.e., the rigid translation of one grain relative to another at the GB interface. When grains deform, GBS is an inevitable process as a result of the relative movements and rearrangements of the grains.[1] GBS is a strongly inhomogeneous mode of deformation localized at a very narrow interface layer, thus creating very strong shear forces. This, together with the fact that the shear strength of a general GB between grains of high misorientation angle (high-angle GB) is much lower than the shear strength of the perfect crystal,[2,3] causes GBS to be a prominent deformation mode. Particularly in nanocrystalline metals, because of their extremely small grain size (less than 100 nm) resulting in a dense network of GBs, GBS assisted through stress-driven GB diffusion for grains smaller than 20 nm[4–6] and through dislocation slip in larger grains[7–9] was found to be a major mode of deformation. During GBS, the load transfer between the sliding surfaces is significantly reduced, and the load is redirected to other places more resistant to sliding such as the triple junctions where three GBs meet and GBS cannot be accommodated. In this way, sliding creates redistribution of the load and the appearance of stress localization in the microstructure that, in the absence of an efficient accommodation mechanism, can lead to void formation and microcracking starting at the triple junctions.[10,11]

This chapter describes a multiscale modeling strategy that is used to study the effect of GBS for stress localization in a polycrystalline microstructure. Although they reveal system behavior at atomic-level resolution, molecular dynamics (MD) calculations of large systems of grains are computationally prohibitive to perform, and a more efficient analysis technique is sought. Finite element method (FEM) analysis is an obvious choice, but the FEM models cannot simulate the deformation mechanisms found within the systems of grains *a priori*. Thus, the FEM models must be tuned in order to reproduce the stress localization observed in the atomistic simulations. For this purpose, a nanocrystalline model of bimodal grain-size distribution was constructed and is convenient for study by both MD and FEM simulations. The MD simulations revealed the behavior of this model with atomic-scale details. Additional MD simulations on a bicrystal model were performed to extract the elastic and yield properties of the GBs presented in the model. These GB properties were then used to tune the FEM model to reproduce closely the behavior of the nanocrystalline model to match with the MD simulation results. The FEM model then served as part of a multiscale modeling strategy to extrapolate the MD-derived information to larger scales to study the failure properties of nanocrystalline metals.

In this chapter, details of the configuration model used in both MD and FEM simulations are discussed first. The specific features of the MD model are then presented. Next the FEM version of the model is discussed, and the results of the FEM simulation are compared with those from the MD simulation. Finally a large-scale FEM simulation of a typical microstructure of 100 grains is analyzed to reveal the stress distribution due to grain boundary sliding in a more general polycrystalline system.

22.2 CONFIGURATION MODEL

To investigate the effect of GBS on stress distribution in a polycrystal of inhomogeneous grain size, a configuration model, presented in Fig. 22.1(a) and (b), was used. This configuration was used in both MD and FEM simulations with the

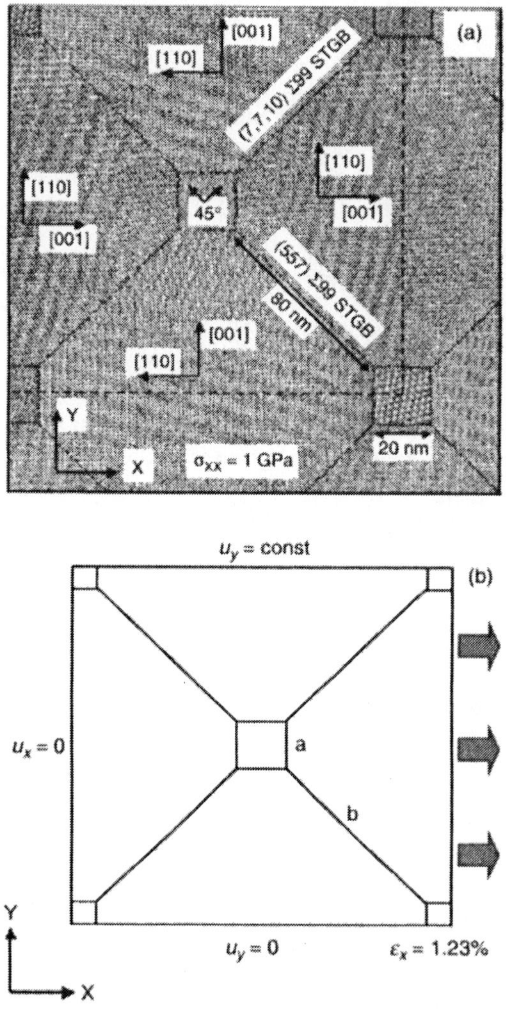

FIGURE 22.1 (a) Snapshot of the thermally equilibrated polycrystalline microstructure designed to study the effect of grain boundary sliding on stress distribution in a polycrystal of bimodal grain size by molecular dynamics simulation. The grain boundaries are shown as darker lines of atoms that were identified to be in a disordered (noncrystalline) local environment. The dashed lines mark the actual system box dimensions, while for better visualization, the microstructure is slightly extended using periodic boundary conditions. (b) Repeating unit used in finite element simulations to model grain structure shown in (a). As there is no characteristic length scale in the finite element model, the dimensions in (b) are relative, characterized by $a:b = 1:4$ and consistent with (a).

additional goal of tuning the parameters of FEM to reproduce as closely as possible the MD simulation. The model is a periodically repeating unit of a four-grain octagon–square configuration that, when replicated in the *x*- and *y*-directions, represents a polycrystal of a bimodal grain-size distribution. The advantages of this configuration are the following: (1) The model allows exploration of the effects of large variations in the ratio $a:b$ between the grain sizes (Fig. 22.1(b)), which for this study was set to 1:4; (2) the model is both simple enough to allow for MD simulation and informative enough to be used in the FEM simulation to compare with the MD results; and (3) the configuration has a four-fold symmetry (against 90° rotation), which further simplifies the analysis. In both MD and FEM simulations, this configuration is deformed by applying stress or strain, respectively, along the *x*-direction. The deformation conditions are chosen such that the grains deform elastically, while the GBs deform plastically by GBS. An interatomic potential representative of aluminum (Al) is used in the MD model, and an isotropic continuum is assumed for the grains in the FE model to isolate the effect of crystal anisotropy on the stress distributions. Though this would result in some inevitable discrepancy in the results, it helped to identify the amount of stress distribution due to GBS alone. Fixed boundary conditions shown in Fig. 22.1(b) that for an isotropic media mimic the periodic boundary conditions in the MD simulation shown in Fig. 22.1(a) were applied in the FE simulation.

22.3 MOLECULAR DYNAMICS MODEL

To investigate the effect of GBS on the stress distribution in a polycrystal of inhomogeneous grain size, atomistic MD simulations were used where the only predefined input quantity was the interactive potential between individual atoms, which, following Newtonian dynamics, defines their motion. The embedded atom method (EAM) many-body potential[12] was fitted to reproduce closely the elastic and thermodynamic properties of a perfect Al crystal and used in the simulation. A textured or columnar microstructure model with periodic boundary conditions in all three dimensions to mimic bulk conditions[13] was selected. While providing a fully three-dimensional (3-D) treatment of the underlying physics, this model makes it possible to simulate relatively large grains, because only a few lattice planes need to be considered in the periodically repeated texture direction. The texture *z*-axis is along a [110] crystallographic orientation, which enables realistic dislocation slip dynamics along six available slip systems in two [111] slip planes in each grain.[14] These slip systems are adequate to accommodate any two-dimensional deformation in the *x-y* plane of the simulated structure. The grain boundaries, formed between the grains in this model, are [110] tilt GBs that have been studied extensively in the past 15 years, both experimentally[15–18] and by simulations.[17–21] In spite of some drawbacks inherent to this model when compared to a fully 3-D structure, as discussed in Reference 7, the model was remarkably successful in predicting the highly unexpected deformation twinning in nanocrystalline Al[22,23] and the transition from partial to perfect slip in nanocrystalline fcc metals.[14,22] Both of these phenomena have since gained solid experimental support.[24–26]

A four-grain octagon-square configuration was used as a test model. A snapshot of the four-grain octagon-square microstructure after thermal equilibration at 100 K is given in Fig. 22.1(a). The configuration represents a periodically repeating unit of large octagonal grains of size $b = 80$ nm encapsulating small square grains of size $a = 20$ nm. The thickness of the system in the texture direction was set equal to 5 (1, 1, 0) atomic planes (1.43 nm) as in the previous columnar models.[13,14,22,23] Within these dimensions, the system contains 1.7 million atoms. By rotating one of the two octagonal grains at 90° relative to the other (the crystallographic orientations are shown in Fig. 22.1), the four 45° inclined diagonal GBs become 90° Σ99 symmetric-tilt GBs (STGBs) for which the atomic structure in Al is known from the literature[18]. The two types of Σ99 STGBs with (5, 5, 7) and (7, 7, 10) GB planes (see Fig. 22.1(a)) are structurally very similar and practically undistinguishable from the point of view of their mechanical properties, which preserves the four-fold symmetry (against 90° rotation) of the structure. In addition, the two small square grains are rotated by 45° relative to the large grains, thus forming high-angle asymmetric tilt GBs[15,16] (Fig. 22.1(a)).

To initiate GB sliding, the microstructure was loaded with a uniform tensile stress of 1 GPa along the x-axis in Fig. 22.1(a). The stress was applied using the Parrinello–Rahman[27] constant stress technique combined with a Nose–Hoover thermostat[28] for constant-temperature simulation. This stress is well below the threshold stress of about 2 GPa[7,14] needed to start nucleation of dislocations from the GBs, which are the only possible dislocation sources in grains of nanometer size.[7,22] Thus the dislocation activity, as a possible accommodation mechanism, was readily suppressed. By running the simulations at a very low temperature of 100 K (a value still high enough to avoid quantum mechanical effects existing in a real structure that cannot be captured by the classical MD technique), GB diffusion, as another possible accommodation mechanism was also eliminated. Low temperature also prevents grain growth, a process that otherwise would be very strong in such a system of a bimodal grain-size distribution. Restricting the dislocation activity and GB diffusion helps to map the MD model nearer to our FEM simulations, which, at the present stage, do not include these mechanisms.

During loading, the system reached an equilibrium strain of 1.23%. The strain was elastic within the grains and plastic at the GBs, thus initiating GB sliding. At this stage, to present the stress distribution in the system, two-dimensional stress maps were created. The two-dimensional resolution of the maps is an area of 6 × 6 lattice parameters (2.43 × 2.43 × 1.43 nm^3 volume enclosing 432 atoms) over which the local stress was averaged. The stress maps were stored every 1 ps over a period of 22 ps after reaching equilibrium (approx. 40,000 MD steps) and then averaged in time to smooth out the fluctuations always present in a system of such small size. These averaging procedures in space and time gave a good estimate of the local stress calculated by the virial expansion technique in Parrinello–Rahman stress calculation.[27] All the stress maps for the three stress components in the x-y plane of deformation (i.e., normal to the texture z-axis), σ_{xx}, σ_{yy}, and σ_{xy}, showed distinctive stress distribution and stress localization that were then compared with an FEM simulation of the same configuration, as will be discussed.

22.4 SHEAR STRENGTH OF A GRAIN BOUNDARY

Knowledge of the shear strength of the GBs presented in the system is crucial for understanding the stress distribution due to GBS. This shear strength is also required as an input for the FEM model discussed later. For this purpose, a separate MD simulation of a bicrystal cubic system, presented in Fig. 22.2, was used. The same interatomic potential[12] as in the octagon–square model was applied.

The two crystals were crystallographically oriented in such a way as to form a (7, 7, 10) Σ99 STGB (Fig. 22.2). This type of GB, together with the very similar (5, 5, 7) Σ99 STGB, was expected to exhibit most of the GBS in the octagon–square model (Fig. 22.1) due to the maximum resolved shear stress on the planes inclined at 45° to the tensile direction. In the simulation, the system box with periodic boundary conditions was allowed to shear in all three directions in addition to expansion or contraction, making it possible to estimate the shear strain directly. The system was loaded with a symmetric shear stress ($\sigma_{xy} = \sigma_{yx}$) to prevent torque forces and to initiate GBS in the direction consistent with the sliding direction between the octagons in the model. In addition, a control simulation was performed on a perfect crystal of the same size and orientation ([5, 5, 7]/[7, 7, 10]/[1, 1, 0]) to obtain the shear strain of a perfect crystal under the same simulation conditions. The perfect crystal strain was then subtracted from the bicrystal strain, and the resulting strain due to the GB alone is presented in Fig. 22.3 as a function of load for three different temperatures.

Following Schiotz et al.,[4] the yield stress was defined as the stress where the strain starts to depart from linearity, a definition convenient for use in MD simulations. For the temperature of $T = 100$ K used in the simulation of the octagon–square model the obtained yield stress is $\sigma_y = 0.2$ GPa. As expected, the yield stress decreases with increasing temperature of the system, becoming 0.14 GPa at 200 K

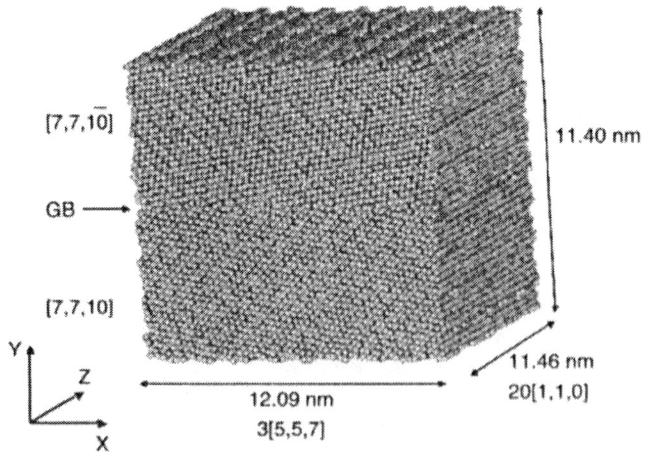

FIGURE 22.2 Bicrystal cubic microstructure used to determine the shear strength of Σ99 symmetric tilt grain boundary formed in the octagon–square system.

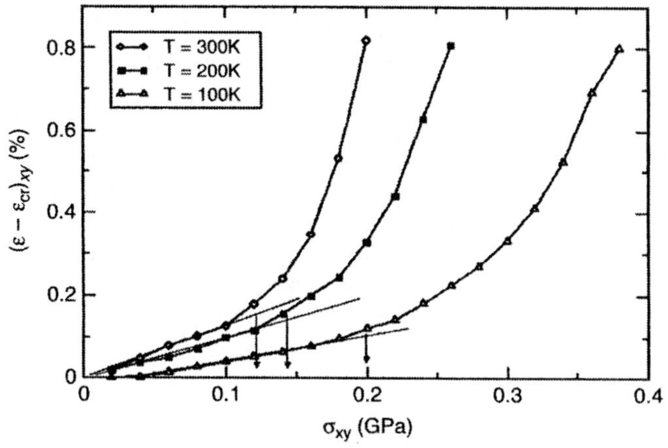

FIGURE 22.3 Grain boundary shear strain versus stress extracted from the bicrystal model. The straight lines tangential to the curves at their origin mark the linear elastic regime. The arrows show the yield stress for each of the three simulated temperatures defined as the stress where the strain starts to depart from linearity.

and 0.12 GPa at 300 K. For comparison, the theoretical shear strength of a perfect Al crystal was estimated at around 2.8 GPa.[29] Thus, the GBs are shown to be more than an order of magnitude weaker to shear than the perfect crystal grains. Applying a tensile load of 1 GPa along the x-axis in the octagon–square model (Fig. 22.1(a)) resulted in a resolved shear stress of 0.5 GPa along the diagonal $\Sigma 99$ GB planes. This stress is more than twice as large as the GB yield stress, as defined here (see Fig. 22.3), but still well below the shear strength of the perfect crystal. This ensures a regime of deformation where the GBs would experience plastic sliding while the crystal grains would deform elastically.

22.5 FEM SIMULATION

For the finite element simulations, a modified version of the FRANC code developed at Cornell University[30] was used and specifically designed to study crack initiation and propagation in metallic polycrystals with explicit representations of grains and grain boundaries. Each of the grains in the model was given isotropic, elastic material properties. The properties used were Young's modulus, $E = 74.8$ GPa, and Poisson's ratio, $\nu = 0.346$. These values were calculated averages[31] from the anisotropic elastic constants of the interatomic potential[12] used in the MD approach. To accurately account for the entire structure, a repeating unit, similar in geometry to the one used in the MD simulation, was chosen (Fig. 22.1(b)). As in the MD model, the ratio of the lengths of the grain boundaries was $a:b = 1:4$. Periodic boundary conditions are necessary to accurately compare results to those obtained from MD simulations. The boundary and loading conditions shown in Fig. 22.1(b) are consistent with periodic boundary conditions for an isotropic material.

In this work, cohesive zone models (CZMs) were used to characterize grain boundary behavior.[32] Cohesive zone models assume cohesive interactions of the material around a grain boundary and permit the appearance of fracture surfaces in a continuum.[33] To simulate the sliding behavior along grain boundaries, two independent cohesive zone models were chosen for the normal and shear components of the traction and displacement. To permit sliding, a perfectly plastic relationship was chosen for the shear model with yield stress of 0.2 GPa; this stress is consistent with the numbers obtained from the MD criterion discussed previously. To restrict opening, the normal CZM was specified as having linear elastic behavior with high stiffness. The analysis was performed for 1.23% applied strain. This applied strain corresponds to the 1 GPa applied stress in the MD simulations.

22.6 RESULTS AND DISCUSSION

Applying a uniform tensile stress of 1 GPa creates resolved shear stresses larger than the GB yield stress but lower than the yield stress of the grains, producing plastic sliding at the GBs and elastic deformation in the grain interiors. GBS reduces the load transfer from one grain to another and redistributes the stress. The σ_{xx}, σ_{yy}, and σ_{xy} stress distributions for the octagon-square model are shown in Fig. 22.4(a–f). The results from the FE simulations presented in Fig. 22.4(a–c) are compared with the stress maps obtained from the MD simulation and presented in Fig. 22.4(d–f). Both models show that the small square grains take most of the load produced from the tensile stress (Fig. 22.4(a) and (d)) and from the Poisson contraction perpendicular to the tensile x-direction (Fig. 22.4(b) and (e)). This contraction creates a strong tensile stress at the sides of the square grains normal to the tensile direction (regions marked (1) in Fig. 22.4(b) and (e)). The close similarity between the MD and FEM stress distribution and the fact that the FEM model was isotropic indicate that this stress distribution is solely due to GBS. Through GBS, the grains become decoupled in their elastic deformation. At the GBs, where only two grains meet, this decoupling creates sliding, but at the triple junctions, where three grains meet, strong incompatibility stresses are produced. This is seen in both FEM and MD results (Fig. 22.4(c) and (f)).

The most direct evidence for the presence of GBS is seen on the MD σ_{xy}–stress map in Fig. 22.4(f). The four diagonal GBs showed strong shear stress in the two opposite directions (with stress values presented in red and blue). Figure 22.5 is a schematic representation of the grain sliding responsible for the observed shear stress distribution. Grains (1) and (3) are displacing toward each other, compressing the small grain (5) in the y-direction, while grains (2) and (4) are displacing apart and stretching the grain (5) in the x-direction. The displacement of the large grains results in sliding at the GBs, schematically shown in Fig. 22.5 as couples of oppositely directed arrows on both sides of the diagonal GBs. The direction of each arrow corresponds to the local displacement of each grain at the GB. This displacement produces strong local shear stress at the diagonal GBs, as seen in Fig. 22.4(c) and (f).

In spite of the overall good match between the stress distribution from MD and FEM simulations, there are also significant mismatches that can be discovered after a careful examination of Fig. 22.4(a–f). For example, in the MD simulation, there is no full symmetry of the stresses produced on the two small grains, and there are

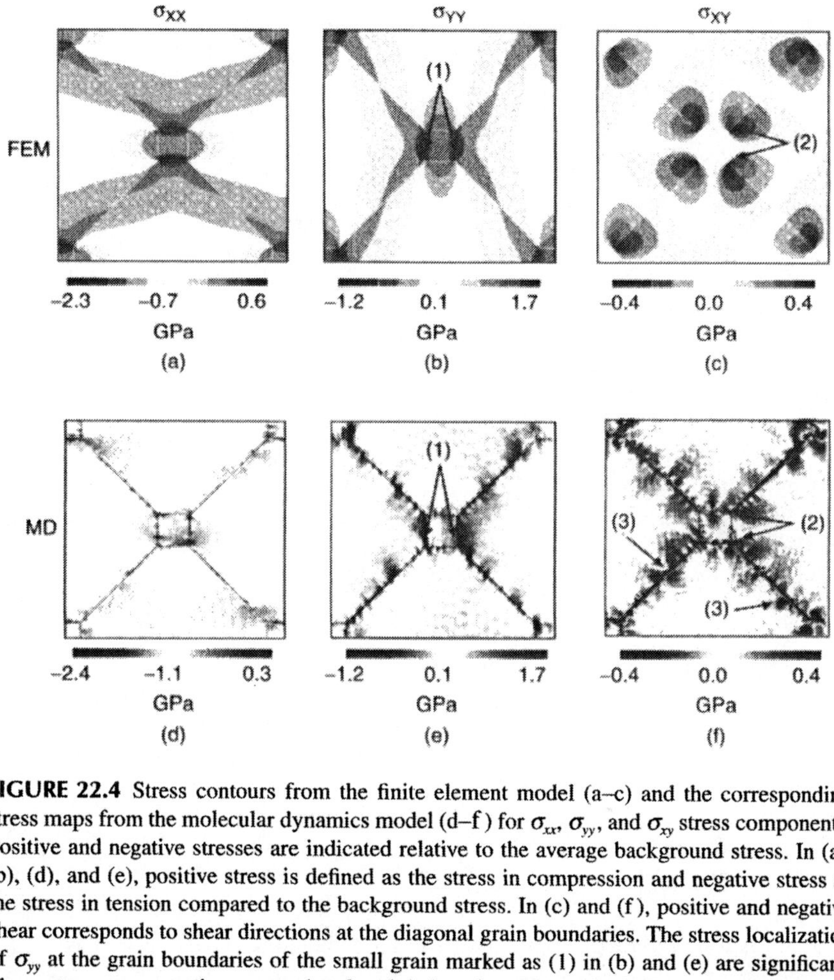

FIGURE 22.4 Stress contours from the finite element model (a–c) and the corresponding stress maps from the molecular dynamics model (d–f) for σ_{xx}, σ_{yy}, and σ_{xy} stress components. Positive and negative stresses are indicated relative to the average background stress. In (a), (b), (d), and (e), positive stress is defined as the stress in compression and negative stress as the stress in tension compared to the background stress. In (c) and (f), positive and negative shear corresponds to shear directions at the diagonal grain boundaries. The stress localization of σ_{yy} at the grain boundaries of the small grain marked as (1) in (b) and (e) are significant. Shear stress concentrations created at the triple junctions are marked (2) in (c) and (f). A few grain boundary dislocations that appeared in the molecular dynamics simulation and which create disturbances in the grain boundary stress field are indicated by (3) in (f).

some deviations in the stresses on the large grains as well (most pronounced in Fig. 22.4(d)). Possibly the main reason for these deviations is the small but noticeable anisotropy in the MD model that results from the interatomic potential. This potential was fitted[12] to reproduce closely the elastic constants including the anisotropy of monocrystalline Al. An anisotropic FEM simulation with the proper periodic boundary conditions is needed to achieve a better comparison.

Another source of mismatch between the two models is the inevitable existence of GB dislocations in the MD model. In a polycrystalline MD model, absolutely perfect GBs are impossible to sustain during loading, even if they were produced initially.

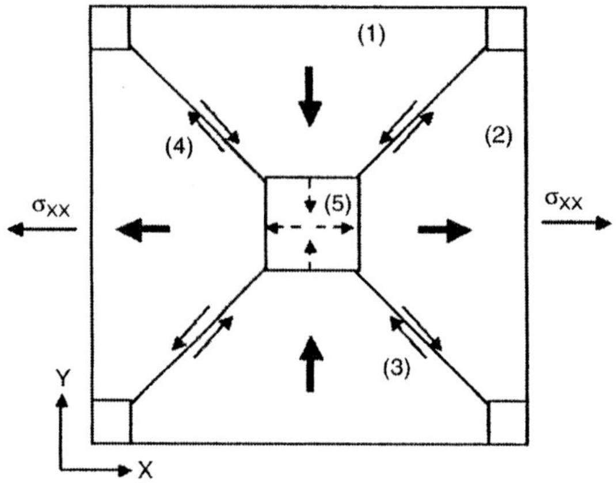

FIGURE 22.5 Schematic representation of the sliding directions of the four octagonal grains (bold arrows) and the inserted stresses on the small square grain in the middle (dashed arrows). The local displacements of the material at the grain boundaries producing grain boundary sliding are presented by couples of opposite pointing arrows on both sides of the boundaries.

The same is true for real materials such as nanocrystalline metals, in which GB dislocations are known to contribute to the deformation.[10] However, at this stage, GB dislocations are not considered in the FEM model. The GB dislocations present in the MD model have long-range elastic fields that can significantly modify the stress distribution in close proximity to the GBs, thus altering it from the FEM result. Such an effect produced by GB dislocations is marked as (3) in Fig. 22.4(f). Also, the phonon waves always existing in finite temperature MD simulation create periodic modulations in the MD stress maps (Fig. 22.4(d–f)) that are nonexistent in the FEM stress distributions (Fig. 22.4(a–c)). Even averaging the stress over a 20-ps time interval was not enough to smooth these fluctuations. One reason may be that the periodic boundary conditions imposed on the system, together with its high configuration symmetry, create stationary waves that cannot be smoothed out over such a short period of time. All these factors make the MD maps shown in Fig. 22.4 much richer and more complicated compared to the substantially more idealized FEM results.

However, the generally close quantitative match between the stress distribution obtained from MD and FEM simulations of the octagon–square configuration (Fig. 22.4) gives confidence that implementing the properly parameterized CZM elements in the FEM model correctly reproduces GBS for a polycrystalline specimen. With this confidence, FEM can now be used to simulate a typical nanocrystalline microstructure. For this purpose, a sample with 100 randomly shaped and sized grains represented by Voronoi polygons[34] was constructed. The grain configuration, along with the boundary and loading conditions, is shown in Fig. 22.6.

For 1.23% applied strain, the σ_{xx} stress distribution is shown in Fig. 22.7. The similarities in grain boundary behavior between the randomly configured model and the idealized octagon–square configuration are highlighted. This larger scale

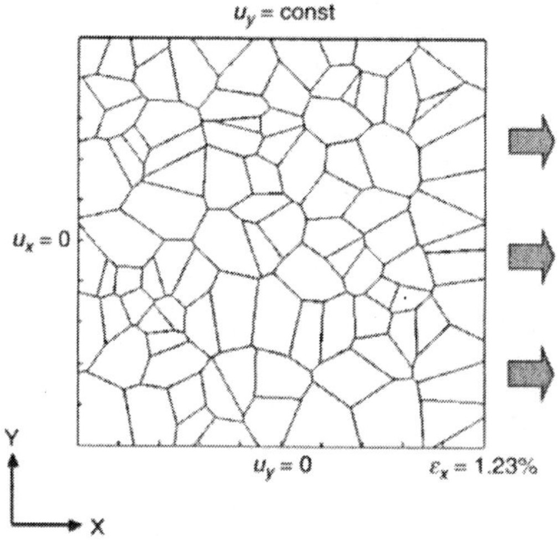

FIGURE 22.6 Material model with 100 randomly shaped and sized grains represented by Voronoi polygons.

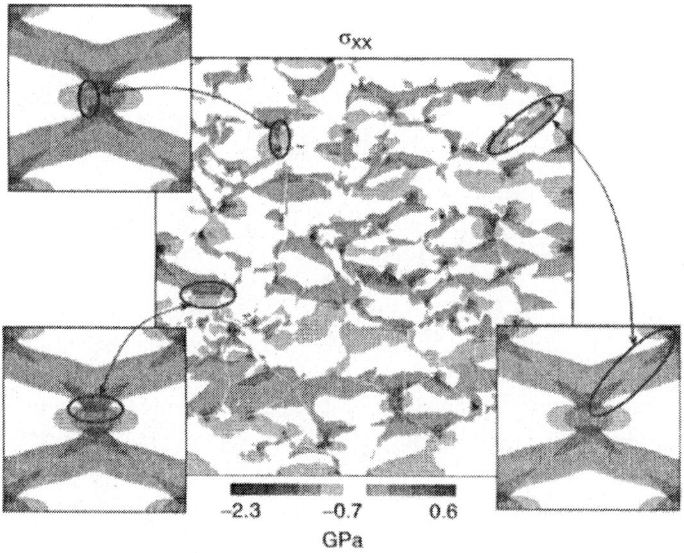

FIGURE 22.7 (Color figure follows p. 12.) Stress contour σ_{xx} for 100-grain material subjected to 1.23% strain with similarities to the idealized model highlighted.

simulation shows that stress localizations due to GBS are common within the nanocrystals, and local stress can exceed several times the background stress, thus marking spots for a possible crack nucleation. It is not surprising that these stress concentrations are mostly localized at the triple junctions where GB sliding cannot be accommodated.[35]

Summarizing the results from this section, it was shown that simulating the behavior of geometrically identical nanocrystalline structures by MD and FEM techniques can serve two purposes. First, it helps to tune the parameters of the larger scale FEM simulation to the results from the lower scale MD simulation, thus incorporating the effects of atomistic mechanisms, present in MD and essential for the deformation at nanoscales, into a continuum model. Second, such a parallel use of two very different techniques can help to identify more unambiguously the underlying mechanism behind a certain phenomenon. Reproducing a stress distribution by isotropic FE simulation similar to the distribution obtained from MD simulation where anisotropy is inherent from the interatomic potential and cannot be turned off proves that GBS played the major role in this case. As a result, even in a material of weak anisotropy, such as Al, substantial stress localization can be produced due to GBS. It should also be noted that the effects of anisotropy and the intragranular dislocation slip are much more limited in the deformation of nanocrystalline materials compared to coarse-grained materials. Thus, neglecting both of these effects in the nanocrystalline FEM model presented here appears to be reasonable.

22.7 CONCLUDING REMARKS

A combined multiscale implementation of two modeling techniques, MD and FEM, has been demonstrated at different length-scale resolutions on two geometrically equivalent nanocrystalline systems. The results, after comparing these two simulations, are very instructive. First, a method was successfully applied to match the parameters in the constitutive equations (in this case the CZM) used in the FEM model with quantities extracted from the MD simulation (such as GB yield stress). By this method, the FEM was tuned to the MD simulation. Second, in an MD simulation where the resolution is at the level of individual atoms, it is usually very difficult to identify unambiguously the underlying mechanism responsible for a given process. By reproducing the MD simulation experiment by a continuum FEM model, where the mechanisms are implemented through constitutive equations and can be easily controlled, the underlying mechanism can be revealed with higher certainty. For example, in the octagon–square model presented here, the stress distribution obtained by MD simulation cannot be attributed readily to GBS, as there is also anisotropy that can produce analogous effects. After reproducing the similar simulation experiment by FEM using isotropic material parameters, but allowing for GBS with the equivalent sliding resistance as in the MD model and comparing the results, it becomes possible to identify what part of the stress distribution is due to GBS alone and what part is due to other sources such as anisotropy and GB dislocations.

It has been also demonstrated that now, because of the advanced computational technology, high-end MD simulations can overlap well with FEM simulations on small but still informative systems to allow direct comparison of the results. This possibility suggests a new way of bridging length and time scales and is an alternative to the direct multiscale models, where the simulated system is divided into regions of different length and time scales that are treated with different simulation techniques simultaneously.[36] Instead of trying to combine the two inherently very different simulation techniques into one multiscale modeling program, it is now possible to use the overlapping region to tune the larger scale model to the lower scale one by direct comparison. This approach has the advantage that it preserves the uniqueness and the integrity of each of the models and avoids creating overcomplicated software codes with very restricted applicability. This approach also avoids potential artifacts of the interface between regions of different length and time scales existing in the direct multiscale models, such as unrealistic phonon reflection or dislocation repulsion from the interface, etc.[36] Thus, it becomes feasible to model much larger polycrystalline systems than could be handled by MD alone.

Finally, the large-scale implementation of the MD-tuned FEM simulation on a typical 100-grain nanocrystalline microstructure showed that GBS, by redistributing the load in the system, can produce stress localization several times higher than the background stress under external uniform load. These stress localizations, under suitable conditions, may become nuclei for microcracking and void formation. These results may help to explain the generally observed much higher brittleness of nanocrystalline metals compared to the coarse-grained metals.[37]

PROBLEMS

Selected problems pertaining to the multiscale combination of discrete MD and continuum FEM analyses are presented below. It is assumed that the reader has a formal knowledge or practical familiarity with either molecular dynamics or finite element methods. Fundamental issues involving the underlying physics or details of the computational methods have been, by necessity, not elaborated in the present chapter but are assumed available to the reader through the listed references or through general literature searches. Solutions to the following problems are aimed at providing the reader with a good understanding of many of the important theoretical issues involved in MD-FEM multiscale analysis methods.

1. Identify the differences between sequential and simultaneous multiscale methods. What are the advantages and disadvantages of each?
2. Discuss the considerations related to determining the smallest length scale for which a continuum model is applicable. Under what conditions do continuum mechanics assumptions break down? List the sources of error associated with the application of continuum mechanics constructs at very small length scales.
3. List several nonphysical artifacts that can be generated in modeling small molecular dynamics systems.
4. Discuss several deformation mechanisms that are explicitly modeled by molecular dynamics analyses that can only be considered in an average sense in finite element analyses.

5. Why is a columnar microstructural configuration needed in molecular dynamics analyses to properly correlate with a two-dimensional finite element analysis? In general, what are some of the advantages and disadvantages of using a two-dimensional model in these solution methods?
6. Discuss the qualitative relationship between the size of a grain and its dominant deformation mechanism. Under what conditions is grain boundary sliding the dominant deformation mechanism of a grain? What deformation mechanisms are suppressed in small grains? Hint: perform a literature search on the Hall–Petch relationship to gain insight to this question.

ACKNOWLEDGMENTS

The work of V. Yamakov and D.R. Phillips was performed at NASA's Langley Research Center and sponsored through cooperative agreement NCC-1-02043 with the National Institute of Aerospace and contract NAS1-00135 with Lockheed Martin Space Operations.

REFERENCES

1. Langdon, T.G., Identifying creep mechanisms at low stresses, *Mater. Sci. Eng. A, 283,* 266–273, 2000.
2. Wolf, D. and Jaszczak, J.A., Tailored elastic behavior of multilayers through controlled interface structure, *J. Computer-Aided Mater. Design, 1,* 111–148, 1993.
3. Adams, J.B., Wolfer, W.G., and Foiles, S.M., Elastic properties of grain boundaries in Cu and their relationship to bulk elastic constants, *Phys. Rev. B, 40,* 9479–9484, 1989.
4. Schiotz, J., DiTolla, F.D., and Jacobsen, K.W., Softening of nanocrystalline metals at very small grain sizes, *Nature, 391,* 561–563, 1998.
5. Van Swygenhoven, H., Polycrystalline materials—grain boundaries and dislocations, *Science, 296,* 66–67, 2002.
6. Yamakov, V., Wolf, D., Phillpot, S.R., and Gleiter, H., Grain-boundary diffusion creep in nanocrystalline palladium by molecular-dynamics simulation, *Acta Mater., 50,* 61–73, 2002.
7. Yamakov, V., Wolf, D., Phillpot, S.R., Mukherjee, A.K., and Gleiter, H., Crossover in Hall–Petch behavior in nanocrystalline materials by molecular-dynamics simulation, *Phil. Mag. Lett., 83,* 385–393, 2003.
8. Yamakov, V., Wolf, D., Phillpot, S.R., Mukherjee, A.K., and Gleiter, H., Deformation-mechanism map for nanocrystalline metals by molecular-dynamics simulation, *Nature Mater., 3,* 43–47, 2003.
9. Schiotz, J. and Jacobsen, K.W., A maximum in the strength of nanocrystalline copper, *Science, 301,* 1357–1359, 2003.
10. Gutkin, M.Y., Ovid'ko, I.A., and Skiba, N.V., Strengthening and softening mechanisms in nanocrystalline materials under superplastic deformation, *Acta Mater, 52,* 1711–1720, 2004.
11. Kumar, K.S., Suresh, S., Chisholm, M.F., Horton, J.A., and Wang, P., Deformation of electrodeposited nanocrystalline nickel, *Acta Mater, 51,* 387–405, 2003.

12. Mishin, Y., Farkas, D., Mehl, M.J., and Papaconstantopoulos, D.A., Interatomic potentials for monoatomic metals from experimental data and *ab initio* calculations, *Phys. Rev. B, 59,* 3393–3407, 1999.
13. Yamakov, V., Wolf, D., Salazar, M., Phillpot, S.R., and Gleiter, H., Length-scale effects in the nucleation of extended lattice dislocations in nanocrystalline Al by molecular-dynamics simulation, *Acta Mater., 49,* 2713–2722, 2001.
14. Yamakov, V., Wolf, D., Phillpot, S.R., and Gleiter, H., Dislocation–dislocation and dislocation–twin reactions in nanocrystalline Al by molecular-dynamics simulation, *Acta Mater., 51,* 4135–4147, 2003.
15. Merkle, K.L., High-resolution electron microscopy of interfaces in FCC materials, *Ultramicroscopy, 37,* 130–152, 1991.
16. Merkle, K.L., Atomic structure of grain boundaries, *J. Phys. Chem. Solids, 55,* 991–1005, 1994.
17. Medlin, D.L., Foiles, S.M., and Cohen, D., A dislocation-based description of grain boundary dissociation: application to a 90° ⟨110⟩ tilt boundary in gold, *Acta Mater., 49,* 3689–3697, 2001.
18. Dahmen, U., Hetherington, J.D., O'Keefe, M.A., Westmacott, K.H., Mills, M.J., Daw, M.S., and Vitek, V., Atomic structure of a Σ99 grain boundary in Al: a comparison between atomic-resolution observation and pair-potential and embedded-atom simulations, *Phil. Mag. Lett., 62,* 327–335, 1990.
19. Wolf, D., Structure–energy correlation for grain boundaries in F.C.C. metals III. Symmetrical tilt boundaries, *Acta Metal., 38,* 781–790, 1990.
20. Rittner, J.D. and Seidman, D.N., ⟨110⟩ Symmetric tilt grain-boundary structures in FCC metals with low stacking-fault energies, *Phys. Rev. B, 54,* 6999–7015, 1996.
21. Nishitani, S.R., Ohgushi, S., Inoue, Y., and Adachi, H., Grain boundary energies of Al simulated by environment-dependent embedded atom method, *Mat. Sci. Eng. A, 309–310,* 490–494, 2001.
22. Yamakov, V., Wolf, D., Phillpot, S.R., Mukherjee, A.K, and Gleiter, H., Dislocation processes in the deformation of nanocrystalline Al by molecular-dynamics simulation, *Nature Mater., 1,* 45–48, 2002.
23. Yamakov, V., Wolf, D., Phillpot, S.R., and Gleiter, H., Deformation twinning in nanocrystalline Al by molecular-dynamics simulation, *Acta Mater., 50,* 5005–5020, 2002.
24. Chen, M., Ma, E., Hemker, K.J., Sheng, H., Wang, Y., and Cheng, X., Deformation twinning in nanocrystalline Al, *Science, 300,* 1275–1277, 2003.
25. Liao, X.Z., Zhou, F., Lavernia, E.J, Srinivasan, S.G., Baskes, M.I., He, D.W., and Zhu, Y.T., Deformation mechanism in nanocrystalline Al: partial dislocation slip, *Appl. Phys. Lett., 83,* 632–634, 2003.
26. Liao, X.Z., Zhou, F., Lavernia, E.J., He, D.W., and Zhu, Y.T., Deformation twins in nanocrystalline Al, *Appl. Phys. Lett., 83,* 5062–5064, 2003.
27. Parrinello, M. and Rahman, A., Polymorphic transitions in single crystals: a new molecular dynamics method, *J. Appl. Phys., 52,* 7182–7190, 1981.
28. Nose, S., A unified formulation of the constant temperature molecular dynamics method, *J. Chem. Phys., 81,* 511–519, 1984.
29. Ogata, S., Li, J., and Yip, S., Ideal pure shear strength of aluminum and copper, *Science, 298,* 807–811, 2002.
30. Iesulauro, E., Decohesion of grain boundaries in statistical representations of aluminum polycrystals, *Cornell University Report, 02–01,* 2002.
31. Hirth, J.P. and Lothe, J., *Theory of Dislocations,* 2nd ed., John Wiley & Sons, New York, 1992.

32. Tvergaard, V. and Hutchinson, J.W., The relation between crack growth resistance and fracture process parameters in elastic-plastic solids, *J. Mechan. Phys. Solids, 40,* 1377–1397, 1992.
33. Klein, P. and Gao, H., Crack nucleation and growth as strain localization in a virtual-bond continuum, *Engin. Fracture Mechan., 61,* 21–48, 1998.
34. O'Rourke, J., *Computational Geometry in C,* 2nd ed., Cambridge University Press, 2001.
35. Shobu, K., Tani, E., and Watanabe, T., Geometrical study on the inhomogeneous deformation of polycrystals due to grain-boundary sliding, *Phil. Mag. A, 74,* 957–964, 1996.
36. Tadmor, E.B., Phillips, R., and Ortiz, M., Mixed atomistic and continuum models of deformation in solids, *Langmuir, 12,* 4529–4534, 1996.
37. Wang, Y., Chen, M., Zhou, F., and Ma, E., High tensile ductility in a nanostructured metal, *Nature, 419,* 912–915, 2002.

23 Modeling of Carbon Nanotube/Polymer Composites

Gregory M. Odegard

CONTENTS

23.1 INTRODUCTION

The research and development of carbon fiber-reinforced polymer composites started in the 1960s. Carbon fibers consist of tightly packed graphene sheets that are aligned parallel to the fiber axis. The principal advantages of using carbon fibers include their high modulus (up to 400 GPa), high strength (up to 3.4 GPa), low density (specific gravity around 2.0), and low cost compared to many other reinforcement materials.[1]

In 1985, Smalley and coworkers discovered the fullerene structure.[2] Fullerenes are cage-like structures of carbon atoms that are composed of hexagonal and pentagonal faces. In 1991, Iijima[3] discovered carbon nanotubes. Nanotubes are elongated

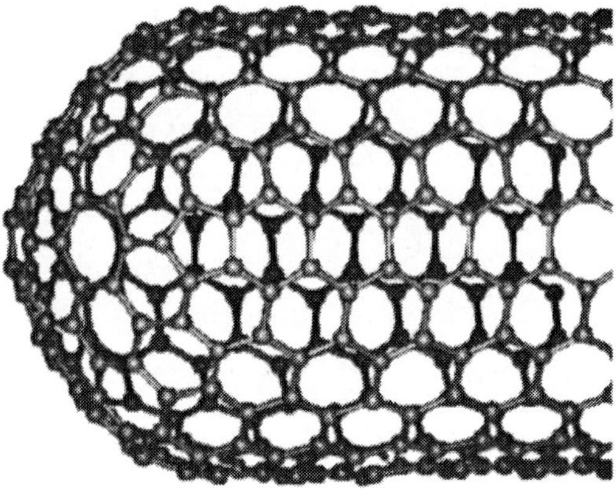

FIGURE 23.1 End section of SWNT.

fullerenes with walls consisting of graphene carbon structures (Fig. 23.1). Nanotubes that consist of a single, closed graphene sheet are referred to as single-wall carbon nanotubes (SWNTs). Nanotubes with multiple layers of closed graphene sheets are referred to as multiwalled carbon nanotubes (MWNTs). The atomic structure of a carbon nanotube is commonly described in terms of the chirality of the nanotube, which is defined by the chiral vector **D**, which joins two equivalent atoms of a graphene sheet. The vector is expressed as

$$\mathbf{D} = \mathrm{n}\boldsymbol{a}_1 + \mathrm{m}\boldsymbol{a}_2 \tag{23.1}$$

where $\boldsymbol{a}_1$ and $\boldsymbol{a}_2$ are the unit vectors indicated in Fig. 23.2, the integers n and m describe the number of steps along the vectors $\boldsymbol{a}_1$ and $\boldsymbol{a}_2$, respectively, and the boldface indicates a vector quantity. Nanotubes have extraordinary mechanical, electrical, and thermal properties.[4,5]

The composite materials community is showing tremendous interest in carbon nanotubes for some of the same reasons carbon fibers generated interest 40 years ago. That is, with a large aspect ratio (up to 3000), extraordinary properties (tensile modulus and strength of up to 1 TPa and 200 GPa, respectively), and a relatively low density (less than carbon fiber density), nanotubes could possibly serve as ideal reinforcements for polymer resins. In addition, for a given volume fraction of reinforcement, a composite reinforced with nanotubes contains a larger surface-to-volume ratio of reinforcement than a composite reinforced with carbon fibers. Therefore, more surface area for load transfer between the polymer matrix and reinforcement exists for carbon nanotubes, which should, in theory, increase the reinforcement efficiency.

Because of the high cost of carbon nanotubes (up to $500 per gram for purified SWNTs), their research and development is quite expensive. In addition, trial-and-error approaches are typically used to optimize the fabrication parameters for these materials in the laboratory, with alternating steps of synthesis and characterization.

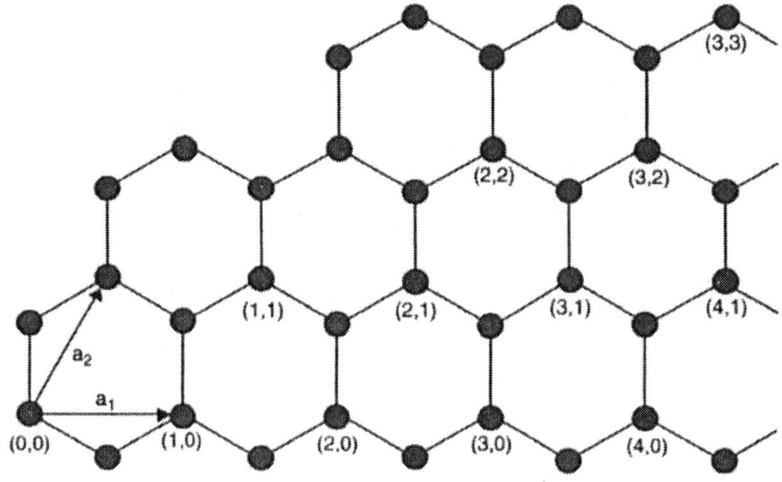

FIGURE 23.2 Unit vector notational for a graphene sheet.

Modeling can be employed to facilitate the development of carbon nanotube materials. Incorporation of modeling leads to lower cost and quicker analysis of a wide range of carbon nanotube/polymer composite materials compared to a purely experimental approach. Two broad classes of modeling tools may be employed: micromechanics and molecular modeling. While the continuum-based micromechanical methods were originally developed for use on traditional, larger scale composite materials, molecular modeling tools were developed for studying various forms of molecular systems.

The objective of this chapter is to describe modeling methods for carbon nanotube/polymer composites. First some of the fundamental characteristics and recent advances of carbon nanotube/polymer interfaces are reviewed, followed by a description of the four principal modeling tools for carbon nanotube/polymer composites: analytical micromechanics, numerical micromechanics, *ab initio* simulations, and molecular dynamics (MD) simulations. Finally examples of modeling techniques are presented for SWNT/polyimide and SWNT/polyethylene composites.

23.2 CARBON NANOTUBE/POLYMER INTERFACE

The ability to transfer mechanical load between a carbon nanotube and the adjacent polymer molecules in a composite material has been a matter of significant discussion in the literature.[6–19] In general, two basic approaches have been proposed to improve the strength of the interaction. The first approach involves forming a strong, nonbonded interaction between the polymer and the nanotube without modifying the nanotube structure. This approach assumes a chemical strengthening of the nonbonded interactions with the nanotube[20] and the improvement of the apparent mechanical connection with the nanotube, for example, by wrapping a large polymer molecule around the nanotube.[21,22] The second approach requires the formation of a chemical covalent bond between the nanotube and polymer directly, also known as functionalization. There is reasonable evidence for the presence of such nanotube/polymer covalent bonds.[23,24] Some studies have indicated that functionalization can occur through chemical bonds added to the nanotube sidewalls.[7–12] Addition of small organic groups to nanotube sidewalls has been reported via *in situ*-generated diazonium compounds[8] and by fluorination of nanotubes[9,10] followed by alkylation.[11] Functionalization can also occur through oxidizing the nanotube sample to induce the formation of hydroxyl of carboxylic acid groups at surface defects on the nanotube.[13–18] This type of functionalization can occur with copolymers,[13–15] proteins,[16] organosilanes,[17] and metal catalysts.[18] It is known that covalent bonding affects the elastic mechanical properties of the nanotube itself because the formation of a chemical bond with carbon atoms in a nanotube interrupts the sp^2 hybridization of the nanotube, thereby forming a site that is closer to an sp^3 hybridized carbon.[25,26]

A recent simulation study by Frankland et al.[27] predicted that for carbon nanotube/polyethylene composites there is at least a one-order-of-magnitude increase in the strength for composites with covalent bonding between the nanotube and adjacent polymer molecules, relative to systems without the covalent bonds. However, the hybridization change may weaken the chemical bonds in the vicinity of the functionalization, and this effect should be manifested as a decrease in the material elastic constants.[20,21,28] Therefore, a trade-off exists between the effects of increased load transfer and decreased nanotube properties because of functionalization. The effects of functionalization are expected to depend on the polymer matrix material.

23.3 MICROMECHANICS

Micromechanical models are used to predict the bulk properties of composite materials as a function of the properties and geometry of the constituent materials (Fig. 23.3). In particular, for fiber-reinforced composites, the fiber properties, shape,

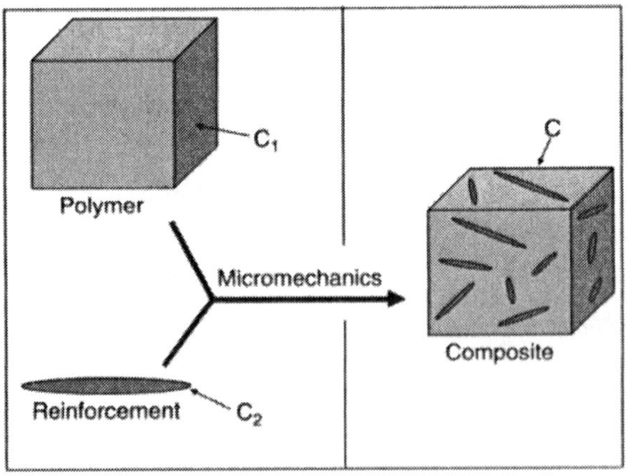

FIGURE 23.3 Micromechanical modeling of a composite.

length, volume fraction, orientation, and the matrix properties are used as input into micromechanical models. Generally in the development and use of micromechanical models, it is assumed that the constituents are continuous, the properties of which are mathematical described with continuum mechanics. Micromechanics models for fiber-reinforced composites can be broken down into two general categories, analytical and numerical.

23.3.1 Analytical Micromechanical Models

Many forms of analytical micromechanical models have been developed over the past 50 years. Important contributions to this effort include the works of Hashin,[29] Hill,[30,31] Hashin and Rosen,[32] Aboudi,[33] Eshelby,[34] Budiansky,[35] Mori and Tanaka,[36] Benveniste,[37] and Halpin and Tsai.[38] Many of these models can be formulated in a consistent mathematical framework.

Using the direct approach[39] for the estimate of overall properties of heterogeneous materials, the volume-averaged stress and strain fields of the composite with a total of N phases are, respectively,

$$\bar{\sigma} = \sum_{r=1}^{N} c_r \bar{\sigma}_r \tag{23.2}$$

$$\bar{\varepsilon} = \sum_{r=1}^{N} c_r \bar{\varepsilon}_r \tag{23.3}$$

where σ is the stress tensor, ε is the strain tensor, c_r is the volume fraction of phase r, the overbar denotes a volume-averaged quantity, the subscript r denotes the phase, and $r = 1$ is the matrix phase. The constitutive equation for each phase is given by

$$\sigma_r = \mathbf{C}_r \varepsilon_r \tag{23.4}$$

where $\mathbf{C}_r$ is the stiffness tensor of phase r. For a composite subjected to homogeneous elastic strain and electric field boundary conditions, ε^0, it can easily be shown that $\bar{\varepsilon} = \varepsilon^0$.[40] The constitutive equation for the composite can be expressed in terms of the volume-averaged fields

$$\bar{\sigma} = \mathbf{C}\bar{\varepsilon} \tag{23.5}$$

The volume-average strain field in phase r is

$$\bar{\varepsilon}_r = \mathbf{A}_r \bar{\varepsilon} \tag{23.6}$$

where $\mathbf{A}_r$ is the concentration tensor of phase r, and

$$\sum_{r=1}^{N} c_r \mathbf{A}_r = \mathbf{I} \tag{23.7}$$

where **I** is the identity tensor. Combining Equation 23.2 and Equation 23.7 yields the modulus of the composite in terms of the constituent moduli

$$\mathbf{C} = \mathbf{C}_1 + \sum_{r=2}^{N} c_r(\mathbf{C}_r - \mathbf{C}_1)\mathbf{A}_r \tag{23.8}$$

Various procedures exist for evaluating the concentration tensor. The simplest approximation is $\mathbf{A}_r^{voi} = \mathbf{I}$, which is the Voight approximation, also referred to as the rule-of-mixtures approach. The next simplest approach is the dilute approximation, for which the interaction between the reinforcement particles is ignored. This results in the dilute concentration tensor:

$$\mathbf{A}_r^{dil} = \left[\mathbf{I} + \mathbf{S}_r\mathbf{C}_1^{-1}(\mathbf{C}_r - \mathbf{C}_1)\right]^{-1} \tag{23.9}$$

where $\mathbf{S}_r$ is the Eshelby tensor for phase r,[34] and is evaluated as a function of the lengths of the principal axes of the reinforcing phase r, a_i^r, and the mechanical properties of the surrounding matrix

$$\mathbf{S}_r = f\left(\mathbf{C}_1, a_1^r, a_2^r, a_3^r\right) \tag{23.10}$$

Numerous forms of the complete expression for Equation 23.10 are given elsewhere.[34,41] For the Mori–Tanaka approach, the concentration tensor is

$$\mathbf{A}_s^{MT} = \mathbf{A}_s^{dil}\left[c_1\mathbf{I} + \sum_{r=2}^{N} c_r\mathbf{A}_r^{dil}\right]^{-1} \tag{23.11}$$

where $\mathbf{A}_r^{dil}$ is the dilute concentration tensor given by Equation 23.9. The Mori–Tanaka approach has been shown to be quite accurate, even at larger reinforcement volume fractions.[42] In the self-consistent scheme, the concentration tensor is

$$\mathbf{A}_r^{SC} = \left[\mathbf{I} + \mathbf{S}_r\mathbf{C}^{-1}(\mathbf{C}_r - \mathbf{C})\right]^{-1} \tag{23.12}$$

where **C** is the unknown modulus of the composite. Since the moduli of the composite appear in both Equation 23.8 and 23.12, iterative schemes or numerical techniques are ultimately required for the prediction of the moduli of composites using the self-consistent method. This approach results in slow and complicated calculations.

The use of analytical micromechanical models for predicting the properties of carbon nanotube/polymer composites has been reported in the literature. Pipes and Hubert[43] developed a model for predicting the properties of arrays of SWNTs that are assembled in helical arrays of circular cross-section. Anisotropic elasticity was used to determine the behavior of a layered cylinder with layers of discontinuous SWNT embedded in a polymeric matrix and with the collimation direction in each layer following a helical path. This model was extended to predict the thermal expansion properties of such a material[44] and to predict the properties of larger scale structures composed of multiple helically wound arrays.[45] Odegard et al.[46] used the Mori–Tanaka approach to predict the mechanical properties of a (6,6) SWNT/polyimide composite with a PmPV nanotube/polyimide interface for various SWNT volume fractions, lengths, and orientations. A similar approach was used by Odegard et al.[28] to predict the properties of a (10,10) SWNT/polyethylene composite with and without chemical functionalization. It was shown that chemical functionalization degraded most of the elastic properties of the composites. The models of Pipes and Hubert[43,45] and Odegard et al.[46] were compared for the same SWNT/polymer composite material[47]. The two approaches showed agreement in predicted Young's moduli for a wide range of SWNT volume fractions. Thostenson and Chou[48] used the Halpin–Tsai approach to predict elastic moduli of MWNT/polystyrene composites as a function of constituent properties, reinforcement geometry, and nanotube structure. They demonstrated that the composite elastic properties were particularly sensitive to the nanotube diameter. Yoon et al.[49] predicted the resonant frequencies and vibration modes of MWNTs embedded in an elastic medium using a multiple-elastic beam model. Lagoudas and Seidel[50] used a variety of micromechanical techniques, including a two-step Mori–Tanaka approach and the composite cylinders approach to examine the properties of carbon nanotube-reinforced epoxy composites. While the clustering of nanotubes in the composite contributed to some reduction in composite properties, nanotube cluster misalignment and poor nanotube–matrix bonding played a much larger role.

23.3.2 Numerical Micromechanical Models

Numerical predictions of bulk properties of carbon nanotube/polymer composites based on the constituent properties and nanotube geometry and volume fraction can be achieved with the finite element method (FEM). If performed correctly, the FEM-based micromechanical approach has the potential to predict bulk properties more accurately than those predicted with analytical micromechanical techniques. However, the gain in accuracy when using FEM micromechanical models comes at a price. Due to the complexity of FEM models and the high costs of FEM software, implementation of FEM-based micromechanical models can be costly and time consuming.

The principal advantage of FEM micromechanical models is the direct determination of the stress and strain distribution in the reinforcement and surrounding matrix. The careful prediction of these fields results in the accurate estimation of the bulk properties. The stress and strain fields are determined by discretizing the heterogeneous continuum into elements (Fig. 23.4), of which elastic solutions exist for applied loads and displacements on nodes. The nodes are placed on the corners and sometimes on the midsides of the element boundaries. As long as the geometry of the elements (mesh) is not too coarse, the overall predicted properties of FEM models can be accurately estimated by solving for the simultaneous stress and strain fields of all of the elements in the model. The pioneering works in the use of FEM-based micromechanical models include those of Adams,[51] Lin et al.,[52] and Dvorak et al.[53]

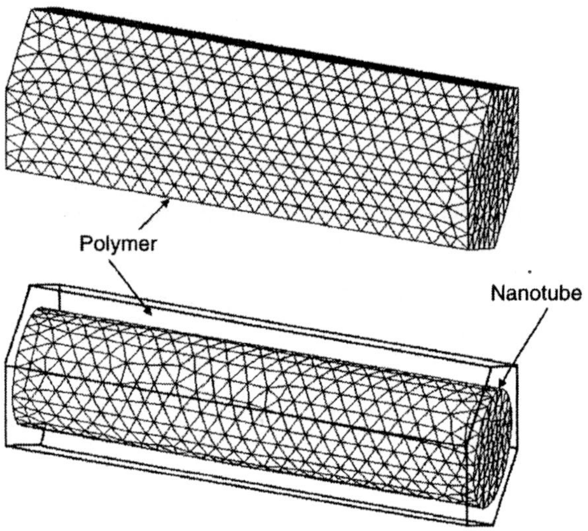

FIGURE 23.4 Finite element model of a carbon nanotube/polymer composite.

Several studies have used FEM-based micromechanical models to predict the properties of carbon nanotube/polymer composites. Liu and Chen[54] and Chen and Liu[55] used various representative volume element (RVE) geometries to evaluate the properties of SWNT/polymer composites. These studies demonstrated the reinforcement potential of carbon nanotubes in nanocomposite materials. Li and Chou[56] used finite elements to study the interfacial shear stress distribution, stress concentration in the matrix in the vicinity of carbon nanotube ends, axial stress profile in the nanotube, and the effect of nanotube aspect ratio on load transfer in a carbon nanotube/polymer composite. Fisher et al.[57] developed a FEM model to investigate the effect of fiber waviness on overall mechanical properties of an effective reinforcing modulus for carbon nanotube/polymer composites. Bradshaw et al.[58] used FEM models to numerically determine the strain concentration tensor of a carbon nanotube/polymer composite with wavy carbon nanotubes. These two studies demonstrated the reduction in mechanical properties that accompanies the increase in nanotube waviness.

23.4 MOLECULAR MODELS

In the past two decades, computational molecular modeling approaches have emerged as important tools that can be used to predict atomic structure, vibrational frequencies, binding energies, heats of reaction, electrical properties, and mechanical properties of organic and inorganic materials. Unlike the continuum-based micromechanical approaches, no assumptions of a material continuum are made. Therefore, these methods are ideal in studying materials in which the RVE is on the scale of nanometers. However, because of the discrete nature of molecular models, only a small volume of atoms may be considered in an analysis of a material, which is a concern in the development of nanostructured materials for structural applications, such as in the aerospace industry. Furthermore, many molecular modeling approaches require long simulation times on relatively powerful computers. For the simulation of carbon nanotube/polymer composites, two main computational molecular modeling techniques are employed: MD and *ab initio* simulations.

23.4.1 *Ab Initio* Simulations

In *ab initio* simulations of atomic systems, an approximation to the solution of the Schrödinger equation is performed. The method can be used for systems containing 1 to 40 atoms, depending on the power of the computational equipment available. Although a wide range of *ab initio* methods have been developed, the most commonly used approach is the Hartree–Fock method. More detailed discussions of *ab initio* simulations can be found elsewhere.[59]

There are many procedures that can be followed in *ab initio* simulations. Three commonly used procedures are single point calculations, geometry optimization calculations, and frequency calculations. In single point calculations, the energy and wave function are calculated for a fixed geometry. This approach is sometimes employed as a preliminary step in a more detailed study on a molecular structure or after a geometry optimization. In geometry optimization calculations, the wave function and energy of a molecule are calculated at a starting geometry and subsequent geometries with lower energy levels. The process is repeated until a minimized energy geometry is achieved. Various procedures exist for establishing the geometries for every calculation step. Frequency calculations are used to predict the frequencies and the infrared and Raman intensities of a molecular system.

Ab initio studies of carbon nanotube/polymer composites in the literature are limited. Mylvaganam and Zhang[60] studied the bonding between polyethylene and carbon nanotubes. The study predicted that the bonding is energetically favorable and that the tubes of smaller diameters have higher binding energies. Jaffe[61] examined the bonding of fluorine atoms and nonplanar carbon geometries corresponding to (10,10), (5,5), and (16,0) carbon nanotubes. It was shown that zigzag nanotubes form more stable fluorination products than armchair tubes of comparable diameter. Bauschlicher[62,63] studied the binding energies of hydrogen to (10,0), (9,0), and (5,5) carbon nanotubes. Hydrogen coverages of 50% and 100% were applied onto the carbon nanotubes. Bauschlicher[64] studied the bonding of hydrogen and fluorine atoms to the sides of (10,0) carbon nanotubes. The fluorine atoms appeared to favor bonding next to existing fluorine atoms, while the addition of two or four hydrogen atoms to the nanotubes was shown to be endothermic.

23.4.2 Molecular Dynamics

Molecular dynamics is a computational technique in which a time evolution of a set of interacting atoms is followed by integrating their equations of motion. The motion of the atoms is described by Newton's law:

$$\mathbf{F}_i = m_i \mathbf{a}_i \tag{23.13}$$

where $\mathbf{F}_i$ is the force vector acting on atom i, and m_i and $\mathbf{a}_i$ are the mass and acceleration vectors, respectively, of atom i in a system with N atoms. The forces acting on each atom are due to the interactions with the other atoms. A trajectory is calculated in a 6N-dimensional phase space (three positions and three momenta components for each atom). Since MD is a statistical mechanics method, a set of configurations distributed according to a statistical distribution function (statistical ensemble) is obtained. Typical MD simulations are performed on systems containing thousands of atoms and for simulation times ranging from picoseconds to nanoseconds.

The physical quantities of the system are represented by averages over configurations distributed according to the chosen statistical ensemble. A trajectory obtained with MD provides such a set of configurations. Therefore the computation of a physical quantity is obtained as an arithmetic average of the instantaneous values. Statistical mechanics is the link between the microscopic behavior and thermodynamics. Thus the atomic system is expected to behave differently for different pressures and temperatures.

The interactions of the particular atom types are described by the total potential energy of the system as a function of the positions of the individual atoms at a particular instant in time:

$$V = V(r_i, \ldots, r_N) \tag{23.14}$$

where r_i represents the coordinates of atom i in a system of N atoms. The potential equation is invariant to coordinate transformations, and is expressed in terms of the relative positions of the atoms with respect to each other, rather than from absolute coordinates. The forces on atom i are determined from the gradient of the potential with respect to the atomic displacements

$$\mathbf{F}_i = -\nabla V(r_1, \ldots, r_N) \tag{23.15}$$

where ∇ is the gradient operator. The total energy of the system is

$$E = K + V \tag{23.16}$$

where K is the instantaneous kinetic energy, which is dependent on the thermodynamic condition of the system.

Numerous options exist for the selection of the potential, V. For systems that consist of only carbon or hydrogen atoms, such as carbon nanotube/polyethylene composites, the Brenner potential[65] has been extensively used.[27,66,67] The basic concept of the Brenner potential is that the strength of a bond between two atoms is not constant, but depends on local conditions. The potential is expressed as

$$V = \sum_{i} \sum_{j(>i)} [V_R(r_{ij}) - B_{ij} V_A(r_{ij})] \tag{23.17}$$

where the summations are performed over the bonds in the system, r_{ij} is the distance between atoms i and j, $V_R(r_{ij})$ and $V_A(r_{ij})$ are repulsive and attractive interaction terms,

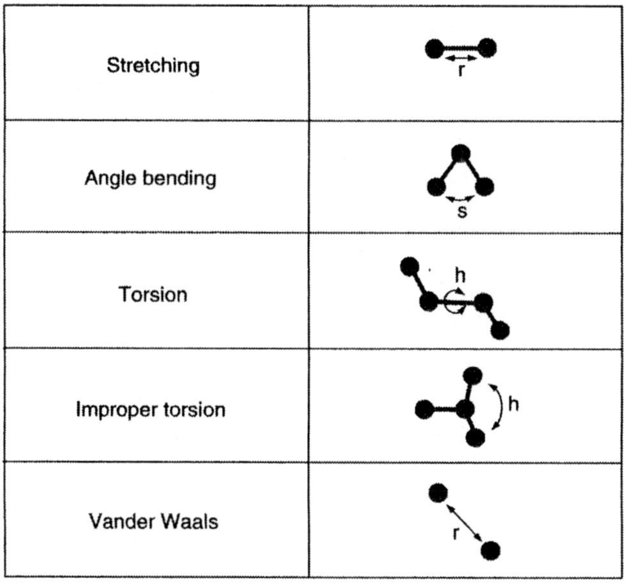

FIGURE 23.5 Molecular mechanics degrees of freedom.

respectively, and B_{ij} represents a many-body coupling between the bond from atom i to atom j and the local environment of atom i. For organic material systems that consist of more than just carbon and hydrogen atoms, a simple yet effective approach is to describe the potential with molecular mechanics.[68–72] With the molecular mechanics approach, the potential energy of the system is described by the sum of the individual energy contributions of each degree of freedom for each of the N atoms (Fig. 23.5). The form of the individual energy contributions is often referred to as a force field. For example, for the force field of Cornell et al.,[68] the potential energy is

$$V = \sum_{bonds} K_r (r - r_{eq})^2 + \sum_{angles} K_\theta (\theta - \theta_{eq})^2 + \sum_{dihedrals} \frac{V_n}{2}[1 + \cos(n\phi - \gamma)] + \sum_{I<J} \left[\frac{A_{IJ}}{r_{IJ}^{12}} - \frac{B_{IJ}}{r_{IJ}^{6}} \right] \quad (23.18)$$

where K_r is the bond stretching force constant, r is the distance between atoms, r_{eq} is the equilibrium spacing between atoms, K_θ is the bond-angle bending force constant, θ is the bond angle, θ_{eq} is the equilibrium bond angle, V_n is the torsion force constant, γ is the phase offset, n is the periodicity of the torsion, A_{IJ} and B_{IJ} are van der Waals force constants between nonbonded atoms I and J, and r_{IJ} is the distance between atoms I and J. In Equation 23.18, the summations occur over all

TABLE 23.1
Force Constants for Aliphatic Carbon Atoms

Bond	Parameters
C–C bond stretch	$K_r = 310.0$ kcal/(mole Å^2)
	$r_{eq} = 1.526$ Å
C–C–C bond angle	$K_\theta = 40.0$ kcal/(mole rad^2)
	$\theta_{eq} = 109.5°$
C–C–C–C torsion	$V_n/2 = 1.4$ (kcal/mole)
	$\gamma = 0°$
	$n = 3$
C–C van der Waals interaction	$R^* = 1.908$ Å
	$\varepsilon = 0.1094$ (kcal/mole)

of the corresponding interactions in the molecular system. The van der Waals term in Equation 23.18 is also known as the Lennard–Jones potential. Examples of force constants are shown in Table 23.1 for aliphatic carbon atoms. More details on MD simulation technique can be found elsewhere.[73]

Numerous studies have addressed the simulation of carbon nanotube/polymer composites using molecular mechanics and MD techniques. Lordi and Yao[22] used molecular mechanics simulations to calculate sliding frictional stresses between carbon nanotubes and a range of polymer substrates. It was determined that frictional forces and helical polymer conformations played minor and major roles respectively in determining the interfacial strength. Frankland et al.[27] used MD simulations to examine the influence of functionalization between a SWNT and polyethylene matrix on shear strengths and critical lengths required for load transfer. The simulations predicted that shear strengths and critical lengths required for load transfer increase and decrease, respectively, when the nanotubes are functionalized to the matrix. Frankland and Harik[74] studied the pull-out of SWNT from a polyethylene matrix to characterize an interfacial friction model. A periodic variation in the nanotube displacements and velocities was observed, which corresponded to the atomic structure of the carbon nanotube. Frankland et al.[66] predicted stress–strain curves of SWNT/polyethylene composites with infinite and finite-length nanotubes. The curves were subsequently compared to micromechanical predictions. Hu and Sinnott[75] and Hu et al.[76] investigated the effects of chemical functionalization of carbon nanotube/polystyrene composite on the composite toughness. The chemical modification of the carbon nanotubes during polyatomic ion beam deposition was detailed. Jaffe and coworkers[77–79] examined the interaction of carbon nanotubes, water droplets, and aqueous solutions. Radial density profiles and radial hydrogen bond distributions were predicted. Liang et al.[80] demonstrated the attractive interaction between (10,10) SWNTs and an epoxy polymer matrix. The results indicated that the aromatic ring structures of the polymer molecules have a tendency to align with the aromatic structures in the SWNTs. Wei et al.[81] showed that the addition of nanotubes to a polyethylene matrix increased the glass transition temperature, thermal expansion, and diffusion coefficients of the polymer. Griebel and Hamaekers[67] determined stress–strain curves of nanotube/polyethylene composites. Good agreement with rule-of-mixture micromechanical models was found. Odegard et al.[82] predicted the mechanical properties of cross-linked nanotubes for various SWNT cross-link densities. While the cross-linking significantly increased the transverse shear moduli of the nanotube arrays by a factor of up to 30, the corresponding degradation of the nanotube caused an overall decrease in the axial properties of 98%.

23.5 EXAMPLE: SWNT/POLYIMIDE COMPOSITE

For this example, the constitutive properties of a nanotube/polyimide composite with a poly(m-phenylenevinylene) (PmPV) interface were predicted based on the method developed by Odegard et al.[46] This method relies on an equivalent continuum modeling technique[83] used to predict the bulk mechanical behaviors of nanostructured materials. In summary, the method consists of four major steps: establishing representative volume elements (RVEs) for the molecular and equivalent continuum models, establishing a constitutive relationship for the equivalent continuum model, deriving and equating potential energies of deformation for both models subjected to identical boundary conditions, and using traditional micromechanics techniques to determine larger scale properties of the composite. Each step of the modeling is described in the next section. Further details of the modeling are discussed elsewhere.[46]

23.5.1 Modeling Procedure

In order to establish the RVE of the molecular model, and thus the equivalent continuum model, the molecular structure was first determined. An MD simulation was used to generate the equilibrium structure of a nanotube/polyimide composite, which consisted of a (6,6) single-wall nanotube and five PmPV oligomers, each ten repeating units in length. The initial structure was constructed by placing the nanotube at the center of the MD cell, and by inserting the PmPV molecules at random, nonoverlapping positions within the MD cell. The molecular structure for a single time increment for an equilibrated system was used as the RVE of the molecular model (Fig. 23.6). The RVE of the equivalent continuum model was chosen to be a solid cylinder of the same dimensions as the molecular model RVE. The equivalent continuum model will be referred to as the effective fiber for the remainder of this example discussion.

The second step in the modeling technique involved establishing a constitutive equation for the effective fiber. It was assumed that the effective fiber had a linear elastic constitutive behavior. Examination of the molecular model (Fig. 23.6) revealed that it was accurately described as having transversely isotropic symmetry, with the plane of isotropy perpendicular to the long axis of the nanotube. There are five independent elastic constants required to determine the entire set of elastic constants for a transversely isotropic material.

In the third step of the modeling approach, the potential energies of deformation for the molecular model and effective fiber were derived and equated for identical

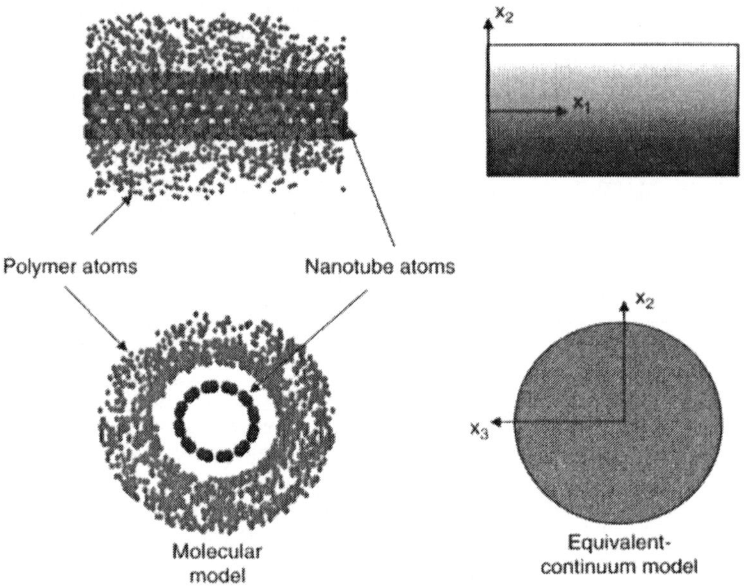

FIGURE 23.6 Equivalent continuum model of the nanotube/PmPV composite.

loading conditions. Five independent elastic constants were determined by employing a technique adapted from the approach of Hashin and Rosen.[32] These elastic constants were the transverse shear modulus, G_T^f, transverse bulk modulus, K_T^f, longitudinal shear modulus, G_L^f, longitudinal Young's modulus, Y_L^f, and the longitudinal elastic stiffness component, C_L^f, where the superscript f denotes effective fiber. The longitudinal properties were associated with the direction parallel to the longitudinal axis of the nanotube, and the transverse properties were associated with the transverse plane of isotropy (Fig. 23.6).

The bonded and nonbonded interactions of the atoms in a polymeric molecular structure were described by using a force field similar to that shown in Equation 23.18. Each of the five elastic constants of the effective fiber was determined from a single boundary condition applied to both molecular and effective fiber models. Therefore, for each applied boundary condition, one elastic constant of the effective fiber was uniquely determined. For this particular study, an intermediate step was used to equate the molecular and effective fiber models. An equivalent truss model was used to approximate the energy of the molecular model. The calculated values of the five independent parameters for the effective fiber are listed in Table 23.2.

Overall constitutive properties of the nanocomposite with randomly oriented and unidirectional nanotubes were determined with the micromechanical-based Mori–Tanaka method[36,37] by using the mechanical properties of the effective fiber and the bulk polymer matrix material (Fig. 23.3). The layer of polymer molecules near the polymer/nanotube interface was included in the effective fiber, and it was assumed that the matrix polymer surrounding the effective fiber had mechanical

TABLE 23.2
Elastic Properties of Effective Fiber

$G_T^f = 4.4$ GPa
$K_T^f = 9.9$ GPa
$G_L^f = 27.0$ GPa
$Y_L^f = 450.4$ GPa
$C_L^f = 457.6$ GPa

properties equal to those of the bulk polymer. Because the bulk polymer molecules and the polymer molecules included in the effective fiber are physically entangled, perfect bonding between the effective fiber and the surrounding polymer matrix was assumed. The bulk polymer matrix material had a Young's modulus and Poisson's ratio of 3.8 GPa and 0.4, respectively.

The maximum nanotube volume fraction that can be obtained with this approach is limited by the maximum effective fiber volume fraction that can be used in the micromechanics analysis. For a hexagonal packing arrangement, the maximum effective fiber volume fraction is 90.7%. While the nanotube and effective fiber lengths are equal, the nanotube volume fraction is 34% of the effective fiber volume fraction if it is assumed that the nanotube volume is a hollow cylinder with a wall thickness equal to the interatomic spacing of graphene sheets (0.34 nm).

23.5.2 Results and Discussion

In this section, the moduli of the effective-fiber/polyimide composite are presented in terms of nanotube length, volume fraction, and orientation. Figure 23.7 is a plot of the calculated longitudinal Young's modulus, E_L, and longitudinal shear modulus, G_L, for the random and aligned composites as a function of nanotube length, for a 1% nanotube volume fraction. These quantities were calculated from the elastic stiffness tensor of the composite, **C**, by using Equation 23.8 for the composite. The results indicate an approximately 55% increase in the shear modulus of the randomly oriented nanotube composite in the range of nanotube lengths between 0 to 200 nm, with a significant change in the slope between nanotube lengths of 50 and 100 nm. Conversely the calculated longitudinal shear modulus for the aligned nanotube composite was constant for the given range of nanotube length. Therefore, under these conditions, increasing the degree of alignment resulted in a decrease in shear modulus.

From Fig. 23.7 it can be seen that alignment of the nanotubes results in nearly a 300% increase in the longitudinal Young's modulus for the composite with nanotubes that are 200 nm long with respect to the composite with short nanotubes. Unlike the case for shear modulus, an increase in the degree of alignment resulted in an increase in the longitudinal Young's modulus. A significant decrease in the slope of the Young's modulus curves occurs between nanotube lengths of 60 to 80 nm.

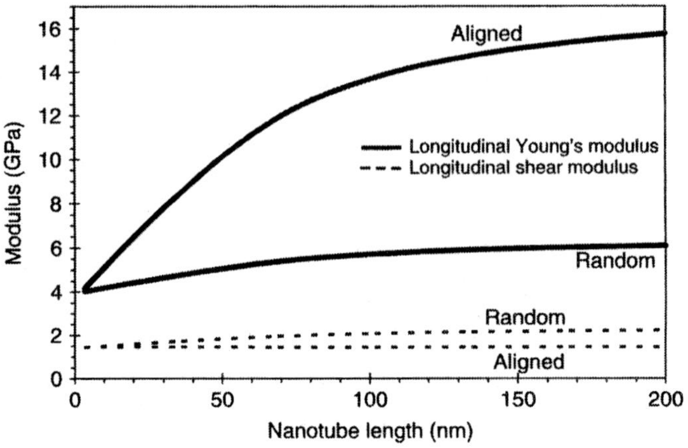

FIGURE 23.7 Modulus of SWNT/polyimide composite material versus nanotube length for a 1% nanotube volume fraction.

The longitudinal Young's modulus of the aligned composite is plotted in Fig. 23.8 as a function of nanotube volume fraction for nanotubes that are 10, 50, and 500 nm long. The Young's modulus increased with an increase in volume fraction, with the most pronounced rate of increase associated with nanotubes of length 50 nm or greater. The dependence of the longitudinal Young's modulus on the nanotube volume fraction became more linear as the nanotube length increased. This dependence is expected because of the well known effect of the increase in load transfer with subsequent increases in reinforcement length and volume fraction.[84]

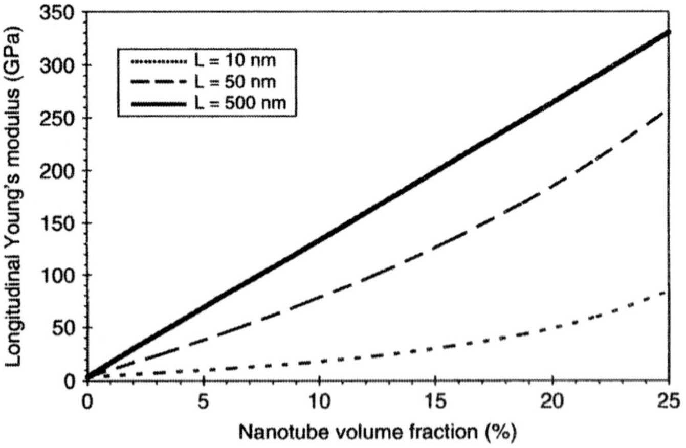

FIGURE 23.8 Longitudinal Young's modulus of aligned SWNT/polyimide composite versus nanotube volume fraction.

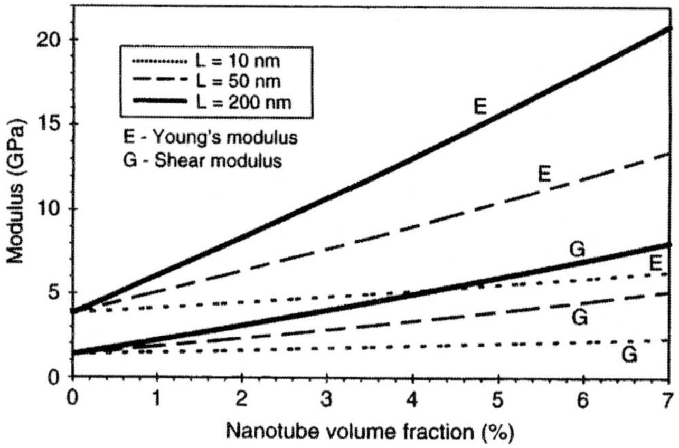

FIGURE 23.9 Modulus of random SWNT/polyimide composite material versus nanotube volume fraction.

For 500-nm-long aligned nanotubes, the longitudinal Young's modulus of a composite with a 25% nanotube volume fraction was about 85 times larger than the Young's modulus of the unreinforced resin.

Figure 23.9 is a plot of the Young's modulus and the shear modulus for the random composite as a function of nanotube volume fraction, for three nanotube lengths. In general, an increase in nanotube volume fraction resulted in increased moduli values. For both the Young's and the shear moduli, increasing the volume fraction for the nanotubes of length near 10 nm provided little to no improvement in stiffness. However, for nanotubes between 50 nm to 200 nm, equivalent stiffness can easily be obtained by trading off a decrease in nanotube length for a small (2× or less) change in volume fraction.

23.6 EXAMPLE: SWNT/POLYETHYLENE COMPOSITE

For this example, stress–strain curves of two unidirectional polyethylene/NT composites were generated from MD simulations.[66] The first composite contained an infinitely long SWNT, and the second composite contained a short SWNT. In the following discussion, details of the composite structure and MD simulations are presented. A detailed description of the computation of stress from molecular force fields follows. Finally the stress–strain curves generated from MD simulation are presented for longitudinal and transverse loading conditions for each of the two composites considered.

23.6.1 MD Simulations

The unidirectional SWNT/polymer composites considered in this example contained long and short SWNTs, as shown in Fig. 23.10. The dashed boxes in Fig. 23.10 enclose a RVE that was simulated by MD. The SWNT composite in Fig. 23.10(a)

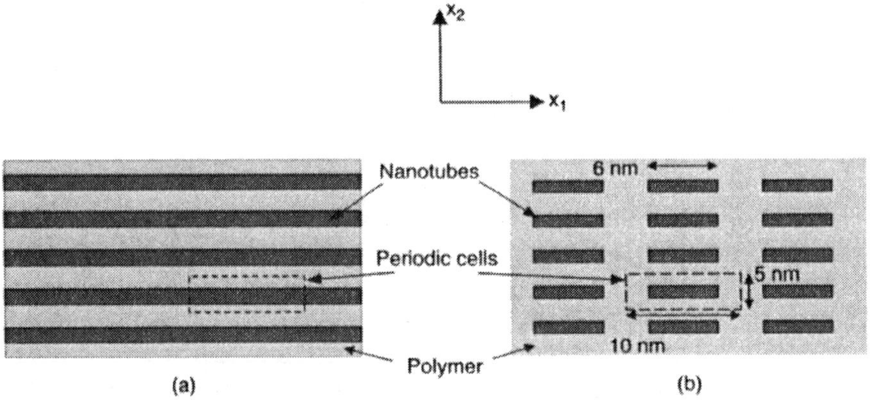

FIGURE 23.10 Schematic of polymer nanocomposites filled with long and short carbon nanotubes.

contained a periodically replicated (10,10) SWNT that spanned the length of the simulation cell. In this composite, the SWNT was embedded in an amorphous polyethylene matrix, which was represented by beads of united atom $-CH_2^-$ units. Specifically the polyethylene matrix had eight chains of 1095 $-CH_2^-$ units. The short-SWNT composite, shown in Fig. 23.10(b), contained a 6-nm capped (10,10) SWNT that was approximately half the length of the simulation cell. The SWNT caps each consisted of one-half of a C_{240} molecule. In this composite, the amorphous polyethylene matrix contained eight chains of 1420 $-CH_2^-$ units. The overall dimensions for the unit cells of each composite in the MD simulation were approximately $5 \times 5 \times 10$ nm. Periodic boundary conditions were used to replicate the cell in all three dimensions. For comparison to the SWNT/polyethylene composites, an equivalent-sized block of amorphous polyethylene without a SWNT was also simulated. For each structure, a polyethylene density of 0.71 g/cm^3 was used.

In the MD simulation, the van der Waals interfacial interaction between the polymer and the SWNT was modeled with the Lennard–Jones potential Equation 23.18). The polyethylene chains were simulated with a molecular mechanics force field adapted from the literature.[66] Specifically the $-CH_2^-$ units of the polyethylene chains were separated by bond lengths of 0.153 nm by using the SHAKE algorithm, a constraint dynamics method that constrains the bond length within a user-defined tolerance.[73] Angle-bending forces were modeled with a harmonic valence angle potential having an equilibrium angle of 112.813° and a barrier of 520 kJ/mol. A torsional potential was used for the torsion angle around the CH_2-CH_2 bond.[66] The Lennard–Jones potential was also used to describe nonbonding interactions between $-CH_2^-$ units in either the same chain or between different chains. The SWNT was modeled with a many-body bond-order potential developed for carbon.[85] This carbon potential was parameterized for C–C bonds of lengths up to 0.17 nm, which was within the magnitude of the strain applied to the composites in the present work. All simulations were carried out at 300 K, with 2 femtosecond time steps.

23.6.2 Stress–Strain Curves from Simulation

Stress–strain curves were generated for the long and short SWNT composites and for the pure polymer via MD simulation. For both composite configurations, the longitudinal (parallel to the SWNT axis) and transverse responses were simulated. The prescription of strain and calculation of stresses is explained in this section.

For each increment of applied deformation, a uniform strain was prescribed on the entire MD model. For the longitudinal and transverse deformations, pure states of strain, ε_{11} and ε_{22}, respectively, were initially applied (Fig. 23.11). The application of strain was accomplished by uniformly expanding the dimensions of the MD cell in the direction of the deformation and rescaling the new coordinates of the atoms to fit within the new dimensions. After this initial deformation, the MD simulation was continued and the atoms were allowed to equilibrate within the new MD cell dimensions. This process was carried out for the subsequent increments of deformation. The applied strain increment in both the longitudinal and transverse directions was 2%, and was applied in two equal increments of 1%. After each 2% increment of strain, the system was relaxed for 2 ps, and then the stress on the system was averaged over an interval of 10 ps. For each composite configuration, six increments of 2% strain were applied up to a total of approximately 12% over a period of 72 ps. The corresponding strain rate was 1.0×10^{10} s^{-1}. This high strain rate is inherent to MD simulation, which includes dynamic information usually on picosecond-to-nanosecond timescales.

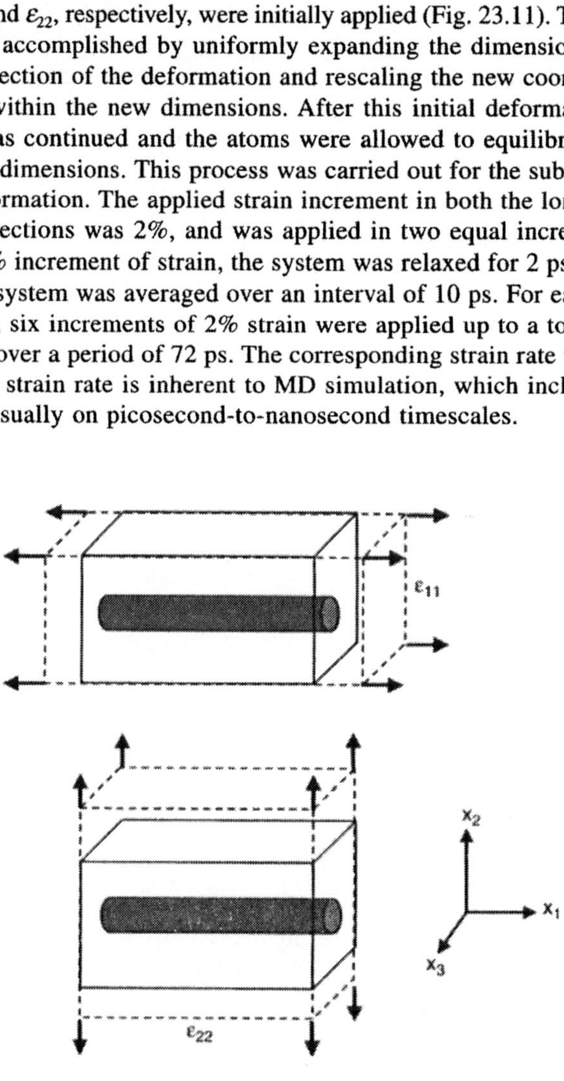

FIGURE 23.11 Definitions of the strains applied to the composites.

In general, the stress in a solid (or a group of interacting particles in the form of a solid) is defined as the change in the internal energy (in the thermodynamic sense) with respect to the strain per unit volume. For example, at the continuum level, the stress tensor, σ_{ij}, for a linear-elastic material is[86]

$$\sigma_{ij} = \frac{1}{V}\left(\frac{\partial E}{\partial \varepsilon_{ij}}\right)_S \tag{23.19}$$

where V is the volume of the solid, E is the total internal energy, ε_{ij} is the strain tensor, and the subscript S denotes constant entropy. When the internal energy is equal to the strain energy of the solid, Hooke's law may be derived from Equation 23.19. Furthermore, if the strain energy is expressed in terms of an applied force acting over the surface area of a solid, then a more familiar form of stress as force per unit area is derived.

At the atomic level, the total internal energy given in Equation 23.19 can be expressed as the summation of the energies of the individual atoms, E^α, that compose the solid:

$$E^\alpha = T^\alpha + U^\alpha = \frac{1}{2}M^\alpha (v^\alpha)^2 + \Phi^\alpha(r) \tag{23.20}$$

where for each atom α, T^α is the kinetic energy, U^α is the potential energy, M^α is the mass, v^α is the magnitude of its velocity, and $\Phi^\alpha(r)$ is the potential energy at the atom location r. Using a Hamiltonian that is based on these individual energy contributions, E^α, it has been shown that the stress contribution, σ_{ij}^α, for a given atom is

$$\sigma_{ij}^\alpha = -\frac{1}{V^\alpha}\left(M^\alpha v_i^\alpha v_j^\alpha + \sum_\beta F_i^{\alpha\beta} r_j^{\alpha\beta}\right) \tag{23.21}$$

where V^α is the atomic volume of atom α, v_i^α is the i-component of the velocity of atom α, v_j^α is the j-component of the velocity of atom α, $F_i^{\alpha\beta}$ is the i-component of the force between atoms α and β obtainable from the derivative of the potential $\Phi(r)$, and $r_j^{\alpha\beta}$ is the j-component of the separation of atoms α and β.[87,88] These parameters are shown in Fig. 23.12 as well.

The stresses that were used to generate the stress-strain curves for the SWNT composites were average atomic stresses for the volume of the model. Therefore, the stress components of each model were calculated for each strain increment by using

$$\sigma_{ij} = -\frac{1}{V}\sum_\alpha\left(M^\alpha v_i^\alpha v_j^\alpha + \sum_\beta F_i^{\alpha\beta} r_j^{\alpha\beta}\right) \tag{23.22}$$

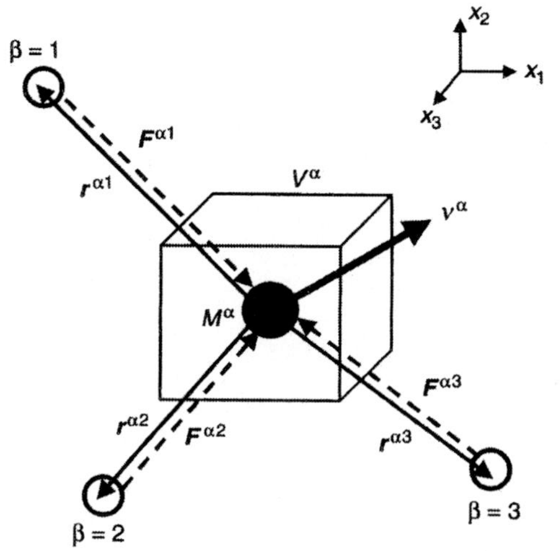

FIGURE 23.12 Parameters used to compute stresses in the simulation.

where V is the volume of the MD model and $V = \sum_{\alpha} V^{\alpha}$. The stress calculated with Equation 23.22 was then averaged over time via the MD simulation.

23.6.3 Results

The stress–strain curve of the long SWNT composite under longitudinal loading is shown in Fig. 23.13. The stress–strain curve of the polymer is also shown in Fig. 23.13. A large increase in the slope of the stress–strain curve is observed upon reinforcement of the polymer material. The stress–strain curve of the short SWNT

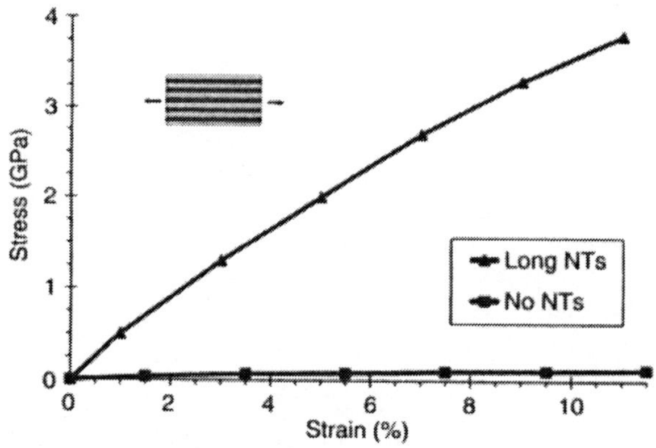

FIGURE 23.13 Longitudinal stress–strain relation of long NT composite and polymer matrix without nanotubes.

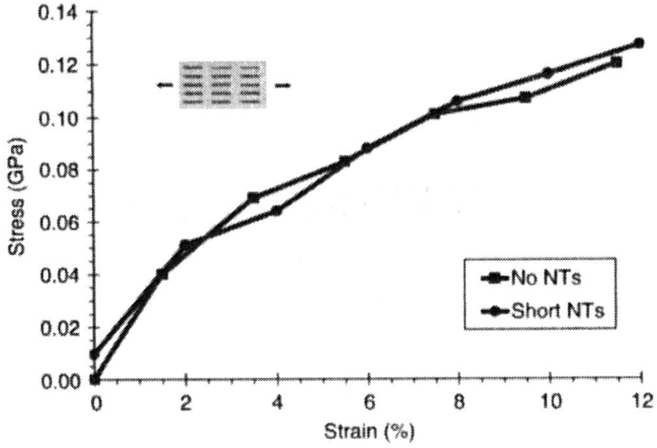

FIGURE 23.14 Longitudinal stress–strain relation for short NT-filled composite compared with stress–strain behavior of the polymer matrix without nanotubes.

subjected to longitudinal loading conditions is plotted with the simulated polymer stress–strain curve in Fig. 23.14. Almost no enhancement relative to the polymer is observed in the stress–strain curve of the composite with the short SWNT. The stress–strain curves of both composites subjected to transverse loading conditions are compared with each other and the polymer in Fig. 23.15. All three MD-generated curves are similar. For the composite with the long SWNT, the stress after loading

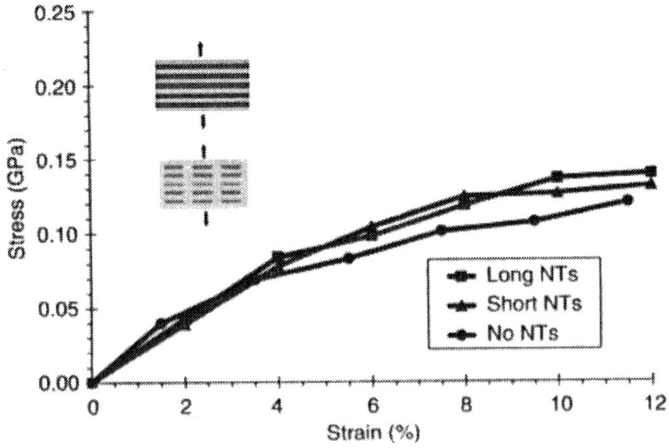

FIGURE 23.15 Transverse stress–strain curves of both long and short NT-filled composites compared with polymer matrices without nanotubes.

in the transverse direction is approximately 30 times lower than the stress levels after loading to the same strain level in the longitudinal direction (Fig. 23.13). This difference in longitudinal and transverse behavior illustrates the anisotropy of the composite.

23.7 SUMMARY AND CONCLUSIONS

This chapter has discussed some of the fundamental concepts of the modeling of carbon nanotube/polymer composites. Four principal modeling tools have been described: analytical micromechanics, numerical micromechanics, *ab initio* simulations, and MD. A comprehensive summary of recent work performed using these four modeling approaches has been presented, along with two in-depth examples of the modeling of carbon nanotube/polyimide and carbon nanotube/polyethylene composites.

The principal difficulty in modeling nanostructured materials is the selection of the correct (or combination) of modeling tools. Micromechanical models are convenient; however, their fundamental assumption is that a continuum exists in the RVE. For a nanostructured material, the RVE is best described as a lattice structure of atoms and molecules. Therefore micromechanical models alone are not expected to accurately predict the behavior of nanotube/polymer composites. Finite element models are more difficult to develop and run than the analytical micromechanical models, but can be more accurate. However, finite element models also assume the presence of a continuum on the nanometer-length scale. *Ab initio* models operate at a first-principles level, and thus predict atomic structures in a very accurate manner. However, the computational difficulty associated with them limits the model size to a very small set of atoms. MD models are capable of modeling larger molecular systems than *ab initio* simulations; however, information about atomic properties is lost in the development of the atomic potentials. A combination of modeling tools can be successfully used to predict the properties of carbon nanotube/polymer composites. The first example (Section 23.5) used MD to establish the atomic structure of the nanotube and polymer molecules in the RVE, and subsequently used micromechanics to predict mechanical properties of the composite at a larger length scale. Therefore, the choice of a modeling tool (or combination of tools) for carbon nanotube/polymer composites should be based on size of the RVE, the computational resources available, and the structure of the material at length scales larger than the RVE.

PROBLEMS

1. How does the use of carbon nanotubes as reinforcement in composite materials improve the mechanical and electrical properties over those of the pure polymer resin?
2. In what ways do nanotubes offer advantages as polymer reinforcements over graphite and glass fibers?

3. What information is lost when continuum mechanics-based models are used to describe the behavior of an atomic structure?
4. What limits do molecular modeling techniques (MD and *ab initio* simulations) have in predicting behavior of heterogeneous materials?

REFERENCES

1. Hull, D. and Clyne, T.W., *An Introduction to Composite Materials,* Cambridge University Press, Cambridge, 1996.
2. Kroto, H.W., Heath, J.R., O'Brien, S.C., Curl, R.F., and Smalley, R.E., C60: Buckminsterfullerene, *Nature, 318,* 162, 1985.
3. Iijima, S., Helical microtubules of graphitic carbon, *Nature, 354,* 56, 1991.
4. Thostenson, E.T., Ren, Z., and Chou, T.W., Advances in the science and technology of carbon nanotubes and their composites: a review, *Compos. Sci. Technol., 61,* 1899, 2001.
5. Lau, K.T. and Hui, D., The revolutionary creation of new advanced materials—carbon nanotube composites, *Compos. Part B, 33,* 263, 2002.
6. Sun, Y.P., Fu, K., Lin, Y., and Huang, W., Functionalized carbon nanotubes: properties and applications, *Accounts Chem. Res., 35,* 1096, 2002.
7. Chen, J., Hamon, M.A., Hu, H., Chen, Y., Rao, A.M., Eklund, P.C., and Haddon, R.C., Solutions properties of single-walled carbon nanotubes, *Science, 282,* 95, 1998.
8. Bahr, J.L. and Tour, J.M., Highly functionalized carbon nanotubes using *in situ* generated diazonium compounds, *Chem. Mater., 13,* 3823, 2001.
9. Michelson, E.T., Huffman, C.B., Rinzler, A.G., Smalley, R.E., Hauge, R.H., and Margrave, J.L., Fluorination of buckytubes, *Chem. Phys. Lett., 296,* 188, 1998.
10. Michelson, E.T., Chiang, I.W., Zimmerman, J.L., Boul, P.J., Lozano, J., Liu, J., Smalley, R.E., Hauge, R.H., and Margrave, J.L., Solvation of fluorinated single-wall carbon nanotubes in alcohol solvents, *J. Phys. Chem. B, 103,* 4318, 1999.
11. Boul, P.J., Liu, J., Michelson, E.T., Huffman, C.B., Ericson, L.M., Chiang, I.W., Smith, K.A., Colbert, D.T., Hauge, R.H., Margrave, J.L., and Smalley, R.E., Reversible sidewall functionalization of buckytubes, *Chem. Phys. Lett., 310,* 367, 1999.
12. Chen, Y., Haddon, R.C., Fang, S., Rao, A.M., Eklund, P.C., Lee, W.H., Dickey, E.C., Grulke, E.A., Pendergrass, J.C., Chavan, A., Haley, B.E., and Smalley, R.E., Chemical attachment of organic functional groups to single-walled carbon nanotube material, *J. Mater. Res., 13,* 2423, 1998.
13. Jin, Z., Sun, X., Xu, G., Goh, S.H., and Ji, W., Non-linear optical properties of some polymer/multi-walled carbon nanotube composites, *Chem. Phys. Lett., 318,* 505, 2000.
14. Hill, D.E., Lin, Y., Rao, A.M., Allard, L.F., and Sun, Y.P., Functionalization of carbon nanotubes with polystyrene, *Macromolecules, 35,* 9466, 2002.
15. Lin, Y., Rao, A.M., Sadanadan, B., Kenik, E.A., and Sun, Y.P., Functionalizing multiple-walled carbon nanotubes with aminopolymers, *J. Phys. Chem. B, 106,* 1294, 2002.
16. Huang, W., Taylor, S., Fu, K., Lin, Y., Zhang, D., Hanks, T.W., Rao, A.M., and Sun, Y.P., Attaching proteins to carbon nanotubes via diimide-activated amidation, *Nano Lett., 2,* 311, 2002.
17. Velasco-Santos, C., Martinez-Hernandez, A.L., Lozada-Cassou, M., Alvarex-Castillo, A., and Castano, V.M., Chemical functionalization of carbon nanotubes through an organosilane, *Nanotechnology, 13,* 495, 2002.

18. Banerjee, S. and Wong, S.S., Structural characterization, optical properties, and improved solubility of carbon nanotubes functionalized with Wilkinson's catalyst, *J. Am. Chem. Soc., 124,* 8940, 2002.
19. Sinnott, S.B., Chemical functionalization of carbon nanotubes, *J. Nanosci. Nanotechnol., 2,* 113, 2002.
20. Chen, R.J., Zhang, Y., Wang, D., and Dai, H., Noncovalent sidewall functionalization of single-walled carbon nanotubes for protein immobilization, *J. Am. Chem. Soc., 123,* 3838, 2001.
21. Star, A., Stoddart, J.F., Steuerman, D., Diehl, M., Boukai, A., Wong, E.W., Yang, X., Chung, S., Choi, H., and Heath, J.R., Preparation and properties of polymer-wrapped single-walled carbon nanotubes, *Angew. Chemie Int. Ed., 40,* 1721, 2001.
22. Lordi, V. and Yao, N., Molecular mechanics of binding in carbon-nanotube–polymer composites, *J. Mater. Res., 15,* 2770, 2000.
23. Jia, Z., Wang, Z., Xu, C., Liang, J., Wei, B., Wu, D., and Zhu, S., Study on poly(methyl methacrylate)/carbon nanotube composites, *Mater. Sci. Engin. A, 271,* 395, 1999.
24. Wagner, H.D., Lourie, O., Feldman, Y., and Tenne, R., Stress-induced fragmentation of multiwall carbon nanotubes in a polymer matrix, *Appl. Phys. Lett., 72,* 188, 1998.
25. Georgakilas, V., Kordatos, K., Prato, M., Guldi, D.M., Holzinger, M., and Hirsch, A., Organic functionalization of carbon nanotubes, *J. Am. Chem. Soc., 124,* 760, 2002.
26. Georgakilas, V., Voulgaris, D., Vazquez, E., Prato, M., Guldi, D.M., Kukovecz, A., and Kuzmany, H., Purification of HiPCO carbon nanotubes via organic functionalization, *J. Am. Chem. Soc., 124,* 14318, 2002.
27. Frankland, S.J.V., Caglar, A., Brenner, D.W., and Griebel, M., Molecular simulation of the influence of chemical cross-links on the shear strength of carbon nanotube–polymer interfaces, *J. Phys. Chem. B, 106,* 3046, 2002.
28. Odegard, G.M., Frankland, S.J.V., and Gates, T.S., The effect of chemical functionalization on mechanical properties of nanotube/polymer composites, in *44th AIAA/ASME/ASCE/AHS/ASC Structures, Structural Dynamics, and Materials Conference,* Norfolk, VA, 2003, p. 1701.
29. Hashin, Z., The elastic moduli of heterogeneous materials, *J. Appl. Mech., 29,* 143, 1962.
30. Hill, R., Theory of mechanical properties of fibre-strengthened materials I. Elastic behavior, *J. Mech. Phys. Solids, 12,* 199, 1964.
31. Hill, R., A self-consistent mechanics of composite materials, *J. Mech. Phys. Solids, 13,* 213, 1965.
32. Hashin, Z. and Rosen, B.W., The elastic moduli of fiber-reinforced materials, *J. Appl. Mech., 31,* 223, 1964.
33. Aboudi, J., *Mechanics of Composite Materials: A Unified Micromechanical Approach,* Elsevier, Amsterdam, 1991.
34. Eshelby, J. D., The determination of the elastic field of an ellipsoidal inclusion, and related problems, *Proc. Roy. Soc. London, Series A, 241,* 376, 1957.
35. Budiansky, B., On the elastic moduli of some heterogeneous materials, *J. Mech. Phys. Solids, 13,* 223, 1965.
36. Mori, T. and Tanaka, K., Average stress in matrix and average elastic energy of materials with misfitting inclusions, *Acta Metal., 21,* 571, 1973.
37. Benveniste, Y., A new approach to the application of Mori–Tanaka's theory in composite materials, *Mech. Mater., 6,* 147, 1987.
38. Halpin, J.C. and Tsai, S.W., Environmental Factors in Composites Design, AFML-TR-67-423.

39. Hill, R., Elastic properties of reinforced solids: Some theoretical principles, *J. Mech. Phys. Solids, 11,* 357, 1963.
40. Hashin, Z., Theory of Fiber Reinforced Materials, NASA CR-1974.
41. Mura, T., *Micromechanics of Defects in Solids,* Martinus Nijhoff, The Hague, 1982.
42. Benedikt, B., Rupnowski, P., and Kumosa, M., Visco-elastic stress distributions and elastic properties in unidirectional composites with large volume fractions of fibers, *Acta Mater., 51,* 3483, 2003.
43. Pipes, R.B. and Hubert, P., Helical carbon nanotube arrays: mechanical properties, *Compos. Sci. Technol., 62,* 419, 2002.
44. Pipes, R.B. and Hubert, P., Helical carbon nanotube arrays: thermal expansion, *Compos. Sci. Technol., 63,* 1571, 2003.
45. Pipes, R.B. and Hubert, P., Scale effects in carbon nanostructures: self-similar analysis, *Nano Lett., 3,* 239, 2003.
46. Odegard, G.M., Gates, T.S., Wise, K.E., Park, C., and Siochi, E.J., Constitutive modeling of nanotube-reinforced polymer composites, *Compos. Sci. Technol., 63,* 1671, 2003.
47. Odegard, G.M., Pipes, R.B., and Hubert, P., Comparison of two models of SWCN polymer composites, *Compos. Sci. Technol., 64,* 1011, 2004.
48. Thostenson, E.T. and Chou, T.W., On the elastic properties of carbon nanotube-based composites: modeling and characterization, *J. Phys. D: Appl. Phys., 36,* 573, 2003.
49. Yoon, J., Ru, C.Q., and Mioduchowski, A., Vibration of an embedded multiwall carbon nanotube, *Compos. Sci. Technol., 63,* 1533, 2003.
50. Lagoudas, D.C. and Seidel, G.D., Effective Elastic Properties of Carbon Nanotube Reinforced Composites, in *45th AIAA/ASME/ASCE/AHS/ASC Structures, Structural Dynamics, and Materials Conference,* Palm Springs, CA, 2004.
51. Adams, D.F., Inelastic analysis of a unidirectional composite subjected to transverse normal loading, *J. Compos. Mater., 4,* 310, 1970.
52. Lin, T.H., Salinas, D., and Ito, Y.M., Elastic-plastic analysis of unidirectional composites, *J. Compos. Mater., 6,* 48, 1972.
53. Dvorak, G.J., Rao, M.S.M., and Tarn, J.Q., Yielding in unidirectional composites under external loads and temperature changes, *J. Compos. Mater., 7,* 1973.
54. Liu, Y.J. and Chen, X.L., Evaluations of the effective material properties of carbon nanotube-based composites using a nanoscale representative volume element, *Mech. Mater., 35,* 69, 2003.
55. Chen, X.L. and Liu, Y.J., Square representative volume elements for evaluating the effective material properties of carbon nanotube-based composites, *Comp. Mater. Sci., 29,* 1, 2004.
56. Li, C.Y. and Chou, T.W., Multiscale modeling of carbon nanotube reinforced polymer composites, *J. Nanosci. Nanotechnol., 3,* 423, 2003.
57. Fisher, F.T., Bradshaw, R.D., and Brinson, L.C., Fiber waviness in nanotube-reinforced polymer composites I. Modulus predictions using effective nanotube properties, *Compos. Sci. Technol., 63,* 1689, 2003.
58. Bradshaw, R.D., Fisher, F.T., and Brinson, L.C., Fiber waviness in nanotube-reinforced polymer composites II. Modeling via numerical approximation of the dilute strain concentration tensor, *Compos. Sci. Technol., 63,* 1705, 2003.
59. Levin, R. N., *Quantum Chemistry,* Prentice Hall, Englewood Cliffs, NJ, 1999.
60. Mylvaganam, K. and Zhang, L.C., Chemical bonding in polyethylene–nanotube composites: a quantum mechanics prediction, *J. Phys. Chem. B, 108,* 5217, 2004.
61. Jaffe, R.L., Quantum chemistry study of fullerene and carbon nanotube fluorination, *J. Phys. Chem. B, 107,* 10378, 2003.

62. Bauschlicher, C.W., High coverages of hydrogen on a (10,10) carbon nanotube, *Nano Lett., 1,* 223, 2001.
63. Bauschlicher, C.W. and So, C.R., High coverages of hydrogen on (10,10), (9,0) and (5,5) carbon nanotubes, *Nano Lett., 2,* 337, 2002.
64. Bauschlicher, C.W., Hydrogen and fluorine binding to the sidewalls of a (10,10) carbon nanotube, *Chem. Phys. Lett., 322,* 237, 2000.
65. Brenner, D.W., Empirical potential for hydrocarbons for use in simulating the chemical vapor deposition of diamond films, *Phys. Rev. B, 42,* 9458, 1990.
66. Frankland, S.J.V., Harik, V.M., Odegard, G.M., Brenner, D.W., and Gates, T.S., The stress–strain behavior of polymer–nanotube composites from molecular dynamics simulation, *Compos. Sci. Technol., 63,* 1655, 2003.
67. Griebel, M. and Hamaekers, J., Molecular dynamics simulations of the elastic moduli of polymer-carbon nanotube composites, *Comp. Meth. Appl. Mech. Engin., 193,* 1773, 2004.
68. Cornell, W.D., Cieplak, P., Bayly, C.I., Gould, I.R., Merz, K.M., Ferguson, D.M., Spellmeyer, D.C., Fox, T., Caldwell, J.W., and Kollman, P.A., A second generation force field for the simulation of proteins, nucleic acids, and organic molecules, *J. Am. Chem. Soc., 117,* 5179, 1995.
69. Allinger, N.L., Yuh, Y.H., and Lii, J.H., Molecular mechanics. The MM3 force field for hydrocarbons, *J. Am. Chem. Soc., 111,* 8551, 1989.
70. Duffy, E.M., Kowalczyk, P.J., and Jorgensen, W.L., Do denaturants interact with aromatic hydrocarbons in water? *J. Am. Chem. Soc., 115,* 9271, 1993.
71. Smith, J.C. and Karplus, M., Empirical force field study of geometries and conformational transitions of some organic molecules, *J. Am. Chem. Soc., 114,* 801, 1992.
72. Rappe, A.K. and Casewit, C.J., *Molecular Mechanics across Chemistry,* University Science Books, Sausalito, CA, 1997.
73. Allen, M.P. and Tildesley, D.J., *Computer Simulation of Liquids,* Oxford University Press, Oxford, 1987.
74. Frankland, S.J.V. and Harik, V.M., Analysis of carbon nanotube pull-out from a polymer matrix, *Surf. Sci., 525,* L103, 2003.
75. Hu, Y.H. and Sinnott, S.B., Molecular dynamics simulations of polyatomic-ion beam deposition-induced chemical modification of carbon nanotube/polymer composites, *J. Mater. Chem., 14,* 719, 2004.
76. Hu, Y., Jang, I., and Sinnott, S.B., Modification of carbon nanotube-polystyrene matrix composites through polyatomic-ion beam deposition: predictions from molecular dynamics simulations, *Compos. Sci. Technol., 63,* 1663, 2003.
77. Werder, T., Walther, J.H., Jaffe, R.L., Halicioglu, T., Noca, F., and Koumoutsakos, P., Molecular dynamics simulation of contact angles of water droplets in carbon nanotubes, *Nano Lett., 1,* 697, 2001.
78. Halicioglu, T. and Jaffe, R.L., Solvent effect on functional groups attached to edges of carbon nanotubes, *Nano Lett., 2,* 573, 2002.
79. Werder, T., Walther, J.H., Jaffe, R.L., Halicioglu, T., and Koumoutsakos, P., On the water-carbon interaction for use in molecular dynamics simulations of graphite and carbon nanotubes, *J. Phys. Chem. B, 107,* 1345, 2003.
80. Liang, Z., Gou, J., Zhang, C., Wang, B., and Kramer, L., Investigation of molecular interactions between (10,10) single-walled nanotube and Epon 862 resin/DETDA curing agent molecules, *Mater. Sci. Engin. A, 365,* 228, 2004.
81. Wei, C.Y., Srivastava, D., and Cho, K.J., Thermal expansion and diffusion coefficients of carbon nanotube–polymer composites, *Nano Letters, 2,* 647, 2002.

82. Odegard, G.M., Frankland, S.J.V., Herzog, M.N., Gates, T.S., and Fay, C.C., Constitutive modeling of crosslinked nanotube materials, in *45th AIAA/ASME/ASCE/AHS/ASC Structures, Structural Dynamics, and Materials Conference*, Palm Springs, CA, 2004.
83. Odegard, G.M., Gates, T.S., Nicholson, L.M., and Wise, K.E., Equivalent-continuum modeling of nano-structured materials, *Compos. Sci. Technol., 62,* 1869, 2002.
84. Fukuda, H. and Takao, Y., Thermoelastic properties of discontinuous fiber composites, in Chou, T.W., *Comprehensive Composite Materials Volume 1: Fiber Reinforcements and General Theory of Composites,* Elsevier, New York, 2000.
85. Brenner, D.W., Shenderova, O.A., Harrison, J.A., Stuart, S.J., Ni, B., and Sinnott, S.B., Second generation reactive empirical bond order (REBO) potential energy expression for hydrocarbons, *J. Phys. C: Condensed Matter, 14,* 783, 2002.
86. Fung, Y.C., *Foundations of Solid Mechanics,* Prentice-Hall, Englewood Cliffs, NJ, 1965.
87. Nielsen, O.H. and Martin, R.M., Quantum-mechanical theory of stress and force, *Phys. Rev. B, 32,* 3780, 1985.
88. Vitek, V. and Egami, T., Atomic level stresses in solids and liquids, *Physica Status Solidi. B: Basic Research, 144,* 145, 1987.

24 Introduction to Nanoscale, Microscale, and Macroscale Heat Transport: Characterization and Bridging of Space and Time Scales

Christianne V.D.R. Anderson and Kumar K. Tamma

CONTENTS

24.1 INTRODUCTION

This chapter serves as an introduction and a prelude to understanding the fundamentals encompassing nanoscale, microscale, and macroscale heat transport. Because of the continued trend in the miniaturization of devices and rapid progress in the design and synthesis of materials and structures leading to the notion of *materials by design,* there is significant interest in improved understanding of heat transport across the broad spectrum of length scales ranging from nanoscale to macroscale regimes and time scales encompassing finite to infinite speeds of heat propagation.

Since the early work on size effects on thermal conductivity in 1938 by Haas and Biermasz and theoretical work by Casimir,[1] and more recent efforts demonstrating that the thermal conductivity of thin film structures is one to two orders of magnitude smaller than its bulk counterpart,[2] from an engineering perspective many researchers have shown a keen interest in providing an improved scientific understanding of thermal transport at small length scales such as the nanoscale regime. Some early works in the 1950s and 1960s by Klemens,[3] Callaway,[4] and Holland[5] emanating from the Boltzmann transport equation (BTE) have indeed provided viable approaches to modeling lattice thermal conductivity based on phonon-scattering processes. Alternately, since the early 1990s, developments emanating from the BTE with relation to the equation of radiation transfer (ERT), namely the equations for phonon radiative transfer (EPRT)[6,7] and others such as Monte Carlo[8–11] and molecular dynamics (MD) simulations,[12–16] have also received considerable research activity. While MD simulations do not require any fitting parameters, they lack the ability to perform large-scale simulations at desired length scales that are of practical interest. Alternately, the BTE-based approaches invariably need fitting parameters(for example, the relaxation time, the acoustic phonon dispersion from property of bulk medium, and the like). On the other hand, these more recent efforts since the 1990s attempt to bring such physics into an engineering perspective and follow the same methods as in Callaway[4] and Holland[5] or improvements thereof and stem serious concerns in employing phenomenological models such as the classical Fourier model and other modifications such as that of Cattaneo–Vernotte.[17,18] Consequently, these more recent efforts have led to microscopic energy transfer models, and although they also need to employ fitting parameters, they appear to show reasonably good fit to experimental data.[6,9,19,20] The primary reasons and arguments for directing research efforts to microscopic energy transfer models and other formulations that have been cited in the literature since the 1990s, for example, in the prediction of thermal conductivity of very thin films, note that (1) the Fourier law overpredicts the actual heat flux,[6] resulting in thermal conductivity results that could be as much as two orders of magnitude greater than experimental reported values for thin dielectric films; (2) the Cattaneo equation lacks the ability to capture the ballistic transport process;[8,19] and (3) for processes that are small in both space and time, the energy balance fails based on the phenomenological Fourier model.[9] Taking some of these arguments at face value, a reference is also made similar to issue (1) by the present authors;[21] however, in this chapter we accurately put into context arguments related to all the preceding issues. This chapter fundamentally places into context and provides challenging answers to the aforementioned arguments existing in the literature, and the reader is encouraged to analyze the underlying details and draw relevant conclusions to understanding thermal transport across a broad range of spatial scales and also temporal scales ranging from finite to infinite speeds of heat propagation.

24.2 SPATIAL AND TEMPORAL REGIMES IN HEAT CONDUCTION

Quantum mechanics help us recognize that solids are composed of several discrete particles (electrons, neutrons, and protons) at the microscopic level. Depending on the type of solid, different mechanisms of heat transport take place in materials in different manners and are called heat carriers. Modeling the spatial and temporal regimes of materials can be broadly subdivided into four categories[22]:

- Macroscale
- Mesoscale
- Microscale
- Nanoscale

The characteristic lengths and times for these regimes are depicted in Table 24.1. In heat conduction, two major areas have been prominent: (1) the *macroscale*, which involves the modeling of macroscopic effects; and (2) the *microscale*, which encompasses all the other regimes in materials modeling (nanoscale effects are implied).

TABLE 24.1
Characteristic Lengths and Times in Materials Modeling

Regime	Characteristic Length	Characteristic Time
Macroscale	$\geq 10^{-3}$ m	$\geq 10^{-3}$ s
Mesoscale	~10^{-4}–10^{-7} m	~10^{-3}–10^{-9} s
Microscale	~10^{-6}–10^{-8} m	~10^{-8}–10^{-11} s
Nanoscale	~10^{-7}–10^{-9} m	~10^{-10}–10^{-14} s

Source: Zhigilei, L.V., *MSE 524: Modeling in Materials Science*, Spring 2002.

The *macroscale* formulation in heat conduction is based on the continuum assumption. It does not consider the size and time dependence of the heat transport. Heat is assumed to be carried by the atoms. The *microscale* formulation in heat conduction considers the physical mechanisms of heat transport through heat carriers where size and time dependence are crucial. This leads to the explanation of several parameters that characterize the microscopic regimes in both temporal and spatial regimes.

24.3 CONSIDERATIONS IN TIME–HEAT CONDUCTION

Besides the notion of temporal scale aspects ranging from the finite nature of heat propagation to infinite speeds of thermal transport, the important time parameters that govern the microscopic heat transport are as follows[9]:

- The thermalization time — The time for the electrons and lattice to reach equilibrium
- The diffusion time — The time the heat information takes to travel through the specimen
- The relaxation time — The time associated with the speed at which a thermal disturbance moves through the specimen
- The heating time — The time that an external source heats the specimen
- The physical process time — The total time duration of interest

When the physical process time is comparable to any of the aforementioned times, the time effect becomes important and the process is considered to be microscopic in time. When the heating time is on the order of the thermalization time, the energy deposition must be considered. The immediate question that naturally arises from this consideration is whether the energy is deposited in the lattice, the electron, or both. Finally, when the heating time is comparable to the diffusion time or the relaxation time, a finite speed of the thermal propagation must be considered.

24.4 CONSIDERATIONS IN SIZE–HEAT CONDUCTION

The important size parameters that govern the heat transport in a device are as follows[9]:

- The mean free path (λ)
- The characteristic dimension of the material (L)

From a physical viewpoint, when the mean free path is much less than the characteristic dimension of the material ($\lambda \ll L$), the heat transport is said to be *macroscopic* and is commonly termed the *purely diffusive limit*. In this limit, there exist enough scattering mechanisms within the film. In dielectrics, these scattering mechanisms help bring the phonons within the film back to equilibrium and help establish a temperature gradient.

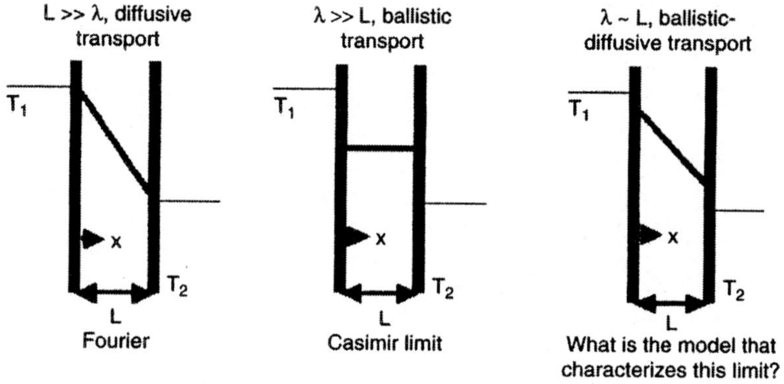

FIGURE 24.1 Comparison of heat transport characterization.

When the mean free path is on the order of or much greater than the characteristic dimension of the material ($\lambda \sim L$ or $\lambda \gg L$), the heat transport is said to be *microscopic,* and the transport presents itself in a partially diffusive ballistic and purely ballistic manner, respectively. As the size of the material is decreased (and in comparison to λ becomes small), so do the scattering mechanisms within the film, and a temperature gradient might not be established at the ballistic limit. As observed at the purely ballistic limit, there exist temperature jumps at the physical boundaries (analogous to slip conditions), and the final temperature profile is characterized by $[(T_{\infty L}^4 + T_{\infty R}^4)/2]^{1/4}$.[23] Since a temperature gradient is not established at the purely ballistic or partially ballistic diffusive limits, then the Fourier law in its purest definition breaks down, thus making it impossible to predict the thermal conductivity.

These limits can be observed in Figure 24.1. The size effects to thermal conductivity were originally observed in 1938 in the experimental works of Haas and Biermasz, and the theoretical explanation soon followed in the works of Casimir.[1] Casimir noted that at the purely ballistic regime a temperature gradient could not be established, and strictly according to the Fourier law, it was not possible to prescribe the thermal conductivity. At the purely ballistic limit, the temperature at the boundaries, and not the temperature gradient within the film, governs the heat transport. Since thermodynamic equilibrium is restored due to the scattering of particles from the boundaries, the heat conduction by these particles is suggested to be similar to conduction by photons and can be analyzed as a radiative transfer model where the heat flux across the specimen can be described by

$$q = \sigma\left(T_1^4 - T_2^4\right) \tag{24.1}$$

where σ is the Stefan–Boltzmann constant for the heat carrier, and T_1 and T_2 are the temperatures of the faces of a thin film. This is commonly referred to as the Casimir limit.

At this juncture, some noteworthy questions naturally arise. What are the underlying issues and approaches for characterization and bridging space and time scales? When do we need to resort to nonclassical approaches and what are the limits of the classical models? Is there a unified theory underlying heat transport that best describes the conduction of heat spanning the various scales, and what are the limitations? From a modeling and simulation viewpoint, what are the simplest but effective formulations that are easy to implement and are not computationally intensive? Several of these issues are addressed in the text to follow and the various methods are also briefly highlighted.

24.5 BOLTZMANN TRANSPORT EQUATION

The heart of heat transport theory is the Boltzmann transport equation (BTE). The BTE determines the status of a particle via its location and velocity. The most general form of the BTE[24] is given as

$$\frac{\partial f}{\partial t} + v \cdot \nabla f + a \cdot \frac{\partial f}{\partial v} = \left(\frac{\partial f}{\partial t}\right)_{\text{scatt}} \tag{24.2}$$

where $f(\vec{r}, T, \omega)$ is the nonequilibrium thermodynamic distribution function, $v(\omega)$ is the phonon velocity (that is, the speed of sound in dielectrics), $a(v)$ is the particle acceleration, and t is time. The first term in Equation 24.2 represents the net rate of particles over time, the second term is the convective inflow of particles in physical space, the third term is the net convective inflow due to acceleration in velocity space, and the term on the right-hand side is the net rate of change of particles inside a control volume due to collisions. The phonon velocity in a dielectric material is fairly constant over a large frequency range, and thus $\partial f/\partial v$ can be neglected. Both energy and temperature gradients tend to disturb the electron distribution, and this tendency is opposed by processes that restore equilibrium, such as the scattering of electrons or phonons by lattice vibrations and crystal defects. As a result, the scattering term is approximated under the relaxation time approximation[25] as

$$\left(\frac{\partial f}{\partial t}\right)_{\text{scatt}} = \frac{\partial (f - f^0)}{\partial t} = \frac{f^0 - f}{\tau} \tag{24.3}$$

where $f^0(\vec{r}, T, \omega)$ is the thermodynamic distribution at equilibrium (Bose–Einstein distribution for boson particles [such as phonons] and Fermi–Dirac distribution for fermion particles [such as electrons]), $\partial f^0/\partial t = 0$, and $\tau(\omega, v)$ is the rate of return to equilibrium and is called the relaxation time.

Kinetic theory is derived from the BTE under the premise that there is *local thermal dynamic equilibrium* (LTE).[26] In cases where LTE is not achieved, one solves the BTE. However, under the presence of a temperature gradient, the LTE is implied, and for simplicity in illustrating the basic concepts we consider the one-dimensional form of the BTE. The diffusion term in Equation 24.2, $\partial f/\partial x \sim \partial f^0/\partial x$, and the $\partial f/\partial x$ term can be approximated as[4]

$$\frac{\partial f}{\partial x} = \frac{df^0}{dT}\frac{dT}{dx} \tag{24.4}$$

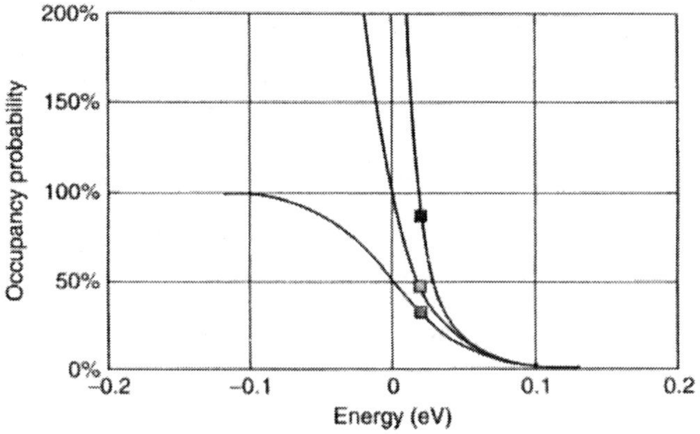

FIGURE 24.2 Occupancy probability for the Fermi–Dirac, Bose–Einstein and Maxwell–Boltzmann distributions. (From Van Zghbroeck, B.J., *Distribution Functions*, 1996.)

The flux of particles is given by[25]

$$q(x) = \int_0^{\omega_D} v_x f(x)\hbar\omega D(\omega)d\omega \tag{24.5}$$

The transient one-dimensional BTE under the relaxation time approximation and the temperature gradient approximation is given by

$$\frac{\partial f}{\partial t} + v_x \frac{df^0}{dT}\frac{dT}{dx} = \frac{f^0 - f(x)}{\tau} \tag{24.6}$$

As the temperature of a material is raised or decreased, these particles interact and collide with each other in order to return the system to thermodynamic equilibrium (the mean time associated with these scattering processes is the relaxation time [τ], and the mean distance associated with these scattering processes is the mean free path [λ]).

24.5.1 One-Temperature Models

Consider the Maxwell–Boltzmann, Bose–Einstein, and Fermi–Dirac distribution functions shown in Fig. 24.2 where the Fermi energy is set to zero. At high frequencies (or energies), the effect of the Fermi–Dirac and the Bose–Einstein statistical functions are eliminated and all distributions converge to a Maxwell–Boltzmann distribution.

Maxwell–Boltzmann distribution function:

$$f(E(\omega)) = f_{MB}(E(\omega)) := \frac{1}{Ae^{\frac{E(\omega)}{T\kappa_B}}} \tag{24.7}$$

Bose–Einstein distribution function:

$$f(E(\omega)) = f_{BE}(E(\omega)) := \frac{1}{Ae^{\frac{E(\omega)}{T\kappa_B}} - 1} \tag{24.8}$$

Fermi–Dirac distribution function:

$$f(E(\omega)) = f_{FD}(E(\omega)) := \frac{1}{Ae^{\frac{(E(\omega)-E_F)}{T\kappa_B}} + 1} \tag{24.9}$$

where A is a normalized constant, κ_B is the Boltzmann's constant, and E_F is the Fermi energy. If the frequency ω is large enough, in the foregoing distribution functions we have

$$f(E(\omega)) \approx 0, \quad \forall\omega \in [\omega_T, \omega_D] \tag{24.10}$$

Therefore, we have

$$\frac{\partial f(E(\omega))}{\partial t} \approx 0, \quad \forall\omega \in [\omega_T, \omega_D] \tag{24.11}$$

At high frequencies, the wavelength of the particle is short and the statistical distinction is unimportant. Multiplying Equation 24.6 by $\upsilon_x \hbar\omega D(\omega)$ and integrating over the two separate frequency ranges $[0,\omega_T]$ and $[\omega_T, \omega_D]$, based on the physics of the actual distribution functions, because of this statistical unimportance, it is reasonable to assume that $\int_{\omega T}^{\omega D} v_x \hbar\omega D(\omega) \frac{\partial f}{\partial t} d\omega$ equals 0 , due to the observation that the distribution function is fairly constant over time for high frequencies. In other words, as the different distribution functions reach their equilibrium positions, most of the change in the distribution functions occurs at the lower frequencies, and the distributions at the high frequency tail are fairly constant over time. Hence, the *df/dt* term is neglected. Also note that as time evolves, the threshold frequency must change as the distributions reach these equilibrium stages. This is a similar analogy to the change of the Fermi energy with temperature (see Reference 27).

Using this fundamental basis, we now highlight a novel C- and F-processes model based on the hypothesis that upon the application of a temperature gradient, there simultaneously coexist both slow processes (at low energies and termed C-processes) and fast processes (at high energies and termed F-processes) associated with the heat carriers and evolving with time as the processes proceed from a finite nature of propagation to a process involving infinite speeds.[17–19] The spanning of space scales is in the sense of characterizing ballistic to diffusive limits, and the spanning of the time scales is in the sense of characterizing finite to infinite speeds of heat propagation. As with BTE-based approaches, certain limitations exist; nonetheless, coupling of the simple yet effective C- and F-processes model with computationally attractive variants of MD or lattice dynamics to avoid the need for fitting parameters may provide extensions to circumvent some of the limitations and provide extensions for multiscale capabilities.

24.5.2 C- and F-Processes Model — A Unified Theory

Following the previous discussion, we now define the total heat flux as

$$q=\int_0^{\omega_T} v_x f\hbar\omega D(\omega)d\omega+\int_{\omega_T}^{\omega_D} v_x f\hbar\omega D(\omega)d\omega=q_C+q_F \tag{24.12}$$

where the postulation is that the integral up to a threshold frequency ω_T involves the slow C-processes and is termed q_C, and that the integral from the threshold to infinity involves the fast F-processes and is termed q_F.

As discussed previously, considering the total heat conduction process, the BTE can be multiplied by $v_x\hbar\omega D(\omega)$ and integrated over the entire frequency range. Also, note that $\int_{\omega_T}^{\omega_D} v_x\hbar\omega D(\omega)\frac{\partial f}{\partial t}d\omega=0$ (not detailed here) due to the observation that the distribution function is constant over time for high frequency; therefore, $df/dt=0$.

Next, introducing the nondimensional heat conduction model number F_T as the following ratio, we have

$$F_T=\frac{\int_{\omega_T}^{\omega_D} v_x^2\tau\frac{df^0}{dT}\hbar\omega D(\omega)d\omega}{\int_0^{\omega_D} v_x^2\tau\frac{df^0}{dT}\hbar\omega D(\omega)d\omega} \tag{24.13}$$

This is physically depicted in Reference 28 and has the interpretation

$$F_T=\frac{K_F}{K_F+K_C}$$

$$=\frac{\text{Conductivity }(K_F)\text{ associated with fast F-Processes}}{\text{Conductivity }(K_F)\text{ associated with fast F-processes}+\text{Conductivity}(K_C)\text{ associated with slow C-processes}} \tag{24.14}$$

Further, the following is defined

$$K=\int_0^{\omega_T} v_x^2\tau\frac{df^0}{dT}\hbar\omega D(\omega)d\omega+\int_{\omega_T}^{\omega_D} v_x^2\tau\frac{df^0}{dT}\hbar\omega D(\omega)d\omega=K_C+K_F \tag{24.15}$$

Upon application of the heat flux of particles given in Equation 24.12 and the thermal conductivity given in Equation 24.15, with the definition from Equation 24.13, the transient BTE finally yields the C- and F-processes heat conduction constitutive model in terms of the heat conduction model number as

$$q=q_F+q_C \tag{24.16}$$

$$q_F=-F_T K\frac{dT}{dx} \tag{24.17}$$

$$q_C+\tau\frac{dq_C}{dt}=-(1-F_T)K\frac{dT}{dx} \tag{24.18}$$

In general, the C- and F-processes heat conduction constitutive model is given by

$$\mathbf{q}_F = -F_T K \nabla T \tag{24.19}$$

$$\mathbf{q}_C + \tau \frac{\partial \mathbf{q}_C}{\partial t} = -(1 - F_T) K \nabla T \tag{24.20}$$

$$\mathbf{q} = \mathbf{q}_F + \mathbf{q}_C \tag{24.21}$$

which explains the present derivation based on fundamental physical principles emanating from the Boltzmann transport equation.

In the previous equations, $\mathbf{q}$ is the total heat flux due to the mechanisms of heat conduction (composed of those associated with each of the Fourier-type fast processes and the Cattaneo-type slow processes), and K is the total conductivity, which is the sum of the Fourier (effective) conductivity, K_F, and the Cattaneo (elastic) conductivity, K_C. Thus, $K = K_F + K_C$.

When $F_T = 1$, the right-hand side of Equation 24.20 is zero and the total heat flux is given by the Fourier law. This implies that the conductivity associated with the slow C-processes $K_C = 0$, or in other words, most of the dominant transport with evolution of time to steady state is via the fast F-processes. On the other hand, when $F_T = 0$, the right-hand side of Equation 24.19 is zero and the total heat flux is given by the Cattaneo law. Consequently, $K_F = 0$ and most of the dominant transport with the evolution of time to steady state is via the slow C-processes. Thus, in the limiting cases of F_T the C- and F-processes model can recover both Cattaneo and Fourier laws.

The combined representation of the C- and F-processes model can be shown by adding Equations 24.17 and 24.18 to yield

$$\mathbf{q} + \tau \frac{d\mathbf{q}_C}{dt} = -K \nabla T \tag{24.22}$$

By substituting the C- and F-processes postulation that the total heat flux is a combination of the flux associated with the low-energy processes and the high-energy processes (see Equation 24.12) where $\mathbf{q}_C = \mathbf{q} - \mathbf{q}_F$, and the high-energy processes are described by a Fourier-like heat flux given by $\mathbf{q}_F = -K_F\, dT/dx$, Equation 24.22 reduces to

$$\mathbf{q} + \tau \frac{d(\mathbf{q} - (-K_F \nabla T))}{dt} = -K \nabla T \tag{24.23}$$

Rearranging Equation 24.23, we obtain the so-called Jeffreys model of heat conduction[29] as

$$\mathbf{q} + \tau \frac{d\mathbf{q}}{dt} = -K \left[\nabla T + \tau \frac{K_F}{K} \frac{d}{dt} (\nabla T) \right] \tag{24.24}$$

where the flux **q**, the thermal conductivity K, and the relaxation time τ are the total contributions, and $\tau K_F/K$ is defined as the retardation time τ_R.[29] By adhering to the notion of the heat conduction model number, it is finally possible to characterize the energy transport of the heat conduction process from transient to steady state. It is noteworthy to mention that the Jeffreys model also reduces to the Cattaneo model when $K_F = 0$. However, when $K_F = K$ it only reduces to a Fourier-like model that contains the elusive relaxation time parameter τ, unlike the C- and F-processes model that identically yields the Fourier model.

By substituting Equations 24.19 through 24.21 into the energy equation, we eliminate the flux and obtain the generalized one-step (GOS) C- and F-processes one-temperature formulation as

$$L_1(T) = f_1(T, \dot{T}, \ddot{T}, T_{xx}, K, \alpha, c_T, S, \dot{S}, \tau, F_T), \quad \text{for } F_T < 1 \tag{24.25}$$

$$L_2(T) = f_2(T, \dot{T}, T_{xx}, K, \alpha, S), \quad \text{for } F_T = 1 \tag{24.26}$$

where T is the total temperature, K is the total thermal conductivity given by kinetic theory as $K = Cv^2\tau/3$, the temperature propagation speed is given by $c_T = \sqrt{K/\rho c\tau}$, the thermal diffusivity is given by $\alpha = K/\rho c$, C is the specific heat of the heat carrier, ρ is the density of the heat carrier, S is the external heat source, and the retardation time is given by $\tau_R = \tau F_T$.

In conjunction with the C- and F-processes model , we argue following the initial work by Klitsner and colleagues[23] on phonon radiative heat transfer where *heat conduction by phonons can be analyzed as radiative transfer* that the phonons are emitted from the surface and that the boundary conditions are developed based on an energy balance at the surface as $q_n = q_{\text{rad}}$, where

$$q_n = \vec{q} \cdot \hat{n} \propto f(C, v, T, T_\infty), \quad \text{for } x = 0, L \tag{24.27}$$

where T represents the surface temperatures of the left and right sides of the film, and T_∞ is the ambient temperature on the right and left of the film. It is assumed that the thermalizing black boundaries (the ambients) are the boundaries of the metal film between which the dielectric film is enclosed.

Finally, after the correct heat flux is computed via the C–F model (which must be a constant across the film due to the radiative equilibrium assumption at steady state), an effective thermal conductivity is obtained. We invoke the following proposition that *the thermal conductivity provided by the C–F model is due to the heat flux within the film based on the established boundary temperatures divided by the temperature gradient based on the imposed ambient temperatures*. Thus,

$$K_{C-F} = \frac{\text{true heat flux within the film due to established boundary temperatures}}{\text{temperature gradient due to the imposed ambient temperatures}} \tag{24.28}$$

24.5.3 Validation of Thermodynamics Second Law for C- and F-Processes Model

The thermodynamics second law is represented in terms of the Clausius–Duhem inequality as

$$\rho\left(\frac{Ds}{Dt} - rT^{-1}\right) \geq -\nabla \cdot (T^{-1}\mathbf{q}) \tag{24.29}$$

where s is the entropy, and r is the heat source.

It can be shown that for a solid,

$$\rho\left(T\frac{Ds}{Dt} - r\right) + \nabla \cdot \mathbf{q} = 0 \tag{24.30}$$

The thermodynamics second law can also be represented in the form

$$\rho(T\dot{s} - r) + \nabla \cdot \mathbf{q} - T^{-1}\mathbf{q} \cdot \nabla T \geq 0 \tag{24.31}$$

Therefore, the thermodynamics second law can be simply decoupled to yield the following two sufficient conditions:

$$\rho(T\dot{s} - r) + \nabla \cdot \mathbf{q} = 0 \tag{24.32}$$

$$\mathbf{q} \cdot \nabla T \leq 0 \tag{24.33}$$

where the first equation represents no viscous dissipation of kinetic energy into thermal energy, and the second condition represents that in heat conduction, the heat flow is from a high-temperature region to a low-temperature region. From the previous section, we showed that the processes in the C- and F-processes heat conduction model belong to two different frequency ranges. Therefore, the F-processes and the C-processes are independent, and either of the processes indeed satisfies the thermodynamics second law independently, namely,

$$\mathbf{q}_F \cdot \nabla T = -F_T k(\nabla T)^2 \leq 0 \tag{24.34}$$

$$\mathbf{q}_C \cdot \nabla T = -\frac{(1 - F_T)k}{\tau} \nabla T \int_{-\infty}^{t} e^{-\frac{t-s}{\tau}} \nabla T(\mathbf{x}, s) ds \leq 0 \tag{24.35}$$

To satisfy the thermodynamics second law, the heat conduction model number F_T has to strictly satisfy the bounds $F_T \in [0,1]$. For other cases such as $F_T < 0$ or $F_T > 1$, the thermodynamics second law will be violated. To be consistent with the second law of thermodynamics, one must ensure $0 \leq k_F$ and $0 \leq k_C$; hence, $0 \leq k = k_F + k_C$. Thus, from its basic underlying definition, the bounds are strictly $F_T \in [0, 1]$. For the case of $F_T < 0$, it leads to $0 < -k_F < k_C$ or $0 < -k_C < k_F$, and for the case of $1 < F_T$, it leads to $0 < k_F < -k_C$ or $0 < k_C < k_F$. For either of the previous two underlying cases, the second law of thermodynamics is violated, and hence it is neither valid nor physically acceptable. Some existing classical models highlighted next can all be readily explained via the C- and F-processes model as described previously.

24.5.3.1 Fourier Model

Under the assumption of a macroscopic continuum formulation, this phenomenological model proposed by Fourier[30] is given by

$$\mathbf{q} = -\mathbf{k}\nabla T \tag{24.36}$$

where **k** is the conductivity tensor, **q** is the heat flux, and T is temperature, which together with the energy equation yields the parabolic one-step (POS) heat conduction equation or the parabolic heat conduction (PHC) equation and is diffusive with the notion of infinite speed of propagation of thermal disturbances.

24.5.3.2 Cattaneo Model

Because of the anomalies associated with the Fourier model and in order to account for a finite temperature propagation speed, the Cattaneo-type model[17,18] is based on the notion of relaxing the heat flux and is given as

$$\tau \frac{\partial \mathbf{q}}{\partial t} = -\mathbf{q} - k\nabla T \tag{24.37}$$

where τ is the relaxation time (Cattaneo originally developed this model for gases).

For the Cattaneo model, the temperature propagation speed is given as

$$c_T = \sqrt{\frac{k}{\rho c \tau}} \tag{24.38}$$

where ρ is the material density, and c is the heat capacity. Together with the energy equation, it yields the hyperbolic one-step (HOS) heat conduction equation or the hyperbolic heat conduction (HHC), which is propagative with the notion of a finite speed of heat propagation.

24.5.3.3 Jeffreys-Type Model

The original Jeffreys model[31,32] was originally proposed for studying the wave propagation in the earth's mantle,[33] and is given as

$$\sigma + \lambda_1 \frac{\partial \sigma}{\partial t} = \eta_0 \left(\dot{\gamma} + \lambda_2 \frac{\partial \dot{\gamma}}{\partial t} \right) \tag{24.39}$$

which relates the stress and the rate of strain tensor $(\sigma, \dot{\gamma})$, and η_0 is the zero shear rate viscosity. In this constitutive model originally proposed in 1929, λ_1 is defined as the *relaxation time* and λ_2 is defined as the *retardation time;* no mention was ever made by Jeffreys relating this to the notion of heat. Based on experiences with viscoelastic fluids, a Jeffreys-type heat flux phenomenological model drawn from the original Jeffreys equation (Equation 24.39) was subsequently proposed and introduced by Joseph and Preziosi[29] for heat conduction, which appears as

$$\mathbf{q} + \tau \frac{\partial \mathbf{q}}{\partial t} = -k\left[\nabla T + K \frac{\partial (\nabla T)}{\partial t}\right] \tag{24.40}$$

where τ is the relaxation time and $K = \tau k_1/k$ is the retardation time. It is important to note that in the original model proposed by Jeffreys,[33] which relates stress and strain rate, or in the corresponding heat flux model introduced by Joseph and Preziosi,[29] which relates the heat flux to the temperature gradient, the retardation time can never exceed the relaxation time because it makes the physical interpretation meaningless if violated. When selecting the retardation time, $K = 0$, the Jeffreys-type model degenerates to the Cattaneo model as

$$\tau \frac{\partial \mathbf{q}}{\partial t} = -\mathbf{q} - k\nabla T \tag{24.41}$$

and when selecting the retardation time equal to the relaxation time, $K = \tau$, the Jeffreys-type model only degenerates to a Fourier-like diffusive model with relaxation as

$$\mathbf{q} + \tau \frac{\partial \mathbf{q}}{\partial t} = -k\left[\nabla T + \tau \frac{\partial (\nabla T)}{\partial t}\right] \tag{24.42}$$

24.6 TWO-TEMPERATURE MODELS

Likewise, there also exists the so-called microscale two-temperature theory (namely, the two-step temperature equations describing nonequilibrium behavior between electrons and phonons). From a microscale viewpoint, the pioneering work of the two-temperature diffusive formulation of Anisimov et al.,[34] who seem to have developed the first parabolic two-step (PTS) energy balance equations, is significant and given as

$$C_e(T_e)\frac{\partial T_e}{\partial t} = -\frac{\partial \mathbf{q}_e}{\partial \mathbf{r}} - G(T_e - T_l) + S(\mathbf{r}, t) \tag{24.43}$$

$$C_l \frac{\partial T_l}{\partial t} = G(T_e - T_l) \tag{24.44}$$

$$\mathbf{q}_e = -k_e \nabla T_e \tag{24.45}$$

where $S(\mathbf{r},t)$ is the internal heating due to a laser pulse and is an integral part of the two-temperature theory, C_e is the electron heat capacity and C_l is lattice heat capacity (with $C = C_e + C_l$), G is the electron-lattice coupling factor whose experimental determination has been challenged,[35,36] and k_e is associated with the electron thermal conductivity in steady state. The electron heat flux $\mathbf{q}_e$ pertains to the Fourier model, and diffusion in the lattice is neglected during the transient duration.

The generalized two-step (GTS) pulse heating model for metals recently proposed by Tamma and Zhou[37] allows not only choices of propagation or diffusion transport mechanisms at the microscale levels but also transitions via the simultaneous introduction of microscale relaxation and retardation times.

We next introduce the physical notion of a *heat conduction model number, F_{T_e}*, associated with the electron heat flux involving F-processes and C-processes as

$$\mathbf{q}_e = -k_e\nabla T_e - \tau\frac{\partial \mathbf{q}_e}{\partial t} - \tau k_{eF}\frac{\partial}{\partial t}(\nabla T_e)$$

$$= -k_e\left[\nabla T_e \tau + \frac{\tau}{k_e}\frac{\partial \mathbf{q}}{\partial t} + \tau F_{T_e}\frac{\partial}{\partial t}(\nabla T_e)\right] \tag{24.46}$$

For microscale heat transport it now has a different physical meaning than in the macroscale formulation, although it has similarities with the associated macroscale formulation and is physically defined in terms of a *heat conduction model number*, $F_{T_e} \in [0,1]$, as

$$F_{T_e} = \frac{k_{eF}}{k_{eF} + k_{eC}} \tag{24.47}$$

where k_{eF} and k_{eC} are contributions of the F-processes and C-processes to the total electron thermal conductivity given by $k_e = k_{eF} + k_{eC}$. Furthermore, the aforementioned constitutive formulation for the two-step theory also has a different physical meaning than that of the macroscale formulation in the following sense.

We consider for microscale heat transport

$$\mathbf{q}_{eF} = -k_{eF}\nabla T_e = -F_{T_e}k_e\nabla T_e$$

$$\mathbf{q}_{eC} + \tau\frac{\partial \mathbf{q}_{eC}}{\partial t} = -k_{eC}\nabla T_e = -\left(1 - F_{T_e}\right)k_e\nabla T_e \tag{24.48}$$

$$\mathbf{q}_e = \mathbf{q}_{eF} + \mathbf{q}_{eC}$$

where $\mathbf{q}_e$ is the total electron heat flux and is composed of the sum of the electron heat flux associated with the F-processes, $\mathbf{q}_{eF}$, and the C-processes, $\mathbf{q}_{eC}$, respectively.

Employing this constitutive model into the Anisimov et al.[34] two-step energy equations yields the generalized microscale electron-lattice two-step model equations with $F_{T_e} \in [0, 1]$ of which the GTS pertains to the range $F_{T_e} \in (0,1)$, which is parabolic (diffusive) in nature:

$$C_e(T_e)\frac{\partial T_e}{\partial t} = -\frac{\partial \mathbf{q}_e}{\partial \mathbf{r}} - G(T_e - T_l) + S$$

$$C_l \frac{\partial T_l}{\partial t} = G(T_e - T_l) \tag{24.49}$$

$$\mathbf{q}_e = -k_e\left[\nabla T_e + \frac{\tau}{k_e}\frac{\partial \mathbf{q}_e}{\partial t} + \tau F_{T_e}\frac{\partial}{\partial t}(\nabla T_e)\right]$$

The preceding GTS model is parabolic (diffusive), and the transmission of information is immediately felt everywhere. It is also to be noted that by approximate mathematical manipulations, the aforementioned equations may be combined to lead to a single equation for the determination of the electron or lattice temperatures, respectively.

24.7 RELAXATION TIME

The first step in solving for the heat transport at the different length scales is to approximate the elusive relaxation time, τ. Accounting for all the scattering mechanisms within a film can approximate the relaxation time. Accurately quantifying and qualifying all the possible scattering mechanisms proves to be the most difficult task in solving the C–F model, or any model derived from BTE for that matter.[3–6,9] This task can be simplified by accounting for all the internal scattering processes within a specimen by assuming that all scattering mechanisms are independent. And, using Matthiessen's rule, which inversely adds all scattering contributions as $1/\tau_i = \Sigma_j\ 1/\tau_{i,j}$ where $\tau_{i,j}$'s are the contributions of the various scattering mechanisms based on i, the modes of polarizations of the phonons, and j represents the different types of scattering rates. These scattering mechanisms contribute to the total resistance to the heat transport.

Many researchers[1,5,38–42] describe all the different scattering mechanisms that have been used for predicting relaxation times for doped and undoped single and polycrystalline dielectric films. In a nutshell, most of the scattering is due to crystal imperfections or interactions with other phonons, and at low temperatures (or for films where the $\lambda \sim L$) the boundary of a crystal. For the results shown here, we assume that these are the main scattering mechanisms

$$\frac{1}{\tau} = \frac{1}{\tau_{\text{defect}}} + \frac{1}{\tau_U} + \frac{1}{\tau_{GB}} \tag{24.50}$$

where the scattering due to defects is given as

$$\tau_{\text{defect}} = \frac{1}{\alpha\sigma_c\eta\nu} \tag{24.51}$$

where α is a constant (usually one), σ_c is the scattering cross-section, η is the level of impurity in the medium (in other words, the number of scattering sites in the medium), and v is the speed of the phonons, which for dielectrics is the speed of sound. The scattering cross-section is approximated as[6]

$$\sigma_c = \pi R^2 \left(\frac{s^4}{s^4 + 1} \right) \tag{24.52}$$

where R is the radius of the lattice imperfection, s is the size parameter given by $s = 2\pi R/\Lambda_{\text{dominant}}$, and $\Lambda_{\text{dominant}}$ is the dominant wavelength of phonons denoted by $hv \sim \Lambda_{\text{dominant}} K_B T$, where K_B is the Boltzmann constant, and h is Planck's constant.

The scattering due to other phonons is accounted for by the U-process scattering as[6]

$$\tau_U = A \frac{T}{\theta_D \omega} e^{\frac{\theta_D}{aT}} \tag{24.53}$$

where A is a nondimensional constant that depends on the atomic mass, the lattice spacing, and the Grüinessen constant; θ_D is the Debye temperature; ω is the frequency; T is the temperature; and a is a parameter representing the effect of the crystal structure.

Finally, the grain boundary scattering is given as[40]

$$\tau_{GB} = \frac{d_g}{v_s} \tag{24.54}$$

where v_s is the average phonon velocity, and d_g is the sample size in the direction perpendicular to heat flow.

24.8 NUMERICAL ILLUSTRATION—TWO-TEMPERATURE MODEL AND PULSE LASER HEATING

For the experiment by Brorson et al.,[43] $C_e = 1.84E + 4$(J/m³K), $C_l = 2.5E + 6$(J/m³K), $\delta = 15.3$ nm, $G = 2.6E + 16$(W/m³K), $J = 10$(J/m²), $k_{eo} = 310$(W/mK), $L = 0.1$ μm, and $R = 0.093$ were employed. Also, $t_p = 96$ fs was used except when the source was that of case (1) as described subsequently; $t_p = 100$ fs was employed for the experimental simulations.

For the selected experiment from Brorson et al.[43] for short-pulse laser heating of thin gold films, the following were investigated:

Case 1: For the source, the actual measured autocorrelation of the laser pulse was first employed.[44]

Case 2: The source was specifically selected as

$$S = 0.94 \left(\frac{1 - R}{t_p \delta} \right) J e^{-\frac{x}{\delta} - 2.77 \left(\frac{t}{t_p} \right)^2} \tag{24.55}$$

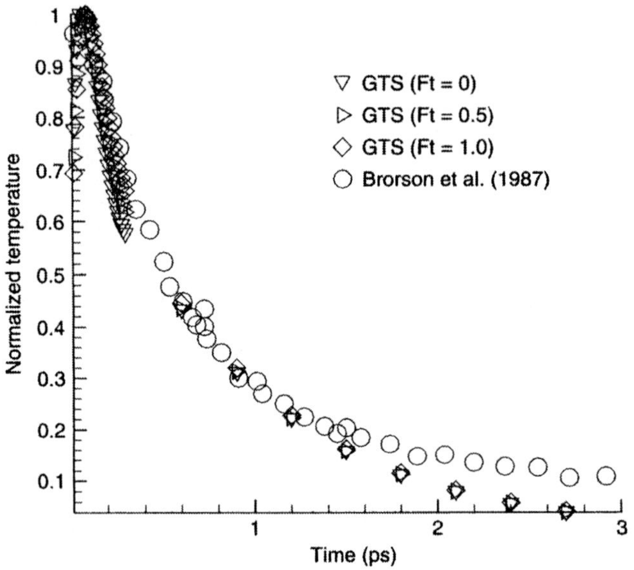

FIGURE 24.3 Comparative results for experiment of Brorson et al. $\tau = \tau(\mathrm{T})$ and heat source. (From Bronson, S.D. et al., *Phys. Rev. Lett.*, 59, 1962, 1987.)

where t_p is the full width at half maximum of the pulse duration and δ is the radiation penetration depth.

Furthermore, for the relaxation parameter, τ, the following were investigated: (1) τ accounts for the temperature dependence with data taken from Pells and Shiga,[45] and (2) $\tau = 0.04$ ps (constant value). In all the situations, temperature dependence of the electron thermal conductivity and heat capacity was assumed.

Employing the GTS, studies were conducted to demonstrate the theoretical consequences alluded to earlier, including efforts to shed further light on the characterization of the heat transport behavior in these experiments. Figure 24.3 shows the comparative results of the Brorson et al.[43] experiments for the the front surface with the GTS ($F_{T_e} = 0$ and permitting $\tau = \tau\,[T_l]$; $F_{T_e} = 1$, which implies $\tau = 0$; and for illustration, $F_{T_e} = 0.5$ with $\tau = \tau\,[T_l]$). The source was selected from the actual measured autocorrelation of the laser probe (Case 1). Although not clearly distinguishable, the cases of $F_{T_e} = 1$ and $F_{T_e} = 0.5$ (arbitrarily selected) seem to be somewhat closer than $F_{T_e} = 0$ at very early times in the transient.

Also based on the theoretical premise that there simultaneously exist fast and slow transport processes, the added objective was to characterize (admittedly, a first and primitive attempt) the *microscale heat conduction model number,* namely the F_{T_e} value, and select this to match closely to the experiment. For example, if $F_{T_e} = 0.1$ was a close match, then $k_{eF} = 0.1k_e = 0.1(k_{eF} + k_{eC})$, which implies that plausibly

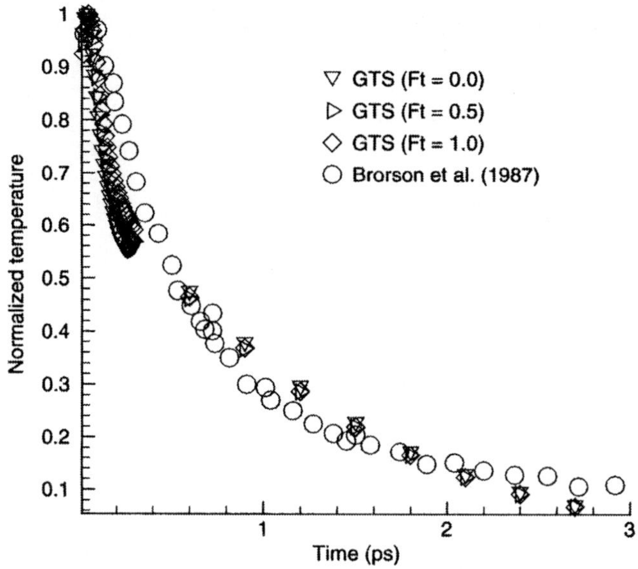

FIGURE 24.4 Comparative results for experiment of Brorson et al. $\tau = \tau(T)$ and heat source, case (2) (From Bronson et al., *Phys. Rev. Lett.*, 59, 1962, 1987.)

90% pertains to transport associated with k_{eC} and 10% pertains to transport associated with k_{eF}. Since total thermal conductivity is customarily measured experimentally, such a plausible challenge was undertaken. However, based on very limited experiments conducted to date, no bounds on errors are available (and the possibility exists for experimental errors), only a quantitative check could be made and no conclusive results or inferences could be drawn.

Figure 24.4 shows the comparative results with the source selected instead of Case 2. The results are in reasonable agreement in the initial part of the transient; however, some in-depth studies drawing comparisons with the measured autocorrelation of the laser pulse need to be conducted.

24.9 NUMERICAL ILLUSTRATION—ONE-TEMPERATURE MODEL AND HEAT CONDUCTION MODEL NUMBER F_T

Consider the following experiment. Since little literature on experimental results describing the transient behavior at the ballistic limit exists to our knowledge, we next present a systematic analysis of the significance of F_T spanning the temporal ballistic limits for the hypothetical case of a diamond film of 1 μm. Considering the film structure in Fig. 24.5(a), assume that the curve shown in Fig. 24.5(b) represents the characteristic thermal conductivity data at room temperature for the diamond film spanning the ballistic to diffusive limits (asymptotically approaching the bulk data provided by Reference 46 as shown). For the 1-μm diamond film, suppose that the total thermal conductivity $K \sim 2100$ W/m/K. Note that the steady-state prediction is independent of F_T. We assume that the correct transient behavior can be characterized by F_T in the C–F model where these responses are depicted in Fig. 24.5(c) with evolution of time to steady state. In the case of the 1-μm diamond film, assume that $F_T = 0.5$ will correctly characterize the transient to steady-state temperature

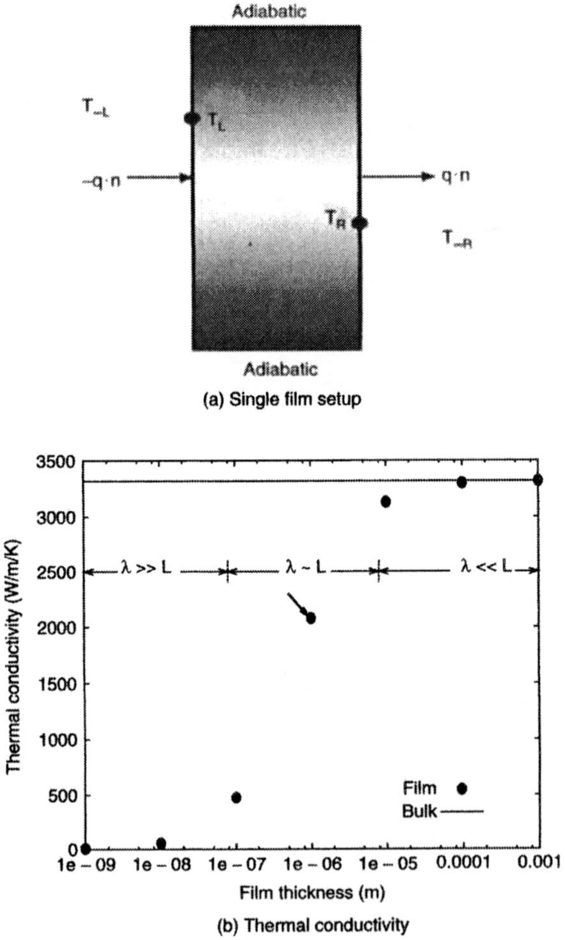

FIGURE 24.5 One temperature model and heat conduction model F_T.

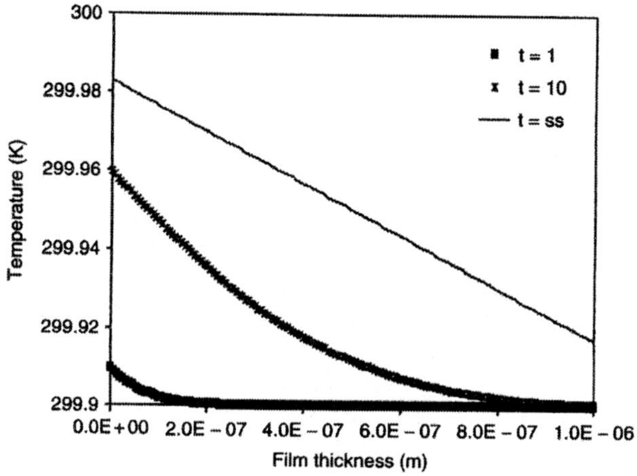

(c) Transient temperature profile

FIGURE 24.5 *(Continued)*

response of the experiment across the film thickness. The information that this hypothetical experiment has provided is that the actual transport behavior for this film that is in the partially diffusive ballistic region (due to the observed temperature jump on boundaries at steady state) also provides a transient response that is 50% wave-like at a finite speed and 50% diffusive at an infinite speed of heat propagation. Since $F_T = 0.5$, $K_F = 0.5(K)$. That is, in obtaining the total conductivity, the conductivity associated with the fast F-processes is half the total conductivity (the other half is due to the conductivity associated with the slow C-processes). Based on the definition of F_T , for the first time it is now possible to characterize the elastic component of the thermal conductivity[29] and provide a fundamental understanding of the role of the fast F-processes and slow C-processes in the propagation of heat transport leading to the thermal conductivity property obtained for a given film.

24.9.1 Spanning Spatial Scales

At steady state, the nondimensional temperature profiles (under the boundary conditions in Equation 24.27, which are of the third kind) of type IIa diamond films spanning lengths from 0.001 to 1000 μm based on the C–F model implementation are shown in Fig. 24.6. At room temperature, the mean free path of diamond films for experiments performed by Anthony et al.[46] is ~0.447 μm (see Table 24.2). At these film thicknesses, the temperature plots spanned both the ballistic ($\lambda \gg L$) and diffusive ($\lambda \ll L$) limits. At steady state, all temperature profile results for various values of $F_T \in [0,1]$ were equal as shown in Fig. 24.6.

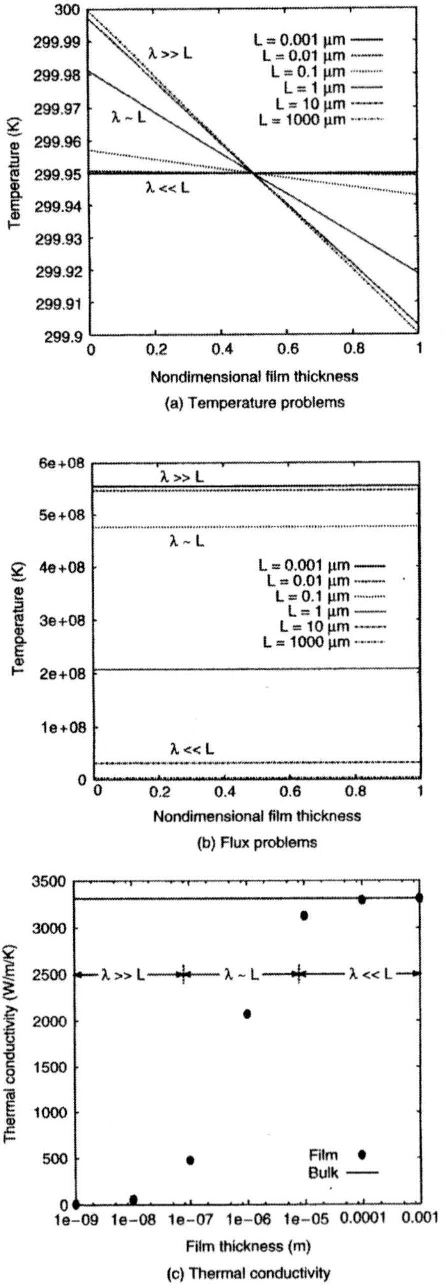

FIGURE 24.6 Steady-state results for several diamond film thicknesses using a C–F model with boundary conditions of the third kind. The nondimensional thickness is l/L where l is the x-axis position and L is the film thickness.

TABLE 24.2
Data Used to Predict the Relaxation Times of Diamond Type IIa Films

Constant	Properties
Bulk K[b]	3320 W/m/K
Lattice constant	3.567 A
Specific heat	517.05 J/kg/K
Mass density	3510 kg/m³
Phonon velocity[c]	12,288 m/s
Constant A (Eq. 24.53)	188.06[a]
Effective mean free path	0.447 μm
η^b (Eq. 24.51)	0.154×10^{26} atoms/m³
a (Eq. 24.53)	1.58
Debye temperature	1860 K

[a] This value is slightly different from the one reported in Reference 6 because we have chosen to compute all parameters based on the phonon velocity set to 12,288 m/s.
[b] Anthony, T.R. et al., *Phys. Rev. B*, 42, 1104–1111, 1990.
[c] Majumdar, A., *J. Heat Transfer*, 115, 7–16, 1993.

24.10 MULTILAYERS AND SUPERLATTICES

We also demonstrate applications relevant to the interface conditions in multilayers and superlattices via the C- and F-processes model based on two schools of thought: (1) the use of contact conduction and contact resistance of layered structures based on geometric, mechanical load, and thermal aspects,[47] and (2) the consideration of phonon transport in nanostructures.[48] In both cases, it is assumed that the interface of dissimilar materials of the multifilm structure has imperfect thermal contact at x_i between layers $(i-1)$ and (i). Via the present C–F model, for method 1 the appropriate interface conditions are given as

$$K^{(i-1)}\left[\frac{\partial T^{(i-1)}}{\partial x}+\tau^{(i-1)}F_T^{(i-1)}\frac{\partial \dot{T}^{(i-1)}}{\partial x}\right]=K^{(i)}\left[\frac{\partial T^{(i)}}{\partial x}+\tau^{(i)}F_T^{(i)}\frac{\partial \dot{T}^{(i)}}{\partial x}\right],\quad \text{at } x_i \tag{24.56}$$

$$K^{(i-1)}\left[\frac{\partial T^{(i-1)}}{\partial x}+\tau^{(i-1)}F_T^{(i-1)}\frac{\partial \dot{T}^{(i-1)}}{\partial x}\right]=\frac{1}{R_{(i-1)-(i)}}\left(T_{(i)}-T_{(j)}\right)+\frac{\tau F_T}{R_{(i-1)-(i)}}\left(\dot{T}_{(i)}-\dot{T}_{(j)}\right) \tag{24.57}$$

for $i = 1, 2, \ldots, n$. As applicable to the C–F model,[47] Equation 24.56 represents the continuity of the interface heat flux, and Equation 24.57 states that the difference between the two surface temperatures is proportional to the heat flux. The proportionality coefficient is the thermal contact resistance $R_{(i-1)-(i)}$ between the two layers, which is a function of forces acting on layer $(i-1)$ due to layer (i) and vice versa.

Usually, this is a parameter obtained by experiments and represents an unknown in this formulation.

Alternatively, for method 2, via the C–F model, the phonon transmission across the interface can be accounted for based on the effect of diffusive scattering as

$$\alpha_{(i-1)-(i)}K^{(i-1)}\left[\frac{\partial T^{(i-1)}}{\partial x}+\tau^{(i-1)}F_T^{(i-1)}\frac{\partial \dot{T}^{(i-1)}}{\partial x}\right]=\alpha_{(i)-(i-1)}K^{(i)}\left[\frac{\partial T^{(i)}}{\partial x}+\tau^{(i)}F_T^{(i)}\frac{\partial \dot{T}^{(i)}}{\partial x}\right],\ \text{at } x_i \tag{24.58}$$

where $\alpha_{(i-1)-(i)}$ is the phonon transmissivity across the rough gap from the $(i-1)$ to the (i) layer. Method 2 formulation does not require direct knowledge of the thermal boundary resistance, and after solving for the unknown temperatures it can be obtained from Reference 48. As an illustration, GaAs/AlAs temperature profiles spanning space scales and the prediction of thermal conductivity are shown in Fig. 24.7. Thus far, we have demonstrated that the C- and F-processes model can (1) explain the heat transport behavior from ballistic limit (Casimir limit: $\lambda \gg L$) to the diffusive limit (Fourier limit: $\lambda \ll L$) for the prediction of the thermal conductivity, (2) span time scales explaining finite to infinite speeds of heat propagation, and (3) overall provide a balanced energy equation with the inclusion of a nondimensional heat conduction model number, which is given as the ratio of the conductivity associated with fast processes to the total conductivity composed of both fast and slow processes. It is simple to implement and circumvents many of the deficiencies existing in the literature.

24.11 EQUATION OF PHONON RADIATIVE TRANSFER (EPRT)

In thin dielectric films, the major heat carriers are phonons. The thermodynamic equilibrium distribution for these carriers follows the Bose–Einstein distribution, which is similar to the distribution of photons. Under this premise the equation of phonon radiative transfer (EPRT) was originally described in Reference 49 based on the correlation between the radiation theory (i.e., the equation of radiative transfer [ERT]) and the transport theory in dielectric thin films. The major assumption made regarding solving the transport properties in very thin films is that the problem can be solved as a one-dimensional gray medium between black walls at specified temperatures under radiative equilibrium. This condition arises when a dielectric film is sandwiched between two metallic films. The following is a detailed derivation of this theory.

The intensity of photons, that is, the radiation emitted in any direction by a wave packet, is given by[24]

$$I(\theta,\phi,\omega,x,t)=\sum_p v(\theta,\phi)f(x,t)\hbar\omega\, D(\omega) \tag{24.59}$$

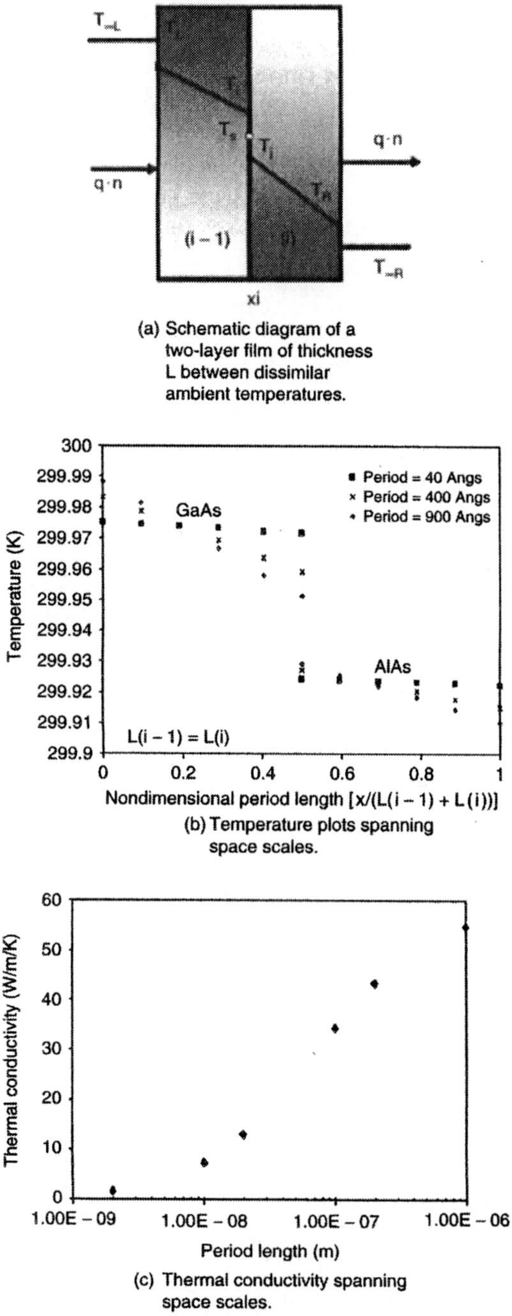

(a) Schematic diagram of a two-layer film of thickness L between dissimilar ambient temperatures.

(b) Temperature plots spanning space scales.

(c) Thermal conductivity spanning space scales.

FIGURE 24.7 Representative C- and F-processes model results for GaAs/AlAs superlattices.

where the summation is over the photon polarizations, $v(\theta, \phi)$ is the velocity vector in the direction of θ and ϕ within a unit solid angle, $\hbar\omega$ is the energy at which the photons propagate, and $D(\omega)$ is the density of states per unit volume. The frequency dependency of the phonon intensity is eliminated by assuming that the film medium acts as a gray body.

For cases where $\lambda \gg L$, the EPRT appears to model microscale aspects encompassing the ballistic to diffusive limits. In these cases, the thermal conductivity for thin films is obtained at steady state due to scattering of phonons from the boundaries, which returns the system to thermodynamic equilibrium. This suggests that the heat conduction by phonons is similar to that of photons and thus can be analyzed as a radiative transfer problem.[26]

Applying the analogy between photons and phonons and multiplying Equation 24.6 by $v_x \hbar\omega D(\omega)$ and using the intensity definition (Equation 24.59), the equation of phonon radiative transfer is obtained as

$$\frac{\partial I(x,\mu)}{\partial t} + v_x \frac{\partial I(x,\mu)}{\partial x} = \frac{I^0(T(x)) - I(x,\mu)}{\tau} \tag{24.60}$$

where the velocity in the direction of the phonon propagation is $v_x = v\mu$, $\mu = \cos\theta$, θ is the angle between the phonon propagation and the x-direction, and I^0 is the equilibrium intensity.

At steady-state, Equation 24.60 reduces to

$$v\mu\tau \frac{\partial I(x,\mu)}{\partial x} + I(x,\mu) = I^0(T(x)) \tag{24.61}$$

Majumdar[6] analyzed the EPRT for both acoustically thin ($\lambda \gg L$) and acoustically thick ($\lambda \ll L$) limits and showed that the EPRT does indeed provide results that span both limiting cases where the Casimir and the Fourier laws are applicable.

24.12 CALLAWAY–HOLLAND'S MODEL

Heat conduction in solids is well understood to follow the *kinetic formula*

$$K = \frac{1}{3} C v_s \lambda \tag{24.62}$$

where K, the thermal conductivity of an ensemble of heat carriers, is given by the total specific heat C, the average speed of the heat carries v_s and the mean free path λ. When the assumption that all phonons have the same energy and velocity is not desired, it is possible to account for the phonon dispersion where C, v, and λ are frequency dependent. Equation 24.62 can be modified by including the summation over all phonon branches (one longitudinal and two transverse) and by integrating over the phonon spectrum width of each branch as presented in the early works of Peierls in 1929,

$$K = \frac{1}{3} \sum_{p} \int C(\omega) v(\omega) \lambda(\omega) d\omega \tag{24.63}$$

The temperature-dependent thermal conductivities of silicon and germanium films can be modeled based on the approximated solutions to the Boltzmann transport equation (BTE) where the frequency-dependent relaxation times represent the phonon-scattering events. The most widely used model for determining the thermal conductivity is a method first proposed by Callaway,[4] which assumes that the N-processes dominate the scattering and the total thermal conductivity can be expressed as a sum of $K = K_1 + K_2$.

The first modified form of Callaway's model, known as Holland's model,[5] assumes that $K_2 = 0$ and includes the three phonon polarizations in K_1:

$$K_1 = K_{TO} + K_{TU} + K_L \tag{24.64}$$

Note that Holland's and the modified Callaway–Holland models are solutions of the steady-state version of the BTE and can only provide the temperature-dependent results of thermal conductivity and therefore only address the issue of microscopic size, i.e., they can only determine thermal conductivity as a function of size.

24.12.1 Comparison of Holland Model and EPRT Results

Figure 24.8(a) depicts the experimental and analytical results obtained by Asheghi and colleagues[42] for crystalline silicon layers 0.42, 0.83, and 1.6 μm thick. In their study, yet another modification to Callaway's temperature-dependent thermal conductivity model is employed, which considers the possibility of specular reflection through the surface. Their results show the peak in the conductivity for the thin films to occur at 70 K. The recommended bulk conductivity reaches a maximum of 5500 W/m/K at 30 K.[42]

Figure 24.8(b) shows the results from the finite element implementation of the steady-state EPRT compared with the experimental and analytical results given by Reference 42. Although the results in Fig. 24.8(b) present values that are higher than those seen in the results of Asheghi and colleagues, they provide a better fit of the data at low temperatures. The current study of the EPRT did not include effects of grain boundary scattering. The addition of the grain boundary scattering reduced the thermal conductivity by adding more sites for the thermal resistance. The results obtained by the EPRT method seem more appropriately used with films of the 1-μm range. Note that at room temperature the mean free paths of silicon and diamond films are 0.0409 and 0.39 μm, respectively. Hence, films thinner than 1 μm will present microscopic ballistic behavior at room temperature.

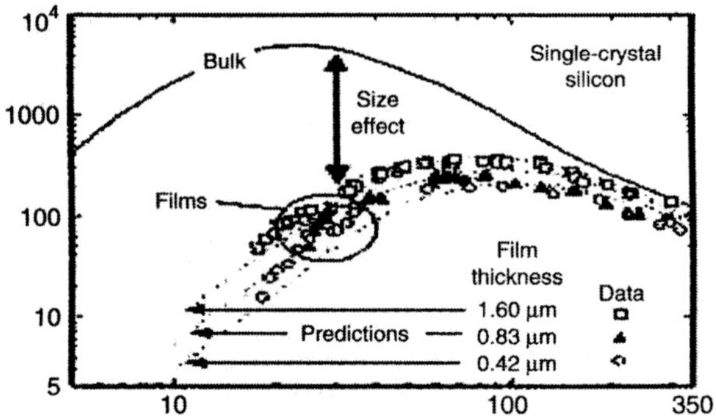

(a) Experimental data and analytical results following Holland's method of crystalline silicon layers of several thickness (from Asheghi *et al.*, 1998 with grain boundary)

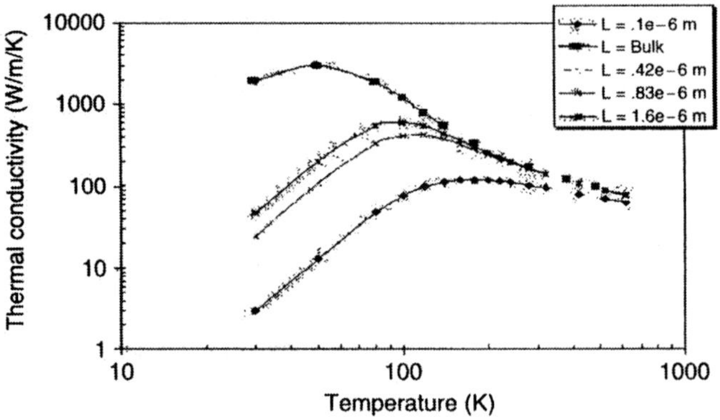

(b) FEM results following the EPRT method for silicon layers of several thickness (without grain boundary)

FIGURE 24.8 Asheghi et al.'s[42] work compared to results using FEM to solve the EPRT in determining temperature-dependent thermal conductivity for thin silicon films.

The results shown with the finite element method (FEM) are based on experimental thermal conductivity data available from work by Anthony and colleagues[46] (Table 24.3).

24.13 MOLECULAR DYNAMICS

Molecular dynamics (MD) is a computer simulation technique in which the time evolution of a set of interacting atoms is followed by integrating their classical equations of motion corresponding to the second law of classical mechanics formulated by Sir Isaac Newton in 1687

$$\vec{F}_i(\vec{r}_N) = m_i\vec{a}_i \tag{24.65}$$

TABLE 24.3
Parameters Used in Solving EPRT for Silicon Film

Parameter	Silicon
v (m/s)	6500
λ (μm)	0.044
A	101.4
σ (W/m²/K^4)	24.8
θ_D K	625

where $\vec{F}$ is the force acting on the atom i at a given time in a system containing N atoms, m_i is the atom mass, and $\vec{a}_i$ is the atom acceleration given by

$$\vec{a}_i = \frac{d^2\vec{r}_i}{dt^2}$$

with $\vec{r}_i$ as the atom position.

However, the motion of molecules is very complicated and the forces involved are nonlinear and depend on the positions of the other atoms. Therefore, a different form of the classical equation of motion is used and follows Hamilton's equation where particles are described by their positions $\vec{r}$ and momenta $\vec{p} = m\vec{v}$. The forces on the particles are obtained from the gradient of the interatomic potential energy surface $U(r_1, r_2, \ldots, r_N)$ as a function of the positions of all the atoms:

$$\vec{F}_i = -\vec{\nabla}_{r_i} U(\vec{r}_1, \vec{r}_2, \ldots, \vec{r}_N) \tag{24.66}$$

where $\vec{\nabla}_{r_i}$ operates on the position r_i of atom i. Any change in the potential energy that results from a displacement of atom i contributes to the force acting on atom i.

In order to develop a molecular dynamics simulation, it is necessary to specify both the atomic structures of the nanotubes where atoms are arranged at specific positions and the interatomic potential, which, for carbon nanotubes, is based on the Tersoff–Brenner[50–52] functional form with parameters fitted from experimental and quantum mechanical data to reflect good structural, mechanical, and thermal properties of nanotubes. A pristine nanotube is illustrated in this work. After the initial conditions and the interaction potential are defined, the equations of motion are numerically integrated and solved using a predictor–corrector algorithm with fixed time steps of 0.2 fs. The molecular dynamics simulation provides information at the microscopic level via the positions and velocities of all atoms as a function of time,

$$\vec{r}_i(t), \vec{v}_i(t)$$

In order to obtain thermal conductivity results from the information provided at the microscopic level, a nonequilibrium molecular dynamics simulation-based direct method[12,13] is used. The direct method emulates the actual experimental situation by setting two regions of hot and cold atoms.

24.13.1 Pristine Nanotube

The results for thermal conductivity shown here are for (10,10) single-wall carbon nanotubes (SWNTs). The theoretical expectation for SWNTs at room temperature (300 K) is either 2000 to 3000 W/m/K[12,14] or about 6000 W/m/K.[15] As mentioned earlier, results reported for the peak value of thermal conductivity in Reference 15 are an order of magnitude larger than the results reported in other work[12,14] and in recent experiments.[16] This is because an extremely small sample of carbon nanotubes was used in that simulation, and extrapolation of the data was probably erroneous. It is noteworthy to mention that in running a MD simulation it was observed that the thermal conductivity is highly dependent on the size of the nanotube and the temperature gradient chosen, as shown in Fig. 24.9. The temperature gradient used for obtaining the thermal conductivity results shown in Fig. 24.9(b), is kept constant at 2.03 K/nm, and the size of the nanotube is varied from 98.4 to 787.0 Å (1600 to 12,800 atoms). As the size of the nanotube is increased, the thermal conductivity appears to asymptotically approach a constant value. This is an interesting finding, which is in agreement with experimental results found in thin films, where the thermal conductivity is size dependent. As the film size approaches a critical value, the thermal conductivity equals that of its bulk counterpart. This concept is worthy of further investigation because the idea of "bulk" nanotube size is still under investigation. Note that, for a given thermal gradient, which can be determined from the experimental conditions, the thermal conductivity as a function of length converges to a constant value within 40 to 80 nm for a (10,10) SWNT.

Figure 24.9(c) shows the change in thermal conductivity as a function of the temperature gradient for a nanotube under periodic boundary conditions with 3200 atoms. Based on the expected room temperature thermal conductivity provided in Reference 14, it appears that it would be more appropriate to choose a temperature difference between 40 and 60 K for pristine nanotubes with 3200 atoms, instead of 80 K as shown in the results of Fig. 24.9(b). This finding, along with the question of what will happen to nanotubes of other sizes, is being further investigated.

24.14 CONCLUDING REMARKS

An introduction that serves as a prelude to an improved understanding of nanoscale, microscale, and macroscale thermal transport was presented. Since the early work in the 1930s, there has been a quest to provide an improved scientific understanding of size effects in thermal conductivity. More notably, with the advent of nanotechnology, the 1990s brought a surge of research interest and activity in the area of thermal transport. Although limitations exist in many of the methods, it is time to take a critical and in-depth look at the various investigations, questions, and concerns raised in the literature. The present chapter highlights some of the recent advances in thermal transport analysis and also provides some challenging solutions

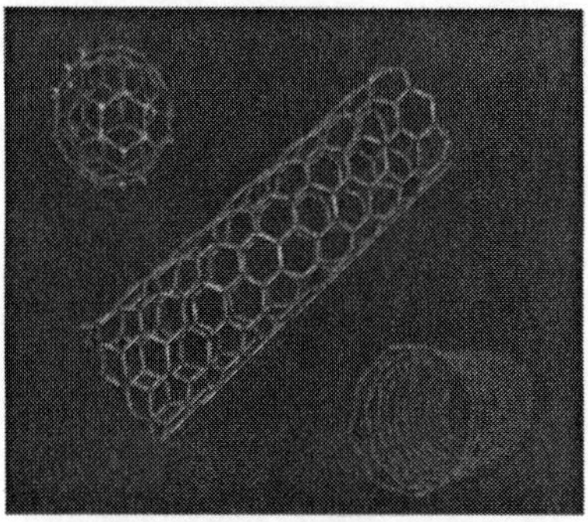

(a) Typical carbon nanotube (picture courtesy of the National Center of Competence in Research (NCCR) 4).

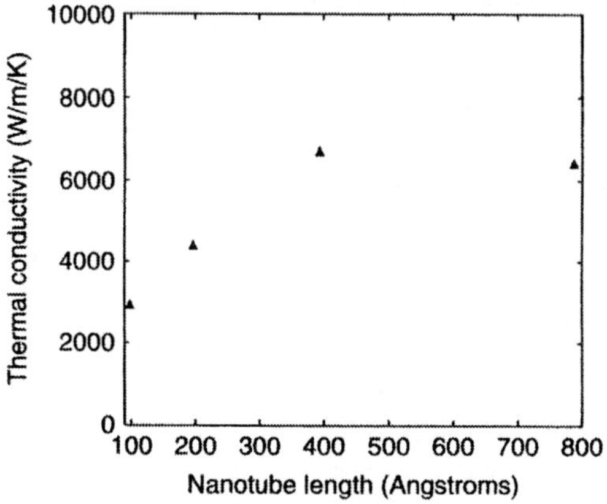

(b) Thermal conductivity versus nanotube length.

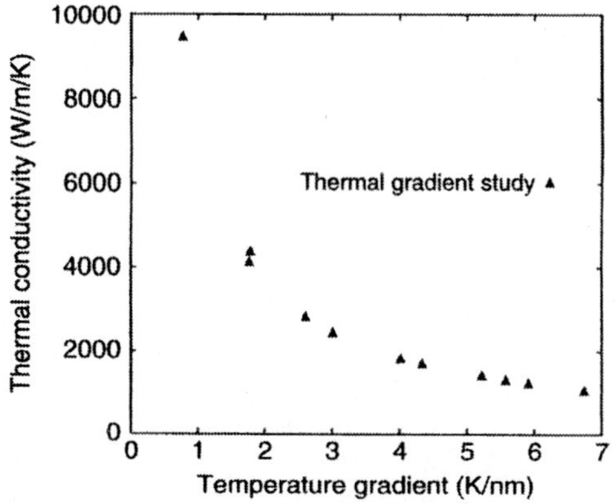

(c) Thermal conductivity versus temperature gradient.

FIGURE 24.9 Thermal conductivity of carbon nanotubes.

to various issues appearing in the literature. It is hoped that this chapter fosters future research in the directions discussed.

PROBLEMS

1. Identify the various methods (classical and nonclassical) existing in the literature for the prediction of thermal conductivity at scales ranging from nanoscale to macroscale regimes.
2. Conduct a thorough literature review and point out the limitations cited in the literature for the classical models. Verify these limitations using examples in the literature. Conduct the same for nonclassical models.
3. Derive the C- and F-processes constitutive model and understand the notion of spanning ballistic to diffusive limits and finite to infinite speeds of heat propagation. Since the method is not computationally intensive, verify some numerical examples such as thin films, superlattices, thermal contact conduction/resistance, and the like.
4. Investigate methods that are not computationally intensive yet do not require fitting parameters. For example, look into coupling the C- and F-processes model with variants that are less cumbersome to molecular dynamics to circumvent potential limitations.
5. Develop multiscale methods suited for modeling and simulation of large-scale applications on modern high-performance computing (HPC) platforms.

ACKNOWLEDGMENTS

The support for this work is in the form of computer grants from the Minnesota Supercomputer Institute and in part by the Army High Performance Computing Research Center under the auspices of the Department of the Army, Army Research Laboratory (DAAD19-01-2-0014). The content does not necessarily reflect the position or the policy of the government, and no official endorsement should be inferred. Special thanks are due to X. Zhou for related technical discussions. Thanks are also due to D. Srivastava for collaborations on the MD simulations. Special thanks are due to Amit Jain for the preparation of the chapter.

REFERENCES

1. Casimir, H.B.G., Note on the conduction of heat in crystals, *Physica*, 5, 495–500, 1938.
2. Lambropoulos, J.C., Jolly, M.R., Amsden, C.A., Gilman, S.E., Sinicropi, M. J., Diakomihalis, D., and Jacobs, S.D., Thermal conductivity of dielectric thin films, *J. Appl. Phys.*, 66, 4230–4242, 1989.
3. Klemens, P.G., *Proc. Roy. Soc. (London)*, A208, 108, 1951.
4. Callaway, J., Model of lattice thermal conductivity at low temperatures, *Phys. Rev.*, 113, 1046–1051, 1959.
5. Holland, M.G., Analysis of lattice thermal conductivity, *Phys. Rev.*, 132, 2461–2471, 1963.
6. Majumdar, A., Microscale heat conduction in dielectric thin films, *ASME J. Heat Transfer*, 115, 7–16, 1993.

7. Chen, G. and Tien, C.L., Thermal conductivity of quantum well structures, *J. Thermophys. Heat Transfer,* 7, 311–318, 1993.
8. Chen, G., Ballistic-diffusive heat-conduction equations, *Phys. Rev. Lett.,* 86, 2297–2300, 2001.
9. Goodson, K. and Flik, M.I., Microscale phonon transport in dielectrics and intrinsic semiconductors, *ASME,* 227, 29–36, 1992.
10. Mazumder, S. and Majumdar, A., Monte Carlo study of phonon transport in solid thin films including dispersion and polarization, *ASME J. Heat Transfer,* 123, 749–759, 2001.
11. Peterson, R.B., Direct simulation of phonon-mediated heat transfer in a Debye crystal, *ASME J. Heat Transfer,* 116, 815, 1994.
12. Osman, M.A. and Srivastava, D., Temperature dependence of the thermal conductivity of single-wall carbon nanotubes, *Nanotechnology,* 12, 21–24, 2001.
13. Muller-Plathe, F., A simple nonequilibrium molecular dynamics method for calculating the thermal conductivity, *J. Chem. Phys.,* 106, 6082–6085, 1997.
14. Che, J., Çağin, T., and Goddard III, W., A., Thermal conductivity of carbon nanotubes, *Nanotechnology,* 11, 65–69, 2000.
15. Berber, S.K., Young-Kyun and Tománek, D., Unusually high thermal conductivity of carbon nanotubes, *Phys. Rev. Lett.,* 84, 4613–4616, 2000.
16. Hone, J., Whitney, M., and Zettl, A., Thermal conductivity of single-walled carbon nanotubes, *Synthetic Metals,* 103, 2498–2499, 1999.
17. Cattaneo, M. C., Sur une frome de l'équation de la chaleur éliminant le paradoxe d'une porpagation instantanée, *C. R. Acad. Sci.,* 247, 431, 1958.
18. Vernotte, P., Les paradoxes de la théorie continue de l'équation de la chaleur, *C. R. Acad. Sci.,* 246, 3154, 1958.
19. Joshi, A.A., and Majumdar, A., Transient ballistic and diffusive phonon heat transport in thin films, *J. Appl. Phys.,* 74, 31–39, 1993.
20. Tien, C.L., Qiu, T.Q., and Norris, P.M., Microscale thermal phenomena in contemporary technology, *J. Thermal Sci. Eng.,* 2, 1–11, 1994.
21. Anderson, C.V.D.R. and Tamma, K.K., An overview of advances in heat conduction models and approaches for prediction of thermal conductivity in thin dielectric films, *Int. J. Heat Fluid Flow,* 14, 12–65, 2003.
22. Zhigilei, L.V., MSE 524: Modeling in Materials Science, Spring 2002.http://www.people.virginia.edu/lz2n/mse524/.
23. Klitsner, T., VanCleve, J.E., Fischer, H.E., and Pohl, R.O., Phonon radiative heat transfer and surface scattering, *Phys. Rev. B,* 38, 7576–7598, 1988.
24. Vincenti, W.G., and Kruger, C.H., *Introduction to Physical Gas Dynamics,* Robert Krieger, Ed., New York, 1977.
25. Kittel, C., *Introduction to Solid State Physics,* 6th ed. John Wiley & Sons, New York, 1996.
26. Ashcroft, N.W., and Mermin, N.D., *Solid State Physics,* W.B., Saunders, Philadelphia, 1976.
27. Lee, J.F., Sears, F.W., and Turcotte, D.L., *Statistical Thermodynamics,* 2nd ed., Addison-Wesley Publishing Co. Inc., Reading, MA, 1973.
28. Zhou, X., Tamma, K.K., and Anderson, C.V.D.R., On a new C-F processes heat conduction constitutive model and the associated dynamic thermoelasticity, *J. Thermal Stresses,* 24, 531–564, 2001.
29. Joseph, D.D. and Preziosi, L., Heat waves, *Rev. Mod. Phys.,* 61, 41–73, 1989.
30. Fourier, J.B.J., *Théorie Analytique De La Chaleur,* Paris, 1822.
31. Bird, R. B., Armstrong, B., and Hassager, O., *Dynamics of Polymeric Liquids,* John Wiley & Sons, 1977.

32. Joseph, D.D., Narain, A., and Riccius, O., Shear-wave speeds and elastic moduli for different liquids, Part I: Theory, *J. Fluid Mech.*, 171, 41, 1986.
33. Jeffreys, H., *The Earth*, University Press, Cambridge, U.K., 1929.
34. Anisimov, S.I., Kapeliovich, B.L., and Perel'man, T.L., Femtosecond electronic heat-transfer dynamics in thin gold film, *Sov. Phys. JETP*, 39, 375, 1974.
35. Corkum, P.B., Brunel, F., and Sherman, N.K., Thermal response of metals to ultrashort-pulse laser excitation, *Phys. Rev. Lett.*, 61, 2886, 1988.
36. Joseph, D.D., and Preziosi, L., Addendum to the paper "Heat Waves," *Rev. Mod. Phys.*, 62, 375, 1990.
37. Tamma, K.K., and Zhou, X., Macroscale and microscale thermal transport and thermo-mechanical interactions: Some noteworthy perspectives, *J. Thermal Stresses*, 21, 405, 1998.
38. Ziman, J.M., *Electrons and Phonons: The Theory of Transport Phenomena in Solids*, Claredon Press, Oxford, 1960.
39. Berman, R., *Thermal Conduction in Solids*, Clarendon Press, Oxford, 1978.
40. Graebner, J.E., Reiss, M.E., Seibles, L., Hartnett, T.M., Miller, R.P., and Robinson, C.J., Phonon scattering in chemical-vapor-deposited diamond, *Phys. Rev. B*, 50, 3702–3713, 1994.
41. Asen-Palmer, M., Bartkowski, K., Gmelin, E., Cardona, M., Zhernov, A.P., Inyushkin, A.V., Taldenkov, A., Ozhogin, V.I., Itoh, K.M., and Haller, E.E., Thermal conductivity of germanium crystals with different isotopic compositions, *Phys. Rev. B*, 56, 9431–9447, 1997.
42. Asheghi, M., Touzelbaev, M.N., Goodson, K.E., Leung, Y.K., and Wong, S.S., Temperature-dependent thermal conductivity of single-crystal silicon layers in soil substrates, *Trans. ASME*, 120, 30–36, 1998.
43. Brorson, S.D., Fujimoto, J.G., and Ippen, E.P., Femtosecond electronic heat transfer dynamics in thin gold film, *Phys. Rev. Lett.*, 59, 1962, 1987.
44. Qiu, T.Q., Juhasz, T., Suarez, C., Bron, W.E., and Tien, C.L., Femtosecond laser heating of multi-layered metals II. Experiments, *Intl. J. Heat Mass Trans.*, 37, 719, 1994.
45. Pells, G.P., and Shiga, M., The optical properties of copper and gold as a function of temperature, *J. Phys. C*, 2, 1835, 1969.
46. Anthony, T.R., Banholzer, W.F., Fleischer, J.F., Wei, L., Kuo, P.K., Thomas, R.L., and Pryor, R.W., Thermal diffusivity of isotropically enriched 12c diamond, *Phys. Rev. B*, 42, 1104–1111, 1990.
47. Blandford, G.E. and Tauchert, T.R., Nonlinear thermoelastic analysis of layered structures, *Finite Elements Anal. Design*, 1, 271–285, 1985.
48. Chen, G., Heat transport in the perpendicular direction of superlattices and periodic thin-film structures, *ASME*, 59, 13–24, 1996.
49. Majumdar, A., Microscale heat conduction in dielectric thin films, *ASME*, 184, 33–42, 1991.
50. Brenner, D.W., Empirical potential for hydrocarbons for use in simulating the chemical vapor deposition of siamond films, *Phys. Rev. B*, 42, 9458–9471, 1990.
51. Tersoff, J., New emperical approach for the structure and energy of covalent systems, *Phys. Rev. B*, 38, 6991–7000, 1988.
52. Tersoff, J., Empirical interatomic potential for silicon with improved elastic properties, *Phys. Rev. B*, 38, 9902–9905, 1988.
53. Van Zegbroeck, B.J., Distribution Functions, 1996, http://ece-www.colorado.edu/~bart/book/distrib.html.
54. Peierls, R.E., Zur theorie der galvanomagnetischen effekte, Zeitschrift fur physik, 53 (1929), 255.

Index

A

B

C

D

E

F

G

H

I

J

K

N

Q

R

S

U

V

Z